GUIDANCE AND CONTROL
2003

CONFERENCE COMMITTEE OFFICERS

Chairperson:	*Ian Gravseth* *Ball Aerospace & Technologies Corp.*
Co-Chairperson:	*Steven Jolly* *Lockheed Martin Astronautics*
Secretary:	Lis Garratt *Ball Aerospace & Technologies Corp.*
Treasurer:	*Marv Odefey* *Left Hand Design Corp.*
Rocky Mtn. AAS Chair:	*Ron Rausch* *Lockheed Martin Astronautics*
Technical Committee Chair:	*Larry Germann* *Left Hand Design Corp..*

ADDITIONAL COMMITTEE MEMBERS

Sol Benson . . . . Left Hand Design Corporation
Jay Brownfield . . . . Honeywell, Inc.
Robert Culp . . . . University of Colorado at Boulder
Ed Euler . . . . Lockheed Martin Astronautics
Joyce Idler . . . . Lockheed Martin Astronautics
Eric Lander . . . . Advanced Solutions, Inc.
Lou Morine . . . . Schafer Corp. (retired)
Stephen Price . . . . Lockheed Martin Astronautics
Charlie Schira . . . . Ball Aerospace & Technology Corp.
Doug Wiemer . . . . Ball Aerospace & Technology Corp.

Reprinted from *Guidance and Control 2003, Volume 113, Advances in the Astronautical Sciences*, Edited by Ian J. Gravseth and Robert D. Culp, 2002. Published for the American Astronautical Society by Univelt, Incorporated, P.O. Box 28130 San Diego, California 92198, U.S.A. Web Site: http://www.univelt.com

AAS PRESIDENT
Dr. Robert E. Lindberg — National Institute of Aerospace

VICE PRESIDENT - PUBLICATIONS
Dr. Robert G. Melton — Pennsylvania State University

EDITORS
Dr. Ian J. Gravseth — Ball Aerospace & Technologies Corporation
Dr. Robert D. Culp — University of Colorado at Boulder

SERIES EDITOR
Robert H. Jacobs — Univelt, Incorporated

Thanks are due Bonita Roach, Univelt, Incorporated, for computer preparation work done on the manuscript.

Front Cover Illustration:

This illustration shows an artist's concept of the Deep Impact Flyby, Impactor, and comet Tempel 1. Deep Impact is scheduled to launch in December of 2004, and will encounter Tempel 1 on July 4th, 2005. Deep Impact will be the first mission to study the interior of a comet (Courtesy of Martha Lageschulte).

Frontispiece:

This picture shows ICESat following its final integration at Ball Aerospace & Technologies Corp. ICESat was successfully launched on January 12th, 2003. ICESat will provide multi-year elevation data needed to determine ice sheet mass balance, cloud and aerosol heights, as well as land topography and vegetation characteristics (Photo: Courtesy of and used with permission of Ball Aerospace).

American Astronautical Society

GUIDANCE AND CONTROL 2003

Volume 113
ADVANCES IN THE ASTRONAUTICAL SCIENCES

Edited by
Ian J. Gravseth
Robert D. Culp

Proceedings of the Annual AAS Rocky Mountain Guidance and Control Conference held February 5-9, 2003, Breckenridge, Colorado, U.S.A.

Published for the American Astronautical Society by
Univelt, Incorporated, P.O. Box 28130, San Diego, California 92198
Web Site: http://www.univelt.com

AAS Publications Office
P.O. Box 28130
San Diego, California 92198

Affiliated with the American Association for the Advancement of Science
Member of the International Astronautical Federation

First Printing 2003

Library of Congress Card No. 57-43769

ISSN 0065-3438

ISBN 0-87703-502-4 (Hard Cover)
ISBN 0-87703-503-2 (Soft Cover)

Published for the American Astronautical Society
by Univelt, Incorporated, P.O. Box 28130, San Diego, California 92198
Web Site: http://www.univelt.com

Printed and Bound in the U.S.A.

FOREWORD

HISTORICAL SUMMARY

The Annual Rocky Mountain Guidance and Control Conference began as an informal exchange of ideas and reports of achievements among local guidance and control specialists. Since most area guidance and control experts participate in the American Astronautical Society, it was natural to gather under the auspices of the Rocky Mountain Section of the AAS.

In 1977, Bud Gates, Don Parsons, and Bob Culp organized the conference formally and began the annual series of meetings the following winter. In March 1978, the Annual Rocky Mountain Guidance and Control Conference met at Keystone, Colorado. It met there for eighteen years. The last eight Conferences have been held at Breckenridge, Colorado.

The Rocky Mountain Section of the Society established a broad-based Conference Committee, the Rocky Mountain Guidance and Control Committee, chaired ex-officio by the next Conference Chairman, to run the Annual Conference. The Conference has been a success from the start. Currently, the meeting, now named the AAS Guidance and Control Conference and sponsored by the national AAS, attracts about 200 of the nation's top specialists in space guidance and control. This year's Conference was the twenty-sixth Conference.

In recent years, the Conference has strongly supported students interested in aerospace guidance and control. The Section, using proceeds from this Conference, every year gives $2000 to two scholarships at the University of Colorado: one in Aerospace Engineering Sciences and one in Electrical and Computer Engineering. The Section has assured the continuation of these scholarships through a $60,000 endowment. The Section supports other space education through grants to K-12 classes throughout the Section at a rate of approximately $10,000 per year. Other grants are made to national organizations to further the ends of space education. All this made possible by this Conference.

In 1983, Don Parsons, Joe Spencer, Ed Dorroh, and Bob Culp, at the request of the National Board of the American Astronautical Society, wrote a charter which established the Annual Conference as the AAS Guidance and Control Conference. The Conference is sponsored by the National AAS, meets annually in Colorado, and is run by the Rocky Mountain Section.

The AAS Guidance and Control Technical Committee, with its national representation, provides oversight to the local conference committee. W. Edwin Dorroh, Jr., was the first chairman of the AAS Guidance and Control Committee; from 1985 through 1995 Bud Gates chaired the committee; from 1995 through 2000, James McQuerry chaired the committee. Since 2000, Larry Germann has chaired this committee, which meets every year at the Conference, and also sometimes at the summer Guidance and Control Meeting or the fall AAS Annual Meeting.

The AAS Guidance and Control Conference hosted by the Rocky Mountain Section in Colorado, continues as the premier conference of its type anywhere. As a National Conference sponsored by the AAS, it promises to be the preferred idea exchange for guidance and control experts for years to come.

In 2003, the twenty-sixth AAS Guidance and Control Conference attracted 178 participants. The Conference Committee extends its congratulations to the 2003 Conference Chair, Ian Gravseth, for an excellent conference and a job well done. His report follows.

Robert D. Culp
University of Colorado
Boulder, Colorado

PREFACE

The 26th annual AAS Rocky Mountain Guidance and Control Conference was held from February 5th-9th, 2003. The conference began with the traditional Wine and Cheese Social and Registration on Wednesday evening.

The technical portion of the conference began on Thursday morning with the *Advances in Guidance and Control* session, which presents innovative, cutting edge ideas for guidance and control hardware and software. The national co-chairs for this session were Julie Thienel of Goddard Spaceflight Center and Jeremy Kasdin of Princeton. Some of the highlights of this session included a paper on a star tracker/earth sensor capable of simultaneous geosynchronous attitude and orbit determination, and another paper discussing an integrated tracker and MEMs gyro capable of providing continuous attitude data even through tracker outages. Other papers were written and presented on precision attitude determination for the GOES-N satellite, closed loop hardware simulation of decentralized formation control, electromagnetic formation flight, precision formation control, and non-linear observers for gyro alignment estimation.

The Thursday evening session on *Autonomous Rendezvous and Proximity Operations* was well received. Karen Frank of Johnson Spaceflight Center and Major Jim Shoemaker of Darpa were the national co-chairs for this session. Jennifer Wagenknecht of JSC filled in for Karen at the conference and helped with the introductions. Several excellent papers were presented, including ones on coupled attitude and orbit dynamics and control in formation flying systems, a combined open-loop and autonomous search and rendezvous navigation system for the CNES/NASA Mars premier orbiter mission, an overview of the spheres autonomous rendezvous and docking laboratory on the international space station, laser-based relative navigation and guidance for space shuttle proximity operations, design, development and testing of the miniature autonomous extravehicular robotic camera (Mimi AERCam) guidance, navigation, and control system, and stochastic optimization of spacecraft rendezvous trajectories in the presence of disturbing forces.

Friday morning's session presented several papers on *Interplanetary Missions*. The national co-chairs for this session were Luisella Giulicchi from ESA, and James Randolph from JPL. Papers from this session discussed autonomous navigation for deep space missions, pointing at Tempel-1 in the presence of a dead impactor, rendezvous and landing on a comet, and positively identifying stars with star tracker data when within a comet's dust cloud.

Friday afternoon's *Storyboards* session is a repeat from preceding years and is a traditional conference favorite. In this forum, vendors present hardware in a hands-on environment. This session allows individuals to present their innovations and achievements in a posterboard format and have direct discussion with conference attendees. Rick Jackson and Joe Vellinga co-chaired this session. Many thanks to the corporate sponsors who help make Storyboards possible every year.

Saturday morning's session *Guidance and Control Design Approaches for Precision Pointing Applications* was another excellent session. The national co-chairs for this session were Brian Kawauchi of Aerospace Corp., and Steve Lee of JPL. Precision pointing papers were presented on the Hubble space telescope, the Thuraya satellite, a space systems Loral geosynchronous satellite, the STS-7 spacecraft controller, the LISA mission, and the Mars Reconnaissance Orbiter mission.

Saturday evening was the banquet which provides an opportunity for the conference attendees and their families to enjoy a nice dinner and an informative and entertaining presentation. Clearly the highlight of the conference was the banquet speech by Bill Gregory. His unfortunately timely talk on piloting STS-67 shuttle mission, which was just one week after the Columbia disaster, was the best we had heard in many years, appealing to both professionals working in the guidance and control community and their kids and spouses. Bill was able to clearly convey to the audience the experience of flying aboard the Space shuttle.

Sunday morning the conference was closed by an outstanding *Recent Experiences* session, co-chaired by Darrell Zimbelman of NASA/Goddard and David Haley from APL. The Recent Experiences session is also offered yearly and gives a forum to share flight successes and failures, along with lessons learned. This year's session had papers on Satellite GN&C anomaly trends, a CMG failure on the International Space Station, the TIMED spacecraft, the MRO spacecraft, and the TOPEX/Poseidon spacecraft.

Conference attendance at this year's conference was back up to 178, indicating a continued interest and support of this conference by the G&C community. The support of the research laboratories, the universities, the many corporations, and the G&C engineers and researches from around the country and the world continues to make this conference a great success.

Sincere appreciation and congratulations are expressed to the conference planning committee, the local session chairs, the session co-chairs, the authors, presenters, the banquet speaker and Beaver Run Resort in Breckenridge, Colorado. I want to express my personal thanks to everyone for making this year's conference a success and I look forward to seeing you at next year's conference, February 4th-8th, 2004 at the Beaver Run Resort.

Ian Gravseth
Conference Chairperson
2003 AAS Guidance and Control Conference

CONTENTS

Section I
ADVANCES IN GUIDANCE AND CONTROL

SESSION I

Joint Session Chairperson:	Julie Thienel NASA Goddard Space Flight Center
Joint Session Charperson:	Jeremy Kasdin Princeton University
Local Session Chairperson:	Dave Igli Lockheed Martin Astronautics

The following paper numbers were not assigned:
AAS 03-009 to -010

MICRO- TECH.- SENSOR FOR ATTITUDE AND ORBIT DETERMINATION

Christopher Kuehl, Markus Melf, Herman Diehl and Eveline Gottzein*

A sensor for autonomous attitude and orbit determination of satellites, based on active pixel technology, is proposed. The system is characterized by a beam splitter, which provides two non-collinear lines-of-sight and attenuation of the Earth's magnitude. Hereby images of a star field, perpendicular to the Earth vector, and the Earth are projected on a single focal plane.

The paper is concerned with extensive functional tests and an initial evaluation of performance based on various algorithms.

Results from three levels of functional tests will be presented:

- Laboratory tests,
- Field tests, in the Bavarian Alps,
- In- Orbit tests, on MITA†, an Italian small satellite platform in circular orbit, altitude 450 km, inclination 87,3 deg.

A laboratory test facility was used to test the algorithm and get a grasp on the system's peculiarities and potentials. It consists of an opto- electronic star simulator and an active pixel camera to provide star images.

The field tests were performed to test the algorithm's behavior in a heavily disturbed environment, and to provide realistic spatially distributed star images.

The In- Orbit tests were performed to prove the system's functionality in space.

All experiments were successful and confirmed the feasibility of the proposed system:

- The laboratory tests showed the system's integrity,
- The field experiment verified the assumptions and expectations with regard to the camera system, the performance of the breadboard and the correct implementation of the laboratory set- up,
- The In- Orbit experiment proved, that it is possible to observe stars and the Earth on a single focal plane.

The experiments in their union served as a vast source of ideas on future developments and as a warning of possible traps, when using a purely optical system for attitude and orbit determination.

---

* Astrium GmbH, Germany.

† Minisatellite Italiano a Technologia Avanzata.

INTRODUCTION

In today's space environment it is desirable to provide the means for autonomous navigation and control of spacecraft. This is caused by several factors which occur in combination as well as separately and can be summarized under reduction of operational efforts and mission critical constraints such as timing. A typical Earth-orbiting and Earth-oriented satellite, for instance, is subject to disturbing torques and forces. The resulting deviations in orientation and position need to be detected and corrected by the spacecraft's control system. (see Fig. 1)

Attitude and Orbit Control Systems (AOCS) are used to position and point the satellite according to its mission and compensate the influence of disturbances.

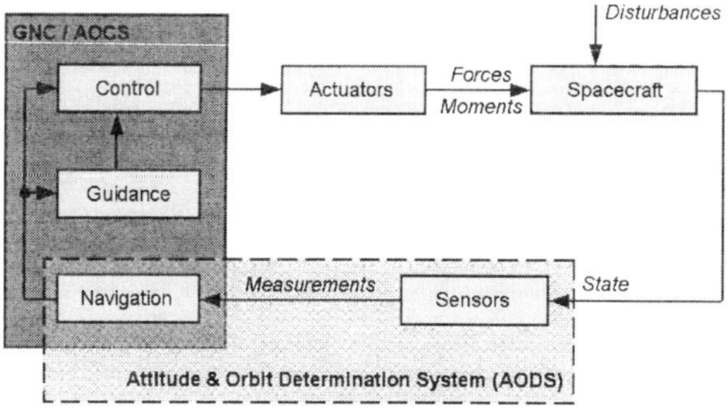

Figure 1: Schematic of AOCS

Current Attitude and Orbit Determination Systems (AODS) for communication satellites are typically based on Earth- and sun-sensors, supplemented by a gyro package, and ground stations for orbit determination. A typical example of a modern communication satellite is the satellite shown in Fig.2.

Typical AOCS components of a geostationary satellite.

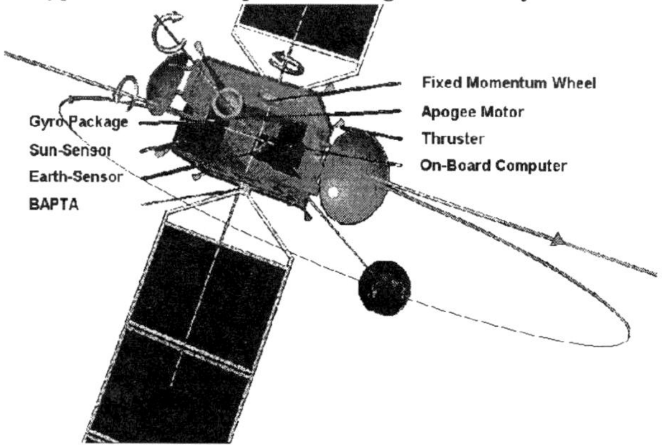

Figure 2: Attitude Sensors of a GEO-Satellite

One functional disadvantage of the particular sensor-configuration in Fig.2 is a periodic occurrence of collinearity of Earth and sun, during which a direct three-axis attitude determination is no longer possible. While autonomous determination would theoretically be possible with the sensors mentioned, it is practically limited due to thermal deformations of the spacecraft which result in varying off-set angles between the sensor-axes. Determination of the calibration-function requires ground based measurements. The need for a time varying calibration-function makes the orbit determination rather difficult and potentially inaccurate. [Juckenhöfel, 2001]

Instead of using ground based measurements the use of GPS in combination with various attitude sensors, e.g. star-sensors, is used in a semi-autonomous system. The nomenclature "semi-autonomous" has been chosen since it relies on the presence of GPS satellites for orbit determination. [e.g. Fichter, 2001]

AUTONOMOUS SENSOR USING EARTH- AND STAR- REFERENCES

A novel optical sensor for attitude and orbit determination has been developed, which provides the means for truly autonomous attitude and orbit determination of a spacecraft. Based on the already well-known idea of sensor-fusion to obtain a satellite's position (e.g. 'Space Sextant' [Magee, 1961]), a new sensor has been developed at Astrium which incorporates the principle of sensor-fusion on a sextant-like sensor, using a single optical detector in the visual regime. For equatorial orbits three-axis-attitude determination in the visual regime using Earth reference and polar stars, observed by a single CID video camera, has been proposed by EDO Corporation [Falbel, 1993].

Our approach is based on the "Active Pixel Sensor" (APS) technology which provides detectors with high dynamic range and high degree of functionality ("windowing"). The compact design reduces the influence of thermal deformation and thus facilitates accurate in-orbit-calibration. The proposed design, in contrast to the 'Space Sextant', developed in the 1960's, uses no moving parts, therefore reducing the risk of mechanical failures. A schematic of the sensor is shown in Fig.3.

The sensor consists of a partially transmissive mirror, coated on a grayfilter to further attenuate the Earth's magnitude, while preserving the stars' brightness. This allows the observation of two non-collinear lines-of-sight (LOS). In the current configuration the mirror is mounted at an angle of 45°, thus resulting in two perpendicular LOS. Depending on the satellite's structure and the chosen orbit other angles might be of interest, but will not be addressed in this report.

The proposed sensor consists of a partially transmissive mirror and a common optics which projects the combined visual information of the two lines-of-sight on a single detector

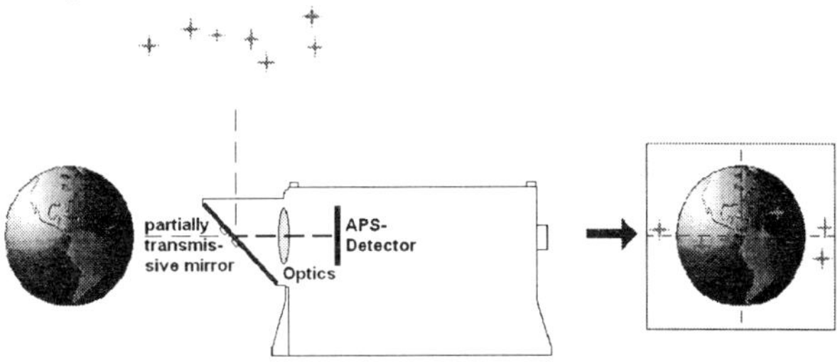

Figure 3: Schematic of Sensor's Principle

The proposed sensor will be able to determine a satellite's attitude and position in all orbits, as long as a determination of the central body's center is possible. Thus, in case of the Earth, a single sensor covers a wide range of orbits ranging from low Earth orbits (LEO) to geostationary orbits (GEO), including inclined and polar orbits, as well as orbits around other planets. Additionally to having a practically universal sensor for all orbits the solution is expected to reduce mass, power and complexity as compared to currently available sensor systems.

ANALYTICAL INVESTIGATIONS AND SENSOR DESIGN

The sensor's design is mainly dependent on the detector, the optical system, the integration constraints on the satellite, the satellite's orbit and on the objects observed. The following table lists the main design aspects, which need to be taken into account for this particular sensor:

- High dynamic range experienced in observation of stars and Earth on one detector
- Observation of Earth under varying illumination conditions (e.g. terminator, horizon-detection at night, sun in Field-of-View)
- Mechanical and optical restrictions and constraints (e.g. obstruction or reflection due to solar panels, bloom caused by propulsion system)
- Baffle- / Filter-/ Mirror-/ Beam Splitter- Design
- Characteristics arising from a large Field-of-View and its influence on the optics
- Limitations due to low fillfactor (currently commercially available APS detectors)
- Provision for In-Orbit and Pre-Flight calibration

The following chapters will deal in detail with some of these points.

Analytical Investigations

Since the point-solution for the satellite's position requires the knowledge of attitude and the central body's center, the investigations presented will address the determination of star-centroids and the Earth's rim. Using the knowledge of the Earth's rim it is possible to determine it's center, as well as its apparent radius.

Reference data sets for these investigations were obtained through laboratory experiments, fiel experiments, METEOSAT-pictures (geostationary weather satellite) and a flight experiment named "Micro Tech Sensor for Attitude and Orbit Measurement Systems" (MTS-AOMS), which was conducted on the Italian micro-satellite platform MITA between 08/2000 and 08/2001.

Centroiding

Tests estimating the system's centroiding performance first of all determined the physical limits. Using the experimental system's parameters it was shown, that a centroiding accuracy of better than $1/10^{th}$ of a pixel can be achieved by observing stars of visual magnitude 4 or brighter, taking into account photon-noise only (see Fig.4).

This implies a constraint on attitude performance, which is directly dependent on the stars' centroiding accuracy. Other dependencies are the stars' distribution on the detector and the number of stars. A numerical full sky test showed that the accuracy is still approximately proportional to the statistically expected inverse of the square-root of the number of observed stars (see Fig.5). The points in Fig.5 resemble a parametric plot with different centroiding-accuracies. The plotted points are obtained by averaging of 400 numerically simulated samples.

Another phenomenon to be taken into account is the non-linear trace of the centroid which is observed during a linear scan of a star over the detector. The pixel-sensitivity used to obtain the theoretical data is approximately hat-shaped, as shown in Fig.6. As seen in Fig.7 a) and b) the calculated position of the star's center thus deviates from the true value in form of an S-shaped curve. This phenomenon can be observed on all detectors which do not show a flat sensitivity within a single pixel.

The curve shows the dependance of centroiding accuracy on the brightness of stars. Assumptions made are given in Table 1.

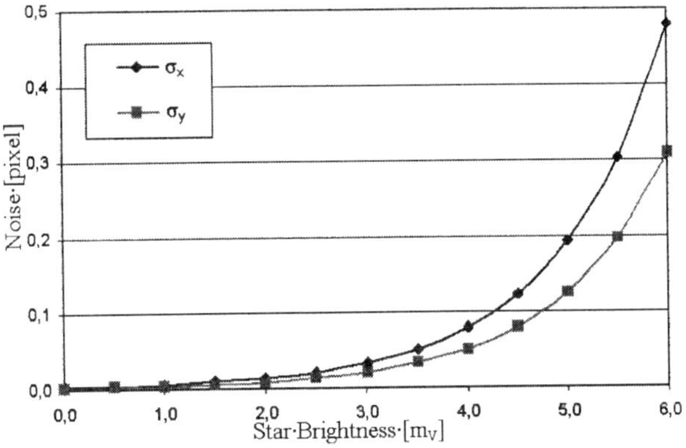

Figure 4: Photon-Noise Induced Inaccuracy

As expected, a numerical full sky test showed an approximately inverse square root behaviour for the increase in accuracy in dependency of the number o observed stars. The deviation from the expected value is best explained by the accuracy dependence on the star distribution on the detector.

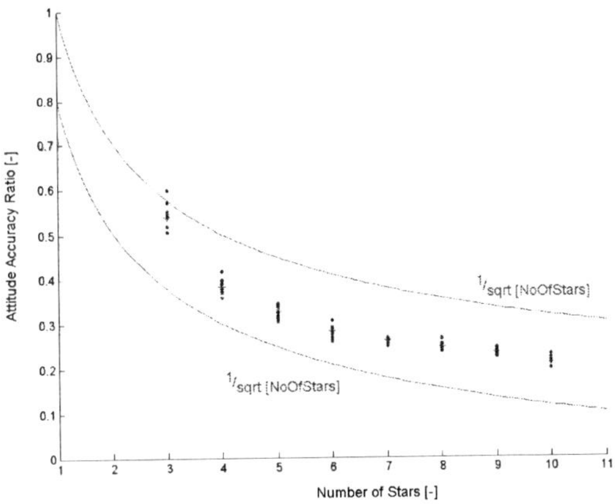

Figure 5: Accuracy Improvement by Increased Number of Observed Stars, Parametric Plot for Different Centroiding Accuracies

The proposed sensitivity function is based on a typical pixel sensitivity while additionally taking into account a fillfactor <100 %. For computational simplicity the non-sensitive part is assumed to be evenly distributed along the pixel- boundary.

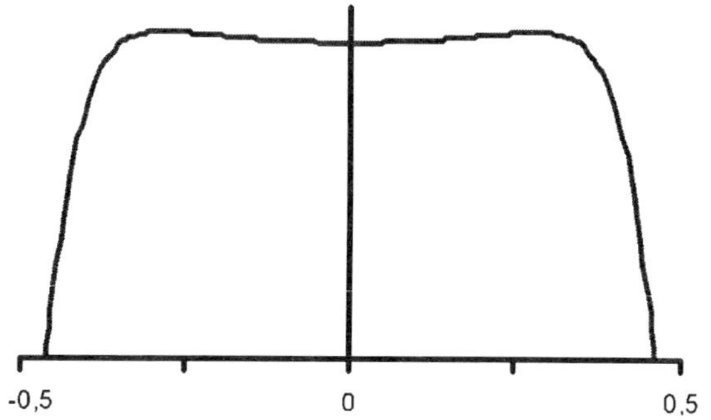

Figure 6: Assumed Sensitivity for a Single Pixel

Laboratory tests were performed using a sensor similar to the one integrated in the flight experiment MITA, having a fillfactor of approx. 72 %. The reference-angle was obtained with a high-accurac rotational table. The star was simulated using a star-simulator with high accuracy in the simulation of stars' centers and spectrum. The test results in Fig.8 a) and b) both exhibit the predicted S-shaped dependency in the FOV's center and a slightly deformed S-shape when observing stars with incidence angles of approx. 7.5 °. It might be of interest to note that while the shown functions are bijective and can therefore be corrected, systems can be found that exhibit ambiguous solutions when trying to determine the inverse of the resulting trace.

The curves shown illustrate the nonlinear relationship between the actual and the perceived position of a star centroid while it is being moved over a pixel. Left is the function for a star in direction of the LOS, right is the function for an observation angle of approx. 12 °.

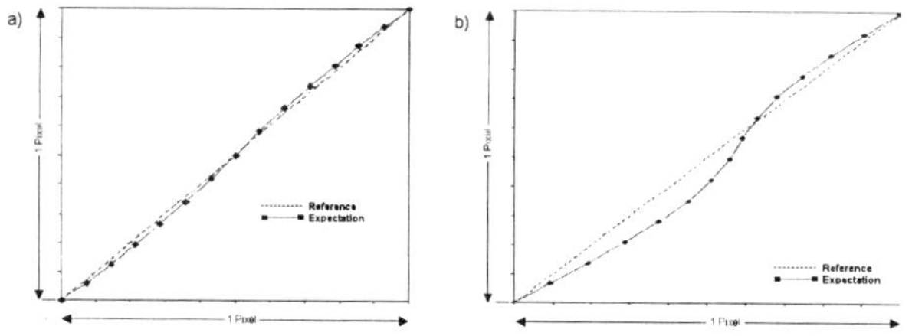

Figure 7: Centroiding Error During Line Scan, Analytical Solution

In addition to the afore-mentioned influence on the attitude accuracy, the centroiding accuracy has another important property: It sets a lower limit to the time necessary to identify the observed stars. In Fig.9 the probability of identification for a set of 5 stars as a function of the inter-star-angle tolerance is shown. Inter-star-angle tolerance in this case is defined as the expected 1σ-centroiding accuracy. The curve-parameters define the 1σ-centroiding accuracy in multiples of 10''. The figure illustrates that starting with a tolerance factor of about 3 the probability of star identification does not change significantly any more. As an example: Choosing a centroiding accuracy factor of 3, resulting in a 1σ- accuracy of 30'', and using an inter-star-angle-tolerance factor of 5, resulting in an inter-star tolerance of 150'', an identification probability level of about 98% is achieved.

The curves shown illustrate the nonlinear relationship between the actual and the perceived position of a star centroid while it is being moved over a pixel. Left is the function for a star in direction of the LOS, right is the function for an observation angle of approx. 7.5 °.

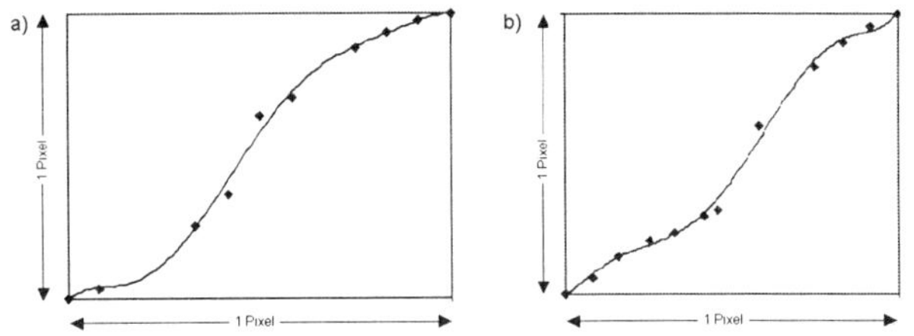

Figure 8: Centroiding Error During Line Scan, Experimental Data

It further can be seen, that a decrease in centroiding accuracy has only a minor impact on identification probability as long as there are 5 stars present.

Fig.10 gives the time necessary for a 5-star identification procedure as a function of the chosen inter-star-angle-tolerance. As can readily be seen, the computation time rises non-linearly. A polynomial fit showed an approximately quadratic dependency. This means, that, e.g. a decrease in centroiding accuracy from 10" to 30" has only an insignificant effect on the star identification probability during acquisition, while using the corresponding tolerance of about 50" and 150", respectively, the computation time for star identification rises by a factor of approx. 8.2.

<u>Parametric Curve:</u>
Parameters given are the factors defining the sensor (centroiding) accuracy in multiples 10''

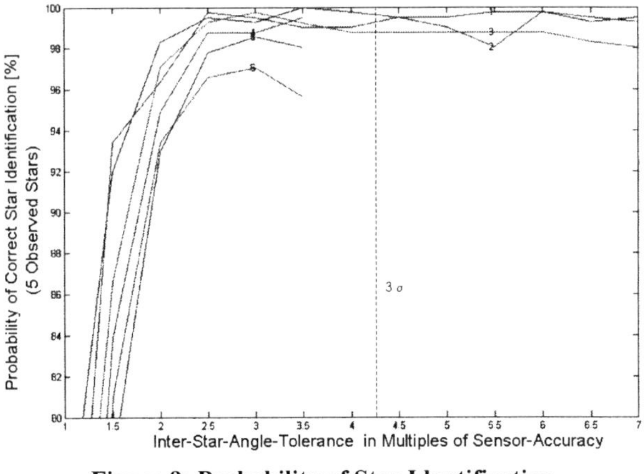

Figure 9: Probability of Star Identification

The graph given shows the average computation time for the identification of a 5 star observation for a given tolerance in arbitrary units. Sample size was approx. 400 observations per tolerance.

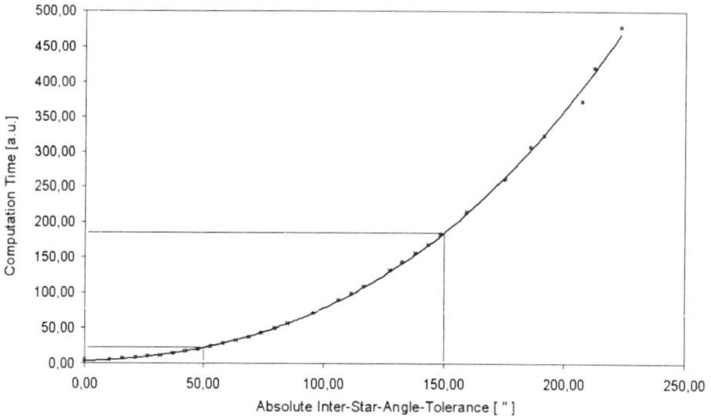

Figure 10: Computation Time as a Function of Tolerance

Earth Center

While an optical way of determining the Earth's center seems to be straightforward, there are several issues to be taken into account when coming up with an appropriate algorithm. First of all, the rim itself is not as easily identified as it might be expected: the terminator, i.e. the day/ night- division, is to be taken into account and, by far more disturbing, cloud constellations have similar local properties as the Earth's rim. So instead of relying on local criterions alone, it is necessary to introduce global criterions. Since the Earth is approximately spherical, the global criterion used is that "enough" of the observed points will be located on a circle within a specified tolerance. The local criterions rely on various features: first of all a threshold must be passed. Secondly, the ratio between "light" and "dark" points in a field around the point examined must satisfy specified conditions. Then the centers of the dark and light areas must be separate enough to ensure distinct regions. Now, since the threshold needs to be adaptive in order to take into account the darker regions where the sunlight is nearly parallel to the Earth's rim, the field's center needs to be corrected along the line connecting the dark and light centers. Thereafter, the check is repeated as often as necessary, while it was shown, that a single iteration generally is sufficient.

It should be explicitly stated that a fixed threshold dramatically decreases the obtainable accuracy due to the darker regions close to the poles, which results in a "rim" which is then closer to the Earth's center.

Instead of checking all points of the picture taken, only distinct lines and columns are to be examined. This reduces the computational effort while still resulting in a sufficient number of points. As additional feature, once a set of points is determined to be part of the Earth's rim, a second algorithm could search for additional points along the expected rim to enhance reliability and accuracy.

The next step is to use the points which were identified as being part of the Earth's rim to determine the Earth's center. This, again can be done in mainly two ways: In a first approach a circle is fitted to the points to determine an optimal circular fit and its corresponding center. The advantage is that robust analytical algorithms are available for this fit. The disadvantage is that the Earth is not a sphere, but an ellipsoid. This results in a systematic error in the determination of the Earth's center. The algorithm would systematically determine the center to be further to the illuminated side of the Earth. The effect is even worse when the points around the pole are not detected due to too low illumination levels. The two situations are shown in Fig.11.

The figure shows the exaggerated deviation of the estimated Earth center from the true one for a circular fit of measurements along the Earth-rim. Deviation (red) is caused by the Earth's ellipticity and is enhanced (yellow) by observation of points close to the equator.

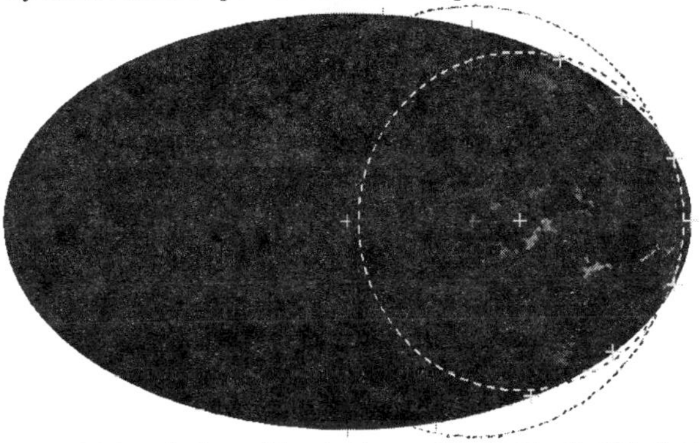

Figure 11: Deviation of Earth Center Due to Earth Ellipticity

In total, as shown in Fig.12, the center will be mis-estimated by approx. ± 0.075 ° (± 10 pixel). While this is a rather high value, it should be pointed out, that the values could be corrected, once a deterministic model is established for the variation throughout the year. Since the data is available, it would simply be needed to be adapted to the algorithm used.

Since the illumination conditions show a daily and yearly variation, the resultant estimate of the Earth's center tends to change. The abscissa is the picture number, where the first and the last picture are taken around midnight, GMT.

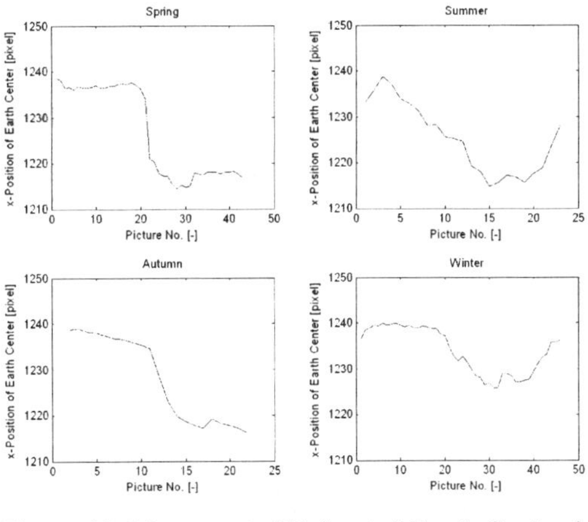

Figure 12: Movement of Estimated Earth Center for Varying Illumination Conditions Throughout a Day's Cycle.

Another, more general idea, would be a parametric fit of an ellipse. The problem associated with that approach is that only a general fit can be solved analytically [Fitzgerald, 1998; Halir, 1998]. A nonlinear optimisation with a cost-function punishing excentricity for a more robust solution can only be solved using iterative methods. Additionally, there are more than one cost function which can be solved for: e.g. algebraic or geometric fits. As an example, an elaborate study on the optimal fitting of measurements to an ellipse has been carried out by the ETH-Zurich, Switzerland [Gander, 1994].

After comparing the circular and elliptical approaches and taking into account today's computational power, it seemed to be more beneficial to use a "semi-analytical" approach, starting with a circular fit and correcting the Earth's center by knowledge of the distribution of points along the rim (e.g. Fig.12).

It should be mentioned, that the approach, as it exploits geometrical features, solves for the geometrical center. This might differ from the gravitational center. In case of orbits around the Moon, for example, an additional correction needs to be performed to correct the position with respect to the gravitational center in order to perform autonomous orbit determination.

The use of Earth sensors in the optical spectrum is further investigated by current research projects on similar topics, like the MOSAIC-Earth sensor [Beard, 1994] and the Sun-Earth-Attitude-Sensor [Mortari, 1999].

Testing

To test the feasibility of such a sensor, several simulations and experiments were performed: The experiments can be classified as laboratory experiments, field experiments and in-orbit tests.

A software has been developed using the Matlab environment which copes with high inaccuracies of star measurements (approx. 300''). This software has been tested on various star-patterns using real star images taken with different cameras. The use of real star data verified the simulation as well as the identification of star fields. The used images were mostly taken by digital cameras with APS- technology. The APS-technology has several advantages compared to the commonly used 'Charge Coupled Devices' (CCDs) and is currently increasingly tested for their use in space applications. A summary o the advantages and disadvantages of APS-detectors is given in Table 2 in the Appendix.

The main features of the three digital cameras used are given in Table 3 in the Appendix. It contains the data for the camera flown on the technological research satellite, MITA, as well as that of two cameras used later-on in field-experiments in the Bavarian Alps and the on-going laboratory experiments. The two newer cameras with a higher resolution and slightly different pixel sizes were both successfully tested in the Bavarian Alps (see Fig.13).

The latest model of the camera's used is shown. Read-out is simplified compared to the older models due to a FireWire connection, allowing laptop use.

Figure 13: Fieldexperiment

The gathered star images were identified off-line, starting with a prelimiary effective focus-length determined by the specifications provided by the manufacturer and a rather high tolerance of about 300" as centroiding-accuracy of the captured stars. The field-experiments provided the necessary star data to validate the star identification scheme, as well as the star-finding-algorithm which extracts stars from the picture. The advantage as opposed to simulated pictures is the actual size of the stars with their real distribution on the pixel field. The drawback is the atmosphere's influence, which leads to disturbances due to scintillating effects. Additionally it was possible to confirm the estimate of th sensor's sensitivity, while again the atmosphere's influence needs to be taken into account. As on example of the field-experiments' results, a star-identification of Ursa Major's tail is shown in Fig.14.

A closer examination of the pictures taken during that campaign showed, that stars down to a magnitude of approx. $m_V = 4.6$ can be detected. Taking into account the atmospheric attenuation and possible cloud coverage in the upper atmosphere, this is sufficiently close to the expected $m_V = 5$.

The picture was taken on the "Maxl-Rainer-Alm" in the Bavarian Alps using the Fillfactory IBIS4b camera. Exposure time was 100 ms. To enhance the visual reception, the intensity was enhanced and the picture inverted.

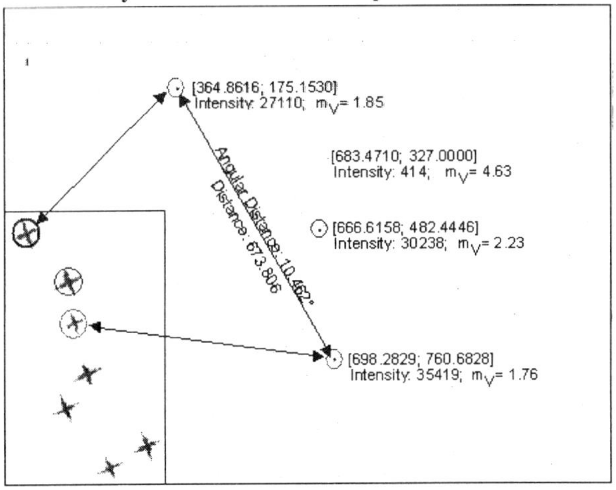

Figure 14: Star Identification on Field-Experiment Picture of Ursa Major

Two approaches were used for the confirmation of a correct star identification:

The first approach was a simulation of the expected image resulting from an identification of the given observation using the 5 brightest stars only. Using this identification the camera's inertial attitude was determined. The attitude in turn was used to simulate the corresponding star field, which was then superimposed on the original picture. Now the stars not used in the identification process can be compared with those which were predicted. This is a very successful technique, as long as enough stars are visible. The advantage to using the full set of stars present is reduction of computation time.

The second approach was using the knowledge of the captured star constellation as an approximate knowledge of the camera orientation. Even though the constellation is in general not fully covered by the camera's field of view (FOV), this knowledge can serve as a reassuring final check.

Using this identification it was then possible to determine several camera parameters which allowed for a higher precision in the centroiding algorithms. The proposed calibration scheme currently is a single-frame algorithm, needing six star separation angles, which is equivalent to a set of fou known stars. The obvious advantage is twofold: Using the preliminary camera-parameters, it is possible to implement an autonomous on-board-calibration of the camera, which can be updated every given time span. This is especially helpful for the star identification algorithm, since it increases the centroiding accuracy which in turn reduces computation time. As an obvious by-product, the number of possible ambiguities is reduced by better knowledge of the measured separation angles.

The laboratory setup consists of an Opto-Electronical Star-Simulator and a lit sphere as Earth-substitute for optical stimulation. The OESS is controlled by "OESS Control"-unit which receives its commands from "Main Computer". The "Main Computer" also performs the processing of the data obtained by the APS camera.

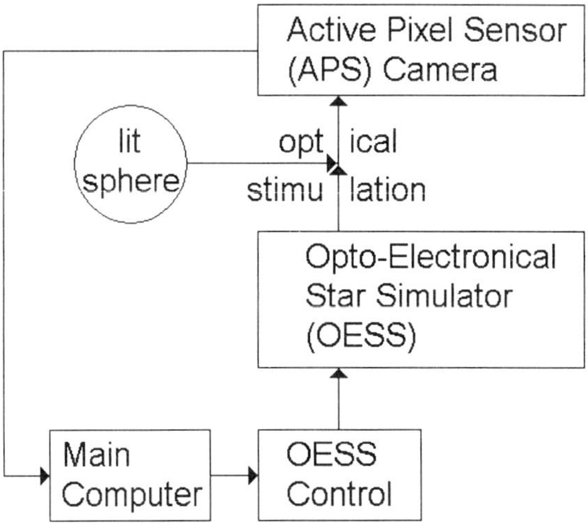

Figure 15: Flowchart of Laboratory Setup

In addition to these experiments, a laboratory setup was built. The laboratory setup, as it is currently configured, is meant for being used to prove the system's functionality. It consists of an Opto-Electronical Star-Simulator (OESS) and an Earth-simulation consisting of a sphere lit by a spot beam with an exit diameter at the source, which is slightly larger than the sphere, as is the case with the sunlit Earth. The control of the OESS is performed by a computer fed with star positions calculated by the main computer, which as well performs the processing of the gathered data. A flowchart of the experimental setup is shown in Fig.15, an actual picture in Fig.16.

The laboratory setup consists of an Opto-Electronical Star-Simulator, a partially transmissive mirror, a sphere as Earth simulation and an APS-camera.

Figure 16: Laboratory Setup

An example of a picture taken by the setup in Fig.16 is shown in Fig.17. Using the setup, it is possible to change the star-pattern according to a simulated orbit. The illumination of the Earth, as well as its position is fixed to simplify the experimental conditions.

Using the laboratory setup it is possible to simultaneously generate observations for two lines of sight. These are then combined using a partially transmissive mirror to be received by a single detector. For visualization the stars have been marked by circles. As in the real case, stars which coincide with the Earth can rarely be detected.

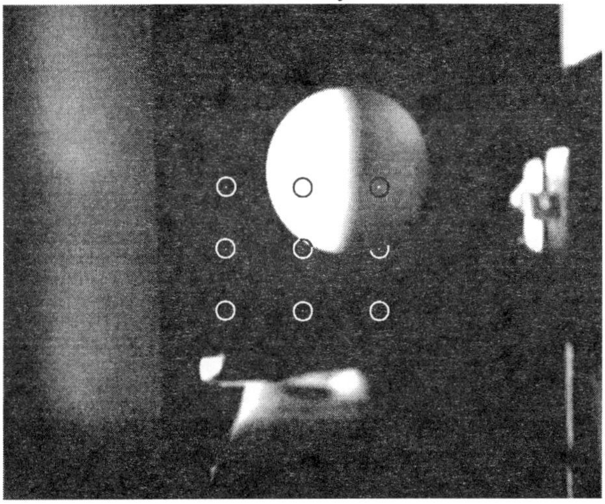

Figure 17: Picture Taken With the APS Camera.
It Shows the Simultaneously Visible Stars- and Earth as Supplied by the Experimental Setup

The On-Board experiment on MITA was designed to investigate micro-system-technology (MST in orbit. The satellite can be seen in Fig.18 the experimental prototype in Fig.19 . It was most valuable by providing data which confirmed the feasibility of the project. Furthermore it showed necessary improvements to guarantee optimal performance.

In the experiment functions of different micro systems technology devices are combined in such a way that they complement each other to form a multifunctional sensor for an attitude and orbit measurement system.

The sensor contains three MST components:
(i) CMOS Active Pixel Sensor,
(ii) Magnetic Field Sensor
and (iii) Angular Rate Sensor.

The CMOS Active Pixel Sensor (APS), by IMEC/ Fillfactory (Leuven, Belgium), is part of a camera in Fig.19 which is used in this experiment in a so-called 'sextant-mode'. This means the star fiel and the Earth horizon are imaged simultaneously on the sensor via a beamsplitter. This in turn is the key to autonomous attitude and orbit determination. To further reduce complexity and mission costs, "off-the-shelf" optics were used.

The Magnetic Field Sensor (MFS), by FhG IMS (Dresden, Germany), measures the Earth magnetic field in two axes.

The moving mass Angular Rate Sensor (ARS), by Bosch (Stuttgart, Germany), is an all-in-silicon sensor with integrated electronics. It measures the angular rate around one axis.

The results concerning the two latter sensors will not be addressed in this publication. Further information on the scope of the experiment can be found in [Melf, 2000].

During the lifetime of the satellite valuable data was gathered, which was post-processed on ground. Its most important message was, that simultaneous observation of the Earth and stars is possible on a single focal plane. Successful star identification and Earth-horizon detection was carried out using the obtained pictures (as an example, see Fig.20).

MITA (Microsatellite Italiano a Technologia Avanzata)
Mass: 170 kg
Dimensions: 1450:400:620 mm (w/o solar array)
Orbit: 450 km, 87.3°
Attitude Control: 3-axis, ±1°
Mission:
Start: Jul. 15th 2000
End: Aug. 15th 2001

Figure 18: MITA

The experimental prototype as flown on MITA contains three MST components:
(i) CMOS Active Pixel Sensor,
(ii) Magnetic Field Sensor
and (iii) Angular Rate Sensor.

Figure 19: Experimental Prototype of MTS-AOMS-Sensor

It was shown that, as with current star trackers, the sun presence in the sensor's FOV should be avoided. The positive result, though, is that the sun presence does not seem to be fatal for the detector. This result is taken from the fact, that during acquisition the sensor was pointed in direction of the su (see Fig.21).

In Fig.20 the Earth can be seen as observed by the sensor. For visualisation, one of the stars present in the picture has been extracted and adapted in intensity.

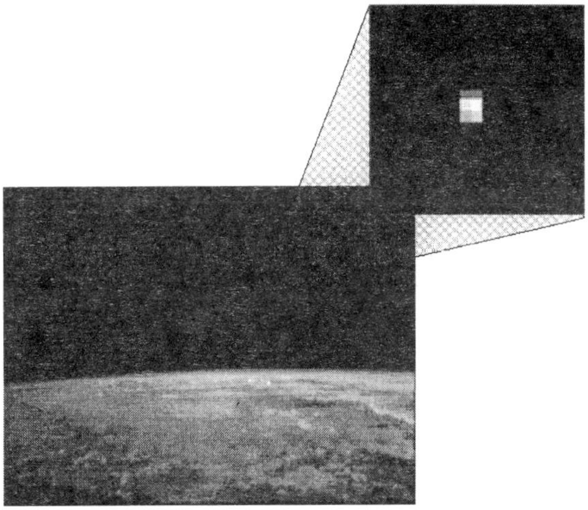

Figure 20: Picture of Earth Taken by the MTS-AOMS-Experiment

In Fig.21 the sun was detected usig the "star-sensor" entrance of the proposed sensor. The sensor survived sun-presence without damage, while during exposure an effect, assumed to be an effect of the readout register, inverted the picture.

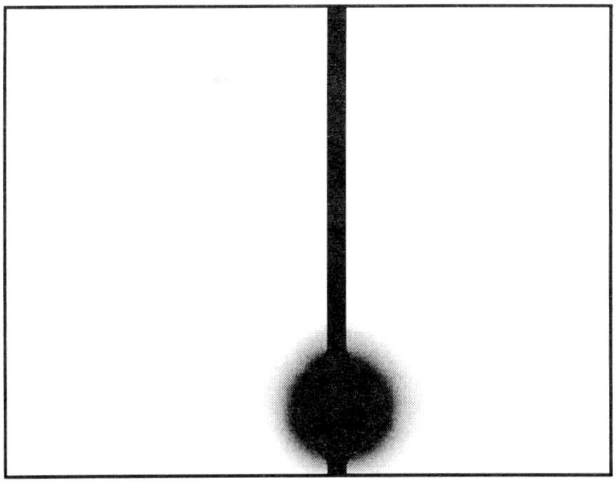

Figure 21: Sun in FOV of "Star Sensor"

The duration of sun-exposure in this case lasted approximately throughout the initial sun-pointing manoeuvre. Thus, the corresponding attitude was maintained for several days. Similar, shorter, occurrences were associated with the Earth-pointing mode in a nearly sun-synchronous orbit and lasted about three minutes. Similarly, the sun-presence in the Earth-sensing direction (attenuated FOV), leads to outages due to overexposure.

Recovery time until new pictures can be evaluated is approx. ten read-out-cycles of the detector, which in normal mode would result in approx. one second.

In the current mechanical sensor prototype, no baffle was being used in the Earth sensing direction. The reason was the integration as a "secondary payload", which resulted in stringent constraints on the sensor's dimensions and therefore a necessary reduction of size. This in turn resulted in th pictures as seen in Fig. 22. Improvement in sensor availability will most-probably arise once an adequate baffle and filter design is applied.

Fig.22 shows the sun at the fringe of the "Earth-FOV". The halo's would still allow an Earth-detection but prevent any attitude information since no stars can be extracted. In this case a baffle would be beneficial.

Figure 22: Sun at the Fringe of "Earth-FOV"

CONCLUSION

The major message from our work and in particular the experimental part is, that autonomous attitude and position determination is possible in the visible range by combination of a star and an Earth-sensor in a single instrument. The core of the instrument is a partially transmissive mirror (beam splitter) by which light from stars and Earth from perpendicular directions are combined on a single focal plane detector.

The sensor can be used on Earth oriented satellites for a wide range of orbit altitudes. Although it recovers quickly after sun exposure, shutters and filters are presently under investigation to prevent incident sunlight from blinding the system e.g. around eclipse times and to facilitate preflight and on-orbit calibration. An advantage of the system is that three-axis-attitude-determination is maintained even during eclipse, while only the Earth-reference needs to be propagated.

The expected accuracies at the current point are:

Inertial Attitude	7.5 "	(1σ) (pitch/yaw)		
	36"	(1σ) (roll)		
Earth-Reference	0.1°	(1σ)		
Position	Point Solution:		O[100km]	(1σ)
	Navigation Solution:		O[10km]	(1σ)

Even though the autonomous attitude and orbit determination using optical sensors might not reach the position accuracy of other systems, e.g. GPS, its advantage is obvious: The system is completely independent of any infrastructure and orbit and can be used with any target planet by a mere change of software.

REFERENCES

1. C. Beard and G. Hayhurst, "The MOSAIC Earth Sensor – A Novel Sensor Development For Geostationary Spacecraft", 2nd ESA International Conference Spacecraft Guidance, Navigation and Control Systems, Noordwijk, The Netherlands, 1994.
2. G. Falbel, "Three Axis Earth/ Star Sensor", United States Patent 5,189,295, 1993.
3. W. Fichter, A. Konrad, P. A. Krauss, M. Mittnacht, P. Ruhland, T. Vogel, "The MosaicAODS Technology for Autonomous Spacecraft GNC System", GNC Conference, ESTEC, Noordwijk, The Netherlands, 2001.
4. Fitzgibbon, M. Pilu, R.B. Fisher, "Direct Least Square Fitting Of Ellipses", IEEE Transactions on Pattern Analysis and Machine Intelligence, May 1999 (Vol. 21, No. 5), pp. 476-480, 1998.
5. W. Gander, G.H. Golub, R. Strebel, „Fitting of Circles and Ellipses Least Square Solution",Technical Report TR-217, ETH Zurich, Switzerland, 1994.
6. R. Halir, J. Flusser, „Numerically Stable Direct Least Squares Fitting Of Ellipses", Proceedings of the 6th International Conference in Central Europe on Computer Graphics and Visualization. WSCG '98. CZ, Plzeò 1998, pp. 125-132, 1998.
7. O. Juckenhöfel, "Einsatz Richtungsmessender Sensoren In der Autonomen Bahnregelun Geostationärer Satelliten", Ph.D.-Thesis, University of Stuttgart, Germany, 2001.
8. M. Melf, S. Manhart, "Micro Systems Technology In-Orbit Experiment: Microtechsensor for Attitude and Orbit Measurement Systems", 3rd Round Table on Micro/ Nano-Technologies for Space, ESTEC, Noordwijk, The Netherlands, 2000.
   and
   http://www.vdivde-it.de/mst/news/0208/mts-tr-007%20-%20final%20report.pdf
9. D. Mortari and S. Gigli, "Earth-Sun Attitude Sensor: Hardware Design and Ground Test", Advances in the Astronautical Sciences, Vol. 103, Part 3, (AAS 99-438), 1999

APPENDIX

Table 1
ASSUMPTIONS MADE IN CREATION OF ANALYTICAL DATA AS SEEN IN FIG.4

Feature	Value
Integration Time	100 ms
Aperture	10 mm
Bandwitdh	500 nm…900 nm
Resulting Photon Flux	3e5 Photons for m_V=0
Quantum Efficiency	50 %
Fill Factor	70 %
Star Size (smear)	3x5 pixel

Table 2
ADVANTAGES AND DRAWBACKS OF APS-DETECTOR AS COMPARED TO CCD

Advantages	Drawbacks
High dynamic range >60 db	Low fill-factor (~ 70%) → new alternative: α-Silicon
Low cross-talk	
Low energy-consumption	Higher noise than CCD → lower S/N-ratio
Simple electrical connection	
Simple read-out electronics	

Table 3
TECHNICAL DATA OF APS DETECTORS

	IBIS1 Flown on MITA	IBIS4b Fieldexperiment	PixeLink Fieldexperiment
Dye Size [mm]	6.5 x 5.3	24.26	165.38
No. of Pixels	386 x 290	1280 x 1024	1280 x 1024
Pixel Size [μm]	14 x 14	7 x 7	7.5 x 7.5
Man.	Fillfactory	Fillfactory	Opdix

AAS 03-002

PRECISION ATTITUDE DETERMINATION FOR GOES N SATELLITE

Yeong-wei Andy Wu, Rongsheng Ken Li, and Andrew D. Robertson*

Since 1998, Boeing Satellite Systems (BSS) has developed an advanced Stellar Inertial Attitude Determination (SIAD) system for the next-generation Geostationary Operational Environmental Satellites (GOES) we're building for NASA. The goals of the next-generation GOES system are to maintain continuous environmental and storm warning systems with an enhanced ground resolution (1.5 km ground resolution at nadir) and to monitor the Earth's surface and space environmental conditions. To achieve high-resolution Image Navigation and Registration (INR) operations, the next-generation GOES satellites require better than 10 micro radians (3 s) attitude knowledge performance. This requirement is far beyond the capability of any previously existing attitude determination method.

The BSS developed SIAD system uses star measurements provided by Ball Aerospace's CT-602 star trackers (3-for 2 redundancies), spacecraft rates measured by Northrop's Hemispheric Inertia Reference Unit (HIRU), and a 6-states Kalman filter implemented in the Spacecraft Control Processor (SCP) to determine spacecraft 3-axis attitude. It has the advantage of two orders of magnitude improvement in accuracy over existing earth/sun sensor-based attitude determination systems.

To meet this never-before achieved 10 micro radian bus pointing for GOES N through Q, we have created many innovative solutions including: a method to account for the star tracker's non-Gaussian, non-white spatial dependent errors in the Kalman filter design to optimize SIAD performance, a 45-degree star tracker boresight orientation to attenuate star tracker high spatial frequency error, a time-matching technique to minimize attitude error between star tracker based attitude and gyro based attitude introduced during spacecraft slews, use of an equalized star catalog to minimize on-board catalog size while enhancing SIAD performance, and a Kalman filter implementation with weighted measurement noise covariance matrices.

A laboratory test bed has been developed and used to fully validate our SIAD design. The test bed consists of a Ball star tracker flight-like engineering model mounted on an optical bench, a flight HIRU mounted on a 3-axis Contraves Motion Table (CMT), an ADI Real Time System (RTS) hosting the flight software on an emulated Spacecraft Controls Processor (ESCP) and other dynamics modelling software on real-time computers. A high resolution CRT scene simulator mounted on the same optical bench is used to project stars seen by the star tracker as the spacecraft moves in space. A collimator placed between the star tracker and the scene simulator produces far-field stars to be observed by the star tracker. By commanding the gimbal motions, the 3-axis CMT produces the desired spacecraft motions, which introduce rates sensed by the HIRU mounted on the CMT.

In this paper we will describe our SIAD design for the GOES N program and present our ground hardware-in-the-loop tests that fully validate our design.

---

* Boeing Satellite Systems, El Segundo, California 90245.

INTRODUCTION

In January 1998, BSS (then Hughes Space and Communications Company) of El Segundo, CA, was awarded the contract from NASA's Goddard Space Flight Center in Greenbelt, MD to design, manufacture, and launch two Geostationary Operational Environmental Satellites, GOES N and GOES O, with two optional satellites, GOES P and GOES Q. The GOES program is funded and operated by the National Oceanic and Atmospheric Administration (NOAA).

The goals of the GOES program are to:

- maintain continuous, reliable operational [CU1], environmental, and storm warning systems to protect life and property,
- monitor the earth's surface and space environmental conditions,
- introduce improved atmospheric and oceanic observations and data dissemination capabilities,
- develop and provide new and improved applications and products for a wide range of federal agencies, state and local governments, and private users.

To accomplish the GOES mission, Space and Ground segments interconnected as shown in Figure 1 are used [1]. The Space segment is comprised of two spacecraft, one at 75^0 West longitude and the other at 135^0 West longitude to provide a full geographic coverage. The GOES N-Q series spacecraft are three-axis, body-stabilized Boeing 601 spacecraft designed to enhance the capability of the GOES system to continuously observe and measure meteorological phenomena in real time. Each spacecraft carries two primary payload instruments:

- *Imager* - a five-channel instrument that senses radiant energy and reflected solar energy from the earth's surface and atmosphere and produces visible and infrared images of Earth's surface, oceans, cloud cover, and severe storm developments;
- *Sounder* - a 19-channel discrete-filter radiometer, designed to sense emitted thermal energy and reflected solar energy from sampled areas of the Earth's surface and atmosphere to provide data for computing vertical profiles of temperature and moisture, surface and cloud-top temperatures, and ozone distribution,

and performs three major functions:

- *Environmental sensing* - Acquire, process, and disseminate the imaging, sounding, and space environment monitoring data;
- *Data collection* – Collect and interrogate data received from earth surface-based Data Collection Platforms (DCPs) and relay such data to the NOAA Command and Data Acquisition (CDA) stations.
- *Data broadcast* – Broadcast the processed environmental sensor data, relay distress signals from aircraft to search and rescue ground stations, and relay emergency weather information to civil emergency managers.

The Ground segment processes the raw Imager and Sounder data received at the NOAA CDA stations with the other data to provide highly accurate, earth-located, calibrated imagery and sounding data in near real time for retransmission via GOES spacecraft to primary end users located throughout the United States.

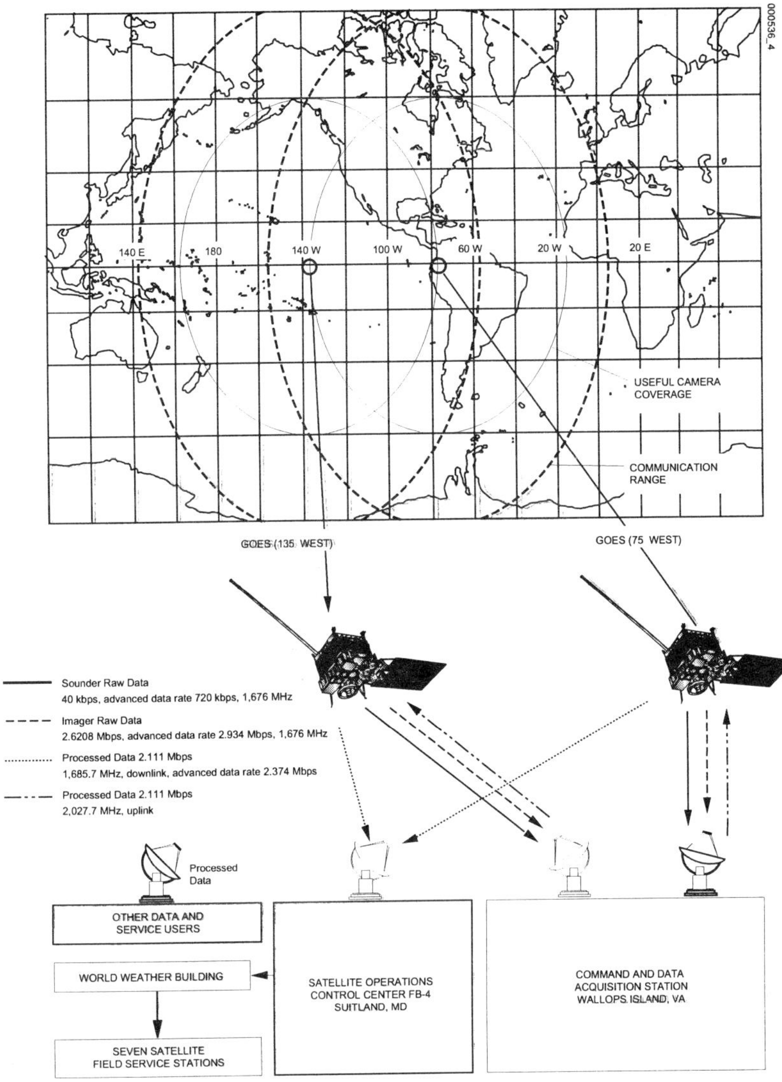

Figure 1 Two nominal GOES satellite stations at 75 and 135 degrees west longitude provide the geographic coverage shown.

Figure 2 shows the GOES N spacecraft in a fully deployed, on-orbit configuration. The Imager and Sounder are colocated with the stellar inertial attitude sensors on a common baseplate supported by structural flexures and attached to the nadir panel. The stellar inertial attitude sensors are comprised of three Ball Aerospace CT 602 star trackers and dual strings Northrop-Grumman's Hemispheric Inertial Reference Unit (HIRU). Reaction wheels are mounted to the aft corners of the spacecraft main body. The single panel solar array provides a clear field of view (FOV) for the Imager and Sounder coolers, maximizing their radiometric performance. The solar array, populated with efficient dual-junction gallium arsenide solar cells, has been optimized to provide ample power while stowed for the spinning transfer orbit and when deployed in the on-orbit configuration. Shown in the figure are also the Solar X-ray Imager (SXI) mounted on an North-South (N-S) gimbaled platform attached to the solar array yoke, and the 8.5 meters magnetometer boom used to sample the Earth's magnetosphere. The SXI, built by Lockheed-Martin, is used to determine when to issue forecasts and alerts of space weather conditions that may interfere with ground and space systems.

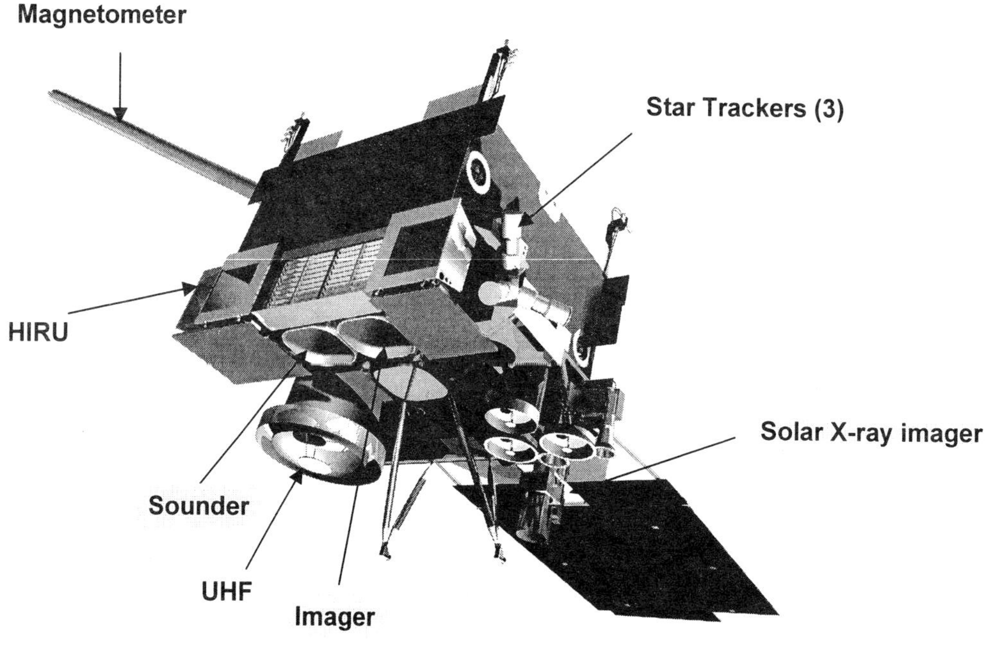

Figure 2 GOES N Spacecraft On-orbit Configuration

Image Navigation and Registration

To continuously obtain precise and clear weather pictures, an Image Navigation and Registration (INR) system providing accurate earth location knowledge and control for each imager and sounder picture element is implemented. Image navigation refers to supplying knowledge of pixel earth locations in earth coordinates (latitude and longitude). Image registration refers to using that knowledge to actively control the pixel earth locations so that they are aligned to a predetermined fixed earth grid [1].

The essential elements of the INR process, originated on GOES I-M [2] are:

- A ground computer that determines orbit and attitude solutions;
- The spacecraft Attitude Control System (ACS) that points the imager and sounder instruments toward earth;
- Instrument (Imager and Sounder) scan mirror servos that scan the imaging detectors across the earth;
- A mirror pointing compensation system in which the ACS generates correction signals to the mirror servos in order to adjust for the effects of orbit and attitude motion.

The basic elements of the INR system design are:

- Orbit and attitude determination is performed periodically in the Spacecraft Support Ground System (SSGS) using spacecraft range and instrument star and landmark measurements. Predictions of long-term orbit and attitude variations are generated in the SSGS and relayed to the spacecraft in parametric form via the Image Motion Compensation (IMC) coefficient set.

- The spacecraft ACS autonomously determines and controls the spacecraft bus attitude to a reference frame aligned to the orbit plane and the nadir direction. The ACS continuously senses bus attitude deviations induced by dynamic disturbances.
- The spacecraft ACS uses the IMC set to generate instrument scan mirror corrections for long-term orbit and attitude variations. It uses the sensed bus attitude deviations to develop mirror corrections for short-term attitude variations (Dynamic Motion Compensation [DMC]). Additionally, based on ground calibration data, the ACS generates mirror corrections for certain instrument fixed pattern errors.
- The instrument scan mirror controllers execute commanded raster scan patterns while continuously applying the scan mirror corrections generated by the ACS.

The INR performance is a function of: instrument servo errors, spacecraft attitude control errors, thermal distortion errors, orbit errors, and scan mirror compensation implementation errors. To support high resolution INR performance (better than 1.5-km (or 42 μrad) ground resolution at nadir), the attitude determination knowledge error (or jitter) shall be less than 10 micro radians, 3 sigma, in both East-West and North-South scan directions. This requirement is far beyond the capability of any previously existing attitude determination method.

SIAD DESIGN DESCRIPTION

Requirements

The SIAD, as one of the key enabling technologies for the GOES N-Q INR design, is used for two purposes: as the attitude determination subsystem for controlling the spacecraft with a pointing accuracy of less than 0.01 [CU2]degree, which clearly is not the driver for SIAD, and as the attitude determination subsystem for controlling the imager and sounder mirrors to dynamically compensate in real time for spacecraft jitters as measured by the SIAD subsystem with an accuracy of 10 micro radians. This technology, named as dynamic motion compensation (DMC), allows a high precision "virtual platform" to be used to support imaging and sounding activities. The requirements levied by the DMC are the primary drivers of SIAD performance.

Design Features

To meet these high performance requirements, the GOES N-Q SIAD design features the following tree of essential design features:

- *SIAD Subsystem Architecture that pushes performance to the limit*

1. 2-for-1 Northrop's HIRU (Hemispherical Inertial Measurement Unit) provide low noise stable inertial reference (3 sigma angle random walk (ARW): 0.001 deg/sqrt(hr), 3 sigma rate random walk (RRW): 0.0036 deg/hour/sqrt(hour)).
2. 3-for-2 Ball CT 602 star-trackers specially tuned to meet GOES's high performance requirement.
3. 45-degree- rotation of star-trackers around its boresight to ensure that the star-tracker high spatial frequency errors are properly attenuated [13].
4. BSS's 702 SCP flight computer with dual 1553 data bus to interface with both the HIRU and the star-trackers.
5. HIRU/Star-trackers and the instruments are mounted on the optical bench with +/-40 micro-radian thermal deformation over life. The optical bench design significantly reduces the

"thrashing effects" due to uncalibrated star-tracker to star-tracker misalignment errors and the variations of the number of stars in the trackers' FOV and their positions.

6. Ground based star-tracker to star-tracker mean misalignment calibration to calibrate out the mean thermal deformation to further reduce the effect of "thrashing" [10].
7. Tight temperature control of the star-tracker environment (+/- 0.5 deg C) to minimize star-tracker temperature dependent errors and boresight shift.
8. Precision time-match between HIRU and Star-tracker data to squeeze the performance to the last micro-radian [12].
9. Directed Field Of View (DFOV) operation of star-tracker and equalized star catalog to facilitate the averaging of the star-tracker low-spatial frequency errors [8,11].

- *Highly optimized Kalman Filter design*

1. 6 state Kalman Filter design optimized against non-Gaussian, non-white star-tracker error [15].
2. Anti-thrashing Kalman Filter design to reduce the "thrashing" effects due to number- of-star variation in the star-tracker FOV [7].
3. Spatial dependent Kalman Filter noise covariance matrix to minimize the star-tracker color dependent error [7].
4. A Simple and efficient method to ensure Kalman Filter numerical stability [9].

Hardware Configuration

Figure 3 shows the hardware configuration of the SIAD subsystem with the HIRU, the CT 602 star trackers, and the Boeing 702 SCP connected by the dual 1553 data bus

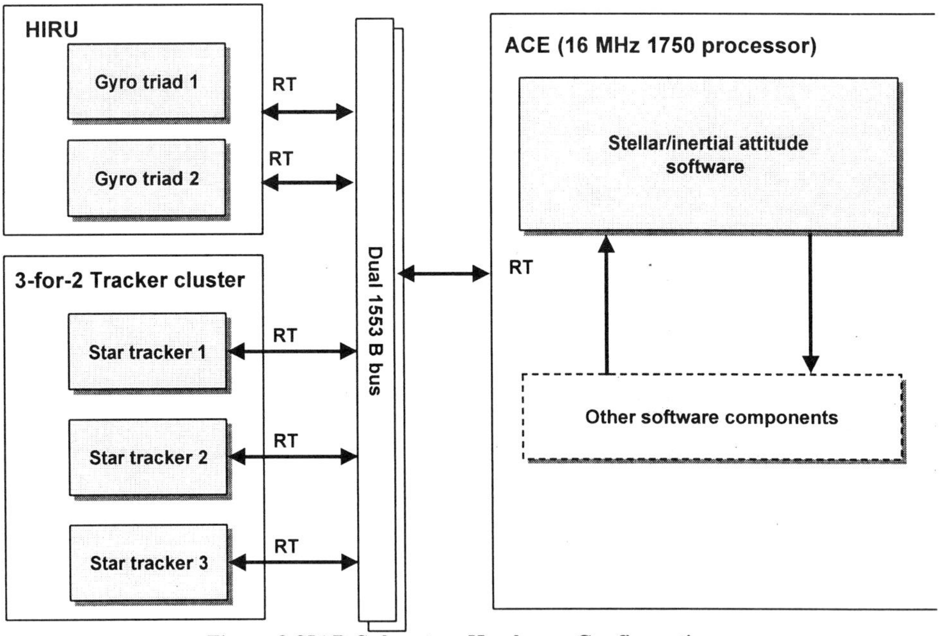

Figure 3 SIAD Subsystem Hardware Configuration

Figure 4 highlights the key characteristics of the IRU.

- **Litton Hemispherical Inertial Reference Unit (HIRU)**
- **Full-time operation**
- **Two redundant sets of 3 orthogonal gyros**
- **With 0.001 deg/sqrt(hr) angle random walk and 0.3 arc sec angle white noise performance (3σ)**

Figure 4 Northrop's (previously Litton's) Hemispherical IRU

The HIRU performance parameters are summarized in the following table.

Additive Errors	Spec (3σ)	Unit
Turn-on bias	2.00	deg/hr
Bias stability	0.015	deg/hr/20 hr
Angle Random Walk (ARW)	0.001	deg/sqrt(hr)
Angle White Noise (AWN)	0.3	arcsec
Multiplicative errors		
Scale factor stability (6 hr)	300	PPM
Scale factor stability (1 year)	1000	PPM
Scale factor stability (Life)	2400	PPM
Scale factor linearity/asymmetry (<1 deg/sec)	300	PPM
Cross-axis multiplicative errors		
Axis alignment stability	175	µrad

Table 1 HIRU performance parameters

Figure 5 summarizes some of the characteristics of Ball CT 602 star tracker for GOES N-Q and Figure 6 is the detailed end-of-life (EOL) error budget (per star position accuracy) for the Ball CT 602.

- Features:
  - 8° x 8° FOV,
  - SI502 CCD, 512x512 full frame
  - 8 Hz update with 4 alternating frame averaging
  - Sensitivity +1 to +6.2 EOL Mi
  - 5 stars tracked simultaneously
  - 30° Sun exclusion angle
  - 25° Earth exclusion angle
  - 25° Moon exclusion angle
- Weight: 13.5 lbs with light shade
- Power: 10 W max, 9.5 W nominal
- Dimensions: 17.8" by 7.6"∅

Figure 5 CT602 Star-tracker for GOES N-Q

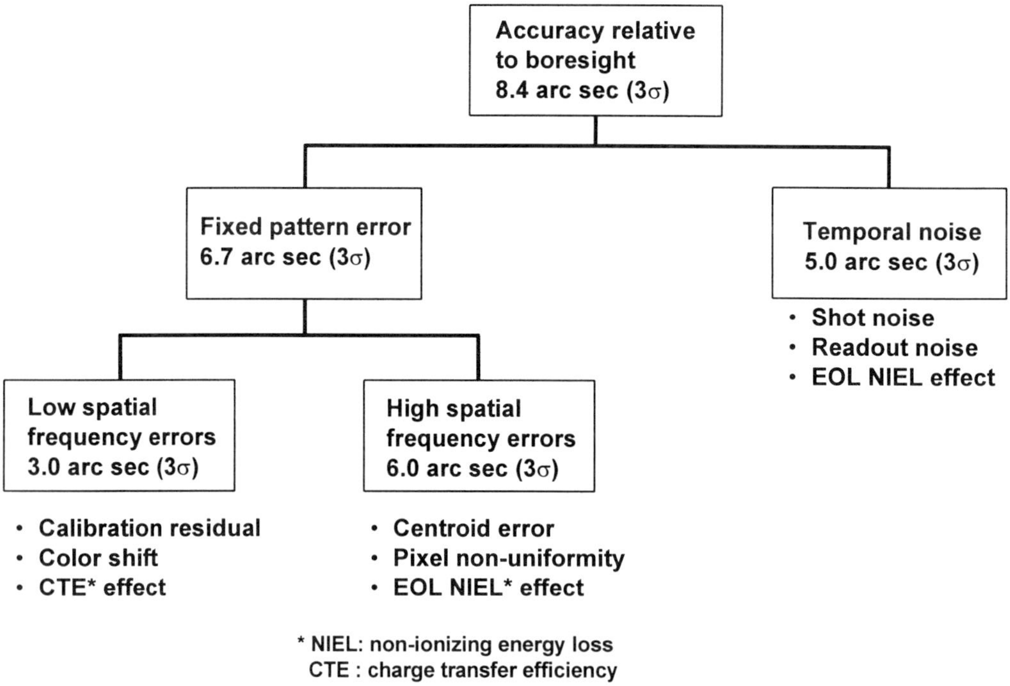

Figure 6 CT 602 Error budget (per star position accuracy)

Software/Algorithm Configuration and Mechanization

Figure 7 shows the software configuration of the SIAD subsystem. The software is partitioned into two "paths", the "inertial path" and the "star-tracker update path". A time-matched buffer is used to ensure that whenever the HIRU derived attitude is compared with the star tracker derived attitude, they are done in a time-matched fashion.

The "inertial path" consists of the HSP (HIRU Signal Processing) CSU (Computer System Unit) and ATD (ATtitude Determination) CSU. The HSP CSU is responsible for requesting and receiving the total angle and time tag from the HIRU. The ATD CSU is responsible for differentiating the total angle into delta angles and integrates them into 3-axis attitude. The ATD CSU also owns a the time-matching buffer (TMB), the attitude and rate states are saved into the TMB every RTI (real time interrupt, approximately 30 Hz)

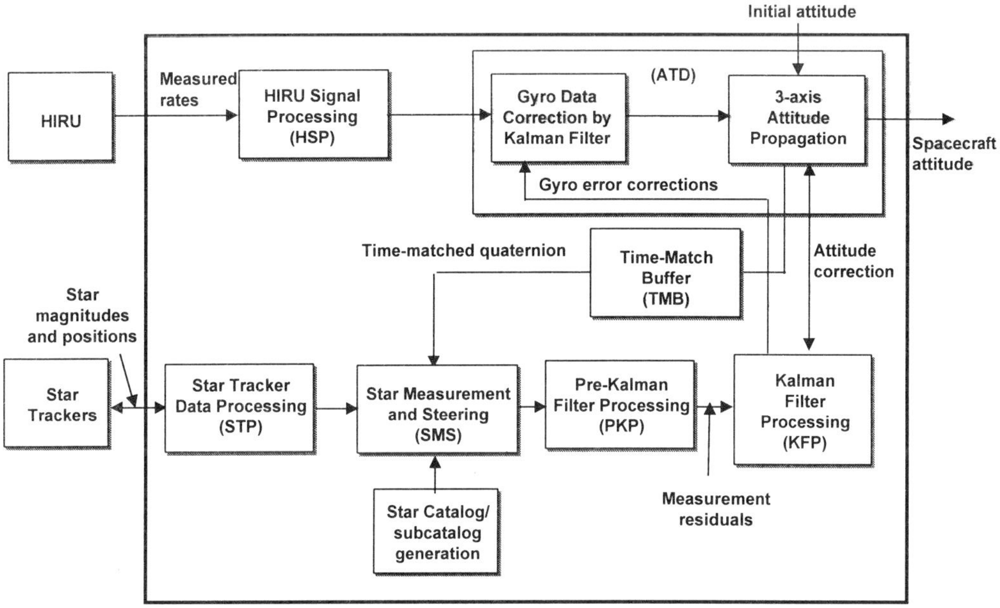

Figure 7 GOES N-Q SIAD subsystem software architecture

The "star-tracker update path" consists of a number of CSUs. The STP (Star Tracker Processing) CSU requests and receives star-tracker data and performs some basic corrections and transformations. The SMS (Star Measurement and Steering) CSUs does the following:

1. It computes predicted star positions using data from the onboard star catalog and the time-matched attitude from ATD retrieved using the star data time tag received from STP,
2. It computes the star measurement residual by differencing the measured star position and the time-matched predicted star position,
3. It identifies the stars by checking their position residuals and star magnitude residuals,
4. It determines which stars shall be tracked based on current attitude and the equalized star catalog

The PKP (Pre-Kalman Filter Processing) CSU corrects the measurement residuals produced by SMS for aberration corrections and also forms the "H_l" matrix (the matrix that maps the attitude error in ECI frames to star position error in star-tracker frame) for each star. The KFP (Kalman

Filter Processing) CSU maintains the Kalman Filter covariance matrices by propagating and updating. It computes the gain matrices for attitude error correction and HIRU bias estimate correction. It also multiplies the gains with the measurement residuals and produces the corrections for attitude and the HIRU bias. The corrections are sent to ATD CSU where they are applied and used.

The Kalman Filter only has 6 states, 3 attitude error states in ECI frame and 3 HIRU bias states in the body frame. Since GOES is a GEO orbit normal steered spacecraft, the body rate is nearly constant when performance is required. The HIRU scale factor and misalignment errors are, consequently, lumped together into the "compound" HIRU bias states. The Kalman Filter is constructed based on the following state transition and measurement equations [16]:

Error Propagation Model:

$$\frac{d}{dt}\begin{bmatrix}\delta\theta \\ \delta b\end{bmatrix} = \begin{bmatrix}0 & C_B^{ECI} \\ 0 & 0\end{bmatrix}\begin{bmatrix}\delta\theta \\ \delta b\end{bmatrix} + \begin{bmatrix}C_{ECI}^{B} ARW \\ RRW\end{bmatrix}$$

White Noise Vector that generates Angle Random Walk

White Noise Vector that generates Rate Random Walk

Measurement Equation for a single star:

$$y = \begin{bmatrix}\delta\theta_h \\ \delta\theta_v\end{bmatrix} = \begin{bmatrix}H_1 & 0_{2x3}\end{bmatrix}\begin{bmatrix}\delta\theta \\ \delta b\end{bmatrix} + V$$

$$H_1 = \begin{bmatrix}\frac{1}{r_{ST}(3)} & 0 & -\frac{r_{ST}(1)}{r_{ST}^2(3)} \\ 0 & \frac{1}{r_{ST}(3)} & -\frac{r_{ST}(2)}{r_{ST}^2(3)}\end{bmatrix} C_B^{ST} C_{ECI}^{B}\left[r_{ECI}\times\right]$$

Geometric Mapping from ECI Frame Attitude Error to Tracker Frame Star Position Error

The KFP CSU has two dispatched tasks. One is dispatched in the foreground at 1 RTI computing the integration of time-varying matrices that are used for covariance propagation. The other is dispatched in the mid-ground once a second performing covariance propagation, gain computation, covariance update and correction computation.

Performance Optimization against Star-Tracker Errors

One of the issues of SIAD design is that the star-tracker errors are not white noises. In fact star tracker errors can typically be classified into:

- Low Spatial Frequency Errors (errors that are dependent on star positions in the FOV with correlation distance on the order of the field of view size),
- High Spatial Frequency Errors (errors that are dependent on star positions in the FOV changing 1 cycle per pixel),
- Temporal Noises (errors that change randomly with time, close to being white noises)

The SIAD performance optimization is based on the following strategy:

1. The star-tracker is rotated around its boresight for 45 degrees so that stars will be moving across the FOV diagonally, exciting the high spatial frequency error along both the horizontal and vertical axes of the star-tracker [13]. As the result, the high spatial frequency error appears in the time domain an error with frequency around 0.1 Hz, significantly higher than low spatial frequency errors that appears to have cycles with period on the order of 30-50 minutes.
2. Similarly, the temporal noise also appears much higher bandwidth than the low spatial frequency errors.
3. As a result, a Kalman Filter that has the capability to attenuate to a certain degree the low spatial frequency errors will be able to attenuate the high spatial and temporal noise to a level that the residual error can almost be neglected.
4. An important measure we have taken to reduce the low spatial frequency error contribution is to use an equalized star catalog [3] and DFOV mode. This allows that the stars commanded, tracked and identified be distributed in the field of view in an equalized manner. Consequently, when the stars are used for attitude determination, the low spatial frequency errors of stars are more likely to be averaged out due to spatial de-correlation.
5. To push the performance to the limit, it is also recognized that part of the low spatial frequency errors are color-induced errors which are proportional to the field angle of the star. A weighting function that is a function of the field angle is used as part of the weighting matrix [7].
6. With the above considerations, an optimization process known as "r-tuning" is formed. The "r-tuning" changes the parameters associated with the Kalman Filter "R" matrix and optimize for the best performance [7].

The "r-tuning" was done in two steps. The first step uses a Matlab based Kalman Filter covariance model. In this model, the low spatial and high spatial frequency errors are modeled as colored noises with parameters representing their time domain signature in GEO orbit. The temporal noises are simply modeled as white noises. The second step uses the high fidelity non-linear simulation to validate the result of the first step optimization and adjust the model parameters if necessary based on the validation result.

Anti-Thrashing Design

The most demanding requirement is actually the 15-minutes registration requirement. To meet this requirement, one needs to minimize fluctuations of SIAD error over time. Apart from the star-tracker errors and HIRU errors, another major contributor to the fluctuation of the SIAD error is the coupling between the star-tracker-to-star-tracker (ST/ST) alignment error and the variation of number of stars observed in each tracker.

To reduce the contribution of the thrashing errors, three measures are taken:

1. The optical bench ensures small thermal deformation between star-trackers.
2. Ground based ST/ST calibration ensures the ST/ST alignment variations are centered around zero, and consequently reduces the dynamic range by half.
3. Anti-thrashing features built into the Kalman Filter [7].

STELLAR INERTIAL TEST BED DESCRIPTION

Figure 8 shows the Stellar Inertial Test Bed (SITB) developed at BSS. The SITB includes a Star Scene Generator (SSG), a 3-axis Contraves Motion Table (CMT), an ADI Real Time System (RTS), an Emulated Spacecraft Control Processor (ESCP), and Ground Support Software (GSS).

Star Scene Generator. The SSG consists of a high-resolution CRT display system, a collimating lens, and software that generates a star-field according to the current spacecraft attitude and star tracker boresight orientation. The high-resolution CRT, built by CELCO, gives a 4000x4000-resolution display across a 7-inch-diameter flat screen. The light is produced by a P11 red phosphor, which has a decay of 10% within 48 microseconds with a 65-nano-seconds rise time. An electronic unit controls the horizontal and vertical positions of spots projected on the CRT with a 16-bit resolution, which produces a position accuracy of better than 0.9 arc-seconds across the 8x8 degrees FOV. The CRT display system is capable of projecting up to 50 spots at a rate of 40 Hz. The average spot size is about 10 arc-seconds, which is small relative to the star tracker pixel size (~ 60 arc-seconds).

3-axis Contraves Motion Table. The 3-axis CMT is used to generate any desired spacecraft motion. The real-time simulation software controls the CMT so that a HIRU mounted on the CMT measures the desired spacecraft rates; rates due to Earth motion are removed via CMT motion. The CMT has 1 arcsec position accuracy, digital rate stability of 0.1% over 1 degree, and 0.001% over 360 degrees, and 0.36 arcsec of short-term jitter.

ADI Real Time System. The dynamic model of the spacecraft is hosted on an ADI real-time computer system running simulation software. The model includes the dynamics, as well as sensor and actuator models. Of particular interest, the real-time computer includes models of two star trackers and of the HIRU, providing the capability to test the SIAD software with any combination of sensors and sensor models. For the test described in this paper, one star tracker model and the HIRU model plus the actual star tracker mounted on the optical bench provide all SIAD inputs.

ESCP. The ESCP faithfully replicates the functionality of the SCP used on-board the GOES spacecraft. The ESCP contains the spacecraft flight software, including the SIAD algorithm. The ESCP interfaces to the real-time computer (and its sensor and actuator models) via a 1553 bus and associated interface electronics. The SIAD estimates drive a simple controller in the ESCP, generating torque commands, which are sent via the 1553 to the real-time computer's model of the reaction wheel actuator.

Ground Support Software/EPOCH Ground Software. The operator controls both the ESCP and the real-time computer from a workstation. Control of the real-time computer is accomplished via ADI's COSIM interface software. The ESCP interface is through an emulated telemetry and control unit (ETC) which interfaces with two operator interface packages: EPOCH and GSS. The EPOCH software provides a user interface that is roughly similar to the ground software used to communicate with the spacecraft on-orbit. Ground Support Software (GSS) interprets star tracker and SIAD telemetry graphically to assist the operator in visualizing the performance of the star trackers. A screen shot from the GSS software is captured in Figure 9.

SIAD TEST SET UP

The SITB provides the capability to test SIAD flight software with a star tracker unit in the loop. Such hardware-in-the-loop testing is common for star tracker-based attitude determination systems (e.g. see References 4 and 5). In the BSS testbed, a high-fidelity star-scene generator (SSG) coupled with a reference star catalog in the simulation provides a simulated-field of view, which the star tracker stares at through a collimating lens. The collimating lens focuses the light from the SSG so that it appears to the star tracker to have originated a great distance away. All optical components are mounted on an optical bench. While hardware-in-the-loop testing on the ground can not reproduce on-orbit operating conditions exactly, the testbed and the experimental procedure were designed to mimic the important aspects of the expected operating conditions. Experimental results – discussed in the next section – include the effects of testbed-noise sources that will not be present on-orbit and thus show less attitude determination accuracy than the on-orbit system is expected to provide.

The imperfect spacing and alignment of the optical components on the optical bench in the test bed must be accounted for via a calibration step prior to any experiment. Table 2 lists error contributions pre- and post-calibration. Calibration produces a mathematical mapping between desired star locations (as commanded from the simulation) and voltages applied to the SSG (producing the stars on the CRT). The mapping is generated by commanding a grid of stars on the SSG (without a map), measuring them via the star tracker, and then calculating the inverse of the transformation between command and measurement. With the resulting map engaged, star position measurement errors are greatly reduced. The alignment does drift over time, however, such that a map correct at the start of a test will no longer completely remove the errors 24 hours later.

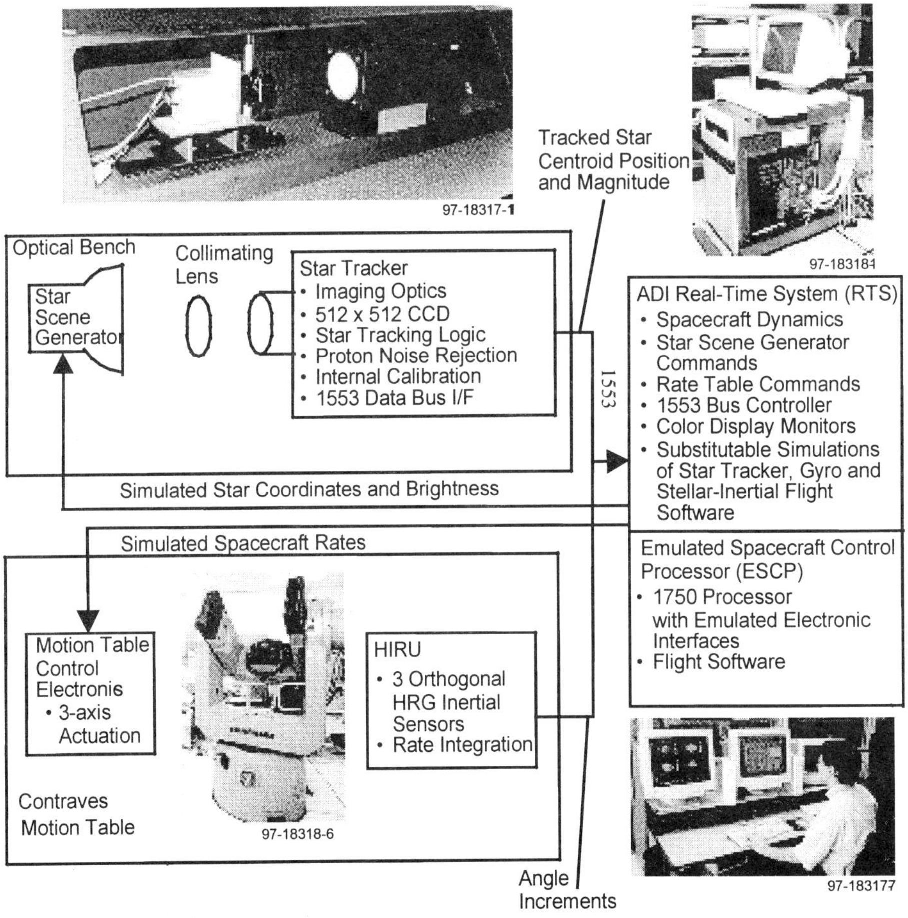

Figure 8. The SIAD testbed provides the capability to test flight software with combinations of simulated and real star tracker and HIRU sensors.

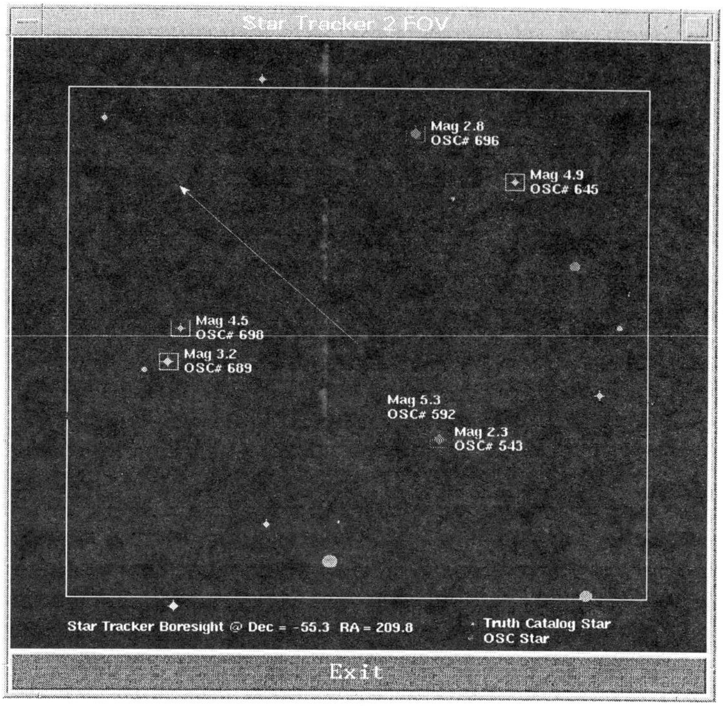

Figure 9. GSS software displays the star tracker's field of view, including tracked stars.

Table 2. Optical test bed errors pre- and post-calibration

Source	Uncorrected Error	Residual Error After Correction
Star Scene Simulator		
Scale Factor	10 asec	< 4 asec
Drift / Repeatability	< 10 asec day-to-day	–
Roll	~ 100 asec	< 4 asec
Distortions across CRT	~ 150 asec	< 4 asec
Effects from Contraves console	5 asec	< 4 asec
Star Scene Display DAC's		
Resolution	1 asec	–
Linearity	< 25 asec	< 1 asec
Analog Noise	< 4 asec	–
Star Tracker		
Mag dependent position errors	< 200 asec	< 4 asec (unimplemented)
High spatial frequency (S-curve)	10 asec	–
Optics		
Alignment	< 50 asec	< 4 asec
Vignetting	< 10 asec < 1 magnitude	< 4 asec < 0.25 mag (projected)

GOES SIAD TEST RESULTS/PERFORMANCE VALIDATION

A 25-hour SIAD validation run was conducted on the SIAD test bed in December 2001. The ESCP was initialized into its nominal on-station configuration: Earth-pointing under reaction wheel control. SIAD performance was monitored during the run using the GSS; simulation data, including attitude estimation errors and SIAD diagnostic data was logged continuously during the run. The test included one Ball CT 602 star tracker (ST1) in the loop pointing to the Northern Hemisphere (55 degrees declination angle), a simulated star tracker (ST3) pointing to the Southern Hemisphere (55 degrees declination angle), and a simulated HIRU providing spacecraft rate measurements. Test bed calibrations were conducted immediately before and after the test; serving to calibrate the test bed initially and to measure how much the optics had changed during the course of the run. Figures 10 (a) and (b) show the SSG to star tracker calibration residuals before and after the test respectively. The results indicate that the test bed introduces an effective star tracker boresight drift of 3.1 arcsec in x-direction and – 4.5 arcsec in y-direction during the 25-hours test. In addition, the effective star tracker low-spatial frequency errors are also increased.

Figure 11 shows the number of stars within the star tracker FOV and the number of stars identified from the flight software implemented in the SCP over the 25-hours run. Figure 12 shows the time responses of spacecraft attitude errors after removing the biases and drifts, which was caused by test bed instability as mentioned previously. The resultant 3-sigma performance based on the 24-hours data (excluding the first hour data from the 25-hours run) is 18.69 micro radians in roll, 22.88 micro radians in pitch, and 15.36 micro radians in yaw. These resultant performances are approximately 3 times larger than the predicted results based on the pure simulation.

Careful analysis has concluded that: (1) the imperfection of the SSG, which caused the star tracker to drop track occasionally on stars projected by the SSG as shown in Figure 11, and (2) 2 to 3 times increase in the effective LSF error for the real tracker are the primary reasons for the observed SIAD performance degradation. Further tests helped us to understand the reasons for dropped star tracks occasionally from the real star tracker, these include: the dim star (dimmer than 5.5) caused star tracker to mark it as a hot-pixel because the star magnitude produced by the SSG was below the star tracker detection threshold, the bright star (brighter than 2.3) was occasionally marked as an invalid data by the star tracker because the blooming of star image on the CRT caused star positions to shift from frame-to-frame, thus triggering the invalid condition, and a non-zero FOV offset caused the real star tracker to lose track earlier compared to the simulated tracker. These effects are all due to the SSG/optical setup of the test bed and won't occur on orbit.

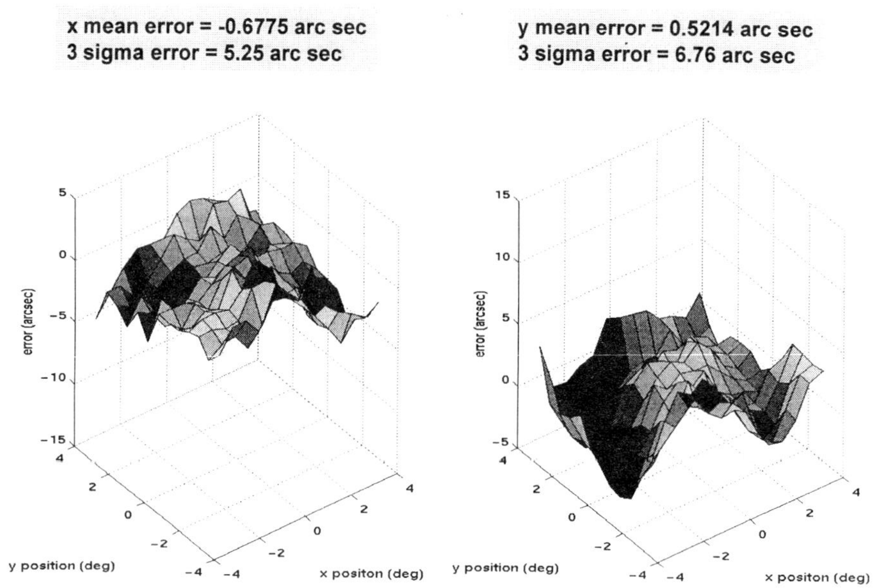

Figure 10 (a) SSG to Star Tracker Calibration Residuals Before the 25-hours Test

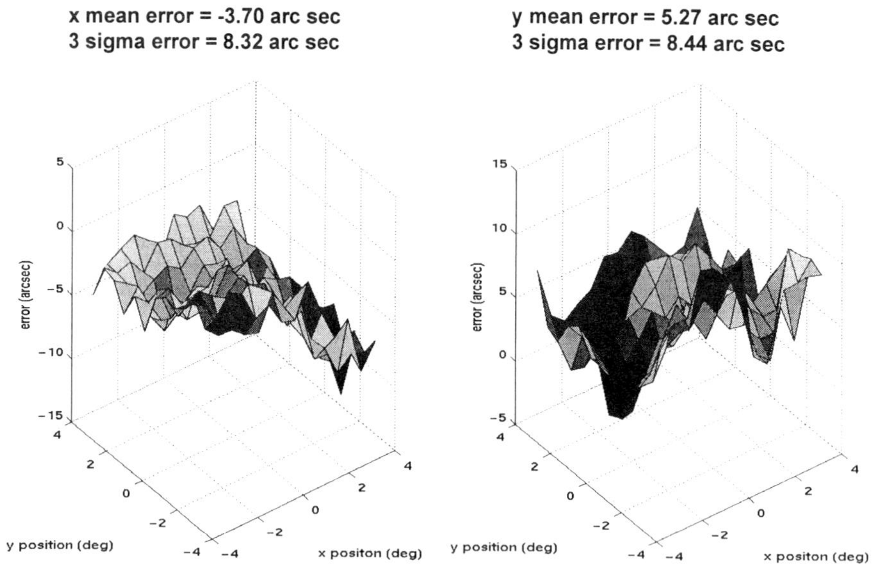

Figure 10 (b) SSG to Star Tracker Calibration Residuals After the 25-hours Test

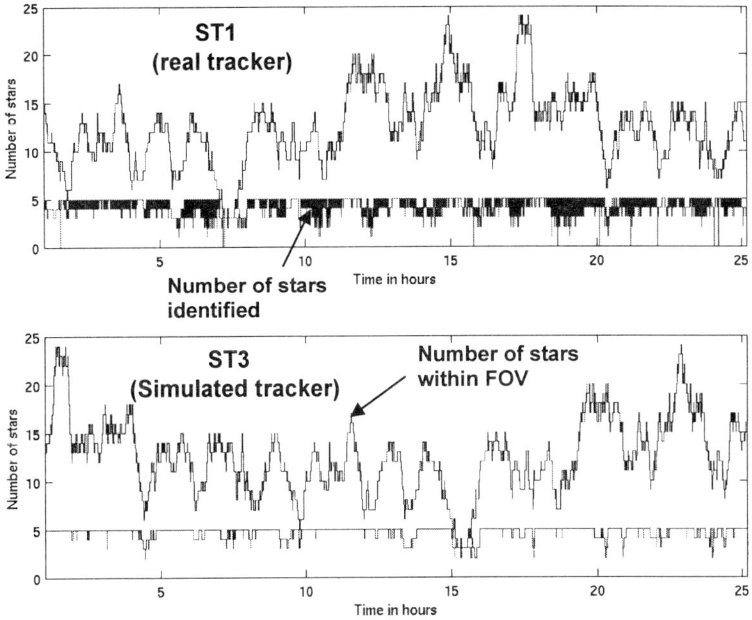

Figure 11 Number of Stars within FOV/Number of Star Identified

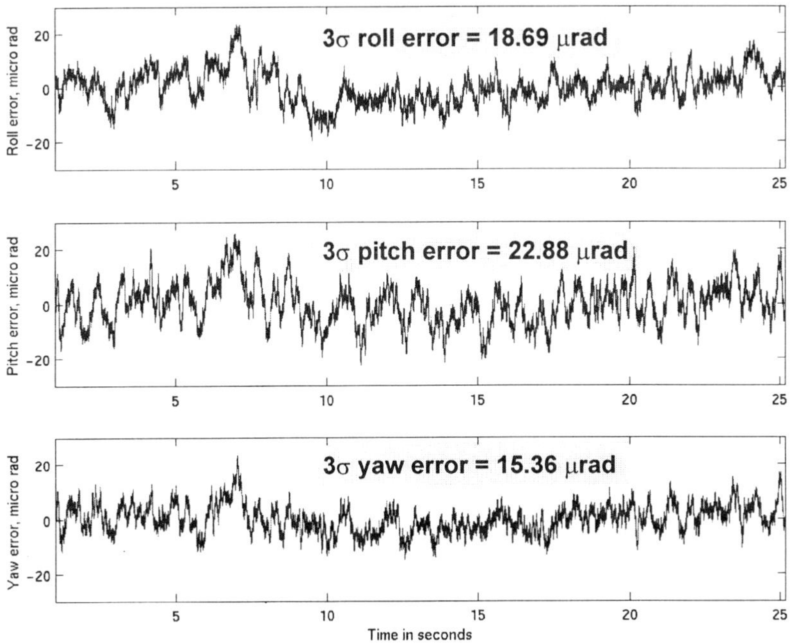

Figure 12 Time Responses of Spacecraft Attitude Errors

CONCLUSIONS

An advanced stellar inertial attitude determination system has been developed for the GOES N spacecraft to enhance the capability of the GOES system to continuously observe and measure meteorological phenomena. Flight qualification testing (FQT) was completed in 2001 for the GOES N spacecraft, which is scheduled to launch in 2004. Design features to meet the stringent 10 micro radian, 3-sigma, attitude determination performance requirement are described in this paper along with a comprehensive stellar inertial test bed that was used to validate the SIAD flight software design. We performed a 25-hour test and analyzed the resultant test data. We demonstrated that the flight-like engineering model star tracker provided stable operation and no glitches were observed over the 25-hours period. We also validated the star tracker model against the real tracker in terms of its interfaces and functionality.

ACKNOWLEDGEMENT

The authors would like to thank Loren Slafer, Doug Hein, David Needelman, Chris Rice, Paul Welton, and John Liu of BSS for their contributions to the GOES N SIAD design and validation, also Landis Markley, Eric Finnegan of NASA's Goddard Space Flight Center, Sandy Ashton, Quang Lam of Swales Aerospace, and Scott Miller of Orbital Sciences Corporation for their valuable discussions during the course of SIAD development at BSS.

REFERENCES

1. Boeing Satellite Systems, "GOES N-Q Data Book", CDRL PM-1-1-03, NASA Goddard Space Flight Center, Greenbelt, Maryland, November 2000.
2. Space Systems/Loral, "GOES I-M Image Navigation and Registration Final Report", SS/L-TR10937A, NASA Goddard Space Flight Center, Greenbelt, Maryland, March 4, 1994.
3. Arun Prakash, Andy Wu, John Liu, and R. Ken Li, "Performance based evaluation of star catalog generation methods," AIAA Guidance, Navigation, and Control Conference, paper #AIAA 2002-4669, 5-8 August 2002, Monterey, California.
4. Giancarlo Rufino and Antonio Moccia, "Laboratory Test System for Performance Evaluation of Advanced Star Sensors", Journal of Guidance, Control and Dynamics, Vol. 25, No 2, March-April 2002.
5. Carl Christian Liebe, "Accuracy Performance of Star Trackers – A Tutorial", IEEE Transactions on Aerospace and Electronic Systems, Vol. 38, No 2, April 2002.
6. US Patent 6,362,464, "System for stellar inertial attitude determination".
7. US Patent 6,356,815, "Stellar attitude-control systems and methods with weighted measurement-noise covariance matrices".
8. US Patent 6,330,988, "System for star catalog equalization to enhance attitude determination".
9. US Patent 6,317,662, "Stable and verifiable state estimation methods and systems with spacecraft applications".
10. "System and method for calibrating inter star-tracker misalignments in a stellar inertial attitude determination system", US pending patent, filed July 26, 2000.
11. US Patent 6,304,822, "Spacecraft attitude control systems with dynamic methods and structures for processing star tracker signals".
12. US Patent 6,285,927, "Spacecraft attitude determination system and Method".
13. US Patent 6,236,939, "Method and apparatus for controlling spacecraft attitude with rotational star trackers".
14. US Patent 6,108,594, "Autonomous attitude acquisition for a stellar inertial attitude determination system".
15. US Patent 6,047,226, "Enhanced stellar attitude determination system".
16. Y. A. Wu and Doug Hein, "Stellar Inertial Attitude Determination for LEO Spacecraft", 35th IEEE Conference on Decision and Control, December 11-13, 1996, pp. 3236-3244.

THE INERTIAL STELLAR COMPASS: A MULTIFUNCTION, LOW POWER, ATTITUDE DETERMINATION TECHNOLOGY BREAKTHROUGH

T. Brady,* S. Buckley*
C. J. Dennehy,† J. Gambino† and A. Maynard†

The Inertial Stellar Compass (ISC) is a miniature, low power, stellar inertial attitude determination system with an accuracy of better than 0.1° (1 sigma) in three axes. The ISC consumes only 3.5 Watts of power and is contained in a 2.5 kg package. With its embedded on-board processor, the ISC provides attitude quaternion information and has Lost-in-Space (LIS) initialization capability. The attitude accuracy and LIS capability are provided by combining a wide field of view Active Pixel Sensor (APS) star camera and Micro-ElectroMechanical System (MEMS) inertial sensor information in an integrated sensor system. The performance and small form factor make the ISC a useful sensor for a wide range of missions. In particular, the ISC represents an enabling, fully integrated, micro-satellite attitude determination system. Other applications include using the ISC as a "single sensor" solution for attitude determination on medium performance spacecraft and as a "bolt on" independent safe-hold sensor or coarse acquisition sensor for many other spacecraft.

NASA's New Millennium Program (NMP) has selected the ISC technology for a Space Technology 6 (ST6) flight validation experiment scheduled for 2004. NMP missions, such as ST6, are intended to validate advanced technologies that have not flown in space in order to reduce the risk associated with their infusion into future NASA missions.

This paper describes the design, operation, and performance of the ISC and outlines the technology validation plan. A number of mission applications for the ISC technology are highlighted, both for the baseline ST6 ISC configuration and more ambitious applications where ISC hardware and software modifications would be required. These applications demonstrate the wide range of Space and Earth Science missions that would benefit from infusion of the ISC technology.

* Systems Engineering and Evaluation Directorate, Charles Stark Draper Laboratory, Inc., 555 Technology Square, Cambridge, Massachusetts 02139.

† Guidance, Navigation, and Control Division, NASA Goddard Space Flight Center (GSFC), Greenbelt, Maryland 20771.

INTRODUCTION

While the attitude needs of various spacecraft differ, the vast majority of them use a combination of sensors to determine how they are oriented in space. Traditionally, the size, weight, power, and integration requirements for many of these attitude determination sensors place a large burden on the spacecraft bus designer and noticeably reduce the amount of payload potential for small and micro-spacecraft. In order to expand the mission capabilities of these small spacecraft, the ability to determine their attitude using a low power and low mass system is of great value.

Draper is developing the ISC to answer the need for a large dynamic range, high accuracy, low power, and real-time attitude determination sensor for micro-spacecraft. The ISC, developed under NASA's New Millennium Program's Space Technology 6 Project, is an enabling technology that utilizes Draper's expertise in MEMS sensor technology, inertial navigation algorithms, and APS imagers. It integrates a wide field-of-view star camera, MEMS gyros, software, and associated processing electronics into a 2.5 kg two-piece attitude sensor that consumes less than 3.5 Watts of power.

The ISC is an innovative attitude determination sensor that combines these emerging MEMS and APS technologies into an integrated package that will be space-validated in 2004. Key ISC performance features include:

- Better than 0.1° (1-sigma) accuracy in each axis
- High-rate maneuver capability (up to 50°/sec)
- Self-initializing (over 99% of the sky)
- Low Mass ~ 2.5 kg
- Low Power ~ 3.5 W

Among the key advantages of the ISC are its ultra-low power, ease of integration with a host spacecraft, and ability to maintain a better than 0.1° accuracy during a high rate (up to 50°/sec) maneuver. As a sensor, the integrated functionality and high rate capability are unique and represent a step forward in spacecraft technology. All of the navigation algorithms and sensor fusion reside within the ISC, thereby eliminating the typical integration of the star camera data with the spacecraft attitude control software. The ISC output, consisting of an Earth Centered Inertial (ECI) attitude quaternion and attitude rates, provides a "bolt on" attitude determination solution.

SYSTEM OVERVIEW

During normal operation, attitude information is provided by the ISC's MEMS gyros. The gyros sense inertial rates that are sampled at a high frequency (320 Hz). The raw gyro data is compensated and processed through a Kalman filter to produce the output quaternion, which is transmitted to the host spacecraft in real time, at a frequency of 5 Hz. The star camera is used periodically (every few minutes) to obtain a camera quaternion that enables the gyro errors to be removed and the inherent drift of the gyros to be calibrated and

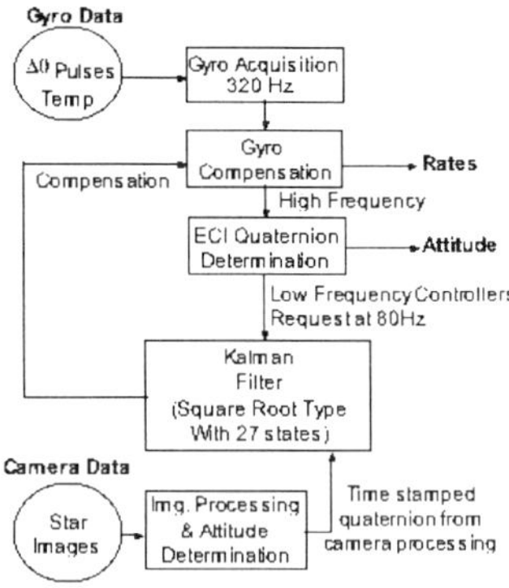

Figure 1: ISC System Data Flow

compensated. A simple system data flow is shown in Figure 1.

The fusion of star camera and MEMS gyro data provides a significant improvement in the performance of both devices. The gyros provide information continuously to the ISC during periods when the spacecraft may be undergoing high slew rates such that the camera cannot obtain good star images (due to unacceptable pixel smearing and periods when the camera is occluded by the Sun, moon, or Earth). The camera provides information for periodic updates to support gyro compensation and system initialization.

To self-initialize the ISC, star field images taken by the ISC's camera electronics are processed to provide a camera quaternion. Stars in the image are identified using a Lost-in-Space attitude determination algorithm developed by Mortari, which analyzes the image against a catalog of 1500 stars to help identify the camera's line of sight without any prior knowledge of the spacecraft's attitude.[1] Once initialized, the gyros are used to maintain attitude knowledge continuously until the next stellar update can be obtained to support gyro compensation. The complementary use of the gyros and camera data help the spacecraft overcome difficulties in providing attitude knowledge during transients, high slew rates (up to 50°/s), or periods of star camera occlusion.

ISC Performance

The ISC maintains an attitude determination accuracy of 0.1° (1 sigma) or better in each axis provided star sightings occur at least once every 150 seconds, the star sightings occur at a host spacecraft body rate <0.25°/sec, the accumulated angle between star sightings is <360° per axis, and the maximum body rate is <50°/sec per axis. The NMP ST6 top-level requirements state that accuracy will be maintained at all times using less than 4.5 Watts of power with an ISC packaged weight of less than 3 kg (excluding mission specific baffle and cabling). Current Best Estimates (CBE) for operational power consumption is 3.5 Watts and 2.5 kg for mass.

An analytical model was constructed to appropriately distribute performance requirements to subassemblies in order to capture major error sources, establish the robustness of the design, and identify the highly sensitive parameters. The overall system accuracy can be met if the camera system can achieve a 75 arcsec accuracy and the gyro subsystem achieves a 300 arcsec accuracy. This apportionment requires the MEMs gyros have an ARW (Angle Random Walk) of less than 0.16 deg/rt-hr, with a bias drift of less than 3.3. deg/hr, and scale factor errors of less than 100 ppm.

Gyro-based attitude accuracy is driven by the ability of the Kalman filter to estimate gyro misalignment, bias instability, and scale factor errors. The stability of the gyro bias in the ISC has been modeled as a first order Markov time-dependent variable with a 45-minute time constant. Gyro test data was reviewed to better estimate the bias stability as both a function of time and temperature. Results indicate that 75% of the bias stability is attributed to

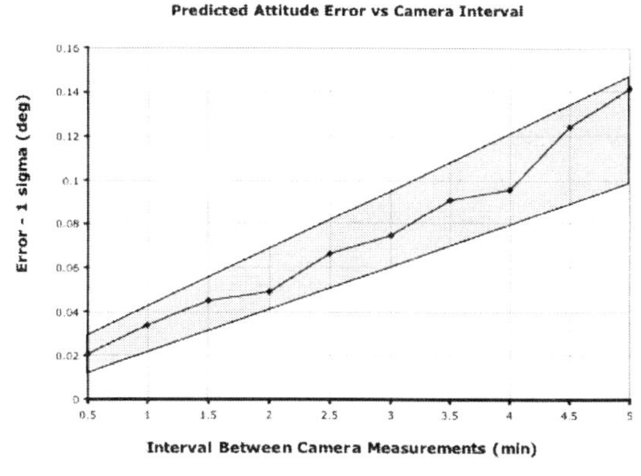

Figure 2: Attitude Error vs. Camera Interval

temperature fluctuations and 25% to the typical time-driven bias instability. The same was done for the gyro scale factor and resulted in a temperature-dependent variation of 75 ppm and a time dependent variation of 25 ppm. An expanded range of gyro thermal transients was investigated given various orbits, solar exposure, and baseplate conductivities. This analysis indicated that the mechanical design provides excellent thermal resistance between the ISC housing and the internal gyro modules. Variations of 0.09 deg-C/min at the camera board and the gyro board occurred given a 10 deg-C/hr rate of change at the ISC baseplate, which satisfies the accuracy requirements.

Camera error sources were quantified and modeled both through analysis and experiment. Pixel level effects such as detector noise, interpolation errors, and pixel smearing were individually modeled and fed into an overall Field-Of-View (FOV) level analysis in the camera model. Lens-detector misalignments, astronomical aberration, line scan aberration, and relative star distributions in a randomly chosen FOV were modeled (and verified through test) to estimate camera accuracy both in typical and worst case situations. Worst case operating conditions included a 40°C focal plane, 0.25°/sec angular slew rate, and the maximum mechanical misalignment tolerance (1/3 pixel in pitch/yaw and 25 microns in focal length). Under these worst-case operating conditions, the camera results predict an accuracy of 55 arc-seconds in roll and 47 arc-seconds in pitch/yaw and estimate the median number of visible stars to be 13. For the worst-case analysis, the error is dominated by the 8 micron mechanical misalignment in the pitch-yaw plane. Typical conditions such as a 20°C focal plane, tight mechanical tolerance (1/6 pixel in pitch/yaw and 10 microns in focal length), and a quasi-steady host spacecraft (0.1°/sec) result in approximately 37 arc-seconds in roll and 18 arc-seconds in pitch/yaw and predict the median number of stars to be 14. In both typical and worst case operating conditions, the percent of trials with greater than 3 stars in the FOV was greater than 99% and satisfied initialization requirements.

ISC camera data is provided as truth to update gyro states within the Kalman filter. The camera update rate is variable and ranges from 30 seconds to 20 minutes. A longer camera update interval results in decreased attitude accuracy. Figure 2 illustrates the predicted accuracies of the ISC as a function of camera update rate. There is approximately a 15% to 20% range in the accuracies, as shown by the overlay, since it is a function of the actual maneuver being performed and the temperature variability at the gyro board. Increased performance can be achieved by shortening the interval between camera updates. At a 30 second update rate, the overall ISC accuracy (~50 arc-seconds in pitch/yaw, ~75 arc-seconds in roll) approaches the accuracy of the camera under typical conditions. Shorter camera update rates (< 60 seconds) increase the software throughput by 25%, the average operational power by 100 mW, and provide for a greater chance of optical interference.

ISC Design

The ISC consists of two separate units (as shown in Figure 3), connected by a cable: the Camera Gyro Assembly (CGA), which contains the sensors, and the Data Processing Assembly (DPA) containing the sensor's embedded computer and power supply electronics. The two-unit design facilitates integration with a host spacecraft. Only the CGA needs to be

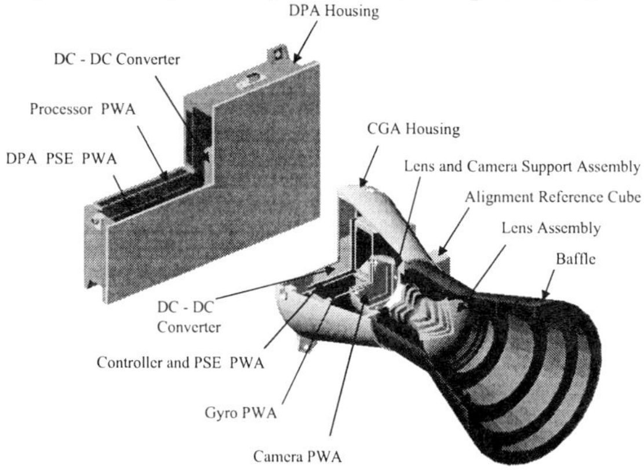

Figure 3: CGA and DPA Cutaway View

precisely aligned with the host spacecraft using the reference cube located on the CGA housing.

The CGA collects raw sensor data upon command and returns the data to the DPA for processing. The CGA and DPA communicate via a serial data link over an interface cable that can extend up to 2 meters. The CGA is 16 cm high (without mission specific baffle) and approximately 17 cm wide at its circular base. CGA mass is estimated at 1.2 kg. The DPA enclosure measures 15 cm x 23 cm x 4 cm with a mass of approximately 1.3 kg.

The CGA includes the lens, mission specific baffle, focal plane, three MEMS single-axis gyros with analog-to-digital conversion electronics, a controller board, and a 28V, triple output, DC/DC converter. The heart of the star camera is a STAR250 512x512 active pixel sensor array with an on-chip 10-bit A/D converter from Fill Factory. Advantages of the APS array over existing CCD technology include good radiation tolerance, no requirement for a thermal electric cooler, no requirement for a high voltage power supply, highly integrated digital control electronics, and low cost standard CMOS fabrication.[2,3] The camera has a command-ready interface to support windowing, various integration times, selectable frame count, and built-in test. The 21° square FOV star camera optic is a custom 35 mm, f1.2 lens manufactured by Zeiss and modified for space flight applications.

The CGA's controller board provides a simple serial interface to the DPA (or any other flight computer) and directs all necessary timing and control needed by the star camera and MEMS gyros. In addition, the controller provides 500 KBytes of SRAM storage to capture a full-frame compressed camera image for transfer to the data processor. Nominally, the image is transferred serially over the 16 MHz interface to the DPA allowing for full 512x512 pixel image transfer at 3.7 Hz. Given a faster interface to a controlling computer, the star camera and controller could support a full 30 frames per second, 512x512 pixel image transfer rate.

Figure 4: ISC focal plane (top); MEMS gyro board (middle); DPA electronics (bottom)

The Draper MEMS gyros sense angular rate by detecting the Coriolis effect on a sense mass. The tiny gyro sensors are etched in silicon using a Draper developed MEMS process. A sense mass is driven into oscillation by electrostatic motors. The mass oscillates in one axis and as the body is rotated, the Coriolis effect causes the sense mass to oscillate out of plane. This change is measured by capacitive plates and is proportional to the rotational rate of the body.

The DPA contains an Atmel ERC32 processor, power supply electronics, and a 28V, single output, DC/DC converter. The DPA interface to a host spacecraft is through a 3-wire, bidirectional, asynchronous RS422 serial port. Input rates are 9600 baud with a variable output data rate up to 38.4K baud. The large downlink capability of the ISC can support raw imagery from the star camera and also highly sampled raw and compensated gyro data transfer from the gyro electronics. The DPA processor can operate up to 32 MHz, but the ISC application clocks it down to 4 MHz to minimize power. Given a 4 MHz operating speed, the DPA can still provide sufficient throughput necessary to maintain accuracy given the infrequent camera updates and their associated image processing . Memory resources include 1 MB of EEPROM (to hold the operational flight code along with a master dark current frame, star catalog, and boot loader), 1.5 MB of EDAC SRAM (to execute operational flight code), and 6 MB of EDAC DRAM (for use as processor scratch pad memory, store up to 8 camera images, and 1096 sets of gyro data).

The ISC structure and its 50 Krad tolerant electronics are designed to accommodate launch loads and space radiation effects of Low Earth Orbit (LEO) and Geosynchronous Earth Orbit (GEO) spacecraft applications. Mechanical analysis was performed to General Environmental Verification Specification (GEVS) levels.

The cornerstone of the ISC design is its low power. As measured from hardware engineering models, CGA power is approximately 2 Watts (including DC/DC inefficiencies) with the camera continuously generating 512x512 pixel images at 3.7 Hz, three MEMS gyros reporting rate information at 320 Hz, and transmission of this data over an LVDS interface. The low power is possible due to the inherently low power requirements of the APS focal plane, MEMS gyros, and a prudent FPGA design. The space-ready DPA operating at 4 MIPS with floating point operations consumes less than 1.5 Watts of power (including DC/DC inefficiencies) due to the inherently low power of the ERC32 processor, 4 MHz operation, and a power saving FPGA design.‡

TECHNOLOGY VALIDATION

The overall goal for flight validation as stated by NMP is to reduce the risk and cost of using new technologies in NASA Space Science missions. Developing high fidelity models and validating them with ground and flight test data will accomplish this goal. The models allow the envelope of applicability to be extended to future applications that are not identical to the current experiment or technology.

Technology validation will be accomplished by a rigorous ground and flight test suite. Draper will conduct a series of analytical measurements, computer simulations, rate-table tests, star simulation tests, and observatory tests to validate the concept on the ground. These tests will characterize the ISC accuracy performance over varying camera update rates, angular slew rates, and temperature ranges with respect to power draw, computational throughput requirements, and memory storage requirements.

Table 1: Key Validation Objectives

Objective	Where	Metric	CBE
Accuracy (1-sigma) in each axis with slewing maneuvers	Ground & Flight	0.1 deg	<0.1 deg as function of camera update rate
Self-Initialize	Ground & Flight	<10 min over >90% of sky	<2 min over >99% of sky
Power	Ground & Flight	<= 4.5 W	~ 3.5 W
Mass	Ground	<= 3 kg	~ 2.5 kg
Space Qualified Component	Ground & Flight	Operates in specified environments	Designed for GEVS

‡ Further reading on the ISC design and operation is available to the reader.[4]

During flight, the ISC testing will validate predicted operation of the MEMS gyros (ARW, scale factor, and bias stability) in a space environment (micro-gravity, overall radiation exposure, micrometeoroids, and vacuum). Predicted imager performance (dim star limit, stray light rejection, attitude accuracy) will also be validated in the relevant space environment. The integrated performance of the MEMS gyros and star imager will be validated in space.

Validation Objectives

The key validation objectives are shown in Table 1. The CBE is given in the last column. Model accuracy and self-initialization CBE were verified through a series of analysis. The power and weight CBE are measured from actual hardware. The environmental testing to GEVS specification will start in 2003.[5]

Ground Validation

Ground validation is divided into six major areas: gyro performance, camera performance, integrated simulation, software tests, observatory tests and environmental tests.

Gyro performance tests will verify that the scale factor error, angle random walk, turn-on bias, time and temperature bias stability, axis alignment and stability, gyro analog/digital noise and quantization can meet the gyro subsystem error requirements of less than 300 arc-seconds.

The camera performance tests will verify that the detector noise, interpolation error, image smearing, lens-detector misalignment, line scan aberration, number of stars and their distribution in the FOV, and other minor error sources (including chromatic aberration and calibration residuals) can meet the camera subsystem error requirements of less than 75 arc-seconds.

The integrated simulation includes a computer simulation of the camera, gyro, and Kalman filter. This will verify that the gyro and camera performance measurements combined in the Kalman filter can produce a single ECI quaternion of 0.1° accuracy under nominal and worst case conditions.

In the engineering model test environment, "canned" images will be collected using the actual ISC camera and electronics. Based on the gyro performance measurements, a gyro simulator will generate high fidelity, repeatable, motion profiles to verify system accuracy under most any conditions as virtually any input parameter can be varied to determine the system response. This environment will also be used to test off-nominal and fault conditions.

Observatory tests will verify overall ISC system accuracy and self-initialization. The flight ISC will be mounted on a 24-inch telescope at the Wallace Observatory in Westford MA. The telescope (and ISC) will be slewed and then positioned for camera updates. The telescope has 2 arc-second knowledge and an ability to slew up to 1.5°/second. The RA/DEC output of the observatory will be recorded, converted to an ECI quaternion, and compared with the ISC quaternion to

Figure 5: MIT's Wallace Observatory located in Westford, MA

determine the overall system accuracy. Mission test cases very similar to those that will be conducted during flight validation with the host vehicle (but limited to the Northern hemisphere) will be established and run at the observatory.

Finally, the flight ISC will undergo an extensive environmental test program including GEVS level random and acoustic vibration, shock, thermal vacuum, and MIL-STD-461 EMI/EMC.

Figure 6: ISC Self-Score Concept

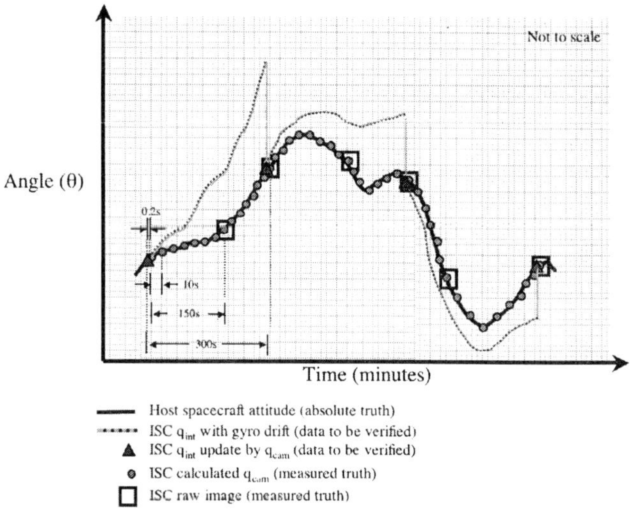

Space Validation

The ISC's performance will be verified in space over the same varying camera update rates and angular slew rates that were conducted during ground tests. The performance in space will be correlated to the accuracy and power measurements taken on the ground. The ISC power consumption and self-initialization time will be monitored over various sky quadrants and angular rates.

For accuracy validation, two of the three sources of truth are produced by the ISC itself. This concept is referred to as self-score. The ISC attitude quaternion to be validated, shown by the yellow dots in Figure 6, is output at 5 Hz and sent to the ground. In the self-score concept, there are two sources of truth attitude for comparison to the ISC quaternion; ground processed raw images and the ISC camera quaternion.

The first source of truth is the raw images captured, downlinked, and processed on the ground. These are shown by the purple squares in Figure 6. The LIS algorithms will independently calculate a camera quaternion from the raw image for a comparison to the ISC quaternion at that point in time.

The second source of truth is the ISC camera quaternion. This is the on-board ISC processed camera data without any gyro data (as shown by the green circles in Figure 6). For validation, the camera will be commanded to acquire an image at short intervals and process the LIS algorithm in the ISC. These short interval camera quaternions will not update the ISC quaternion, except at the camera update interval, and will be forwarded to the ground.

The third source of truth is the host spacecraft attitude for which a time-tagged independent attitude representation will be generated and sent in telemetry to the ground. This data will be correlated to the data generated onboard the host spacecraft.

Draper will develop a report summarizing the ground and flight validation results as required by the NMP for ST6. In the report, numerous details of the mission will be documented along with an overall assessment of the ISC's technical performance.

NEXT GENERATION ISC PERFORMANCE ENHANCEMENTS

Several enhancements have already been identified that will substantially improve the performance and functionality of the ISC. These represent the next generation of integrated attitude and navigation hardware. The benefits and costs of each will be examined to assess which are worth developing and incorporating into the next generation ISC sensor.

Improving MEMS gyro performance will improve the system performance. The development of the Draper MEMS gyros is ongoing and performance continues to improve as the design evolves. Draper is a leader in MEMS gyro development and is constantly improving both the sensor and electronics design. Next generation MEMS gyros currently in development will be fabricated with an upper sense plate (USP) to reduce noise.[6] This improvement is expected to reduce the angle random walk to half of its current value resulting in significant system level improvements in performance.

APS detector designs continue to evolve as well with improvements such as larger arrays, better process uniformity, and improved quantum efficiency. Larger arrays such as the Fill Factory 1024x1024 pixel array are already available.[7] Larger arrays can improve the resolution of the detector. As CMOS processes evolve, feature uniformity improves and feature size decreases. As feature size decreases, quantum efficiency is improved as features occlude less of the APS pixel well.

Another worthwhile design improvement is to provide the ability to interface two CGAs to a single DPA. This feature was not part of the baseline design, but would be of value to space missions, especially when higher performance is required. On average, star tracker boresight performance is worse than the performance across the field of view. For this reason, two trackers are used for most missions. Allowing two CGAs to interface to a single DPA would eliminate a DPA and reduce mass and power requirements further.

Improving the mechanical alignment stability of the ISC to the host spacecraft will also improve performance. For example, a kinematic mount could be used to obtain better mounting accuracy and reduce the amount of alignment error between the ISC and the host spacecraft.

Novel electronics packaging techniques are being explored to further reduce size and add functionality to the ISC. Draper recognizes that as the feature size of these devices continues to shrink, new and different ways of packaging are required. Ball Grid Array, chip-scale packaging, Chip on Board, and Flip Chip methods are ways that would provide more efficient interconnection and therefore increase the overall ISC packaging efficiency.[8] Envisioned is an integrated DPA module that combines the ERC32 microprocessor, memory, and glue logic in a stacked module that is estimated to be 2 inches on a side.

Incorporating MEMS accelerometers would provide a full MEMS-based Inertial Measurement Unit (IMU) and introduce a basic navigation capability to the ISC. Draper, as well as being a leader in MEMS gyro development, has also developed high performance MEMS accelerometers that have been proven in extensive and severe environmental testing. These accelerometers could be integrated alongside the gyros in the CGA. Beyond the addition of a miniature MEMS accelerometer triad to the ISC, one could also add a miniature GPS receiver to ISC to provide a full navigational capability in LEO mission orbits. These additional hardware elements could be integrated into the ISC with minimal size, power, and mass increases through a combination of Ultra-Low Power (ULP) electronics along with advanced packaging/interconnect techniques.

Algorithms that identify and track the stars are constantly improving and could result in faster acquisition and initialization times. The specific Mortari algorithm and its implementation in the ISC will be tuned based on test results of the validation system for better performance and to minimize the associated resources.

TECHNOLOGY INFUSION

Technology infusion is the process of providing a new and useful technology to NASA missions that benefit from the innovation. The ST6 ISC flight experiment not only provides the mechanism to verify and benchmark pre-launch performance predictions, but also serves as a showcase for technology possibilities early enough in the ISC product development cycle to easily adapt to changing requirements.

The foundation of the infusion process is fostering an understanding and interest in the ISC's capabilities. This motivates the need to initiate and maintain a continuing contact with the potential mission customers and stakeholders to assure a clear understanding of technology needs and infusion opportunities. This involves seeking out candidate missions and fine-tuning the product to be as widely beneficial as practical. It is expected that the ISC will offer designers an attractive alternative for challenging applications where power, mass, and volume constraints preclude the use of the traditional star tracker and inertial sensing components. It is envisioned that many of NASA's future missions, ranging from satellites in LEO to deep space interplanetary probes, could benefit from the infusion of the versatile ISC technology in some manner.

The infusion process begins with a survey of missions currently in formulation and longer-term "Roadmap" missions to establish both the attitude determination requirements and the hardware resource constraints. A critical element of this survey process will be technical interaction with NASA Code S (Space Science) Theme Technologists to understand their theme-unique technology needs. The NASA Earth Science Technology Office (ESTO) would also be solicited to understand the Code Y (Earth Science) platform technology needs for attitude determination. Also, the Integrated Mission Design Centers at both GSFC and JPL, where collaborative end-to-end mission conceptual design studies are performed, will also serve as rich sources of requirements and constraints data.[9] An analysis of this information should identify specific missions where the ISC technology is most applicable, providing the biggest "payoff".

When the ISC technology appears to be applicable to a mission, direct contact will be made with that mission's Program Office, formulation team, and/or lead GN&C engineer to explore the opportunity for potential infusion. In some cases the ISC technology may be a direct "as is" fit, whereas in other cases it may be close to meeting mission requirements.

As described above, several useful design enhancements have been identified that will substantially improve the performance and functionality of the ISC. For those missions where the ISC is close to meeting the requirements, this set of potential design enhancements will be explored. Specific enhancements to the ST6 ISC baseline design will be studied and evaluated to determine the cost versus benefit trade space. Key trade parameters would focus on its size, mass, and power and would drive the decision on which modifications are worth pursuing for a specific mission application.

In addition to efforts to seek out missions that would benefit from ISC, information about ISC will be published. The NMP website will be updated with the most up to date information regarding the ISC capabilities and development status. Contact information will be provided for those interested in learning more about the vast potential of the ISC. Also, the ISC will be added to the technology item database in the Integrated Mission Design Centers at both GSFC and JPL so that this innovative technology is available to the spacecraft designers conducting end-to-end mission conceptual design studies.

Certainly, following successful ST6 on-orbit validation and subsequent commercialization of the technology into an affordable product form, one can readily foresee a path for rapid adoption of the ISC by NASA, DoD, and Industry as a uniquely enabling technology to perform the micro-satellite attitude determination function. While that particular ISC use will address a very critical and specific need, it is difficult to imagine the infusion of this technology being limited to that single micro-satellite application. While it is unreasonable to predict that the ISC will be a panacea solution suitable for all next generation missions, there is clearly a wide-ranging set of future GN&C problems for which the ISC technology is relevant and may represent part or all of the solution. It is entirely possible that the ISC technology, in either its initial ST6 configuration or in a slightly modified form, will be employed to address some of the GN&C design challenges posed by: highly dynamic vehicles, tumbling and/or "lost in space" spacecraft, semi-autonomous (tele-operated) satellite servicing platforms, autonomous rendezvous vehicles, reusable launch vehicles, formation flying platforms, deep space probes, and planetary exploration vehicles (surface rovers, sample return craft, aerial platforms, and balloons).

MISSION APPLICATIONS

The success of NASA's next generation Earth and Space Science missions will be critically dependent on the development, validation, and infusion of spacecraft avionics that are not only highly integrated, power efficient, and minimally packaged, but also flexible and versatile enough to satisfy multi-mission requirements. The imminent introduction of the ISC into the spacecraft designer's inventory will directly address this need for advanced technology avionics and will herald a breakthrough in how the function of medium (arc-minute class) accuracy attitude determination will be implemented in future space missions. It is perhaps the inherent multifunction capability of the ISC that is of greatest relevance to future space missions. Technologists at Draper and NASA understood the powerful advantages to be gained in power, mass, and volume by combining the traditionally separate gyro (inertial) sensing and stellar sensing functions into one unique ISC package. The APS star sensor and the MEMS gyro technology combination described above provides the innovative technology "push" for ISC. Conversely, the primary customer "pull" for ISC is derived from NASA's need to implement affordable, next generation, micro-spacecraft capabilities. In short, the ISC will not only support the on-going general evolution towards smaller, lower power spacecraft sensors, but also will directly focus on and address the technology needs for the on-going micro-satellite revolution.

It is envisioned that the ISC will serve as both an *Enhancing* and an *Enabling* technology for many future Space Science and Earth Science missions. Enabling technologies are those that provide the presently unavailable capabilities necessary for a mission's implementation and are vital to both NASA's intermediate and long term missions. Enhancing technologies typically provide significant mission performance improvements, mitigations of critical mission risks, and/or significant increases in mission critical resources (e.g., cost, power, and mass).

A number of potential general mission applications for the ISC technology are described below. Mission applications directly addressed by the baseline ST6 ISC configuration will be described, then ambitious mission applications, where hardware and software modifications to the baseline ISC would be required, are subsequently discussed.

The Microsat Application

One of NASA's top technology priorities focuses on economical, highly autonomous, micro-satellites for multi-spacecraft distributed missions. The ISC represents an enabling, fully integrated, micro-satellite attitude determination system that directly addresses this NASA priority. Preliminary analysis of a three-axis stabilized 50kg micro-satellite conceptual design indicates the critical need for the ISC technology to meet stringent power, mass, and volume requirements while simultaneously satisfying a 0.03-0.05 degree attitude determination requirement. In this conceptual design, the ISC is a fully compatible external sensor interfaced to the Multi-Function GN&C System (MFGS) currently being developed at NASA/GSFC.[10] The versatile MFGS unit represents the successful coalescence of several hardware and software technology innovations into one single, highly integrated, compact, low power, and low cost unit that simultaneously provides attitude determination solutions and navigation solutions with accuracies that satisfy many future GSFC micro-satellite mission requirements.

The Satellite "Single Sensor" Application

Mission designers could employ the ST6 configuration ISC unit as a significant enhancing technology for a number of moderate (< 0.1^{o}) non-microsatellite applications. In these cases, the ISC would serve as a compact and low power "Single Sensor" unit that could replace typical attitude sensors and associated attitude determination flight software.

For example, one could foresee the use of the ISC on some GEO communications or observation satellites replacing the typical attitude sensor complement of horizon crossing indicators, Earth sensors, and Sun sensors. Mass savings in this particular geosynchronous application would directly translate to additional station-keeping propellant or payload capacity.

The ISC seems uniquely suited to fill a medium (arc-minute class) accuracy attitude sensor technology niche between relatively low cost/low performance COTS Sun Sensors/Magnetometers and relatively high cost/high power COTS Star Trackers with arc-second performance.

The Satellite Safe-Hold/Coarse Acquisition Application

Identifying and implementing simple, reliable, independent, and affordable (in terms of cost, mass, and power) methods for autonomous satellite safing and protection has long been a significant challenge for designers. Employing one or as many as three (if an attitude voting scheme is desired) "bolt on" ISC's to provide a fully independent safe-hold attitude determination system for high end (very costly) one-of-a-kind observatories, such as the James Webb Space Telescope (JWST), may yield a simple solution to a complex problem. Specific implementations would vary, but in general, this application entails one or more of the ISC units being mounted on the observatory to mitigate the risk of mission loss for a small relative cost. ISC represents an enhancing technology in this application. The low mass and small volume of the ISC should preclude any major accommodation issues on a large observatory. Additionally, the performance of the baseline ST6 ISC configuration can easily satisfy the typically modest safe-hold attitude determination requirements (on the order of a degree or less) to maintain the observatory in a power-safe and thermal-safe attitude.

One notional safe-hold operational concept calls for the ISC units to be powered off at launch. They would then be autonomously powered-on upon indication of observatory separation from its launch vehicle. In this state, the outputs of the ISC's MEMS gyros alone could provide sufficient angular rate polarity feedback to stabilize the observatory if in a tumbling state due to high tip-off rates at separation. The ISC can also provide a coarse attitude acquisition function prior to handing over attitude determination to the observatory's precision attitude sensors or, in some cases, the payload's Fine Guidance Sensor (FGS). The ISC units would subsequently be powered-off once the nominal observatory attitude state is achieved. They would be left in an unpowered state during normal mission operations and would only be activated when a safe-hold event is triggered by the observatory's FDC logic. The low power draw of the ISC would be very compatible with the initial phase of safe-hold operations when only very limited battery power may be available. The outputs of the MEMS gyros alone could provide sufficient angular rate polarity signal feedback to stabilize the observatory if in a tumbling state following an on-orbit anomaly. Once the anomaly is resolved, the ISC could perform the coarse attitude determination function supporting the transition back to normal mission operations.

MODIFIED ISC CONFIGURATION MISSION APPLICATIONS

In addition to the mission applications discussed above, there are a number of other potential uses of the ISC technology that can be conceived. The Draper/NASA team has conceptually identified, but not studied in any significant detail, the following potential next-generation ISC technology infusion targets.

The GEC Mission Application

Planned for launch in 2009, the Geospace Electrodynamics Connections (GEC) mission will be an element of the NASA's Sun-Earth Connection (SEC) Space Science theme.[11] GEC will make in-situ measurements in the Ionosphere-Thermosphere region to improve models of how the Earth's upper atmosphere electrodynamically couples to the magnetosphere. In this discussion, it is important to note that the GEC spacecraft with an on-orbit mass of approximately 375 kg is not a micro-satellite class vehicle. In fact, a considerable amount of propellant will be carried on GEC to perform "deep-dipping" campaigns down to perigee orbital altitudes, as low as 130 km, where the mission's primary science data is taken. The amount of science data collected overall is a function of the number of dipping campaigns, which itself is a function of the amount of propellant. The current baseline GEC Attitude Determination and Control Subsystem (ADCS) design incorporates a COTS star tracker unit, an Inertial Reference Unit (IRU), sun sensors, and customized Kalman filter flight software algorithms hosted on the spacecraft's central processor to satisfy the attitude determination requirement of 0.01 degree. The performance of the baseline ST6 ISC technology is not currently predicted to satisfy the GEC attitude knowledge requirement of 0.01 degree but that level of required performance could be achieved with technology enhancements. The ISC could, with a minimum of modifications, serve as the "Single Sensor" replacing the individual COTS Star Tracker and the IRU.

It is estimated that approximately 1.5 kg of mass and 26 Watts of power can be saved by employing a single ISC unit rather than the separate COTS Star Tracker and IRU. The mass savings afforded by using the ISC to perform the attitude determination function could be allocated for additional propellant, thus more "deep-dipping", and thus more GEC science data. That is one advantage of using ISC on GEC. The other has to do with power and it looks to be even more significant than the mass advantage.

The design of the GEC Electrical Power Subsystem (EPS) poses significant challenges. The ADCS is the largest power consumer of all the spacecraft bus subsystems. Controlling the spacecraft while "deep-dipping" at perigee may demand as much as 100 Watts of power for the ADCS. This is about twice the ADCS power requirement while operating in the non-perigee portions of the mission orbit. The large perigee ADCS power draw is primarily driven by the need for the reaction wheels to generate attitude control torques to counteract the significant aerodynamic disturbances acting on the spacecraft during perigee passage. The complicating factor for the GEC EPS designers is the added difficulty in generating electric power during perigee passage. Using the ISC significantly lowers the ADCS power requirement (by approximately 26 Watts) during perigee passage thus reducing the demand on the EPS and/or making more power available for the GEC science payload.

Lastly, there is another, more subtle, but significant, advantage of using the ISC technology on a mission such as GEC. As described in the System Overview above, the ISC contains all the necessary flight software algorithms embedded in its dedicated internal processor. The fact that customized Kalman filter flight software attitude determination algorithms would not have to be designed, developed, tested, and integrated into the spacecraft central processor should serve to reduce mission costs and risk.

Although several implementation details remain to be studied, a strong compatibility of the ISC technology with the GEC mission requirements has emerged from the preliminary top-level considerations. The GEC represents a very solid potential technology infusion target for ISC that merits additional consideration.

The Autonomous Visual Relative Navigation Sensor Application

In order to accomplish its ambitious mission goals for the next two decades there is a heightened priority within NASA to develop and infuse the multiple technologies needed to autonomously perform both the functions of: 1) multiple, distributed spacecraft formation flying and 2) spacecraft rendezvous, proximity operations, and docking. Several missions currently being formulated at NASA rely on the capability to fly multiple distributed spacecraft to enable revolutionary methods of Space and Earth Science observation.[12] Likewise, NASA foresees several important applications of the capabilities for spacecraft rendezvous, proximity operations, and docking such as on-orbit deployment, inspection, assembly, servicing, re-fueling, payload retrieval/transfer (e.g., sample canisters and/or experiment packages), and re-configuration at various orbital locations.

The ultimate success of the planned formation flying and rendezvous missions will depend on the development of advanced electro-optical sensors (and algorithms) for performing vision-based relative navigation. The ISC technology is a natural candidate for this type of application and could be readily adapted with a minimum of camera and flight software modifications.

The ISC could be a coarse acquisition visual sensor to determine the relative location of multiple formation spacecraft immediately following deployment from the launch vehicle dispenser. Although there are a number of passive and active techniques that could be used in this rendezvous application, it was assumed in this concept that each individual spacecraft is equipped with a uniquely identifiable array of small, perhaps actively modulated, Light Emitting Diode (LED) beacons that can be sensed by the ISC detector. In this specific application, the ISC would first be used to provide coarse navigation outputs of sufficient accuracy to simply ensure collision avoidance between the spacecraft. Simple relative navigation algorithms would need to be embedded on the ISC processor.

In a similar manner, the ISC technology lends itself to the rendezvous problem of first acquiring and subsequently sensing the relative position and attitude of the orbiting target spacecraft by the maneuvering chase spacecraft. In this application, vision-based feedback from the ISC is used in an integrated orbital/attitude control loop on the maneuvering chase spacecraft. It is conceivable that the ISC could potentially serve as the relative navigation sensor for all phases of the rendezvous and docking process: long range acquisition and tracking as well as close range relative orientation and position determination.

Long range initial acquisition of the target vehicle could be performed passively by the ISC by detecting the target as a moving point source against the inertially fixed star background. The appropriate relative navigation filter algorithms would need to be hosted on the ISC internal processor to accomplish this task.

Close range maneuvering of the chase spacecraft towards terminal engagement with the target could then be performed using the attitude and position derived from object feature (model matching) recognition processing algorithms using a series of ISC-generated video image inputs. Potentially these image processing algorithms could be hosted on the internal ISC processor, but that would be dependent on tailoring their size and complexity.

Once close enough to conduct proximity operations, the multipurpose ISC can serve as a high bandwidth video camera system to conduct detailed inspections of the target spacecraft prior to performing any docking, servicing, re-fueling, or payload retrieval operations. Conceivably two ISC's could be employed to generate stereo images of the target vehicle.

The Nanosatellite Core Bus Application

A novel and promising application is to use the lightweight, highly compact, and low power ISC as the enabling technology core of a 10-15 kg nano-satellite class bus architecture. It would inherently satisfy the attitude determination requirement and the remainder of the vehicle could be packaged around it by employing other advanced technologies for miniature, highly integrated, power efficient electronics. MEMS technology could be exploited in this architecture to implement many of the subsystem functions creating flexible and versatile design for multi-mission applications.[13] The ISC camera system could conceivably serve as the imaging payload in a two CGA/one DPA hardware configuration. With some hardware enhancements to the ISC's internal processor, it could serve as the nano-satellite central flight processor taking on the additional computational tasks required to perform the payload, command/telemetry, communications, power management, I/O, and attitude control functions. Additional interfaces would be necessary for subsystem control elements such as nano-wheel or MEMS thruster attitude control actuators.

CONCLUSION

An attitude determination sensor has been introduced and shown to be useful to a broad range of spacecraft as either a primary attitude sensor or as a safe-hold sensor. For moderate performance micro-satellites, ISC provides an enabling technology. The ISC promises benefits of ultra-low power and low mass by integrating state of the art complimentary sensors and an embedded processor in a compact, efficient package. An additional benefit of ISC is simplified spacecraft integration.

As a "bolt-on" smart sensor, the ISC reduces complexity by tightly integrating the gyros and APS star camera within the CGA to simplify alignment and calibration. Integration costs are reduced since the DPA performs all of the attitude determination and filtering with a simple serial interface to provide the spacecraft with attitude information in the form of an attitude quaternion and attitude rates. To further reduce size and power impact on the spacecraft, the ISC design also offers the possibility of using its data processor as the spacecraft's primary processor to run both the ISC and spacecraft software.

Preliminary analysis indicates that a multitude of spacecraft will benefit from the ISC technology infusion process, pending specific design trades. One objective of this paper was to stimulate ideas for infusion and adoption of the ISC technology in the minds of future mission designers. It is exciting to contemplate the various mission applications this breakthrough ISC technology might support in the near to mid-range future.

ACKNOWLEDGEMENT

The NASA research described here is being carried out at Draper Laboratory under a contract with the California Institute of Technology's Jet Propulsion Laboratory.

The authors would like to acknowledge the advice, insight, encouragement, and support received from the following individuals at NASA and JPL: Chuck Gay, Bryant Cramer, Frank Bauer, Mike Ryschkewitsch, Timothy VanSant, Chris Stevens, Jack Stocky, and Art Chmielewski.

Bill Wyman, the Draper ISC Program Manager, deserves special recognition for his tireless support, dedication, and insight. In addition, the ISC program would not be possible without the dedicated and talented engineering staff at Draper, whom is a privilege to work with.

The GSFC authors would especially like to posthumously acknowledge the significant engineering contributions to the ST6 ISC project by their dear friend and mentor, Tom Budney. Until his untimely, and greatly lamented passing, Mr. Budney served as GSFC's Chief Engineer for GN&C.

REFERENCES

1. MORTARI, D., JUNKINS, J., and SAMAAN, M. "Lost-in-Space Pyramid Algorithm for Robust Star Pattern Recognition," *24th annual AAS Guidance and Control Conference, Breckenridge,* CO, 31 January - 4 February 2001.

2. OGIERS, W. "Identification of CMOS Imager Applications in Space", DOC P60280-MS-RP-01, May 1997.

3. MARSTON, N. "Solid-state imaging: A Critique of the CMOS sensor", Ph.D. Dissertation, The University of Edinburgh, November 1998.

4. BRADY, T., et al. "The Inertial Stellar Compass: A New Direction in Spacecraft Attitude Determination", *16th Annual AIAA/USU Conference on Small Satellites*, Logan, Utah, 12 – 15 August 2002.

5. "General Environmental Verification Specification for STS and ELV Payload, Subsystems, and Components", Rev A, GSFC, June 1996.

6. ANDERSON, R., HANSON, D., and KOUREPENIS, A. "Evolution of Low-Cost MEMS Inertial System Technologies", *International Technical Meeting of the Satellite Division of the Institute of Navigation (ION GPS),* Salt Lake City, UT, 9 – 11 September 2001.

7. STAR1000 CMOS Image Sensor Datasheets, www.fillfactory.com

8. http://www.atmel.com/atmel/products/prod47.htm

9. http://idc.gsfc.nasa.gov

10. GAMBINO, J., and DENNEHY, N. "A Multi-Function Guidance, Navigation and Control System for Future Earth and Space Science Missions", AAS Paper 02-062, *Proceedings of the 25th Annual AAS Guidance and Control Conference*, 6-10 February 2002.

11. http://stp.gsfc.nasa.gov/missions/gec/gec.htm

12. LEITNER, J., BAUER, F., FOLTA, D., CARPENTER, R., and MOREAU, M. "Formation Flight in Space", GPS World, February 2002.

13. DENNEHY, N., HATTIS, P., JOHNSON, W., and SOCHA, M. "MEMS-Based GN&C Sensors and Actuators for Micro/Nano Satellites", AAS Paper 00-065, *Proceedings of the 23rd Annual AAS Guidance and Control Conference*, February 2000.

ACRONYMS

ADS	Attitude Determination System
APS	Active Pixel Sensor
ARW	Angle Random Walk
ASIC	Application Specific Integrated Circuit
CBE	Current Best Estimate
CCD	Charge Coupled Device
CGA	Camera Gyro Assembly
CMOS	Complementary Metal Oxide Semiconductor
DPA	Data Processing Assembly
ECI	Earth Centered Inertial
ISC	Inertial Stellar Compass
FDC	Fault Detection and Correction
FOV	Field of View
FPGA	Field Programmable Gate Array
GEC	Geospace Electrodynamics Connection
GN&C	Guidance, Navigation, and Control
GSFC	Goddard Space Flight Center
LIS	Lost-in-Space
LVDS	Low Voltage Differential Signal
MIPS	Million Instructions Per Second
NASA	National Aeronautics and Space Administration
PPM	Parts Per Million
NMP	New Millenium Program
GEO	Geosynchronous Earth Orbit
GEVS	General Environmental Verification Specification
GPS	Global Positioning System
LEO	Low Earth Orbit
MEMS	Micro-ElectroMechanical System
MFGS	Multifunction GN&C System
RA/DEC	Right Ascension and Declination

AAS 03-004

A CLOSED-LOOP HARDWARE SIMULATION OF DECENTRALIZED SATELLITE FORMATION CONTROL

Takuji Ebinuma* and E. Glenn Lightsey†

A closed-loop simulation system has been developed to demonstrate autonomous satellite formation control using the Global Positioning System (GPS). The developed system is capable of realistic simulations of GPS-based guidance, navigation, and control in real-time and enables research into orbit control strategies for autonomous formation flying of multiple spacecraft. A sample formation flying scenario employing a decentralized controller architecture has been investigated in this paper. The technical issues surrounding the decentralized architecture implementation, such as data flow and timing requirements, are also investigated. It is expected that the decentralized architecture promotes agility, adaptability, scalability, and affordability of future multiple spacecraft platforms.

INTRODUCTION

In recent years, there has been significant interest in the use of formation flying spacecraft for a variety of earth and space science missions [1]. Formation flying may provide smaller and cheaper satellites that, working together, have more capability than larger and more expensive satellites. Several decentralized architectures have been proposed for autonomous establishment and maintenance of satellite formations. In such architectures, each satellite cooperatively maintains the shape of the formation without a central supervisor, and processes only local measurement information. The Global Positioning System (GPS) sensors are ideally suited to provide such local position and velocity measurements to the individual satellite.

An investigation of the feasibility of a decentralized approach to satellite formations was originally presented by Carpenter [2]. He extended a decentralized linear-quadratic-Gaussian (LQG) framework proposed by Speyer [3] in a fashion similar to a linearized Kalman filter (LKF) which processed GPS position fix solutions. The new decentralized LQG architecture was demonstrated in a numerical simulation for an ideal scenario that is similar to the TechSat-21 mission [4] that has been proposed by NASA and the U.S. Air Force.

Another decentralized architecture was proposed by Park et al. [5] using carrier differential-phase GPS (CDGPS). Busse et al. [6] recently demonstrated the decentralized CDGPS architecture in a hardware-in-the-loop (HWIL) simulation on the Formation Flying Test-Bed (FFTB) at NASA's Goddard Space Flight Center (GSFC), which features four 16-channel GPS signal generators.

Although the early experiments represented a step forward by utilizing GPS signal simulators for a spacecraft formation flying simulation, only open-loop performance was considered. Open-loop

* Research Engineer, Center for Space Research, The University of Texas at Austin, 3925 West Braker Lane, Suite 200, Austin, Texas 78759. E-mail: tak@csr.utexas.edu.

† Assistant Professor, Department of Aerospace Engineering and Engineering Mechanics, The University of Texas at Austin, Austin, Texas 78712. E-mail: lightsey@csr.utexas.edu.

tests can provide some information about the basic performance of a system, but they are limited to scripted trajectories that do not allow for control of the vehicles based on the sensor measurements. Satellite formation operations require precise relative navigation, but this alone is insufficient to allow autonomous on-orbit operations. The guidance algorithm relies on the navigation state and must operate in concert with the navigation system in a real-time mode.

In this research, hardware experimentation has been extended to include closed-loop integrated guidance and navigation of spacecraft formations using GPS receivers and an inter-spacecraft communication system. A hardware closed-loop simulation employing the decentralized LQG architecture proposed by Carpenter is presented in this paper. The technical issues surrounding the integration of guidance and navigation systems in a real-time HWIL test are also investigated to ensure that no practical issues (such as timing mismatch) are overlooked.

DECENTRALIZED CONTROL

A brief introduction to decentralized control is presented in this section. More complete mathematical definitions are presented by Carpenter [2]. Subsequently, the subscript i denotes time epoch, and the superscript j denotes local information of each satellite in the formation.

Consider the solution to the discrete decentralized LQG control problem obtained by minimizing the cost function

$$J = E\left\{\sum_{i=1}^{N}\left(\boldsymbol{x}_i^T \boldsymbol{W}_i \boldsymbol{x}_i + \sum_{j=1}^{K}(\boldsymbol{u}_i^j)^T \boldsymbol{V}_i^j \boldsymbol{u}_i^j\right)\right\} \tag{1}$$

over control vectors $\boldsymbol{u}_i^j$, subject to the measurement $\boldsymbol{y}_i^j$ given by

$$\boldsymbol{y}_i^j = \boldsymbol{H}^j \boldsymbol{x}_i + \boldsymbol{v}_i^j \ , \tag{2}$$

and the state $\boldsymbol{x}_i$ given by

$$\boldsymbol{x}_{i+1} = \boldsymbol{\Phi}_{i+1}\boldsymbol{x}_i + \sum_{j=1}^{K}\boldsymbol{\Lambda}^j \boldsymbol{u}_i^j + \boldsymbol{w}_i \ , \tag{3}$$

where $\boldsymbol{W}_i$ and $\boldsymbol{V}_i^j$ are weight matrices for the state and the control, respectively, $\boldsymbol{\Phi}_i$ is the state transition matrix, $\boldsymbol{H}^j$ and $\boldsymbol{\Lambda}^j$ are the measurement and control matrices of spacecraft j, respectively, K is the number of spacecraft in the formation, N is the number of epochs, and $\boldsymbol{v}_i^j$ and $\boldsymbol{w}_i$ are white noise vectors, such that

$$E\left\{\boldsymbol{v}_i^j\right\} = \boldsymbol{0} \ , \quad E\left\{\boldsymbol{v}_i^j(\boldsymbol{v}_k^j)^T\right\} = \boldsymbol{R}_i^j \delta_{ik} \ , \quad E\left\{\boldsymbol{v}_i^j(\boldsymbol{v}_i^l)^T\right\} = \boldsymbol{0} \ , \tag{4}$$

$$E\left\{\boldsymbol{w}_i\right\} = \boldsymbol{0} \ , \quad E\left\{\boldsymbol{w}_i(\boldsymbol{w}_k)^T\right\} = \boldsymbol{Q}_i \delta_{ik} \ . \tag{5}$$

Speyer [3] decomposed the state space into a control-dependent partition $\boldsymbol{x}_i^C$ and a data-dependent partition $\boldsymbol{x}_i^D$ as follows:

$$\boldsymbol{x}_i = \boldsymbol{x}_i^C + \boldsymbol{x}_i^D \ . \tag{6}$$

A local LKF algorithm operating only on the local mesurements of spacecraft j can be described as

$$\boldsymbol{x}_i^C = \boldsymbol{\Phi}_i \boldsymbol{x}_{i-1}^C + \sum_{j=1}^{K} \boldsymbol{\Lambda}^j \boldsymbol{u}_{i-1}^j \tag{7}$$

$$\bar{\boldsymbol{x}}_i^{Dj} = \boldsymbol{\Phi}_i \hat{\boldsymbol{x}}_{i-1}^{Dj} \tag{8}$$

$$\bar{\boldsymbol{P}}_i^j = \boldsymbol{\Phi}_i \hat{\boldsymbol{P}}_{i-1}^j \boldsymbol{\Phi}_i^T + \boldsymbol{Q}_i \tag{9}$$

$$\boldsymbol{K}_i^j = \bar{\boldsymbol{P}}_i^j (\boldsymbol{H}^j)^T \left[\boldsymbol{H}^j \bar{\boldsymbol{P}}_i^j (\boldsymbol{H}^j)^T + \boldsymbol{R}_i^j \right]^{-1} \tag{10}$$

$$\hat{\boldsymbol{x}}_i^{Dj} = \bar{\boldsymbol{x}}_i^{Dj} + \boldsymbol{K}_i^j \left[\boldsymbol{y}_i^j - \boldsymbol{H}^j (\boldsymbol{x}_i^C + \bar{\boldsymbol{x}}_i^{Dj}) \right] \tag{11}$$

$$\hat{\boldsymbol{P}}_i^j = (\boldsymbol{I} - \boldsymbol{K}_i^j \boldsymbol{H}^j) \bar{\boldsymbol{P}}_i^j (\boldsymbol{I} - \boldsymbol{K}_i^j \boldsymbol{H}^j)^T + \boldsymbol{K}_i^j \boldsymbol{R}_i^j (\boldsymbol{K}_i^j)^T , \tag{12}$$

where $\boldsymbol{K}_i^j$ and $\boldsymbol{P}_i^j$ are the local Kalman gain and local state error covariance matrices, respectively. Even though this decentralized approach does not require a global Kalman filter, it does require the global state error covariance matrix $\boldsymbol{P}_i$ to be maintained locally via

$$\bar{\boldsymbol{P}}_i = \boldsymbol{\Phi}_i \hat{\boldsymbol{P}}_{i-1} \boldsymbol{\Phi}_i^T + \boldsymbol{Q}_i \tag{13}$$

$$\hat{\boldsymbol{P}}_i = \left[\bar{\boldsymbol{P}}_i^{-1} + \sum_{j=1}^{K} s_i^j (\boldsymbol{H}^j)^T (\boldsymbol{R}_i^j)^{-1} \boldsymbol{H}^j \right]^{-1} . \tag{14}$$

A significant practical issue with the local reconstruction of the globally optimal covariance matrix via Eq. (14) is that the measurement updates may not occur at all local filters uniformly. This issue requires an additional transmission of a semaphore $s_i^j \in \{0, 1\}$ from each satellite j to all the other satellites every time a local measurement update is successfully executed.

The global estimate $\hat{\boldsymbol{x}}_i^D$ is then obtained by a linear combination of the local estimates $\hat{\boldsymbol{x}}_i^{Dj}$ and an additional data vector $\boldsymbol{h}_i^j$, which contains non-local information, via

$$\hat{\boldsymbol{x}}_i^D = \sum_{j=1}^{K} \left(\hat{\boldsymbol{P}}_i (\hat{\boldsymbol{P}}_i^j)^{-1} \hat{\boldsymbol{x}}_i^{Dj} + \boldsymbol{h}_i^j \right) , \tag{15}$$

where

$$\boldsymbol{h}_i^j = \boldsymbol{F}_i \boldsymbol{h}_{i-1}^j + \boldsymbol{G}_i^j \bar{\boldsymbol{x}}_i^{Dj} \; ; \; \boldsymbol{h}_0^j = \boldsymbol{0} \tag{16}$$

$$\boldsymbol{F}_i = \hat{\boldsymbol{P}}_i (\bar{\boldsymbol{P}}_i)^{-1} \boldsymbol{\Phi}_i \tag{17}$$

$$\boldsymbol{G}_i^j = \boldsymbol{F}_i \hat{\boldsymbol{P}}_{i-1} (\bar{\boldsymbol{P}}_{i-1}^j)^{-1} \boldsymbol{\Phi}_i^{-1} - \hat{\boldsymbol{P}}_i (\bar{\boldsymbol{P}}_i^j)^{-1} . \tag{18}$$

The main advantage of this decentralized algorithm is that each spacecraft need not reconstruct the globally optimal state estimate $\hat{\boldsymbol{x}}_i^D$ via Eq. (15) to compute the globally optimized control. By exchanging an additional vector

$$\boldsymbol{\alpha}_i^{lj} = (\boldsymbol{\Lambda}^l)^T \boldsymbol{S}_{i+1} \boldsymbol{\Phi}_i \left[\hat{\boldsymbol{P}}_i (\hat{\boldsymbol{P}}_i^j)^{-1} \hat{\boldsymbol{x}}_i^{Dj} + \boldsymbol{h}_i^j \right] , \quad l = 1, 2, \ldots, K \tag{19}$$

between the satellites l and j, the globally optimal control $\boldsymbol{u}_i^j$ can be obtained locally via

$$\boldsymbol{u}_i^j = - \left[\boldsymbol{V}_i^j + (\boldsymbol{\Lambda}^j)^T \boldsymbol{S}_{i+1} \boldsymbol{\Lambda}^j \right]^{-1} \left\{ (\boldsymbol{\Lambda}^j)^T \boldsymbol{S}_{i+1} \boldsymbol{x}_i^C + \sum_{l=1}^{K} \boldsymbol{\alpha}_i^{jl} \right\} , \tag{20}$$

where the controller Riccati matrix $\boldsymbol{S}_i$ can be found via a backward sweep of

$$\boldsymbol{S}_i = \boldsymbol{\Phi}_i^T \boldsymbol{S}_{i+1} \boldsymbol{\Phi}_i - \sum_{j=1}^{K} \left\{ (\boldsymbol{L}_i^j)^T \left[\boldsymbol{V}_i^j + (\boldsymbol{\Lambda}^j)^T \boldsymbol{S}_{i+1} \boldsymbol{\Lambda}^j \right] \boldsymbol{L}_i^j \right\} + \boldsymbol{W}_i \tag{21}$$

$$\boldsymbol{L}_i^j = - \left[\boldsymbol{V}_i^j + (\boldsymbol{\Lambda}^j)^T \boldsymbol{S}_{i+1} \boldsymbol{\Lambda}^j \right]^{-1} (\boldsymbol{\Lambda}^j)^T \boldsymbol{S}_{i+1} \boldsymbol{\Phi}_i , \tag{22}$$

and may be stored on-board each satellite as a table-lookup for use in Eq. (20).

Note that $\boldsymbol{\alpha}_i^{lj}$ has the dimension of the local controller, which is almost always smaller than that of the state. Since the dimension of the controller is the minimum data exchange rate required to execute optimal control, the inter-spacecraft link budget is much cheaper than passing raw measurements for a global Kalman filter. This can be a significant advantage when the number of spacecraft in the formation is large.

System Dynamics

In order to demonstrate the application of the decentralized algorithm to a spacecraft formation, an example is presented. Each satellite is assumed to be orbiting the Earth in a near-circular orbit and to remain in the vicinity of the formation origin. For an intuitive interpretation of the relative motion of the satellites with respect to the formation origin, it is suitable to express the differential position vector $\boldsymbol{r}$ and velocity vector $\boldsymbol{v}$ in a reference frame aligned with the local radial (x), along-track (y), and cross-track (z) directions. The corresponding unit vectors of this local reference frame are defined by

$$\boldsymbol{e}_x = \frac{\boldsymbol{r}_\mathrm{R}}{|\boldsymbol{r}_\mathrm{R}|}\ , \ \boldsymbol{e}_y = \frac{\boldsymbol{r}_\mathrm{R}\times\boldsymbol{v}_\mathrm{R}}{|\boldsymbol{r}_\mathrm{R}\times\boldsymbol{v}_\mathrm{R}|}\ , \ \boldsymbol{e}_z = \boldsymbol{e}_y \times \boldsymbol{e}_x\ , \tag{23}$$

where $\boldsymbol{r}_\mathrm{R}$ is the inertial position of the formation origin, and the inertial velocity

$$\boldsymbol{v}_\mathrm{R} = \dot{\boldsymbol{r}}_\mathrm{R} + \boldsymbol{\omega}_\oplus \times \boldsymbol{r}_\mathrm{R} \tag{24}$$

is obtained by correcting the derivative of the WGS84 position for the angular velocity $\boldsymbol{\omega}_\oplus$ of the Earth's rotation.

The motion of each satellite in the local reference frame is then described by the Clohessy-Wiltshire (CW) equations. For the time interval Δt, the state transition matrix is given by

$$\boldsymbol{\Phi}^j = \begin{bmatrix} 4-3C & 0 & 0 & S/n & 2(1-C)/n & 0 \\ 6(S-n\Delta t) & 1 & 0 & 2(C-1)/n & (4S-3n\Delta t)/n & 0 \\ 0 & 0 & C & 0 & 0 & S/n \\ 3nS & 0 & 0 & C & 2S & 0 \\ 6n(C-1) & 0 & 0 & -2S & 4C-3 & 0 \\ 0 & 0 & -nS & 0 & 0 & C \end{bmatrix}\ , \tag{25}$$

where n is the instantaneous frame rotation rate

$$n = (\boldsymbol{e}_y^T)/|\boldsymbol{r}_\mathrm{R}|\ , \tag{26}$$

and

$$S = \sin n\Delta t\ ,\ C = \cos n\Delta t\ . \tag{27}$$

It is also assumed that each satellite has small thrusters that can apply an impulsive velocity vector

$$\boldsymbol{u}^j = \begin{bmatrix} u_x^j & u_y^j & u_z^j \end{bmatrix}^T\ . \tag{28}$$

Then the state equation defined in Eq. (3) gives

$$\begin{bmatrix} \boldsymbol{r}^1 \\ \boldsymbol{v}^1 \\ \boldsymbol{r}^2 \\ \boldsymbol{v}^2 \\ \vdots \\ \boldsymbol{r}^K \\ \boldsymbol{v}^K \end{bmatrix}_{i+1} = \begin{bmatrix} \boldsymbol{\Phi}^1 & \boldsymbol{0} & \cdots & \boldsymbol{0} \\ \boldsymbol{0} & \boldsymbol{\Phi}^2 & & \\ \vdots & & \ddots & \\ \boldsymbol{0} & & & \boldsymbol{\Phi}^K \end{bmatrix}_{i+1} \begin{bmatrix} \boldsymbol{r}^1 \\ \boldsymbol{v}^1 \\ \boldsymbol{r}^2 \\ \boldsymbol{v}^2 \\ \vdots \\ \boldsymbol{r}^K \\ \boldsymbol{v}^K \end{bmatrix}_i + \begin{bmatrix} \boldsymbol{0} \\ \boldsymbol{I} \\ \boldsymbol{0} \\ \boldsymbol{0} \\ \vdots \\ \boldsymbol{0} \\ \boldsymbol{0} \end{bmatrix} \boldsymbol{u}_i^1 + \begin{bmatrix} \boldsymbol{0} \\ \boldsymbol{0} \\ \boldsymbol{0} \\ \boldsymbol{I} \\ \vdots \\ \boldsymbol{0} \\ \boldsymbol{0} \end{bmatrix} \boldsymbol{u}_i^2 + \cdots + \begin{bmatrix} \boldsymbol{0} \\ \boldsymbol{0} \\ \boldsymbol{0} \\ \boldsymbol{0} \\ \vdots \\ \boldsymbol{0} \\ \boldsymbol{I} \end{bmatrix} \boldsymbol{u}_i^K + \begin{bmatrix} \boldsymbol{w}^1 \\ \boldsymbol{w}^2 \\ \vdots \\ \boldsymbol{w}^K \end{bmatrix}_i\ . \tag{29}$$

Measurement Models

In this research, it is assumed that the navigation system onboard each satellite in the formation consists of a GPS receiver and proximity sensors.

The GPS receivers are one of the primary sensors for Earth orbiting spacecraft navigation systems. Most GPS receivers are capable of providing a position solution $\boldsymbol{r}^j$ in the WGS84 frame. The differential position measurement $\boldsymbol{y}^j_{\text{GPS}}$ in the reference frame is then given by

$$\boldsymbol{y}^j_{\text{GPS}} = (\boldsymbol{e}_x, \boldsymbol{e}_y, \boldsymbol{e}_z)^T(\boldsymbol{r}^j - \boldsymbol{r}_{\text{R}}) \; . \tag{30}$$

There is much work currently going into the development of various proximity sensors for the formation flying missions. One of the examples is the VISNAV sensor [7], which is a vision-based system processing measurements from an optical sensor to estimate the sensor location (and orientation) with respect to the known light sources (beacons). The navigation system can also be extended to support pseudolite augmentation systems [8] for GPS-based relative navigation. Although there are diverse measurement types, this paper assumes that the local proximity sensors are capable of providing the relative positions of all the other satellites in the formation given by

$$\boldsymbol{y}^{lj}_{\text{REL}} = (\boldsymbol{e}_x, \boldsymbol{e}_y, \boldsymbol{e}_z)^T(\boldsymbol{r}^l - \boldsymbol{r}^j) \; . \tag{31}$$

The measurement vector in Eq. (2) then becomes

$$\boldsymbol{y}^j_i = \left[(\boldsymbol{y}^j_{\text{GPS}})^T \quad (\boldsymbol{y}^{1j}_{\text{REL}})^T \quad \cdots \quad (\boldsymbol{y}^{(j-1)j}_{\text{REL}})^T \quad (\boldsymbol{y}^{(j+1)j}_{\text{REL}})^T \quad \cdots \quad (\boldsymbol{y}^{Kj}_{\text{REL}})^T \right]^T_i \; . \tag{32}$$

HARDWARE SETUP

The decentralized controller performance is investigated in closed-loop HWIL simulations using two actual space-capable GPS receivers. Figure 1 shows the hardware setup, including: a GPS signal simulator, two GPS receivers, flight computers, and an external controller.

The core component of the facility is the Spirent STR4760 GPS signal simulator, which is capable of simultaneously simulating L1 signals for two vehicles on up to 16 channels each. Each receiver is connected to one of the RF outputs of the simulator via a coaxial cable and provides its position fix solution to the flight computer which runs the decentralized controller. In addition to the STR4760 system, the closed-loop simulation requires the STR4762 remote control option that allows the simulations to be controlled by an external computer in real-time. The external controller runs two satellite trajectory propagators simultaneously and provides position, velocity, and acceleration vectors of the two satellites to the STR4760 system every 100 ms. The propagators are also capable of accepting control vectors from the flight computers for real-time trajectory updates. In this research, the external controller is also modified to provide numerically simulated relative measurements defined in Eq. (31) to the flight computers.

GPS Receivers

The GPS receivers (Figure 2) used in this study are Zarlink's GPS Orion receivers [9], which have been modified for space applications. The original receiver provides C/A code tracking on 12 channels at the L1 frequency. To support user specific software adaptations for the GPS receiver, the GPS Architect development kit was made available by Mitel Semiconductor [10].

For use on low Earth orbit satellites and other space applications, numerous software modifications and enhancements have been made to the original firmware of the Orion receiver [11]. These modifications include the fixes related to the implicit assumption of a low speed vehicle in the Doppler prediction and the time tagging error of the raw measurements. Aside from these fixes, an open-loop Doppler and visibility prediction algorithm has been added to the receiver code to ensure robust tracking and rapid signal acquisition under the conditions of a high-dynamic space vehicle [12].

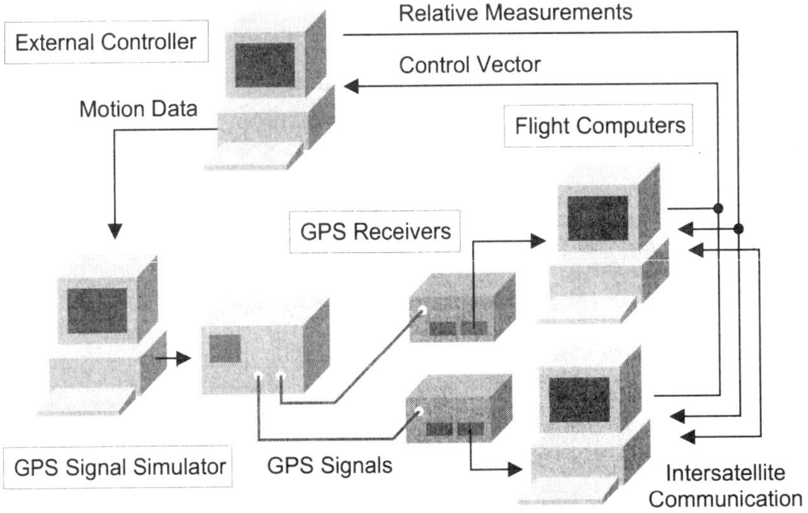

Figure 1: Closed-Loop GPS Test Facility

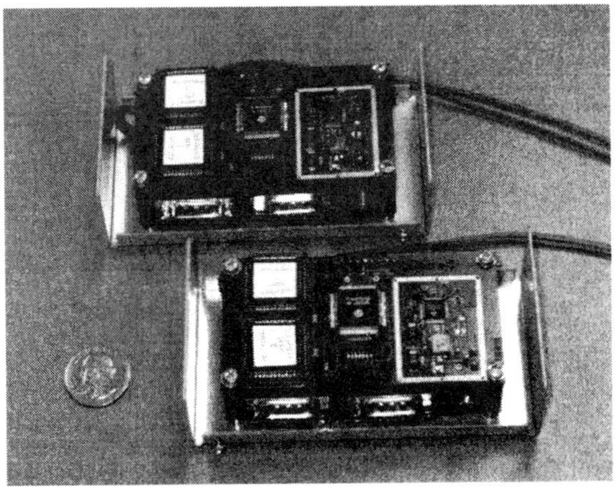

Figure 2: GPS Orion Receivers

Also, an active alignment of measurement epochs and navigation solutions to the integer second of GPS system time has been implemented, which ensures synchronized measurements among multiple independent receivers (typically better than 100 ns in the absence of multipath). To improve the overall navigation performance of the Orion receiver, integrated carrier phase measurements have been made available using a 3rd order phase lock loop (PLL) assisted by a 2nd order frequency lock loop (FLL) [13]. The loop provides accurate tracking and stable acquisition over a wide range of dynamic conditions. Raw measurement accuracies obtained in signal simulator tests are better than 1 m for C/A code pseudorange, 1 mm for L1 carrier phase, and 10 cm/s for L1 Doppler measurements in the absence of environmental error sources such as multipath [14].

Decentralized Controller Implementation

As an example, the sequence of events of the decentralized controller on satellite 1 in a two-satellite formation gives the following result:

1. Given $\{\boldsymbol{r}_0^1, \boldsymbol{v}_0^1, \bar{\boldsymbol{P}}_0^1\}$ and $\{\boldsymbol{r}_0^2, \boldsymbol{v}_0^2, \bar{\boldsymbol{P}}_0^2\}$ at the epoch $i = 0$, initialize the controller

$$\boldsymbol{x}_0^C = \begin{bmatrix} \boldsymbol{r}_0^1 \\ \boldsymbol{v}_0^1 \\ \boldsymbol{r}_0^2 \\ \boldsymbol{v}_0^2 \end{bmatrix} , \quad \hat{\boldsymbol{x}}_0^{D1} = \begin{bmatrix} \mathbf{0} \\ \mathbf{0} \\ \mathbf{0} \\ \mathbf{0} \end{bmatrix} , \quad \boldsymbol{h}_0^1 = \begin{bmatrix} \mathbf{0} \\ \mathbf{0} \\ \mathbf{0} \\ \mathbf{0} \end{bmatrix} ,$$

$$\hat{\boldsymbol{P}}_0 = \begin{bmatrix} \bar{\boldsymbol{P}}_0^1 & \mathbf{0} \\ \mathbf{0} & \bar{\boldsymbol{P}}_0^2 \end{bmatrix} , \quad \hat{\boldsymbol{P}}_0^1 = \begin{bmatrix} \bar{\boldsymbol{P}}_0^1 & \mathbf{0} \\ \mathbf{0} & \bar{\boldsymbol{P}}_0^2 \end{bmatrix} ,$$

   and transmit the local control vector $\boldsymbol{u}_0^1$ if available.

2. Receive the local control vector $\boldsymbol{u}_0^2$ from satellite 2, then propagate the control dependent part of the state state vector via

$$\boldsymbol{x}_1^C = \begin{bmatrix} \boldsymbol{\Phi}^1 & \mathbf{0} \\ \mathbf{0} & \boldsymbol{\Phi}^2 \end{bmatrix} \boldsymbol{x}_0^C + \begin{bmatrix} \mathbf{0} \\ \boldsymbol{I} \\ \mathbf{0} \\ \mathbf{0} \end{bmatrix} \boldsymbol{u}_0^1 + \begin{bmatrix} \mathbf{0} \\ \mathbf{0} \\ \mathbf{0} \\ \boldsymbol{I} \end{bmatrix} \boldsymbol{u}_0^2 ,$$

   and the data dependent part via

$$\bar{\boldsymbol{x}}_1^{D1} = \begin{bmatrix} \boldsymbol{\Phi}^1 & \mathbf{0} \\ \mathbf{0} & \boldsymbol{\Phi}^2 \end{bmatrix} \hat{\boldsymbol{x}}_0^{D1} .$$

   Also propagate the local and global covariance matrices via

$$\bar{\boldsymbol{P}}_1^1 = \begin{bmatrix} \boldsymbol{\Phi}^1 & \mathbf{0} \\ \mathbf{0} & \boldsymbol{\Phi}^2 \end{bmatrix} \hat{\boldsymbol{P}}_0^1 \begin{bmatrix} \boldsymbol{\Phi}^1 & \mathbf{0} \\ \mathbf{0} & \boldsymbol{\Phi}^2 \end{bmatrix}^T + \boldsymbol{Q}_1 ,$$

$$\bar{\boldsymbol{P}}_1 = \begin{bmatrix} \boldsymbol{\Phi}^1 & \mathbf{0} \\ \mathbf{0} & \boldsymbol{\Phi}^2 \end{bmatrix} \hat{\boldsymbol{P}}_0 \begin{bmatrix} \boldsymbol{\Phi}^1 & \mathbf{0} \\ \mathbf{0} & \boldsymbol{\Phi}^2 \end{bmatrix}^T + \boldsymbol{Q}_1 .$$

3. Given the local measurement

$$\boldsymbol{y}_1^1 = \begin{bmatrix} \boldsymbol{y}_{GPS}^1 \\ \boldsymbol{y}_{REL}^{12} \end{bmatrix} ,$$

   update the data dependent part of the local state vector via

$$\hat{\boldsymbol{x}}_1^{D1} = \bar{\boldsymbol{x}}_1^{D1} + \boldsymbol{K}_1^1 \left[\boldsymbol{y}_1^1 - \boldsymbol{H}^1 (\boldsymbol{x}_1^C + \bar{\boldsymbol{x}}_1^{D1}) \right] ,$$

and the local state error covariance matrix via

$$\hat{\boldsymbol{P}}_1^1 = (\boldsymbol{I} - \boldsymbol{K}_1^1\boldsymbol{H}^1)\bar{\boldsymbol{P}}_1^1(\boldsymbol{I} - \boldsymbol{K}_1^1\boldsymbol{H}^1)^T + \boldsymbol{K}_1^1\boldsymbol{R}_1^1(\boldsymbol{K}_1^1)^T ,$$

where

$$\boldsymbol{H}^1 = \begin{bmatrix} \boldsymbol{I} & 0 & 0 & 0 \\ -\boldsymbol{I} & 0 & \boldsymbol{I} & 0 \end{bmatrix} .$$

If the local measurement update is successfully executed, transmit a semaphore $s_1^1 = 1$ to satellite 2. Otherwise, $s_1^1 = 0$.

4. After the reception of the semaphore s_1^2 from the satellite 2, update the global state error covariance matrix via

$$\hat{\boldsymbol{P}}_1 = \left[\bar{\boldsymbol{P}}_1^{-1} + s_1^1(\boldsymbol{H}^1)^T(\boldsymbol{R}_1^1)^{-1}\boldsymbol{H}^1 + s_1^2(\boldsymbol{H}^2)^T(\boldsymbol{R}_1^2)^{-1}\boldsymbol{H}^2\right]^{-1} ,$$

where

$$\boldsymbol{H}^2 = \begin{bmatrix} 0 & 0 & \boldsymbol{I} & 0 \\ \boldsymbol{I} & 0 & -\boldsymbol{I} & 0 \end{bmatrix} .$$

5. Update the data vector $\boldsymbol{h}^1$ via Eq. (16) and obtain $\boldsymbol{\alpha}_1^{11}$ and $\boldsymbol{\alpha}_1^{21}$ via Eq. (19) using

$$\boldsymbol{\Lambda}^1 = \begin{bmatrix} 0 \\ \boldsymbol{I} \\ 0 \\ 0 \end{bmatrix} , \text{ and } \boldsymbol{\Lambda}^2 = \begin{bmatrix} 0 \\ 0 \\ 0 \\ \boldsymbol{I} \end{bmatrix} ,$$

respectively. Then transmit $\boldsymbol{\alpha}_1^{21}$ to satellite 2.

6. From the locally available $\boldsymbol{\alpha}_1^{11}$ and $\boldsymbol{\alpha}_1^{12}$ received from satellite 2, the globally optimal control vector of satellite 1 can be obtained via

$$\boldsymbol{u}_1^1 = -\left[\boldsymbol{V}_1^1 + (\boldsymbol{\Lambda}^1)^T\boldsymbol{S}_1\boldsymbol{\Lambda}^1\right]^{-1}\left\{(\boldsymbol{\Lambda}^1)^T\boldsymbol{S}_1\boldsymbol{x}_1^C + \boldsymbol{\alpha}_1^{11} + \boldsymbol{\alpha}_1^{12}\right\} ,$$

where the controller Riccati matrix $\boldsymbol{S}_1$ is read from a pre-computed memory source. Then transmit the control vector to satellite 2.

7. Update the current epoch i, and go to 2.

In the current implementation, the receipt of the output string from the GPS receiver is used as a trigger for the control process. Although the output string trigger provides enough synchronization for this experiment, more precise alignment between the navigation and control systems can be achieved by utilizing the 1 pulse-per-second (1PPS) signal from the GPS Orion receiver, which is aligned with a UTC or GPS system integer second [15].

CLOSED-LOOP SIMULATION RESULTS

This section presents preliminary closed-loop simulation results of the decentralized control algorithm for satellite formations. Figure 3 depicts the nominal trajectories of the two satellites (◇) selected for this experiment. This example mission could be of interest for synthetic aperture imaging or optical interferometry. Alfriend et al. [16] describes this type of satellite formation in more detail. The orbits of the satellites in the formation differ slightly from the circular orbit of the formation center, so that they suborbit around the center in an elliptical epicycle.

Since the primary goal of this research is not to design a definitive control low for formation flying missions, but to develop and test the capability of demonstrating technologies for such missions, the simulation scenario in this paper is a rather simplified example. The propagator model consists of the

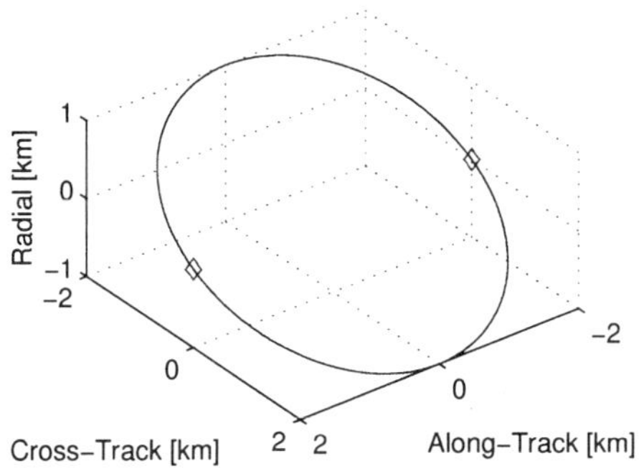

Figure 3: Example Formation Flying Mission

J2 gravitational perturbation and atmospheric drag, and the physical properties of the two satellites are identical. The commanded maneuvers are perfectly executed, and the GPS measurements are bias-free (no broadcast ephemeris, ionosperic delay, and multipath errors). Table 1 summarizes the simulator and controller parameters used in this research. More extensive design and analysis of the distributed control system can be considered as future work.

Table 1: Simulator and Controller Parameters

Start date and time	6-Nov-2001, 16:30:00
GPS YUMA almanac	Week 1138
Initial hub orbit radius, eccentricity, and inclination	400 km, 0, 56 deg
Formation horizontal radius	2 km
Number of Satellites	2
Measurement rate	1/60 Hz
Maneuver rate	1/60 Hz
GPS position fix measurement noise (x, y, z)	3, 1, 1 m (1σ)
Proximity sensor measurement noise (x, y, z)	0.1, 0.1, 0.1 m (1σ)
Initial position uncertainties (x, y, z)	100, 300, 300 m (1σ)
Initial velocity uncertainties (x, y, z)	0.1, 0.3, 0.3 m/s (1σ)
Process noise intensity	9.81e-6 m/sec$^{3/2}$
Cost function position, velocity weight	4e-2, 4e-8
Cost function control weight	1e3

Figures 4 and 5 show the resulting formation trajectories and tracking errors, respectively. It is verified that the decentralized controller can be successfully employed in a real-time HWIL test to maintain the shape of the formation. Figure 6 depicts the local state errors of satellite 1, while Figure 7 shows the reconstructed global state errors. The absolute state accuracy strongly depends on the GPS position fix solutions and less on the relative state measurements. A local positioning accuracy of 0.8 m is achievable in both the along-track and cross-track directions, while the accuracy of the radial component is slightly worse (about 2.5 m) because of the less favorable vertical dilution of precision (DOP). Although the proximity sensor measurements do not affect the global absolute state accuracy very much, they improve the relative position state as shown in Figure 8. The reconstructed global absolute states of the two satellites provide the relative positioning accuracy of 0.08 m, which is similar to the simulated proximity sensor measurement noise defined in Table 1. This indicates that the decentralized controller properly processed the relative measurements.

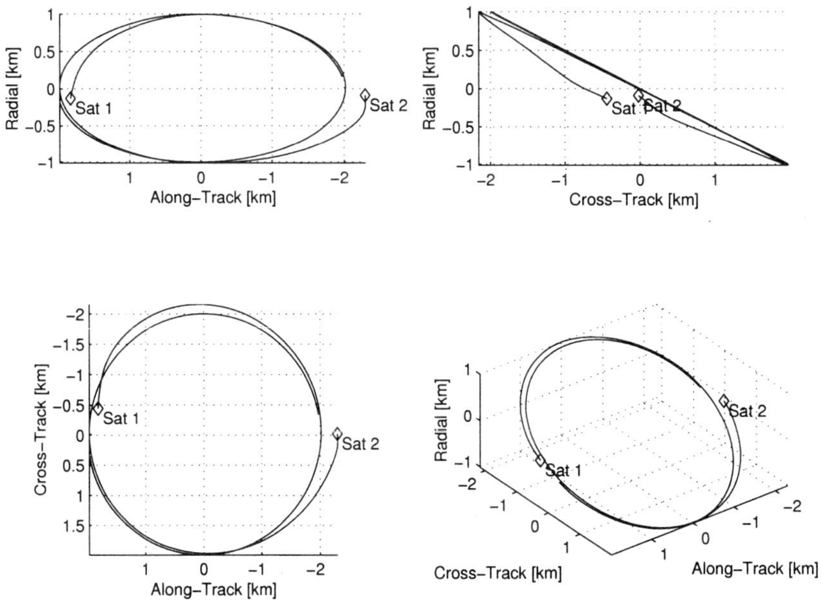

Figure 4: Satellite Formation Flying Result

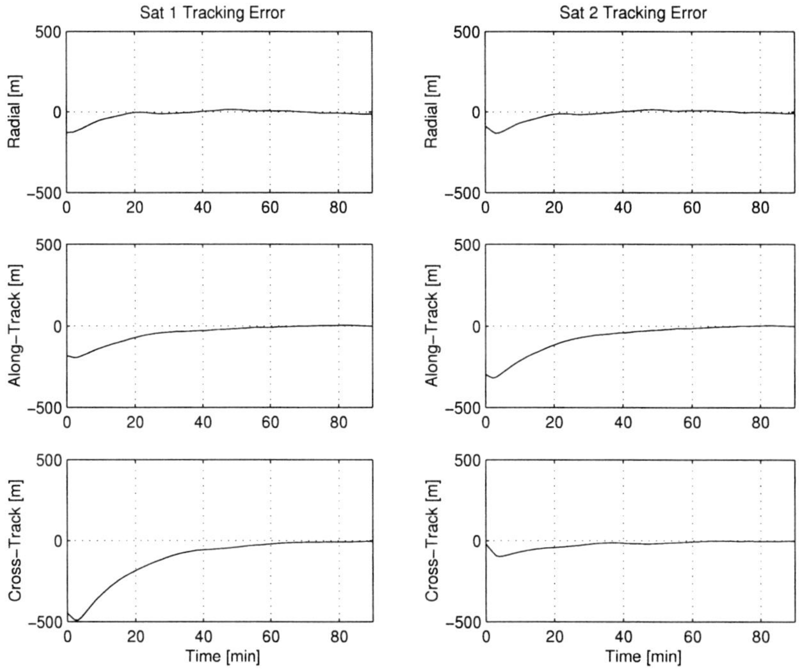

Figure 5: Tracking Errors

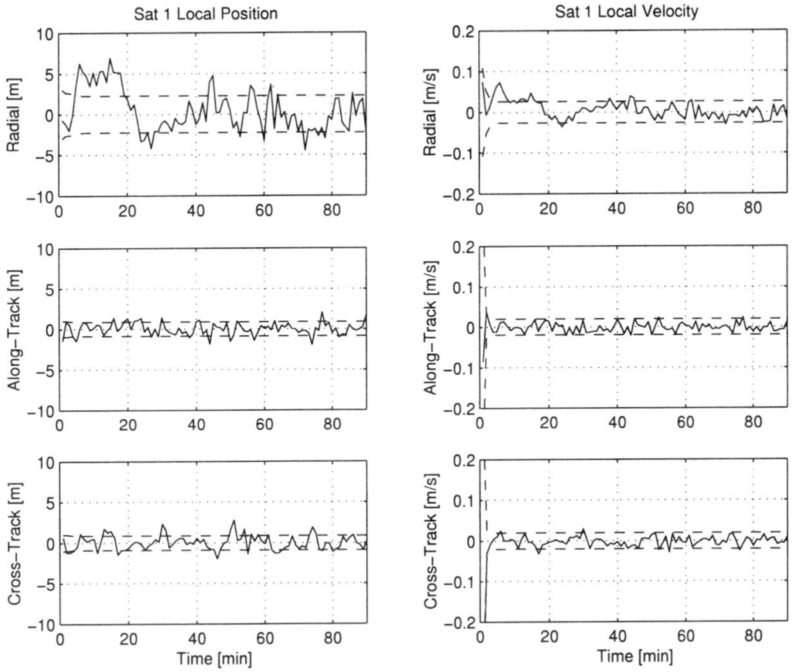

Figure 6: Local State Errors of Satellite 1

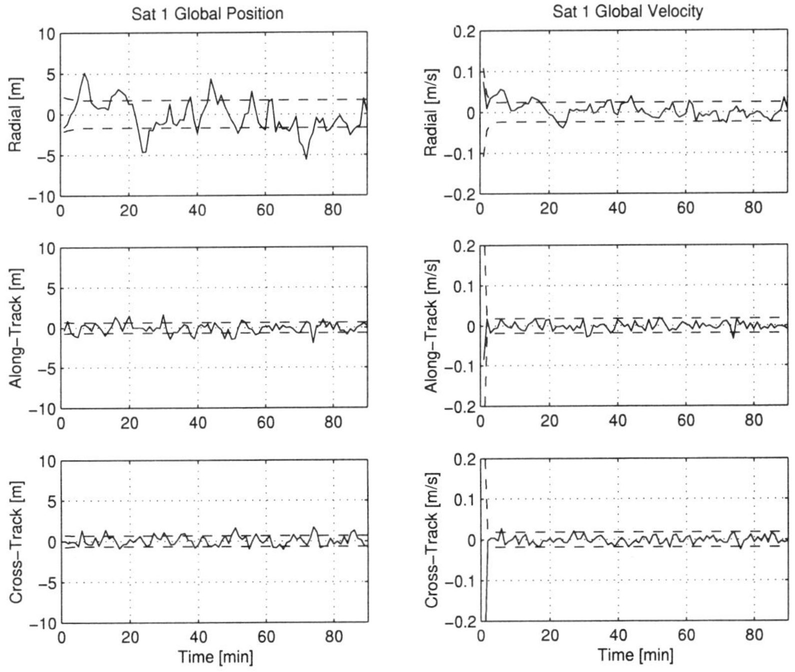

Figure 7: Reconstructed Global State Errors of Satellite 1

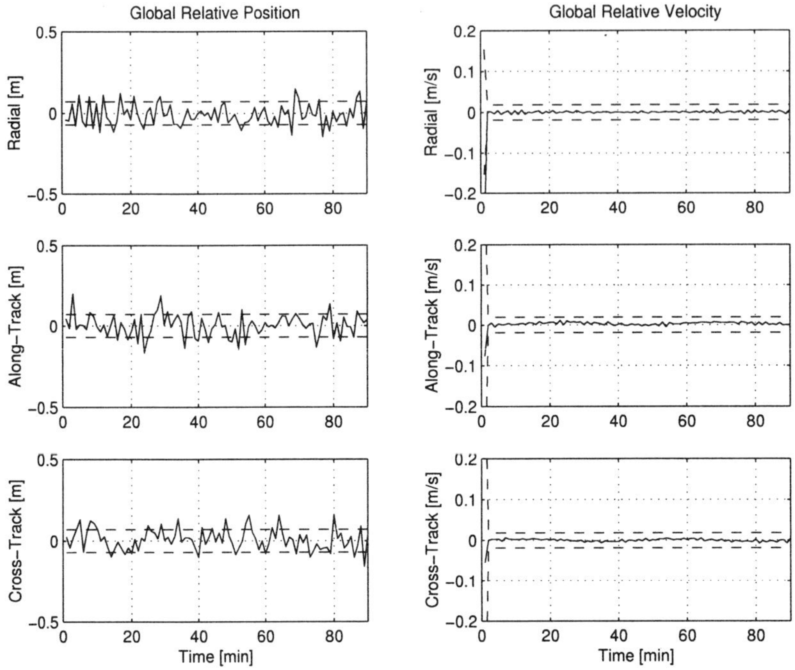

Figure 8: Reconstructed Global Relative State Errors

CONCLUSIONS

A unique closed-loop simulation facility has been developed for design and analysis of GPS-based navigation and control architectures for formation flying satellites. The preliminary results from the decentralized controller implementation show that the designed system is capable of supporting hardware and software technology development for future distributed spacecraft operations.

Although the current simulation setup is capable of simulating only two satellite formations due to local hardware limitations, the system is currently being extended to allow multiple spacecraft formation flying simulations by connecting remotely located GPS simulation systems together through the internet. The inter-spacecraft communication system will then be modeled as an IP network. The developed closed-loop capability has been transferred to the formation flying testbed at NASA Goddard Space Flight Center, which is capable of simulating the GPS signals for up to four spacecraft.

ACKNOWLEDGEMENTS

The authors are grateful to Oliver Montenbruck at German Space Operation Center (GSOC) for his effort to modify the Orion GPS receiver software to work efficiently in the low Earth orbit dynamics environment. This research was completed under NASA Goddard Space Flight Center (GSFC) contract NAG5-11278. The authors would like to thank Russell Carpenter of NASA GSFC for his support.

REFERENCES

[1] Bauer, F. H., Hartman, K., How, J. P., Bristow, J., Weidow, D., and Busse, F., "Enabling Spacecraft Formation Flying through Spaceborne GPS and Enhanced Automation Technologies," Proceedings of ION GPS 1999, pp. 369-384, 1999.

[2] Carpenter, J. R., "A Preliminary Investigation of Decentralized Control of Satellite Formations," 2000 IEEE Aerospace Conference Proceedings, Vol. 7, pp. 63-74, 2000.

[3] Speyer, J. L., "Computation and Transmission Requirements for a Decentralized Linear-Quadratic-Gaussian Control Problem," IEEE Transactions on Automatic Control, AC-24 (2), pp. 266-269, 1979.

[4] Burns, R., MacLaughlin, C. A., Leitner, J., and Martin, M., "TechSat-21: Formation Design, Control, and Simulation," 2000 IEEE Aerospace Conference Proceedings, Vol. 7, pp. 19-25, 2000.

[5] Park, C., How, J. P., and Capots, L., "Sensing Technologies for Formation Flying Spacecraft in LEO using CDGPS and an Inter-Spacecraft Communication System," Proceedings of ION GPS 2000, pp. 1595-1607, 2000.

[6] Busse, F. D., How, J. P., Simpson, J., and Leitner, J., "For-Vehicle Formation Flying Hardware Simulation Results," ION GPS 2002, Portland, Oregon, Sep. 24-27, 2002.

[7] Junkins, J. L., Hughes, D. C., Wazni, K. P., and Pariypong, V., "Vision-Based Navigation for Rendezvous, Docking and Proximity Operations," 22nd Annual AAS Guidance and Control Conference, AAS 99-021, 1999.

[8] Wawrzyniak, G., Lightsey E. G., Key, K. W., Bell, J. W., Majka, M., and Provence, R. S., "Ground Experimentation of a Pseudolite-Only Method for the Relative Positioning of Two Spacecraft," ION GPS 2001, Salt Lake City, Utah, Sep. 11-14, 2001.

[9] Zarlink, "GPS Orion 12 Channel GPS Reference Design," Zarlink Semiconductor, AN4808, Issue 2.0, October 2001.

[10] Mitel, "GPS Architect 12 Channel GPS Development System," Mitel Semiconductor, DS4605, Issue 2.5, March 1997.

[11] Montenbruck, O., Markgraf, M., Leung, S., and Gill, E., "A GPS Receiver for Space Applications," ION GPS 2001, Salt Lake City, UT, Sep. 12-14, 2001.

[12] Leung, S., Montenbruck, O., and Bruninga, B., "Hot Start of GPS Receiver for LEO Microsatellites," NAVITECH 2001, Noordwijk, Dec. 10-12, 2001.

[13] Kaplan, E. D. (editor), "Understanding GPS: Principles and Applications," Artech House, Boston, 1996.

[14] Montenbruck, O., and Holt, G., "Spaceborne GPS Receiver Performance Testing," DLR-GSOC TN 02-04, 2002.

[15] Leung, S., "Pulse Per Second (PPS) Implementation for the Mitel GPS Architect Receiver and Mitel Orion Receiver," DLR-GSOC TN 01-02, 2001.

[16] Alfriend, K. T., Schaub, H., and Gim D. W., "Gravitational Perturbations, Nonlinearity and Circular Orbit Assumption Effects on Formation Flying Control Strategies," AAS Rocky Mountain Guidance & Control Conference, AAS 00-012, Breckenridge, CO, Feb. 3-7, 2000.

ELECTROMAGNETIC FORMATION FLIGHT

Raymond J. Sedwick[*] and Samuel A. Schweighart[†]

The ultimate limiter of mission lifetime for current space systems is onboard propellant supply. Spacecraft clusters that are insensitive to their center of mass location can lift this constraint by using internally generated electromagnetic forces between vehicles as an alternative formation control. Such systems include rendezvous and docking applications and sparsely distributed apertures for interferometry. The dipole nature of the electromagnetic force allows for full control of the relative degrees of freedom, position and orientation, provided reaction wheels or control moment gyros are used for angular momentum storage or control. This paper briefly presents the background research in this area at the MIT Space Systems Laboratory, and develops the theory behind electromagnetic formation flight under the far field approximation, giving examples of the 2-D and 3-D dynamics of multiple spacecraft clusters. Discussion of recent work on determining dipole configurations is included.

INTRODUCTION

Recent research at the MIT Space Systems Laboratory (SSL) has examined the use of steerable electromagnetic dipoles and reaction wheels as a means of controlling the relative positions and orientations of a cluster of formation flying spacecraft. This technology was independently conceived by Hashimoto, et. al. at the Institute of Space and Astronautical Science in Japan[1]. The motivation for this effort stems from two factors that make the use of thrusters for this purpose undesirable. One is the inherent limit on operational lifetime, caused by the need for a non-replenishable resource such as propellant. High specific impulse electric propulsion has gone a long way to extend this operational lifetime for reasonable propellant fractions, but eventually all propellant based systems will exhaust their onboard supply. A second consideration is for systems that do optical sensing, specifically near the infrared. Propellant exhaust can affect these systems in the short term by causing a hot plume of gas that can obscure the sensors, and in the long term by directly coating the optics with effluents. Using only electromagnets and reaction wheels means that the only consumable is power, which through the use of solar arrays is a renewable resource.

There are also two main drawbacks to electromagnetic formation flight (EMFF). The first is that only the relative degrees of freedom can be actuated, meaning that there is no mechanism to move the center of mass of the system. However, most applications where multiple spacecraft comprise a cluster are insensitive to the absolute position of the cluster as a whole. Examples include sparse aperture interferometry and rendezvous/docking. If moving the center of mass of the system is required, then a hybrid system could be constructed using both EMFF and thrusters. This may extend the operational lifetime of the system beyond using thrusters alone, but not indefinitely, since propellant is still necessary. The second drawback is the limited spatial extent over which EMFF can be effectively applied. As will be discussed shortly, the force law between two dipoles drops off as separation to the fourth power, meaning that doubling the separation will reduce the force by a factor of 16.

---

* Principal Research Scientist, MIT Space Systems Laboratory, 77 Massachusetts Avenue, Cambridge, Massachusetts 02139. E-mail: sedwick@mit.edu.

† Doctoral Candidate, MIT Space Systems Laboratory, 77 Massachusetts Avenue, Cambridge, Massachusetts 02139. E-mail: schwghrt@mit.edu.

To date, the majority of research in the SSL has focused on system trades[2], with some feasibility simulations of maneuvers that would be of particular interest to NASA missions such as Terrestrial Planet Finder (TPF)[3] and the Submillimeter Probe of the Evolution of Cosmic Structure (SPECS)[4,5]. In the system trades, the use of High Temperature Superconducting (HTS) wire is considered an enabler for this technology, and a testbed to validate the feasibility of such a system is currently under construction. Trades that baseline the current state of the art in this technology show performance that is competitive, and in some cases superior to thruster based systems, and future enhancements in HTS wire current density will favor EMFF even more. In the meantime, successful implementation of EMFF requires an efficient way of calculating the dipole strengths and orientations that are necessary (or at least sufficient) to provide the desired motion of each spacecraft in the cluster. This paper discusses the dynamics of such systems, giving specific examples in both 2-D and 3-D, and addresses the problem of finding a solution to the dipole configuration for a given prescribed motion.

ELECTROMAGNETIC DIPOLE INTERACTION

An assumption that has been used quite liberally to date is that the interaction between electromagnetic coils can be considered to be an ideal dipole-dipole interaction. The validity of this assumption increases with separation, and as a rule of thumb can be applied with negligible error if the separation distance of the coils is about an order of magnitude larger than their radii. This is known as the 'far-field' approximation. In this regime, the magnetic field created by a dipole (A) at some point in space (B) is given by

$$\vec{B}_A = \frac{\mu_o}{4\pi}\left[3\frac{(\vec{m}_A \cdot \vec{r}_B)}{r_B^5}\vec{r}_B - \frac{\vec{m}_A}{r_B^3}\right] = \frac{\mu_0}{4\pi}\left(\frac{m_A}{{r_B}^3}\right)\left[3(\hat{m}_A \cdot \hat{r}_B)\hat{r}_B - \hat{m}_A\right] \tag{1}$$

where μ_o is the permeability of free space, $\vec{m}_A$ is the magnetic moment and $\vec{r}_B$ is the position vector from dipole A to point B. The field strength is seen to drop off as separation distance cubed. For coils of wire, the dipole moment magnitude is expressed as $m = NiA$, where N is the number of turns of wire, i is the current, A is the cross sectional area enclosed, and the dipole direction is normal to A, with i flowing in a right handed sense around the positive normal. Introducing a second dipole (B), the potential energy of the second dipole in the field generated by dipole A is given as

$$U_B = -\vec{m}_B \cdot \vec{B}_A \tag{2}$$

and it can be seen that the minimum energy configuration is when the dipole is aligned with the local field. From this potential, the force on dipole B can be found by taking the gradient with respect to the position of this dipole, resulting in

$$\vec{F}_B = -\nabla U_B = \vec{m}_B \cdot \nabla \vec{B}_A \tag{3}$$

Similarly, the torque on dipole B can be found by taking the gradient of the potential relative to the orientation of this dipole, giving after some manipulation

$$\vec{T}_B = \vec{m}_B \times \vec{B}_A \tag{4}$$

Carrying out these operations explicitly, we find the force and torque relations can be expressed as

$$\begin{aligned}\vec{F}_B &= \frac{3\mu_o}{4\pi}\left(\frac{m_A m_B}{r_B^4}\right)\left\{[5(\hat{m}_A \cdot \hat{r}_B)(\hat{m}_B \cdot \hat{r}_B) - (\hat{m}_A \cdot \hat{m}_B)]\hat{r}_B - (\hat{m}_B \cdot \hat{r}_B)\hat{m}_A - (\hat{m}_A \cdot \hat{r}_B)\hat{m}_B\right\} \\ \vec{T}_B &= \frac{\mu_o}{4\pi}\left(\frac{m_A m_B}{r_B^3}\right)\left[3(\hat{m}_A \cdot \hat{m}_B)(\hat{m}_B \times \hat{r}_B) - (\hat{m}_B \times \hat{m}_A)\right]\end{aligned} \tag{5}$$

and several noteworthy results can be identified. First we see that although the force drops off as the separation distance to the fourth power, the torque only drops off as separation distance cubed. It is for this reason that magnetic torque rods can be used against the Earth's magnetic field (principally a dipole) rather effectively on a spacecraft to control orientation. However, since the force on the spacecraft in LEO from this interaction is some six orders of magnitude smaller than the torque, it is ineffective in controlling the mass center of the vehicle. The second item of note is that by swapping the subscripts A and B in the force expression, and noting that $\vec{r}_B = -\vec{r}_A$, the force on dipole A is seen to be equal and opposite to that on dipole B, conserving linear momentum, which is of course to be expected. However, performing this same operation on the expression for torque does not result in the same conclusion. The reason is that the torques generated are themselves not equal and opposite, since non-radial forces can be generated between the dipole pairs. The true statement of angular momentum conservation comes from summing the total torque on the system as

$$\sum \vec{T} = \vec{T}_A + \vec{T}_B + \vec{r}_B \times \vec{F}_B \tag{6}$$

and evaluation of this sum shows that the total torque on the system is indeed zero. By placing reaction wheels on each vehicle, the kinematics that result from the application of each of the torques in Eq. (6) can be completely controlled, allowing each of the relative degrees of freedom to be controlled independently.

A more intuitive feel for the forces and torques generated in a single plane can be achieved by using an electrostatic analogy, and looking at the two specific orientations shown in Figure 1. In the far field approximation, the electrostatic and electromagnetic dipole interactions are the same, so we can imagine that the north pole of the magnet is a positive charge, and the south pole is negative, and examine the radial forces that each monopole exerts on the other.

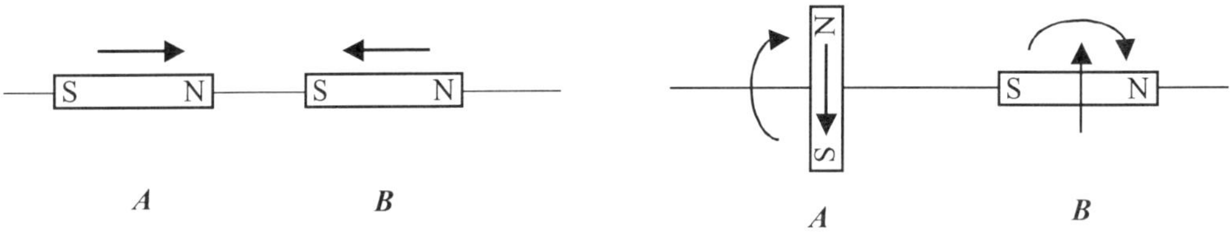

Figure 1. Two Dipole Configurations that Provide Different Forces and Torques

Looking at the left side of Figure 1, the north (positive) pole of A attracts the south (negative) pole of B, but repels the north pole of B. Since the south pole is closer than the north, the net effect is an attraction between the dipoles. The opposite line of reasoning applies for the south pole of A (repelling B), but since it is farther away the effect is less and an attraction results. In the configuration on the right side, the north pole of A not only attracts the south pole of B to the left, but also up. Likewise, the south pole of A pushes the south pole of B to the right, as well as up. By symmetry, we would expect that the left and right forces would cancel, but a net upward force on the south pole of B remains. Similarly, a net downward force on the north pole of B would exist, although with a lower strength, since this pole is farther away from A. The result is then a net upward force, as well as a torque tending to rotate dipole B in a clockwise manner. Repeating the analysis shows that dipole A is forced down, but also rotated clockwise. The torque generated on A in this configuration is twice that generated on B, due to the moment arms presented by each. This factor of two is not obvious from the picture, but can be easily shown. Although both dipoles are torqued in a clockwise manner, the downward force on A and equal but opposite upward force on B produce the counterclockwise torque on the system that maintains angular momentum conservation.

ELECTROMAGNETIC COIL PERFORMANCE

Before developing the dynamics associated with electromagnetic formation flight (EMFF), it is instructive to discuss some of the necessary operating conditions and their implications. The scaling of an EMFF system can be most easily seen by starting with Eq. (5), and evaluating the force expression for the dipole configuration given on the left side of Figure 1. Assuming both dipoles have the same strength, the equation for the purely axial force between the two dipoles is then given as

$$F = \frac{3\mu_o}{2\pi}\frac{(NiA)^2}{r^4} = \frac{3\pi}{2}\mu_o(Ni)^2\left(\frac{R}{r}\right)^4 \tag{7}$$

where the expression for the dipole moment in terms of area and current has been used, and a circular coil of N turns and radius R is assumed. For this far-field approximation to be valid the ratio of R/r must be at most 0.1 – 0.2, and under this assumption, the second term in parenthesis would evaluate to 10^{-4}-10^{-3}. The permeability in MKS units is $4\pi x10^{-7}$ and for the sake of argument we will assume that formation flight would require thrust levels in the milli-Newton range, although fine positioning could be much smaller, and aggressive maneuvering could be much larger. The number of amp-turns that must be present is then on the order of 10^5.

For resistive materials, the current density is limited by the cross-section of the conductor, since the ohmic power generated scales as the volume of the material, whereas the radiative cooling scales as the surface area. Equating these per unit length of the coil (assumed circular in cross section), the number of amp-turns can be related to the total conducting cross section as

$$Ni = \left(\frac{2\sqrt{\pi}\sigma T^4}{\rho_r}\right)^{1/2} A^{3/4} \approx 6x10^5 A^{3/4} \qquad \text{(copper at } 150^o C) \tag{8}$$

where σ is the Stefan-Boltzmann constant, T is the temperature and ρ_r is the resistivity. The numerical value is for copper at 150°C, since for multiple windings this represents the temperature at which most magnet wire insulation will melt. For the value of $Ni = 10^5$ given above, this implies a cross section of 850 cm^2. This coil would most likely dominate the mass of the spacecraft, since even a 2-meter diameter coil with this cross section would have a mass of over 4500 kg.

In light of this result, it makes sense to consider the applicability of High Temperature Superconductors. Since ohmic heating is no longer a contributing factor, much smaller cross sections of conductor can be used. Experimentally, it has been shown that some high temperature superconductors can carry a current density as high as 10^6 amps/cm^2, dropping to around half that value in the presence of very high (8 Tesla) magnetic fields[6]. With these current densities, the cross-section would be less than a square centimeter. There is clearly considerable technological overhead associated with achieving large coils of HTS, such as cryogenic cooling and the control of large currents, however it appears that the mass overhead associated with the technology is still quite attractive, and worth exploring. Currently, a testbed is being fabricated at the MIT SSL, which uses HTS wire provided by American Superconductor of Westborough Massachusetts. The wire must be cooled to a maximum temperature of 110 K before it will transition to superconducting, but the use of liquid nitrogen (77 K) provides a straightforward way in the laboratory to achieve this condition. The margin also allows for coils to be made, since the 110 K limit applies to only a single wire, and if wires are conducting side-by-side, a lower temperature is necessary to maintain superconductivity in the presence of the B-fields induced by other wires. Although the current density in this HTS wire is only about 13 kA/cm^2, it represents the state of the art in the technology and still provides a significant improvements over copper wire.

ELECTROMAGNETIC FORMATION FLIGHT DYNAMICS

One of the practical difficulties that exists with EMFF is the coupling among the dynamics of each vehicle in the cluster. With thruster based systems, an error in the position or orientation of a spacecraft can be adjusted by that spacecraft alone, without affecting the others. However with the force on each vehicle determined by (and affecting) each of the other vehicles, maneuvers must be determined on a more global scale. Two questions that arise are 1) what are the allowable motions of the system, and 2) what distribution of dipoles (and reaction wheels or CMGs) is necessary and/or sufficient to provide these motions?

We will start by considering the motion in a plane. Figure 2 shows a cluster of 4 spacecraft, with arrows indicating the direction that each dipole points on its vehicle. To allow the vehicle to steer its dipole independently of its orientation, two orthogonal coils are placed with a common origin, and with their normals tangent to the plane of motion. In the same far-field limit that the dipole approximation applies, two orthogonal coils will independently produce components of a steerable dipole field. This can be seen by taking Eq. (5), and assuming that two dipoles ($\vec{m}'_A$ and $\vec{m}''_A$) are collocated and operate independently on vehicle B. Summing the forces that act on B, and recollecting terms will show that there is an effective dipole $\vec{m}_A = \vec{m}'_A + \vec{m}''_A$ present on vehicle A. The reason they must be collocated can also be seen, since an offset of the centers of the coils will create higher terms that are not consistent with the dipole formulation, and in fact represent higher order poles. The relative strength of these poles will of course become less significant as the separation distance increases.

If there are n vehicles, there are then $2n$ values that can be set (the components of each dipole) to determine the forces on each of the spacecraft. There are also $2n$ force components that must be satisfied, and both the force and dipole components can be evaluated in any convenient frame, such as the inertial frame shown in Figure 2. However, since only internal forces can be generated between the vehicles, they must come as equal and opposite pairs, was shown earlier. Therefore, only $2n$-2 components of force can be arbitrarily chosen, with the remaining two set to the values necessary to conserve linear momentum. A consequence of this is that there are only $2n$-2 constraints on the system, leaving one dipole to be freely chosen, while still allowing all forces to be achieved. This same situation occurs in three dimensions, since the third component of the 'last' force in the summation is also constrained to set the sum to zero, again freeing one component of a dipole in the 3rd dimension.

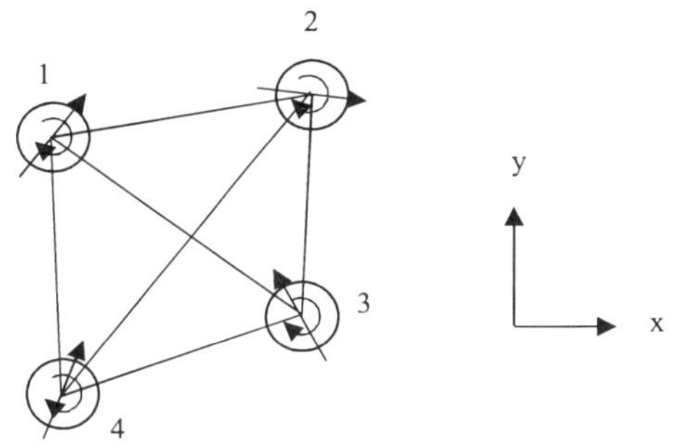

Figure 2. Arbitrary Cluster of Vehicles with Steerable Dipoles

If the system of equations were linear, this would be the final word. However, the force equations are quadratic in the dipole strength, and dominated by cross-terms, and as such we cannot know a priori how many additional solutions may be present. This problem will be re-visited later.

The freedom of multiple solutions is convenient, since it allows for consideration of other system parameters. For example, a given configuration of dipoles that produce the desired force profile will also by necessity create torques on each of the vehicles, as shown in Eq. (5). It is very unlikely that these torques are going to correspond to the angular accelerations that are desired of each vehicle. Therefore, some type of inertial torquing devices, such as reaction wheels or CMGs are necessary to counter these torques. Having some freedom to choose the dipole configuration may allow a more even distribution of angular momentum among the vehicles, thus keeping any one vehicle from reaching a saturation point. However, in the examples to follow, a natural geometric constraint on the free dipole is chosen for simplicity.

There is one subtle point to the ability to select at least 2 of the $2n$ parameters arbitrarily. Referring back to Eq. (5), one sees that in the force equation, the direction of the force is determined by a superposition of terms that are aligned either with the radial vector between the dipoles or with either of the dipoles themselves. Therefore, the ability to arbitrarily set the force direction relies on having these three vectors span the space where the force is to be defined. A degenerate case exists when each of the dipoles is aligned with the radial vector ("— —"), as in the left side of Figure 2. Here, all three vectors are collinear, and none present a basis vector for extending orthogonally. Therefore there is no way to produce an off-axis force in this configuration. An alternative example is when both vectors are arranged vertically ("| | "). In this configuration the radial vector is orthogonal to both dipole vectors, so it appears at first as though this configuration could support any force. However repeating the force Eq. (5) as Eq. (9) below, one sees that having both dipoles orthogonal to the radial vector eliminates them from the equation entirely.

$$\vec{F}_B = \frac{3\mu_o}{4\pi}\left(\frac{m_A m_B}{r_B^4}\right)\{[5(\hat{m}_A \cdot \hat{r}_B)(\hat{m}_B \cdot \hat{r}_B) - (\hat{m}_A \cdot \hat{m}_B)]\hat{r}_B - (\hat{m}_B \cdot \hat{r}_B)\hat{m}_A - (\hat{m}_A \cdot \hat{r}_B)\hat{m}_B \qquad (9)$$

This demonstrates that there are degeneracies associated with selecting dipole directions. It is therefore not allowable to select the two dipole directions as the free parameters, leaving the two dipole strengths as the parameters to be chosen. With only two spacecraft, an additional reason for this constraint is that the two magnitudes are really only a single parameter, since they appear as a product. This however is unique to the two spacecraft case, as additional vehicles allow a mapping of the n dipole strengths into n dipole strength products. The above constraint still holds, however, and the free parameters are to be chosen as a magnitude and a direction, or simply one complete dipole.

Examples of 2-D Vehicle Dynamics

Two Vehicle Case. In 2-D, a two-spacecraft maneuver is easily calculated, since selecting one of the dipoles results in a linear equation that determines the second one exactly. Initially, we will start with a simple system of two spacecraft at rest. Figure 3 shows the result of a spin-up simulation with two vehicles. The gray lines do not indicate a physical connection, but they are included to show the relative positions of both vehicles at different positions in time. The arrows point in the direction of the magnetic dipole axis, along the north pole. Initially their dipoles are in a "| –" configuration along the horizontal axis. This is a similar configuration to that of the right side of Figure 2, except the vertical dipole is inverted.

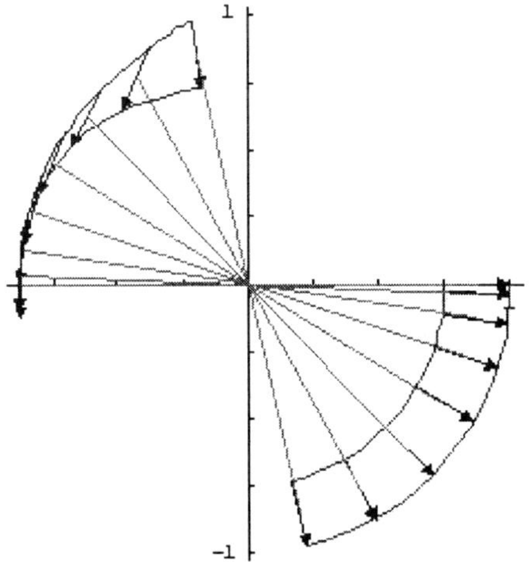

Figure 3. Simulation of Two Spacecraft Spin-up Maneuver

As discussed before, in this configuration there is no radial force between the vehicles, only a tangential force. As the spacecraft begin to rotate, the dipoles are steered so that an attractive component of their force is in the radial direction in order to counteract the increasing centrifugal acceleration. The torque applied to each vehicle must be absorbed through the use of reaction wheels. At the end of the maneuver, the vehicles' dipoles are completely aligned and only produce and attractive radial force to counteract the centrifugal acceleration. The geometric constraint that has been chosen is to maintain one of the dipoles radially from the center of rotation, allowing the other to rotate as necessary. In this particular maneuver, the strength of the free dipole was also held constant.

Three Collinear Vehicle Case. An extension of this maneuver is to consider the spin-up of three collinear vehicles. Figure 4 shows the three spacecraft initially at rest on the horizontal axis. The left simulation shows the three vehicles initially in a "— | —" configuration, while the simulation on the right shows the vehicles initially in a "| — |" configuration.

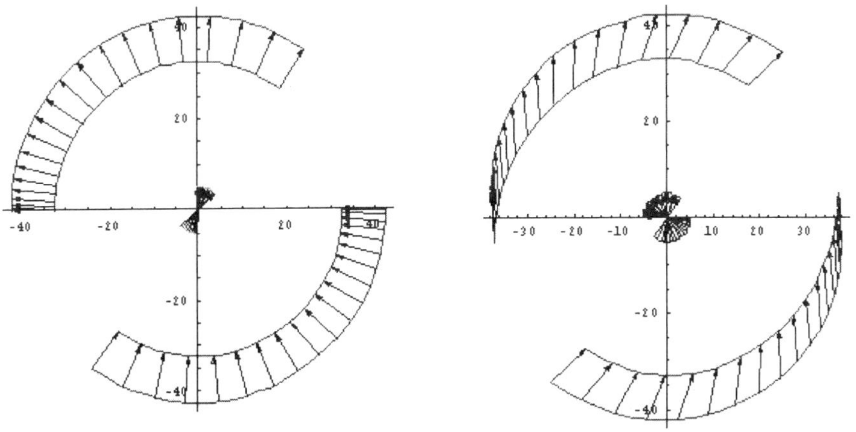

Figure 4. Simulation of a Three Spacecraft Spin-up Maneuver

The two different configurations do not differ in the total system angular momentum, but they do differ greatly in how the angular momentum is distributed among the vehicles. It was mentioned earlier that the torque applied to the dipole that is perpendicular to the line between the two vehicles has twice the torque applied to it than the other vehicle. In the left configuration, the center vehicle has twice the torque applied to it from the left vehicle than the left vehicle receives. It also has twice the torque applied to it from the right vehicle than the right vehicle receives, and both of these contributions are in the same direction. The ratios of applied torques are therefore 1:4:1. The outer vehicles' reaction wheels must absorb approximately 16% of the total angular momentum each, while the center vehicle must absorb about 68% of the total angular momentum. In the right configuration, the outer vehicles each receive twice as much torque as they apply to the center vehicle. Again the torques applied to the center vehicle are in the same direction, so the ratios of applied torques are now 1:1:1, and each vehicle's reaction wheel absorbs the same amount of angular momentum.

If the spacecraft formation is made up of identical vehicles, the right configuration would be preferable because each vehicle would have the same reaction wheel system and all vehicles would absorb the same amount of angular momentum. However, if the center vehicle had a larger reaction wheel, and the outer vehicles had less massive reaction wheels, the total angular momentum of the system and the force necessary to hold the outer vehicles in place would be less in the final spinning condition. One would then reason that the total system mass could then be reduced. This is true up to the point that the reaction wheel and coil masses become 'small' in comparison to the other subsystems. In either case, the choice between identical vehicles or a more massive central vehicle occurs at the system level, and must consider other factors such as manufacturing cost and loss of redundancy by having a cluster of different vehicles.

TPF Configuration: Five Co-linear Spacecraft. While the spin-up of two and three spacecraft formations is fairly straight forward, more complex formations can also be controlled. One design of the Terrestrial Planet Finder (TPF)[3] mission utilizes five co-linear spacecraft. Figure 5 shows five vehicles initially at rest on the horizontal axis. The dipoles are aligned in a "| – | – |" fashion with each vertical dipole nearly perpendicular to each horizontal. Once again the grey lines do not indicate a physical connection, but an imaginary axis between the vehicles at different times during the simulation. While the outer (**A**) and inner (**B**) vehicles have perpendicular dipoles, the center (**C**) vehicle's dipole is initially offset. The reason for this is to counteract the attraction force between the two inner spacecraft that have their axes aligned. With multiple spacecraft formations, the control becomes increasingly complex because the number of interactions between *n* vehicles increases an *n(n-1)*.

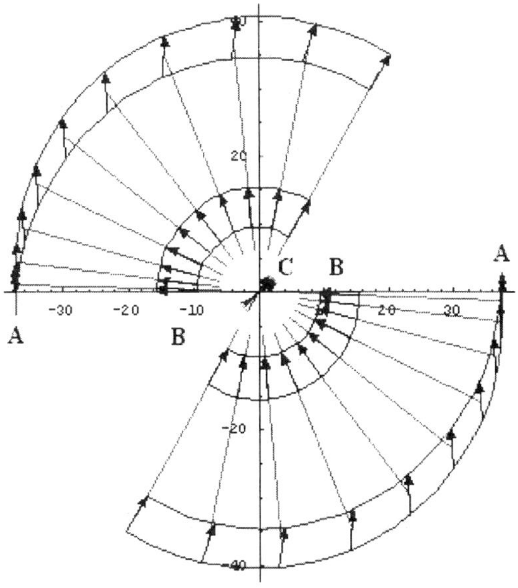

Figure 5. Simulation of Five Vehicle Spin-up Maneuver

Just as with the three collinear spacecraft spin-up problem, there are also multiple solutions to the five vehicle spin-up problem. One could easily imagine a dipole order in a "– | – | – " configuration, and such solutions have been found. Also, solutions where the center or inner vehicles produce an outward radial force have been found. In these configurations the outer vehicles must counteract this force in addition to the centrifugal acceleration. These solutions, while not optimal in terms of magnet sizing, do allow for different angular momentum distributions, and demonstrate the increasing complexity of solutions for multiple spacecraft systems.

Equilateral Triangle Configuration. While all of the previous configurations have been exclusively linear systems, this is certainly not a requirement. An equilateral triangle configuration, which may have application to interferometry with a Golay Configuration, can also be spun-up from rest. See Figure 6 below.

The configuration shown begins rotating with the dipoles located at the 3, 7 and 11 O'clock positions, and each has both a tangential component as well as a radial component. As the formation spins up, the dipoles once again change their orientation in order to counteract the increasing centrifugal acceleration. In the linear formations, the dipoles always align themselves with the radial vector in the steady state. In the equilateral triangle configuration they end up perpendicular to the radial vector. The final steady state configuration is shown with gray radial lines.

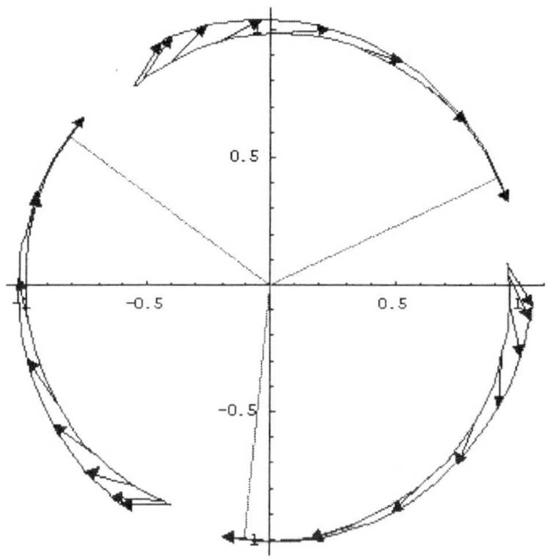

Figure 6. Simulation of Equilateral Triangle Spin-up Maneuver

Examples of 3-D Vehicle Dynamics

Helical Path. The true test of the EMFF concept is the ability to guide the spacecraft through arbitrary, and complex 3-D maneuvers. The ones shown in this section are not for any specific purpose, other than to demonstrate this capability. One potential issue that arises in 3-D motion that was not present in 2-D is the gyroscopic effect caused by the reaction wheels. Thus far, no specific treatment of the angular momentum management has occurred, since in 2-D it is dynamically trivial to apply the necessary torques for the spacecraft to resist rotation. The only caveat is to insure that the reaction wheel has sufficient angular momentum storage that it can complete the maneuver without saturating. In 3-D, the potentially large angular momentum vector present due to the spinning reaction wheels will come into play if they are rigidly mounted to the spacecraft, and a change in attitude is prescribed. However, since the dipoles are assumed steerable, we will only consider cases in the current work where attitude is to be held fixed throughout the maneuver. This situation is likely in the case of interferometric imaging of a distant target. The first case considered is shown below in Figure 7.

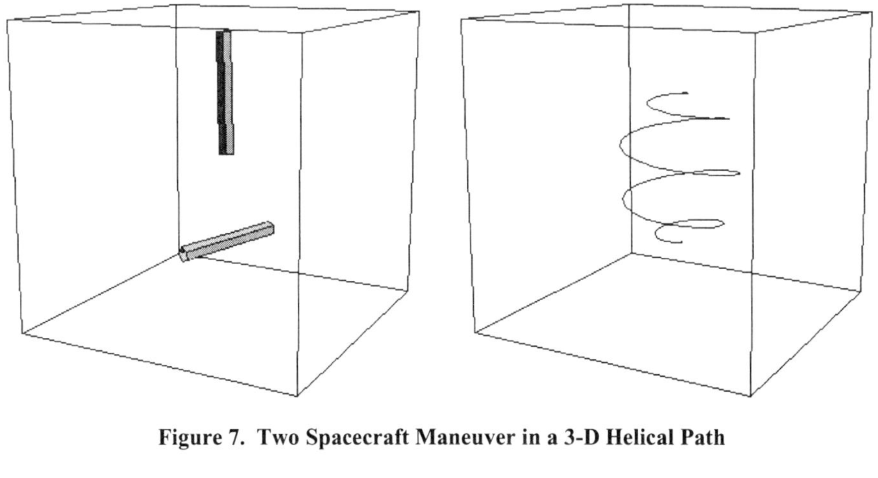

Figure 7. Two Spacecraft Maneuver in a 3-D Helical Path

In this maneuver the dipoles start out with their centers aligned along the vertical axis. The bottom dipole is horizontal, and the top dipole is aligned with the vertical axis. The top dipole follows the helical trajectory shown on the right of Figure 7. As the top dipole follows the path, its dipole direction is constrained to always point toward the center of mass of the system, whereas the orientation of the second one is determined to create the required forces. Because there are only two dipoles, the path of the second one is clearly the complement of the one shown, with opposite helicity.

Another, slightly more complex maneuver is shown below in Figure 8. Here we have a four-dipole maneuver, with two dipoles diametrically opposed on a horizontal circular path, and two dipoles diametrically opposed on a vertical circular path. The shading indicates matched sets of dipoles at an instant in time. The lightest dipoles are the initial locations, progressing through time to the darkest shading. The light dipole at the highest point of the vertical path is the one that is constrained, again chosen to always point toward the center of mass of the system. The other three-dipole orientations are calculated based on the required forces. The phasing of the dipoles around their paths is such that the four dipoles always remain in a plane.

Although it cannot be seen in this still image, the animation of this maneuver shows the first case of multiple solutions. The intersection of solutions appears to occur when the dipoles traveling in the horizontal circle cross the intersection points of the two paths. At this point, it appears that the gradient search technique locks on to the alternate solution. A method for dealing with the presence of multiple solutions is discussed in the next section.

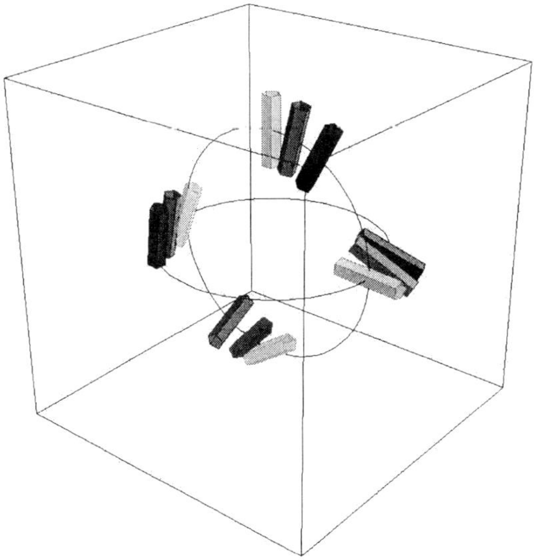

Figure 8. Four Dipole Maneuver in Orthogonal Planes

SOLVING FOR THE DIPOLE CONFIGURATION

It was mentioned earlier that the nonlinearity of the problem gives rise to multiple solutions, even after fixing one of the dipoles. Although this can be beneficial form the standpoint of limiting angular momentum build-up, or perhaps to optimize some other parameter of interest, it can present problems when trying chart a continuous solution of the dipole orientations over a complex maneuver. Returning yet again to Eq. (9), we can write the general solution for the force generated at dipole j, as a summation of contributions over all other dipoles i, resulting in

$$\vec{F}_j = \sum_{i \neq j} \frac{3\mu_o}{4\pi r_{ij}^4} \left\{ 5(\vec{m}_i \cdot \frac{\vec{r}_{ij}}{r_{ij}})(\vec{m}_j \cdot \frac{\vec{r}_{ij}}{r_{ij}})\frac{\vec{r}_{ij}}{r_{ij}} - (\vec{m}_i \cdot \vec{m}_j)\frac{\vec{r}_{ij}}{r_{ij}} - (\vec{m}_j \cdot \frac{\vec{r}_{ij}}{r_{ij}})\vec{m}_i - (\vec{m}_i \cdot \frac{\vec{r}_{ij}}{r_{ij}})\vec{m}_j \right\} \tag{10}$$

where the magnitudes have been re-introduced into the unit vectors of Eq. (9). In *d* dimensional space, this represents *d*(*n*-1) independent, multivariate polynomial equations, in terms of the *dn* component variables. A Cartesian inertial frame can be used to project the components, producing the variable set m_{jx}, m_{jy} and m_{jz} for $1 \leq j \leq n$. If spacecraft trajectories have been chosen which conserve linear momentum (as they must), then differentiating these positions twice and multiplying by the spacecraft masses will result in the desired forces that appear on the left side of Eq. (10). For the solutions presented in this paper, one of the dipoles was fixed by an obvious (yet arbitrary) geometric constraint at each time step (usually aligned with the center of mass of the system), and a gradient search was then performed to determine a complete dipole configuration that satisfied the equations.

This search requires the calculation of the Jacobian of Eq. (10), with derivatives of the *d*(*n*-1) components of force taken with respect to the *dn* components of the *n* dipoles, and the position vectors held constant. The components are related by

$$\Delta F_{(j\alpha)} = Jac(m,r)_{(j\alpha)(i\beta)} \Delta m_{(i\beta)} \tag{11}$$

where $(i\alpha)$ acts as a single index, which indexes each spatial component (α) of the resultant force on each dipole (*j*), and likewise $(i\beta)$ indexes each spatial component (β) of each dipole moment (*i*). Summation over a repeated set of indices within a product is implied. Inverting the Jacobian then allows corrections to the dipole components to be calculated in terms of errors in the forces resulting from a previous guess of the dipole moments and exact knowledge of the desired positions. Because one of the forces is not linearly independent this equation cannot be inverted as it stands. Instead, the *d* columns associated with the specified dipole moment components are separated from the rest, and *d* rows are removed that correspond to the force components that will be determined by setting the sum to zero. The new equation after inverting then becomes

$$\Delta m^*_{(i\beta)} = \left[Jac(m,r)^{**}_{(j\alpha)(i\beta)} \right]^{-1} \left(\Delta F^*_{(j\alpha)} - Jac(m,r)^{d*} \Delta m^*_{(n\beta)} \right) \tag{12}$$

where * indicates removal of *d* rows, ** indicates an additional removal of *d* columns, and the *d** indicates the *d* columns that were removed from the Jacobian. Since there is no error in the components of the one specified dipole, $\Delta m^*_{(n\beta)} = 0$, and that term drops from the right side. For each successive time step, the solution for the previous one was used to seed the search, and it was found in the maneuver shown in Figure 8 that a transition from one solution path to another can occur, resulting in a quick change in direction or jerk of one or more dipoles. This has practical implications, since the inductance of the coils and the design voltage of the power system will limit the rate at which the current can be changed within the coil. Ideally, a smooth, more slowly changing dipole moment profile is desired.

For control purposes, this same approach is useful to provide feedback to the system. However in addition to finding the Jacobian with respect to the dipole moments, an additional Jacobian with respect to errors in position is also desired. In this case

$$\Delta F_{(j\alpha)} = Jac(m,r)_{(j\alpha)(i\beta)} \Delta m_{(i\beta)} + Jac2(m,r)_{(j\alpha)(i\beta)} \Delta r_{(i\beta)} \tag{13}$$

and an modified version of Eq. (12) is given by

$$\Delta m^*_{(i\beta)} = \left[Jac(m,r)^{**}_{(j\alpha)(i\beta)} \right]^{-1} \left(\Delta F^*_{(j\alpha)} - Jac(m,r)^{d*} \Delta m^*_{(n\beta)} - Jac2(m,r)^*_{(j\alpha)(i\beta)} \Delta r_{(i\beta)} \right) \tag{14}$$

If, at a given point in time, the position of any of the satellites is off-nominal, the proper selection of $\Delta r_{(i\beta)}$ and $\Delta F^*_{(j\alpha)}$ can be chosen to smoothly bring the system back into the proper configuration. Note that the 'proper selection' of these quantities is not determined by Eq. (14), however if determined otherwise, this equation shows how they translate into changes in the dipole moments.

This approach also lends itself as an alternate method of finding a time history of the dipoles that will satisfy the desired motion. Instead of considering ΔF and Δr to be errors to a nominal trajectory at a particular point n time, they can be considered changes from one time step to the next. Since it has been assumed that the desired trajectory is known for all times, then these quantities are known ahead of time, and Eq. (14) can be used to forward integrate the solution from a known starting point. However, the same problem exists with regard to the existence of multiple solutions. Although, as stated earlier, multiple solutions are desirable to allow for other system considerations to be addressed, the desire is to know of them ahead of time, and be able to choose freely between them, rather than come across them unexpectedly.

One approach to this is through the use of Gröbner Bases[7]. For a set of multivariate polynomials, the set of Gröbner Basis polynomials share all the same roots as the original polynomials, however additional roots may exist. The appealing property of these bases is that each basis polynomial has one additional variable eliminated in comparison to another basis polynomial, with the lowest order basis appearing in only a single variable. This one basis may be solved directly for its roots, and the solutions propagated through the other bases, each time with the resulting polynomial in only a single variable. In this way, all solutions are obtained, and the extraneous solutions are eliminated by evaluating them within the original polynomials. The idea is similar to heuristic substitution of one polynomial into another, until only one variable remains, however the method of determining the Gröbner basis set is systematic, and many symbolic manipulators such as Maple® and Mathematica® have routines in place for doing so. This approach to identifying all possible solutions is still under development at the time of this writing.

CONCLUSIONS

In this paper we have introduced the physics behind Electromagnetic Formation Flight in the far-field approximation, and briefly discussed the performance that can be expected when High Temperature Superconducting wire is used as the dipole source. The use of the current state-of-the-art in HTS wire technology already trades well on the basis of mass alone to an equivalent system using electric propulsion, and improvements in current density by as much as two orders of magnitude are theoretically possible. The focus of this work has been on the practical aspects of formation flight dynamics in 2-D and 3-D. Conservation of momentum allows for one dipole moment, both magnitude and direction, to be arbitrarily chosen, and as the number of spacecraft increases, the nonlinearity of the system provides even more freedom. This is a double-edged sword. On the one hand, the existence of multiple solutions allows for considerations of other system parameters, such as angular momentum storage to be addressed. However, these multiple solutions mean that gradient search methods or forward integration can result in somewhat discontinuous jumps from one solution to the next. An approach to this problem using Gröbner bases has been outlined, and is currently being researched.

ACKNOWLEDGMENTS

This work was sponsored in part by the NASA Institute for Advanced Concepts, and also in part by the National Reconnaissance Office Director's Innovation Initiative Program.

REFERENCES

[1] Hashimoto, T., S. Sakai, K. Ninomiya, K. Maeda and T. Saitoh, "*Formation Flight Control Using Super-Conducting Magnets*", International Symposium on Formation Flying Missions and Technologies, Toulouse, France, Oct. 2002.

[2] Kong, E.M.C., "*Spacecraft Formation Flight Exploiting Potential Fields*", PhD Thesis, Dept. of Aero/Astro, MIT, 2002.

[3] Beichman, C. A., Woolf, N. J., and Lindensmith, C. A., ed., *The Terrestrial Planed Finder (TPF): A NASA Origins Program to Search for Habitable Planets,* JPL Publication 99-003, 1999.

[4] Sedwick, R.J. and S.A. Schweighart, "*Propellantless Spin-Up and Reorientation of Tethered or Electromagnetically Coupled Spacecraft*", SPIE Highly Innovative Space Telescope Concepts Conference, Waikoloa, HI, Aug 22-28, 2002.

[5] Farley, R.E. and D.A. Quinn, "*Tethered Formation Configurations: Meeting the Scientific Objectives of Large Aperture and Interferometric Science*", AIAA 2001-4770.

[6] http://www.aip.org/enews/physnews/1995/split/pnu227-2.htm, Physics News Update, The American Institute of Physics Bulletin of Physics News, Number 227, May 22, 1995, Phillip F. Schewe and Ben Stein.

[7] Boege, W., R. Gebauer, and H. Kredel, "Some Examples for Solving Systems of Algebraic Equations by Calculating Gröbner bases.", *J. Symbolic Computation*, 1:83-98, 1986.

AAS 03-006

A COMPARISON OF TRAJECTORY OPTIMIZATION METHODS FOR THE IMPULSIVE MINIMUM FUEL RENDEZVOUS PROBLEM

Steven P. Hughes,* Laurie M. Mailhe† and Jose J. Guzman‡

In this paper we present a comparison of trajectory optimization approaches for the minimum fuel rendezvous problem. Both indirect and direct methods are compared for a variety of test cases. The indirect approach is based on primer vector theory. The direct approaches are implemented numerically and include Sequential Quadratic Programming (SQP), Quasi-Newton, and Nelder-Meade Simplex. Several cost function parameterizations are considered for the direct approach. We choose one direct approach that appears to be the most flexible. Both the direct and indirect methods are applied to a variety of test cases which are chosen to demonstrate the performance of each method in different flight regimes. The first test case is a simple circular-to-circular coplanar rendezvous. The second test case is an elliptic-to-elliptic line of apsides rotation. The final test case is an orbit phasing maneuver sequence in a highly elliptic orbit. For each test case we present a comparison of the performance of all methods we consider in this paper.

INTRODUCTION

The minimum fuel rendezvous problem has received considerable attention in the literature and numerous approaches to both posing and solving the problem have been developed. The basic objective is to find a minimum solution to a two-point boundary value problem (TPBVP) that has multiple feasible solutions. The primary difference between approaches is the choice of independent variables and how the TPBVP is posed in terms of these variables. Methods also differ on how constraints are handled. In the next few paragraphs we give a brief overview of the methods considered in this paper. We also discuss the test problems we used to allow a comparison of each method we consider for different flight regimes.

This paper deals with the problem of optimizing trajectories where the maneuvers can be modeled impulsively. In general, two approaches are utilized for optimization: direct and indirect.[1] Both approaches will be compared in this investigation. A survey of the rendezvous problem is provided by Jezewski et al.[2]

A direct optimization method tries to minimize the cost function by directly varying the *control (independent) variables.* The methods employed for the optimization process are mathematical programming techniques.[3] The mathematical programming techniques utilized in this paper include Sequential Quadratic Programming (SQP), Quasi-Newton, and Nelder-Meade Simplex. (Other techniques, such as Genetic Algorithms, are under current investigation but are not discussed here). For direct optimization

* Aerospace Engineer, Flight Dynamics Analysis Branch, NASA Goddard Space Flight Center, Greenbelt, Maryland 20771. E-mail: steven.hughes@gsfc.nasa.gov.

† Mission Analyst, a.i. solutions, Inc., Lanham, Maryland 20706. E-mail: mailhe@ai-solutions.com.

‡ Mission Analyst, a.i. solutions, Inc., Lanham, Maryland 20706. E-mail: guzman@ai-solutions.com.

the choice of independent variables and constraints is extremely important. For example, suppose that there are n impulses in the trajectory of interest. Then, two possible choices for independent variables are (1) the actual ΔV's, or (2) the impulse locations and times. In this paper, we find the option of using the impulse locations and times more effective. This is consistent with previous work by Brusch.[4] Hence, we utilize a Lambert-type approach with implicit position continuity. Thus, as the impulse locations change, new arcs are computed utilizing either Lambert arcs (in the two-body model) or differentially corrected arcs (allows for more complex models than the two-body model). The result of computing the lambert arcs is a path continuous trajectory with velocity discontinuities. The velocity discontinuities are then utilized to formulate the cost function. Other approaches are possible but include the addition of position continuity as constraints.[5]

An indirect optimization method tries to minimize the cost function by considering its variations, thus it involves Calculus of Variations (COV). It leads to two-point boundary value problems in the *co-state variables*. Once the co-state variables are obtained, the optimal control variables are usually readily available. Primer vector theory can be considered to be a byproduct of applying the Calculus of Variations to the problem of minimizing the fuel usage of impulsive trajectories. Perhaps more importantly, primer vector theory can be utilized to "find" the optimal number of impulses. A survey of primer vector theory is presented by Guzman[6] *et al.*

The parameterization of the cost function chosen in this work ensures that for every cost function evaluation, the rendezvous constraints are satisfied implicitly. Hence, from the point-of-the-view of a numerical optimizer, the problem is unconstrained. Several numerical optimization routines are considered. Three of the routines are products of The Mathworks and are available in their Optimization Toolbox. The Mathworks' routines that we employ are *fmincon*, *fminunc*, and *fminsearch*. The *fmincon* function is an SQP algorithm, *fminunc* is a Quasi-Newton algorithm, and *fminsearch* is a Nelder-Meade Simplex algorithm. For a detailed discussion of The Mathworks' routines and their implementations, we refer the reader to The Mathworks Optimization Toolbox[7] documentation. We also use *SNOPT*, an SQP algorithm developed by Gill[8] *et. al.* For details on *SNOPT* we refer the reader to the user's guide for *SNOPT*[8] 5.3. For the indirect method we develop software in MATLAB that uses a fully automated algorithm which implements the primer vector theory first developed by Lawden.[9]

We employ three test cases to compare the performance of each trajectory optimization method in different flight regimes. The first test case is a simple circular-to-circular, coplanar rendezvous. The second test case is an elliptic-to-elliptic, line of apsides rotation. The final test case is a phasing maneuver in a highly elliptic orbit. For each test case we generate numerous initial guesses. The initial guesses are generated using simple intuition. The times and locations of the initial and final maneuvers are chosen based on experience. Given the initial and final burn locations we generate a two-burn rendezvous sequence. Multiple maneuver sequences are generated by placing small maneuvers, equally spaced in time, along two maneuver initial guesses. Sequences of two, three, and four burns are considered. The performance of all methods for the test cases described above is presented in the final section. It is difficult to provide a single metric that can conclude which method is the best. Hence we have provided a variety of statistics to illustrate the performance of all of the trajectory optimization approaches we consider. It should be noted that we have not considered the rate of convergence to be a measure of importance. Since all methods are acceptably fast, we use metrics based on the total ΔV of the converged solution to be the primary measures of performance.

To solve the minimum fuel rendezvous problem we must first develop a mathematical model. In the next section we discuss the distinction between an orbit transfer and an orbit rendezvous and develop a model for the rendezvous problem.

PROBLEM STATEMENT

The minimum fuel rendezvous problem can be posed in numerous ways. The primary difference between approaches is the choice of independent variables, how the boundary value problem is posed in terms of the independent variables, and how the constraint functions are handled. It is useful at this point to make a clear distinction between a rendezvous problem and a transfer problem. In an orbit transfer problem we are only concerned with finding a maneuver sequence that will take a spacecraft from some position in its initial orbit, to some position in the desired final orbit. For a transfer, the initial maneuver can occur anywhere along the initial orbit, and the final maneuver can occur anywhere along the final orbit, without concern for the orbit phasing. However, for a rendezvous problem, we have the additional constraints determined by the time and place of the spacecraft in the initial orbit, and the time and place of the target spacecraft in its orbit. In this section we develop a mathematical model of the rendezvous problem. In later sections the model is used to develop optimization algorithms to find minimum fuel solutions.

Figure 1 is a diagram to aid in illustrating the general impulsive rendezvous problem. The arc denoted $\mathcal{P}_o$ represents the path of the spacecraft in its initial orbit. Similarly, the arc denoted $\mathcal{P}_f$ represents the path of path of the spacecraft in its final orbit. Throughout the paper a superscript '+' is used to denote a condition immediately prior to an impulsive maneuver, and a superscript '–' is used to denote a condition immediately following an impulsive maneuver. According to these conventions, we can write the conditions for the ith burn as,

$$\mathbf{r}_i^+ = \mathbf{r}_i^- \tag{1}$$

$$t_i^+ = t_i^- \tag{2}$$

$$\Delta\mathbf{V}_i = \mathbf{V}_i^- - \mathbf{V}_i^+ \tag{3}$$

where $\mathbf{r}_i^+$ and $\mathbf{r}_i^-$ are the positions immediately before and after the ith burn respectively, t_i^+ and t_i^- are the times immediately before and after the ith burn respectively, and $\mathbf{V}_i^+$ and $\mathbf{V}_i^-$ are the velocity vectors immediately before and after the ith burn respectively. The constraints in Eqs.(1) and (2) are a result of the definition of an impulsive maneuver.

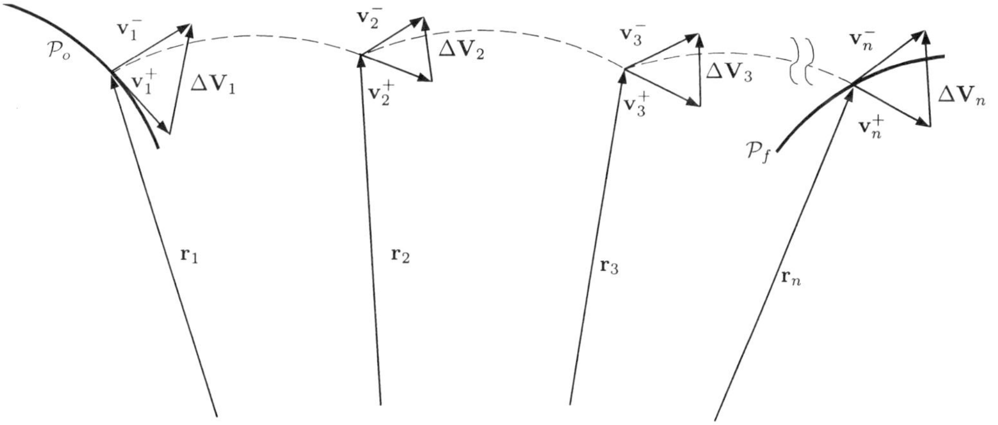

Figure 1: Impulsive Orbit Transfer Diagram

There are additional constraints that must be satisfied at the boundaries, and at the interior impulses which are imposed by the orbit dynamics. We assume that the dynamics for the entire maneuver sequence are given by

$$\ddot{\mathbf{r}} = -\mu\frac{\mathbf{r}}{r^3} \tag{4}$$

Let $\mathbf{r}_f = \mathbf{f}(\mathbf{r}_o, \mathbf{v}_o, t_o, t_f)$ and $\mathbf{v}_f = \mathbf{g}(\mathbf{r}_o, \mathbf{v}_o, t_o, t_f)$ be the solution to Eq. (4), with the initial conditions $(\mathbf{r}_o, \mathbf{v}_o, t_o)$ and the final time t_f. For an interior impulse $i \neq 1$, $i \neq n$, and $n > 2$, the additional constraints that must be satisfied are given by

$$\mathbf{r}_i^+ = \mathbf{f}(\mathbf{r}_{i-1}, \mathbf{v}_{i-1}, t_{i-1}, t_i) \tag{5}$$

$$\mathbf{v}_i^+ = \mathbf{g}(\mathbf{r}_{i-1}, \mathbf{v}_{i-1}, t_{i-1}, t_i) \tag{6}$$

The constraints that must be satisfied at the initial boundary are given by

$$\mathbf{r}_1^+ = \mathbf{f}(\mathbf{r}_o, \mathbf{v}_o, t_o, t_1) \tag{7}$$

$$\mathbf{v}_1^+ = \mathbf{g}(\mathbf{r}_o, \mathbf{v}_o, t_o, t_1) \tag{8}$$

where $(\mathbf{r}_o, \mathbf{v}_o, t_o)$ are the position and velocity of the initial orbit at the reference epoch t_o, and t_1 is the time of the first maneuver. A similar set of constraints apply at the final boundary and are give by

$$\mathbf{r}_n^- = \mathbf{f}(\mathbf{r}_f, \mathbf{v}_f, t_f, t_n) \tag{9}$$

$$\mathbf{v}_n^- = \mathbf{g}(\mathbf{r}_f, \mathbf{v}_f, t_f, t_n) \tag{10}$$

where $(\mathbf{r}_f, \mathbf{v}_f, t_f)$ are the position and velocity of the initial orbit at the reference epoch t_f, and t_n is the time of the last maneuver. Without loss of generality, we can appropriately formulate the optimal control problem to ensure that the constraints $\mathbf{r}_i^+ = \mathbf{r}_i^-$ and $t_i^+ = t_i^-$ are satisfied implicitly. For the remainder of this work we drop the superscripts on $\mathbf{r}$ and t, and assume

$$\mathbf{r}_i = \mathbf{r}_i^+ = \mathbf{r}_i^- \tag{11}$$

$$t_i = t_i^+ = r_i^- \tag{12}$$

In summary, the problem is to find the set $(\mathbf{r}_i, \mathbf{v}_i^+, \mathbf{v}_i^-, t_i)$, where $i = 1, 2...n$ and $n \geq 2$, that satisfies the constraints

$$\mathbf{r}_1 = \mathbf{f}(\mathbf{r}_o, \mathbf{v}_o, t_o, t_1) \tag{13}$$

$$\mathbf{v}_1^- = \mathbf{g}(\mathbf{r}_o, \mathbf{v}_o, t_o, t_1) \tag{14}$$

$$\mathbf{r}_i = \mathbf{f}(\mathbf{r}_{i-1}, \mathbf{v}_{i-1}, t_{i-1}, t_i) \tag{15}$$

$$\mathbf{v}_i^- = \mathbf{g}(\mathbf{r}_{i-1}, \mathbf{v}_{i-1}, t_{i-1}, t_i) \tag{16}$$

$$\mathbf{r}_n = \mathbf{f}(\mathbf{r}_f, \mathbf{v}_f, t_f, t_n) \tag{17}$$

$$\mathbf{v}_n^+ = \mathbf{g}(\mathbf{r}_f, \mathbf{v}_f, t_f, t_n) \tag{18}$$

and that is a minimum solution to the function

$$J = \sum_{i=1}^{n} \|\Delta \mathbf{V}_i\| \tag{19}$$

where n is the number of maneuvers, the set $(\mathbf{r}_o, \mathbf{v}_o, t_o)$ defines the initial orbit and the set $(\mathbf{r}_f, \mathbf{v}_f, t_f)$ defines the final orbit.

There are two main methods available for solving the optimal control problem, including direct and indirect approaches. Within each method there are numerous possible implementations. Often methods are actually a combination of indirect and direct techniques. For direct methods, the parameterization of the problem depends on the type of optimizers available. Indirect methods are sensitive to the constraints and often require complete reformulation if new constraints are imposed. In the next two sections we discuss several approaches for solving the minimum fuel rendezvous problem using both indirect and direct methods.

DIRECT APPROACH

The success of a direct method is intimately dependent on the choice of independent variables and the constraint function formulation. In this section we discuss three parameterizations for solving the minimum fuel rendezvous problem using direct methods. There are numerous possible parameterizations and comparing all approaches is beyond the scope of this work. We categorize the parameterizations we consider into two groups called the Feasible Iterate Approach and the Infeasible Iterate Approach. We present a brief comparison of the parameterizations, and the most promising method is chosen for further development. For this work we choose a Feasible Iterate Approach for reasons explained later. This section is concluded with a discussion of several numerical optimization packages used in this work.

Objective Function Parameterizations

There are numerous approaches to solving the minimum fuel rendezvous problem using direct methods. For convenience we break down the direct approaches considered here into two categories called the Feasible Iterate Approach and the Infeasible Iterate Approach. In the Feasible Iterate Approach, each evaluation of the objective function is also a feasible solution that satisfies the rendezvous conditions. Hence, from the point of view of the optimizer, the problem is unconstrained. The Feasible Iterate Approach requires that for each objective function evaluation a TBPVP is solved. The Feasible Iterate Approach also requires a careful selection of independent variables for the objective function parameterization. In the Infeasible Iterate Approach, there is less restriction on the choice of independent variables, however, the constraints must be defined carefully, and the optimizer must be able to handle nonlinear constraints. For continuity, we present the discussion of alternative parameterizations in Appendix 1. Here we present the parameterization chosen for this work.

Our choice of independent variables is as follows:

$$\text{Given: } (\mathbf{r}_o, \mathbf{v}_o, t_o) \text{ and } (\mathbf{r}_f, \mathbf{v}_f, t_f) \tag{20}$$

Choose the independent variables:

$$t_i \qquad i = 1, 2 \ldots n \tag{21}$$

$$\mathbf{r}_j \qquad j = 2, 3 \ldots . n - 1 \tag{22}$$

where t_i are the times of the maneuvers, and $\mathbf{r}_j$ are the positions of the intermediate burns. Given the independent variables defined in Eqs.(21) and Eqs.(22), we must develop an algorithm to determine the total ΔV of the maneuver sequence. The boundary conditions $\mathbf{r}_1$, $\mathbf{v}_1^+$, $\mathbf{r}_n$, and $\mathbf{v}_n^-$ are determined using Eqs.(13),(14),(17),and (18), where t_1 and t_n are known from Eq.(21). After solving for the boundary conditions we know the times and positions of all the maneuvers. Therefore, we can solve for the velocities before and after each maneuver, by solving Lambert's problem for each of the $n-1$ trajectory segments. There are numerous well-known approaches to solving Lambert's problem. We choose a method developed by Howell and Pernicka[10] and further developed by Guzman.[11] The method developed by Howell is chosen because it can solve Lambert's problem with perturbations included, as well as when there is more than one significant gravitational body. Perhaps more importantly, Howell demonstrated that using the method described below, we can provide analytic approximations for the gradient of the total ΔV with respect to the independent variables chosen in Eqs. (21) and Eqs.(22). The algorithm is described as follows. Given an initial position $\mathbf{r}_{i-1}$, and an initial velocity $\mathbf{v}_{i-1}$ both at time t_{i-1}, find $\delta\mathbf{v}_{i-1}$ applied at time t_{i-1} so that we achieve $\mathbf{r}_i$ at t_i. Figure 2 illustrates the problem. The dark black arc denotes the path a spacecraft would follow if no $\delta\mathbf{v}_{i-1}$ is applied. For this arc, the final position denoted by $\mathbf{r}_a$ is not equal to the desired final position $\mathbf{r}_i$. Hence $\delta\mathbf{r}_i \neq 0$. The dashed arc denotes the trajectory that is the solution to Lambert's problem. For this arc $\mathbf{r}_i = \mathbf{r}_a$, or $\delta\mathbf{r}_i = 0$. To solve for $\delta\mathbf{v}_{i-1}$ such that $\delta\mathbf{r}_i = 0$ first define $\mathbf{x}$ as

$$\mathbf{x} = \begin{bmatrix} \mathbf{r} \\ \mathbf{v} \end{bmatrix} \tag{23}$$

then $\delta\mathbf{x}_i$ is given by

$$\delta\mathbf{x}_i = \mathbf{\Phi}\left(t_i, t_{i-1}\right)\delta\mathbf{x}_{i-1} \tag{24}$$

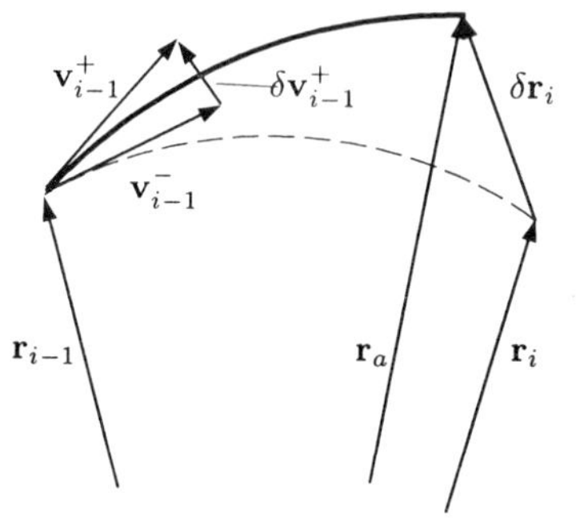

Figure 2: Illustration of One Trajectory Arc

where t_i and t_{i-1} are fixed. We can write $\mathbf{\Phi}(t_i, t_{i-1})$ as

$$\mathbf{\Phi}(t_i, t_{i-1}) = \begin{bmatrix} \mathbf{A}_{i,i-1} & \mathbf{B}_{i,i-1} \\ \mathbf{C}_{i,i-1} & \mathbf{D}_{i,i-1} \end{bmatrix} \tag{25}$$

Using the fact that $\delta\mathbf{r}_{i-1} = 0$ we can write

$$\delta\mathbf{r}_i = \mathbf{B}_{i,i-1}\delta\mathbf{v}_{i-1} \tag{26}$$

Solving for $\delta\mathbf{v}_{i-1}$ we see that

$$\delta\mathbf{v}_{i-1} = \mathbf{B}_{i,i-1}^{-}\delta\mathbf{r}_i \tag{27}$$

From inspection of Fig. 2 we can write

$$\mathbf{r}_i + \delta\mathbf{r}_i = \mathbf{r}_a \tag{28}$$

Solving for $\delta\mathbf{r}_i$ we have

$$\delta\mathbf{r}_i = \mathbf{r}_a - \mathbf{r}_i \tag{29}$$

Substituting Eq.(29) into Eq.(27) yields

$$\delta\mathbf{v}_{i-1} = \mathbf{B}_{i,i-1}^{-}(\mathbf{r}_a - \mathbf{r}_i) \tag{30}$$

We can solve for $\delta\mathbf{v}_{i-1}$ such that $\delta\mathbf{r}_i$ is zero by iterating on Eq.(30) until $(\mathbf{r}_a - \mathbf{r}_i)$ meets a user defined tolerance. It is important to note that this approach assumes that $\mathbf{B}_{i,i-1}^{-}$ exists. For cases when $\mathbf{B}_{i,i-1}^{-}$ is not invertible we use a method by Gedeon[12] to solve the TPBVP. We solve all $n-1$ trajectory arcs using the algorithm defined above, then the ΔV is calculated using Eq.(19). Note that although the algorithm above can be extended to provide a gradient approximation, we have used finite differencing for gradients in this work. Analytic approximations for the gradient are a topic of current research.

There are several other concerns to address to completely define the objective function parameterization. We also must choose appropriate time and coordinate systems to express the independent variables given in Eqs. (21) and (22). There are several issues in choosing a time parameterization that must be considered. The first issue is to select appropriate units, the second is to select an appropriate reference epoch. We choose the units of seconds for time, and reference each time t_i to the reference epoch t_0 given in Eq. (20). The positions in Eq. (22) are expressed in cartesian coordinates in the Mean of J2000 Earth Equatorial system and the units are in *km*.

With the parameterization of the cost function described above, the rendezvous conditions are satisfied implicitly for every cost function evaluation. Hence, the problem is essentially an unconstrained optimization problem and we are free to use unconstrained as well as constrained optimization techniques to find a minimum ΔV solution. It is exceedingly rare that both TPBVP methods described above fail to converge. In the case that neither method converges we must "inform" the optimizer. This is discussed in a subsequent section. In the next subsection we discuss the numerical optimization packages we use in this work.

Direct Optimizers

There are numerous optimization packages available that solve unconstrained optimization problems. We investigate the performance of four routines for the parameterization of the minimum fuel rendezvous problem chosen above. Three of the routines are products of *The Mathworks* and are available in their *Optimization Toolbox*.[7] The first routine is an SQP algorithm called *fmincon*. The second routine is a quasi-Newton algorithm called *fminunc*. The third routine is a Nelder-Meade simplex algorithm and is called *fminsearch*. For a detailed discussion of the optimization routines developed by *The Mathworks* we refer the reader to the *Optimization Toolbox Documentation*.[7] The fourth numerical optimization routine we employ is an SQP routine developed by Gill[8] *et al.* called *SNOPT*. For a detailed discussion of *SNOPT* we refer the reader to the *SNOPT*[8] 5.3 documentation.

Numerical Issues

To avoid numerical difficulties the independent variables and the objective function values are normalized to be on the order of one. Secondly, all derivatives are calculated using finite differencing. Providing analytic derivatives for gradient based methods is a topic of current research. The final numerical issue occurs when both TPBVP solvers fail to converge. This is exceedingly rare. However when it occurs we must "inform" the optimizer. *SNOPT* has a built in capability to step back from a poorly conditioned state vector. Hence, for *SNOPT*, if both TPBVP solvers fail, the appropriate message is sent to *SNOPT* and *SNOPT* steps away from the poorly conditioned state. However, *The Mathworks* routines do not have this capability yet. As a temporary solution, when using a *Mathworks'* routine, in the event of a poorly conditioned state vector we pass back a crude approximation of the total ΔV.

To solve the minimum fuel rendezvous problem using direct methods we must choose an appropriate parameterization of the problem, and an appropriate numerical optimization routine. In this section we discussed some choices involved in choosing a specific parameterization. We chose one method that performed well in a preliminary comparison. We also discussed several numerical optimization routines we employed and some numerical issues one must deal with to ensure adequate performance. In the next section we discuss an indirect approach to solve the minimum fuel rendezvous problem.

INDIRECT APPROACH

In this section, we present a review of the primer vector theory as well as the main challenges involved in its implementation.[13] Primer vector theory[14] is an optimization technique based on calculus of variations. The theory has several appealing features including the indication of when and where to add an impulse to a non-optimal trajectory, and a visual assessment of the optimality of neighboring solutions.

Primer Vector Theory

Primer vector theory provides a set of first order necessary conditions that a trajectory must meet to be locally optimal. The necessary conditions, first derived by Lawden, are expressed in terms of the primer vector, which is defined as the adjoint to the velocity vector in the variational Hamiltonian formulation.[14] If any of Lawden's conditions are violated, the rendezvous trajectory is not optimal and we can use the primer vector history to obtain information on how to improve its ΔV cost. In his initial work, Lawden solved a fixed-time rendezvous. His theory was extended by Lion and Handelsman and later, by Jezewski and Rozendaal, to solve the n-impulse optimal rendezvous problem.[15,16] A more detailed derivation of the primer vector theory is provided by Hiday.[17] For the rendezvous problem considered in this paper, we can express the primer vector equations using calculus of variations theory. The initial trajectory or first-guess is labeled as the reference trajectory. Since primer vector is a first-order theory based on local variations, it will converge to locally optimal neighboring trajectories of the reference trajectory. Therefore, the optimal solution is highly dependent on the reference trajectory but it will also depend on other design parameters discussed later in this section. The primer vector obeys the following equation, also known as the second order canonical form of the Euler-Lagrange equation,

$$\ddot{\mathbf{p}} = \mu\,(-\frac{\mathbf{p}}{r^3} + 3\frac{\mathbf{r}}{r^5}(\mathbf{p}\cdot\mathbf{r})) \tag{31}$$

where $\mathbf{r}$ is the satellite position vector on the reference trajectory, $\mathbf{p}$ is the primer vector and μ is the earth's gravitational constant. The satellite position is found using Eq. (4). As evident in Eq. (31), the primer vector state can not be derived simultaneously with the spacecraft state. The spacecraft trajectory must be propagated first for the primer vector history to be computed. Using calculus of variations, Lawden derived four necessary conditions expressed in terms of the primer vector defined by Eq. (31) for an optimal rendezvous trajectory:

1. The primer vector and its first derivative are continuous for the entire history.
2. Between impulses, $||\mathbf{p}|| = p <= 1$. Impulses occur when $p = 1$.
3. At an impulse, $\mathbf{p} = \hat{\mathbf{p}} = \hat{\mathbf{u}}_T^*$, where $\hat{\mathbf{u}}_T^*$ is the optimal thrust direction.
4. At all interior impulses (not at the initial or final times) $\mathbf{p}\cdot\dot{\mathbf{p}} = 0$. This condition has implications on the slope of the primer vector magnitude since

$$\frac{d||\mathbf{p}||}{dt} = \frac{d}{dt}(\mathbf{p}\cdot\mathbf{p})^{1/2} = \frac{\dot{\mathbf{p}}\cdot\mathbf{p}}{||\mathbf{p}||}.$$

Therefore, $\frac{d||\mathbf{p}||}{dt} = 0$ at the intermediate impulses. Also, for convenience, let $\dot{p} = \frac{d||\mathbf{p}||}{dt}$.

These conditions are necessary (but not sufficient) for an optimal trajectory. In this paper, a trajectory that meets the above conditions will be called an optimal trajectory. When solving a rendezvous problem using primer vector theory, we first need to evaluate the primer vector history along the reference trajectory or initial guess. Solving for the primer vector history is equivalent to solving a TPBVP. Let's first assume that we are solving a two-burn rendezvous problem. We know from Lawden's conditions that, at an impulse, the primer vector is in the direction of the thrust vector. Thus, we can express the initial and final primer vector as :

$$\mathbf{p}_o = \frac{\Delta\mathbf{V}_o}{|\Delta\mathbf{V}_o|}, \tag{32}$$

$$\mathbf{p}_f = \frac{\Delta\mathbf{V}_f}{|\Delta\mathbf{V}_f|}. \tag{33}$$

The initial and final primer vectors and the time-of-flight $(t_f - t_o)$ are known. However, to propagate Eq. (31), we need the complete primer initial state $(\mathbf{p}_o, \dot{\mathbf{p}}_o)$. To obtain $\dot{\mathbf{p}}_o$, we can either use a shooting method (TPBVP solver) or use the state-transition-matrix (STM) formulation from Eq. (34) below.

$$\begin{pmatrix} \mathbf{p} \\ \dot{\mathbf{p}} \end{pmatrix} = \mathbf{\Phi}(t,t_o)\cdot\begin{pmatrix} \mathbf{p}_o \\ \dot{\mathbf{p}}_o \end{pmatrix} = \begin{bmatrix} \mathbf{A} & \mathbf{B} \\ \mathbf{C} & \mathbf{D} \end{bmatrix}\cdot\begin{pmatrix} \mathbf{p}_o \\ \dot{\mathbf{p}}_o \end{pmatrix} \tag{34}$$

where $\mathbf{A}$, $\mathbf{B}$, $\mathbf{C}$ and $\mathbf{D}$ are 3x3 matrices, partitions of $\mathbf{\Phi}(t,t_o)$. In general, using the STM is faster and although not as accurate as the TPBVP solver, it is sufficient. Using Eq. (34), the initial derivative of the primer vector can be expressed as:

$$\dot{\mathbf{p}}_o = \mathbf{B}^{-1}\cdot(\mathbf{p}_f - \mathbf{A}\cdot\mathbf{p}_o) \tag{35}$$

It is worthwhile noting that if $\mathbf{B}^{-1}$ is singular, $\dot{\mathbf{p}}_o$ cannot be computed using Eq. (35). In the two-body problem, there are three known singularities.[18] Eq. (35) can be generalized for a n-impulse trajectory by examining each individual two-burn transfer. Once the primer vector history is computed, we can determine the optimality of the trajectory using Lawden's necessary conditions. If any of Lawden's conditions are violated, the rendezvous trajectory is not optimal and we can use the primer vector history to obtain information on how to improve its ΔV cost. Improving non-optimal trajectories using primer vector theory is the main contribution of Lion and Handelsman.[15] Essentially, first order variations of the total ΔV are considered for different perturbed trajectories.[15–17] For a non-optimal primer vector history, two types of actions are possible to lower the ΔV cost: (I) moving the initial or final impulse and (II) adding and/or moving an interior impulse. If the epoch of the endpoints is unconstrained, an action of type I is performed whenever the initial and/or the final primer vector magnitude slope is different from zero. The general expression for time variations at both endpoints is given by Hiday[17] as

$$\delta J = -\dot{p}_o||\Delta \mathbf{v}_o||dt_o - \dot{p}_f||\Delta \mathbf{v}_f||dt_f, \tag{36}$$

where the initial and final time variations are given by dt_o and dt_f respectively. Furthermore, the initial and final primer magnitude slopes are given by $\dot{p}_o$ and $\dot{p}_f$, respectively. To have a lower cost, δJ must be less than zero. If the following terminology is defined, 1) initial coast ($dt_o > 0$), 2) early departure ($dt_o < 0$), 3) final coast ($dt_f < 0$), and, 4) late arrival ($dt_f > 0$); then, the following four cases cover all the non-zero primer slope combinations,

- If $\dot{p}_o > 0$ and $\dot{p}_f < 0 \Rightarrow$ Apply initial coast and final coast.
- If $\dot{p}_o > 0$ and $\dot{p}_f > 0 \Rightarrow$ Apply initial coast and late arrival.
- If $\dot{p}_o < 0$ and $\dot{p}_f < 0 \Rightarrow$ Apply early departure and final coast.
- If $\dot{p}_o < 0$ and $\dot{p}_f > 0 \Rightarrow$ Apply early departure and late arrival.

See Hiday[17] for details. Of course, if the slopes are zero, no further improvement is achieved by varying the endpoints times. If the primer vector magnitude goes above one during a coasting phase on a given segment, an impulse is added to the trajectory (action type II) to improve the overall ΔV budget. This impulse is examined by considering a neighboring path with an additional impulse. The two-impulse arc is perturbed with the addition of a mid-impulse at some time, t_m, and position $\mathbf{r}_m + \delta\mathbf{r}_m$. (The position $\mathbf{r}_m$ is the position along the unperturbed path at t_m.) The initial and the final times and positions, $(t_o,\ \mathbf{r}_o)$ and $(t_f,\ \mathbf{r}_f)$, are fixed. The cost function variation is defined as $\delta J = J' - J$, where J' is the cost on the perturbed trajectory. Then, considering first order terms only, it can be shown that

$$\delta J = c(1 - \mathbf{p}_m^T \hat{\eta}), \tag{37}$$

where the following are mid-impulse parameters: c is the magnitude of the impulse, $\mathbf{p}_m$ is the primer vector at t_m and $\hat{\eta}$ is an unit vector in the direction of the impulse. Then, for an improvement in cost, $\delta J < 0$. Thus, using the definition of a dot product, if $||\mathbf{p}_m|| > 1$ at any time, a third impulse is beneficial. Furthermore, the greatest decrease in the cost function will be achieved if the impulse is applied at the maximum of $||\mathbf{p}_m||$ at time t_m and in the direction of $\hat{\eta}$. The position along the perturbed path and the magnitude of the impulse are yet to be determined. Consider finding the position of the impulse. Utilizing the fact that the position across an impulse is continuous, using the STM before and after the impulse, and utilizing the information obtained from Eq. (37), gives the following variational equation,

$$\delta\mathbf{r}_m = c\mathbf{A}^{-1}\frac{\mathbf{p}_m}{||\mathbf{p}_m||}, \tag{38}$$

where

$$\begin{aligned}\mathbf{A} &= \mathbf{D}(t_m, t_f)\mathbf{B}(t_m, t_f)^{-1} \\ &\quad -\mathbf{D}(t_m, t_o)\mathbf{B}(t_m, t_o)^{-1}.\end{aligned}$$

This equation, of course, is valid only for non-singular $\mathbf{A}$ and determines the position of the mid-impulse. To estimate the magnitude, c, of the impulse, the expression for the cost on the perturbed path, J', can be expressed as a function of c and minimized. The mid-impulse should decrease the cost but might not produce an optimal trajectory in the sense of Lawden. The subsequent optimization of the three-impulse trajectory is presented next. For a three-impulse arc, if the *time and position* of the mid-impulse are allowed to vary, it can be shown that,

$$\delta J = (\Delta \dot{\mathbf{p}}_m)^T \, d\mathbf{r}_m + (\Delta H_m) \, dt_m \tag{39}$$

where $\Delta \dot{\mathbf{p}}_m$ and ΔH_m are, respectively, the primer vector and the Hamiltonian differences at the impulse. If the trajectory is indeed optimal, the first cost variation must vanish. Thus, it is required that, $\Delta \dot{\mathbf{p}}_m = \mathbf{0}$ and, $\Delta H_m = 0$. It can be shown that $\Delta H_m = \dot{\mathbf{p}}_m^{-T} \mathbf{v}_m^- - \dot{\mathbf{p}}_m^{+T} \mathbf{v}_m^+$. In an interesting "mix", a direct optimization method can be applied to vary the mid-impulse time and position to meet the above conditions. In this investigation, following Hiday,[17] the Broyden-Fletcher-Goldfarb-Shanno (BFGS) algorithm is used.

Primer Vector Implementation

The primer vector theory described above was implemented in MATLAB in a tool we call PVAT. PVAT has a fully automated algorithm, which iterates following the primer vector principles to optimize a non-optimal reference trajectory. Most of the time multiple actions for improvements are possible. However, there is yet no mathematical theory that determines the optimal sequencing of the actions and, in general, different sequencing will lead to different neighboring solutions. One way of solving this problem is to combine Eqs. (36) and (39) to form a unique cost function gradient. This gradient is then used in the BFGS optimization algorithm to simultaneously move the endpoints times and the midcourse impulses positions and times. A separate check for the addition of an impulse is then required. In this paper, we choose to implement a sequential algorithm where the endpoint times and the midcourse impulses are varied with separate PVAT iterations. Each time the underlying trajectory is changed, the algorithm recomputes the primer vector data and re-iterates until no improvement can be made.

A conceptual flowchart of PVAT is shown in Fig. 3. First, the initial and final endpoint gradients are evaluated to check whether their values are greater than zero within some specified tolerance ϵ_t. Note that only one endpoint is moved at a time for a given iteration and the priority is given to the boundary with the highest gradient value. To determine the departure/arrival state which corresponds to a zero initial slope, we use a bisection method. Both endpoint algorithms being identical, only the initial boundary is discussed here. Whenever the primer magnitude slope, $\dot{p}_o$, is positive the departure state is propagated forward, otherwise it is propagated backward. To quickly recompute the trajectory data, we use a Lambert solver. The primer magnitude slope is then updated for the new trajectory and the process is repeated until the value of the slope is below the tolerance. Whenever there are internal impulses, the norm of the internal gradient is evaluated. The midcourse burn(s) are moved whenever the norm exceeds a specified tolerance ϵ_m. To move the midcourse impulses, we use the BFGS minimization technique. This variant of the Quasi-Newton method uses an approximation of the Hessian matrix instead of its direct evaluation. The approximation formula is called a BFGS update. Finally, if the primer vector is above some maximum value, $p_{threshold}$, during a coasting arc, an impulse is added to the trajectory. Note that theoretically the threshold value is one, but for numerical implementation, the value is typically chosen to be slightly higher than one. The value of the perturbation impulse is computed using an estimate derived by Jezewski.[16] However, we need to ensure that its magnitude remains small for the theory to be valid. In this paper, we implemented two different algorithms labeled PVAT1 and PVAT2 respectively. For PVAT1, the algorithm was not permitted to add a burn in spite of indication of potential improvements by the primer vector history. The second implementation permits the addition of burns to the rendezvous trajectory to take full advantage of the primer vector theory. However, for practical reasons, we limited the number of burns to a maximum of six. For this first version of the code, the equations where not non-dimensionalized which means that the choices of the

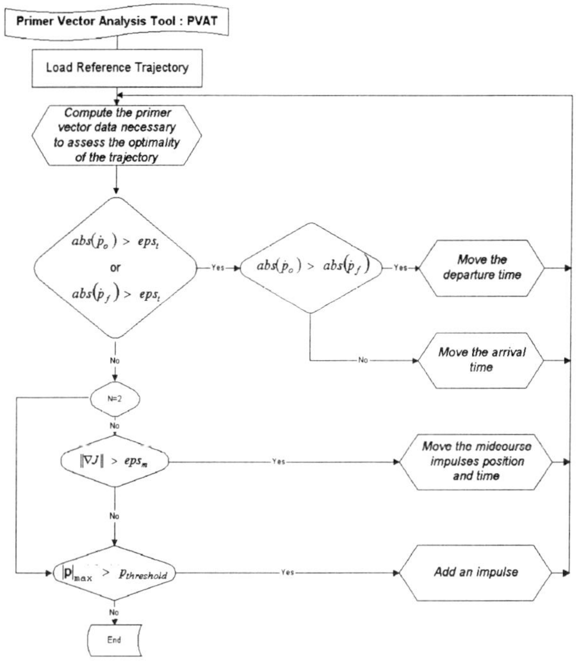

Figure 3: **PVAT** Conceptual Flowchart

various tolerances are specific to the problem solved and that the final optimal solution will depend on their values.

TEST CASES

In order to compare the performance of each optimization method we utilize three test cases. The test cases are chosen to investigate the performance of the each method in different flight regimes. The first test case is a low-Earth, circular-to-circular, coplanar rendezvous. The second test case is an elliptic-to-elliptic line of apsides rotation. The third case is a highly elliptic orbit phasing sequence. For each test case there are numerous initial guesses. To generate different initial guesses for a particular test case we first define an initial and final orbit. Next, we vary the true anomaly and the epoch for the initial and final maneuvers. Given the times and positions of the initial and final maneuvers for a particular initial guess, we solve Lambert's problem to yield a two burn solution. A simple algorithm is used to place small maneuvers along the two-burn trajectory arc to generate multiple maneuver solutions. In the next three subsections we discuss the three test cases in more detail.

Case One

Test Case One is a simple circular-to-circular coplanar transfer. The optimal solution is simply the Hohmann ΔV. The orbital elements for this test case are shown in Eqs.(40) and (41). The format for the orbital elements is [$a\ e\ i\ \omega\ \Omega\ \nu$] unless specified otherwise. Distance is measured in *km*, and angles are measure in *deg*.

$$\text{Initial Orbit:} \quad oe_o = [7000\ 0\ 0\ 0\ 0\ 0]\,, \qquad T = 0 \tag{40}$$

$$\text{Final Orbit:} \quad oe_f = [7500\ 0\ 0\ 0\ 0\ M_f]\,, \qquad T = T_f \tag{41}$$

where M is a mean anomaly. The ΔV for a Hohmann Transfer for Case One is 255.8 m/s. A set of initial guesses for case one is generated by varying M_f and T_f. We vary M_f from 90° to 270° in increments of 20°. For each value of M_f we choose two values for T_f. For each pair of M_f and T_f we solve Lambert's problem to obtain a two-maneuver rendezvous for the initial to the final orbit. The values of T_f are chosen so that the ΔV for the initial guess is less than 1.5 km/s. Note this is considerably higher than the known optimal solution. For each two burn solution we add small interior burns to create similar three and four burn solutions. In summary, there are nine different values for M_f, two different values of T_f for each M_f, and a two, three and four burn solution for each pair of M_f and T_f. Hence there are fifty four initial guesses for Case One. A figure showing a sample of the initial guesses for Case One is shown in Figure 4.

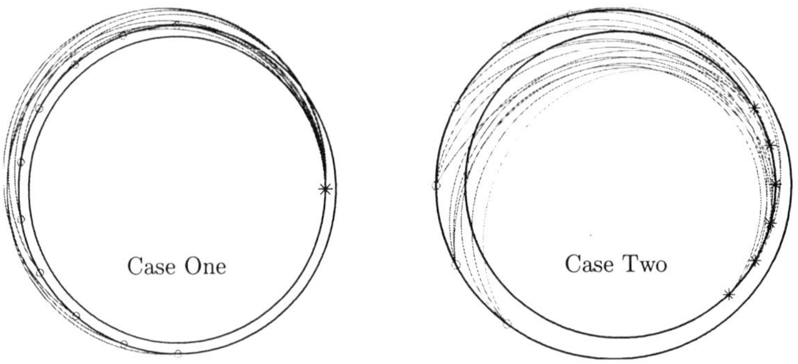

Figure 4: Sample of Initial Guesses for Cases One and Two

Case Two

The second test case involves a combined maneuver that changes both the semimajor axis and the inclination of the orbit. The orbital elements for Case Two are shown in Eqs. (42) and (43). The eccentricity of both orbits is 0.3. The semimajor axis for the initial orbit is 7000 km, and the semimajor axis for the final orbit is 8000 km. The initial orbit is equatorial, and has an argument of periapsis of 0°. Hence the five degree plane change is a rotation about the line of apsides. We generate 180 sets of initial conditions by varying the position and epoch of the initial maneuver, ν_o and T_o respectively, as well as the position and epoch of the final maneuver, ν_f and T_f. We vary ν_o from 300° to 40° in increments of 20°. We vary ν_f from 140° to 220° in increments of 20°. These ranges are chosen because we know from intuition and experience that the initial maneuver should occur near periapis, and the final maneuver should occur near apoapsis. For each possible pair of ν_o and ν_f we choose two times of flight and compute two transfer trajectories that result in a two-burn maneuver sequence with a total ΔV of less than 2 km/s. Three and four burn maneuver sequences are generated from each two burn solution by putting small maneuvers, equally spaced in time along the initial two burn sequence. Hence, there are thirty combinations of ν_o and ν_f, two times of flight for each pair of ν_o and ν_f, and two, three and four burn sequences for each time of flight resulting in one hundred and eighty initial guesses. A representative sample of the initial guesses for Case Two are shown in Figure 4. Positions of the initial maneuvers are labeled by black asterisks. Positions of the final maneuvers are marked by red circles.

$$\text{Initial Orbit:} \quad oe_o = [7000\ 0.3\ 0\ 0\ 0\ \nu_o], \quad T = T_o \tag{42}$$

$$\text{Final Orbit:} \quad oe_f = [8000\ 0.3\ 5.0\ 0\ 0\ \nu_f], \quad T = T_f \tag{43}$$

Case Three

Case Three is a phasing maneuver sequence in a highly elliptic orbit. The orbital elements for the initial and final orbits are given in Eqs.(44) and (45) where E_o and E_f are the eccentric anomalies of the initial orbit and final orbit at times T_o and T_f respectively. To generate a set of initial guesses for Case Three we vary E_o from 0° to 345° in increments of 15°. For each value of E_o we calculate two values of E_f such that there is a 30 minute separation in time between the initial and final orbits. The first value of E_f, for a given E_o, is calculated so that the final position is thirty minutes ahead of the initial position. The second value of E_f, for a given E_o, is calculated so that the final position is thirty minutes behind the initial position. For each E_f and E_o pair we generate a two maneuver sequence by solving Lambert's problem for an E_f and E_o pair knowing that Δt is thirty minutes. Three and four burn cases are generated by inserting small maneuvers along the two burn sequence. For Case Three there are a total of 96 initial guesses.

$$\text{Initial Orbit:} \quad oe_o = [42000\ 0.8\ 10\ 0.0\ 0.0\ E_o], \quad T = T_o \tag{44}$$

$$\text{Final Orbit:} \quad oe_f = [42000\ 0.8\ 10\ 0.0\ 0.0\ E_f], \quad T = T_f \tag{45}$$

The test cases developed above are intended to allow a comparison of the methods for rendezvous sequences in different flight regimes. In the next section we present results that compare the converged solutions for the six methods investigated in this paper.

RESULTS

Comparing the performance of the trajectory optimization methods studied in this work is non trivial. To begin, we must develop some metrics that allow us to compare the solutions provided by the different techniques. The ultimate goal is to provide an analyst with some insight into which technique to employ for a given problem. So ultimately, we want to know which technique is most likely to converge to the lowest ΔV cost. However, this is not the only important statistic. We are also interested in determining the relative performance of different methods. For example, if one approach converges to a known optimum 70% of the time, and a second approach converges to a known optimum 25% of the time, the second approach may still be useful for several reasons. Firstly, for a given problem the hypothetical method 1 will not converge to the known optimal 30% of the time. Hence we will need another approach. Secondly, the hypothetical method 2 might converge to a slightly higher solution in terms of the ΔV cost, but it might provide a solution that is better when other mission constraints are taken into account.

We employ two statistics to capture the relative performance of the different trajectory optimization techniques. The first statistic is simply the percentage of initial guesses where a particular method yields the lowest ΔV cost. We denote this statistic $\mathcal{S}_1$. For example, there are 54 initial guesses for Case One. If *fmincon* converges to the lowest ΔV cost of all the techniques for 10 of the initial guesses, then $\mathcal{S}_1$ for *fmincon* is 17.54%. According to this definition, the sum of the $\mathcal{S}_1$ statistic for for all of the techniques for a particular test case is 100%.

The second statistic we employ is intended to provide more insight into the relative performance of the different techniques. For example, if *SNOPT* converges to a solution of 260 m/s for a particular initial guess, and PVAT1 converges to a solution of 261 m/s for the same initial guess, then although PVAT1 has a higher solution, it still performed well relative to *SNOPT*. This is not captured in the $\mathcal{S}_1$ statistic. We define a second statistic, $\mathcal{S}_2$, and define it as the percentage of initial guesses that converge to within 5% of the lowest solution yielded by the six techniques for the particular initial guess. An example makes this clear. Suppose we are comparing solutions from *fminunc* and *fminsearch* for a set of three initial guesses. Suppose *fminunc* converges to 100 m/s for the first initial guess, and *fminsearch* converges to 102 m/s for the first initial guess. For the second initial guess *fminunc* converges to 104 m/s and *fminsearch* converges to 120 m/s. For the third initial guess *fminunc* and *fminsearch* converge to 110, and 101 m/s respectively. For this set of solutions $\mathcal{S}_2$ for *fminsearch* is 66.66%, which means that 66.66% of the time *fminsearch* converged to within the 5% of the lowest solution of all the methods.

Both the $\mathcal{S}_1$ and $\mathcal{S}_2$ statistic can be applied to the results in several ways. For example, if only four-burn initial guesses are used in generating an $\mathcal{S}_2$ statistic, we add an additional subscript and call the statistic $\mathcal{S}_{2_4}$. If all of the initial guesses are used for determining an $\mathcal{S}_2$ statistic, there is no additional subscript. This convention is also used for the $\mathcal{S}_1$ statistic.

In the next three subsections we present a performance comparison of the six techniques for each of the three test cases. We use both the statistics described above, as well as graphical methods to illustrate how the techniques compare with one another.

Case One

Recall that Case One is a simple circular-to-circular coplanar rendezvous problem. Hence the optimal solution is simply the Hohmann ΔV, which for the orbits defined in Eqs. (40) and (41) is 255.8 m/s. A figure illustrating the results for all six optimization techniques we investigate is shown in Fig. 5. The top subplot contains the solutions for all two burn maneuver sequences. The middle and bottom subplots contain the solutions for all three and four burn solutions respectively. On each subplot the x-axis is the initial guess number. The initial guesses have been numbered in order of increasing ΔV. This is done to enable one to draw conclusions as to the performance of each method as the ΔV of initial guess increases. Furthermore, the initial guesses between the subplots are similar. For example, the initial guess number one in the two-maneuver subplot is similar to initial guess number one in the three-maneuver subplot. The difference being only that a small mid-course maneuver is placed along the two-maneuver sequence to create a three maneuver sequence. The same is true for initial guess one of the four-maneuver subplot. On all subplots the y-axis is the ΔV in km/s. Solutions for all direct methods are plotted in black. Solutions for indirect methods are plotted in red. Statistics that aid in comparing the results for each of method are found in Table1. By inspection of Fig. 5 we see that the direct methods outperformed the indirect methods. For all but one initial guess, the direct methods always converged to within a few m/s of the known optimal solution for two-maneuver sequences. For three and four burn sequences, *fminsearch* consistently had the poorest performance of the direct methods. In comparisons between PVAT1 and PVAT2, PVAT2 performed better. Examining the $\mathcal{S}_1$ statistics for Case One we see *SNOPT* found the lowest solution most often at 35.09% of the time. The method with the second best performance was *fmincon*, which found the lowest solution 31.58 % of the time. Examining the $\mathcal{S}_2$ statistics we see that both SQP methods converged to within 5% of the lowest solutioin at least 93% of the time. It is also important to note that although other methods performed poorly in comparison to SQP, about 33% of the time the lowest solution was found by a method other than SQP. This suggests that although SPQ is the single best performing method for Case One, it is not safe to rely only on SQP as an optimization approach. It is also important to note that although SPQ found the lowest solution most often, solutions from other methods were often close to the solution found by SQP methods.

Table 1: Statistics: Case One Results

	$\mathcal{S}_1$	$\mathcal{S}_{1_2}$	$\mathcal{S}_{1_3}$	$\mathcal{S}_{1_4}$	$\mathcal{S}_2$	$\mathcal{S}_{2_2}$	$\mathcal{S}_{2_3}$	$\mathcal{S}_{2_4}$
fmincon	**31.58**	21.05	15.79	57.89	**92.98**	100	78.95	100
SNOPT	**35.09**	57.89	42.11	5.263	**94.74**	100	89.47	94.74
fminunc	**10.53**	5.263	5.263	21.05	**78.95**	94.74	52.63	89.47
fminsearch	**7.018**	10.53	0.00	10.53	**43.86**	100	21.05	10.53
PVAT1	**3.509**	0.00	5.263	5.263	**71.93**	94.74	42.11	78.95
PVAT2	**12.28**	5.263	31.58	0.00	**66.67**	57.89	57.89	84.21

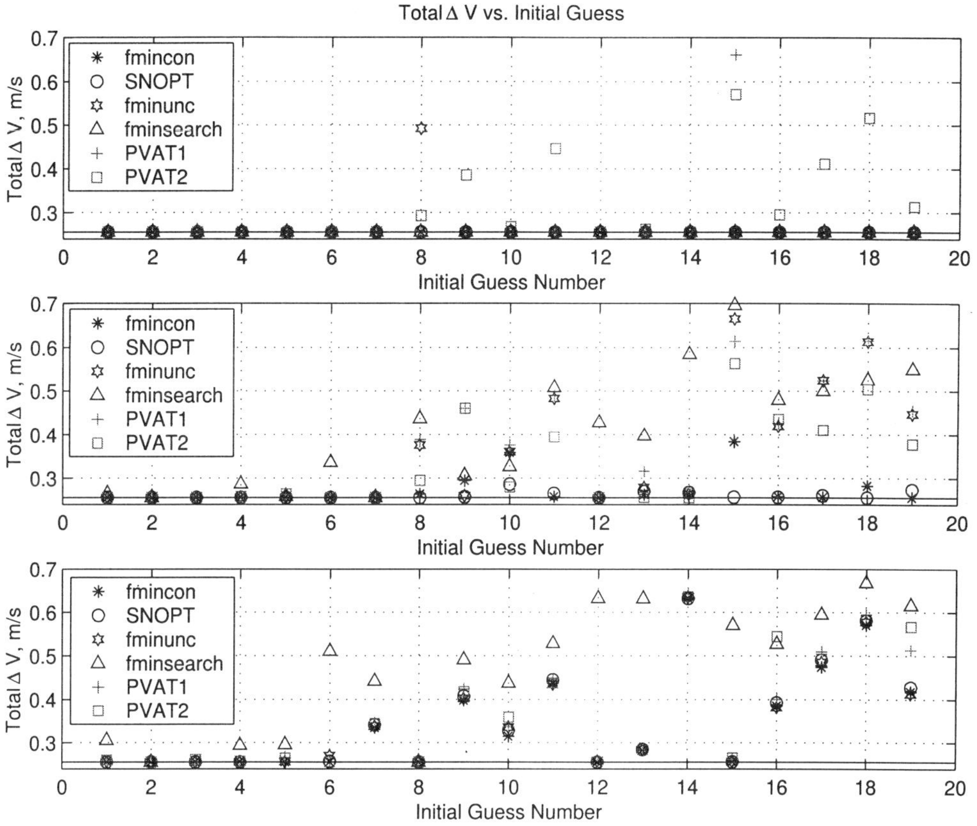

Figure 5: Results for Case One

Case Two

The maneuver sequence in Case Two involves raising the semi-major axis and rotating about the line of apsides, as well as satisfying the rendezvous conditions. Hence, two burn solutions will often perform poorly. Figure 6 shows all of the converged solutions for Case Two. The results are plotted using conventions identical to the results for Case One and we refer the reader to the previous subsection for details. Table 2 summarizes the selected statistics for each method for Case Two. For the two-burn sequences seen in the first subplot, PVAT2 yielded the lowest ΔV about 48% of the time and converged 71% of time within 5% of the lowest solution. PVAT2 higher performance can be explained by its unique ability compared to the other methods to add maneuvers to lower the ΔV cost. For the two-burn sequences all methods, with the exception of PVAT2, frequently converged to the same local minimum. Note that PVAT1 did not perform as well as the direct methods for the two-burn sequences as it converged only 29% of the time within 5% of the best solution. For three-burn sequences, *SNOPT* and *fmincon* yielded to the lowest ΔV about 42% of the time and converged about 90% and 81% of the time, respectively, within 5% of the lowest solution. For this set of initial guesses, PVAT2 yielded the worst results. Finally, for the four-burn sequence, *fmincon* out-performed all the other methods. *fmincon* found the lowest solution 58% of the time and 93% of its converged solutions were within 5% of the lowest ΔV cost. *fminunc* is the second best with about 16% and 64% for $\mathcal{S}_1$ and $\mathcal{S}_2$ respectively. *SNOPT*, *fminsearch* and PVAT1 have comparable statistics for this initial guess sequence. PVAT2 performed poorly compared to the other methods. Overall, we can see that, for Case Two, when there are more than two-impulses, *fmincon* appears to be consistently better. However, for the two-burn sequence,

P.VAT2 dominates over the other method because of its ability to add impulses to the trajectory.

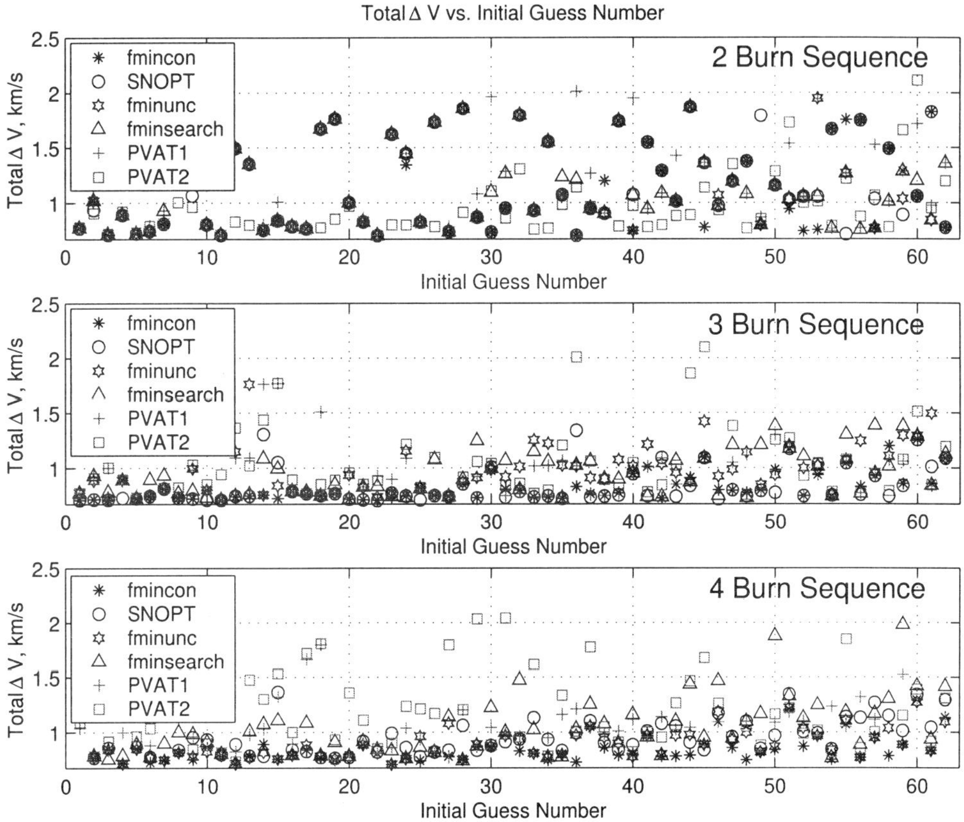

Figure 6: Results for Case Two

Table 2: Statistics Case Two Results

	$\mathcal{S}_1$	$\mathcal{S}_{1_2}$	$\mathcal{S}_{1_3}$	$\mathcal{S}_{1_4}$	$\mathcal{S}_2$	$\mathcal{S}_{2_2}$	$\mathcal{S}_{2_3}$	$\mathcal{S}_{2_4}$
fmincon	**31.58**	9.677	41.94	58.06	**75.81**	53.23	80.65	93.55
SNOPT	**24.73**	22.58	41.94	9.677	**56.99**	48.39	90.32	32.26
fminunc	**6.452**	1.613	1.613	16.13	**50.00**	48.39	37.1	64.52
fminsearch	**8.065**	8.065	8.065	8.065	**39.25**	43.55	45.16	29.03
PVAT1	**7.527**	9.677	4.839	8.065	**34.41**	29.03	41.94	32.26
PVAT2	**16.67**	48.39	1.613	0.00	**33.33**	70.97	17.74	11.29

Case Three

Recall that Case Three involves a phasing maneuver in a highly eccentric orbit which implies that the initial guess is going to be critical in defining which optimal solution is achievable. This sequence is a pure phasing maneuver where the number of revolutions is not to exceed one. Because of the high eccentricity of the orbit, moving the initial and final epochs of the rendezvous will in some cases change the trajectory dramatically making it more difficult to set the proper tolerances for the finite differencing to compute accurate gradients. Figure 7 shows all of the converged solutions for Case Three. Table 3 summarizes the selected statistics for each method for Case Three. For both the three-burn and four-burn sequences seen in the first subplot, all the indirect methods performed very poorly. *fmincon* outperformed all other methods with about 69% and 66.7% of the lowest solutions for the three-burn sequence and the four-burn sequence respectively. For the three and four burn sequences, $\mathcal{S}_2$ for *fmincon* was 89.6% and 91.67% respectively. *fminsearch* and *SNOPT* were the direct methods with the lowest performance rate for both burn sequences. Note that *fminunc* converged 62.5% of the time within 5% of the best solution but was only the best method 10.42% of the time for the three-burn case. However, for the four-burn sequence, *fminunc* statistics improved to a convergence rate of 91.67% within 5% of the best solution and it found the lowest ΔV 22.92% of the time. Overall, *fmincon* and *fminunc* seem to be the preferable methods for solving a Case Three type of rendezvous. All the other methods performed very poorly.

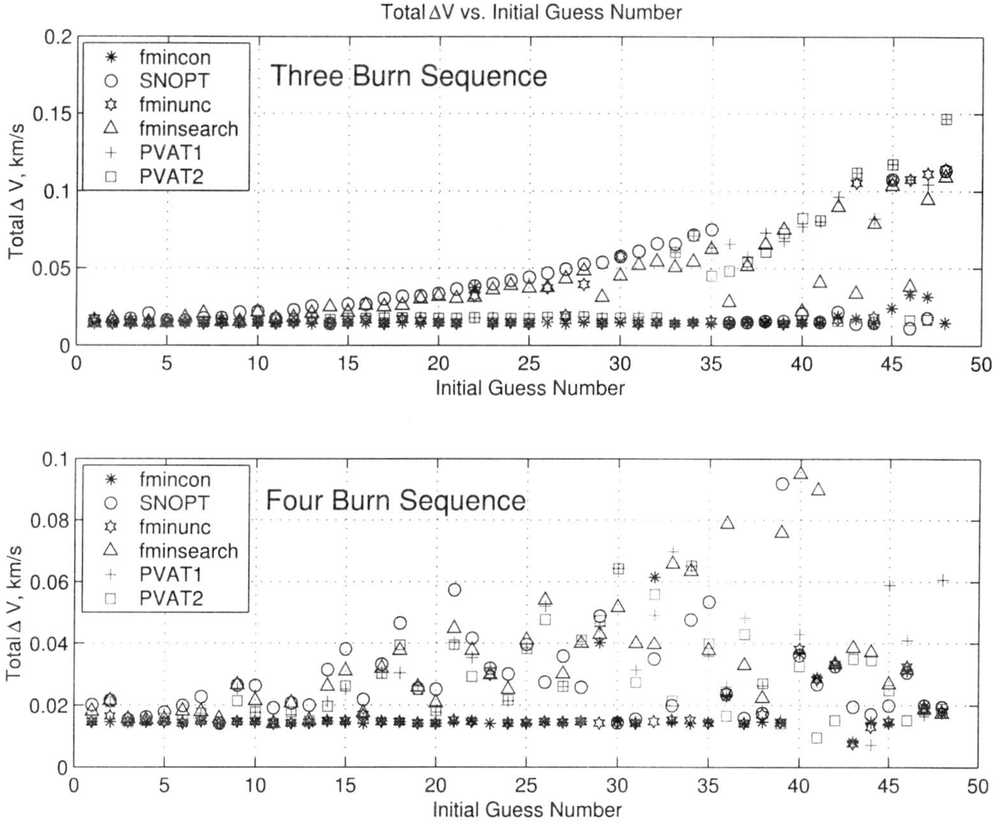

Figure 7: Results for Case Three

Table 3: Statistics: Case Three Results

	$\mathcal{S}_1$	$\mathcal{S}_{1_3}$	$\mathcal{S}_{1_4}$	$\mathcal{S}_2$	$\mathcal{S}_{2_3}$	$\mathcal{S}_{2_4}$
fmincon	**65.63**	72.92	58.33	**84.38**	89.58	79.17
SNOPT	**5.208**	8.333	2.083	**8.333**	12.5	4.167
fminunc	**14.58**	10.42	18.75	**67.71**	60.42	75.00
fminsearch	**3.125**	2.083	4.167	**5.208**	4.167	6.25
PVAT1	**3.125**	2.083	4.167	**14.58**	14.58	14.58
PVAT2	**8.333**	4.167	12.5	**21.88**	20.83	22.92

CONCLUSIONS AND FUTURE WORK

The minimum fuel rendezvous problem has received extensive attention in the literature. There are two primary methods to solving the problem: direct and indirect. However, many methods are actually a hybrid between a direct and indirect approach.

In this work we compared several approaches for finding optimal rendezvous solutions. We compared several parameterizations for an objective function for a direct approach, and chose the one that appeared the best after some preliminary comparisons. We applied four numerical optimization routines to the direct method parameterization. It is important to note that finite differencing was used to calculate gradients for the gradient-based direct methods. Providing analytic gradients is a topic of current research and will likely improve both the rate of convergence, and the final values of the final converged solutions.

We also developed two indirect approaches. The indirect approaches were based on primer vector theory. The first approach does not allow additional impulses higher than the number of impulses contained in the initial guess. The second approach will include an additional impulse if the optimality conditions suggest that it will result in a ΔV reduction.

Ideally, from a software maintenance point of view, it is desirable for one method to always excel over competing methods. However, in this work we found that no single method was always the best. Yet, there are certain methods that outperform rival methods a significant amount of time. In general, in comparing direct approaches, SQP outperformed other methods such as Quasi-Newton, and Nelder-Meade Simplex. However, these results are intimately dependent on the parameterization of the objective function. For alternative parameterizations, SQP might not necessarily outperform other methods.

In comparisons between the direct and indirect approaches we implemented, we found that the direct approaches tended to find lower solutions more often. However, there were a significant number of cases where indirect methods outperformed direct approaches. Therefore, we cannot simply discard the indirect approaches in favor of the direct approaches.

REFERENCES

[1] J. Betts, "Survey of Numerical Methods for Trajectory Optimization," *Journal of Guidance, Control and Dynamics*, Vol. 21, March-April 1998, pp. 193–207.

[2] D. Jezewski, J. Brazzel, E. Prust, B. Brown, T. Mulder, and D. Wissinger, "A Survey of Rendezvous Trajectory Planning," *AAS/AIAA Astrodynamics Specialists Conference*, Durango, Colorado, August 1991. AAS 91-505.

[3] S. Rao, *Engineering Optimization: Theory and Practice*. Wiley-Interscience, third ed., 1996.

[4] R. Brusch, "Constrained Impulsive Trajectory Optimization for Orbit-to-Orbit Transfer," *Journal of Guidance and Control*, Vol. 2, May-June 1979, pp. 204–212.

[5] R. Serban, W. Koon, M. Lo, J. Marsden, L. Petzold, S. Ross, and R. Wilson, "Optimal Control for Halo Orbit Missions," *IFAC Workshop on Langrangian and Hamiltonian Methods for Nonlinear Control*, Princeton, New Jersey, Princeton University, March 16-18 2000.

[6] J. Guzmán, L. Mailhe, C. Schiff, S. Hughes, and D. Folta, "Primer Vector Optimization: Survey of Theory, New Analysis and Applications," *53rd International Astronautical Congress*, Houston, Texas, October 2002. IAC-02-A.6.09.

[7] T. MathWorks, "Optimization Toolbox Documentation," *http://www.mathworks.com.*

[8] P. E. Gill, W. Murray, and M. A. Saunders, "User's Guide for SNOPT 5.3: A Fortran Package for Large-Scale NonLinear Programming," December 1998.

[9] D. F. Lawden, *Optimal Trajectories for Space Navigation.* London: Butterworths, 1963.

[10] K. C. Howell and H. J. Pernicka, "Numerical Determination of Lissajous Trajectories in The Restricted Three-Body Problem," *Celestial Mechanics*, Vol. 41, 1988, pp. 107–124.

[11] J. J. Guzman, *Spacecraft Trajectory Design in the Context of a Coherent Restricted Four-Body Problem.* Ph.D. Dissertation, Purdue University, May 2001.

[12] G. Gedeon, "A Practical Note on the Use of Lambert's Equation," *AIAA Journal*, Vol. 3, No. 1, 1965, pp. 149–150.

[13] C. S. J. Guzman, L. Mailhe, S. Hughes, D. Folta, and S. Schiff, "Primer Vector Optimization: Survey of Theory and Some Applications," IAF Conference, Houston, October 2002.

[14] D. F. Lawden, *Impulsive Transfer Between Elliptic Orbits*, pp. 323–351. Optimization Techniques, New York: Academy Press, 1962. edited by G. Leitman.

[15] P. M. Lion and M. Handelsman, "Primer Vector on Fixed-Time Impulsive Trajectories," *AIAA*, Vol. 6, January 1968, pp. 127–132.

[16] D. J. Jezewski and H. Rozendaal, "An Efficient Method for Calculating Optimal Free-Space N-Impulse Trajectories," *AIAA Journal*, December 1967, pp. 2160–2165.

[17] L. A. Hiday, *Optimal Transfers Between Libration-Point Orbits in the Elliptical Restricted Three-Body Problem.* Ph.D. Dissertation, Purdue University, August 1992.

[18] R. G. Stern, "Singularities in The Analytic Solution of the Linearized Variational Equations of Elliptical Motion," *AIAA*, July, 1 1964, pp. 1–13.

APPENDIX 1

There are numerous considerations to take into account when selecting an objective function parameterization for a direct method. In this work we break down the types of parameterizations into two categories called the Feasible Iterate Approach, and the Infeasible Iterate Approach. In the Feasible Iterate Approach the cost function is parameterized in such a way that the rendezvous constraints are satisfied implicitly for each cost function evaluation. Using a Feasible Iterate Approach allows one to choose between both constrained and unconstrained numerical optimization packages. However, we must provide a robust way to solve the TPBVP. In the Infeasible Iterate Approach the rendezvous constraints are not necessarily satisfied for each cost function evaluation. The rendezvous conditions are satisfied upon convergence of the numerical optimization routine. The strength of the Infeasible Iterate Approach is that we do not have to provide a robust TPBVP algorithm. However, it is necessary that we use only constrained optimization packages. Hence, in the Feasible Iterate Approach we are in a sense applying a change of variables to convert a constrained problem into an unconstrained problem. In this appendix we present one possible parameterization for each category to illustrate some differences in the methods. One possible parameterization of the Infeasible Iterate Approach is

$$\text{Given: } (\mathbf{r}_o, \mathbf{v}_o, t_o) \text{ and } (\mathbf{r}_f, \mathbf{v}_f, t_f) \tag{46}$$

Choose the independent variables:

$$t_i \qquad i = 1, 2...n \tag{47}$$

$$\mathbf{r}_j \qquad j = 1, 2, 3....n \tag{48}$$

require the optimizer to satisfy the following constraints

$$\mathbf{r}_1 - \mathbf{f}(\mathbf{r}_o, \mathbf{v}_o, t_o, t_1) = 0 \quad (49)$$

$$\mathbf{r}_n - \mathbf{f}(\mathbf{r}_f, \mathbf{v}_f, t_f, t_n) = 0 \quad (50)$$

Given Eqs. (47) and (48), the entire maneuver sequence is defined. By solving Lambert's problem for each trajectory segment we can calculate the total ΔV.

Although many parameterizations of the Feasible Iterate Approach are possible, we only present one here. One possible parameterization is

$$\text{Given: } (\mathbf{r}_o, \mathbf{v}_o, t_o) \text{ and } (\mathbf{r}_f, \mathbf{v}_f, t_f) \quad (51)$$

Choose the independent variables:

$$t_i \quad i = 1, 2, ..., n \quad (52)$$

$$\Delta \mathbf{V}_j \quad j = 1, 2, ..., n \quad 2 \quad (53)$$

For the this parameterization the entire maneuver sequence is determined and we can solve for the total ΔV and satisfy the boundary conditions simultaneously. We first determine $\mathbf{r}_1$ and $\mathbf{v}_1^-$ from Eqs.(13) and (14). The next step is to solve $n-2$ initial value problems by iterating on the following algorithm

for $i = 1$ to $n-2$

$$\mathbf{v}_i^+ = \mathbf{v}_i^- + \Delta V_i \quad (54)$$

$$\mathbf{r}_{i+1} = \mathbf{f}(\mathbf{r}_i, \mathbf{v}_i^+, t_i, t_{i+1}) \quad (55)$$

$$\mathbf{v}_{i+1}^- = \mathbf{g}(\mathbf{r}_i, \mathbf{v}_i^+, t_i, t_{i+1}) \quad (56)$$

end

Finally we ensure the rendezvous conditions are satisfied by solving Lambert's problem for $\mathbf{r}_{n-1}$, $\mathbf{r}_n$, and $\Delta t = t_n - t_{n-1}$. With the solution for Lambert's problem we can solve for ΔV_{n-1} and ΔV_n and we then solve Eq. (19) for the total ΔV. There are many more possible choices for independent variables that fall under the Feasible Iterate Approach. Presenting all of the possibilities is beyond the scope of this work. However, it is worth mentioning that other Feasible Iterate Approaches are likely to be a hybrid of the method described by Eqs.(20), (21) and (22) or the method described by Eqs.(51), (52) and (53).

Choosing a specific parameterization for the objective function is nontrivial. Each of the parameterizations discussed above have some strengths and some weaknesses. Although a detailed comparison of the Feasible Iterate Approach and the Infeasible Iterate Approach is beyond the scope of this work, preliminary comparisons suggest that the Feasible Iterate Approach performs better. This is expected because it is often better to convert a constrained problem to an equivalent unconstrained problem if an appropriate change of variables is possible. Therefore we have chosen to consider only Feasible Iterate parameterizations. Choosing a specific parameterization of the Feasible Iterate Approach is also nontrivial. For this paper we choose to parameterize the objective function using the independent variables given in Eqs.(20), (21) and (22). The justification for choosing this parameterization is discussed in a previous section.

AAS 03-007

A NONLINEAR, SIX-DEGREE OF FREEDOM, PRECISION FORMATION CONTROL ALGORITHM, BASED ON RESTRICTED THREE BODY DYNAMICS

Richard J. Luquette* and Robert M. Sanner†

Precision Formation Flying is an enabling technology for a variety of proposed space-based observatories, including the Micro-Arcsecond X-ray Imaging Mission (MAXIM), the associated MAXIM pathfinder mission, Stellar Imager and the Terrestrial Planet Finder (TPF). An essential element of the technology is the control algorithm. This paper discusses the development of a nonlinear, six-degree of freedom (6DOF) control algorithm for maintaining the relative position and attitude of a spacecraft within a formation. The translation dynamics are based on the equations of motion for the general restricted three body problem. The control law guarantees the tracking error convergences to zero, based on a Lyapunov analysis. The simulation, modelled after the MAXIM Pathfinder mission, maintains the relative position and attitude of a Follower spacecraft with respect to a Leader spacecraft, stationed near the L2 libration point in the Sun-Earth system.

INTRODUCTION

Distributed Spacecraft Systems (DSS) represent the future architecture for constructing very large, space-based observatories.[1] Widely varying science goals lead to an equal assortment of design concepts and requirements. Stellar Imager,[2] the Terrestrial Planet Finder (TPF),[3] the Micro-Arcsecond X-ray Imaging Mission (MAXIM),[4] and the associated MAXIM Pathfinder, represent a subset of missions requiring precision formation flying. The spacecraft are expected to be stationed in the vicinity of the Earth-Sun L2 point, or in a heliocentric, Earth drift-away orbit. In both cases the Restricted Three Body Problem (RTBP) provides a natural context for design and analysis of the formation control law. Mission success will depend on the development of many supporting technologies.

One essential technology element is the algorithm for both relative position and attitude control. Traditionally, spacecraft control algorithms treat the orbital dynamics and attitude dynamics as uncoupled. However, precision formation flying requires continuous low thrust to achieve a sub-millimeter mission design criteria. Continuous thruster action introduces a coupling action between orbital trajectory control and attitude control. With proper design the thrusters can serve as actuators to simultaneously control the spacecraft orbit and attitude trajectories. Implementation of this control strategy requires a six-degree of freedom (6DOF) control law. This paper proposes a

* Aerospace Engineer, Flight Dynamics Analysis Branch, NASA Goddard Space Flight Center, Greenbelt, Maryland 20771. E-mail: richard.j.luquette@nasa.gov.

† Associate Professor, University of Maryland, Aerospace Engineering Department, College Park, Maryland. E-mail: rmsanner@eng.umd.edu.

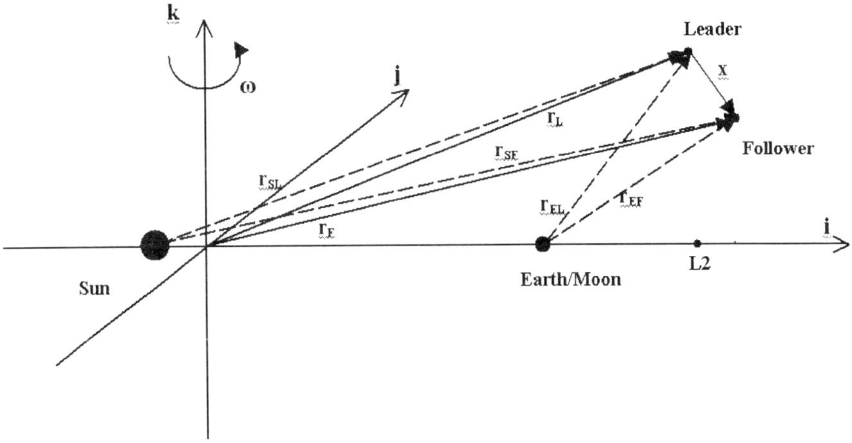

Figure 1: Two Spacecraft Formation Orbiting in the Earth/Moon - Sun Rotating Frame

6DOF nonlinear control law to solve this problem, building on previous work of the authors, which focused on the 3DOF, orbital control problem.[5]

Refer to the Notation section at the end of this paper for definitions of mathematical symbols not provided as part of the discussion.

PROBLEM STATEMENT

Consider a two spacecraft formation stationed in the vicinity of L2 in the Earth/Moon - Sun rotating frame, as shown in Figure 1. The spacecraft designations are Leader and Follower. The Earth-Moon system is modelled as a combined mass located at the system center of mass. The position and relative motion of the Earth/Moon system and the Sun define the rotating frame. The Leader maintains a planned ballistic trajectory with periodic station-keeping maneuvers, and a predetermined attitude trajectory. Attitude control on the Leader is accomplished with reaction wheels to avoid perturbing its orbit trajectory. The Follower tracks a specified separation trajectory, $\mathbf{x_d}$, relative to the Leader, and a predetermined attitude trajectory, $\mathbf{q_d}$. Measurement data provides the relative position and velocity between the two spacecraft. Each spacecraft is equipped to measure attitude, referenced to an inertial coordinate frame. The Follower is equipped with thrusters to serve as actuators for both translation and attitude maneuvers.

The problem:

- Design a control algorithm for the Follower to track a desired trajectory, $\mathbf{x_d}$ and $\mathbf{q_d}$.
- Demonstrate the tracking error approaches zero, at least asymptotically.
- Characterize the stability properties, i.e. local versus global stability.

SYSTEM DYNAMICS AND KINEMATICS

This section provides a brief review of both the dynamics and kinematics of translation and rotation for a spacecraft.

Translation

The following analysis is extracted from Ref. 5. Given a spacecraft stationed near the Sun-Earth/Moon L2 point, the principle environmental forces are gravity and solar pressure. These forces, combined with thruster action, drive the spacecraft dynamics. The principal gravitational sources are the Sun and the Earth-Moon system. In comparison, the mutual gravitational interaction between the spacecraft is insignificant. Based on the reference vectors shown in Figure 1, the Leader dynamics (per unit mass) are given by Eq. (1). The Follower dynamics per unit mass are given by Eq. (2).

$$\ddot{\mathbf{r}}_{\mathrm{L}} = -\mu_{em}\frac{\mathbf{r}_{\mathrm{EL}}}{||\mathbf{r}_{\mathrm{EL}}||_2^3} - \mu_s\frac{\mathbf{r}_{\mathrm{SL}}}{||\mathbf{r}_{\mathrm{SL}}||_2^3} + \mathbf{f}_{\mathrm{solar,L}} + \mathbf{f}_{\mathrm{pert,L}} + \mathbf{u}_{\mathrm{thrust,L}} \quad (1)$$

$$\ddot{\mathbf{r}}_{\mathrm{F}} = -\mu_{em}\frac{\mathbf{r}_{\mathrm{EF}}}{||\mathbf{r}_{\mathrm{EF}}||_2^3} - \mu_s\frac{\mathbf{r}_{\mathrm{SF}}}{||\mathbf{r}_{\mathrm{SF}}||_2^3} + \mathbf{f}_{\mathrm{solar,F}} + \mathbf{f}_{\mathrm{pert,F}} + \mathbf{u}_{\mathrm{thrust,F}} \quad (2)$$

The relative motion of the Follower with respect to the Leader is computed as the difference of Eqs. (1) and (2).

$$\begin{aligned}
\ddot{\mathbf{x}} &= \ddot{\mathbf{r}}_{\mathrm{F}} - \ddot{\mathbf{r}}_{\mathrm{L}} \\
&= -\mu_{em}\frac{\mathbf{r}_{\mathrm{EF}}}{||\mathbf{r}_{\mathrm{EF}}||_2^3} - \mu_s\frac{\mathbf{r}_{\mathrm{SF}}}{||\mathbf{r}_{\mathrm{SF}}||_2^3} + \mathbf{f}_{\mathrm{solar,F}} + \mathbf{f}_{\mathrm{pert,F}} + \mathbf{u}_{\mathrm{thrust,F}} \\
&\quad -\{-\mu_{em}\frac{\mathbf{r}_{\mathrm{EL}}}{||\mathbf{r}_{\mathrm{EL}}||_2^3} - \mu_s\frac{\mathbf{r}_{\mathrm{SL}}}{||\mathbf{r}_{\mathrm{SL}}||_2^3} + \mathbf{f}_{\mathrm{solar,L}} + \mathbf{f}_{\mathrm{pert,L}} + \mathbf{u}_{\mathrm{thrust,L}}\} \\
&= -\mu_{em}\{\frac{\mathbf{r}_{\mathrm{EF}}}{||\mathbf{r}_{\mathrm{EF}}||_2^3} - \frac{\mathbf{r}_{\mathrm{EL}}}{||\mathbf{r}_{\mathrm{EL}}||_2^3}\} - \mu_s\{\frac{\mathbf{r}_{\mathrm{SF}}}{||\mathbf{r}_{\mathrm{SF}}||_2^3} - \frac{\mathbf{r}_{\mathrm{SL}}}{||\mathbf{r}_{\mathrm{SL}}||_2^3}\} \quad (3) \\
&\quad + \Delta\mathbf{f}_{\mathrm{solar}} + \Delta\mathbf{f}_{\mathrm{pert}} + \mathbf{u}_{\mathrm{thrust,F}} - \mathbf{u}_{\mathrm{thrust,L}} \\
&= -\{\frac{\mu_{em}}{||\mathbf{r}_{\mathrm{EF}}||_2^3} + \frac{\mu_s}{||\mathbf{r}_{\mathrm{SF}}||_2^3}\}\mathbf{x} - \mu_{em}\{\frac{1}{||\mathbf{r}_{\mathrm{EF}}||_2^3} - \frac{1}{||\mathbf{r}_{\mathrm{EL}}||_2^3}\}\mathbf{r}_{\mathrm{EL}} \\
&\quad -\mu_s\{\frac{1}{||\mathbf{r}_{\mathrm{SF}}||_2^3} - \frac{1}{||\mathbf{r}_{\mathrm{SL}}||_2^3}\}\mathbf{r}_{\mathrm{SL}} + \Delta\mathbf{f}_{\mathrm{solar}} + \Delta\mathbf{f}_{\mathrm{pert}} + \mathbf{u}_{\mathrm{thrust,F}} - \mathbf{u}_{\mathrm{thrust,L}}
\end{aligned}$$

The terms are arranged to resolve the main gravitational forces into three components along the vectors, $\mathbf{x}$, $\mathbf{r}_{\mathrm{EL}}$, and $\mathbf{r}_{\mathrm{SL}}$. The typical formation under consideration represents a space-based observatory. Therefore, the desired relative position, $\mathbf{x_d}$, is assumed constant in magnitude and direction with respect to inertial space, except during reorientation maneuvers. On a short time scale, defined by the controller response, the spacecraft position vectors, expressed in the rotating coordinate frame, ${}^{r}\mathbf{r}_{\mathrm{EL}}$, and ${}^{r}\mathbf{r}_{\mathrm{SL}}$, remain essentially constant in magnitude and direction, Figure 1. This allows Eq. (3) to be represented as:

$$\ddot{\mathbf{x}} = -\mathbf{C}_1 * \mathbf{I}_3 * \mathring{\mathbf{x}} - \mathbf{C}_2 * \mathbf{A}_{\mathrm{ri}}^{\mathrm{T}} *^{\mathrm{r}} \mathring{\mathbf{r}}_{\mathrm{EL}} - \mathbf{C}_3 * \mathbf{A}_{\mathrm{ri}}^{\mathrm{T}} *^{\mathrm{r}} \mathring{\mathbf{r}}_{\mathrm{SL}} + \Delta\mathbf{f}_{\mathrm{solar}} + \Delta\mathbf{f}_{\mathrm{pert}} + \mathbf{u}_{\mathrm{thrust,F}} - \mathbf{u}_{\mathrm{thrust,L}} \quad (4)$$

The terms, $\mathbf{C}_1$, $\mathbf{C}_2$ and $\mathbf{C}_3$, are "constant" coefficients. The unit vectors, $\mathring{\mathbf{x}}$, ${}^{r}\mathring{\mathbf{r}}_{\mathrm{EL}}$ and ${}^{r}\mathring{\mathbf{r}}_{\mathrm{SL}}$, are "constant" in their respective frames. Therefore, the gravitational component along a fixed $\mathbf{x}$, is treated as constant in inertial space. The other components are considered constant in the rotating frame of the RTBP. Theoretically, the magnitude and direction of these terms could be computed based on the position of the spacecraft relative to the Earth, Sun and other gravitational bodies, allowing direct compensation for the gravitational gradient between spacecraft in the control design. However, in practice precise position data is not available for spacecraft stationed at L2, or a heliocentric drift away orbit. Therefore, knowledge of the magnitude and direction of these terms is considered unknown. So, with $\mathbf{\Theta_1} = \mathbf{C_1} * \mathring{\mathbf{x}}$ and $\mathbf{\Theta_2} = \mathbf{C_2} *^{r} \mathring{\mathbf{r}}_{\mathbf{EL}} - \mathbf{C_3} *^{r} \mathring{\mathbf{r}}_{\mathbf{SL}}$ representing the unknown constant vectors in the inertial and RTBP frames, respectively, Eq. (3) is expressed as:

$$\ddot{\mathbf{x}} = -\mathbf{I_3} * \mathbf{\Theta_1} - \mathbf{A}_{\mathbf{ri}}^{\mathbf{T}} * \mathbf{\Theta_2} + \mathbf{\Delta f}_{\mathbf{solar}} + \mathbf{\Delta f}_{\mathbf{pert}} + \mathbf{u}_{\mathbf{thrust,F}} - \mathbf{u}_{\mathbf{thrust,L}} \quad (5)$$

Further, the perturbing forces due to solar pressure and other gravitational sources are also modelled as unknown with constant components, one inertially fixed, the other fixed in the RTBP frame. Solar pressure acts along the spacecraft to sun line, which is fixed in the RTBP rotating frame. As with the Sun and Earth/Moon, the contribution of other gravitational sources can be resolved into two components, one in each of the inertial and RTBP frames. These components also remain essentially constant under the stated assumptions. Therefore, the terms, $\mathbf{\Delta f}_{\mathbf{solar}}$ and $\mathbf{\Delta f}_{\mathbf{pert}}$, are absorbed into the terms, $\mathbf{\Theta_1}$ and $\mathbf{\Theta_2}$. Then, Eq. (5) is rewritten as:

$$\ddot{\mathbf{x}} = -\mathbf{I_3} * \mathbf{\Theta_1} - \mathbf{A}_{\mathbf{ri}}^{\mathbf{T}} * \mathbf{\Theta_2} + \mathbf{u}_{\mathbf{thrust,F}} - \mathbf{u}_{\mathbf{thrust,L}} \quad (6)$$

Note: The rotation matrix, $\mathbf{A_{ri}}$, is the coordinate transformation from the inertial to the RTBP frame. Its computation is based on planetary ephemeris data. Therefore, it is easily determined, based on knowledge of the ephemeris time.

Rotation

The equation for rotational dynamics of a rigid spacecraft without reaction wheels is given by:

$$\mathbf{H_R} * \dot{\omega} - S([\mathbf{H_R} * \omega]) * \omega = \tau \quad (7)$$

where: $S([\mathbf{H_R} * \omega])$ is the skew symmetric matrix formed by the vector, $[\mathbf{H_R} * \omega]$.

In general, the rotational kinematics are given by:

$$\dot{\mathrm{q}} = \frac{1}{2} \begin{bmatrix} (\eta * I_3 + S(\varepsilon)) \\ -\varepsilon^T \end{bmatrix} * \omega \quad (8)$$

CONTROL LAW DESIGN

The proposed control law design combines algorithms for control of translation from Ref. 5, and rotation from Ref. 6.

Translation

The control design is based on nonlinear adaptive theory developed for robotic applications.[7] Under the assumption that Eq. (6) represents the relative dynamics, Ref. 5 presents an adaptive control strategy that provides globally stable, perfect tracking of a desired smooth trajectory for a spacecraft with a known mass. The analysis assumes perfect knowledge of the motion of the Follower with respect to the Leader. Terms used in the following discussion are defined as:

$\mathbf{m_F}$ – Mass of Follower
$\hat{\mathbf{m}}_\mathbf{F}$ – Estimated Mass of Follower
$\mathbf{x_d}$ – Desired spacecraft separation vector
$\mathbf{\Gamma_T}$ – $\left[\ \mathbf{I_3} \mid \mathbf{A}_{\mathbf{ri}}^{\mathbf{T}}\ \right]$
$\mathbf{I_3}$ – Identity Matrix, 3x3
$\mathbf{A_{ri}}$ – Rotation Matrix from Inertial to Earth/Moon - Sun Rotating Frame

Θ_T – $\begin{bmatrix} \Theta_1 \\ \Theta_2 \end{bmatrix}$
$\hat{\Theta}_T$ – Estimate of Θ_T
$\tilde{\Theta}_T$ – $(\hat{\Theta}_T - \Theta_T)$
$\dot{x}^r$ – $(\dot{x}_d - \Lambda_T * (x - x_d))$, reference velocity
s_T – $(\dot{x} - \dot{x}_d) + \Lambda_T * (x - x_d) = \dot{x} - \dot{x}^r$, error metric
α – Design parameter, adaptive gain, positive constant
K_T – Design parameter, symmetric, uniformly positive definite matrix
Λ_T – Design parameter, constant, symmetric, positive definite matrix

The proposed control law for differential thrust per unit mass is:

$$u_{thrust,F} - u_{thrust,L} = \ddot{x}^r + \left[I_3 \mid A_{ri}^T \right] \begin{bmatrix} \hat{\Theta}_1 \\ \hat{\Theta}_2 \end{bmatrix} - K_T * s_T = \ddot{x}^r + \Gamma_T * \hat{\Theta}_T - K_T * s_T \quad (9)$$

$$\dot{\hat{\Theta}}_T = -\gamma * \Gamma_T^T * s_T \quad (10)$$

For a spacecraft of known mass, Eqs. (6) and (9) combine to form the expression for the tracking error dynamics:

$$\begin{aligned} \ddot{x} &= -I_3 * \Theta_1 - A_{ir} * \Theta_2 + \ddot{x}^r + \Gamma_T * \hat{\Theta}_T - K_T * s_T \\ \dot{s}_T &= \Gamma_T * (\hat{\Theta}_T - \Theta_T) - K_T * s_T \\ \dot{s}_T &= \Gamma_T * \tilde{\Theta}_T - K_T * s_T \end{aligned} \quad (11)$$

The error dynamics, Eq. (11), combined with the adaptive rule, Eq. (10), results in asymptotic stability of the tracking error. The proof is provided by Ref. 5.

With the additional restriction that the desired acceleration, $\ddot{x}_d = 0$, the above algorithm is still valid for a spacecraft with uncertain mass. For a spacecraft of unknown/estimated mass, Eqs. (6) and (9) combine to form the expression for the tracking error dynamics:

$$\ddot{x} = -\Gamma_T * \Theta_T + \frac{\hat{m}_F}{m_F} * (\ddot{x}^r + \Gamma_T * \hat{\Theta}_T - K_T * s'_T) \quad (12)$$

Where the error metric is redefined as, $s'_T = (\dot{x} - \dot{x}_d) + \sigma * \Lambda * (x - x_d)$, with $\sigma = \frac{\hat{m}_F}{m_F}$. Further, define $\tilde{\Theta}_T = \sigma * \hat{\Theta}_T - \Theta_T$. Then Eq. (12) is rewritten as:

$$\begin{aligned} (\ddot{x} - \ddot{x}_d) + \sigma * \Lambda * (\dot{x} - \dot{x}_d) &= \Gamma_T * (\sigma * \hat{\Theta}_T - \Theta_T) - (1 - \sigma) * \ddot{x}_d - K_T * s'_T \\ \dot{s}'_T &= \Gamma_T * \tilde{\Theta}_T - K_T * s'_T - (1 - \sigma) * \ddot{x}_d \end{aligned} \quad (13)$$

Under the stated assumption, $\ddot{x}_d = 0$, the error dynamics are simplified as:

$$\dot{s}'_T = \Gamma_T * \tilde{\Theta}_T - K_T * s'_T \quad (14)$$

Coupled with the following modified adaptive rule the equations assume the same form as Eqs. (9) and (10).

$$\dot{\hat{\Theta}}_T = -\gamma * \Gamma_T^T * s'_T \quad (15)$$

Asymptotic stability of the tracking error is directly inferred from the Lyapunov proof presented in Ref. 5. Convergence of $\tilde{\Theta}_T \longrightarrow 0$, is not guaranteed.

As a final note, the modified adaptive rule and error metric, $\mathbf{s}'_{\mathrm{T}}$, facilitate the proof of asymptotic stability. As the value of $\boldsymbol{\sigma}$ is unknown, it cannot be directly factored into the algorithm design. However, the control law provided in Eq. (9), combined with the adaptive rule, Eq. (10), are sufficient to generate the desired result. Effectively, the term, $(\boldsymbol{\sigma} * \boldsymbol{\Lambda})$, is treated as a single design parameter for defining $\mathbf{s}'_{\mathrm{T}}$.

Rotation

Ref. 6 presents a passivity-based attitude control algorithm for a rigid spacecraft. The formulation for a spacecraft with gas jet actuators is applied to this design. Terms used in the following discussion are defined as:

ω_d	– Desired Spacecraft Angular Rate
$\mathbf{q}_{\mathrm{d}} = [\varepsilon_{\mathrm{d}}\ \eta_{\mathrm{d}}]^{\mathrm{T}}$	– Desired Spacecraft Attitude Quaternion
$\tilde{\mathrm{q}}$	– $[\bar{\varepsilon}\ \bar{\eta}]^{\mathrm{T}} = \mathbf{q} \bigotimes \mathbf{q}_{\mathrm{d}}^{-1}$, error quaternion
ω^r	– $(\omega_d - \boldsymbol{\lambda}_{\mathrm{T}} * \bar{\varepsilon})$, reference angular rate
$\mathbf{s}_{\mathrm{R}}$	– $(\omega - \omega^r)$, angular rate error metric
$\mathbf{K}_{\mathrm{R}}$	– Design parameter, symmetric, uniformly positive definite matrix
$\boldsymbol{\Lambda}_R$	– Design parameter, constant, symmetric, positive definite matrix

Based on the rotational dynamics, Eq. (7), the control law is defined as:

$$\tau = \mathbf{H}_{\mathrm{R}} * \dot{\omega}^r - S([\mathbf{H}_{\mathrm{R}} * \omega]) * \omega^r - \mathbf{K}_{\mathrm{R}} * \mathbf{s}_{\mathrm{R}} \tag{16}$$

Evaluation of the expression requires knowledge of $\mathbf{q}$ and ω, plus definition of $\mathbf{q}_{\mathrm{d}}$, ω_d and $\dot{\omega}_d$. Implementation of this algorithm requires knowledge of the spacecraft moment of inertia, $\mathbf{H}_{\mathrm{R}}$, which is limited by measurement errors and mass property variations. Therefore, an adaptive control strategy, designed to estimate the mass properties, is preferred. This is accomplished by reformulating Eq. (16) as:

$$\tau = \boldsymbol{\Gamma}_{\mathrm{R}}(\mathbf{q}, \omega, \omega_r, \dot{\omega}^r) * \hat{\boldsymbol{\Theta}}_{\mathrm{R}} - \mathbf{K}_{\mathrm{R}} * \mathbf{s}_{\mathrm{R}} \tag{17}$$

Where:

$$\boldsymbol{\Gamma}_{\mathrm{R}}(\mathbf{q}, \omega, \omega_r, \dot{\omega}^r) * \boldsymbol{\Theta}_{\mathrm{R}} = \mathbf{H}_{\mathrm{R}} * \dot{\omega}^r - S([\mathbf{H}_{\mathrm{R}} * \omega]) * \omega^r \tag{18}$$

The matrix, $\boldsymbol{\Gamma}_{\mathrm{R}}(\mathbf{q}, \omega, \omega_r, \dot{\omega}^r)$, contains the known rigid body dynamics and kinematics. The vector, $\boldsymbol{\Theta}_{\mathrm{R}}$, contains the mass properties, considered constant and unknown. The control, Eq. (17), combined with the adaptive rule, Eq. (19), provides global convergence of the tracking error to zero.[6] However, the estimate of $\boldsymbol{\Theta}_{\mathrm{R}}$ will not necessarily converge to its true value.

$$\dot{\hat{\boldsymbol{\Theta}}}_{\mathrm{R}} = -\boldsymbol{\Lambda}_R * \boldsymbol{\Gamma}_{\mathrm{R}}(\mathbf{q}, \omega, \omega_r, \dot{\omega}^r)^{\mathrm{T}} * \mathbf{s}_{\mathrm{R}} \tag{19}$$

Combined

Simultaneous implementation of the above control algorithms requires coordinated thruster action to generate the desired translation and rotation control inputs. The relationship between the thruster output and the net control inputs is expressed as:

$$^{b}\mathbf{U} = \left[\begin{array}{c} \mathbf{R}(\mathbf{q}) * (\mathbf{u}_{\mathbf{thrust,F}} - \mathbf{u}_{\mathbf{thrust,L}}) \\ \tau \end{array} \right] = \mathbf{B} * \mathbf{F} \tag{20}$$

where:

$^{b}\mathbf{U}$ – Combined, translation and rotation control in body coordinates
$\mathbf{R}(\mathbf{q})$ – Transformation from inertial to body coordinates
$\mathbf{B}$ – Control sensitivity matrix
$\mathbf{F}$ – Thruster output, $[\mathbf{f}_1, \mathbf{f}_2, ..., \mathbf{f}_n]^T$

With proper placement of the thrusters the control sensitivity matrix, $\mathbf{B}$, will have a pseudo inverse. Assuming perfect knowledge of the sensitivity matrix, the thruster commands are computed as:

$$\mathbf{F} = \mathbf{B}^T(\mathbf{B} * \mathbf{B}^T)^{-1} * {}^{b}\mathbf{U} \tag{21}$$

This implementation for computing thruster commands generates the desired thrust and torque for translation and rotation control. Therefore, the convergence and stability properties of the previously discussed algorithms are retained.

SIMULATION

The performance of the control design is demonstrated through a simulation based on the MAXIM Pathfinder mission. The spacecraft are initialized in an orbit about the L2 point in the Earth/Moon-Sun system. The leader spacecraft remains on a ballistic trajectory, $\mathbf{u}_{\mathbf{thrust,L}} = \mathbf{0}$. The scenario starts with the Follower at the same position and attitude as the Leader, i.e. initial deployment. The Follower tracks a commanded trajectory to a 200 km separation, Figure 2, while simultaneously tracking a 90 degree slew maneuver, Figure 3. The simulation is implemented in MATLAB©.

The thruster model is comprised or three sets of four thrusters. Each set provides thrust along a single body axis, and torque along a perpendicular body axis, such that thrust and torque authority are available along each of the three body axes.

The dynamics model includes the gravitational influence of the Earth, Moon, Sun and all the other planets. The orbital equations of motion are adapted from the discussion of the N-Body problem found in Ref. 8. Solar pressure is modelled as a function of the radial distance from the sun, but otherwise constant. The cross-sectional area/surface reflectivity of the spacecraft is modelled as attitude independent.

All elements of the parameters, $\hat{\mathbf{\Theta}}_T$ and $\hat{\mathbf{\Theta}}_R$, are initialized to zero to demonstrate the perfomance of the adaptive mechanism. In practice, these parameters would be initialized to the best estimate of their true values.

The successful controller performance is evidenced by the tracking error, shown in Figures 4 and 5. The controller gains are not designed for optimal performance. The sum of the instantaneous output of the thrusters is provided in Figure 6

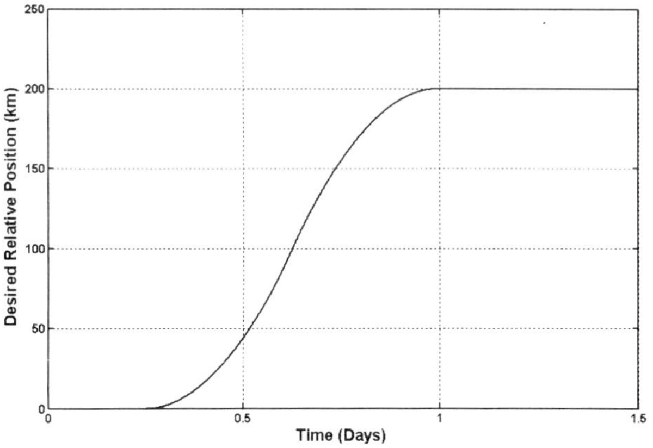

Figure 2: Desired Relative Separation between Follower and Leader

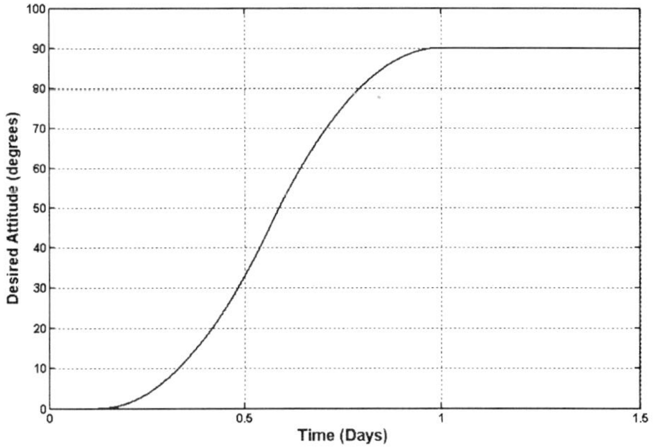

Figure 3: Euler Angle of Desired Attitude, $\mathbf{q_d}$, of Follower with Respect to Inertial

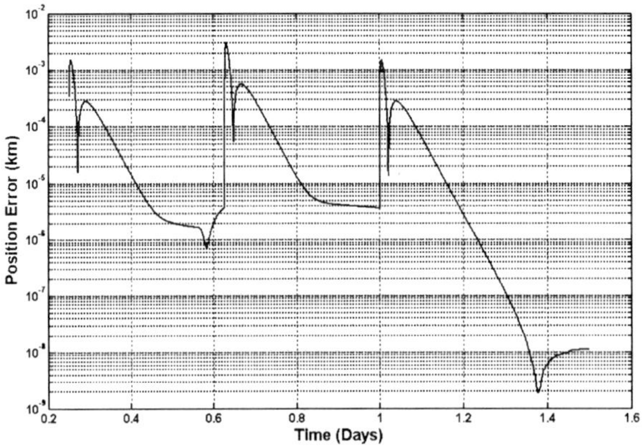

Figure 4: Relative Position Trajectory Tracking Error, $(\mathbf{x} - \mathbf{x_d})$

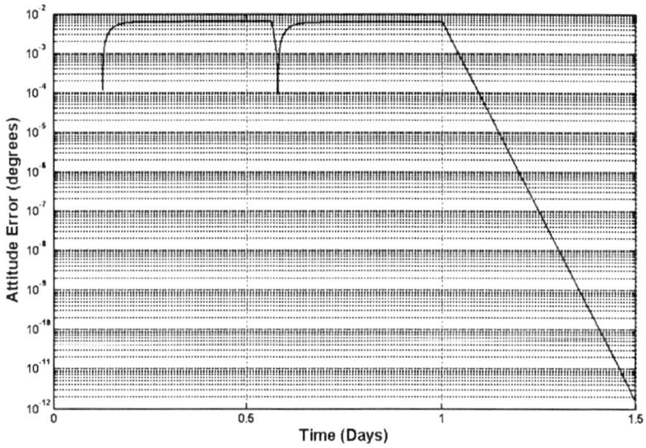

Figure 5: Euler Angle of Quaternion Trajectory Tracking Error, $(\mathbf{q} \otimes \mathbf{q}_{\mathbf{d}}^{-1})$

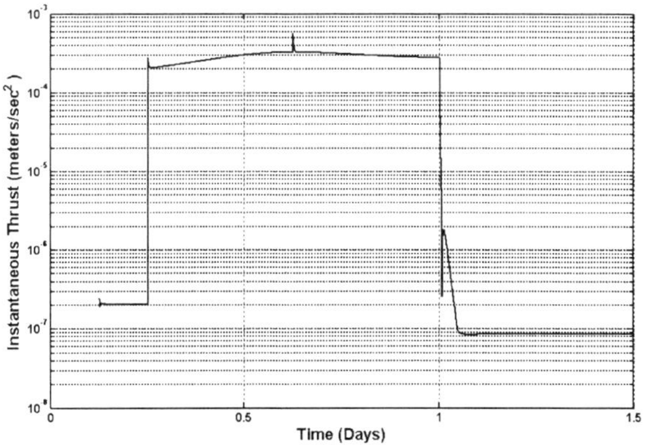

Figure 6: Combined Output of All Thrusters

CONCLUSIONS AND FUTURE WORK

This paper presents a successful adaptive, nonlinear 6DOF formation control strategy for spacecraft experiencing slow changes in environmental forces, typical for an orbit about L2. Adaptation is employed to compensate for unknown spacecraft mass properties. Stellar Imager, Terrestrial Planet Finder (TPF), MAXIM and MAXIM Pathfinder are typical missions with potential for employing this strategy.

The development assumes the availability of perfect measurement data, and ideal thruster performance. Future work will consider issues associated with corrupted measurement data, issues associated with thruster alignment and performance, and coupled stability issues with an observer/estimator.

NOTATION

Symbols for additional terms are defined in the following table. All vectors are resolved in inertial coordinates, unless otherwise stated.

$\mathbf{f}_{\text{solar}}$	– Force exerted on spacecraft due to solar pressure
$\mathbf{f}_{\text{pert}}$	– Force exerted on spacecraft due to other perturbations
$\mathbf{u}_{\text{thrust}}$	– Force exerted on spacecraft due to thrusters
μ_{em}	– Gravitational parameter of the Earth/Moon
μ_s	– Gravitational parameter of the Sun
$\mathbf{H}_{\text{R}}$	– Spacecraft Moment of Inertia
ω	– Spacecraft Angular Rate
τ	– External Torque on Spacecraft
$\mathrm{q} = [\varepsilon\ \eta]^{\mathrm{T}}$	– Spacecraft Attitude Quaternion
ε	– Vector Component of Spacecraft Attitude Quaternion
η	– Scalar Component of Spacecraft Attitude Quaternion

REFERENCES

1. Leitner, J., Bauer, F., Folta, D., Moreau, M., Carpenter, R., and How, J., "Formation in Flight Distributed Spacecraft Systems Develop New GPS Capabilities," *GPS World*, Feb. 2002.

2. "The Stellar Imager (SI) Homepage," http://hires.gsfc.nasa.gov/%7Esi/.

3. "The Terrestrial Planet Finder (TPF) Homepage," http://planetquest.jpl.nasa.gov/TPF/tpf_index.html.

4. "Offical Web Site for the MicroArcsecond X-Ray Imaging Mission (MAXIM)," http://maxim.gsfc.nasa.gov/.

5. Luquette, R. J. and Sanner, R. M., "A Nonlinear Approach to Spacecraft Formation Control in the Vicinity of a Collinear Point," *Proceedings of the AAS/AIAA Astrodynamics Specialist Conference*, July 2001, Paper No. AAS 01-330.

6. Egeland, O. and Godhavn, J. M., "Passivity-Based Adaptive Attitude Control of a Rigid Spacecraft," *IEEE Trans. on Automatic Control*, April 1994.

7. Slotine, J. and Li, W., *Applied Nonlinear Control*, Prentice Hall, Inc., New Jersey, 1991.

8. Bate, R. R., D., M. D., and E., W. J., *Fundamentals of Astrodynamics*, Dover Publications, Inc., New York, 1971.

AAS 03-008

A NONLINEAR OBSERVER FOR GYRO ALIGNMENT ESTIMATION

Julie Thienel[*] and Robert M. Sanner[†]

A nonlinear observer for gyro alignment estimation is presented. The observer is composed of two error terms, an attitude error and an alignment error. The observer is globally stable with exponential convergence of the attitude errors. The gyro alignment estimate converges to the true alignment when the system is completely observable.

INTRODUCTION

Combined observer-controller designs for the attitude control of rigid flight vehicles are a subject of active research[1]. Successful design of such architectures is complicated by the fact that there is, in general, no separation principle for nonlinear systems. In contrast to linear systems, "certainty equivalence" substitution of the states from an exponentially converging observer into a nominally stabilizing, state feedback control law does not necessarily guarantee stable closed-loop operation for the coupled systems[2,3].

One version of this problem, in particular, is the task of forcing the attitude of a rigid vehicle to asymptotically track a (time-varying) reference attitude using feedback from rate sensors with alignment errors. Alignment errors are typically estimated using a least-squares approach, or an extended Kalman filter[4-8]. Here, in order to determine the misalignment errors, we propose utilizing an angular velocity observer similar to the observers in refs.1 and 9, using the estimated misalignment state in the nonlinear observer presented in ref. 10. In the analysis below we demonstrate that the resulting system provides stable estimation of the alignment states.

The main proof proceeds in two steps. First, we extend the analysis of ref. 1 to the case of gyro misalignment first for a spacecraft with a constant angular velocity. Given the necessary observability conditions, we demonstrate that the alignment estimate converges to the true alignment. We extend the analysis for the case of a time varying angular velocity. The proof in ref. 1 uses a Lyapunov argument to obtain a similar result, but under the assumption that the alignment errors are small; this restriction is removed here.

* Flight Dynamics Analysis Branch, NASA Goddard Space Flight Center, Code 572, Greenbelt, Maryland 20771. E-mail: Julie.K.Thienel@nasa.gov.

† Department of Aerospace Engineering, University of Maryland, College Park, Maryland 20742. E-mail: rmsanner@eng.umd.edu.

DEFINITIONS

The attitude of a spacecraft can be represented by a quaternion, consisting of a unit rotation vector **e**, known as the Euler axis, and a rotation ϕ about this axis, given as

$$\mathbf{q} = \begin{bmatrix} \mathbf{e}\sin\frac{\phi}{2} \\ \cos\frac{\phi}{2} \end{bmatrix} = \begin{bmatrix} \varepsilon \\ \eta \end{bmatrix}$$

where **q** is the quaternion, partitioned into a vector part, ε and a scalar part, η. Typically, in spacecraft attitude applications, the quaternion represents the rotation from an inertial coordinate system to the spacecraft body coordinate system. Note that $\|\mathbf{q}\| = 1$ by definition. The rotation matrix for a specific attitude can be computed as[11]

$$R(\mathbf{q}) = (\eta^2 - \varepsilon^T\varepsilon)I + 2\varepsilon\,\varepsilon^T - 2\eta S(\varepsilon) \qquad (1)$$

where $S(\varepsilon)$ is a matrix representation of the vector cross product operation.

$$S(\varepsilon) = \begin{bmatrix} 0 & -\varepsilon_z & \varepsilon_y \\ \varepsilon_z & 0 & -\varepsilon_x \\ -\varepsilon_y & \varepsilon_x & 0 \end{bmatrix}$$

A rotation between coordinate frames is computed as[12]

$$\tilde{\mathbf{q}} = \begin{bmatrix} \tilde{\varepsilon} \\ \tilde{\eta} \end{bmatrix} = \mathbf{q}_1 \otimes \mathbf{q}_2^{-1} = \begin{bmatrix} \eta_2 I - S(\varepsilon_2) & -\varepsilon_2 \\ \varepsilon_2^T & \eta_2 \end{bmatrix} \begin{bmatrix} \varepsilon_1 \\ \eta_1 \end{bmatrix} \qquad (2)$$

Where $\tilde{\mathbf{q}}$ defines the rotation from the frame defined by $\mathbf{q}_2$ to the frame defined by $\mathbf{q}_1$. Note that $\|\tilde{\varepsilon}\| = 0,\ \tilde{\eta} = \pm 1$ indicates that frame 2 is aligned with frame 1.

The kinematic equation for the quaternion is given as

$$\dot{\mathbf{q}}(t) = \begin{bmatrix} \dot{\varepsilon} \\ \dot{\eta} \end{bmatrix} = \tfrac{1}{2} Q(\mathbf{q}(t)) R_g(\mathbf{q}_g)\omega_g(t) \qquad (3)$$

where

$$Q(\mathbf{q}(t)) = \begin{bmatrix} \eta(t)I + S(\varepsilon(t)) \\ -\varepsilon(t)^T \end{bmatrix} = \begin{bmatrix} Q_1(\mathbf{q}(t)) \\ -\varepsilon(t)^T \end{bmatrix} \qquad (4)$$

The kinematic equation for an attitude matrix is[11]

$$\dot{R}(\mathbf{q}(t)) = -S(R_g(\mathbf{q}_g)\omega_g(t))R(\mathbf{q}(t)) \qquad (5)$$

$\omega_g(t)$ is the spacecraft angular velocity with respect to inertial space, in gyro coordinates, transformed to body coordinates with the alignment matrix $R_g(\mathbf{q}_g)$. $R_g(\mathbf{q}_g)$ is computed from (1) above with the constant gyro alignment quaternion given as

$$\mathbf{q}_g = \begin{bmatrix} \varepsilon_g \\ \eta_g \end{bmatrix}$$

The angular velocity is typically measured by a gyro, which can be corrupted by various errors, such as bias, alignment, scale factor errors, and noise. In this work we consider the case of alignment errors, where the matrix $R_g(\mathbf{q}_g)$ is unknown.

NONLINEAR OBSERVER

Following the development in Ref. 10, observers for the attitude and gyro alignment quaternions are given as

$$\dot{\hat{\mathbf{q}}}_o(t) = \tfrac{1}{2}Q(\hat{\mathbf{q}}_o(t))\tilde{R}_o^T(\tilde{\mathbf{q}}_o(t))(\hat{R}_g(\hat{\mathbf{q}}_g(t))\omega_g(t) + k\tilde{\varepsilon}_o(t)\operatorname{sign}(\tilde{\eta}_o(t))) \tag{6a}$$

$$\dot{\hat{\mathbf{q}}}_g(t) = \tfrac{1}{2}Q(\hat{\mathbf{q}}_g(t))\hat{\omega}_g(t) \tag{6b}$$

where $\hat{\mathbf{q}}_o$ represents the transformation from inertial coordinates to the attitude observer coordinates, and $\hat{R}_g(\hat{\mathbf{q}}_g)\omega_g$ is the estimated angular velocity, transformed to observer coordinates with $\tilde{R}_o^T(\tilde{\mathbf{q}}_o)$. The attitude error quaternion, $\tilde{\mathbf{q}}_o$, is computed as in (2) with $\mathbf{q}$ and $\hat{\mathbf{q}}_o$, respectively. In Eq.(6b), $\hat{\mathbf{q}}_g$ represents the transformation from gyro coordinates to an intermediate coordinate frame, denoted as $\hat{\mathbf{b}}$, and $\hat{\omega}_g$ is the angular velocity of the $\hat{\mathbf{b}}$ frame with respect to the gyro frame, which is yet to be specified.

Taking the derivative of $\tilde{\mathbf{q}}_o$, and substituting in Eq. (3) and Eq. (6a) results in the following kinematic equation for the attitude error quaternion

$$\dot{\tilde{\mathbf{q}}}_o(t) = \tfrac{1}{2}Q(\tilde{\mathbf{q}}_o(t))(R_g(\mathbf{q}_g)\omega_g(t) - \hat{R}_g(\hat{\mathbf{q}}_g(t))\omega_g(t) - k\tilde{\varepsilon}_o(t)\operatorname{sign}(\tilde{\eta}_o(t))) \tag{7}$$

Define the difference in the two angular velocity terms in Eq. (7) as

$$\tilde{\mathbf{m}}(t) = R_g(\mathbf{q}_g)\omega_g(t) - \hat{R}_g(\hat{\mathbf{q}}_g(t))\omega_g(t) \tag{8}$$

First, consider the case where ω_g is constant. As in Ref. 10, a Lyapunov function is given as

$$V_o(t) = \tfrac{1}{2}\tilde{\varepsilon}_o^T(t)\tilde{\varepsilon}_o(t) + \begin{cases} \tfrac{1}{2}(\tilde{\eta}_o(t)-1)^2 & (\tilde{\eta}_o \geq 0) \\ \tfrac{1}{2}(\tilde{\eta}_o(t)+1)^2 & (\tilde{\eta}_o < 0) \end{cases} + \tfrac{1}{2}\tilde{\mathbf{m}}^T(t)\tilde{\mathbf{m}}(t) \tag{9}$$

The derivative of $V_o(t)$ is

$$\dot{V}_o(t) = \tilde{\varepsilon}_o^T(t)\dot{\tilde{\varepsilon}}_o(t) + \begin{cases} (\tilde{\eta}_o(t) - 1)\dot{\tilde{\eta}}_o(t)(\tilde{\eta}_o \geq 0) \\ (\tilde{\eta}_o(t) + 1)\dot{\tilde{\eta}}_o(t)(\tilde{\eta}_o < 0) \end{cases} + \tilde{\mathbf{m}}^T(t)\dot{\tilde{\mathbf{m}}}(t) \tag{10}$$

Noting that $\tilde{\varepsilon}_o^T(t)\dot{\tilde{\varepsilon}}_o(t) + \tilde{\eta}_o(t)\dot{\tilde{\eta}}_o(t) = 0$ and substituting $\dot{\tilde{\eta}}_o(t)$ from Eq. (7) into Eq. (10) yields

$$\dot{V}_o(t) = -\tfrac{k}{2}\tilde{\varepsilon}_o^T(t)\tilde{\varepsilon}_o(t) + \tfrac{1}{2}\tilde{\varepsilon}_o^T(t)\tilde{\mathbf{m}}(t)\mathrm{sign}(\tilde{\eta}_o(t)) + \tilde{\mathbf{m}}^T(t)\dot{\tilde{\mathbf{m}}}(t) \tag{11}$$

If $\dot{\tilde{\mathbf{m}}}(t)$ is defined as

$$\dot{\tilde{\mathbf{m}}} = -\tfrac{1}{2}\tilde{\varepsilon}_o\mathrm{sign}(\tilde{\eta}_o) \tag{12}$$

The derivative of the Lyapunov becomes

$$\dot{V}(t) = -\tfrac{k}{2}\tilde{\varepsilon}_o^T(t)\tilde{\varepsilon}_o(t) \tag{13}$$

Given that $V_o(t)$ in Eq. (9) is lower bounded, and Eq. (10) shows that $\dot{V}_o(t) \leq 0$, $\tilde{\varepsilon}_o$ and $\tilde{\mathbf{m}}$ are bounded. $V_o(t)$ is a continuous, twice differentiable function, with $\ddot{V}_o(t)$ bounded. Applying Barbalat's lemma[2] shows that, in fact, $\|\tilde{\varepsilon}_o(t)\| \to 0$.

In order to determine a form for $\hat{\omega}_g(t)$ in Eq. (6b), the derivative of $\tilde{\mathbf{m}}$, as given in Eq. (8), is computed as

$$\dot{\tilde{\mathbf{m}}}(t) = -\dot{\hat{R}}_g(\hat{\mathbf{q}}_g(t))\omega_g = S(\hat{\omega}_g(t))\hat{R}_g(\hat{\mathbf{q}}_g(t))\omega_g \tag{14}$$

Equating Eq. (12) and Eq. (14)

$$\tfrac{1}{2}\tilde{\varepsilon}_o(t)\mathrm{sign}(\tilde{\eta}_o(t)) = S(\hat{R}_g(\hat{\mathbf{q}}_g(t))\omega_g)\hat{\omega}_g(t) \tag{15}$$

Since $\hat{\omega}_g(t)$ is an observer design variable, it is designed to be perpendicular to $\hat{R}_g(\hat{\mathbf{q}}_g(t))\omega_g$ (any parallel component does not appear in Eq. (15)). This results in

$$\hat{\omega}_g(t) = \tfrac{1}{2}\mathrm{sign}(\tilde{\eta}_o(t))S(\tilde{\varepsilon}_o(t))\hat{R}_g(\hat{\mathbf{q}}_g(t))\omega_g$$

The above Lyapunov analysis shows that the errors $\tilde{\varepsilon}_o$ and $\tilde{\mathbf{m}}$ are stable and bounded. Therefore, as in Ref. 10, the system can be analyzed as a linear time varying system. Let

$$\begin{bmatrix} \dot{\tilde{\varepsilon}}_o(t) \\ \dot{\tilde{\mathbf{m}}}(t) \end{bmatrix} = A(t) \begin{bmatrix} \tilde{\varepsilon}_o(t) \\ \tilde{\mathbf{m}}(t) \end{bmatrix}$$

$$\text{with } A(t) = \begin{bmatrix} -\tfrac{1}{2}k\,\text{sign}(\tilde{\eta}_o(t))(\tilde{\eta}_o(t)I + S(\tilde{\varepsilon}_o(t)) & -\tfrac{1}{2}(\tilde{\eta}_o(t)I + S(\tilde{\varepsilon}_o(t)) \\ -\tfrac{1}{2}\text{sign}(\tilde{\eta}_o(t)) & 0 \end{bmatrix}$$

As shown in Ref. 10, this system is exponentially stable. In other words, both $\|\tilde{\varepsilon}_o(t)\|$ and $\|\tilde{\mathbf{m}}(t)\|$ converge to zero exponentially fast.

The convergence of $\tilde{\mathbf{m}}(t)$ does not guarantee that the gyro alignment estimate $\hat{\mathbf{q}}_g(t)$ converges to the true gyro alignment, $\mathbf{q}_g$. Rewriting Eq. (7) as

$$\tilde{\mathbf{m}}(t) = (I - \tilde{R}_g^T(\tilde{\mathbf{q}}_g(t))\boldsymbol{\omega}$$

Given that $(I - \tilde{R}_g^T)\tilde{\varepsilon}_g = 0$, the error, $\tilde{\mathbf{m}}(t)$, is zero when the angular velocity term, $\boldsymbol{\omega}(t)$ is parallel to $\tilde{\varepsilon}_g(t)$, or, ideally, when $\|\tilde{\varepsilon}_g(t)\| = 0$ indicating that the estimated alignment quaternion is equal to the true alignment quaternion. Given that the convergence of the attitude error is exponential for a constant angular velocity, a series of discrete attitude changes provides observability of the gyro alignment errors, allowing convergence of $\hat{\mathbf{q}}_g$ to $\mathbf{q}_g$.

Next, consider next the case where $\boldsymbol{\omega}$ is not constant. In this case, the derivative of $\tilde{\mathbf{m}}(t)$, given in Eq. (8), becomes

$$\dot{\tilde{\mathbf{m}}}(t) = (I - \tilde{R}_g(\tilde{\mathbf{q}}_g(t))\dot{\boldsymbol{\omega}}(t) - \frac{d}{dt}(\tilde{R}_g(\tilde{\mathbf{q}}_g(t))\boldsymbol{\omega}(t)) \tag{16}$$

Inserting (16) into (11), along with (15) results in

$$\dot{V}_o(t) = -\tfrac{k}{2}\tilde{\varepsilon}_o^T(t)\tilde{\varepsilon}_o(t) + \tilde{\mathbf{m}}^T(t)\dot{\tilde{\mathbf{m}}}(t) = -\tfrac{k}{2}\tilde{\varepsilon}_o^T(t)\tilde{\varepsilon}_o(t) + \boldsymbol{\omega}^T(t)(I - \tilde{R}_g(\tilde{\mathbf{q}}_g(t))(I - \tilde{R}_g^T(\tilde{\mathbf{q}}_g(t))\dot{\boldsymbol{\omega}}(t)$$

Using Eq. (1) to expand $\tilde{R}_g(\tilde{\mathbf{q}}_g(t))$ in terms of the quaternion components gives

$$\dot{V}_o(t) = -\tfrac{k}{2}\tilde{\varepsilon}_o^T(t)\tilde{\varepsilon}_o(t) + \tilde{\varepsilon}_g^T(t)(\boldsymbol{\omega}^T(t)\dot{\boldsymbol{\omega}}(t) - \boldsymbol{\omega}(t)\dot{\boldsymbol{\omega}}^T(t))\tilde{\varepsilon}_g(t) \tag{17}$$

If $\boldsymbol{\omega}^T(t)\dot{\boldsymbol{\omega}}(t) \le 0$ due to $\boldsymbol{\omega}(t)$ changing direction and/or decreasing in magnitude, (17) can be rewritten as

$$\dot{V}_o(t) \le -\tfrac{k}{2}\|\tilde{\varepsilon}_o(t)\|_2^2 - 8\|\tilde{\mathbf{m}}(t)\|_2^2 \tag{18}$$

using $\|I - R\|_2 = 2\|\varepsilon\|_2$. Since $\dot{V}_o(t) \leq 0$, $\|\tilde{\varepsilon}_o(t)\|$ and $\|\tilde{\mathbf{m}}(t)\|$ converge to zero exponentially fast. Again, this does not guarantee that $\|\tilde{\varepsilon}_g(t)\|$ converges to zero. (The second term on the right of (17) can also evaluate to zero. The convergence then follows from the argument above.)

Integrating (18) results in

$$V_o(t) - V_o(t_o) \leq -\tfrac{k}{2}\int_{t_o}^{t}\|\tilde{\varepsilon}_o(\tau)\|^2 d\tau + \int_{t_o}^{t}\tilde{\varepsilon}_g^T(\tau)(\omega^T(\tau)\dot{\omega}(\tau) - \omega(\tau)\dot{\omega}^T(\tau))\tilde{\varepsilon}_g(\tau)d\tau \qquad (19)$$

Assuming that $\omega^T(t)\dot{\omega}(t) \leq 0$, the integral terms are finite since the integrands converge to zero exponentially fast. If, however, $\omega(t)$ satisfies a 'persistency of excitation' condition, such that

$$-\alpha_2 I \geq \int_{t}^{t+\delta}(\omega^T(t)\dot{\omega}(t) - \omega(t)\dot{\omega}^T(t))d\tau \geq -\alpha_1 I$$

for positive constants α_1, α_2, and δ, the terms in (19) can only remain finite if $\|\tilde{\varepsilon}_g(t)\|$ converges to zero exponentially fast. Intuitively, the alignment is estimated only when $\omega(t)$ changes direction.

SIMULATION RESULTS

The simulation consists of the above observer equations, programmed into Matlab. Several scenarios are tested. The first considers a constant angular velocity, changed discretely at 40 second intervals about the body axes. The initial quaternions are

$$\mathbf{q}_g^T = [1 \;\; 0 \;\; 0 \;\; 0],\; \hat{\mathbf{q}}_g^T = [0 \;\; 0 \;\; 0 \;\; 1],\; \mathbf{q}^T = [0 \;\; 1 \;\; 0 \;\; 0],\; \hat{\mathbf{q}}^T = [0 \;\; 0 \;\; 0 \;\; 1]$$

The initial angular velocity is $\omega_g^T = [1 \;\; 1 \;\; 1]$ rad/sec. Figure 1 shows that the alignment errors converge to zero.

The next tests consider a time varying angular velocity, $\omega(t) = \omega_0 e^{-k_1 t}$. The initial quaternions and ω_0 are as given above. Figure 2 shows that the alignment errors converge to a constant, not to zero.

Next, the angular velocity is given as $\omega(t) = A(t)\omega_0 e^{-k_1 t}$, where A(t) is a rotation matrix, computed from the (arbitrary) 3-1-2 Euler sequence of [0.3t;-0.1t;0.2t]. The initial quaternions and ω_0 are as given above. Figure 3 shows that the alignment errors converge to zero. This test is repeated with random initial quaternions. Figure 4 shows, again, that the alignment errors converge to zero.

Finally, the angular velocity is given as $\omega(t) = A(t)\omega_0$, with A(t) and ω_0 as above. Figures 5 and 6 show the alignment errors given the initial quaternions above, and random initial quaternions, respectively. In both cases, the alignment errors converge to zero.

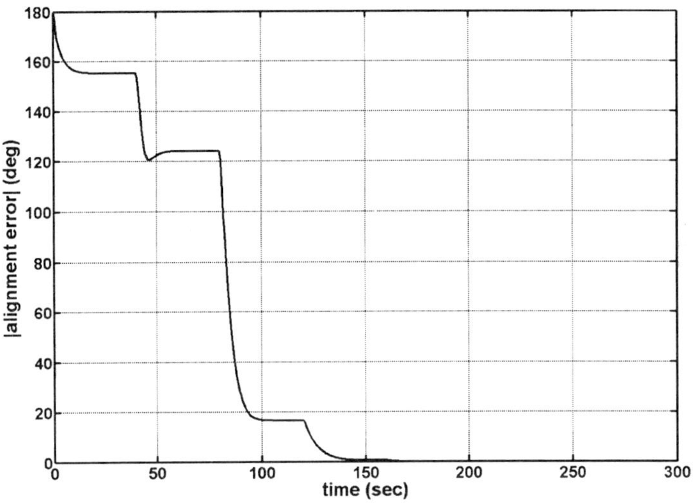

Figure 1. Magnitude of Alignment Errors Given Constant Angular Velocity, With Discrete Changes at 40 Second Intervals

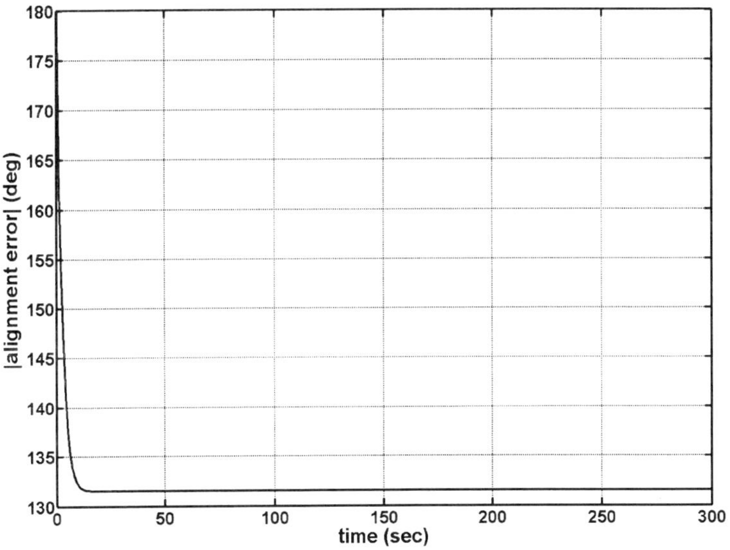

Figure 2. Magnitude of Alignment Errors Given an Exponentially Decaying Magnitude

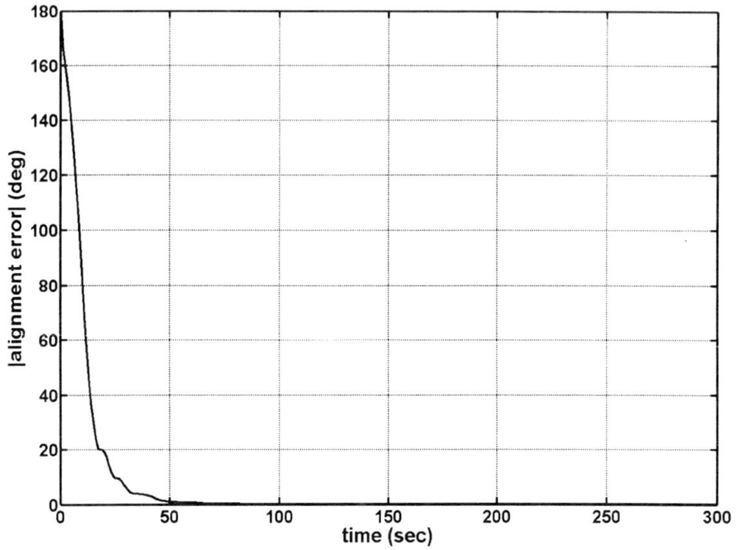

Figure 3. Magnitude of Alignment Errors Given an Exponentially Decaying/Sinusoidal Angular Velocity

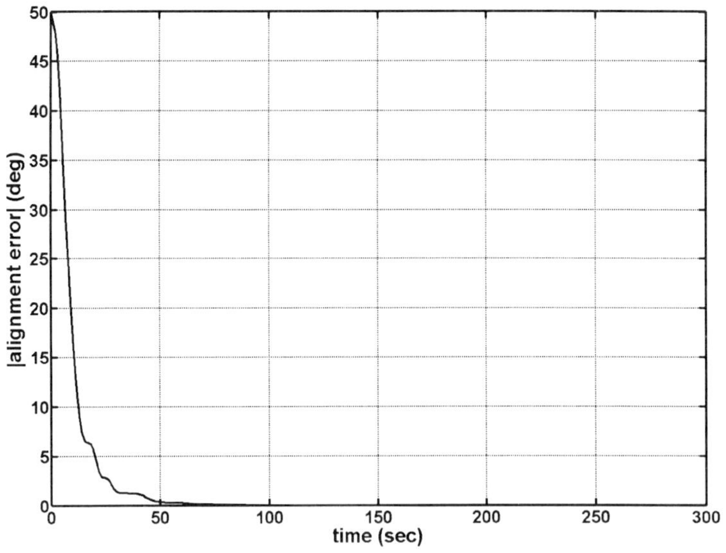

Figure 4. Magnitude of Alignment Errors Given an Exponentially Decaying/Sinusoidal Angular Velocity with Random Initial Quaternions

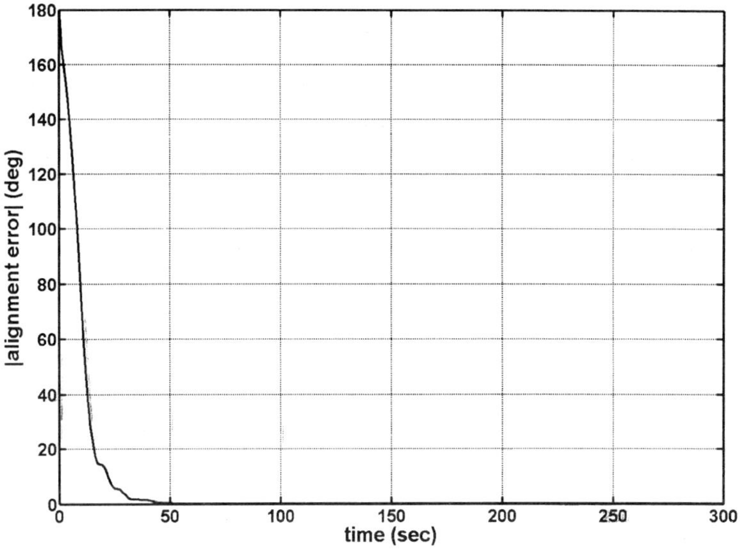

Figure 5. Magnitude of Alignment Errors Given a Sinusoidal Angular Velocity

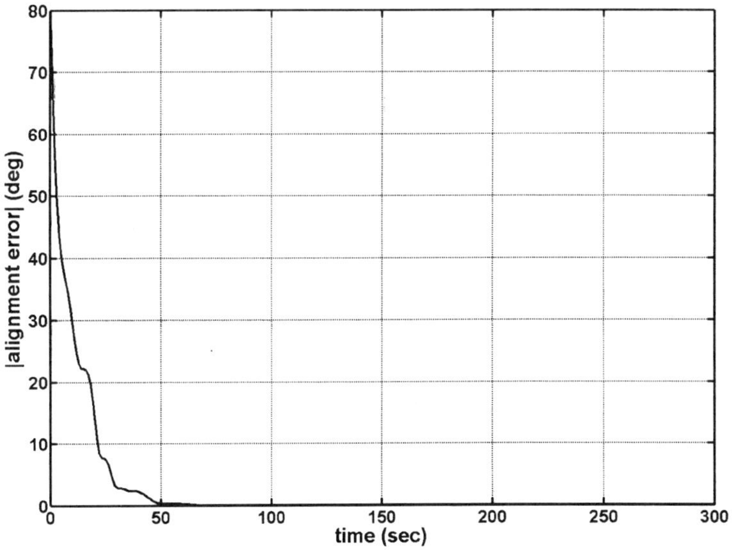

Figure 6. Magnitude of Alignment Errors Given a Sinusoidal Angular Velocity with Random Initial Quaternions

CONCLUSIONS

A nonlinear observer for gyro alignment estimation is developed. The observer is proven, via a Lyapunov argument, to be exponentially stable given a constant angular velocity. The estimated alignment converges to the true alignment when the angular velocity is discretely changed, providing the necessary observability. Given a time varying angular velocity, the estimated alignment converges to the true alignment, when the angular velocity vector changes direction, again to provide the necessary observability.

REFERENCES

1. H. Nijmeijer and T.I. Fossen, *Lecture Notes in Control and Information Sciences, New Directions in Nonlinear Observer Design,* Springer-Verlag, 1999, pp. 135-159.
2. H. Khalil, *Nonlinear Systems*, Prentice-Hall, Inc., 1996, pp. 115-116 and 626-629.
3. M. Krstic, I. Kanellakopoulos, and P. Kokotovic, *Nonlinear and Adaptive Control Design*, John Wiley & Sons, Inc., 1995, p. 491.
4. J.A. Hashmall, M. Radomski, and J. Sedlak, "On-Orbit Calibration of Satellite Gyroscopes", AIAA/AAS Astrodynamics Specialist Conference, Denver, CO, August 13-17, 2000.
5. G.Welter, J. Boia, M. Gakenheimer, E. Kimmer, D. Channell, and L. Hallock, "Variations on the Davenport Gyroscope Calibration Algorithm, Flight Mechanics/Estimation Theory Symposium, NASA Goddard Space Flight Center, Greenbelt, MD, May 14-16, 1996.
6. J. Deutschmann (Thienel) and I.Y. Bar-Itzhack, "Estimating Attitude, Trajectory, and Gyro Biases in an Extended Kalman Filter Using Earth Magnetic Field Data from the Rossi X-Ray Timing Explorer", Flight Mechanics Symposium, NASA Goddard Space Flight Center, Greenbelt, MD, May 19-21, 1997.
7. F.L. Markley and R.G.Reynolds, "Analytic Steady-State Accuracy of a Spacecraft Attitude Estimator", *AIAA Journal of Guidance, Control, and Dynamics*, Vol. 23, No. 6, November-December, 2000, pp. 1065-1067.
8. R.L. Farrenkopf, "Analytic Steady-State Accuracy Solutions for Two Common Spacecraft Attitude Estimators", *AIAA Journal of Guidance and Control,* Vol. 1, No. 4, July-August, 1978, pp. 282-284.
9. Salcudean, "A Globally Convergent Angular Velocity Observer for Rigid Body Motion", *IEEE Transactions on Automatic Control*, Vol. 36, No. 12, December, 1991.
10. J.K. Thienel and R.M. Sanner, "A Coupled Nonlinear Spacecraft Attitude Observer/Controller With an Unknown Constant Gyro Bias", 40th IEEE Conference on Decision and Control, Orlando, FL, December 4-7, 2001.
11. B. Wie, Space Vehicle Dynamics and Control, AIAA Education Series, 1998, pp. 318-323.
12. R.M. Sanner, "Adaptive Attitude Control Using Fixed and Dynamically Structured Neural Networks", AIAA Guidance, Navigation and Control Conference, San Diego, CA, July 29-31, 1996.

Section II
AUTONOMOUS RENDEZVOUS AND PROXIMITY OPERATIONS

SESSION II

Joint Session Chairperson: Jennifer D. Wagenknecht
NASA Johnson Space Center

Joint Session Chairperson: Major Jim Shoemaker
DARPA

Local Session Chairperson: Bill Frazier
Ball Aerospace & Technologies Corp.

The following paper numbers were not assigned:
AAS 03-019 to -020

AAS 03-011

COUPLED ATTITUDE AND ORBIT DYNAMICS AND CONTROL IN FORMATION FLYING SYSTEMS

Yunjun Xu* **and Norman Fitz-Coy***

In this paper, generalized equations of motion for a leader/follower satellite pair are derived. The equations are generalized to include non-circular reference orbits by regularizing the governing equations to be true-anomaly dependent rather than time dependent. In addition, the vehicles are considered to have finite dimensions and are pointed to a specific location on the Earth (i.e., their attitude and orbit dynamics are coupled). An Euler angle attitude representation was chosen and each vehicle is assumed gravity gradient stabilized (i.e., each satellite's z-axis points to its own nadir). Finally, an attitude controller based on the sliding mode concept was developed and implemented to control the attitude of the follower relative to the leader. For the close-loop scenario, both vehicles are required to point to the same position on the surface of the earth (i.e., the nadir of the leader). Simulation results illustrate the capability/performance of the controller developed.

INTRODUCTION

Formation flying systems can range from global constellations of satellites offering extended service coverage to smaller clusters of highly coordinated vehicles that perform distributed sensing[1]. The concept of formation flying or distributed space systems dates back to the early 1980's with a paper by Stachnik[21,22] et al. Recently, these formation flying systems have received renew interest by several government agencies. There are numerous proposed missions that consist of multiple conventional, mini, micro, and/or nano satellites in specific formations. These formation flying systems are designed to perform a unified mission and therefore would replace very larger, complex, monolithic platform.

Formation flying systems are attractive for many reasons. Besides the difficulty and cost associated with placing massive payloads into space, large space structures are usually very flexible and require active vibration control strategies. These control strategies can be very complex and expensive. Formation flying systems typically will not require these active vibration control systems and therefore provide an advantage in this regards; however, they do require positional (and attitude) control for each individual spacecraft in the formation. Thus, regardless of the system being used (monolithic or distributed), accurate models are necessary in order to develop appropriate model based control strategies.

* Department of Mechanical & Aerospace Engineering, University of Florida, Gainesville, Florida 32611.

Whereas the monolithic platforms are typically modeled as discrete elements with strong physical bonds (i.e., a finite element model of a continuum), the formation flying system can be modeled with discrete elements with weak physical bonds (gravitational attractions). A widely used model for the formation flying system is the classic Clohessy-Wiltshire (C-W) relative dynamics equations[2-8]. Taking the advantage of the decoupling between the in-plane and the out-of-plane motions, Vassar and Sedwick used these linear C-W equations to synthesize and analyze closed-form controller for formation system[3, 5]. Nonlinear relative position model are also available and Queiroz and Queirox [9, 10] used such models to analyze adaptive control methodologies. Most of the existing models are restricted to circular reference orbits.

A recent paper by Inalhan[15] examined the relative position motion for noncircular (i.e., elliptic) reference orbits. The emphasis of this study was the development of T-periodic solutions that return the vehicles to their initial relative states at the end of one orbital period. This model also had limitations; for example, it can only be used in the case where the reference orbit is elliptic; it cannot be applied to the circular case. Another example of the elliptic case was based on the FreeFlyer software package developed by AI Solutions, Inc. Using this software, Schiff[11] simulated relative motions for elliptic orbits with eccentricities less than 0.4.

There have been many proposed orbit control strategies for formation systems. Among them, and the earliest and most widely used method, is the linear quadratic regulator based on the C-W model[3, 4, 7]. Others span the range from PD controllers[16] to Lyapunov-based adaptive controllers[9, 17, 18]. The common factors among all these control strategies are (i) they address circular reference orbits and (ii) they assume a decoupled orbit and attitude problem. In practice, there will be scenarios in the formation flying systems where the satellite's orbit and attitude are coupled. These are the scenarios considered in this paper.

GENERALIZED POSITION MODEL

The model is based on the leader-follower pair shown in Figure 1. It is assumed that each vehicle is experiencing two-body motion with respect to the Earth and thus their motions are governed by

$$\ddot{\underline{R}}_L = \frac{\mu_\oplus}{R_L^3}\underline{R}_L + \underline{u}_L \text{ and } \ddot{\underline{R}}_F = \frac{\mu_\oplus}{R_F^3}\underline{R}_F + \underline{u}_F$$

where $\vec{R}_L$ and $\vec{R}_F$ represent the position of the leader and follower, respectively. $\mu_\oplus$ is the Earth's gravitational parameter, and $\vec{u}_L$ and $\vec{u}_F$ represent all other accelerations imposed on the leader and follower, respectively.

A formation flying system can be described in terms of the relative position of the follower with respect to the leader. Let $\vec{\rho}$ represent this quantity. Then, the formation flying system is governed by

$$\ddot{\underline{\rho}} = \frac{\mu}{R_L^3}[\underline{R}_L - (\frac{R_L}{R_F})^3 \underline{R}_F] + (\underline{u}_F - \underline{u}_L) \quad (1)$$

It is customary to coordinatize $\vec{\rho}$ in the leader's coordinate frame, which is the local vertical local horizontal (LVLH) coordinate frame shown in Figure 2. Assuming the leader is in a fixed orbital plane, then the LVLH coordinate frame has an angular velocity of ${}^L\vec{\omega}_L = \dot{f}_L \hat{\underline{k}}$ and angular acceleration ${}^L\dot{\vec{\omega}}_L = \ddot{f}_L \hat{\underline{k}}$ where $\hat{\underline{k}}$ is a unit vector parallel to the leader's angular momentum vector and $\dot{f}_L$ and $\ddot{f}_L$ are the derivatives of the true anomaly. Expressing Eq. (1) in the LVLH frame yields

$${}^L\overset{\circ\circ}{\underline{\rho}} = -{}^L\overset{\circ}{\underline{\omega}}_L \times {}^L\underline{\rho} - 2\,{}^L\underline{\omega}_L \times {}^L\overset{\circ}{\underline{\rho}} - {}^L\underline{\omega}_L \times ({}^L\underline{\omega}_L \times {}^L\underline{\rho}) + \frac{\mu}{R_L^3}[{}^L\underline{R}_L - (\frac{R_L}{R_F})^3\, {}^L\underline{R}_F] + ({}^L\underline{u}_F - {}^L\underline{u}_L) \quad (2)$$

The quantities in Eq. (2) are as follows: $\overset{\circ}{()}$ denotes the time derivative with respect to the leader's LVLH frame; the left superscript L designates that the vectors are coordinatized in the leader's LVLH coordinate frame; the right subscript denotes the satellite of interest (i.e., L for leader and F for follower). Furthermore, the quantities ${}^L\overset{\circ}{\vec{\rho}}$ and ${}^L\overset{\circ\circ}{\vec{\rho}}$ represent the relative velocity and relative acceleration of the follower with respect to the leader coordinatized in the leader's frame.

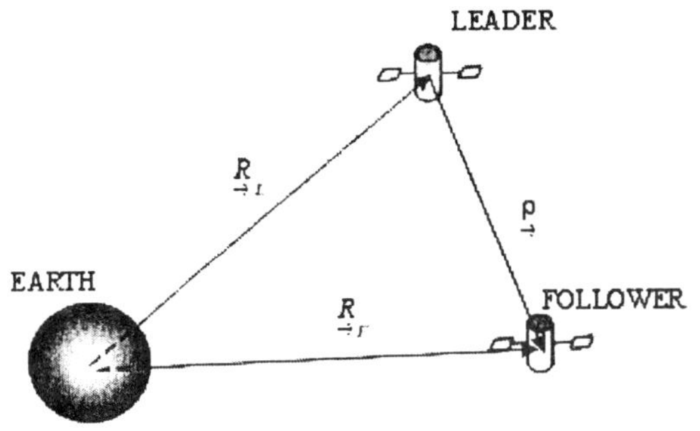

Figure 1 Leader-Follower Pair

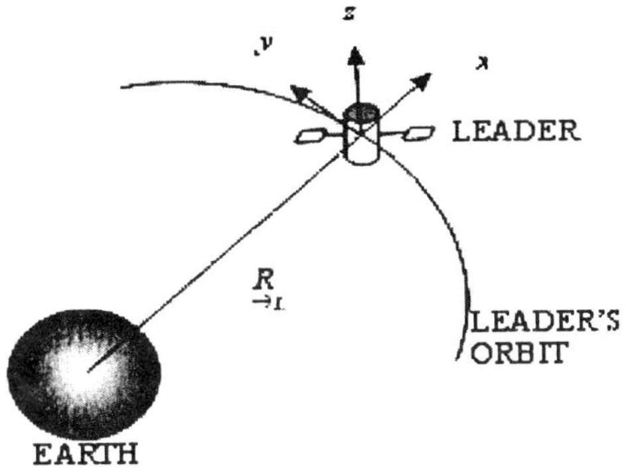

Figure 2 LVLH Coordinate System

Equation (2) represents a set of time dependent equations that governs the motion of the leader-follower formation system described above. For cases in which the leader's orbit is circular, the LVLH coordinate frame has a constant angular velocity (i.e., ${}^L\dot{\vec{\omega}}_L = \underline{0}$) and Eq. (2) completely describes the motion and thus can be implemented in its current form. The same is not true for cases in which the leader's orbit is elliptic. In these noncircular cases, neither the angular velocity nor the angular acceleration of the

LVLH is constant, and thus additional expressions are required to describe these additional time dependencies.

Assuming that only radial accelerations are imposed on the leader, then its specific angular momentum is conserved and this information can be used to develop the additional time dependent relationships for $\dot{f}_L$ and $\ddot{f}_L$. These are

$$\dot{f} = \frac{h}{R_L^2} \text{ and } \ddot{f} = -2\frac{\mu_{\oplus} e}{R_L^3}\sin f_L \tag{3}$$

where h represents the magnitude of the leader's specific angular momentum and e is the eccentricity of its orbit.

The use of the definitions in Eq. (3) along with Eq. (2) results in a complete set of governing equations for the formation flying system. It is worth noting that if $e = 0$, then $\ddot{f} = 0$ and the circular case is recovered. These expressions are amenable to computer implementation; however, in their current form, they are not computational efficient. Computational efficiency can be obtained by a changing from a time-dependence to a true anomaly-dependence. Thus a complete and computationally efficient governing equation for the formation system is

$$\begin{aligned} {}^L\underline{\rho}'' = \frac{1}{\dot{f}_L^2}\{&-\ddot{f}_L\,{}^L\underline{\rho}' - {}^L\dot{\underline{\omega}}_L \times {}^L\underline{\rho} - 2\dot{f}_L\,{}^L\underline{\omega}_L \times {}^L\underline{\rho}' - {}^L\underline{\omega}_L \times ({}^L\underline{\omega}_L \times {}^L\underline{\rho}) \\ &+ \frac{\mu}{R_L^3}[{}^L\underline{R}_L - (\frac{R_L}{R_F})^3\,{}^L\underline{R}_F] + ({}^L\underline{u}_F - {}^L\underline{u}_L)\} \end{aligned} \tag{4}$$

Symbolically, Eq. (4) contains terms that appear to be time dependent (i.e., $\dot{f}_L$ and $\ddot{f}_L$), but as noted in Eq. (3), these terms can be expressed explicitly in terms of the true anomaly f_L. In order to implement Eq. (4) the quantity ${}^L\underline{R}_F$ must be evaluated. This is accomplished with the definition of ${}^L\underline{\rho}$; that is ${}^L\underline{R}_F = {}^L\underline{R}_L + {}^L\underline{\rho}$.

COUPLED ATTITUDE MODEL

The scenarios addressed in this paper consider satellites in orbits with different inclinations but collectively point to the same location. This results in a coupling between the orbit and the pointing (attitude) of the satellites. An example of this scenario is the joint EO-1/LandSat-7 mission in which both satellites collectively observe the same location on the surface of the Earth.

The geometry of the coupled attitude and orbit scenario is shown in Figure 3. Each satellite is assumed gravity gradient stabilized with its z-body axis pointing to its own nadir. Assuming the satellites collectively point to the nadir of the leader, then the follower satellite must maneuver through angles θ_1 and θ_2 as shown in Figure 3. The following points in Figure 3 are of interest in determining these two angles: N_L and N_F are the nadir of the leader and follower, respectively; A is intersection of the longitude of the follower's nadir and the leader's orbit plane; D is intersection of the longitude of the follower's nadir and the reference plane.

From the geometry shown in Figure 3, θ_1 is the angle between vectors $\underline{FO}$ and $\underline{FA}$, and θ_2 is the angle between vectors $\underline{FA}$ and $\underline{FN_L}$; that is,

$$\theta_1 = \begin{cases} -\cos^{-1}[\dfrac{{}^{ECI}\underline{FA} \bullet {}^{ECI}\underline{FO}}{\|\underline{FA}\|\|\underline{FO}\|}] & if\ 0 \le \alpha_F < \pi \\ \cos^{-1}[\dfrac{{}^{ECI}\underline{FA} \bullet {}^{ECI}\underline{FO}}{\|\underline{FA}\|\|\underline{FO}\|}] & if\ \pi \le \alpha_F < 2\pi \end{cases} \tag{5}$$

$$\theta_2 = \cos^{-1}[\frac{{}^{ECI}\underline{FA} \bullet {}^{ECI}\underline{FN_L}}{|\underline{FA}||\underline{FN_L}|}] \tag{6}$$

The vectors in Eqs. (5) and (6) are expressed in the Earth Centered Inertial (ECI) frame as

$${}^{ECI}\underline{FO} = -\underline{\underline{C}}(3,-\Omega_L)\underline{\underline{C}}(1,-i_L)\underline{\underline{C}}(3,-\alpha_L)\begin{bmatrix} R_L + \rho_x \\ \rho_y \\ \rho_z \end{bmatrix}$$

$${}^{ECI}\underline{OA} = \underline{\underline{C}}(3,-\Omega_L)\underline{\underline{C}}(1,-i_L)\underline{\underline{C}}(3,-\alpha_F)\begin{bmatrix} R_e \\ 0 \\ 0 \end{bmatrix}$$

$${}^{ECI}\underline{FA} = -({}^{ECI}\underline{FO} + {}^{ECI}\underline{OA})$$

$${}^{ECI}\underline{ON_L} = \underline{\underline{C}}(3,-\Omega_L)\underline{\underline{C}}(1,-i_L)\underline{\underline{C}}(3,-\alpha_L)\begin{bmatrix} R_e \\ 0 \\ 0 \end{bmatrix}$$

$${}^{ECI}\underline{FN_L} = {}^{ECI}\underline{ON_L} + {}^{ECI}\underline{FO}$$

where $\underline{\underline{C}}(i,\phi)$ is the fundamental rotation matrix about axis i through angle ϕ. Angles i_L Ω_L, and α_L are, respectively, the inclination, longitude of the ascending node, and latitude at epoch of the leader satellite. The angle α_F is defined in TABLE 1 and Figure

4. Its value is computed using spherical trigonometry as show in Eq. 7. Angles s_F and i_F represent the true longitude and inclination of the follower.

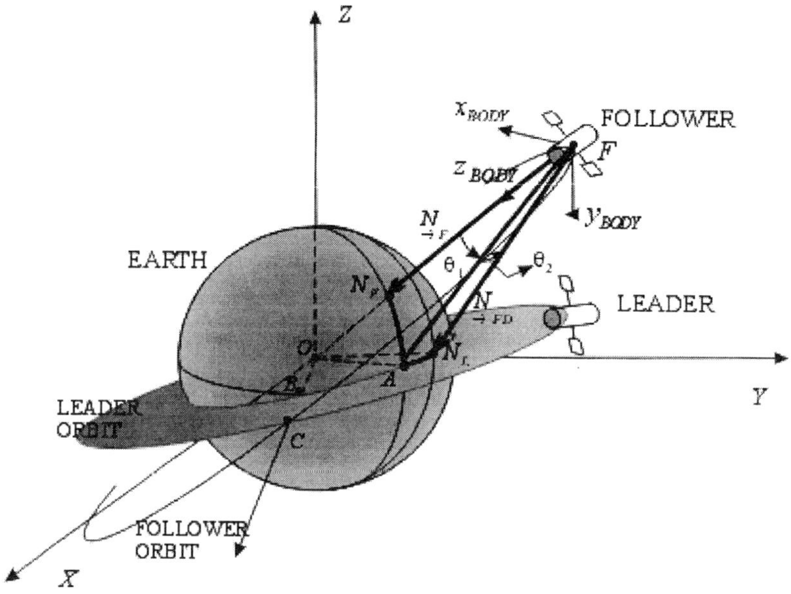

Figure 3 Geometry of Coupled Attitude and Orbit Scenario

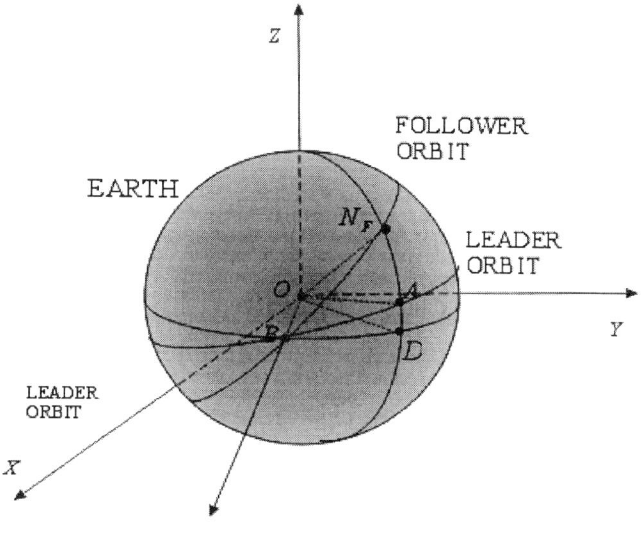

Figure 4 Additional Geometry Definition

TABLE 1 Definition of Angles in Figure 4

Angles	Defined Between	Restriction
L_F	ON_F and OA	$L_F = L_{F1} - L_{F2}$
L_{F1}	ON_F and OD	$-i_F \le L_{F1} \le i_F$
L_{F2}	OD and OA	$-i_L \le L_{F2} \le i_L$
s_F	ON_F and OB	$0 \le s_F \le 2\pi$
α_F	OA and OB	$0 \le \alpha_F \le 2\pi$
α_{F1}	OD and OB	$0 \le \alpha_{F1} \le 2\pi$

$$L_{F1} = \sin^{-1}(\sin s_F \sin i_F)$$

$$\alpha_{F1} = \begin{cases} \Omega_F + \cos^{-1}(\cos s_F / \cos L_{F1}) & if\ 0^o \le s_F < \pi \\ \Omega_F + 2\pi - \cos^{-1}(\cos s_F / \cos L_{F1}) & if\ \pi \le s_F < 2\pi \end{cases}$$

$$L_{F2} = \tan^{-1}(\sin \alpha_{F1} \tan i_L)$$

$$\alpha_F = \begin{cases} \cos^{-1}(\cos L_{F2} \bullet \cos \alpha_{F1}) & if\ 0^o \le s_F < \pi \\ 2\pi - \cos^{-1}(\cos L_{F2} \bullet \cos \alpha_{F1}) & if\ \pi \le s_F < 2\pi \end{cases} \tag{7}$$

The attitude of each satellite is governed by Euler's equation

$$\underline{\underline{J}}\,\dot{\omega}_{\rightarrow BODY} + \widetilde{\omega_{\rightarrow BODY}\,\underline{\underline{J}}\,\omega_{\rightarrow BODY}} = T_{\rightarrow BODY} \tag{8}$$

where $\underline{\underline{J}}$ is the centroidal inertia dyadic and $\widetilde{(\ )}$ is a skew-symmetric matrix representing the cross product. The angular velocity can be expressed in terms of the Euler rates as $\omega_{\rightarrow BODY} = \underline{\underline{S}}\,\dot{\theta}_{\rightarrow}$. Substituting this definition of $\omega_{\rightarrow BODY}$ and its derivative into Eq. (8) yield a second order differential equation for the Eular angles.

$$\ddot{\underline{\theta}} = \underline{\underline{A}}^{-1}[-\underline{\underline{J}}\,\underline{\underline{\dot{S}}}\,\dot{\underline{\theta}} - \widetilde{\underline{\underline{S}}\dot{\underline{\theta}}}\,\underline{\underline{J}}\,\underline{\underline{S}}\,\dot{\underline{\theta}} + \underline{T}_{BODY}] \tag{9}$$

where $\underline{\underline{A}} = \underline{\underline{J}}\,\underline{\underline{S}}$. This form of the attitude equation is amenable for controller design which is addressed in the next section.

SLIDING MODE CONTROLLER FOR ATTITUDE

Let $\underline{x} = [\theta_1, \theta_2, \theta_3]^T$ be the system state, then the attitude of the formation flying system can be expressed as

$$\underline{x}^{(n_i)} = \underline{f}(\underline{x}, \underline{x}^{(1)}, \ldots, \underline{x}^{(n_i-1)}) + \underline{\underline{B}}(\underline{x}, \underline{x}^{(1)}, \ldots, \underline{x}^{(n_i-1)})\underline{u} \tag{10}$$

where $x^{(j)}$ represents the jth derivative. Let $\underline{\underline{B}} = \left[b_{ij}\right]_{m\times m}$ be the input matrix and $\underline{f} = [f_i]_{m\times 1}$ be the dynamics, where $m = 3$. $\hat{f}_i$ and $\hat{b}_{ij}$ are assumed to be the predicted system model. In the examples described in this paper, the system model is assumed exact. Therefore, $\left|\hat{f}_i - f_i\right| \le F_i = 0$, $\underline{\underline{B}} = (\underline{\underline{I}} + \underline{\underline{\Delta}})\hat{\underline{\underline{B}}}$, and $\left|\Delta_{ij}\right| \le D_{ij} = 0$. Furthermore, matrices $\underline{\underline{B}}$ and $\hat{\underline{\underline{B}}}$ are full rank, i.e., torque commands can be applied in all three directions of the body axes.

Let's define the sliding surface to be

$$s_i = (\frac{d}{dt} + \lambda_i)^{(n_i - 1)} \tilde{x}_i = x_i^{(n_i - 1)} - x_{ri}^{(n_i - 1)} \tag{11}$$

Here $\tilde{\underline{x}} = \underline{x} - \underline{x}_{DESIRED}$ is the tracking error, and let $x_{ri}^{(n_i-1)} = x_{di}^{(n_i-1)} - \lambda_i^{n_i - 1}\tilde{x}_i$. Then the control is given by Eq. (12), in which, $\underline{x}_r^{(n-1)} = \left[x_{r1}^{(n_1-1)} \quad x_{r2}^{(n_2-1)} \quad \cdots \quad x_{rm}^{(n_m-1)}\right]^T$ and $\underline{k}\,\mathrm{sgn}(\underline{s}) \triangleq [k_i\,\mathrm{sgn}(s_i)]_{m\times 1}$. Furthermore, k_i should be chosen according to Eq. (13). Based on the stability analysis, $\lambda_i > 0$ and $\eta_i > 0$[20]. In sliding mode control, parameter k_i affects $\tilde{\underline{x}}$ and $\dot{\tilde{\underline{x}}}$, whereas, λ_i affects $\tilde{\underline{x}}$.

$$\hat{\underline{u}} = \hat{\underline{\underline{B}}}^{-1} [\underline{x}_r^{(n-1)} - \hat{\underline{f}} - \underline{k}\,\mathrm{sgn}(\underline{s})] \tag{12}$$

$$(1 - D_{ii})k_i + \sum_{j \ne i} D_{ij} k_j = F_i + \sum_{j=1}^{n} D_{ij}\left|x_{ri}^{(n_i-1)} - \hat{f}_j\right| + \eta_i \tag{13}$$

SIMULATION RESULTS

An algorithm was developed in Matlab to implement the developed models. The algorithm was debugged and validated using special test cases and by comparing its output to those of existing software packages (STK by Analytic Graphics). The validated algorithm was then used to investigate the coupled formation system described in this paper. Results from 3 test cases are presented.

In all the cases discussed below, principal axes were selected for each satellite. Typical microsatellite mass and inertias were used ($\underline{\underline{J}} = [433, 0, 0; 0, 129, 0; 0, 0, 429] Kg \bullet m^2$). Also, the attitude controller was bandwidth limited to 1 Hz and the maximum available torque about any principal axis was limited to $2 N \bullet m$.

Case 1: Open-loop attitude for circular leader and follower orbits

The leader is in the equatorial circular orbit with a $500\,km$ altitude. The follower is also in a $500\,km$ circular orbit with 0.1^o inclination. Initially, $u_{oL} = 0^o$ and $u_{oF} = -0.1^o$. These conditions were chosen to keep the satellites within 20 km of each other. The misalignment angles θ_1 and θ_2 are shown in Figure 5 and Figure 6. Both angles display some periodic behavior with θ_1 oscillating about zero and θ_2 about some fixed bias. This behavior is clearly supported by the geometry of the scenario.

Case 2: Open-loop attitude for circular leader and an elliptic follower

The leader's orbit is unchanged from Case 1 above. The follower's orbit is modified to have an eccentricity of 0.07; the period of both vehicles are identical. The results for error angles θ_1 and θ_2 are shown in Figure 7 and Figure 8. The sinusoids shown in Figure 5 and Figure 6 are no longer visible since the speed of the follower is not constant. As the follower goes through periapsis, the motion is fastest; hence, the changes in θ_1 and θ_2 occur more rapidly at these locations. From both Cases 1 and 2, it is obvious that controllers are need to drive these two error angles to zeros.

Case 3: Closed-loop for circular leader and follower orbits

A sliding mode attitude controller was designed to compensate for the θ_1 and θ_2 misalignment of the follower satellite. The following controller parameters were developed through trail and error: $\underset{\rightarrow}{k} = [50, 80, 0.01]^T$ and $\underset{\rightarrow}{\lambda} = [5, 3, 0.01]^T$. Figure 9 - Figure 11 demonstrate that the controller was capable of maintaining the desired pointing of the follower satellite. The required torques are shown in Figure 12 - Figure 14 and are well within the allowable torques.

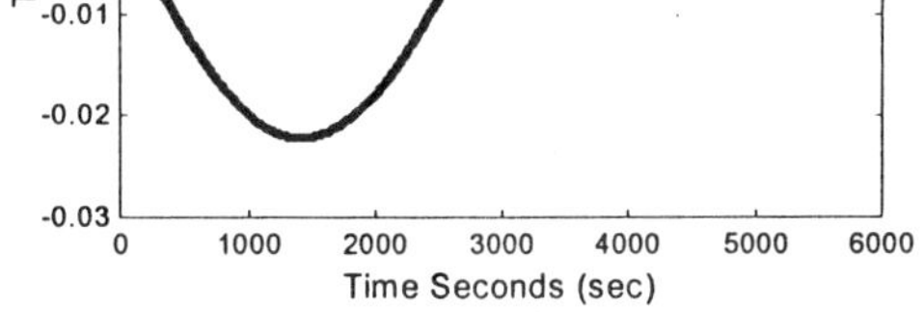

Figure 5 θ_1 in open loop in case 1

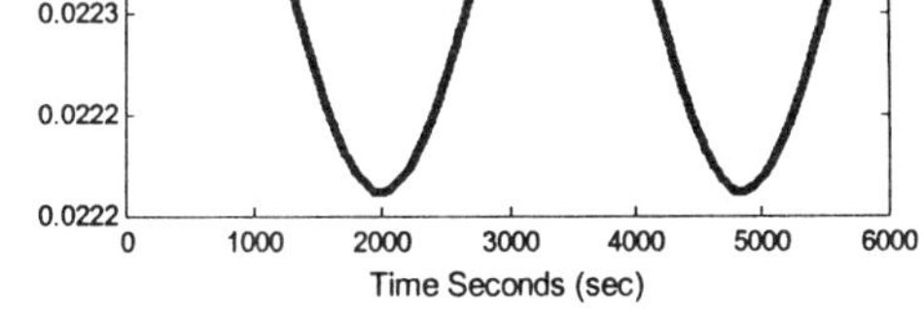

Figure 6 θ_2 in open loop in case 1

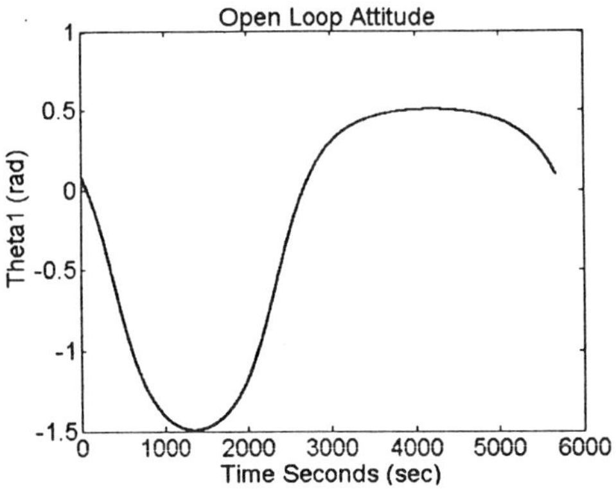

Figure 7 θ_1 in open loop in case 2

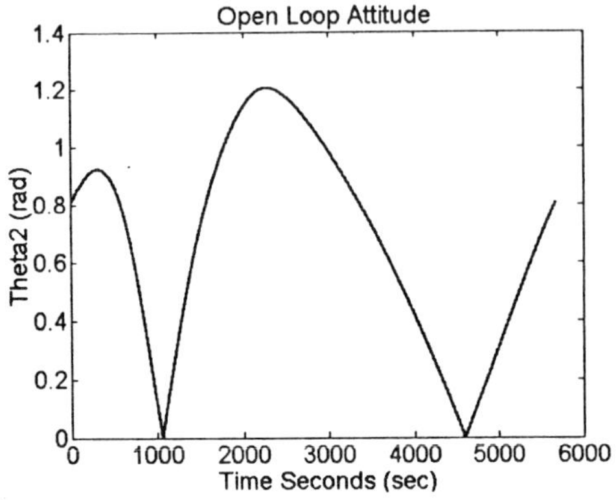

Figure 8 θ_2 in open loop in case 2

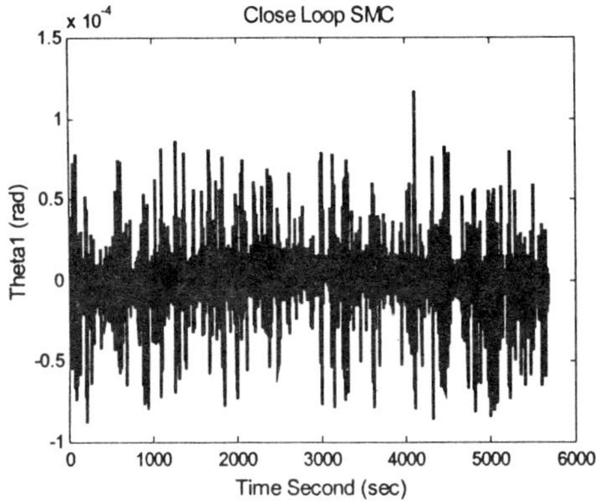

Figure 9 θ_1 in Closed Loop

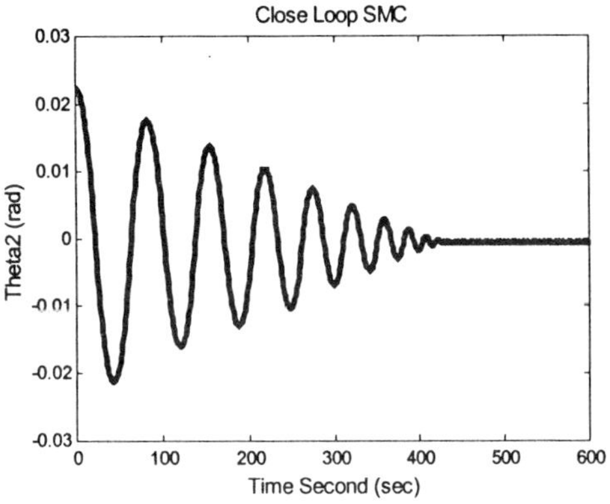

Figure 10 θ_2 in Closed Loop

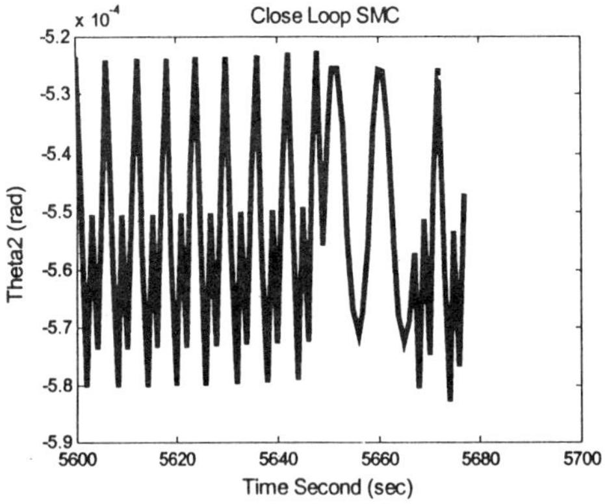

Figure 11 θ_2 in Closed Loop

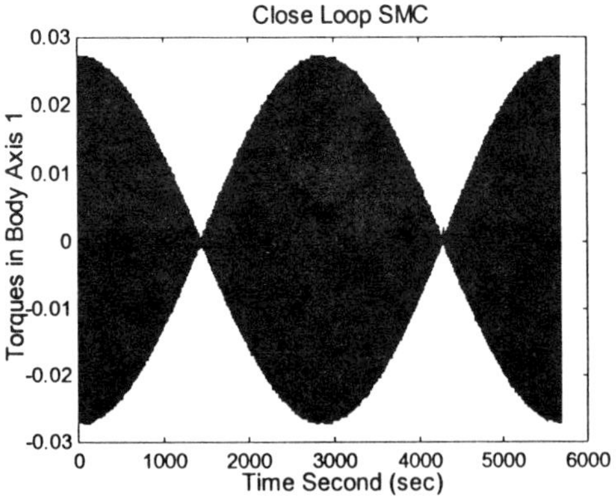

Figure 12 Torque 1 Needed in Closed Loop

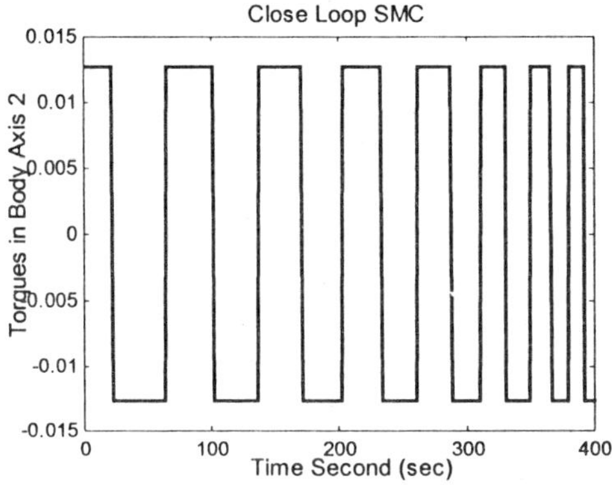

Figure 13 Torque 2 Needed in Closed Loop

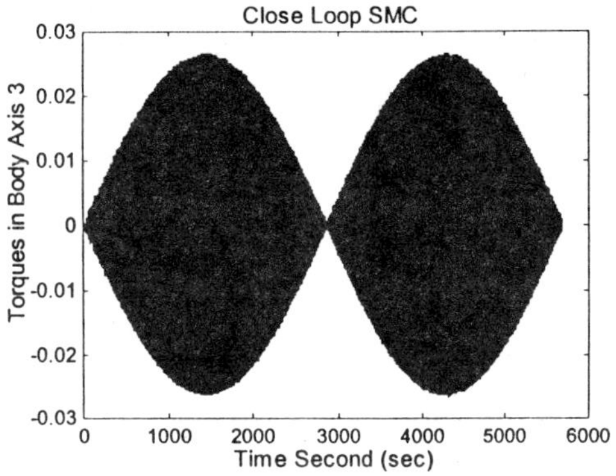

Figure 14 Torque 3 Needed in Closed Loop

CONCLUSION AND FUTURE RESEARCH

In this paper, a generalized dynamic model capable of modeling all closed loop orbits is presented. In addition, mission scenarios that result in a coupling between the attitude and orbit are presented and discussed. An attitude control based on the sliding mode technique was designed and implemented in the coupled system. Typical results were presented and the attitude controller was shown to provide pointing accuracies on the order of arc seconds without actually saturating the actuators. However, this accomplishment was achieved at the cost of significant actuator chattering.

Future work involves the development of techniques to mitigate the chattering phenomenon. Techniques involving boundary layer control are currently being investigated.

ACKNOWLEDGEMENTS

The authors wish to acknowledge the contributions of Dr. Paul Mason. His contributions and suggestions are greatly appreciated.

REFERENCE

1. **http://nmp.jpl.nasa.gov/index_flash.html**. Last update, 12/19/02.
2. B. Morton and N. Nicholas, "Collective Management of Satellite Clusters," *AIAA-99-4267*.
3. R.H. Vassar and R.B. Sherwood, "Formation Keeping for a pair of Satellites in a circular Orbit," *J. of Guidance, Control, and Dynamics*, Vol. 8, No. 2, March-April 1985.
4. D.C. Redding and N.J. Adams, "Linear-Quadratic Station Keeping for the STS Orbiter," *J. of Guidance, Control, and Dynamics*, Vol. 12, No. 2, March-April 1989.
5. R.J. Sedwick, D.W. Miller and E.M.C Kong, "Mitigation of Differential Perturbations in Formation Flying Satellite Clusters," *J. of the Astronautical Sciences*, Vol. 47, No. 3 and 4, July-December 1999, pp. 309-331.
6. H. Schaub, J.L. Junkins, and R.D. Robinett, "New Penalty Functions and Optimal Control Formulation for Spacecraft Attitude Control Problems," *J. of Guidance, Control, and Dynamics*, Vol. 20 No. 3, Issue May 1997.

7. Q. Yan, V. Kapila, A.G. Sparks, "Pulse-based periodic control for spacecraft formation flying," *American Control Conference, 2000. Proceedings of the 2000*, volume:1, 2000, Page(s): 374 -378 vol.1.
8. S.M. Veres, S.B. Gabriel, E. Rogers, D.Q. Mayne, "Analysis of formation flying control of a pair of nano-satellites," *Decision and Control, 2001. Proceedings of the 40th IEEE Conference on*, Volume: 2, 2001, Page(s): 1095 -1100 vol.2.
9. D. Queiroz, V. Kapila, and Q. Yan, "Adaptive Nonlinear Control of Satellite Formation Flying," *AIAA-99-4270.*
10. Q. Queirox, M.S. Yan, Q. Yang, G. Kapila, "Global output feedback tracking control of spacecraft formation flying with parametric uncertainty," *Decision and Control, 1999. Proceedings of the 38th IEEE Conference on*, Volume:1, 1999, Page(s): 584 -589 vol.1.
11. C. Schiff, R. Rohrbaugh, J. Bristow, "Formation flying in elliptical orbits," *Aerospace Conference Proceedings, 2000 IEEE* , Volume: 7, 2000, Page(s): 37 -47 vol.7.
12. K.R. Hartman, C.J. Gramling, T. Lee, D.A. Kelbel, and A.C. Long, "Relative Navigation For SpaceCraft Formation Flying," *AAS-98-353.*
13. C. Kitts, and F. Pranajaya, "Emerald: An experimental Mission in Robust Distributed Space Systems," *13th AIAA/USU Conference on Small Satellites*, SSC-99-VI-5.
14. J.E. Prussing, B.A. Conway, *Orbital Mechanics*, New York, Oxford, Oxford University Press, 1993.
15. G. Inalhan, M. Tillerson, and J.P. How, "Relative Dynamics and Control of Spacecraft Formations in Eccentric Orbits," *J. of Guidance, Control, and Dynamics*, Vol. 25 No. 1, January-February 2002.
16. A. Robertson, G. Inalhan, and J.P. How, "Formation Control Strategies for a Separated Spacecraft Interferometer," *Proceedings of the American Control Conference*, San Diego, California, June, 1999.
17. J. Ahmed, V.T. Coppola, and D.S. Bernstein, "Adaptive Asymptotic Tracking of Spacecraft Attitude Motion with Inertia Matrix Identification," *J. of Guidance, Control, and Dynamics*, Vol. 21 No. 5, Issue Sep 1998.
18. J. Lawton, R.W. Beard, and F.Y. Hadaegh, "An adaptive Control Approach to Satellite Formation Flying with Relative Distance Constraints," *AACC*, 0-7803-4990-6/99.
19. http://www.math.uncc.edu/~droyster/math31-81/notes/hyprgeom/node5.html. Last update: 01/21/01.
20. S. Li, *Applied Nonlinear Control,* Prentice *Hall,* Englewood Cliffs, New Jersey 07652.
21. R. Stachnik, K. Ashlin, and S. Hamilton, "Space Station-SAMSI: A Spacecraft Array for Michelson Spatial Interferometry," *Bulletin of AmericanAstronomical Society*, Vol. 16, No. 3, 1984, pp. 818–827.
22. R. Stachnik, P. Melroy, E.F. McCormack, D. Arnold, and D.Y. Gezari, "Multiple Spacecraft Michelson Stellar Interferometry," *Proceedings of the Society of Photo-Optical Instrumentation Engineers, instrumentation in Astronomy V*, Vol. 445, 1984, pp. 358–369.

AAS 03-012

A COMBINED OPEN-LOOP AND AUTONOMOUS SEARCH AND RENDEZVOUS NAVIGATION SYSTEM FOR THE CNES/NASA MARS PREMIER ORBITER MISSION

J. E. Riedel,* J. Guinn,* M. Delpech,† J. B. Dubois,† D. Geller‡ and P. Kachmar‡

The CNES/NASA Mars Premier Orbiter Mission, at the time of this writing scheduled for launch in 2007, will carry a technology demonstration package that will validate methods of searching for and rendezvousing with a carried target representative of a possible future Mars Sample Return (MSR) Mission. The target will carry a radio beacon and be optically reflective, enabling a suite of radio and optical sensors to locate, track and subsequently autonomously rendezvous with the released target. The low fidelity battery-powered oscillator on the target will provide the capability for an initial crude detection of the small target with an unknown orbit, and through subsequent multi-spacecraft links through the Electra radio transceiver instruments aboard Premiere, and MRO or GMO, provide a high quality navigation capability. Long-range optical observations with the narrow-angle (1.5degree FOV) Navigation Camera provide a completely redundant and robust search capability, and subsequently tighter navigation performance. Though the search processes are open-loop Earth-based navigation operations, when the target is within 10km of the orbiter, the ground-based orbit determination fix will be transferred to the onboard autonomous rendezvous navigation system. Here, with limited onboard compute power available, and a very restricted budget, a completely autonomous onboard rendezvous system is being built to use the narrow-angle MRO-based Opnav Camera supplemented with a MER-based wide-angle camera. Though autonomous, the onboard algorithms have been greatly simplified, and make extensive and maximum use of ground-based computations in order to limit the scope of onboard software. Nevertheless, highly robust autonomous navigation performance is achieved. The key to the simplifications are a complete separation of the attitude-control and navigation problems, made possible by highly capable momentum wheels aboard Premiere, and the very high navigation performance of the Opnav Camera. This paper will describe the search methods to be used using the radio and optical observations, and will provide a description of the methods and performance of the autonomous onboard rendezvous navigation system.

Introduction

The CNES Premier Mars '07 Mission (Ref. 1) mission-design has multiple objectives. First and foremost is the objective to deliver four Netlander science stations to the surface of Mars, and provide at least a year of science-telemetry relay for them (Ref 2). The second principal objective is to demonstrate rendezvous technologies applicable to a future International Mars-Sample-Return (MSR) mission. The third main objective is to perform orbital Mars science. Focusing on the second prime objective, the rendezvous technology demonstration objective is to experimentally validate certain hardware, rendezvous algorithm, and mission design and operations technologies. The rendezvous system uses one-way radiometric and optical observables for distant

* Jet Propulsion Laboratory, California Institute of Technology, 4800 Oak Grove Drive, Pasadena, California 91109.

† Centre National d'Etudes Spatiales (CNES), 18, Avenue Edouard Belin, F-31401 Toulouse Cedex 4, France.

‡ Charles Stark Draper Laboratory, 555 Technology Square, Cambridge, Massachusetts 02139.

(greater than a few kilometers) measurements and optical exclusively for close observations. The radio observables are processed on the ground, whereas the optical observables are processed on the ground and autonomously onboard during critical periods of the demonstration. The mission phases of MSR and the Premier Rendezvous Demonstration are divided into five parts: Intermediate, Terminal Parking, Terminal Rendezvous, Capture and Search; additionally, the Premier technology demonstration has the obvious required phase of Target Release. Fig. 1 describes these phases (except for Target Release), and contrasts them with how they might be implemented in an actual MSR mission.

The hardware configuration of the Rendezvous system consists of two cameras, a narrow angle camera, essentially a duplicate of the Mars Reconnaissance Orbiter (MRO) Optical Navigation camera (Ref 3), and a wide-angle camera, perhaps a version of the Mars Exploration Rover (MER) lander camera. The system also makes use of the Electra UHF-band radio navigation and communication subsystem on Premier (Ref 4), and at least one other spacecraft. All computations involved in the autonomous onboard rendezvous navigation system are handled by the spacecraft main processor, as the computational load of the system is intentionally minimized to obviate the need for an additional computer.

Phases of the Rendezvous Mission

The Search phase of the rendezvous technology demonstration is not a "search" in the sense that knowledge of the Orbital Sample (OS) is totally, or even partly, lost. Instead, the ability to perform an "MSR-like" search is demonstrated, where radio and optical observables are obtained sufficient to validate hardware and actual MSR search procedures. This entails taking radio ob-

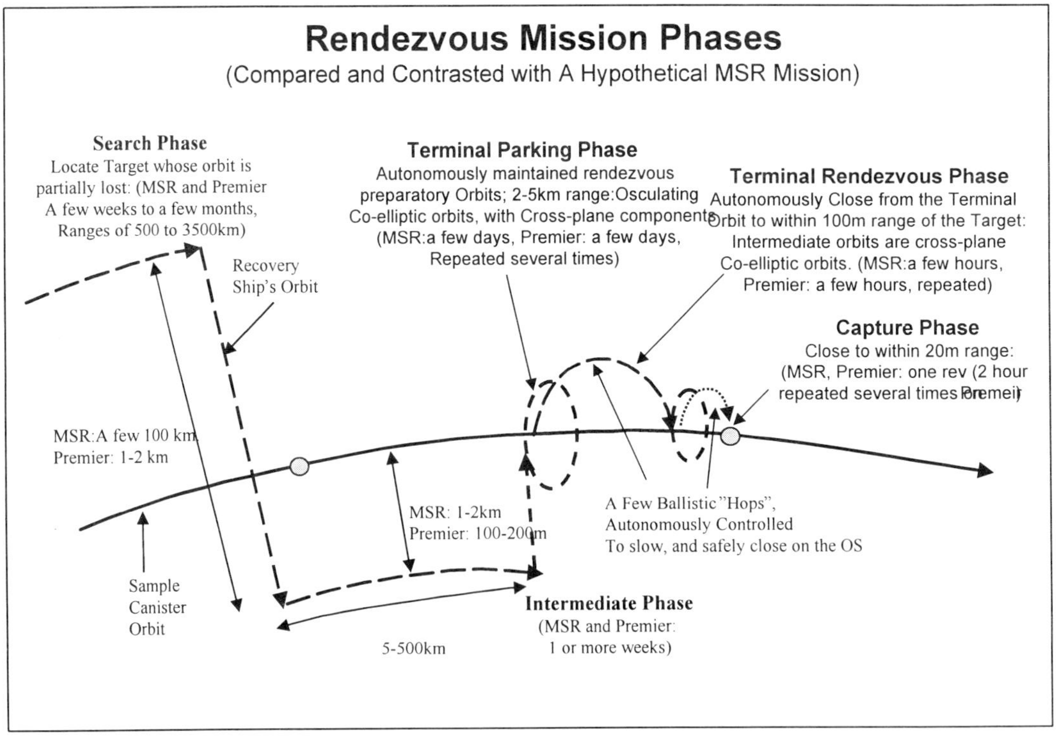

Fig. 1: Schematic of Premier Rendezvous Demonstration Mission Design Compared to an MSR mission.

servations of the OS from multiple receivers (e.g. Premier and MRO or GMO) and taking optical observables at varying ranges and phase angles. In the actual MSR mission it is assumed that total orbit knowledge is not lost, but that only Mean Anomaly is unknown, reflecting the tacit assumption that observations of the OS are obtainable within a few days after it is placed into orbit. Spacecraft/OS ranges during the search phase will span 500 to 4000 km. The optimum search range is from 2000 to 3000 km; in these ranges the s/c will point the body-mounted camera toward a position along the OS uncertainty annulus (see Fig. 2). The spacecraft will also slew the camera to match the apparent motion of the OS in the field, where the apparent motion assumed is that at the center of the annulus where the search mosaic is being taken. Such motion compensation enhances the signal strength of the OS image, at the expense of star images. Optical image processing will entail identification and location of all signals from the camera field that exceed a certain threshold, and will include images of stars and, when captured, the OS. These locations will be correlated on the ground, using multiple occurrences of a bright signal that shows positions and drift rates consistent with a likely OS orbit as the means of identifying the OS signals. In conjunction with the optical measurements, one-way Doppler measurements will be taken with the Electra instrument onboard the carrier spacecraft and simultaneously on at least one other spacecraft in Mars orbit. These radiometric observables will be downlinked, and processed on the ground. The search phase is expected to take only one or two weeks in the real MSR, but may occupy one or two months for the technical demonstration. The purpose for the latter apparently long duration is to achieve large range changes with minimal fuel expenditure, and to provide a maximum opportunity in the middle of intensive rendezvous operations to provide other experimenters an opportunity to have dominant use of spacecraft resources, and to provide an opportunity for the rendezvous team to apply experience learned from the first set of rendezvous events, which preceded the Search experiment, to the second set, which will follow it.

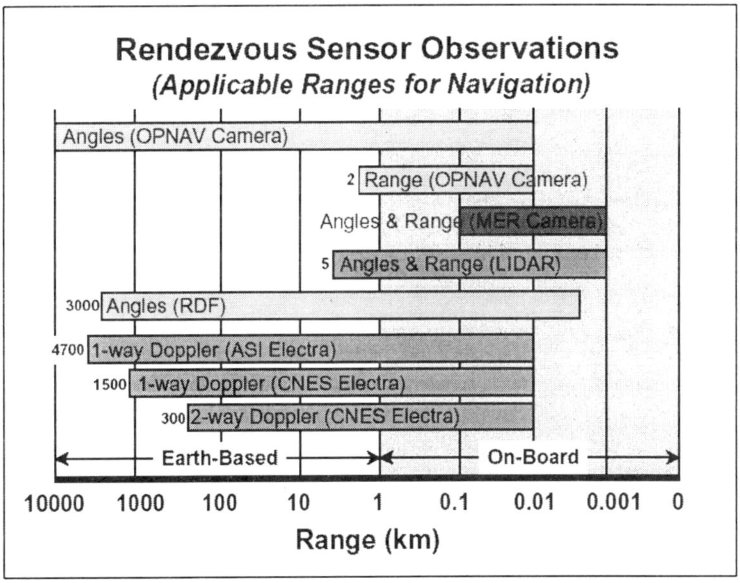

Figure 2: Search Phase Data Taking Geometries and Strategy

In the Intermediate phase of the mission, the spacecraft/OS range is reduced and/or increased to position the spacecraft in anticipation of the following phase. The navigation during Intermediate Phase will be performed primarily optically, although Doppler observables of the OS from the Premier orbiter will be taken and processed on the ground. The power of the radiometrics in the intermediate phase will not be as high as that of the optical. Figure 3 shows a comparison of the applicability at different ranges of various sensors that were considered for use in the demonstration, including a ranging LIDAR and a Radio-Direction-Finder (RDF). The orbit determination, using onboard processed images, is performed on the ground; although in this phase, initial trials of the autonomous closed-loop navigation system will be made. Control of the spacecraft trajectory will be via maneuvers computed on the ground and uplinked. Spacecraft to OS ranges vary from 5 to 500km in this phase. The Intermediate phases of both an MSR and the rendezvous technology demonstration will likely cover a week or so in time.

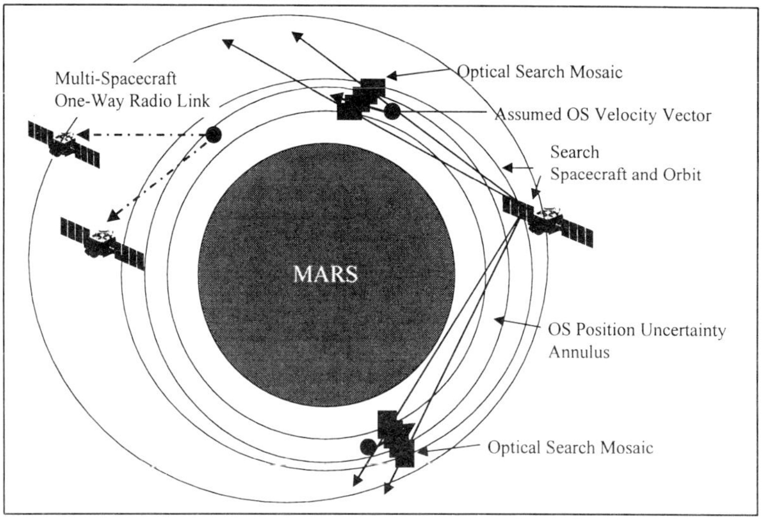

Figure 3: Comparison of Rendezvous Sensors

The Terminal Parking Phase of the mission consists of a series of "parking orbits" that the orbiter enters after executing a pair of range closing maneuvers. The characteristics of these orbits are such as to minimize the chance of OS/orbiter impacts if control of the orbiter is lost at anytime. Radial and out-of plane components of the orbiter's velocity are added to slightly perturb the eccentricity and node or inclination of what would otherwise be an identical - except for Mean-Anomaly - orbit to create relative off-standing "football orbits." These orbits, if allowed to propagate indefinitely provide high degrees of impact safety, due to their low probability of intersecting the orbit of the OS. The Terminal Parking phase will use optical data for orbit determination, and will include large periods of time where the autonomous system is invoked. The OS/spacecraft range of this phase varies from 2 to 5km. The Terminal Parking Phase of a real MSR is likely to take less than a week, as the mission prepares for Terminal Rendezvous and Capture, but in the case of the technology demonstration, with ten or more rendezvous "sorties" planned, the mission is likely to spend many weeks in Terminal Parking Phase. With the autonomous tracking and control systems operational, this fact will likely not lead to excessive operational costs.

The Terminal Rendezvous phase is always operated autonomously, and brings the OS/spacecraft relative range to 100m from 2 to 5km. For the Premier mission, the Netlander tracking orbit where the rendezvous demonstration will take place is approximately at 550km altitude, with an approximately 2 hour period, and nearly noon/midnight nodes, local time, so, in the approximately 12 hours of activities comprising the Terminal Rendezvous phase several eclipses are experienced. In this orbital configuration, an illumination source may be necessary to keep the OS successfully tracked during the eclipses. The light would become operational within a range of about 250m.

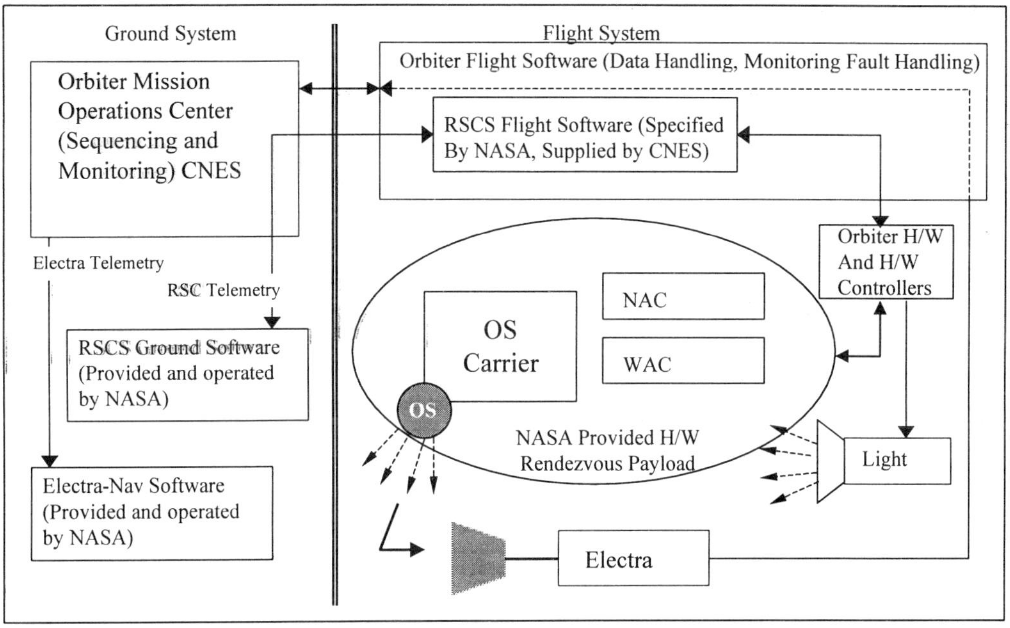

Figure 4: Rendezvous and Sample Capture System Components

The currently configured rendezvous experiment carries no capture mechanism, so the aspects of a rendezvous and capture having to do with mechanically contacting and "trapping" the OS will obviously not be tested. However, all aspects of the capture before that moment will be tested, including driving the orbiter along a slow relative orbit in very close proximity to the OS and slow (e.g. 2m range at 2cm/sec relative rate) with sufficient accuracy that a reasonably designed capture mechanism could capture the OS. So, even though there is not an actual physical capture of the OS, it is felt that these close proximity operations demonstrate the majority of the non-mechanical "capture" technologies and techniques. The Capture Phase will last less than two hours, and probably less than an hour, as the range from the orbiter to the OS is reduced from 100m to 2m.

The Rendezvous and Sample Capture System

The Premier Rendezvous and Sample Capture System (RSCS) is a combined ground and flight software system that, when combined with the optical sensing instruments onboard, provides a means of manual "ground-in-the-loop" tracking and maneuvering relative to the OS. It also provides automatic "closed-loop" tracking and maneuvering relative to the OS, and in the latter case, can accomplish close-proximity operations including a simulation of a "capture." Components of the system are shown in Fig.4.

The design of the RSCS is such as to maximize use of ground computation ability and therefore minimize onboard computation. This unequal division of responsibility between ground and flight systems is made possible by making maximum use of knowledge about the expected geometry of any rendezvous operations. By predicting position and timing of the rendezvous trajectory, and always constraining the Orbiter/OS relative position to be near this nominal trajectory, virtually all-numerical components of a closed-loop rendezvous navigation system can be pre-computed before the fact, uplinked to the spacecraft, and the rendezvous set into motion at the pre-determined time. Though there is substantial loss of generality in such a design, a Mars Sample Return (MSR) mission, for which the RSCS is intended to demonstrate certain technologies, is not a general problem; the recovery events will be very few, and each such event can be very carefully tailored, and be operationally intensive if need be. Such a limited and restricted system also minimizes costs of flight software development, and reduces testing costs due to the very simplified onboard algorithms, and greatly simplified operational scenarios.

The RSCS is a very limited closed-loop system, because it can take no action without a command, either directly, or indirectly from the Command Sequence. The operation proceeds in the following steps, that are sequentially repeated, A) Target Tracking B) Picture Processing, C) Orbit Determination and Ephemeris Update, D) Maneuver Calculation, E) Maneuver Execution.

Though part of the overall Rendezvous Mission Demonstration, the taking and processing (on the ground) of radiometric data through the Electra instrument is not part of the onboard autonomous rendezvous system, the RSCS.

Target Tracking

It is necessary for the Premier orbiter to track the OS, and other objects such as Phobos or Deimos, and certainly Mars (e.g. the nadir point.) The RSC will use and extend this basic AOCS (Attitude and Orbit Control System) Tracking Subsystem (AOCST) by supplying this subsystem with nominal ephemeris files and offsets from those files. This interface will allow the orbiter to put the specified target in the field of view (FOV) of the specific instrument.

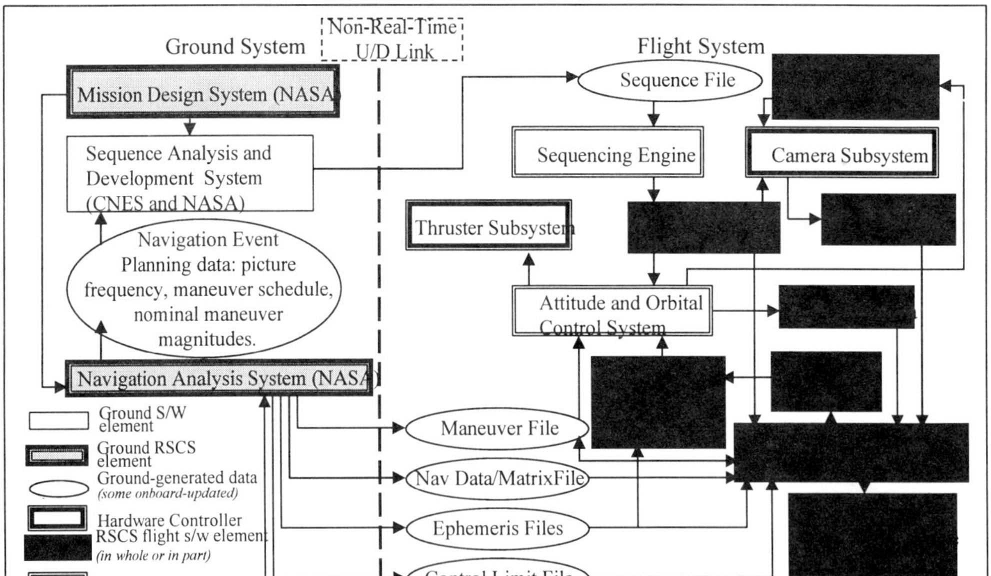

Figure 5: Rendezvous and Sample Capture Architecture and Block Diagram

Picture Processing

All picture data used by the RSCS is created by commands in the ground-generated Command-Sequence. When a "Take-Picture" command is issued by the Command Sequence the camera will shutter the picture and then (if the picture is designated as a "Nav" picture) a message is sent to the RSCS. Upon receipt of this "Picture Ready" Message, the RSCS will process the picture. The onboard RSCS picture processing is limited to simple geometric calculations of locations of objects in the field-of-view, using the (corrected) nominal ephemerides, and then searching in sub-sets of the frame for "bright" objects. If close enough to the OS, the RSCS image processor will measure the maximum width of the OS image, and from this measurement infer a range measurement.

Orbit Determination and Ephemeris Update

Using the extracted information from the picture-position of target image and its range - and that from previous pictures, the RSCS will compute a least-squares-estimate of the Orbiter and OS position and velocity offset vectors; making use of uplinked navigation data on the Nav-Data File. This estimate requires a small number of matrix and vector multiplications and additions with no numerical integrations or other extensive computations are required. After the ephemeris offsets are computed, they are transmitted to the AOCST, to provide current tracking information.

Maneuver Calculation

After a number of pictures have been processed (via commands issued by the Command Sequence), the Command Sequence issues a "Calculate_Maneuver" command to the RSCS. This command causes the RSCS to first determine if the currently estimated state offsets are large enough to warrant a correction (by comparing against a tabular schedule of allowable errors given in the Control Limits File) and if it is, multiply its current estimated Orbiter and OS relative offset vector by a state-transition matrix read from the Nav-Data File, to compute the parameters for two maneuvers that will return the Orbiter to the nominal Orbiter-OS relative trajectory by the completion of the second of the two maneuvers. The parameters of these maneuvers are then written to the Maneuver File.

Maneuver Execution

Finally, at a time consistent with the Maneuver File, the Command Sequence issues a command to the AOCS to execute a maneuver. To do this, the AOCS will read the Maneuver File for the maneuver parameters, specifically, the inertial delta-velocity vector. This process is repeated throughout the Terminal Phase, and until Capture is achieved.

RSCS System Interfaces and Architecture

The onboard software architecture alluded to in the above discussion of the RSCS is displayed in Fig. 5 as a block diagram of the system and its interaction with other onboard systems. The software elements highlighted in color are the only elements that contain RSCS computations. Data necessary for onboard computations is transmitted to onboard RSCS elements through files. Data transfers onboard may be by file or other means. Only one of the RSCS data files is modified during rendezvous operations and that is the Maneuver File. All activity of the RSCS is triggered either by a direct sequence command (e.g. "Plan-Maneuver") or is an indirect result of a command to take a picture. The latter command causes a signal to be sent to the RSCS when the picture file has been transferred into the CPU, and image processing commences.

ID	Name	Size	Occurrence/Frequency
M1	Covariance	6x6	Associated with every Nav picture time
M2	Observable DataPartials	3x6	Associated with every Nav picture time
M3	Maneuver K-Matrix	6x6	Associated with every maneuver time
M4	State-Transition-Matrix	6x6	Frequency ≤ 1/10mins, may be rev-cyclic
M5	STM-Time-Derivative	6x6	Frequency ≤ 1/10mins, may be rev-cyclic

Table 1: Definition and Description of Pre-computed Navigation Quantities

RSCS Onboard Computation Description

As noted earlier, the computations in the RSCS maximally depend upon pre-computed quantities. These are largely matrices that have reference epochs, and refer to some particular pre-computed spacecraft and OS ephemerides. In the discussion that follows, reference is made to five of these quantities, and their meaning is summarized in Table 1.

Target Tracking

The RSCS must communicate information to the AOCST about the location of targets, in particular, the OS. Before any orbit determination (OD) functions have been performed, the means of data transfer is simply through the nominal Ephemeris Files. These files are in the form of Chebyshev polynomials describing position, with the velocity to be computed from position differences. The order of the polynomials to be used and the period of time over which the polynomial applies are adjustable parameters.

After OD functions begin, the nominal ephemerides need to be updated by the state corrections (offsets). The OD function will produce both Orbiter and OS state corrections. These state corrections are referenced to a specific time epoch. In order for the AOCST to have current geometry, RSCS will supply through the Ephemeris Offset Data, the state corrections ΔX as well as state-transition matrices (M4), and their time-derivatives (M5), allowing AOCST to correct the nominal ephemeredes to the current time. If t is the time of the desired position, and t_m is the epoch of M4 and M5, then the updated state $X_{corrected}$ is given by:

$$X_{corrected} = X_{nominal} + M4 \cdot \Delta X + (M5 \cdot \Delta X) \cdot (t - t_m)$$

Having computed the corrected Orbiter and OS states, the relative inertial vector is computed by differencing these positions. The targeting direction is given by this Orbiter-to-OS inertial vector. In the search phase of the rendezvous experiment, when an accurate position to the OS is not available, the inertial attitudes of the search pictures will be explicitly specified in the command sequence, based on search planning done by the ground portion of the RSCS system.

Image Processing

There are two types of image processing required for the RSC: Search Phase image processing, where the position of the target is unknown, and all other phases, where the position of the target is predicted with certainty to be in the field of view of the camera, or even a small portion of the field. These two types of processing are shown schematically below.

For those cases where knowledge of the target's position in the field is required, the computation is accomplished as follows. The predicted target position in the camera field of view is the inertial s/c to target vector projected and mapped into camera coordinates. If ***SC2OS*** is the inertial predicted Orbiter-to-OS vector, φ is the angular extent of one camera pixel, and **TITV** is the 3x3 ortho-normal transformation from inertial to video (or camera) coordinates, where coordinates 1 and 2 are the horizontal and vertical axes of the camera frame, and coordinate 3 is the normal axis, then the predicted pixel, line position of the OS in the camera field of view is given by

$$\overline{V} = TITV \cdot SC2OS'$$

$$\rho = norm(\overline{V})$$

$$\begin{bmatrix} pixel\_position \\ line\_position \end{bmatrix} = \frac{1}{\varphi\rho}\begin{bmatrix} V_1 \\ V_2 \end{bmatrix} + \begin{bmatrix} pixel\_center \\ line\_center \end{bmatrix}$$

Camera frame line and pixel coordinates are trivially converted into meters in the "camera plane" or "plane of sky" at the target range. For both the search and the targeting image processing, the processing begins by identifying all illuminated pixels above a specified brightness. For the search phase it is sufficient to return this data to the ground for further processing. For targeted image processing, a sub-portion of the picture is searched, and the largest region of illuminated contiguous pixels above threshold is chosen as the target.

A simple "center-of-brightness" calculation is sufficient to locate the image, and the widest extent of the image will be measure of the image diameter, alternatively, the diameter of the object can be inferred to higher accuracy by measuring the number of illuminated pixels associated with the target image. From the diameter of the target (when the target image is large enough) the distance to the target is computed. Then, the three quantities -horizontal position, vertical position and range - form the image "observables" for the subsequent Orbit Determination function, and the nominal positions and range form the "modeled" or "computed" values. Differencing the computed values from the observed values forms the set of "residuals" for this picture, which is a 3-vector ***R***

An additional mode of image processing requires location of the bright object in the field (i.e. the target) to allow extraction of dimmer objects (e.g. stars). Such processing is necessary for Mars-approach optical navigation images, and for high-precision OS navigation frames. Such image processing is accomplished by including, in the command to shutter the image, pre-computed pixel/line vectors from the target to the desired image locations where data is desired for extraction.

Orbit Determination

With the observable quantities – horizontal x, vertical y, and range (all in meters) -extracted from the pictures, the RSCS will update the estimated state, and deliver the updated state offsets to the AOCST. The estimated state is computed using data partials (M2) and covariance (M1) from the

Nav-Data File. M2 represents the Observable partial derivatives (partial of the OS position in the camera with respect to the epoch-state of the OS), and M1 is the estimated covariance associated with each observation, and includes the effect of all of the previously scheduled pictures taken in a data-arc of pre-determined length, assuming a predetermined de-weighting decay coefficient. Over a series of pictures, a least squares estimate of the OS (Target) state will be computed by forming the product of the covariance, the observation partials, and the computed residuals $\mathbf{R}_i$ of the Target position in the camera frame. This amounts to a current state Kalman filter. As each picture is processed, the anomalous values of $\Delta \mathbf{X}_i$ are tested for (based on preset limiting parameters) and such pictures are rejected.

$$R_i = \Delta(x, y, range)$$
$$\Delta \overline{X}_i = M1_{i(6x6)} \cdot M2_{i(6x3)} \cdot R_i'$$

As part of the OD process, the position of the Orbiter must be corrected as well. In the case of the orbiter, corrections are based on the Non-Grav File, produced by AOCS. The Non-Grav file provides accumulated translational velocity from all thruster firings. The current updated state of the Orbiter is maintained by incrementally mapping the ephemeris offsets from time-point to time-point, and using the incremented velocity change from the Non-Grav File:

$$\Delta(pos, vel) = (\Delta position_t, \{\Delta velocity_t + \Delta v_{non-grav}\})$$
$$\Delta X_{t+\Delta t} = [M5_{6x6} \cdot \Delta(pos_{1x3}, vel_{1x3})'] \cdot \Delta t$$

When the estimation of the OS ephemeris offset is complete, and the mapping of the Orbiter position is complete, $\Delta \mathbf{X}_i$ **for** both is sent to AOCST with the state transition matrix M4, its reference epoch time, and its rate of change matrix M5, as discussed above.

Maneuver Computations

With the estimated Orbiter state offsets complete, it is possible to compute the magnitude and direction of a pair of maneuvers that will bring the Orbiter back onto the nominal trajectory. This process can be repeated as necessary to keep the trajectory "in trim," where the frequency of such maneuvers is determined a priori, and the opportunities to perform them is imbedded in the Command Sequence. This process is shown schematically in the following figure. It must be emphasized that all maneuvers are planned ahead, and initial values of some maneuvers are pre-computed on the ground. These are the deterministic maneuvers that define the trajectory, e.g. obtaining or leaving a "football" orbit, landing or leaving "v-bar". Other maneuvers are pre-planned as well, and are entered into the Maneuver File with "null" a priori values. When the RSCS is asked to compute a maneuver, it reads the Maneuver File to determine the next pair of maneuver opportunities, and simultaneously reads the Nav-Data/Matrix-File for the K-Matrix (M3) for this pair of maneuvers. Then, having previously obtained the current best estimated and determined positions of the OS and the Orbiter, the Spacecraft to OS vector is computed, ***SC2OS***, and the difference vector from the nominal to the estimated vector, ***ΔSC2OS***, and the pair of maneuvers can be computed, the values of which are added to the current values on the Maneuver File, and that file is then updated with the sum:

$$M3_{6x6} = \begin{bmatrix} M3_{11} & M3_{12} \\ M3_{21} & M3_{22} \end{bmatrix}$$
$$\Delta \overline{v}_1 = [\Delta SC2OS_{1:3} \cdot M3_{11} + \Delta SC2OS_{4:6} \cdot M3_{21}] \cdot M3_{21}^{-1}$$
$$\Delta \overline{v}_2 = [\Delta SC2OS_{1:3} \cdot M3_{12} + (\Delta SC2OS_{4:6} - \Delta \overline{v}_1) \cdot M3_{22}]$$

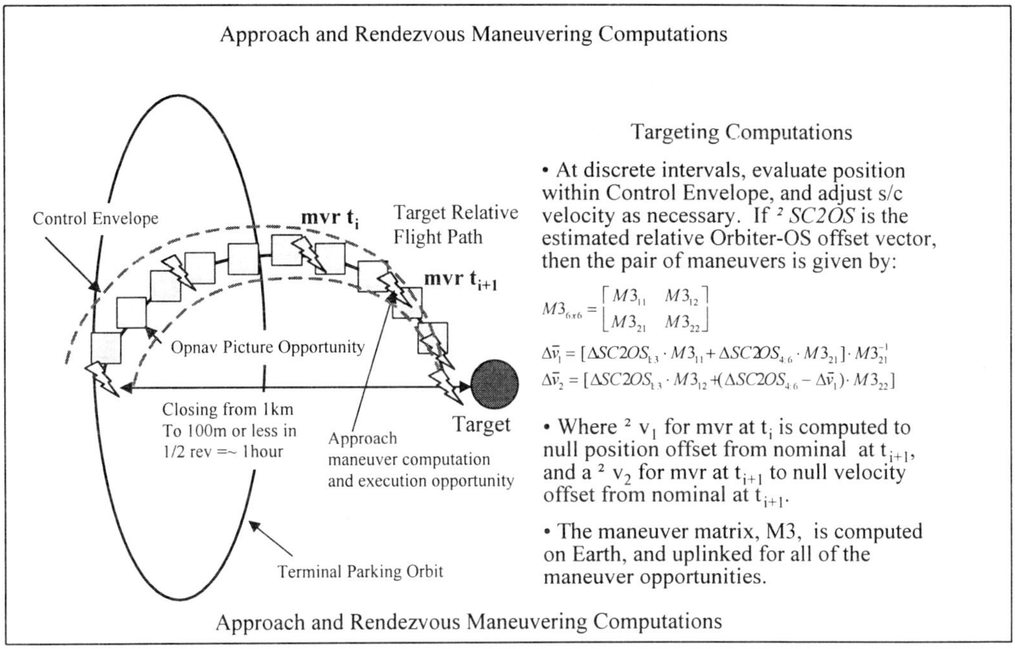

Figure 6: Diagrammatic representation of Approach and Rendezvous Maneuvering Computations

Operational Considerations

The operation of the RSCS is completely deterministic, even though it is an autonomous system, exhibiting only statistically predictable behavior, and responding to only statistically predictable inputs. Deterministic operation of the system is possible because the method of the navigation and trajectory control is to always drive the orbit of the spacecraft back to the nominal pre-computed, and therefore perfectly known, trajectory. This is the basis of being able to pre-compute all navigational quantities, as the deviations from the nominal are never allowed to get larger than what would result in non-linear estimation effects. With the operation of the system being deterministic, it is possible to plan all of the rendezvous activities before the rendezvous sorties, with a known time of "capture" pre-determined. All picture-taking events, all maneuver computations, and all maneuver executions can be pre-planned and placed in the operational sequences. Table 2 presents a summary of the commands used by the system.

Another aspect of the operation of the RSCS is the provision of the data files. Long before the execution of a rendezvous sortie, the desired trajectories for the spacecraft and the OS are computed, based on reasonable expectations of locations of both at the time of the desired rendezvous. Control of the relative orbit of the spacecraft might well be managed by the ground to a point near the nominal trajectories before the autonomous system is invoked. In the meantime, the uplink files are prepared, including the nominal ephemerides, a file containing the nominal deterministic maneuvers, and locations to record computed statistical maneuvers (the Maneuver File), and the file containing the pre-computed matrices of navigational data (the Navigation Data File). These will be uplinked to the spacecraft well before the start of a rendezvous event. It is not the case that data in the Nav Data File need represent every picture to be taken. For long periods of time, perhaps days, the RSCS will maintain the spacecraft orbit relative to the OS. For these periods, one or a few orbits worth of matrix/data, used repetitively will be adequate, and will be cyclically reused.

COMMAND NAME	DESCRIPTION	FREQUENCY of USE
CAM_POINT	Request the AOCST to point camera at target, specifying particular camera.	One request at beginning of tracking session.
PICT_ACQ	Take a picture and deliver to the RSCS, the command will trigger OD calculations	One command per picture
PLAN_MAV	Plan a maneuver based on current OD, and store result on the Maneuver File.	One command per maneuver opportunity.
SET_DFILE	Assignment of data file to the RSCS, specify one of several files.	Assignments all occur before autonomous activity begins.
RESET_DATA	Null currently estimated state biases.	At start of rendezvous event.

Table 2: Rendezvous System Commands

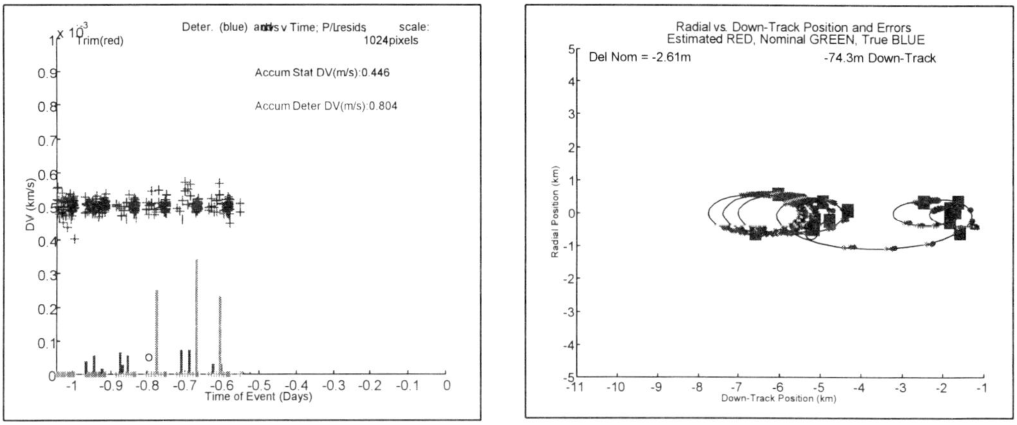

Figure 7a, b: First half-day of operations of a representative rendezvous sortie; picture and maneuver timeline, residuals and magnitudes; downtrack v. radial trajectory residual and maneuver display.

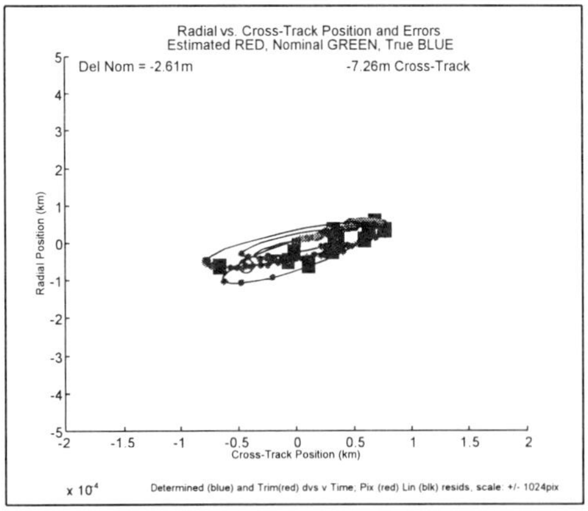

Figure 7c: radial v. crosstrack trajectory residual and maneuver display.

System Simulation

A computer simulation of the system has been constructed, as a proof of concept demonstration. All aspects of the RSCS were simulated with realistic noise models, with the exception of the image processing. Random observational errors were introduced to "ideal" optical observables. Another limitation of this early simulation is that modeling of the Wide-Angle camera has not yet been introduced into this simulation, and the system relies on the Narrow Angle camera alone. No translational dynamics of the spacecraft due to turns necessary to track the OS were considered. None of these limitations are believed to cast substantial doubts

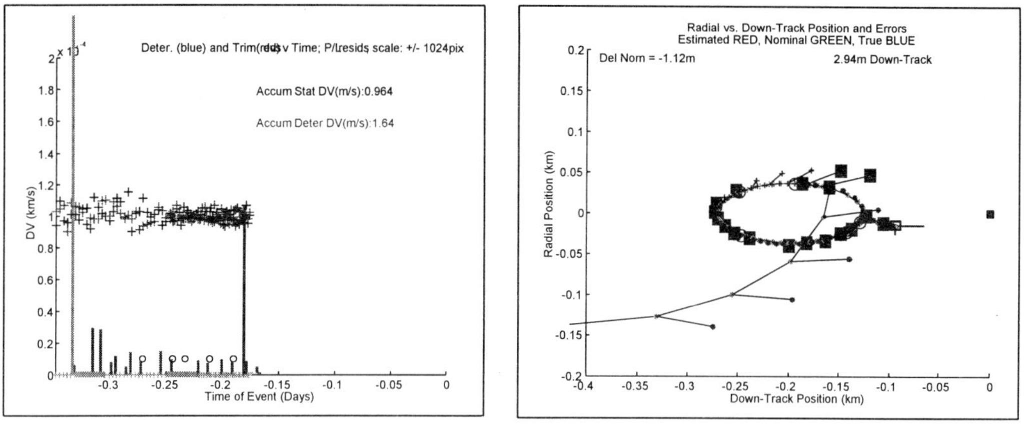

Figure 8a, b: Middle hours of operations of a representative rendezvous sortie; picture and maneuver timeline, residuals and magnitudes; downtrack v. radial trajectory residual and maneuver display.

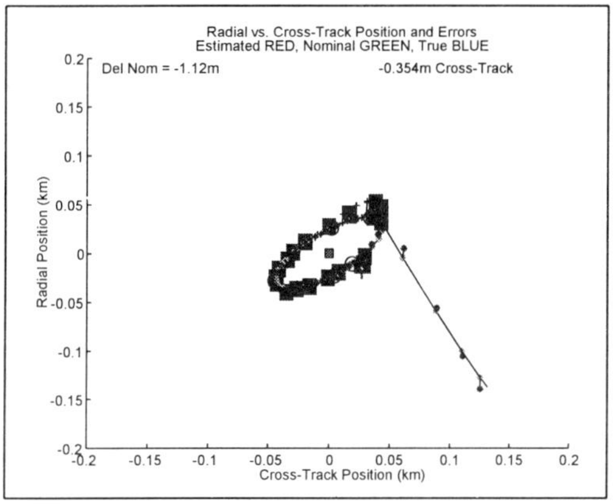

Figure 8c: radial v. crosstrack trajectory residual and maneuver display.

upon the overall feasibility of the technical approach, but such factors will be included in future simulations.

Fig. 7 shows the first approximately 10 hours of operations of a rendezvous sortie. Fig. 7a presents a timeline of events, including picture occurrences (crosses on time-axis), picture residuals (crosses across the horizontal mid-line, with a scale of +/- 1 NAC frame) and maneuver occurrences and magnitudes; dashed bars represent deterministic burns, and solid bars represent statistical burns. Accumulated magnitudes are shown from the start of the sortie, and represent an absolute-value sum across the three axes, giving measure of total delta-v consumed. Figs. 7b, and c display the trajectory of the OS relative to the spacecraft, in radial-downtrack and out-of-plane coordinates; radial vs. downtrack and radial vs. crosstrack components respectively. Maneuvers are shown on the trajectory trace as diamonds. Deviations from the nominal are noted at each observation point by a line and differently colored marker. The deviation from the desired nominal trajectory at the end of this time period is shown at the top of the plot. Figure 7c shows identical information in radial verses crosstrack directions. The projected position of the spacecraft is shown as a square.

Figs. 8 displays a period of several hours in the central portion of the sortie timeline. Figs 8a-c show the information from this time as was described for Figs. 7. Figs. 9 displays the same plots for the final 14 minutes of the rendezvous sortie, in the same format. The trajectory geometry plots present the final miss-targeting of the simulation. In this particular instantiation of the statistical noise sources, the trajectory error was less than 10cm in all directions.

Rough characterization of the performance of the rendezvous system has been made using the simulation. Success rates are high (where success is determined by an accurate delivery of the OS along a trajectory consistent with a hypothetical capture mechanism), but are clearly dependent upon the initial assumptions, which, for this instantiation are given in Table 3, along with general characteristics of the filter.

Table 3: Characteristics and Assumptions of the Rendezvous Navigation Filter	
6 State (Spacecraft Position, Velocity) Kalman Filter	Maneuver execution errors, 1.5% relative, 1mm/sec bias
Exponential "Knowledge Decay" time of 4 hours	Camera pointing biases unmodeled
1mrad pointing knowledge, 0.5mrad pointing noise	Control requirement: 1% of spacecraft/OS range
Diameter measurement noise of 0.5pixels	Illumination of OS is required inside of 250m range
Increased Data Frequency Prior to entering Eclipse	Narrow Angle Camera data only
Initial orbit errors: 1m, 5cm/sec spherical	Maneuvers forbidden immediately before eclipse

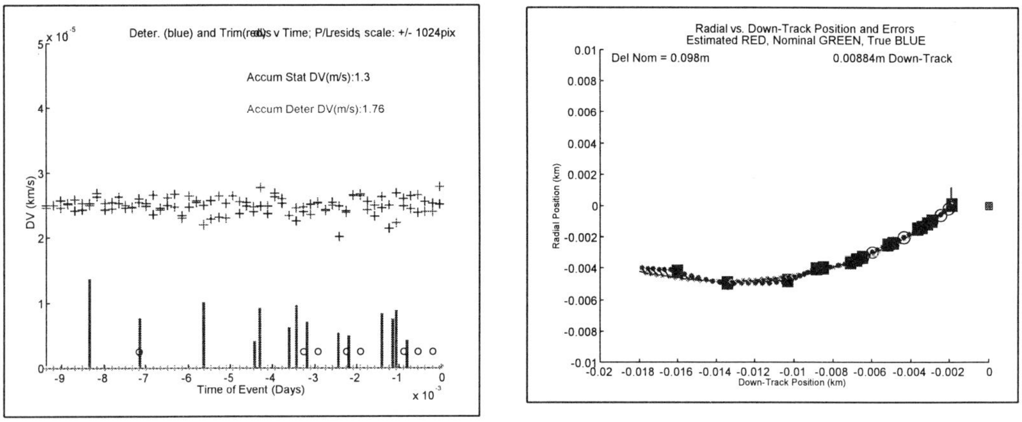

Figure 9a, b: Middle hours of operations of a representative rendezvous sortie; picture and maneuver timeline, residuals and magnitudes; downtrack v. radial trajectory residual and maneuver display.

Important aspects of the operations of the system were discovered with the simulation; two of these are the need for illumination of the OS inside of 250m range, and a determination of the size of the imaging mosaics necessary when the OS comes out of eclipse when the range is greater than 250m. Such mosaics are necessitated by loss of orbital information during eclipse. Further refinements of the simulation will include adding observations from the wide-angle camera, and adding the translational delta-v effects due to turns.

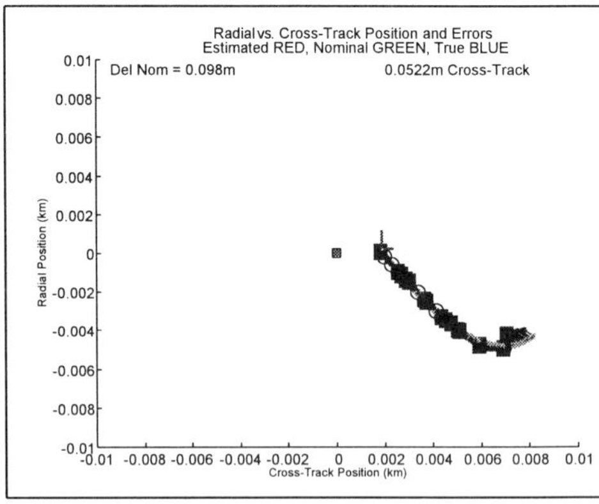

Figure 9c: radial v. crosstrack trajectory residual and maneuver display.

The Radio-Based Search Experiment

Overview

During a real MSR mission, the search phase of the mission will likely make use of several redundant data types, to absolutely guarantee success of initial contact with the OS. One very robust and adaptable instrument for this purpose is a one-way radio beacon. Depending upon the accuracy of the transmitting beacon, good to very good navigation accuracy can be achieved with a single receiver, and even with a relatively low-accuracy transmitter, extremely good results can be achieved with multiple orbiting receivers. The radio receiver, unlike the optical sensor, can also provide an extremely robust "proximity detection" capability, which, if the receiver is positioned in an orbit of substantially different altitude, can provide a very useful initial determination of OS orbital period and mean anomaly, without the need for a wide sky search. The following discussion will show results from analysis of the one-way radio beacon link during the MSR search phase, and which is intended to be part of the Premier Rendezvous Technology Demonstration.

Proximity Detection Capability

Evaluation of the CNES Electra onboard radio received signal strength can provide a crude OS detection capability. At search distances closest approach times can be inferred and used to assess the OS orbital period and mean anomaly. Signal power at closer distances can be helpful in determining pointing directions needed for recovery from situations where the OS is not in the expected field-of-view of the cameras. If multiple orbiters with UHF-band receive capability are available, simple direction finding can also be performed.

Radiometric Orbit Determination

For search and intermediate rendezvous operations one-way Doppler measurements, from the OS UHF-band radio beacon, can be produced by the CNES Electra onboard radio. For search distances larger than 1500 km, the Doppler observations are computed from a Fourier reconstruction of an "open-loop" or full spectrum recording of the received beacon signal. At shorter distances, the measurements are stored and transmitted to Earth where in all cases the OS orbit is reconstructed. Taken alone, the CNES one-way Doppler measurements provide OS location information at the 100-200km levels. The poor OS radio frequency stability (oscillator expected to be 10^{-7} s/s Allan deviation over 10-100 seconds) limits accuracy. Figure 10 shows the bounding performance resulting from 10^{-7} and 10^{-14} oscillators on the OS. Collecting one-way measurements simultaneously with another orbiting spacecraft, in this case the proposed Italian Space Agency (ASI) G. Marconi Orbiter (GMO) also equipped with Electra, allows the simultaneously observed OS frequency instabilities to be cancelled by forming differenced one-way Doppler measurements. Figure 10 additionally presents the performance of these differenced Doppler combinations. For comparison, the relative range is also shown between the CNES orbiter and the OS (bold red line). Thus, simultaneous Doppler collected by the CNES and ASI orbiters with oscillators stable at least to 10^{-10} can provide reconstructed knowledge at the 10 km level. The CNES and ASI orbiters as well as future NASA Mars orbiters have baseline oscillator stabilities of 10^{-12}. While not autonomous, this Earth-based processing radiometric technique is valuable in assisting initialization of autonomous terminal rendezvous and, if necessary, rapid recovery from terminal rendezvous abort conditions.

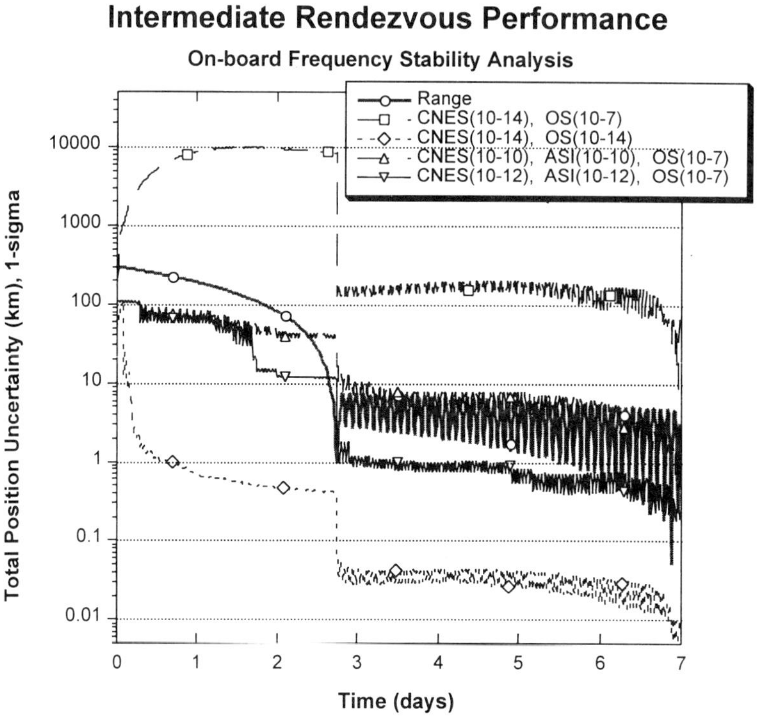

Figure 10: Radio-based covariance analysis results for various OS Beacon and orbiter clock stabilities

Acknowledgements

The research described in this paper was carried out, in part, by the Jet Propulsion Laboratory, California Institute of Technology, under a contract with the National Aeronautics and Space Administration. The authors would like to recognize the contributions of several individuals who provided technical support in the development of the materials presented in this paper. They are: Rob Bailey, Todd Ely, Richard Kornfeld, Steve Synnott, Peter Xaypraseuth, all of JPL, Don Pearson of JSC, Geoff Huntington at Draper Laboratory, and Mark Vincent of Raytheon Corporation.

References

1. Thien Lam-Trong , "Versatility of the CNES 2007 mission to Mars", IAF-02-Q.3.2.01

2. Olivier Marsal & al., "The NetLander geophysical network on the surface of Mars: general mission description and technical design status", IAF-01-Q.3.b.01

3. Stephen. P. Synnott, Jet Propulsion Laboratory, private communication

4. Ely, T., "Electra 2-Way Doppler Measurement Model including Error Budget" Electra JPL IOM 03-18, January 15, 2003 (Internal JPL Document).

OVERVIEW OF THE SPHERES AUTONOMOUS RENDEZVOUS AND DOCKING LABORATORY ON THE INTERNATIONAL SPACE STATION

David W. Miller,* Edmund M. C. Kong† and Alvar Saenz-Otero‡

Currently scheduled for launch to the International Space Station in July 2003, the SPHERES testbed presents a unique opportunity to develop and test autonomous rendezvous and docking algorithms in a zero gravity environment. The shirt-sleeve environment makes the testbed a low risk, low cost and cost effective to testing more aggressive maneuvers than that of a full flight program. For this first flight to the ISS, three types of experiments have been scheduled for ten two hour test sessions: system identification to provide the most updated dynamic models; autonomous rendezvous and docking experiments using two Spheres and formation flight experiments to test maneuvers representative of NASA's Terretrial Planet Finder mission. The last section of the paper is geared towards evolving the SPHERES testbed to further develop staged control algorithms for separated spacecraft interferometry missions and the possibility of performing experiments outside of the ISS using the SPHERES testbed (Exo-SPHERES).

INTRODUCTION

The ability to perform autonomous rendezvous and docking of satellites opens up the possibilities of servicing, upgrading and replenishing of existing satellites, which lead to operational flexibility, injection of new technology and extension of spacecraft lifetime. To this end, DARPA has initiated the Orbital Express program to develop and demonstrate the technologies required to perform such tasks. One key component to the program is the development of autonomous rendezvous and docking algorithms.

The SPHERES testbed provides an unique opportunity to demonstrate and validate rendezvous and docking algorithms in a zero gravity environment. Slated for launch to the International Space Station (ISS) in July 2003, the SPHERES testbed will consist of three self-sufficient volleyball-sized spacecraft operating in the shirt-sleeve environment of the ISS. This shirt-sleeve environment, which allows interaction with the crew member, makes the SPHERES testbed a low risk, low cost and cost effective testbed to develop and test autonomous rendezvous and docking algorithms. Unlike full flight programs, more aggressive maneuvers can be tested onboard the SPHERES testbed without the fear wasting precious resources such as propellant or worse still, damaging flight hardware. Once the algorithms can be demonstrated to be robust and reliable, the incorporation of these algorithms onto the significantly more expensive spacecraft represents a smaller incremental risk to the overall program.

These high risk, yet high payoff control algorithms are applicable to systems requiring coordinated motion of multiple satellites in a micro-gravity environment, including, but not limited to, sparse

---

* Associate Professor, MIT Space Systems Laboratory, Cambridge, Massachusetts 02139. E-mail: millerd@mit.edu.

† Research Scientist, MIT Space Systems Laboratory, Cambridge, Massachusetts 02139. E-mail: mckong@mit.edu.

‡ Research Assistant, MIT Space Systems Laboratory, Cambridge, Massachusetts 02139. E-mail: alvarso@mit.edu.

aperture and interferometry. Other missions such as the Air Force's TechSat 21, and NASA's Terrestrial Planet Finder (TPF), can utilize these algorithms to achieve the capabilities of a single large spacecraft with multiple small separated spacecraft.

In this paper, an overview of the SPHERES testbed is first presented. With the SPHERES testbed currently being manifested for launch to the ISS in July 2003, the experiments currently planned for the testbed are then discussed. Finally, the possibility of using the SPHERES testbed as a stepping stone to mature formation flight technologies to support NASA's TPF mission is explored.

THE SPHERES TESTBED

Figure 1 An assembled SPHERES unit

The SPHERES testbed consists of three self-contained free-flyers (Figure 1), a laptop computer, and five small beacons. It is designed specifically for operation in the shirtsleeve environments of the SSL laboratory, KC 135 reduced gravity airplane, and ISS. The simplicity and hands-on nature of the testbed allow for easy re-configuration and replenishment of consumable, resulting in low experiment risk and cost. The KC 135 and ISS environments provide 3-D environments to test algorithms that can be directly applied to real satellites. The additional laboratory environment at the MIT SSL enables 2-D experiments to be performed prior to testing on the KC 135 or ISS, reducing even further the cost and risk to develop and verify algorithms.

The SPHERES testbed is designed to produce results traceable to proposed formation flying missions. The individual self-contained satellites have the ability to maneuver in six degrees of freedom, to communicate with each other (satellite to satellite: STS) and with the laptop control station (satellite to ground: STG), and to identify their position with respect to each other and to the experiment reference frame. The laptop control station is used to collect and store data as well as to upload control algorithms to the satellites. Currently, one to three satellites may be used, depending on the algorithm being tested. The software architecture allows additional SPHERES to be added to the array, if desired. Figure 2 shows an operational concept for the SPHERES testbed, with inter-satellite and satellite-to-laptop interactions illustrated. In this section, the overall design of the SPHERES testbed and of its subsystems are further discussed.

Testbed Design

During the early design phases of the SPHERES, a clear set of hardware requirements was not available. The team had to determine a set of requirements that ensure future formation flight and docking control algorithms will run in the testbed and provide significant results. These requirements include the precision, accuracy, and operational ranges of sensors and actuators, processing power, untethered operational lifetime, and communications bandwidth. Table 1 summarizes the

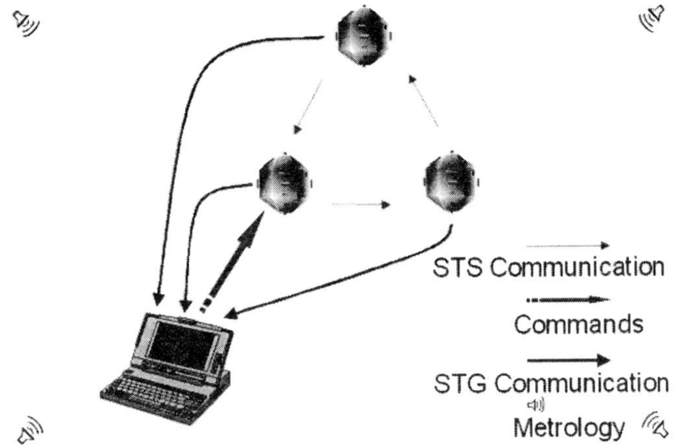

Figure 2 Testbed Operational Concept [1]

quantitative requirements derived from trade analysis. This set of requirements ensures the data feedback precision necessary to conduct useful experiments.

The SPHERES units are identical since they constitute the general equipment of the laboratory. Yet, each unit can easily be outfitted with external components that change their dynamics and/or add operational elements. Therefore, the SPHERES testbed can simulate the operation of both homogeneous (all identical units) and heterogeneous (units with different physical or operational components) systems.

The testbed equipment was designed to fit in a Space Shuttle middeck locker, with room for expendables such as propellant tanks and batteries. This way, the laboratory can integrate seamlessly into existing space available in the ISS environment. Subsequent hardware upgrades would either fit in the same locker, or take only a fraction of another locker space. Therefore, at all times, the SPHERES testbed will occupy only minimal space in the ISS, while providing a multitude of research options. The locker space constraints limit the satellite diameter to 0.25 m, about that of a volleyball, and the total volume of the system, including consumable, to $0.057m^2$.

Table 1 SPHERES REQUIREMENTS [2]

Item	Requirement
Translation (1m start to stop)	5 secs
Rotation (360° start to stop)	5 secs
Translation accuracy	5 mm
Rotation accuracy	2.5^o
Propulsion lifetime	20 secs
Power lifetime	90 min
Mass (all units with consumable)	24.5 kg
Processing Power	23 MFLOPS
Communications Data Rate	12 kbps

Subsystems Descriptions

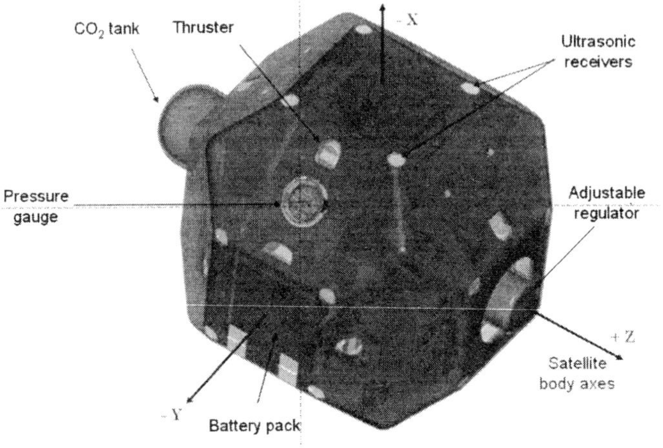

Figure 3 CAD model of the SPHERES unit [3]

Each SPHERES unit contains on-board power, propulsion, processing, communication, and position and attitude determination subsystems. Figure 3 shows a computer aided drawing (CAD) model of an assembled SPHERES unit with key features being highlighted. Physical properties and other specifications for the satellite units are listed in Table 2. The following sections describe each sub-system.

Table 2 SPHERES UNIT SPECIFICATIONS [4]

Items	Values
Diameter	0.25 m
Mass	4.38 kg
Max Linear Acceleration	0.17 ms^{-2}
Max Angular Acceleration	3.5 rad s^{-2}
Battery Life	120 min
Communications Data Rate	18 kbps
Average Power	15 W
Metrology Resolution - Linear	5 mm
Metrology Resolution - Angle	2°
Propellant Tank Life	~10 min

Structure - The SPHERES backbone structure is made from 6 laser cut rings held in place by 6 brackets and 12 cross members as shown in Figure 4. These components were made from aluminum for its cost effectiveness in providing rigidity and stiffness to the structure. All other components are attached to this aluminum structure through the various mounting points.

Propulsion - The satellites are propelled by a cold gas thruster system which uses carbon dioxide as propellant. The CO_2 propellant is stored in mixture of liquid and gas form at 860 psig, without the need for a cryogenic system. A regulator reduces the pressure to 35 psig; the operating pressure

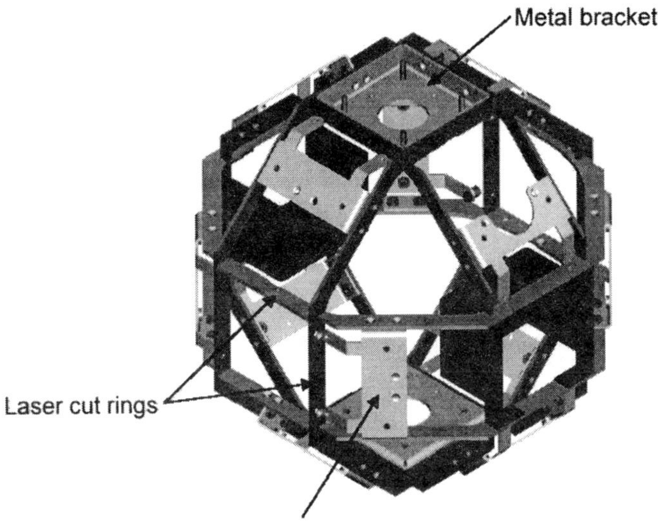

Figure 4 SPHERES unit backbone structure [3]

of the testbed. A teflon tubing system distributes the gas to twelve thruster assemblies. The thrusters are positioned (Figure 5) so as to provide controllability in six degrees of freedom, enabling both attitude and station keeping control. Each thruster assembly consists of a solenoid-actuated micro-valve with machined nozzles optimized for the desired thrust of 0.1 N. The propulsion system may be easily replenished by replacing the spent propellant tank.

The system is replenishable, ensuring that extended investigations can take place. A single tank provides approximately 10 minutes of active operation, using non-optimal trajectories and with environmental disturbances (such as air currents inside the ISS). Ten minutes allows for most control experiments to return substantial information. After a test, a tank can be left in the system partially full, for use at a later time, or can be replaced with a new tank. In both cases, a subsequent test can be done either immediately or the units can be stored for use at a subsequent date.

Position and Attitude Determination - The Position and Attitude Determination System (PADS) has local and global elements that work together to provide metrology information to the satellites

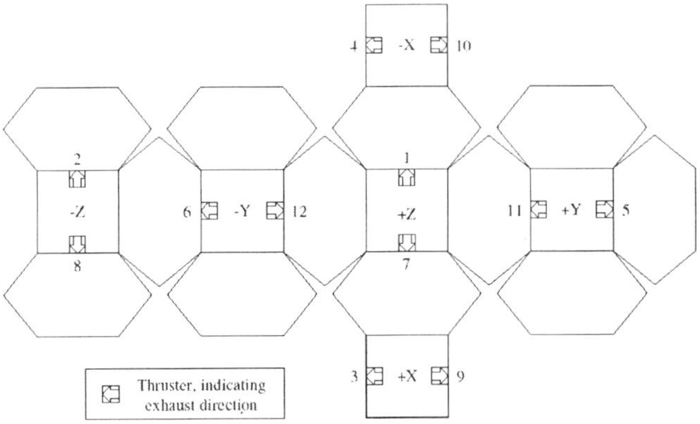

Figure 5 Thruster locations on the SPHERES unit [5]

in real-time. While the global and local elements are capable of independent operation, the readings of both systems are combined during nominal operations to produce continuously updated state information at 1 kHz.

The local PADS element provides inertial measurements at 1 kHz, and the global element measurements update those estimates at 1 Hz. The local element consists of three single-axis rate gyros and three single-axis accelerometers. The global element is a GPS-like ranging system that uses ultrasonic time-of-flight measurements from transmitters placed at known locations in the testbed's reference frame to the ultrasonic receivers distributed on the surface of each satellite. These time of flight measurements are converted to ranges and then used to derive position and attitude with respect to the reference frame. Figure 6 illustrates the time-of-flight range measurements from the various ultrasound beacons to the ultrasound receivers on the SPHERES unit.

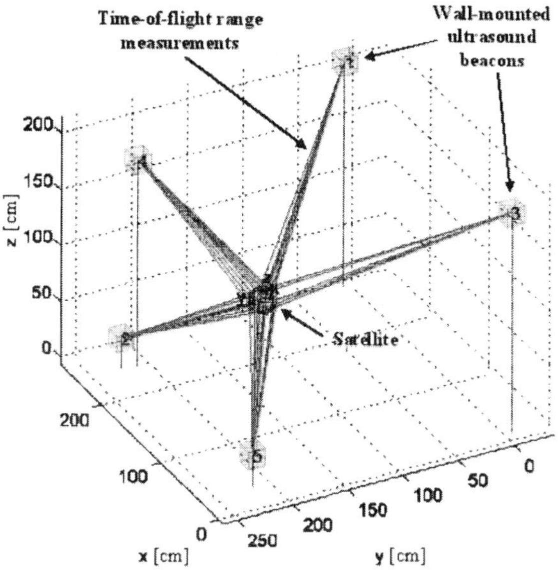

Figure 6 SPHERES global metrology system [5]

The software sub-system provides further advantages. In order to ensure software re-configuration and support multiple investigators, software filters can be easily added. These filters can process the PADS data in real-time to simulate different types of metrology systems. For example, a software filter could add a drift factor to the attitude data of the state vector, in order to simulate a system with only a local (IMU) attitude determination sub-system, while maintaining the position data untouched to simulate a one antenna GPS system.

In addition to PADS, each SPHERES unit is also equipped with a ultra-sonic beacon. The purpose of this beacon is to give inter-sphere ranging and bearing angle information without using the global PADS. The beacon is located at the center of the -X face of the SPHERES unit. To avoid conflict with the global beacons, the beacons onboard the SPHERES unit are normally programmed to emit ultrasonic pulses after pulses from the global beacons are transmitted.

Power and Avionics - A Texas Instruments C6701 Digital Signal Processor (DSP) provides the computational power. DSP processors provide multiple features that ensure real-time operation. Further, the DSP processors include all support functions of a standard processor, allowing it to control the whole unit. The ability of the C6701 to provide between 167 MFLOPS up to 1

GFLOPS, provides significant processing power to prevent being the limiting factor in the performance of the system. The processor is supported with 16 MB of RAM and 512 KB of FLASH memory. The FLASH memory allows software re-configuration at all levels, ensuring that multiple investigators are supported while the system is in the ISS.

The power system utilized in the Space Station consists of packs of AA alkaline batteries. The packs provide each unit with approximately two hours of operation; once a pack is consumed, it can be easily replaced. The power system supports repeatability, extended investigations, and a low-risk environment.

Software - Figure 7 presents the different components of the software architecture. The main components are the controller interrupt, propulsion interrupt, and background processes. The other shown functions are parts of the software architecture that may be exchanged between the different sections of software in order to program the SPHERES testbed with a specific algorithm. This example placement shows how elements of the SPHERES satellite software can work as independent units. The propulsion interrupt operates at 1 kHz, the controller interrupt operates at 50 Hz, and the background processes run freely.

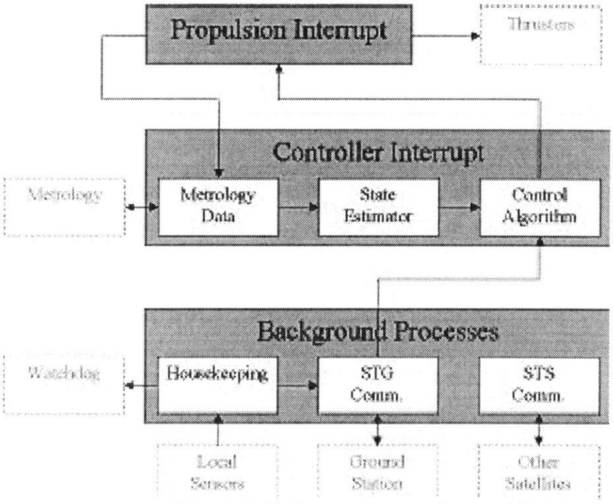

Figure 7 Software system design [1]

The controller interrupt is considered the main section of the software. The controller interrupt process determines the system state, runs the control algorithm, and determines the necessary output in terms of thruster commands and communication packets. The propulsion interrupt provides the interface to the thrusters, and its rate determines the granularity of the thruster actuation. The interrupt runs at 1 kHz, allowing a pulse width resolution of 1 ms (although the solenoids place a hardware restriction of 5 ms minimum pulse width). The background processes are not time dependent; they do not need to run at a specific rate. These processes run freely in the background, whenever neither of the interrupts is being processed.

The SPHERES software presents very few restrictions. While the propulsion interrupt module should not be changed, basically any other software module can be re-configured. An investigator can determine any desired control frequency and the placement of PADS, communications, and housekeeping functions.

All the SPHERES code is written in standard ASCII C language, which is widely used in the industry. This provides several advantages: first, the language is likely to already be in use by possible investigators; second, it allows the creation of multiple simulations that use the SPHERES code directly; third, it allows direct control of all the features available in the unit; fourth, an API (application programming interface) is available in case a user does not want direct control of all the hardware; fifth, the code can be easily ported to C++ or other object-oriented languages once the industry moves in that direction. Therefore, overall, the SPHERES software provides re-configuration, upgradeability, generic and specific functions, and supports multiple investigators.

Communications - Each SPHERES unit uses two separate frequency communications channels with a data rate of 19.2 kbps. One channel is used for satellite-to-satellite (STS) communications; the other channel enables satellite-to-ground (STG) communications. Both channels are bi-directional; however, the communication hardware is half-duplex, meaning that only one unit can transmit at a time. This requires the implementation of a communications protocol to ensure that only one unit at a time will attempt to communicate, but also that all units are allowed to communicate when necessary. The developed system utilizes a time division multi-access (TDMA) network that uses packeted data. The TDMA protocol ensures that only one unit transmits data at a time. The packet format includes the addition of a header with origin, destination, length, and packet type information, plus a tail that contains checksum information. This type of packet allows binary data to be transferred easily, and allows for error detection at the receiving end.

The choice of two communications channels closely models the expected operations of future formation flying missions, creating a physical end-to-end simulation of a real mission. A high-bandwidth, low-power (short distance) communications link sends data between the units while in space. This link is used to send information in real-time, which is needed by the controllers in order to maintain formation. Due to the autonomy of the units, this channel does not require the power to establish communications with the ground, since only high-level commands/tasks will be sent from ground.

Ground communications links with formation flying missions are expected to be high-power and high-bandwidth, but have limited visibility and high latency. To correctly simulate this operation, the SPHERES testbed has a dedicated ground channel. For example, a separated spacecraft telescope may only be in communications with ground during the perigee of its trajectory, at which point it must send a burst of high-quality images to ground. While farther from earth, the ground link will not be utilized. The ground channel of SPHERES, though, is not limited in any way. The channel also serves to download the necessary telemetry information to the "ground" laptop computer to ensure that the researchers have the necessary information feedback. In a method similar to that proposed for the metrology system, software filters can be added to simulate a specific test of the program. For example, a software filter could ensure that commands are only sent every 5 minutes, rather than continuously, simulating a five-minute orbital period.

The laptop represents the ground station. A hardware wireless adapter connects to the serial port of the laptop to establish communications with the units. The use of a standard serial port ensures that the communications sub-system is brand-independent. That is, any laptop or PC computer can establish communications with the SPHERES units. The interface program is compatible with the PC operating system used by NASA in its ISS computers.

Laptop Interface - The SPHERES interface program contains all the necessary software to upload programs to a SPHERES unit, save telemetry information, present the user with the status of the units, start or abort a program, and reset the units remotely. Two different user interfaces are under development. A "research" interface, used by investigators on the ground or KC-135 testbeds, puts emphasis on the availability of telemetry information in real-time. In this way, the investigators can immediately see the telemetry data to determine correct operation of the testbed with better precision than simple observation. This facilitates the iterative research process by providing information to make changes as fast as necessary.

An "operations" GUI is used in the ISS and KC-135 tests. This GUI presents only the interface to operate the testbed; the telemetry data is still recorded, but not shown. In order to ensure good human observability and manipulation, this interface accounts for the fact that the operator is not the investigator. Therefore, this interface emphasizes the operation and status controls of the test-bed. By not distracting the operators with data that they don't necessarily understand or care about, the interface ensures that the test-iterations are each carried out as fast and easily as possible.

SCIENCE/OPERATIONS PLAN

A total of 10 two hour test sessions of crew time has been allocated for the SPHERES testbed onboard the ISS. Each test session will be recorded with two video cameras and the video feed will be made available within 4 days of the tests. Live video is available for the first two test session for the purpose of hardware check outs and setting up of the testbed onboard the ISS. SPHERES data will be downlinked within 24 hours of each test session and will be made available to Guest Scientists within 24 hours after that. In this section, key experiments that are to be carried out on the SPHERES testbed are outlined. This is followed by a brief discussion of the overall Guest Scientist Program.

Experiments

The SPHERES program is currently sponsored by DARPA and NASA. DARPA's interest in the program is to use the testbed to develop and test rendezvous and docking algorithms to possibly support the Orbital Express program. Sponsorship from NASA comes from three different organizations, Jet Propulsion Laboratory (JPL), Goddard Space Flight Center (GSFC) and Ames Research Center (ARC). JPL's interest is to develop formation flight algorithms to support NASA's TPF program, while GSFC's interest is to develop fundamental multiple spacecraft control algorithms. NASA ARC is interested developing on-line mass and inertia property identification onboard the SPHERES testbed. To satisfy the interests from the various funding sources, the following experiments, broken down based upon the number of SPHERES unit, are to be tested on SPHERES testbed.

Single Sphere Experiments - To provide an accurate dynamic model of the SPHERES testbed, the first experiments to be performed on the ISS is to determine the inertia properties of the SPHERES units. With the mass of the CO_2 propellant representing about 4% of the total mass of the SPHERES units, both the center of mass and the inertia properties of the SPHERES units are bound to change as propellant is expelled. As such, an on-line mass and inertia properties identification algorithm will be developed and implemented. One implementation is to use the gyroscope based algorithm developed by NASA Ames [6].

Since the CO_2 propellant is stored partially in liquid form in the tank, the effect of fuel slosh on the SPHERES units may become a concern. Using the accelerometers onboard the SPHERES units, it may be possible to detect and compensate for this disturbance. Even though the extent of this fuel slosh may be small, it may present a major problem when one considers precise formation flight experiments. Laser locks between the SPHERES units may be affected if one were to consider mounting high precision optics onboard the SPHERES units. Further studies will be performed to determine if the fuel slosh effect can even be detected with the SPHERES testbed.

Other single Spheres experiments include development of attitude control, position hold, trajectory tracking, de-tumbling and autonomy experiments. The objective of these experiments is to fine

Table 3 Docking scenarios [7]

	STS Comm	Servicing Vehicle	Target Vehicle
Cooperative	Yes	Full Control Authority	Full Control Authority
Semi-cooperative	Yes	Full Control Authority	Station Keeping Only
Un-cooperative	Yes	Full Control Authority	Free Drift
Non-cooperative	No	Full Control Authority	Free Drift

tune the controllers to ensure that the simple maneuvers can be carried out without fail. Once the appropriate controllers are determined, the goal is then to move on to two SPHERES experiments where more complicated maneuvers or operations can be tested.

Two Sphere (Rendezvous and Docking) Experiments - With DARPA being the major sponsor of the SPHERES program, at least 3/4 of all resources will be dedicated to the development and testing of rendezvous and docking algorithms. The types of docking scenarios that are to be tested with the SPHERES testbed is tabulated in Table 3.

In the *Cooperative* scenario, both the servicing and target vehicles have full control authority. State information will be passed between the two vehicles and they will work together to achieve docking. The *Semi-cooperative* scenario differs from the first scenario by allowing the target vehicle to only station-keep. However, full communication is still allowed between the two spacecraft. In the *Un-cooperative* docking scenario, it is assumed that the target vehicle has no control authority. The servicing vehicle will need to get state information from the target vehicle and determine the path to dock with it. The last scenario, *Non-cooperative*, is the most challenginng scenario. In this case, the target vehicle is considered non-operational. The servicing vehicle has to first determine the where the target vehicle is by using its onboard beacon and then determine the path to dock with it. In this case, no state information will be transferred between the vehicles. However, in order to carry out this experiment, the beacon on the target vehicle must be allowed to response to the infra-red pinging from the servicing vehicle.

In each of these docking scenarios, different variations can also be considered. In the case of *Cooperative* docking, the simplest is to face the docking ports of the two SPHERES units toward each other and initiate a slow collision docking. Additional challenges can involve 'throwing out' the servicing vehicle, as one would when being deployed from a payload fairing, thus requiring the servicing vehicle to first perform a de-tumbling maneuver prior to docking to the target vehicle. In the *Un-cooperative* scenario, one can also consider the target vehicle stuck in a nutating spin and the servicing vehicle has to dock with it. To make it more interesting, one can also put solar array panels on the target vehicle thus requiring the servicing vehicle to avoid collision with the panels with executing the docking maneuver.

Three Spheres (Formation Flight) Experiments - NASA JPL interests in the SPHERES testbed lies in the development of formation flight algorithm in support of NASA's TPF mission. Even though this first SPHERES experiment on the ISS do not have any of the fine optical control elements, the SPHERES testbed can be used to develop coarse formation flight control algorithms. Maneuvers representative of the a TPF formation flight system will be demonstrated with the SPHERES units. The maneuvers considered include:

- "Lost in space" acquisition - where a SPHERES unit executes an "orange-peel" trajectory to find the other unit using its onboard beacon
- Formation hold - where the three SPHERES units are held fix relative to each other in the presence of disturbance
- Array resize - increasing the relative separation of the SPHERES units while maintaining relative angular separation among them
- Array rotation - rotating the entire array from rest while maintaining relative angular separation

The ultimate goal of the formation flight experiments is to test the entire sequence of maneuvers with all three SPHERES units. To achieve this goal, all maneuvers will be tested with only one SPHERES unit prior to increasing to two and finally three SPHERES units. The purpose is to gain confidence in the algorithms by testing them on a SPHERES unit using as little resources as possible. Safety to the crew member must also be considered and by demonstrating the algorithms on a single SPHERES unit will allow us to predict the controller performance in the more complicated scenarios.

Technologies - Experiments on the SPHERES testbed can also be divided in terms of the technologies that are required to enable rendezvous, docking and formation flight of multiple satellite systems. The technologies of interest are the development of Metrology, Control and Autonomy algorithms.

The metrology class algorithms are made available through the standard interface. At the lowest level, the fusion of various sensors in the SPHERES unit to estimate the state of the spacecraft will be investigated. Since the global estimate of the SPHERES unit states can be determined, sensor outputs that would correspond to any metrology suite can be simulated.

Control class algorithms can be divided into two main groups. The first is the close-loop control algorithms that takes the state estimates provided by the Metrology class algorithms to determine the control actuations to reject disturbances. Controls topology ranging from distributed to centralized architectures can be tested. The second group is related to path planning algorithms. From the initial state estimates of the SPHERES unit, the control algorithms are required to determine the optimal (time or fuel) paths for the SPHERES units to achieve the desired goals.

At the highest level, autonomy algorithms are used to determine the mode of operation (type of docking and phase of approach), as well as to monitor the health status and to coordinate the execution of the plan generated by the Control class algorithms. Autonomy algorithms are the ones implementing the appropriate Control algorithms during the maneuver.

Guest Scientist Program

A main feature of a laboratory is the ability to conduct multiple research investigations in it. This is not only a feature, but an advantage. By allowing multiple programs to use the facilities, the cost of the laboratory is greatly reduced, since it is shared among the programs. Yet, in order to ensure that multiple projects can take place in a laboratory, a good program must be established. The SPHERES project has a Guest Scientist Program (GSP) to allow multiple investigators to use the SPHERES testbed, once deployed in the ISS, regardless of their location.

The program inherits many requirements from the laboratory design philosophy: feedback to the guest scientist must be precise, repeatable, and reliable; the guest scientist needs to know the generic equipment available to them, as well as the possibilities of using specific equipment; the program must be long enough to support extended investigations; and it must facilitate the iterative process. Therefore, the Guest Scientist Program was designed with the goal to create a true relationship between the operators at the MIT SSL, and the guest scientist elsewhere.

In order to ensure the success of the GSP, several steps have been created (Figure 8). The initial communications with a guest scientist include delivering the description of the SPHERES testbed, useful numerical data, a complete set of libraries to interface with the SPHERES hardware, and also a non-real-time simulation of the SPHERES testbed itself. The simulation allows the guest scientists to fully develop their algorithms in C code. Although not as precise as running an experiment in the testbed, the simulation is able to confirm that the code can compile correctly, is fully compliant with the SPHERES software environment, and can help detect initial instability problems.

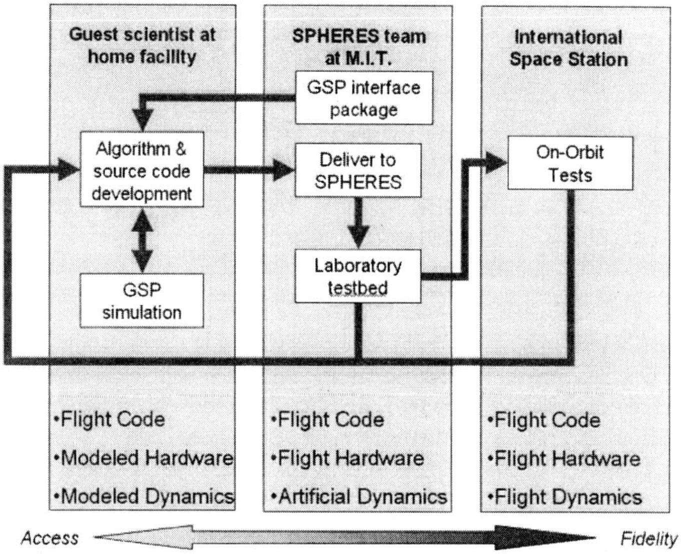

Figure 8 Overview of the Guest Scientist Program

The guest scientist can then send the C code to be tested in SSL SPHERES ground laboratory testbed. The SPHERES ground laboratory testbed, at the MIT SSL, can be operated at a very low operational cost. Therefore, after the code has been tested in the simulator, it can be run in the ground testbed. Results from these experiments will be similar to those expected from the ISS tests, limited to 2-D (3-DOF) data. Any problems discovered in the one-g testbed help lower the cost of the project by reducing the occurrence of errors in the ISS; due to the fast turn-around time of running tests in the ground testbed, this step adds a guarantee at a very low cost.

After the code has been validated in the ground testbed, the algorithms can be tested in the ISS. This process is expected to have a turn-around time of approximately one to two weeks, once SPHERES is operational in the ISS. The algorithms will be uploaded to the ISS via NASA channels, and placed in the ISS laptop computers. They are then uploaded to the SPHERES units during the next experiment set. The tests are run by the astronauts in the ISS, who save the data and then download it to ground. Data analysis occurs a few days after the test, with the ability to restart the iteration also within a few days.

By providing the guest scientist with multiple stages of feedback, the SPHERES GSP ensures that the projects facilitates the iterative research process even when the guest scientist is not on site. Further, since the project is to remain for extended periods of time at both the ground station and the ISS, extended investigations are supported. Finally, SPHERES allows the guest scientists to obtain the data they desire at different stages of development.

SPHERES AS A FORMATION FLIGHT TESTBED

Besides developing autonomous rendezvous and docking algorithms, future formation flight missions, such as interferometry missions, can stand to benefit from the development of the SPHERES testbed. Interferometry, and associated technology, suffers from a lack of flight heritage. Large system-level flight demonstrations suffer from an attempt to validate too many high risk technologies in a single step. Associated cost growth and schedule slippage lead to excessive conservatism. This is in direct conflict with the need to mature high risk yet high payoff technologies in a cost-effective and timely manner.

This section proposes a phased approach to formation flight technology maturation (SPHERES, SPHERES-Reflight, Exo-SPHERES) on the ISS where key technologies can be tackled in manageable amounts and in risk tolerant ways. SPHERES will test TPF array maneuvers, collision avoidance, and "lost-in-space" capture inside the ISS Node starting in July 2003, in addition to developing the rendezvous and docking algorithms. SPHERES-Reflight is proposed to augment SPHERES with laser metrology pointing control inside the ISS-Node as early as 2004. Exo-SPHERES takes this laboratory concept outside the ISS to further mature this technology at larger vehicle separations and in a more representative environment as early as 2006. ISS is selected as the base of operations because it can provide replenishment of consumables, free infrastructure (power, communications, etc.) and frequent access for repair and upgrade. The SPHERES laboratory philosophy is "if we cannot emulate space in the laboratory, let's emulate the laboratory in space." If interferometry and formation flight are our future for telescopes in space, it is time to start working in space.

Why a Phased Facility on ISS?

Control, navigation, and autonomy software is a system-level technology because it spans many sub-systems. This causes its development to be postponed to later stages of mission development when problems have substantial programmatic impact. Maturation and validation traditionally requires system-level flight demonstrations which are expensive and, by necessity, quite conservative.

Maturation and validation through ground-based testbeds and simulations is problematic. Simulations are "doomed-to-succeed" and all aspects of implementation are seldom fully captured. Hardware testbeds are constrained by gravity effects such as suspensions, constrained degrees of freedom, gravity sag and pre-loading, etc. This can often lead to either a lack of real hardware experience or the expenditure of significant resources towards solving hardware issues that are not relevant to the actual flight system.

There is a need to develop, mature, and validate command and control software under systems-level, space-environment conditions in a cost-effective, risk-tolerant manner. The ISS is the only facility that allows such development in a phased approach where reasonable amounts of technical risk are tackled at a time. Inside, ISS provides infrastructure such as video coverage, limited in-flight maintenance, high bandwidth (up)downlink, power, and replenishment of consumables. Operating inside the Node or US Laboratory also relaxes radiation and thermal tolerance requirements on hardware, allowing use of COTS parts, thereby maximizing the amount of limited resources that can be focused on advancing the technology. Operating outside ISS still provides frequent access, replenishment and upgrade but adds larger vehicle separations and access to an even more representative environment.

Such a facility can be used to conduct many tests. Data can be downlinked and new algorithms uplinked in the matter of days. New peripheral hardware can be launched and attached to the SPHERES for testing thereby increasing the test complexity/maturity in phases commensurate with the maturity of the software technology. This phased approach allows more rapid maturation of potentially high risk yet high payoff technologies because work can start now.

Phased Implementation

SPHERES-Reflight could occur as soon as programmatic support becomes available. The goal is to test optical metrology alignment and control on formation flown satellites in 3-D. The optical capture and control problem is challenged most by the 4π steradian angle uncertainty thereby motivating test in 3-D rather than on a flat floor. Three possible scenarios are envisioned.

First, the development of pointing and possibly phasing opto-mechanisms could be as soon as possible. Once available, they could be launched to ISS (assume March 2004). The SPHERES that will be launched in July 2003 have expansion ports that allow small payloads/instruments to be attached by the crew. This allows reflight to only entail launch of the pointing optics and extension of SPHERES' time on orbit. Second, new SPHERES could be built and launched to the ISS along with the new instruments. This entails a modestly higher expense and launch mass but enables reflight to be decoupled from first flight. The third would involve refitting the SPHERES hardware with the new pointing optics upon return of the hardware to the ground.

The goal behind Exo-SPHERES is to test formation flight in the actual space environment present in LEO. They would be launched under soft stowage and functionally testbed inside ISS. During a routine EVA, they would be placed on an external mounting location (e.g., Express Pallet). At a later time, tests would be initiated by the crew. They would be released, flown by the crew to a safe standoff position (~ 100m), autonomously flown to a test area roughly one kilometer behind ISS, tested and returned. Multiple tests can be conducted with downlink of data and uplink of new algorithms. The Exo-SPHERES would also contain expansion ports which would allow optical payloads (metrology and interferometer optics) to be attached at a later time.

SUMMARY

The SPHERES testbed fulfills all the characteristics of a successful laboratory environment, both as a whole and by each of its subsystems. The testbed provides a risk-tolerant environment capable of testing unknown algorithms and recovering from failures at low costs compared to that of full missions. By diminishing the cost of running multiple tests, the testbed facilitates the iterative research process, allowing researchers to be more aggressive in their designs. The program provides human observability and manipulation by both the guest scientists in a ground station, and operators in the ISS. Further, setup and operation of the testbed is simple enough that the environment can be repeated or changed in a controlled manner. The Guest Scientist Program allows multiple investigators to use the facility, regardless of location, further reducing the cost of new missions.

The design and component selection of each sub-system ensures the system is a physical end-to-end simulation of the intended programs. The units create a general equipment testbed which allows hardware reconfiguration of individual units for specific tests. The metrology and propulsion system provide the data feedback precision necessary to demonstrate new algorithms. Since all the expendables (propellant and battery power) are easily replenishable, the testbed allows for extended investigation in a risk-tolerant environment with direct human observation and manipulation. The communications system ensures that data are available both between the units and in the ground station. The data can be analyzed both in real-time and post-test. The software is reconfigurable and upgradeable, ensuring that the units can be used in multiple projects and with emerging technologies.

Experiments that are currently planned onboard the SPHERES testbed on the ISS beginning July 2003 include autonomous rendezvous and docking, system identification and formation flight experiments. System identification experiments are required to ensure the most updated dynamic models are available to the onboard controller. Docking experiments ranging from fully *cooperative* controlled collision to *un-cooperative* docking have been discussed. Formation flight algorithms representative of NASA's TPF mission will also be developed and tested onboard the ISS. Since these experiments are distributed over 10 two hour sessions, all these algorithms can be developed, tested, refined and re-tested over the entire SPHERES tenure on the ISS. This makes implementation of the Guest Scientist Program a viable option to involve other investigators into the SPHERES program.

The viability of evolving the SPHERES testbed for future formation flight experiments have also been discussed. The formation flight goals of this first SPHERES flight is geared towards development of coarse spacecraft control. The SPHERES-reflight, where optics are placed onto the Spheres expansion ports, will allow the development of staged control formation flight algorithms, a key technology required to perform separated spacecraft interferometry. Exo-Spheres will then take the testbed outside of the ISS to perform experiments at a scale larger than that allowed in the ISS, while still using the ISS as a base to operate the testbed.

ACKNOWLEDGEMENTS

This work was made possible in part by the support of DARPA Tactical Technology Office for SPHERES Autonomous Rendezvous Docking - Contract #29601-97-K-0010 under the supervision of Major James R. Shoemaker, PhD.

REFERENCES

[1] Alvar Saenz Otero, David W. Miller, and Mark Hilstad, SPHERES: Development of an ISS Laboratory for Formation Flight and Docking Research, IEEEAC paper #081, Big Sky, MT, March 2002.

[2] David W Miller, et al, "SPHERES Design Document", MIT SSL internal document , v2.0, November 1999.

[3] SPHERES Critical Design Review, http://ssl.mit.edu/spheres/Docs/,MIT SSL internal document, February 2002.

[4] SPHERES Fact Sheet, http://ssl.mit.edu/spheres/Docs/, MIT SSL internal document, December 2002.

[5] Hilstad, M., A Multi-Vehicle Testbed and Interface Framework for the Development and Verification of Separated Spacecraft Control Algorithms, SM. Thesis, Department of Aeronautics and Astronautics, Massachusetts Institute of Technology, June 2002.

[6] Wilson, E., Lages, C., and Mah, R., "On-line, gyro-based, mass-property identification for thruster-controlled spacecraft using recursive least squares," Proceedings of the 45th IEEE International Mid-west Symposium on Circuits and Systems, Tulsa, Oklahoma, Aug 2002.

[7] SPHERES Science Critical design Review, http://ssl.mit.edu/spheres/Docs/, MIT SSL internal document, November 2002.

AAS 03-014

LASER-BASED RELATIVE NAVIGATION AND GUIDANCE FOR SPACE SHUTTLE PROXIMITY OPERATIONS

Fred D. Clark,* Peter T. Spehar,* Jack P. Brazzel, Jr.† and Heather D. Hinkel‡

American manned spacecraft have used visual piloting techniques in the terminal phase of rendezvous during the Gemini, Apollo, Skylab, and Space Shuttle programs. In the last several years, space shuttle astronauts have used the Rendezvous and Proximity Operations Program (RPOP), running on a laptop computer, as a guidance and navigation aid during proximity operations. By processing measurements to the target satellite taken by a laser sensor, RPOP provides the shuttle crew with a more accurate relative position and velocity than from any other source. The inclusion of guidance algorithms allows RPOP to determine delta-velocities to fly very efficient, repeatable trajectories. This paper will focus on the guidance and navigation algorithms in RPOP, as well as results from simulation and flight. Although developed for shuttle proximity operations, the RPOP algorithms have potential applicability to an automated vehicle.

INTRODUCTION

The Space Shuttle crew has used the Rendezvous and Proximity Operations Program (RPOP) as a situational awareness tool for every rendezvous and docking flight since 1993 (the last 40 of the 47 shuttle rendezvous flights). RPOP's primary objective is to provide information necessary to assist the Orbiter pilot in meeting the operational flight constraints during approach to, and separation from, the International Space Station (ISS) and other vehicles. RPOP runs on a laptop computer in the Orbiter cockpit and provides the crew with information not available within the embedded Orbiter systems. Although RPOP could reside in the main Orbiter onboard computers, it is on the laptop because it is much cheaper to develop, certify, and maintain. RPOP receives raw sensor data and computes and displays relative navigation and guidance trajectory information to the crew both digitally and graphically. RPOP is also monitored in real time by personnel on the ground.

As the Orbiter approaches the ISS, several different measurement sources are used for navigation, including ground tracking, star tracker, radar, and lasers. RPOP simultaneously displays the solution from each navigation source, which allows the crew to evaluate and compare navigation solutions. At ranges less than 5000 feet, the most accurate navigation source available is based on laser data. The Trajectory Control Sensor (TCS) laser is located in the Orbiter's payload bay. The TCS takes measurements to a reflector

* Lockheed Martin Space Operations, 2400 NASA Road 1, Houston, Texas 77058.

† Boeing, HS&S, 13100 Space Center Blvd., Houston, Texas 77059.

‡ NASA Johnson Space Center, 2101 NASA Road 1, Houston, Texas 77058.

as well as accelerometer and attitude data from the Orbiter's inertial measurement unit (IMU) to compute very accurate estimates of relative position and velocity. The relative position and velocity from the TCS Kalman filter is the primary navigation information used by the Orbiter pilot during the manually-piloted proximity operations phase of flight.

In the last few years, several guidance and targeting algorithms have been developed and implemented in RPOP to augment the pilot's decision making during manually-piloted proximity operations. The purpose of the guidance algorithms is to improve trajectory control, reduce propellant usage, reduce piloting variability, and enable the pilot to more easily fly a tightly controlled trajectory. The guidance algorithms compute delta-velocities, which are then transformed into Orbiter translational hand-controller (THC) commands. RPOP displays the THC commands as recommendations to the crew.

RPOP's TCS Kalman filter, together with the Glideslope Approach, +V-bar Acquisition, and +R-bar acquisition algorithms have enhanced the pilot's ability to fly precise trajectories and have improved the probability of mission success. RPOP also includes out-of-plane guidance, but it is not discussed in this paper.

RPOP KALMAN FILTER DESCRIPTION

The Kalman filter in RPOP, which processes TCS measurements, is linear in its dynamics model, but the measurement equations are nonlinear. The state vector is comprised of three components for the relative position and three components for the relative velocity. The filter equations can be summarized as follows (ref. 1, 2):

TABLE I.- KALMAN FILTER NOTATION

$\mathbf{x}$	relative state vector
Φ	state transition matrix
$\mathbf{P}$	state covariance matrix
R	measurement variance
$\mathbf{Q}$	state process noise matrix
$\mathbf{H}$	partial derivative of measurement with respect to state
z	measurement
$h(\mathbf{x})$	*a priori* estimated measurement
$\mathbf{K}$	Kalman gain vector

State propagation: $\hat{\mathbf{x}}_{k+1}(-) = \Phi\hat{\mathbf{x}}_k(+) + \mathbf{G}\mathbf{a}_k$

Covariance propagation: $\mathbf{P}_{k+1} = \Phi\mathbf{P}_k\Phi^{\mathrm{T}} + \mathbf{Q}_k$

Gain Computation: $\mathbf{K}_k = \mathbf{P}_k(-)\mathbf{H}_k^{\mathrm{T}}[\mathbf{H}_k\mathbf{P}_k(-)\mathbf{H}_k^{\mathrm{T}} + \mathbf{R}_k]^{-1}$

State update: $\hat{\mathbf{x}}_{k+1}(+) = \hat{\mathbf{x}}_k(-) + \mathbf{K}_k[z_k - h(\hat{\mathbf{x}}_k(-))]$

Covariance update: $\mathbf{P}_k(+) = (\mathbf{I} - \mathbf{K}_k\mathbf{H}_k)\mathbf{P}_k(-)$

Here $\hat{\mathbf{x}}_k(-)$, $\hat{\mathbf{x}}_k(+)$ are the estimated state vectors at step k before and after the measurement, Φ is the state transition matrix, $\mathbf{G}$ is the matrix that maps thrust acceleration ($\mathbf{a}_k$) into the state vector, $\mathbf{P}_k$ is the error covariance matrix, $\mathbf{Q}_k$ is the process noise matrix, $\mathbf{K}_k$ is the gain vector, $\mathbf{H}_k$ is the measurement partial vector, $\mathbf{R}_k$ is the measurement covariance matrix, z_k is the measurement, and $h(\hat{\mathbf{x}}_k(-))$ is the estimated measurement.

The state $\hat{\mathbf{x}}_k$ is a 6 × 1 vector of estimated position and velocity components of the Orbiter relative to a target, in a rotating, target-centered local vertical-local horizontal (LVLH) frame. The state transition matrix (Φ) and the matrix, $\mathbf{G}$, are taken from the Clohessy-Wiltshire (C-W) equations. The process noise matrix is

$$\mathbf{Q} = K\begin{bmatrix} \frac{\Delta t^3}{3}\mathbf{I} & \frac{\Delta t^2}{2}\mathbf{I} \\ \frac{\Delta t^2}{2}\mathbf{I} & \Delta t\,\mathbf{I} \end{bmatrix}$$

where $\mathbf{I}$ denotes the 3 × 3 identity matrix and Δt is the propagation time. The value K is one of two specified constants, depending on whether the change in accumulated sensed velocity from the IMU exceeds a threshold.

TCS measurements of range, range rate, azimuth, and elevation are processed sequentially as scalar measurements. TCS has two modes. At ranges greater than about 1000 ft, it tracks in pulse mode. At ranges less than about 1000 ft, it tracks in continuous wave (CW) mode. The CW mode is much more accurate, especially for measurement of range and range rate. RPOP's Kalman filter automatically adjusts the filter variances appropriately depending on the current TCS tracking mode.

The RPOP Kalman filter can be configured with four additional states to estimate the bias in the measurements of range, range rate, azimuth, and elevation. We have found, however, that little performance improvement is gained by inclusion of these states. Therefore, the operational filter uses only six states.

The RPOP filter accounts for the $\Delta\mathbf{v}$ during the navigation propagation cycle. The accumulated sensed velocity (in Mean-of-1950 inertial coordinates) from the Orbiter's IMU, and the M50-to-Orbiter body quaternion are obtained from the Orbiter's main flight computers. The transformation from M50 coordinates to LVLH coordinates is computed. The IMU sensed $\Delta\mathbf{v}$ is only included in state propagation if the magnitude of the $\Delta\mathbf{v}$ is greater than a threshold to prevent incorporation of IMU bias.

TCS MEASUREMENT EDITING AND REFLECTOR IDENTIFICATION

The RPOP filter employs a standard residual editing capability to reject measurements that are seriously in error. In actual practice, the TCS rarely produces such a measurement.

There are multiple reflectors on the ISS. Since TCS is capable of tracking only one reflector at a time, TCS sometimes instantaneously changes the reflector that it is tracking, causing a large *apparent* error. When a range, azimuth, or elevation measurement is rejected, the reflector identification algorithm is called to determine whether it appears that TCS has switched the reflector that it is tracking. The change in the measurement is compared to all possible changes that would occur if the TCS switched reflectors. If a match is found, the measurement is reprocessed through the filter using the new reflector location; otherwise, the measurement is rejected. Before reprocessing, the first three elements of the diagonal of the covariance matrix are increased by a specified constant. This ensures that the measurements will not be rejected when they are reprocessed.

FILTER PERFORMANCE

TCS tracked the Russian space station Mir at ranges as great as 7000 ft. Because of reflector obscuration issues, TCS typically tracks the International Space Station (ISS) at ranges less than about 5000 ft.

After STS-79 we computed a Best Estimated Trajectory (B.E.T.). We compared the estimates from the Kalman filter to the B.E.T. values. Figure 2 shows the difference in the range rate computed by the Kalman filter and the range rate derived from the STS-79 B.E.T. as a function of range. The superiority of TCS measurement precision in CW mode is immediately evident from figure 2. The data shown between the two vertical lines at a range of about 1000 ft contains instances of both CW and pulse-mode tracking. After the right vertical line, the TCS remained in CW mode for the remainder of the approach to Mir.

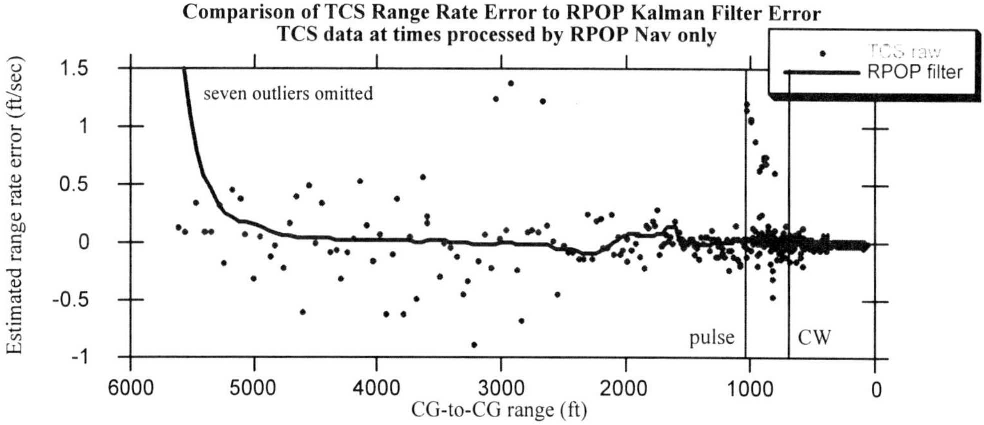

Figure 1.- Comparison of raw TCS range rate error and RPOP filter range rate error; STS-79

Figure 3 shows the RPOP Kalman filter residuals that were obtained on STS-112, which docked with the ISS in October 2002. The solid lines shown in these plots depicts the Kalman filter's estimates of the standard deviation (1σ) of the residuals. The vertical lines at about 154,500 sec mission-elapsed time (MET) were caused by TCS changing the reflector that it was tracking. When this occurs, additional process noise is added to ensure measurement incorporation when a reflector switch is identified, resulting in the one-sigma estimate being off the scale.

Also evident in figure 3 is the difference in the variances specified in the filter depending on whether the TCS is currently tracking in pulse mode or CW mode. This is especially obvious in the range rate plot where the estimated 1σ lines are essentially covered by the residuals.

The transient in the range and elevation residuals that occurred slightly after 153,000 sec (MET) were caused by the midcourse 4 (MC4) maneuver.

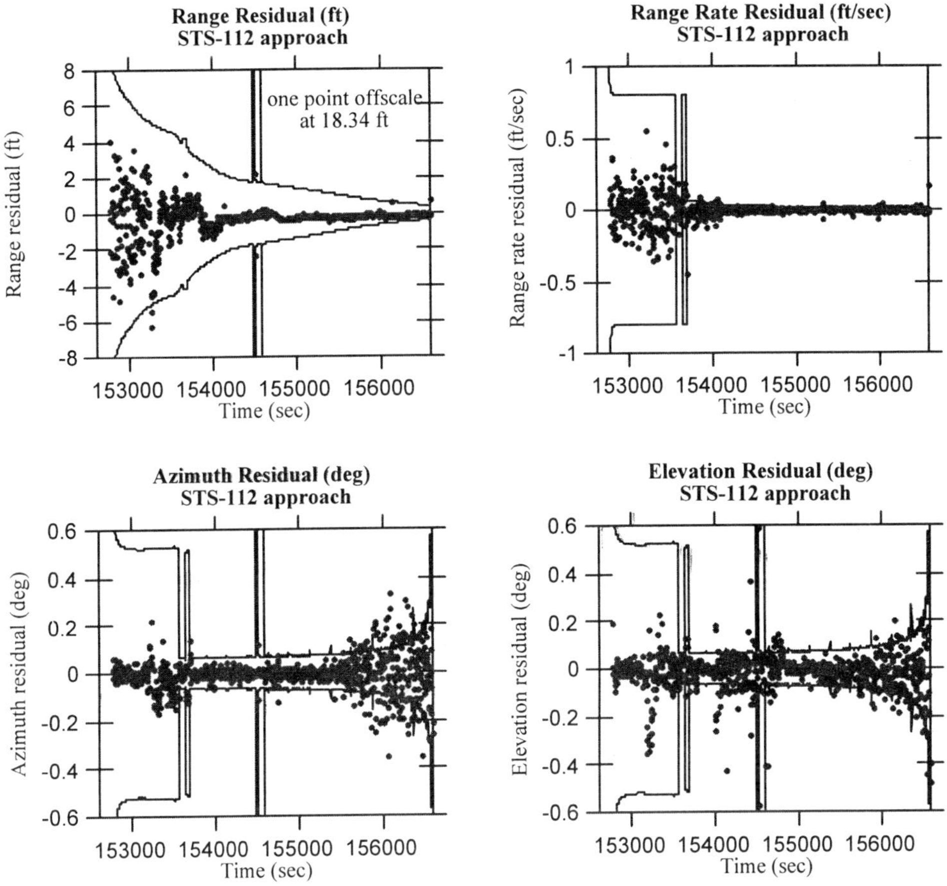

Figure 2.- Measurement residuals; STS-112

RPOP GUIDANCE ALGORITHMS

The manual phase of a space shuttle rendezvous is shown in figure 1. After the MC4 maneuver the Orbiter nominally reaches the R-bar 600 ft below the target at a range rate of –0.8 ft/sec. Then the Twice Orbital Rate V-bar Approach (TORVA) is begun, which takes the Orbiter to the vicinity of the V-bar approximately 350 ft from the target. The V-bar approach portion is usually flown as a glideslope approach at an angle between zero and five degrees above the V-bar.

The glideslope approach, V-bar acquisition, and R-bar acquisition algorithms are described in subsequent subsections in that order.

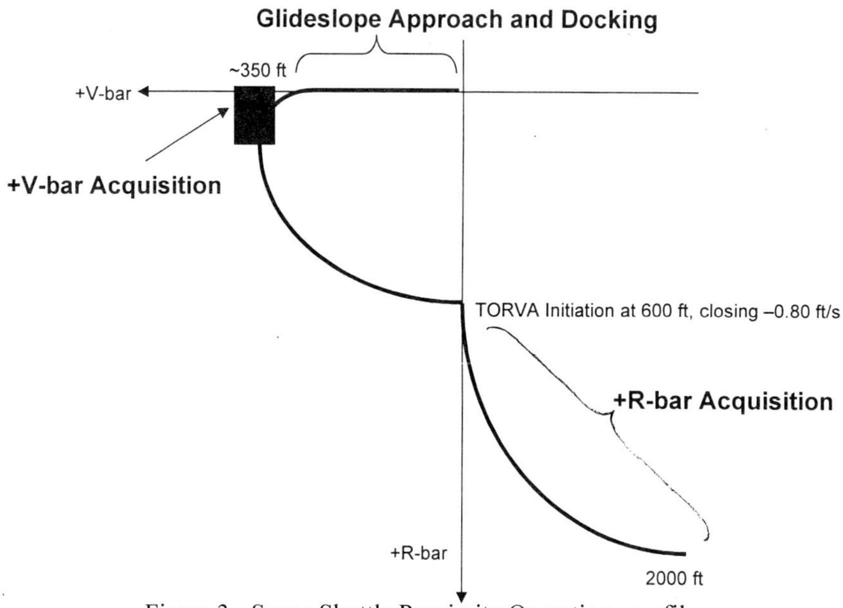

Figure 3.- Space Shuttle Proximity Operations profile

THE GLIDESLOPE TRAJECTORY

Before discussing glideslope approach targeting, we present some background on glideslope trajectories. We would like to determine the range and range rate as a function of time along a glideslope trajectory.

We assume that thrusting is applied continuously. If an approach can be done using only the aft jets, plume impingement of the target is minimized. Ideally, we should be able to translate along a glideslope path with range rate slowly falling (in absolute value), caused by the braking effect of the Orbiter's canted tail jets.

After transforming the differential equations of relative motion to polar coordinates r (range) and θ (constant glideslope angle), and assuming only tail jets are used to control the approach, we may obtain the differential equation[1]

$$\ddot{r} + (2\omega \tan\alpha)\dot{r} - [3\omega^2 \sin\theta\,(\tan\alpha \cos\theta + \sin\theta)]\,r = 3\omega^2(\tan\alpha \cos\theta + \sin\theta)\,z_{offset}$$

where ω denotes the orbital angular velocity, α is the canting angle of the Orbiter's tail jets, and z_{offset} is the vertical distance between the CGs of the two vehicles at docking. From the solution to this equation we may obtain the range rate required at a given range to accomplish the approach firing tail jets.[2]

In reality, it is not possible for the space shuttle Orbiter to follow the glideslope trajectory precisely, because it cannot thrust continuously. The Orbiter's minimal delta-velocity is about 0.02 ft/sec. But, the Orbiter's deviation from a glideslope trajectory can be made surprisingly small.

---

[1] A derivation of the homogeneous form of this differential equation is given by Pearson (ref. 3).

[2] A separate solution is obtained for the ±V-bar cases to avoid a singularity.

The canting angle of the Orbiter's aft jets (α) is about 10 deg. Translational and rotational cross-coupling tends to cause an undesired acceleration toward the target. To account for this, RPOP is configured with the jet cant angle smaller than 10 deg, resulting in guidance assuming the tail jets produce a smaller braking component from each firing.

The previous development can be easily extended to the use of nose-firing jets for a tail-forward R-bar approach, a tail-up V-bar approach, or for separation, but we will not discuss these here.

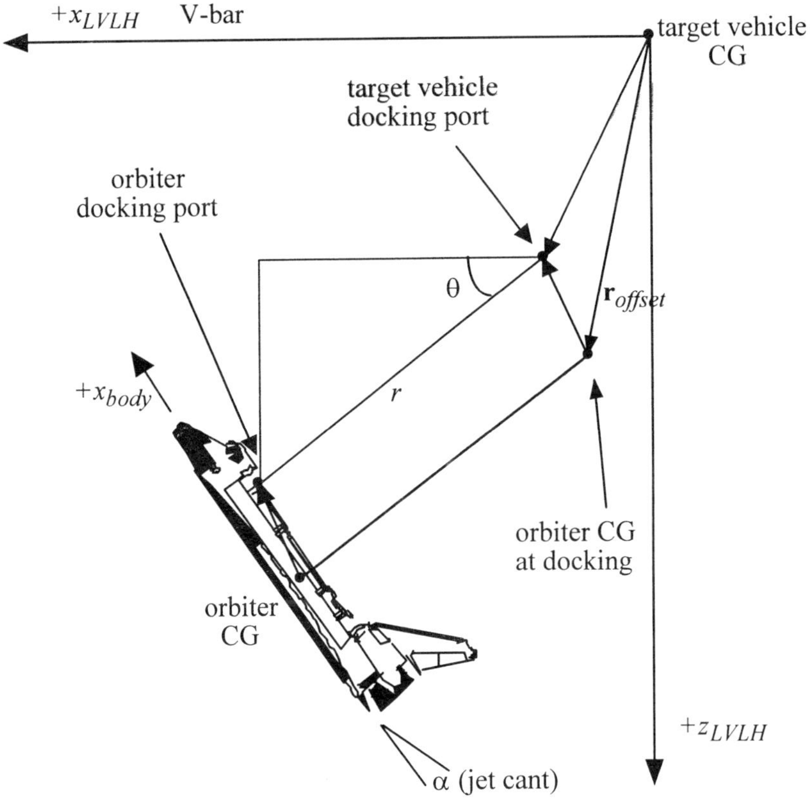

Figure 4.- Glideslope geometry

THE GLIDESLOPE TARGETING ALGORITHM

An algorithm can be derived to target to the glideslope line, leaving the arrival point on the glideslope line unconstrained, and forcing the range rate after the maneuver to equal the desired range rate at the current range. Such an algorithm for V-bar approaches (and acute angles from the V-bar) was developed by William L. Jackson (formerly of JSC) and Quan Le (of JSC) during the mid-1980s, assuming ideal rectilinear motion (ref. 4). The authors of this paper have extended this glideslope approach algorithm using the Clohessy-Wiltshire equations in a glideslope coordinate system. The advantage of this formulation is that it is more accurate, and a single set of targeting equations can be used for any linear in-plane approach.

The transfer time is fixed for a given maneuver, but it is generally specified as a smaller value the closer we are to the target. The smaller the time between maneuvers, the more precisely the in-plane trajectory is controlled. RPOP guidance computes the transfer time from a piecewise linear function of range based on values that are initialization constants.

For a V-bar approach, the glideslope angle (θ) is defined to be zero. In the case of an R-bar approach, the glideslope angle is defined to be 90 deg. Figure 5 shows the results of simulated V-bar and R-bar approaches.

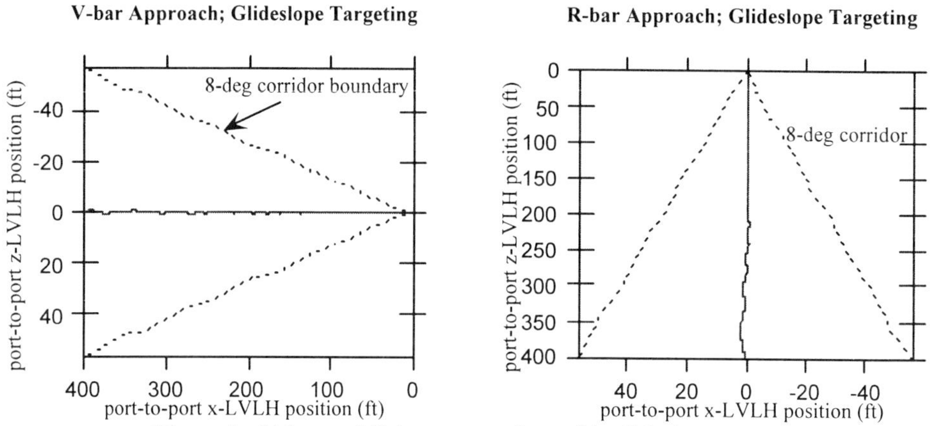

Figure 5.- V-bar and R-bar approaches with glideslope targeting

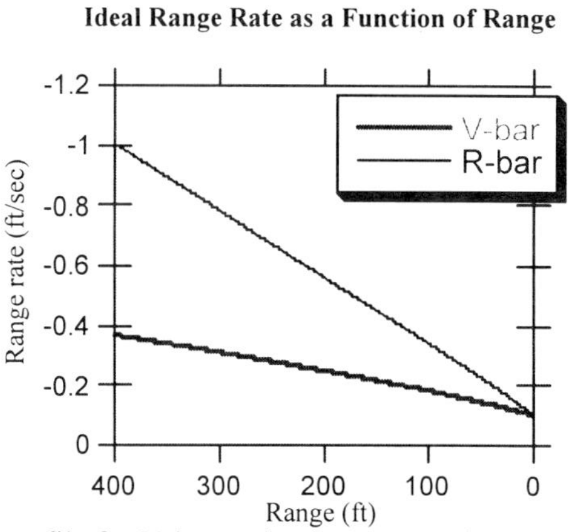

Figure 6.- Range rate profile for V-bar and R-bar approaches; final range rate = –0.1 ft/sec

Figure 6 shows the range rate profile for V-bar and R-bar approaches for a desired range rate of –0.1 ft/sec at docking. Notice that at a given range the range rate is higher for the R-bar approach than for the V-bar approach. This is because natural braking reduces range rate at every point along the R-bar.

Figures 7 through 11 are from man-in-the-loop simulation runs with and without RPOP guidance. In each of these figures the graphs on the left display the results without RPOP guidance, whereas those on the right display the results with RPOP guidance. Figure 7 shows that the trajectory dispersions are much smaller in the cases run with RPOP guidance.

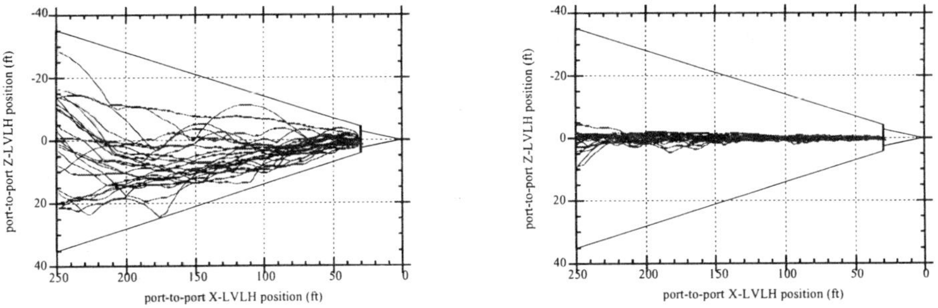

Figure 7.- In-plane relative motion from 250 ft to docking, without and with RPOP guidance

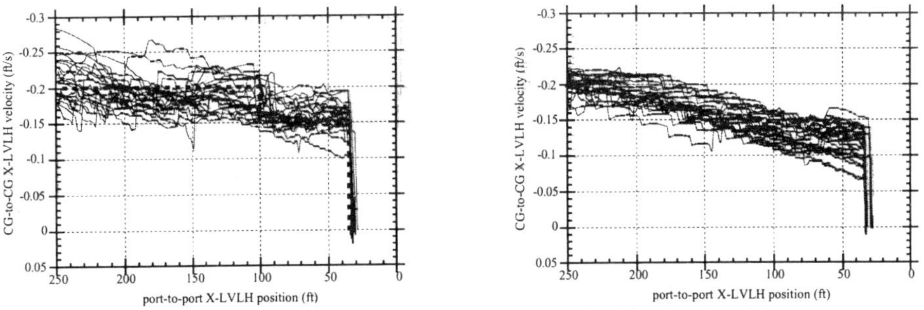

Figure 8.- Range rate vs. range, without and with RPOP guidance

Figure 8 shows that the variability of range rate as a function of range is also smaller when run with RPOP guidance than without it.

Figure 9 displays the view from the centerline camera at the end of the simulations. As one would expect, there is less variability in the cases run with RPOP guidance.

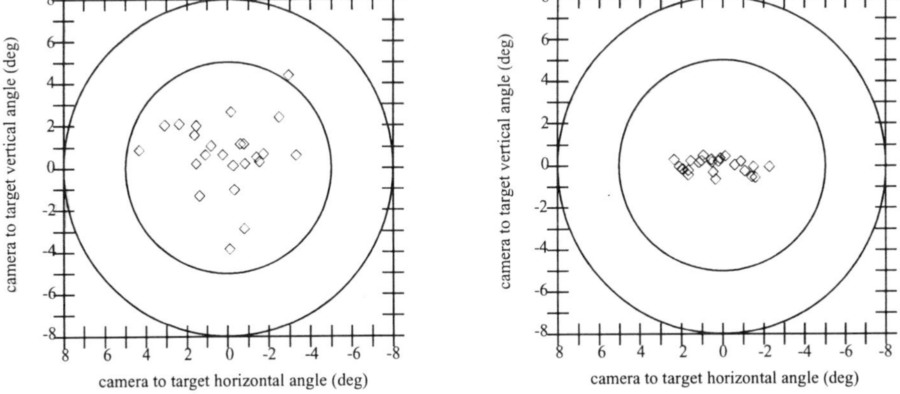

Figure 9.- Final position in centerline camera view, without and with RPOP guidance

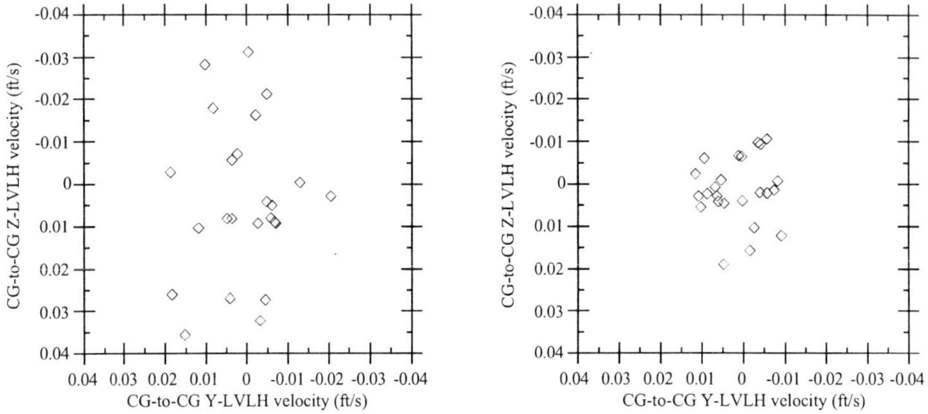

Figure 10.- Final lateral velocities, without and with RPOP guidance

Figure 10 shows the final lateral velocities. Once again, RPOP guidance reduces the variability.

Figure 11 shows propellant consumption without and with RPOP guidance. The mean fuel usage is clearly less while using RPOP guidance—close to the theoretical minimum. Moreover, the variability is also very much smaller than when guidance is not used. The standard deviation of propellant usage with RPOP guidance is only 2 lb.

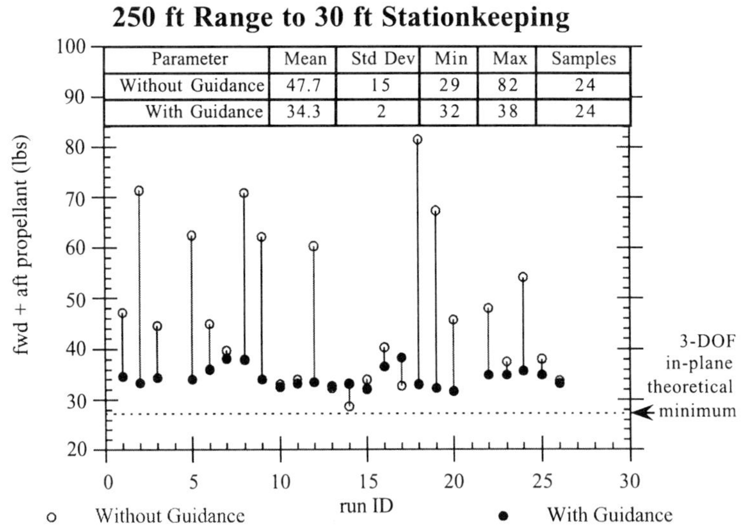

Parameter	Mean	Std Dev	Min	Max	Samples
Without Guidance	47.7	15	29	82	24
With Guidance	34.3	2	32	38	24

Figure 11.- Propellant comparison without and with RPOP guidance

V-BAR-ACQUISITION GUIDANCE

Current procedures call for the Orbiter to leave the R-bar at about 600 ft below the target with approximately 1 ft/sec downrange velocity and range rate of approximately –0.8 ft/sec. This results in the Orbiter reaching the V-bar with the trajectory nearly orthogonal to the V-bar. To reach the V-bar more efficiently, we developed an algorithm that puts the docking port of the Orbiter on a trajectory which is tangent to the approach line. (The approach line is parallel to the V-bar for the case that the docking port is not at the CG of the target.) We also desire that when the Orbiter reaches the approach line it have precisely the range rate that is required by the glideslope controller.

Figure 12 shows three trajectories. Each of these begins on the V-bar with the range rate required by the glideslope controller. If we propagate backward in time, we obtain the coasting flight trajectories shown in the figure. If we were to start on the ending points of the back-propagated trajectories and propagate forward in time, we would arrive at the approach line. Notice that the locus of points at which the slope on the trajectory is vertical appears to form a straight line. In fact, it can be proved that the locus of points is exactly a straight line so long as the range rate at the approach line is a linear function of range and the CGs of the vehicles are aligned. We call this locus of points the *V-bar acquisition line*.

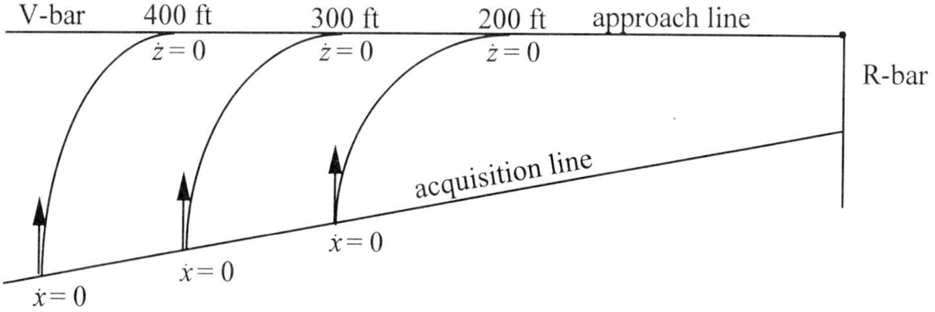

Figure 12.- Acquisition line

In the case that the CG of the Orbiter is on the V-bar when the docking ports are aligned, the transfer time between the acquisition line and the approach line is constant for any point on the acquisition line. For the general case, we may still compute very accurate estimates of the transfer time. We can also determine the required velocity at the acquisition line to intercept the approach line tangentially. Remarkably, the required radial velocity for this maneuver is surprisingly simple:

$$\dot{z} = -3\,\omega z \tan \omega t$$

where z is the radial coordinate on the acquisition line, ω is the orbital angular velocity, and t is the transfer time.

As the Orbiter approaches the acquisition line, the predicted point that the current trajectory will intersect the acquisition line is computed each second. (This requires iteration, but it converges rapidly.) If the trajectory does not intersect the acquisition line, there is no solution.

Figure 13 shows a set of trajectories acquiring the V-bar without and with RPOP V-bar acquisition guidance. Note how much smaller the trajectory dispersions are at V-bar arrival. Figure 14 shows the amount of fuel used without and with the use of RPOP V-bar acquisition guidance. Clearly, the mean propellant used is less when using guidance. The variability is not reduced as much as in the glideslope approach phase (see figure 11), because there is much more variability in the pre-V-bar acquisition initial conditions than in the post-V-bar arrival initial conditions.

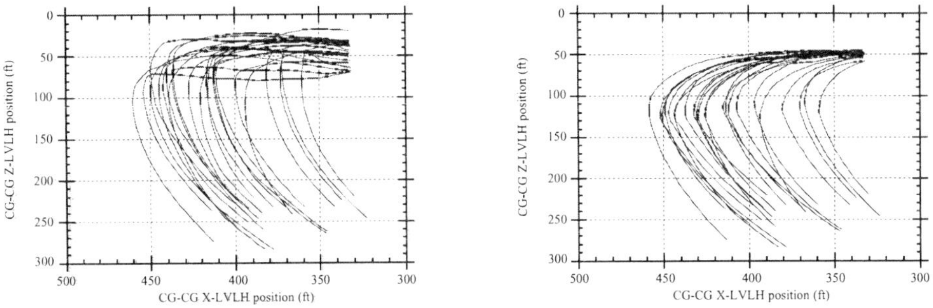

Figure 13.- V-bar acquisition maneuver without and with RPOP guidance

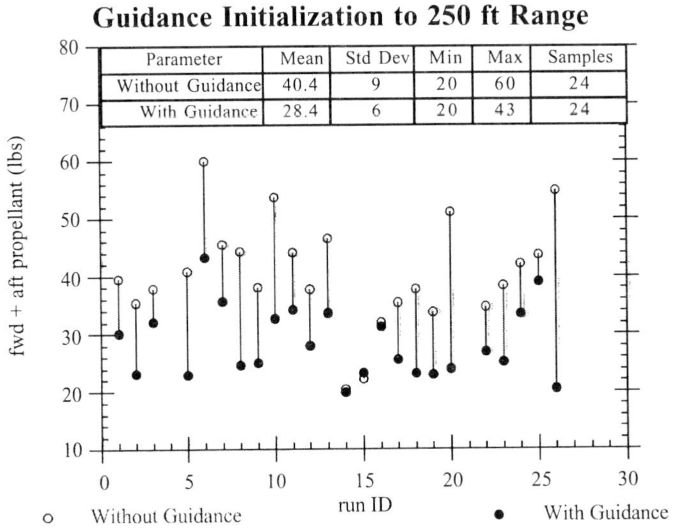

Figure 14.- Propellant comparison without and with RPOP V-bar acquisition guidance

R-BAR ACQUISITION GUIDANCE

The midcourse 4 (MC4) point of the current shuttle rendezvous profile is at $x = -900$ ft, $z = 1800$ ft, where x and z denote the downrange direction and opposite the radial direction in LVLH coordinates. The MC4 maneuver is targeted to $(x, z) = (0, 600)$ ft with velocity $(0, -0.8)$ ft/sec. The initiation of the subsequent Twice Orbital Rate V-bar Approach (TORVA) assumes that these conditions have been achieved. Analysis of data from several flights shows that there is often an undesired component of velocity in the horizontal direction at R-bar arrival. R-bar acquisition targeting is designed to assist the commander in reaching the desired range rate of –0.8 ft/sec at a range of 600 ft on the R-bar.

There is a unique reference trajectory between the nominal MC4 point and the desired condition at (0, 600). If the Orbiter is on the reference trajectory and a burn solution is computed with the transfer time necessary to traverse the reference, the velocity at arrival must also be correct, even if it is not explicitly constrained.

After the MC4 maneuver, ideally the Orbiter is on the reference trajectory. In reality, because of trajectory dispersions, navigation errors, and maneuver trim errors, it is unlikely that the Orbiter will be on the reference trajectory.

R-bar acquisition guidance works as follows: For the current *z*-coordinate, the position in *x* on the reference with the same vertical offset from the target is computed. Knowing the current *x*-LVLH position, and the *x*-position on the reference, we compute a deviation (x_{dev}). We use point-to-point C-W targeting to target to a point x_{dev} from the final point.

This algorithm is executed cyclically. After a few seconds, the new position will be closer to the reference. The next targeting calculation aims for a point closer to the final condition. This slowly "pulls" the trajectory toward the reference. This is shown graphically in figure 15, where for purposes of illustration two maneuvers are shown. In practice several maneuvers would be executed. (But, if the commander desired to do so, he or she could execute a two-impulse solution using R-bar acquisition guidance by waiting until arrival on the reference trajectory to execute the second maneuver.)

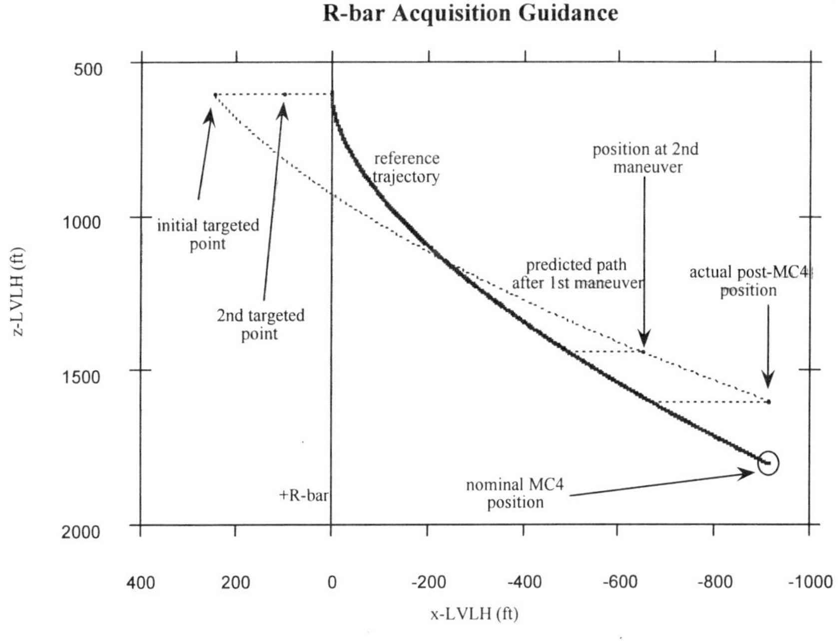

Figure 15.- Iterative use of C-W targeting for R-bar acquisition

R-bar acquisition guidance has been in RPOP since STS-112. The left graph of Figure 16 shows the MC4-to-R-bar acquisition trajectories that were flown on ten shuttle flights (STS flights 92, 97, 98, 100, 101, 102, 104, 105, 106, and 108). The graph on the right shows the trajectories from man-in-the-loop simulations flown with RPOP guidance, and whose initial conditions were taken from the flight data. Clearly, use of R-bar acquisition guidance has potential for reducing trajectory dispersions.

Currently, R-bar acquisition guidance is certified for use for nominal operations. It has potential applicability to a mission for which the rendezvous radar has failed. We hope to extend R-bar acquisition guidance for this scenario in the near future.

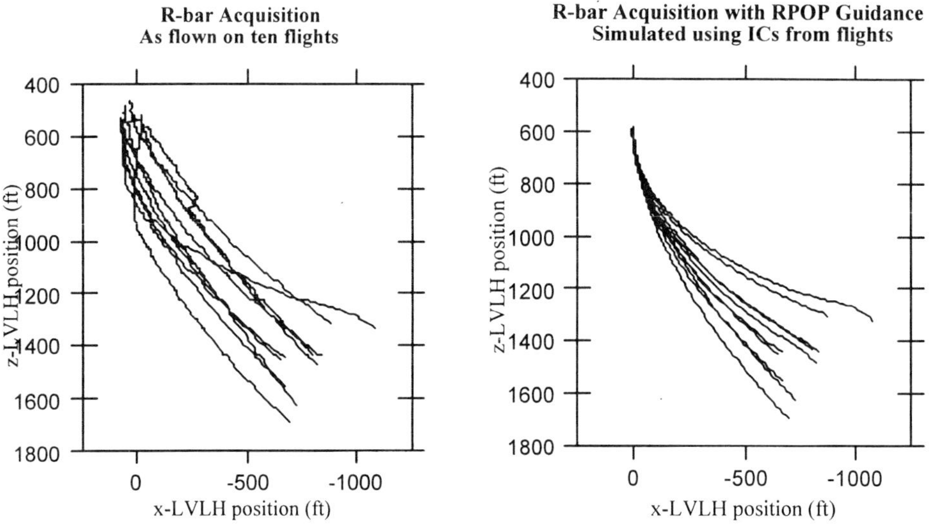

Figure 16.- R-bar acquisition trajectories without and with RPOP guidance

RPOP USER INTERFACE

A picture of the RPOP display is shown in figure 17. The relative state vector in local vertical-local horizontal (LVLH) coordinates is displayed in the top-center of the screen, with raw TCS measurements displayed to the right. If desired, TCS Kalman filter residuals are displayed in the lower left of the screen (not shown in figure 17).

The floating window in the left of figure 17 shows the translational hand controller (THC) recommendations computed by RPOP guidance. An "up" arrow denotes a +x-body axis THC deflection. The number of THC deflections in shuttle digital autopilot (DAP) modes A and B are also shown, with the currently selected mode in bold. A digital and graphical representation of the time remaining until the maneuver is shown in the left of this window.

In the center of the screen is a depiction of the current position of the Orbiter relative to the target. This can be displayed in CG-to-CG coordinates or docking port-to-docking port coordinates. Out-of-plane motion may also be displayed. The numbers 1, 2, 3, ..., 9 depict the prediction of future relative motion if no maneuvers were done, whereas the "+" signs show the predicted trajectory if the current maneuver recommendations are executed.

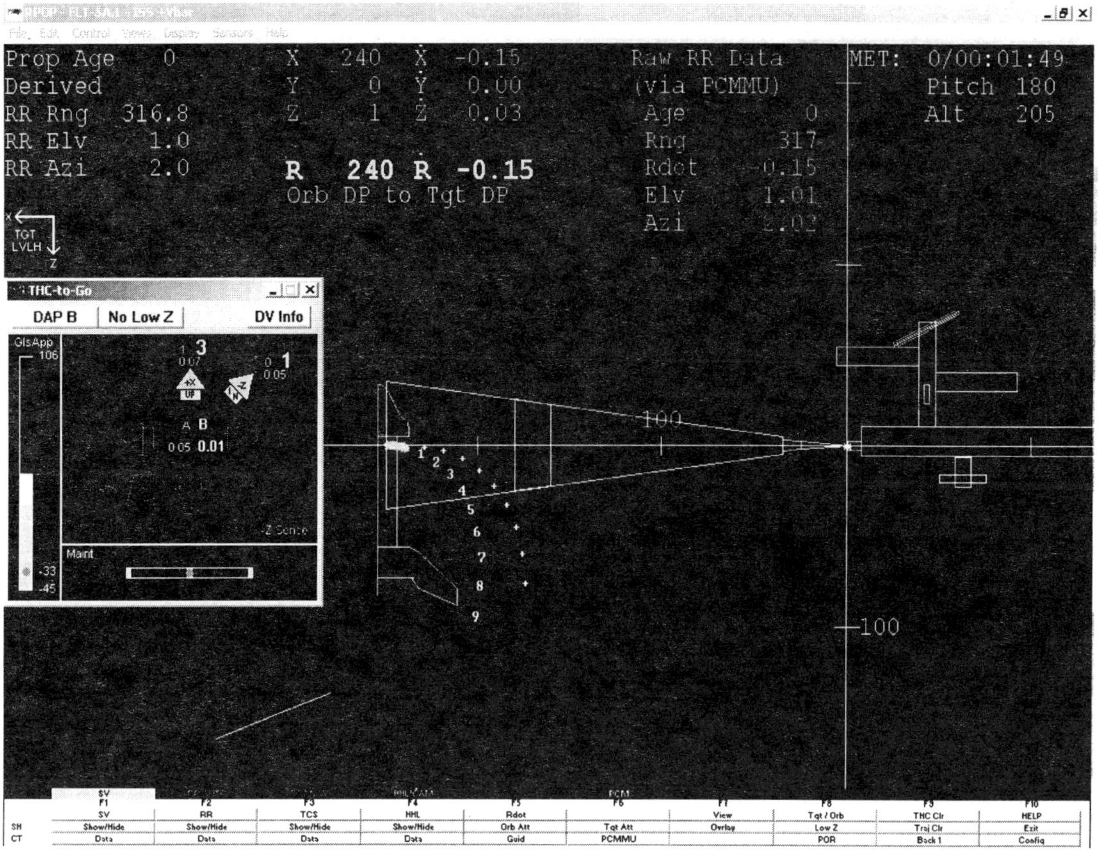

Figure 17. RPOP display

CONCLUSION

The combination of the TCS and RPOP provides the space shuttle crew with the most accurate relative state vector available. Navigation and proximity operations guidance algorithms assist the crew in reducing propellant consumption and reducing trajectory dispersions, while meeting operational constraints.

The guidance and navigation algorithms are not limited to use in manned applications; they have applicability to automated rendezvous and docking, as well. We have already demonstrated a simulation of automated proximity operations by feeding the RPOP maneuver recommendations directly to a simulation. In addition, the value of the algorithms has already been proven in flight.

REFERENCES

1. Gelb, A., Applied Optimal Estimation, M.I.T. Press, 1974.
2. Maybeck, P. S., Stochastic Estimation and Control, Vol. I, Academic Press, 1982.
3. Pearson, D. J., "The Glideslope Approach," (AAS 89-162), *Orbital Mechanics and Mission Design, Advances in the Astronautical Sciences,* Vol. 69, pp. 109–123, Proceedings of an AAS/NASA international symposium, April 1989.
4. Le, T. Q., "Prox Ops Control," NASA Johnson Space Center memo EH2-86M-151, June 30, 1986.

AAS 03-015

DESIGN, DEVELOPMENT AND TESTING OF THE MINIATURE AUTONOMOUS EXTRAVEHICULAR ROBOTIC CAMERA (MINI AERCAM) GUIDANCE, NAVIGATION, AND CONTROL SYSTEM

Jennifer Wagenknecht, Steven Fredrickson, Tom Manning and Brandon Jones*

Engineers at NASA Johnson Space Center have designed, developed, and tested a nanosatellite-class free-flyer intended for future external inspection and remote viewing of human spaceflight activities. The technology demonstration system, known as the Miniature Autonomous Extravehicular Robotic Camera (Mini AERCam), has been integrated into the approximate form and function of a flight system. The primary focus has been to develop a system capable of providing external views of the International Space Station. The Mini AERCam system is spherical-shaped and less than eight inches in diameter. It has a full suite of guidance, navigation, and control hardware and software, and is equipped with two digital video cameras and a high resolution still image camera. The vehicle is designed for either remotely piloted operations or supervised autonomous operations. Tests have been performed in both a six degree-of-freedom closed-loop orbital simulation and on an air-bearing table. The Mini AERCam system could also be used as a test platform for evaluating algorithms and relative navigation for autonomous proximity operations and docking around the Space Shuttle Orbiter or the ISS.

INTRODUCTION

Engineers at NASA Johnson Space Center have designed, developed, and tested a nanosatellite-class free-flyer intended for future external inspection and remote viewing of human spaceflight activities. The technology demonstration system, known as the Miniature Autonomous Extravehicular Robotic Camera (Mini AERCam), has been integrated into the approximate form and function of a flight system.

Scope

This paper provides an overview of the Mini AERCam vehicle design, and a detailed description of the GN&C system design, development, and testing.

---

* Authors Jennifer Wagenknecht and Steven Fredrickson are with the NASA Johnson Space Center, and authors Tom Manning and Brandon Jones are with Titan Corporation. All correspondence should be sent to Jennifer Wagenknecht, NASA Johnson Space Center, Mail Code EG2, Houston, Texas 77058.
E-mail: jennifer.d.wagenknecht@nasa.gov. The Mini AERCam Public website is http://aercam.nasa.gov/.

Background

The predecessor to Mini AERCam, called AERCam "Sprint," was flight tested on STS-87 in December 1997. As shown in Figure 1, Sprint was released in the Space Shuttle Orbiter Payload Bay by astronaut Winston Scott during an extravehicular activity (EVA). Astronaut Steve Lindsay controlled the free-flyer using a combination rotational and translational hand controller from inside the Orbiter aft cockpit. Lindsay was able to see Sprint at all times through the Orbiter windows, and also monitored the video on an internal display. The flight test on STS-87 lasted about an hour and fifteen minutes and was highly successful.

Figure 1 AERCam Sprint Release on STS-87

Sprint was approximately fourteen inches in diameter and weighed about thirty-five pounds. The propulsion system consisted of twelve cold gas thrusters with nitrogen as the propellant. Sprint had angular rate gyros for providing angular rate feedback and automatic attitude hold. The video camera could be switched between two focal lengths for a regular or "telephoto" view. A light was available for illumination. Sprint had a soft covering for impact protection.

From 1997 to 1998, JSC engineers designed, developed, and integrated a ground-test version of AERCam for the "AERCam Integrated Ground Demonstration (IGD)." The AERCam IGD, conducted on an indoor air-bearing table, was intended to demonstrate the availability of advanced technologies that could be used to enhance AERCam. The GN&C technology advancements were a significant departure from the limited GN&C capabilities on Sprint. Then in 1998, a crew evaluation was conducted in JSC's Virtual Reality Laboratory for the purpose of evaluating the suitability of Sprint for use as an operational vehicle outside the International Space Station. The study by nine astronaut participants yielded a set of recommendations for how to improve safety, situational awareness, and handling qualities for a future version of AERCam[1].

In 2000, Mini AERCam development began with the goal of reducing free-flyer size while simultaneously adding capabilities that would contribute to overall system controllability, reliability, and utility. For situational awareness, it was desired to add an additional camera for an orthogonal view, as well as relative navigation so that crewmembers would know the location of the free-flyer with respect to the complex ISS structure. To improve the handling qualities and reduce crew workload, engineers would incorporate automatic position hold and point-to-point maneuvering. By 2002, a fully functional prototype had been designed, developed, integrated and tested as described below in the following sections.

Challenges for the Mini AERCam Design Team

The Mini AERCam design team was challenged by NASA management to design a version of Sprint, with advanced capabilities and within a diameter of eight inches. The team was also directed to select advanced technologies that were at the near edge of the technology envelope, such that if they were not available yet, they would be mature in time for a flight experiment within a few years. Finally, the design team was asked to design the ground demonstration system without making significant compromises such that each subsystem was only one step away from a flight configuration. These challenges were met as described in the following sections.

MINI AERCAM VEHICLE SYSTEMS

The Mini AERCam free-flyer is spherical-shaped and approximately 7.5 inches in diameter. Figure 2 shows a size comparison between AERCam Sprint and Mini AERCam. Mini AERCam achieves its 'nanosatellite' size by a combination of dense packaging and miniaturized components.

Figure 2 AERCam Sprint and Mini AERCam

On Mini AERCam, two digital cameras are aligned with the +X direction of the vehicle. One of these cameras provides National Television System Committee (NTSC)-quality color video, and the other camera can be used for high-resolution still images, when selected. A third color video camera is positioned in the +Y direction for an orthogonal view. The Mini AERCam prototype is designed to have two custom GPS antennas, one on top and one on the bottom. An array of Light Emitting Diodes (LEDs) provides illumination. A communications antenna is located near the top GPS antenna.

As shown in Figure 3, the vehicle is designed with a central ring that houses the power and propulsion system. The batteries are lithium-ion and provide at least four hours of operational time. The propulsion system is designed for cold-gas xenon, which packs more densely than nitrogen, but is compatible with low-cost nitrogen in the current ground test configuration. Attitude and position control are achieved with the use of twelve thrusters, distributed across four thruster pods around the central ring. The batteries are rechargeable and a port is provided for refueling.

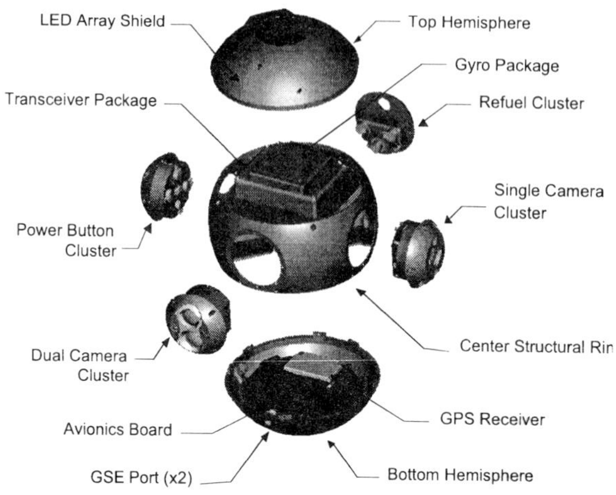

Figure 3 Mini AERCam Exploded View

The Mini AERCam navigation system includes a Global Positioning System (GPS) receiver for position and velocity determination, and Draper Micro-Electro-Mechanical System (MEMS) gyros for angular rate sensing. In order to compute precise relative navigation using GPS, the parent vehicle (e.g. the ISS) would also be equipped with a GPS receiver. The current Mini AERCam GPS receiver is a Thales Navigation Systems (formerly Ashtech) G-12 High-Dynamics and Missile Applications (HDMA) GPS receiver.

The Mini AERCam avionics system includes an avionics processor board, a video compression and imaging system, wireless ethernet (802.11b standard) for communications, and LED illuminators. The avionics processor board is a 22-layer board, and houses a PowerPC 740 which runs at 266 MHz, with 64 Mbytes of RAM. The board includes a field programmable gate array (FPGA) which implements most of the hardware interface functions. Data buses include RS-232, I^2C for instrumentation, and Low Voltage Differential Serial (LVDS). The vision subsystem consists of three complementary metal oxide semiconductor (CMOS) imagers, an onboard video compression system, and an off-board video decompression and display system. The video compression system performs wavelet compression in both hardware and custom software. The video is transmitted as part of a single multiplexed data stream.

The vehicle is designed for either remotely piloted operations or supervised autonomous operations. Manual operations are performed with separate rotational and translational hand controllers. The operator may also input commands through the Control Station, which also displays system status and engineering data. Multiple Mini AERCam camera views are available to the operator. For the ISS application, a Situational Awareness display provides a high fidelity graphical representation of the Mini AERCam location with respect to the ISS.

NAVIGATION DESIGN

This section contains descriptions of the Translational Navigation and the Attitude Navigation.

Translational Navigation

The translational navigation subsystem is the cornerstone of the Mini AERCam GN&C capabilities. AERCam Sprint's capabilities were significantly limited by the absence of any relative navigation information that could be used for operator situational awareness and on-board autonomous functionality. Thus, precise translational navigation (sub-meter accuracy) was a high priority for providing advanced capabilities on Mini AERCam.

The Mini AERCam translational navigation system uses precise relative GPS for estimating relative position and velocity between the free-flyer and parent vehicle (in this case, the ISS). Precise relative GPS is based on differencing the pseudorange and carrier phase measurements between two vehicles' GPS receivers. Throughout this paper, the GPS receiver that would be on-board the ISS is referred to as the "ISS GPS," not to be confused with the current operational ISS GPS, which is a Trimble Force 19 housed in a Honeywell Space Integrated GPS/INS (SIGI). For our implementation and ground testing, we selected a Commercial-Off-the-Shelf (COTS) GPS receiver. It is envisioned that for Mini AERCam operations around the ISS, a receiver like the Ashtech G-12 would be integrated with either an existing ISS GPS antenna or with an added antenna, in order to provide an acceptable level of functionality for precise relative GPS.

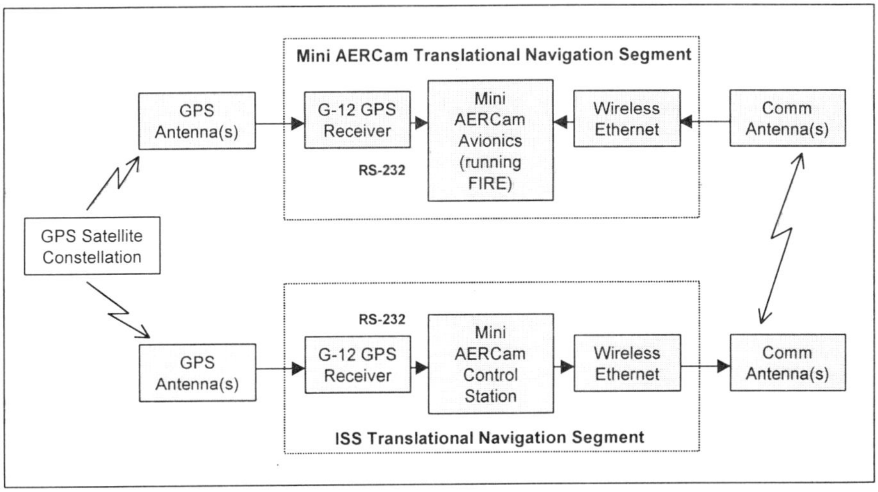

Figure 4 Translational Navigation System Components

Figure 4 shows the Mini AERCam translational navigation system components. In our current prototype configuration, the ISS GPS data is collected at the Mini AERCam control station and transmitted to the free-flyer using a wireless Ethernet transceiver. The data is then processed on-board the free-flyer, along with the free-flyer GPS measurements, by the avionics software.

The GPS measurements are processed and filtered using JSC's Flexible Inertial/Relative Estimator (FIRE) software, which includes a Kalman filter. FIRE was originally based on the Generic Kalman Filter (GKF) algorithms that are used for many NASA navigation applications. FIRE was initiated in 1998 in cooperation with the United States Air Force Research Laboratory

(AFRL), as part of a risk mitigation activity for the XSS-11 micro-satellite program. FIRE development continued when the Mini AERCam project was initiated in 2000. The suite of FIRE software includes not just a Kalman filter, but also the real-time data collection and synchronization software. The FIRE navigation software that runs on the free-flyer is hand-coded in ANSI C and runs under VxWorks.

On-board the free-flyer, FIRE samples the GPS receiver data and processes it. Three Kalman filters run simultaneously on-board: one filter estimates the WGS-84 position and velocity of the parent vehicle based on its GPS data; another filter estimates the WGS-84 position and velocity of the free-flyer based on its GPS data; and the third filter estimates the relative position and velocity of the free-flyer in the WGS-84 frame. FIRE provides the position and velocity estimates at 1 Hz. It also generates detailed performance information for display and recording for post-flight analysis[2].

Another version of FIRE also runs on the Control Station laptop computer, under Microsoft Windows. The Control Station version samples data from the parent vehicle GPS receiver, processes the data, and writes it to a Common Data Area on the laptop for packaging into the communications data structure. The data is transmitted to the free-flyer across the wireless Ethernet interface.

Attitude Navigation

The Mini AERCam attitude navigation system uses angular rate measurements from a three-axis Draper MEMS gyros to estimate inertial attitude and attitude rate for the free-flyer. The MEMS gyro package outputs digital data at a rate of 300 Hz, which is averaged down to a 25 Hz rate. The body axis roll, pitch, and yaw attitude rate measurements are converted to quaternions and then integrated to maintain an estimate of inertial attitude. In order to estimate relative attitude of the free-flyer with respect to the ISS, the inertial attitude is transformed to other reference frames, including the Mini AERCam LVLH-to-body frame, J2000 Inertial-to-Mini AERCam LVLH, Body-to-J2000, Body-to-ISS LVLH, J2000-to-ISS LVLH, and Body-to-an arbitrary inertial frame.

The quaternions are integrated using a first-order Euler integrator. Higher-order integrators were considered, but are not used in order to reduce system lag and error accumulation over time. Rate measurement noise is reduced by using a weighted average rate before the quaternions are passed to the integrators. The weighted average rate is formed by combining gyro measured rates with a feed-forward rate estimation based on previously commanded thruster on-times and the previous body rates. Body rates with respect to the vehicle LVLH are also produced by correcting for the orbital rotation rate, which is computed using the inertial states from the translation navigation and transformed to the body frame. After the quaternions have been updated, the Euler angles are extrapolated in pitch-yaw-roll order and output for pilot telemetry.

The attitude navigation uses a slightly different process when producing the quaternion used for inertial attitude hold. Instead of always holding with respect to an established inertial frame, like J2000 or M50, the body-to-arbitrary inertial transformation is reset with each attitude hold command, and then continues propagation like the other quaternions. This ensures that errors accumulated over time are dropped from the estimate used for inertial attitude holds.

Prior to each air-bearing table test, the free-flyer attitude is initialized from an external source. Additionally, an external rate reference, typically the zero rate of a stationary free-flyer, is used to reset the zero rate point for the gyros and to minimize accumulation of bias errors in the attitude. Because of the performance levels of this generation of MEMS gyro, the free-flyer attitude is designed to be updated from an external source, when such information is available. For typical air-bearing tests lasting one-half hour, the gyro accuracy has been adequate. For longer tests, attitude updates have been required.

GUIDANCE AND CONTROL DESIGN

This section begins with a description of the thruster configuration and thruster selection logic. Next, the attitude guidance and attitude control are described. Finally, a description is provided for the translational guidance and translational control.

Thruster Selection

Figure 5 shows the numbered thrusters and the plume direction for each of the twelve thrusters. The input to the thruster selection logic is a set of commanded thrust durations (in integer milliseconds) for all rotational and translational axes, in order to control all six degrees of freedom. Each thrust duration input is either positive or negative depending on the direction of commanded motion.

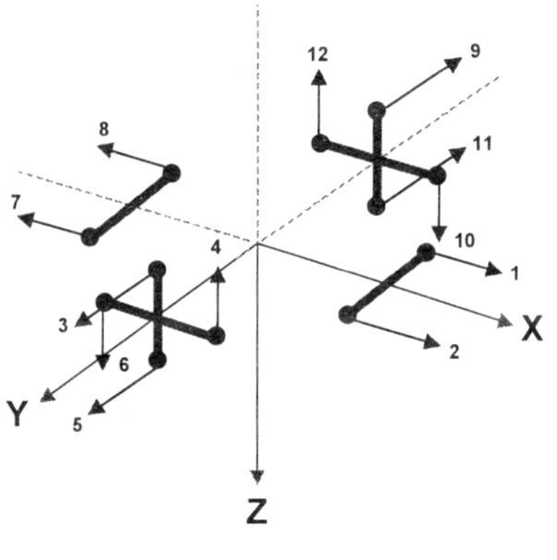

Figure 5 Mini AERCam Thruster Configuration

The thruster configuration shown in Figure 5 is designed such that the same thrusters are used for both a rotation and a translation command. For example, a positive X command would use thrusters 7 and 8, while a positive yaw command would use thrusters 2 and 8. This creates a "contention" between rotation and translation commands. The full set of thruster commands and thrusters (jets) used is shown in Figure 6 below, where the translational axes are identified by X, Y, and Z, and the rotational axes Roll, Pitch, and Yaw are represented by R, P, and W, respectively. The integer indices in the table are sums computed within the jet selection software based on the combination of rotation and translation commands.

	1	2	3	4
	Jet #1	Jet #2	Jet #7	Jet #8
1 -X -W	1	0	0	0
2 -X	1	1	0	0
3 -X +W	0	1	0	0
4 -W	1	0	1	0
5 Unused	0	0	0	0
6 +W	0	1	0	1
7 +X -W	0	0	1	0
8 +X	0	0	1	1
9 +X+W	0	0	0	1

	1	2	3	4
	Jet #3	Jet #5	Jet #9	Jet #11
1 -Y -R	1	0	0	0
2 -Y	1	1	0	0
3 -Y +R	0	1	0	0
4 -R	1	0	0	1
5 Unused	0	0	0	0
6 +R	0	1	1	0
7 +Y -R	0	0	0	1
8 +Y	0	0	1	1
9 +Y+R	0	0	1	0

	1	2	3	4
	Jet #4	Jet #6	Jet #10	Jet #12
1 -Z -P	1	1	0	0
2 -Z	0	1	1	0
3 -Z +P	0	0	1	1
4 -P	1	1	0	0
5 Unused	0	0	0	0
6 +P	0	0	1	1
7 +Z -P	1	1	0	0
8 +Z	1	0	0	1
9 +Z+P	0	0	1	1

Figure 6 Thruster Selection Tables

The contention between rotation and translation is handled through the thruster selection logic that is segmented into X/Yaw, Y/Roll, and Z/Pitch control systems. X/Yaw contention is addressed by turning on only one thruster that simultaneously honors both commands. As an example, for simultaneous positive X (usually thrusters 7 and 8) and positive Yaw (usually thrusters 2 and 8) commands, only thruster 8 is turned on, yielding a simultaneous X and Yaw motion. Y/Roll contention is addressed in the same fashion. The only issue with this approach is that the expected amount of instantaneous thrust will not be achieved when contention occurs. Instead of an anticipated number of milliseconds from two thrusters, only one thruster fires in support of the command on each axis. However, the Mini AERCam thruster selection logic queues the unsupported thrust duration and direction and applies it on the next available cycle, thus providing the full, expected thrust integral. Z/Pitch contention is discussed further in the Translational Control section.

Attitude Guidance

The Attitude Guidance subsystem on Mini AERCam is designed to accommodate either manual (crew-piloted) or autonomous attitude and attitude rate commands. The Attitude Guidance system accepts inputs from either source and computes errors that are then passed on to the Attitude Control subsystem. All guidance functions utilize quaternion mathematics for attitude and body rate error determination. The quaternion convention chosen for Mini AERCam is right-handed quaternions with the scalar in the first element. The current modes supported in the Attitude Guidance are described in Table 1 below.

Table 1

ATTITUDE GUIDANCE MODES

Attitude Guidance Mode	Description
Drift	No attitude or attitude rate guidance
Decel	Null attitude rates
Limit Cycle	Maintain attitude to within specified deadbands
Attitude Maneuver	Execute commanded attitude
Discrete Rate	Execute commanded attitude rate

The attitude guidance modes are Drift mode, Decel mode, Limit Cycle mode, Attitude Maneuver mode, and Discrete Rate mode. All modes can be executed with respect to either the inertial or LVLH reference frames. The pilot makes the selection between inertial and LVLH on the Control Station before any mode is executed. Drift mode is used when the AERCam is completely under the pilot's control and no autonomous capabilities are desired. Decel mode is used to null the angular rates to within a predetermined threshold. In Decel mode, the attitude errors are set to zero since the goal is simply to null rates and not to track a particular attitude. After Decel mode execution is complete, the free-flyer automatically enters Limit Cycle mode and holds the attitude in the desired frame until the pilot issues another command. The Attitude Maneuver mode is used to issue a command to rotate the vehicle to a pilot-selected attitude. Once the vehicle arrives at the commanded attitude it enters Limit Cycle mode. Discrete rate mode increments the commanded body rate by a specified discrete amount with each rotation command issued by the pilot.

Attitude Control System

The Attitude Control System (ACS) uses a phase plane for control of each independent body axis. It is designed to work in such a manner as to center the vehicle in the phase plane rapidly. The phase plane contains switching lines as shown in Figure 7. When the attitude error exceeds a deadband or a rate error exceeds the rate limit, a burn time is computed to force the errors to the burn-to-line (1). If the burn-to-line is not achieved after one cycle then a continuation is commanded (2). The burn-to-line is set at a high enough rate that the vehicle will cross the rate error axis quickly. Once the vehicle crosses the rate error axis, a trim burn is commanded (4) with a duration to force it to the center of the phase plane. If the needed duration is longer than a cycle, then a continuation burn will be commanded (5).

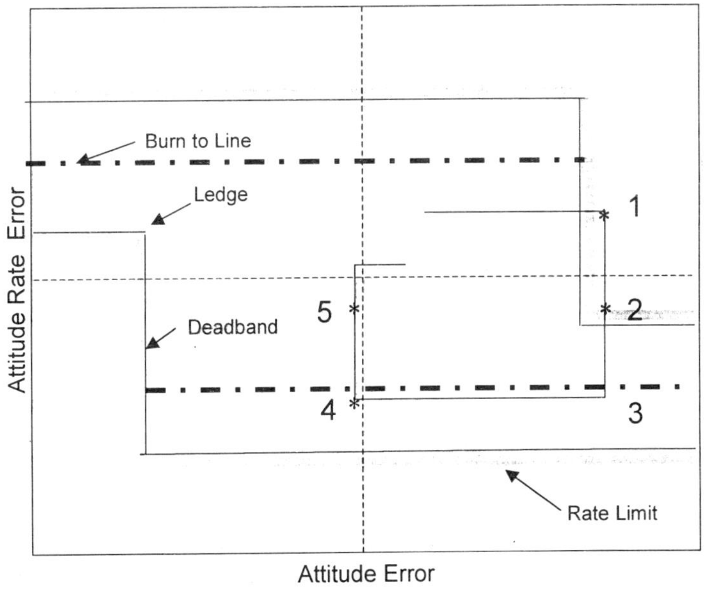

Figure 7 Attitude Control System Phase Plane Diagram

Achieving and remaining in the center of the attitude phase plane means the vehicle is aligned as commanded. The center of the phase plane is always the goal. Drifting away from the center is the result of undesired accelerations, the most significant of which is due to thruster misalignment for translation commands.

The thruster valve drive electronics are designed to allow thrusters to be commanded at a granularity of one milli-second intervals. However, the control cycle runs at 25 Hz (40 milli-second increments). Thus, the software interface for thruster control is designed so that guidance can separately command thruster on-times (burn duration) and delay times (when to initiate the burn). This makes it possible to command the thrusters to stay on until a desired switching line is crossed. The time at which the switching line is expected to be crossed is computed by estimating the minimum acceleration that the thrusters will provide.

The ACS does not send a command longer than 40 ms. Otherwise, command queue problems arise. The unique nature of this control system allows the ACS to run at 25 hz while retaining the thruster granularity of a 1000 hz system. In this way, the control system can run at a low frequency and maintain control authority.

The most important characteristic of the phase plane control is the trim burn. A "perform_trim_burn" flag is set whenever any phase plane burn region logic is invoked. A trim burn is commanded whenever the rate axis is crossed inside of the phase plane and the perform_trim_burn flag is on. The trim burn continues to be commanded until the rate errors are within the trim tolerance. Once the trim burn is accomplished, the perform_trim_burn flag is set to zero. While the Pitch perform_trim_burn flag is on, the flight software refuses to support any Z body translation commands, either from the Hand Controller or Automatic Translational Controller. It will, however, queue Z hand controller commands and support them once the vehicle Pitch is centered (i.e. the trim burn is complete.). This logic is necessary to support a thruster chattering problem with Z/Pitch contention encountered as a result of the Mini AERCam thruster configuration. By waiting until the pitch axis is centered, a more pure Z pulse is obtained.

Translational Guidance

Like the Attitude Guidance subsystem, the Translational Guidance subsystem on Mini AERCam is designed to accommodate either manual (crew-piloted) or autonomous commands. The Translational Guidance system accepts inputs from either source and computes errors that are then passed on to the Translational Control subsystem. The current modes supported in the Translational Guidance are described in Table 2 below.

Table 2

TRANSLATIONAL GUIDANCE MODES

Translational Guidance Mode	Description
Drift	No translational commands
Brake	Null relative velocity between free-flyer and parent vehicle
Station Keep	Maintain relative position between free-flyer and parent vehicle to within specified deadbands
Commanded Maneuver	Execute commanded relative position or velocity maneuver

The first three translational guidance modes are Drift mode, Brake mode, and Station Keep mode. In Drift mode, all of the position and velocity control errors are zeroed. Brake mode is used to support a station-keeping command from the pilot. When the pilot commands the vehicle to stop (or station-keep), the translational guidance first enters Brake mode until the relative velocity slows to within a specified limit. In Brake mode, the position errors are zeroed and the velocity errors are set equal to the most recent LVLH velocity estimate provided by FIRE. Once the velocity is within a specified limit, the Translational Guidance enters Station Keep mode. In this mode, the position errors are the difference between the LVLH position estimate upon entering Station Keeping mode and the current LVLH position estimate. The velocity errors are set to the current LVLH velocity estimate. The Brake and Station Keep algorithms are designed to quickly bring the free-flyer to a relative position hold with respect to the ISS.

The fourth translational mode is Commanded (Translational) Maneuver mode. This mode allows the pilot to select any point in the vicinity of the target vehicle and have the Mini AERCam maneuver to that point with a specified LVLH speed limit. A targeting algorithm computes the commanded position as a function of time over the length of the trajectory. As the maneuver continues, the Translational Guidance computes the current commanded position as determined by the targeting logic and subtracts the estimated state to obtain errors. Those errors are then output to the Translational Control. Once a commanded maneuver is complete, the vehicle enters Station Keep mode.

If the Flight Control system at any time receives a manual translation command from the Hand Controller, it will immediately mode the Translational Guidance into Drift mode. There are two Translational Hand Controller modes currently supported on Mini AERCam. In ***Pulse Mode***, a specified pulse duration is commanded every time the hand controller is deflected about any axis. In the current implementation of ***Pulse Mode***, the rotation and translation pulse durations can be independently set. This helps match rotational response with translational response. The second mode is ***Accel Mode***. In ***Accel Mode***, the commanded thrusters remain actively firing as long as the hand controller is deflected in any direction.

As a future enhancement, GN&C will provide the ability to end a maneuver with a commanded velocity. This will provide a useful interface for waypoint navigation with multiple waypoints. This type of commanded maneuver is completely independent of the Rotational Guidance Mode.

Translational Control

The Translational Control subsystem accepts as inputs the errors from the Translational Guidance subsystem and the LVLH-to-Body quaternion estimate from the Rotational Navigation subsystem. The translational errors are then converted from the LVLH frame to the body frame. There are two methods of Translational Control. The first uses a phase plane controller similar to the attitude controller on the Space Shuttle Orbiter. Like the Mini AERCam attitude phase plane, the translation phase plane uses the anticipated thruster accelerations to compute a burn time that forces the vehicle back toward the center of the phase plane in a fuel-efficient manner. Again the center of the phase plane represents the commanded position and velocity.

Figure 8 shows the translational control phase plane. The switching lines on the phase plane are intended to parallel thruster accelerations with the intercept on the position error axis equating to the deadband. If the vehicle is in Region I of the phase plane outside of the ledge then the shortest command to burn to Region IV is issued. If the vehicle is within Region III then the shortest command to place the vehicle within Region II is issued. Any command that lasts longer than one duty cycle (1 sec.) is truncated to one duty cycle. Any command that is shorter than the minimum impulse (4 msec.) is truncated to 0.

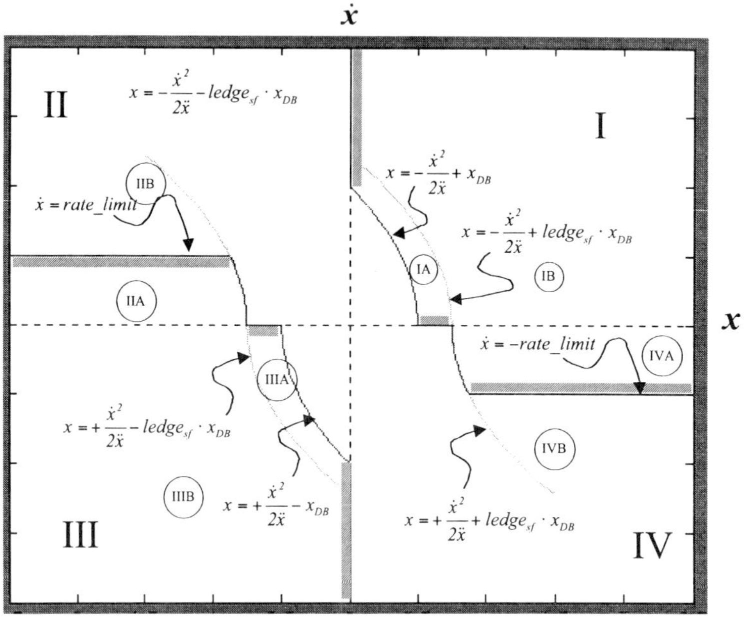

Figure 8 Translational Control Phase Plane Diagram

The second translational control method uses a Linear Quadratic Regulator (LQR) controller developed for Mini AERCam by Charles Stark Draper Laboratories. This controller takes into account the expected relative motion environment using Clohessy-Wiltshire equations, and controls the vehicle within an ellipse for fuel optimality.

Z/Pitch Control

Z/Pitch control is handled differently from other control modes due to a contention problem between Z and pitch which causes thruster chatter, as mentioned earlier in the Thruster Selection section. This is due to the way that the thrusters are configured on the vehicle. The types of conflict and the management strategy for each are described in Table 3.

Table 3

Z/PITCH CONTENTION DESCRIPTION

Z Command Source	Pitch Command Source of Conflict	Conflict Management Strategy
Z hand controller	Pitch hand controller	Support pitch first, and then support Z
Z hand controller	Pitch phase plane	Support pitch phase plane until the trim burn is complete, and then support Z hand controller
Z automatic translation	Pitch hand controller	Support pitch hand controller commands first and postpone Z translation control*
Z automatic translation	Pitch automatic attitude	Support pitch automatic translation commands and postpone Z automatic translation control *

*It is assumed that the translation controller errors will grow, but the system will respond appropriately when the vehicle finishes its Pitch maneuver

GN&C SOFTWARE DEVELOPMENT

Software Architecture

The GN&C software architecture is designed to be compatible with the overall Mini AERCam software architecture. The Mini AERCam system executive, which schedules all tasks, runs at a high rate of 25 Hz and a low rate of 1 Hz. The backbone of the Mini AERCam software is the Common Data Area (CDA), which manages all data that is passed between software subsystems. Rate sensor data is read directly from the FPGA and exported to the CDA for debugging and recording, while raw GPS data is queued in the CDA. All GN&C telemetry also passes through the CDA. The only exception is the thruster commands; which are directly issued by GN&C to the FPGA and the commands never pass through the CDA. The GN&C software runs at a high rate of 25 Hz and a low rate of 1 Hz.

The Mini AERCam is controlled remotely from the Control Station, which is a laptop computer running Microsoft Windows. This computer has RS-232 serial interfaces to two hand controllers (separate rotational and translational) and to the "parent vehicle" GPS receiver. Data is transmitted between the Control Station and the free-flyer using a wireless Ethernet transceiver.

Software Development

Several commercial-off-the-shelf software packages are used on the Mini AERCam project, as well as custom software developed at the Johnson Space Center. The software packages and products are listed in Table 4 below. Trick, developed by Titan Corporation at the Johnson Space Center, is a C-language development environment that makes it easy to "plug-and-play" with an extensive set of models and algorithms. The JSC Virtual Reality Laboratory's Dynamic Onboard Ubiquitous Graphics (DOUG) software package provides high-fidelity graphical models that are updated for each ISS increment and Shuttle flight. The models and graphics are made available to the Mini AERCam project for use in the orbital simulation.

Table 4

SOFTWARE DEVELOPMENT AND SIMULATION TOOLS

Function	Product
Operating system	WindRiver's VxWorks
Communications	Real Time Innovations Network Data Delivery System (NDDS)
Compiler	WindRiver's Diab
GN&C algorithms and software	MathWorks $Matrix_x$
Data processing and plotting	MathWorks Matlab
Real-time data displays	National Instruments LabWindows
Orbital simulation development environment	Titan Corporation's "Trick"
High-fidelity orbital simulation graphics	JSC Virtual Reality Laboratory's Dynamic Onboard Ubiquitous Graphics (DOUG)

Mini AERCam GN&C software includes both hand-coded and autocoded C code. The translational navigation software, including FIRE, is entirely hand-coded in C. The guidance and control software is designed in $Matrix_X$'s SystemBuild, which is a graphical programming environment, and then C code is auto-generated using $Matrix_X$. $Matrix_X$ autocode has a legacy within the AERCam program. $Matrix_X$ autocode was first used on AERCam Sprint[3]; the entire motion control system was built and generated using $Matrix_X$. The auto-generated C code was then integrated with the hand-coded system executive and other routines. Similarly, the guidance and control software for the AERCam IGD vehicle was also generated using $Matrix_X$. The navigation software for the AERCam IGD was mostly hand-coded. The biggest difference between the software for AERCam Sprint and the AERCam IGD vehicles is that the AERCam Sprint processor (a simple microcontroller, the Intel 80C196K) was so limited that the autocode had to be post-processed by hand to improve efficiency. This was not the case for the AERCam IGD, where the processor was a 166 MHz Pentium processor.

$Matrix_X$ SystemBuild made it easy to reuse much of the guidance and control logic from the AERCam IGD for Mini AERCam. The most significant change is that the AERCam IGD had a set of two-dimensional algorithms that were designed for a terrestrial air-bearing table, and the Mini AERCam algorithms are designed to work on-orbit in three dimensions.

The core G&C software for Mini AERCam is primarily developed in $Matrix_X$ SystemBuild. First, unit test cases are developed for each algorithm, and then the system is designed. After design, the unit test cases are executed. Once an algorithm passes the unit tests, the software is autocoded in C and compiled for the "development board" test bed, which houses a real-time embedded processor similar to the actual vehicle Avionics processor board but without the avionics interfaces. The development board test bed software (under VxWorks) is then run through a battery of functional tests. After this, it is integrated with the FIRE software and ported to the Simulation Avionics Processor Board (APB) Test Bed where it is integrated with the Mini AERCam system executive software. In this arrangement the software is then run through a series of tests to insure that the demonstration mission will run. Finally, the GN&C software is ported to the Air Bearing Test Bed where it is again run through a series of maneuvers designed for demonstration. These maneuvers test out the hand controller interface as well as the position and attitude control logic.

The differences between the flight software version of FIRE and the test version of FIRE are minimal. The test version includes all of the serial port interfaces to the GPS receivers. These interfaces are not part of the flight software version. Once ported to flight software, FIRE interfaces to a Common Data Area and the system executive software handles the hardware interfaces in a way that is invisible to FIRE. Each version of FIRE is initially developed to run under the Solaris Operating system. It is then ported to the development board testbed to run under VxWorks. Here the device interfaces change minimally. Once tested on the development board it is then ported to the flight software version and tested on the Sim APB.

TEST FACILITIES

Test Facility Description

The Mini AERCam GN&C testing has four levels. As described in the Software Development section, the first level of testing is simple unit testing either in SystemBuild or using hand written C code. All of the software is originally developed using some variety of UNIX as the development platform, including Solaris and Apple workstations. Once developed and tested on the development platform, the software is ported to the development board system. Once it passes a series of tests in that environment, it is then ported to the Orbital Simulation. Again it must pass a series of tests in that environment before it is finally installed on the Air Bearing Table Test Bed.

Orbital Simulation Testbed

The orbital simulation testbed is designed for testing guidance, navigation, and control capabilities in a high fidelity simulation with much of the Mini AERCam avionics in the loop. In the orbital simulation, the free-flyer can perform rotational and translational maneuvers, including automatic attitude hold, automatic translation hold, and point-to-point maneuvering. The MEMS gyros and thrusters are emulated in hardware, and a high-fidelity GPS signal generator provides Radio Frequency (RF) signals to the GPS receivers. This is illustrated in Figure 9.

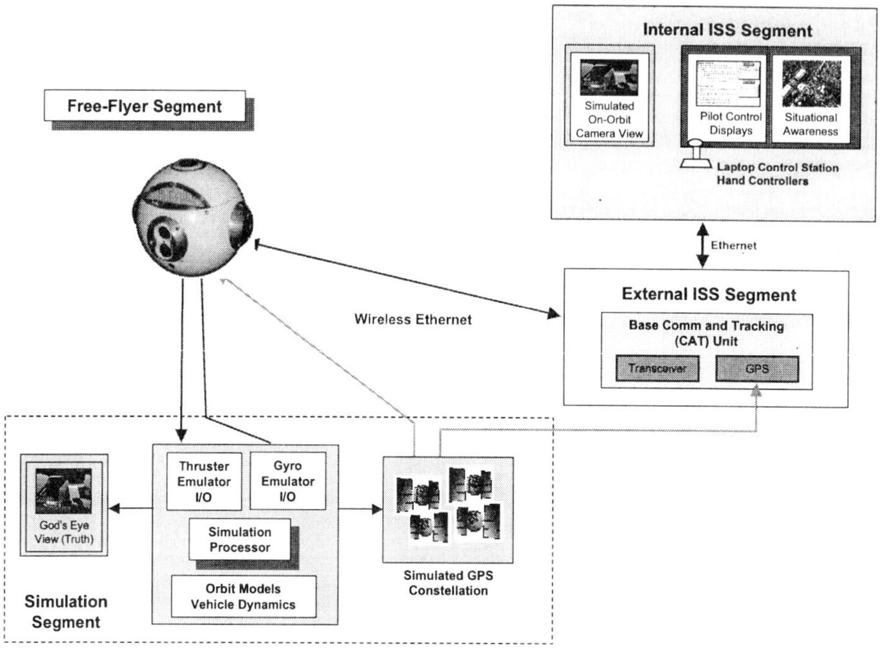

Figure 9 Orbital Simulation Testbed Configuration

In the orbital simulation, the free-flyer is controlled using separate rotational and translational hand controllers and a control station. In place of live video, the images as seen from the free-flyer are simulated with the DOUG high-fidelity graphics. Engineering data and state information are displayed on the control station. In addition, a "God's Eye View" is generated for the Situational Awareness display, which provides the operator with a graphical view of the Mini AERCam location with respect to ISS structure. Figure 10 shows an example scene from the high-fidelity graphics that are used for the orbital simulation. The blackness of space has been changed to white for viewing clarity.

Figure 10 "God's Eye View" for the Orbital Simulation

The simulation avionics processor board system uses a full-fidelity avionics processor like that in the free-flyer prototype. Having a duplicate avionics processor board allows orbital simulation and air-bearing table tests to be conducted in parallel with no resource conflicts. The orbital simulation configuration is similar to the development board testbed, but with higher fidelity. For the orbital simulation, the Trick simulation communicates with a Spirent Federal Systems GPS simulator through a user datagram protocol (UDP) socket connection to an interface PC. The Trick simulation sends vehicle state information for both the free-flyer and ISS to the interface PC. The interface PC packages the data in a format that the GPS signal generator can interpret, and ships the information over an IEEE-488 connection. The signal generator then uses that information to generate the radio frequency (RF) signals, which are transmitted via RF cables to both the Mini AERCam and the ISS GPS receivers. The Mini AERCam Sim APB is connected serially to the G-12 GPS receiver. The control station laptop provides an interface to the ISS GPS receiver and packages the receiver data, along with operator commands, to send across the wireless Ethernet interface to the free-flyer.

In this test bed, data is logged and displayed using LabWindows in real-time by tapping into the telemetry stream. Since Mini AERCam uses Ethernet for communications, the engineering data can be displayed on a separate computer on the local area network. The logged data is also post-processed and analyzed using Matlab.

Air Bearing Table Test bed

An air-bearing table test facility complements the orbital simulation, with an emphasis on more of the hardware components. A special sled, shown in Figure 11, produces a cushion of air to float the free-flyer on an air-bearing table, which simulates microgravity. The vehicle is then able to translate in two directions and rotate about one axis, using the vehicle's own thrusters, propelled with nitrogen for the ground demonstration.

Figure 11 Mini AERCam on Air-Bearing Table Sled

The free-flyer is controlled on the air-bearing table using separate rotational and translational hand controllers, and a control station. Live video and images from the free-flyer are provided on three video displays. Engineering data and state information are displayed on the control station. On the air-bearing table, the free-flyer can perform rotational and translational maneuvers, including automatic attitude hold, automatic translation hold, and point-to-point maneuvering.

Since GPS constellation signals are not available inside the laboratory, a commercial-off-the-shelf (COTS) image-based navigation system called Visualeyez is used for position determination. Visualeyez provides indoor relative navigation using an LED tracking system. The LEDs are mounted on the free-flyer sled. The state (position only) observations from the Visualeyez system are transmitted to the free-flyer on the air-bearing table from off-board, and then are processed through a version of FIRE that is similar to the orbital version but configured for indoor, terrestrial operations. Visualeyez can also be used to initialize the attitude at the beginning of each air-bearing table test.

When executed on the table, FIRE utilizes a separate set of filter tunings than for the orbital simulation, and absolute positions are not calculated. The GN&C main software uses a separate set of initial values for parameters on the air-bearing table that account for the change in conditions, i.e. no orbital rate, mass property changes, etc. Rotation navigation does not compute some quaternions, since inertial frames like J2000 are no longer relevant, and a different translation controller is used for the table.

The air bearing test bed for the Mini-AERCam provides a fully integrated three degree-of-freedom test platform for hardware/software interaction and other hardware systems in a controlled environment. It utilizes all hardware in the loop, including the rate gyros, thrusters, sensors, video systems, and communications. It also allows for demonstrating, profiling and debugging these systems. Testing a fully integrated vehicle also allows the developers to test hardware interfaces, thermal effects, and other hardware systems. One primary example is testing of the video system, in order to evaluate how picture quality is affected by bandwidth changes.

GN&C CAPABILITIES

At the beginning of the project, the Mini AERCam project management set a series of incremental milestones for demonstrations of capabilities, as shown in Table 5. These milestones were achieved by using a spiral development process. Early milestones for the orbital simulation and air-bearing table showed basic manual tele-operation. Later milestones were chosen to demonstrate fully autonomous maneuvers in both the orbital simulation and on the air-bearing table.

Table 5

INCREMENTAL DEMONSTRATION MILESTONES

Orbital Simulation Demonstration Milestones	Air-Bearing Table Demonstration Milestones
1: Manual 6-DOF Control	1: Manual 3-DOF Control
2: Manual 6-DOF Control with Automatic Attitude Hold (Inertial and Relative)	2: Manual 3-DOF Control with Automatic Attitude Hold
3: Relative Navigation with GPS Hardware in the Loop	3: Relative Navigation with Indoor Navigation System
4: Automatic Translation Hold (ATH) with GPS Hardware in the Loop	4: Automatic Translation Hold (ATH) with Indoor Navigation System
5: Automatic Point to Point Maneuvering in 6-DOF	5: Automatic Point to Point Maneuvering in 3-DOF

In the hardware-in-the-loop orbital simulation, our precise relative navigation typically provides accuracies better than 0.5 meters per axis. On the air-bearing table, the Visualeyez system provides the free-flyer with better than centimeter-level navigation accuracy. This has allows us to demonstrate that the current Mini AERCam GN&C system provides the basis for most of the advanced capabilities that were listed as recommendations in Reference 1. The Mini AERCam team continues to add additional functionality, with a near-term focus on adding automatic path-planning, contingency response algorithms, decision-making capabilities, and enhanced displays and controls.

CONCLUSIONS

The current Mini AERCam prototype has demonstrated GN&C capabilities that will allow Mini AERCam to be of significant benefit to NASA's human spaceflight program for external visual inspections of space vehicles. In addition to the primary ISS viewing mission, these capabilities allow Mini AERCam to serve as a test platform for evaluating algorithms and relative navigation for autonomous proximity operations and docking around the Space Shuttle Orbiter or the ISS, or for evaluating candidate sensors and technologies. In the future, small free-flyers like Mini AERCam may serve many functions around the ISS, such as visual or non-visual inspections, communications relay, or viewing for human and robotic extravehicular activities, both in low earth orbit and during exploration missions.

ACKNOWLEDGEMENTS

The authors wish to acknowledge other Mini AERCam GN&C team members for their significant contributions to the project over many years: Mike Lisano (now at JPL) for beginning the FIRE work with GKF; Kevin Key for continuously improving FIRE and making it work well in the real-time orbital simulation; Scott Cryan for GPS hardware testing and evaluations; Sharada Vitalpur for navigation system support and task leadership; Tim Straube for attitude navigation and MEMS gyro contributions, as well as GN&C task leadership and support; Rodolfo Gonzalez for GN&C design, development, and testing all the way back to the AERCam Sprint project; and David Hasan for quality software development, particularly for GPS data handling.

REFERENCES

1. NASA Johnson Space Center Document JSC-39400, "Autonomous Extravehicular Robotic Camera (AERCam) Simplified Vehicle Concept Crew Evaluation Test Report," November 1998.

2. Key, K.W. and Vitalpur, S., "Performance of the Flexible Inertial Relative Estimator (FIRE) for Terrestrial and On-Orbit Relative Navigation," Institute of Navigation Global Positioning Systems Conference, Sept. 2002.

3. Wagenknecht, J. D., and González, R. A., "Guidance, Navigation, and Control (GN&C) for the Autonomous Extravehicular Robotic Camera (AERCam) Free-Flyer Vehicles," AIAA ISS Service Vehicles Conference, April 25-28, 1999.

AAS 03-016

AUTONOMOUS MISSION MANAGER FOR RENDEZVOUS, INSPECTION AND MATING

Douglas J. Zimpfer*

To meet cost and safety objectives, space missions that involve proximity operations between two vehicles require a high level of autonomy to successfully complete their missions. The need for autonomy is primarily driven by the need to conduct complex operations outside of communication windows, and the communication time delays inherent in space missions. Autonomy also supports the goals of both NASA and the DOD to make space operations more routine, and lower operational costs by reducing the requirement for ground personnel.

NASA and the DoD have several programs underway that require a much higher level of autonomy for space vehicles. NASA's Space Launch Initiative (SLI) program has ambitious goals of reducing costs by a factor of 10 and improving safety by a factor of 100. Additionally, other programs are investigating the applications of microsats and servicing vehicles for various military and commercial servicing and inspection missions. A common element in space operations for many NASA and DOD missions is the ability to rendezvous, inspect and/or dock with another spacecraft. For NASA, this is needed to meet the primary SLI design reference mission of International Space Station re-supply, while the other missions require this capability to perform spacecraft servicing, inspection of a failed spacecraft and boost or deorbit of satellites.

A common aspect for each of these programs is an Autonomous Mission Manager that provides highly autonomous planning, execution and monitoring of the rendezvous, inspection and docking operations. This paper provides an overview of the Autonomous Mission Manager (AMM) design being incorporated into many of these technology programs. This AMM provides a highly scalable level of autonomous operations, ranging from automatic execution of ground-derived plans to highly autonomous on-board planning to meet ground developed mission goals. The AMM provides the capability to automatically execute the plans and monitor the system performance. In the event of system dispersions or failures the AMM can modify plans or abort to assure overall system safety.

This paper describes the design and functionality of Draper's AMM framework, presents concept of operations associated with the use of the AMM, and outlines the relevant features of the flight demonstrations.

---

* Program Manager Space Transportation, Charles Stark Draper Laboratory, 2200 Space Park Drive, Suite 210, Houston, Texas 77058. E-mail: zimpfer@draper.com.

INTRODUCTION

Future space missions are expected to require low-cost, highly reliable launch vehicles and orbital vehicles (be it a Space Maneuvering Vehicle or Orbital Space Plane) that can complete a wide range of complex space missions. These space missions will consist of deploying payloads, servicing space stations or satellites, inspecting cooperative and un-cooperative targets and performing formation flight. The ability to meet these key mission objectives routinely will significantly increase space capabilities, e.g., by allowing increased maneuverability of earth monitoring or surveillance satellites, extended duration of science or monitoring spacecraft and re-supply of the International Space Station. Each of these key new space capabilities is highly dependent on the ability of the orbital vehicle to perform rendezvous, inspection and/or mating. Few space missions in the future will not involve these key mission capabilities as the number of integrated space assets increases.

To provide rendezvous, inspection and mating capabilities within the cost and safety objectives for the increased number of missions proposed, a high level of autonomy will be required to successfully complete the missions. The need for autonomy is primarily driven by the need to complete these complex operations outside the sphere of human influence due to unmanned operation of many future orbital vehicles and the limited communication coverage that can be guaranteed for all types of future space vehicles. Autonomy also supports the goals of both NASA and the DOD to make space operations more routine, and lower operational costs by reducing the requirement for ground personnel. Finally, once verified more trusted automatic operations would increase the safety of these complex operations.

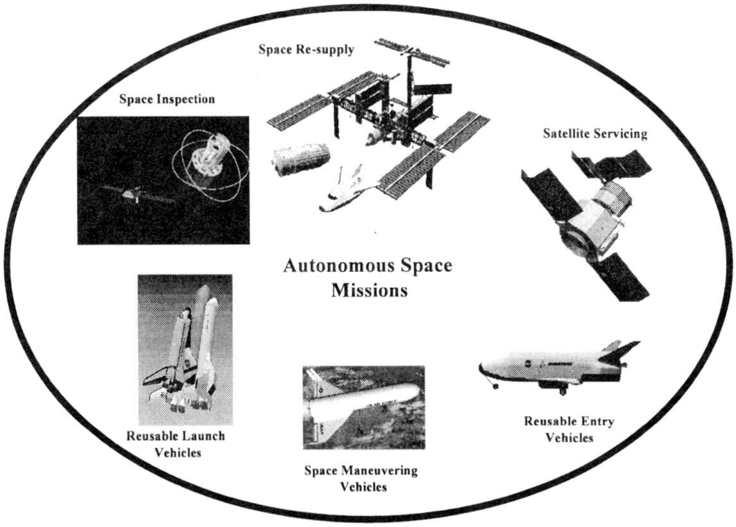

Figure 1 Autonomous Space Missions

To develop and demonstrate the increased automation required for these missions, NASA and the DOD have initiated several technology programs (see Figure 1). NASA's Space

Launch Initiative set goals of improving safety by a factor of 100 and reducing costs by a factor of 10 [1]. NASA is also exploring through its Orbital Space Plane (OSP) and Alternate Access to Station (AAS) the capability for crew escape from and resupply of the International Space Station. All of these development programs are developing key technologies to enable rendezvous and mating, including automatic guidance, navigation and control, autonomous mission management, mating mechanisms, sensors and servicing interfaces. The variety of these missions require a wide range of solutions to meet the needs of the specific missions, e.g., operations in the presence of a manned space station, satellite servicing and inspection of spacecraft.

One aspect common to all of these missions (and to a wide variety of other space missions as well) is the need for an Autonomous Mission Manager (AMM) that provides highly autonomous mission planning, execution and monitoring of the mission activities. To be used across a variety of programs, an AMM must not only be able to automate many of the critical functions, it must be flexible and scalable. It must also be able to integrate with the variety of other key technologies or capabilities required for autonomous rendezvous, inspection and mating. This paper will present such a mission manager being developed to support the on-going space programs.

The first section of the paper will provide a background for autonomous systems. It will describe standard definitions for the level of autonomy, key aspects of any autonomous system and some important elements in the implementation of an AMM. The next section will present the All Domain Planning and Execution Technology (ADEPT) developed at the Draper Laboratory. The key process elements of ADEPT will be presented along with a summary of the standard software implementation. The following section will summarize the application of autonomy to the autonomous rendezvous, inspection and mating programs. Finally, some conclusions will be presented.

AUTONOMOUS SYSTEM BACKGROUND

The initial item of all discussions on autonomy is the definition of an autonomous system. Is it automatic? Is it automated? Does it possess intelligent autonomy? To avoid these questions it is best to discuss autonomous systems based on a sliding scale of autonomy. Figure 2 provides an overview of this sliding scale developed by Sheridan [2]. Some systems developed and their place on the scale is shown in Figure 3. At the lowest end of the scale are fully human controlled systems, while at the highest end are systems that accept simple objectives or goals from the operator and then autonomously develop plans and execute them to achieve the goals. In the middle is scripted autonomy where humans complete detailed planning ahead of time and then the autonomous system automatically executes these scripts. Although the figure clearly shows that the level of autonomy has increased over time, it should be noted that a high level of autonomy is not always the correct answer for the proposed system. An example of this is the International Space Station, which uses the Timeliner executor to automatically execute timelines developed by ground operators [3]. Given the high degree of interaction required with the astronauts to complete activities aboard the station, this level of autonomy works well.

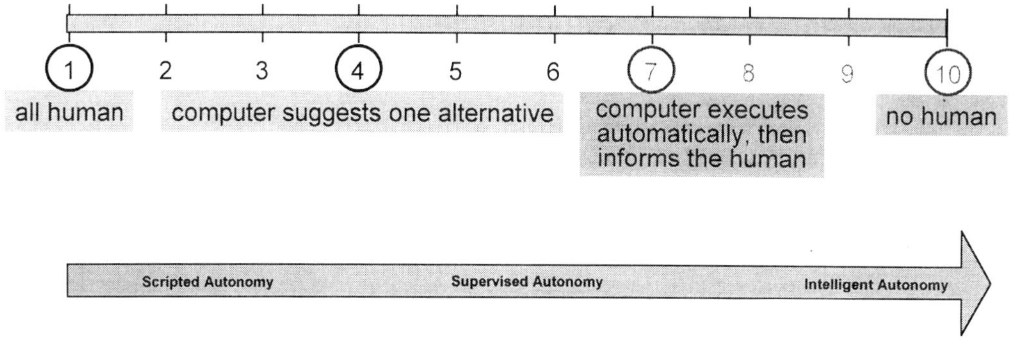

Figure 2 Level of Autonomy

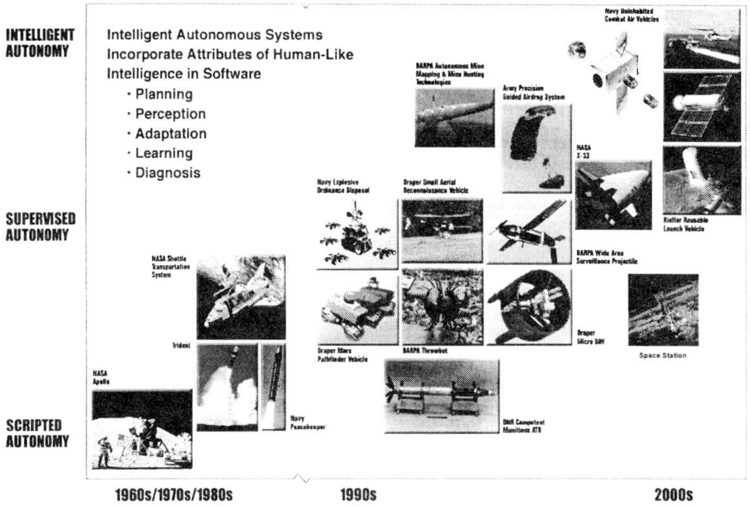

Figure 3 Autonomous Vehicle History

A second key element desired is to provide a distributed, scaleable, intelligent autonomous system [4]. That is the system should be distributed to allow the implementation on multiple processors or between agents (e.g., distributing elements between a ground segment and an onboard segment). Scaleable implies that the system should provide the capability to scale the level of autonomy and to scale the level of operator supervision and direction. The implementation of a framework for autonomy should also provide the ability to scale to different domains and to more complex or less complex problems. Finally, intelligent is used here to state the embedded system should be able to assume roles traditionally performed by human operators (such as planning, perception, adaptation, execution, monitoring and diagnosis).

Finally, planning, execution, monitoring and diagnosis are key elements of the autonomous system and correlate with the sense-think-act paradigm of intelligence and the military's Observe-Orient-Decide-Act (OODA) loop.

Planning – The planning function decides the plan to be executed by the system. It must generate and select an acceptable plan to meet the goals or objectives provided by the operator within the defined constraints. The level of autonomy in the plan is determined by the degree of detail in the goals and objectives defined. The lower the level of objective defined by the operator the less autonomous the system.

Execution – The execution function acts on the mission plan. This function is often tightly integrated with the sub-system sequencer or executive. Unmanned space vehicles will all have a high degree of execution.

Monitoring – The monitoring function observes the system behavior and identifies when problems have occurred. Often sub-system monitoring would be performed by Vehicle Health Management and this information is provided to the mission manager.

Diagnosis – The diagnosis function orients the plan to resolve issues observed by the monitoring function. It will determine if the current plan adequately meets the current objectives given the performance observed by the monitor. It can also determine if current performance indicates new opportunities to meet additional goals with a replan. The diagnosis function will identify when replanning is desired or necessary.

Ideally, an autonomous system framework would provide distributed, scaleable, intelligent capabilities.

ALL DOMAIN PLANNING AND EXECUTION TECHNOLOGY (ADEPT)

Draper has designed the ADEPT architecture to meet the autonomous system challenges described in the previous section [5]. ADEPT bases the development of an autonomous system on the standard aspects of systems engineering. As such it relies on the decomposition of highly complex problems into tractable sub-problems. The decomposition of the problem is then implemented with a standard building block, called a Planning and Execution Module (PEM). At each level of the problem varying levels of autonomy can be employed and these levels can be adjusted to the appropriate level for the specific mission. Additionally, over the life of the mission "trusted autonomy" is achieved as the level of autonomy can be scaled as additional elements of the system are validated. The ADEPT architecture provides a hierarchical, closed-loop, integrated mission manager for real-time vehicle operations.

Figure 4 summarizes the standard ADEPT development process. It begins with the standard systems engineering tasks of developing a concept of operations and functional requirements for the overall system. These provide the basis for the development of the AMM.

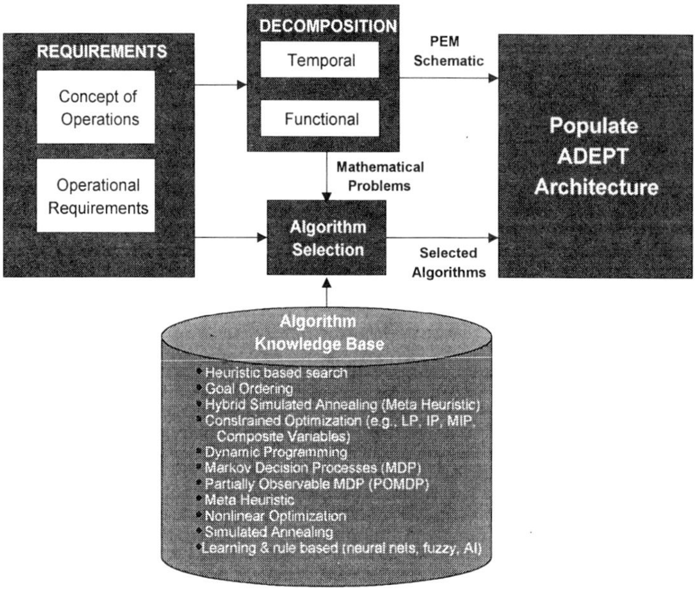

Figure 4 ADEPT Systems Engineering Approach

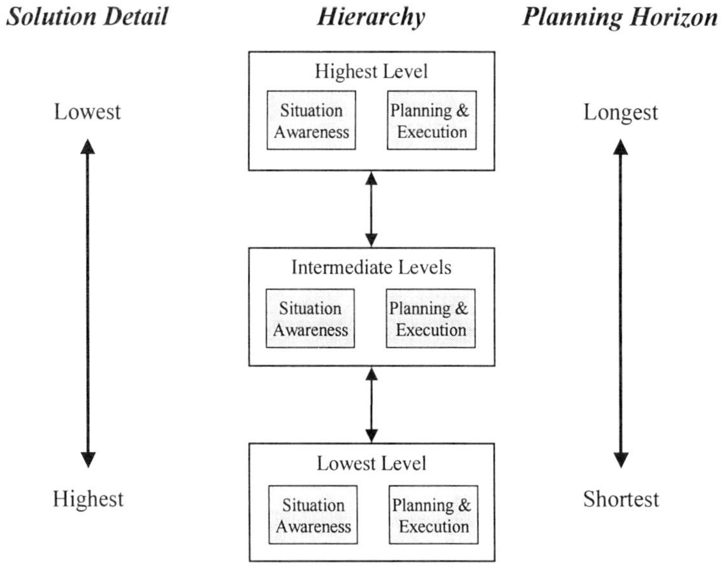

Figure 5 Characteristics of Solutions at Various Levels of the Hierarchy

Completing a temporal decomposition of the problem into tractable sub-problems develops the AMM. As shown in Figure 5, at the highest level of the decomposition long-horizon, low detail problems are posed, while at the lowest level short-horizon highly detailed problems are solved. The ability to decompose the planning, execution and monitoring problems in this framework is key to creating a usable autonomous system. The temporal decomposition results in two key products. The first is the PEM schematic which describes each of the hierarchical functions that can be used to create a mission and the second is a series of use case statements, which describe how the PEMs can be assembled to create a specific mission plan. The lowest level of the PEM schematic will represent the plans provided to the core level system (typically the GN&C, sub-system sequencers, etc.) and will correspond the modes and sequences of the system to be controlled. An example PEM schematic for an SLI visiting vehicle to the ISS is provided later.

The next step in the design process is the functional decomposition. As stated each element of the AMM is developed using a standard PEM building block shown in Figure 6. Each PEM performs planning, execution, monitoring and diagnosis for the function of the system that it represents. The purpose of the functional decomposition is to define the details for completing these common functions and the interfaces with superior and subordinate PEMs.

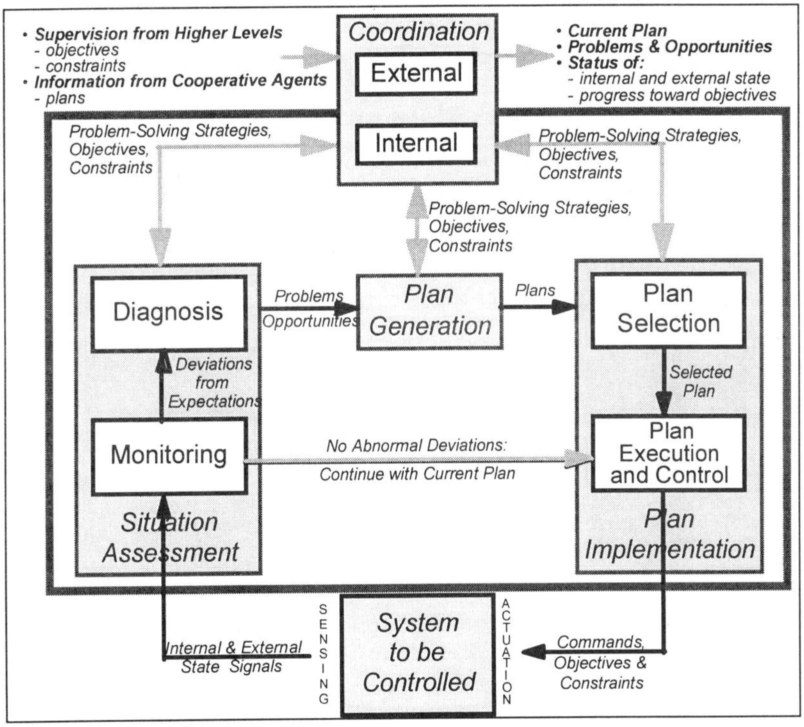

Figure 6 Standard Planning and Execution Module (PEM)

The final step in the process is the population of the PEMs with required planning, execution, monitoring and diagnosis algorithms. These algorithms are specific to the details of the problem to be solved and are chosen from a wide variety of technologies used in optimization, planning and scheduling domains. Ultimately, the algorithms implemented within the planning, execution, monitoring and diagnosis framework determine the level of autonomy. As an example, limiting the planning function of a specific PEM to simply select among pre-defined sequences will provide a highly scripted response, using heuristics increases the plan flexibility, while allowing the planning function to solve an optimization problem would result in a highly adaptive response (but also increasing the complexity of the solution).

As stated the ADEPT implementation framework relies on the standard PEM building block. In addition to provide the standard class libraries for the PEMs, the framework provides the other key aspects required for the implementation of an autonomous system (see Figure 7), including interlevel communication, event/activity scheduling, sequencing and moding, and an operator interface.

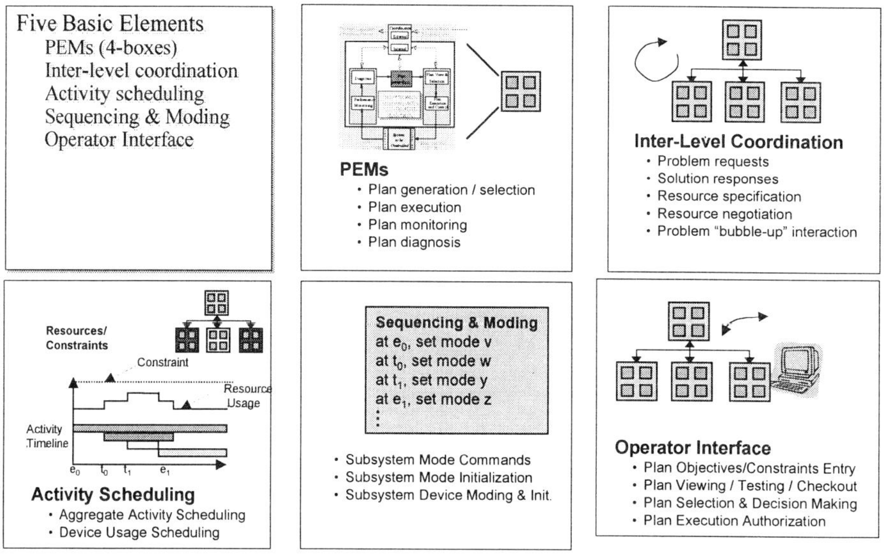

Figure 7 Key Elements of an Autonomous System Framework

AUTONOMOUS RENDEZVOUS, INSPECTION AND MATING MISSIONS

The application of autonomous technologies will be critical to the success of future space applications. To provide routine, safe and efficient operations for satellite servicing, military space operations and ISS visiting vehicle operations, autonomous rendezvous, inspection and mating will be required. Currently there are several programs on-going to develop and demonstrate the technologies required for autonomous rendezvous, inspection and mating. The goals of these programs are to develop and demonstrate the GN&C, Autonomous Mission Management (AMM), sensors, servicing interfaces and mating mechanisms required to make routine space operations a reality. Additionally, studies and research have been started to supply these technologies for future missions such as Mars Sample Return and military space maneuvering vehicle. The following subsections will summarize these programs and describe the application of AMM technologies to meet the program needs.

NASA Space Transportation

The NASA Space Integrated Space Transporation Plan has several areas that are exploring the application of autonomous systems to meet its safety and cost goals. These include the SLI Next Generation RLV, the OSP and the AAS programs. The most ambitious of these efforts was the Spacecraft Mission Assessment and Replanning Tool (SMART) being led by the Flight Mechanics Division at the Johnson Space Center. This effort is developing a prototype of an autonomous mission manager for a next generation RLV, including ascent, orbit, entry and aborts. As part of this effort the initial functional requirements were developed and a prototype ADEPT AMM was demonstrated for a three-stage ascent vehicle (see Figure 8) launch abort [6]. Additionally, the Marshall Space Flight Center has been developing and demonstrating enabling guidance technologies [7].

The SMART activities also developed the design required for an OSP as the crewed stage in the overall system. This included the definition of the rendezvous, inspection and mating AMM for a visiting vehicle system. Figure 9 provides an example of the PEM schematic out-lining the temporal decomposition of the rendezvous, inspection and mating activities. These preliminary designs demonstrated the feasibility of applying ADEPT to automation of the ISS visiting vehicle operations.

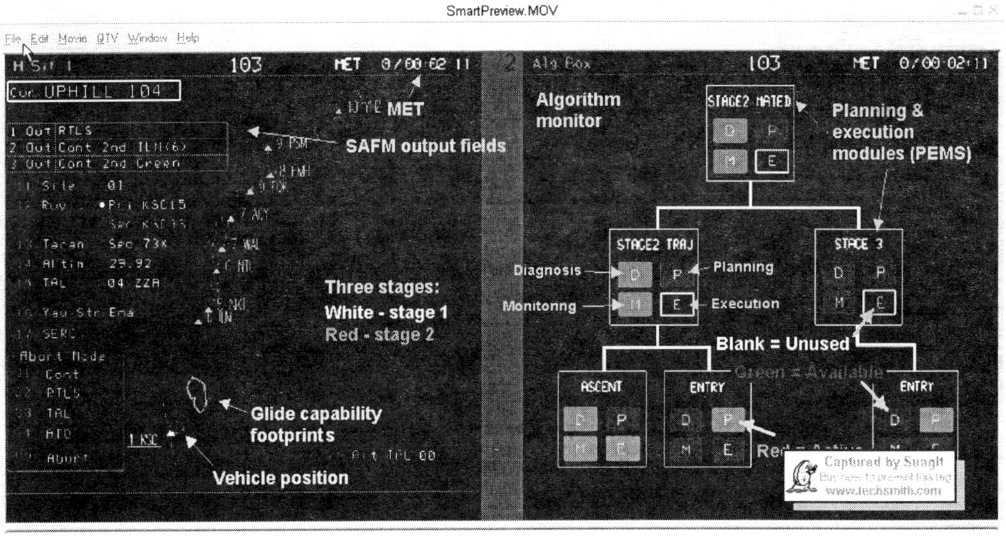

Figure 8 Demonstration of Autonomous Mission Manager for Aborts

A key difference in the applications developed under SMART from many other AMM development efforts was the application to a manned system. The ISS is a unique target for rendezvous, inspection and mating, as it is the only manned target vehicle currently proposed. This increases the safety requirements beyond those for mission success to those required for human safety. This requirement is expected to increase the human supervision required by NASA of the flight operations. The ADEPT architecture was shown to provide a level of supervision tailored to the ISS operations. Several other programs are developing ISS visiting vehicle capabilities, including the SLI Demonstration of Autonomous Rendezvous Technology (DART), the ESA Automated Transfer Vehicle (ATV) and the NASDA H-2 Transfer Vehicle (HTV), but each of these programs is utilizing a limited scripted mission manager.

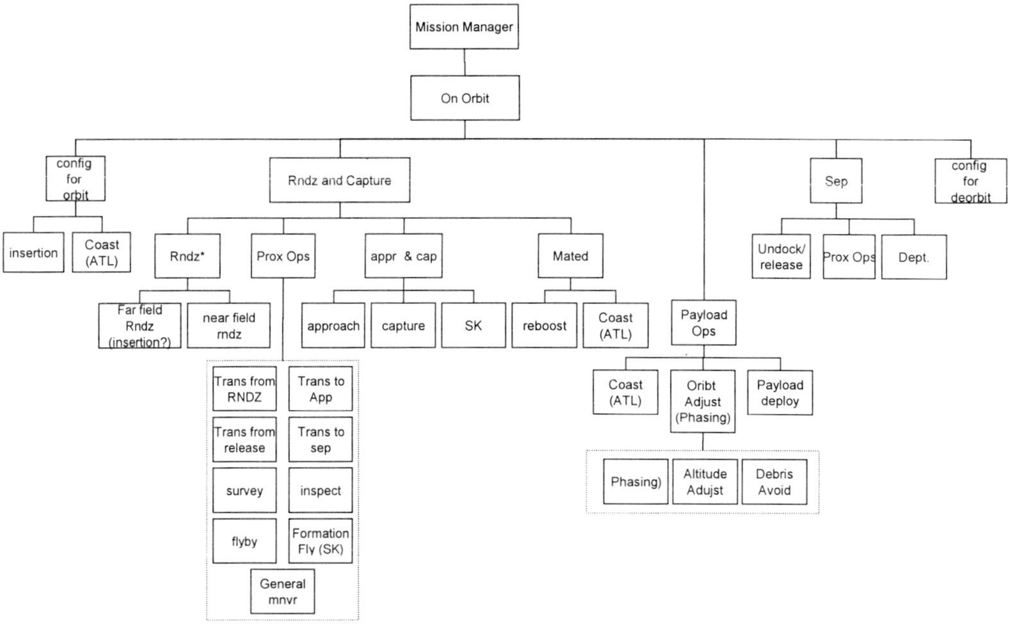

Figure 9 On-orbit, Rendezvous, Inspection and Mating PEM Schematic

Satellite Servicing and Inspection

Many organizations are investigating the technologies required to perform satellite servicing and inspection. As stated, significant benefits can be made in the utilization of space if inspection and servicing can be made routine. For example, spacecraft left in unusable orbits could be boosted to higher orbits or deorbited, satellites that have failed could be inspected to troubleshoot the problem and then be serviced with replacement units. Finally, fuel re-supply would significantly increase the capability of satellites to maneuver to perform their missions.

To make these missions routine, rendezvous, docking and inspection operations must become both more flexible and more autonomous. The goals of these programs should be to make these operations analogous to pizza delivery. The satellite needing service simply sends a command to the servicing provider on what is needed and where/when to provide it; the servicing system then provides an autonomous servicing capability to deliver the necessary space commodity (fuel or replacement unit).

An autonomous mission manager for rendezvous, docking and servicing operations would be a key enabler to create an operational servicing infrastructure. To make servicing a truly usable capability it is likely that a series of servicing vehicles would be required and only an increased level of autonomy over current operations would prove cost effective. An ideal system would provide the capability for a limited operator team to determine the satellite servicing objectives, e.g., which satellites need servicing and what servicing they need. The mission planning system would then provide a series of servicing missions that could be employed to meet these objectives. Once approved by the operations team, including coordination with the satellite operators, mission objectives could be sent to each servicing vehicle. These objectives would then be used to autonomously create a detailed rendezvous, dock, and servicing timeline or mission plan.

The ADEPT hierarchical architecture lends itself well to meeting these mission needs. A ground system would be incorporated to perform the autonomous mission planning and monitoring functions for the overall servicing fleet, while an onboard segment would be developed to autonomously plan and execute the rendezvous, docking and servicing mission of each individual servicing vehicle.

ADDITIONAL AUTONOMOUS SPACE APPLICATIONS

The NASA Advanced Information Systems Technologies (AIST) program goals are to identify, develop and demonstrate the technologies required to enable earth observation by reducing the cost, risk, size and development time of earth science enterprise systems and increasing the accessibility and utility of the science data. The AIST program objective is to provide these technologies to an autonomous and flexible sensor constellation of earth observing satellites that can be reconfigured based on the contemporary scientific problems at hand. To meet these goals the Draper Laboratory has been leading an effort to apply ADEPT to the operations and control of an earth observing sensor web. Autonomous command and control operations are required to provide efficient and safe operation of a large web of complex earth sensor. The proposed ground operational system would plan and control the operations of the overall sensor web, planning specific earth observation and maneuvering satellites to efficiently maximize the science applications. A prototype of a ground operation system has been developed and demonstrated under Phase I of this program [8].

The future military space vehicles will require autonomous space operations to meet the demanding frequency and complexity of the mission proposed by the DOD. The demanding launch operations of up to 170 flights/week and the near-real time requirement for targeting and direction of these missions will require a high level of automation in the planning and execution of coordinated multiple vehicle flight operations. Recent research efforts [9, 10] have begun to develop the optimization technologies that could be incorporated in to the mission planning and execution of the DOD global strike and responsive space lift missions.

The NASA exploration missions also benefit significantly from the application of autonomous vehicle operations. An example is the Mars Sample Return (MSR) Mission planned for 2012. The Jet Propulsion Laboratory (JPL) has made significant progress in developing and demonstrating autonomous rover operations [11]. These efforts have demonstrated the feasibility and benefits of automating the path planning and obstacle avoidance in performing rover operations. Additionally, the rendezvous the sample object with its earth return system will require autonomous operation to meet the time critical mating operation in the presence of long command lags from the earth control stations. Recently, Draper Laboratory demonstrated the autonomous execution of these operations for an MSR mission [12].

CONCLUSIONS

Almost every future space mission will require or benefit significantly from the application of autonomous flight operations. These include launch, servicing, inspection, and exploration and earth observation vehicles. A key function in many of these programs is the ability to make rendezvous, inspection and mating with another space vehicle routine and safe. The ADEPT technologies and software to provide a framework for the implementation of autonomous mission management to space vehicles were presented. This framework has been shown to be flexible to meet varying vehicle domains, functional operations and operator interaction requirements. The application of ADEPT to various mission management systems and programs has been summarized.

ACKNOWLEDGEMENTS

The author would like to acknowledge that the technologies and programs presented were the efforts of many individuals at Draper, NASA and others. The work on SMART was conducted under NASA contract NAS9-01069. The AIST efforts were conducted under NASA contract NAS3-00163. The MSR rendezvous and capture efforts were conducted under NASA contract 1215138.

REFERENCES

1. 2nd Generation Reusable Launch Vehicle Program Level 1 Requirements, NASA Marshall Space Flight Center, MSFC-RQMT-3221, September 2001.
2. Parasuraman, Raja; Sheridan, Thomas; and Wickens, Christopher, "A Model for Types and Levels of Human Interaction with Automation", IEEE Transactions on Systems, Man, and Cybernetics-Part A: Systems and Humans, Vol. 30, No. 3.
3. R. Brown, E. Braunstein, R. Brunet, R. Grace, T. Vu, J. Busa and B. Dwyer, "Timeliner: Automating Procedures on the ISS", Space Ops 2002, October 2002.
4. M. Adams, "Functional Analysis/Decomposition of Closed-loop, Real-time Work Processes", AGARD Lecture Series #200: Knowledge-Based Functions in Aerospace Systems, November 1995.
5. M. Ricard and S. Koelitz, "The ADEPT Framework for Intelligent Autonomy", VKI Lecture Series on Intelligent Systems for Aeronautics, von Karman Institute, Brussels, Belgium, May 2002.
6. H. Hu, "SMART Risk Reduction Review", Presentation to NASA Unique Risk Reduction Review Team, October 2002.
7. J. Hanson, "A Plan for Advanced Guidance and Control Technology for 2nd Generation Reusable Launch Vehicles", AIAA Guidance, Navigation and Control Conference and Exhibit, August 2002.
8. M. Abramson, D. Carter, S. Koelitz, M. Ricard, and P. Scheidler, "Real-time Optimized Near Earth Observation Autonomous Planning", NASA's Earth Science Technology Conference 2002, June 2002.
9. Rao and K. Clarke, "Performance Optimization of Maneuvering Re-entry Vehicle Using a Legendre PsuedoSpectral Method", AIAA Atmospheric Flight Mechanics Conference and Exhibit, August 2002.
10. Ngo and D. Doman, "Footprint Determination for Reusable Launch Vehicles Experiencing Control Effector Failures", AIAA Guidance, Navigation and Control Conference and Exhibit, August 2002.
11. T. Estlin, F. Fisher, D. Gaines, C. Chouinard, S. Schaffer, and I. Nesnas, "Continuous Planning and Execution for an Autonomous Rover", 3rd International NASA Workshop on Planning and Scheduling, October 2002.
12. D. Geller, MARS SAMPLE RETURN MISSION TERMINAL ORBIT RENDEZVOUS ANALYSIS, FY01 Final Report, September 2001.

AAS 03-017

STOCHASTIC OPTIMIZATION OF SPACECRAFT RENDEZVOUS TRAJECTORIES

Bryce Carpenter[*] and Bill Jackson[†]

Many recent and proposed satellite missions involve the rendezvous of two or more satellites in different orbits. The relative motion of two orbiting bodies can be described by the traditional Clohessy-Wiltshire (CW) equations. However, the CW equations rapidly decrease in accuracy when disturbing forces are present. The purpose of this paper is to use the CW equations to initialize a Genetic Algorithm (GA) that will calculate a rendezvous trajectory in the presence of disturbing forces. The GA uses a stochastic process to find optimal variations of the CW equations. Disturbing forces in this study include high fidelity Earth gravity forces, drag, solar radiation, and third body effects. A statistical analysis of the GA solution is included.

INTRODUCTION

The need for performing high-accuracy rendezvous maneuvers is becoming increasingly important. This is due to the emergence of new missions that consist of multiple spacecraft flying in precision formation (e.g., Starlight, Techsat-21, and others), as well as the nascent interest in on-orbit servicing of satellites (e.g., Orbital Express).

There is only one closed-form solution to the rendezvous problem in widespread use today, which was presented by Clohessy and Wiltshire in 1959 (Ref.1). The Clohessy-Wiltshire (CW) equations are useful for calculating either the relative position of two space vehicles as a function of time or the relative velocity required to achieve a desired position. Unfortunately, some simplifying assumptions are required in the use of CW equations (including Earth point mass gravity model). While these equations are useful for preliminary mission planning, the simplifying assumptions can lead to errors in actual use.

The purpose of this study is to develop an optimization tool that can be used in conjunction with the traditional CW equations to plan high-accuracy rendezvous maneuvers in the presence of perturbing forces. This work is important because it allows existing, well-understood mathematics to be adapted for use in a real-world

* Ball Aerospace & Technologies Corporation, 1600 Commerce Street, MS T-1, Boulder, Colorado 80301. E-mail: bcarpent@ball.com.

† Ball Aerospace & Technologies Corporation, 1600 Commerce Street, MS T-1, Boulder, Colorado 80301. E-mail: bjackson@ball.com.

application, without being hindered by simplifying assumptions and constraints. In this paper the perturbing forces will include high-fidelity Earth gravity model forces, differential drag forces, solar radiation forces, and Sun and Moon third body forces.

This study demonstrates the use of the CW equations to provide an initial estimate for a Genetic Algorithm (GA) solution of the rendezvous problem. The GA can accommodate all the disturbances previously mentioned. The approach presented herein will improve upon the performance of the CW equations by minimizing range error. Other parameters, such as total ΔV, could be optimized if desired.

DETAILED PROBLEM DESCRIPTION

CW Coordinate System

The CW coordinate system is defined as follows. Let the origin of the CW coordinate system be coincident with a reference vehicle. Let the fundamental plane of the CW coordinate system be the orbital plane of the reference vehicle and let the fundamental vector be defined in the direction from the center of the Earth to the reference vehicle. Figure 1 shows the geometry of the CW coordinate system.

If the position vectors of both vehicles are known in inertial space, then the relative range vector in the CW coordinate system is given by

$$(1) \qquad \vec{R}_{CW} = \begin{bmatrix} r_{man} - r_{ref} \\ \Delta f r_{ref} \\ \Delta z \end{bmatrix}$$

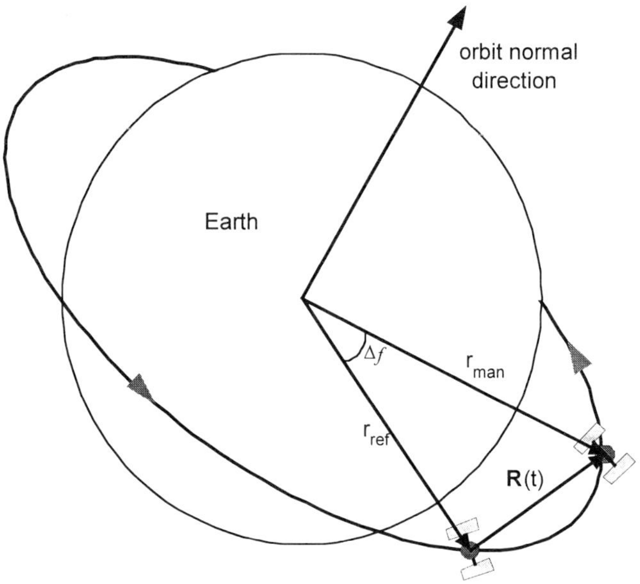

Figure 1. CW Coordinate System Geometry

where the subscripts ref and man refer to the reference vehicle and the maneuverable vehicle. The first term of the relative range is simply the difference in position magnitude of the two vehicles. The second term is a curvi-linear term that contains the angular separation. The third term is the cross track term as measured from the reference vehicle orbital plane along the orbit normal direction. Thus, the relative range vector is a cylindrical representation of the inertial cartesian relative range vector.

Then, if the relative position vector at the final time is known (e.g., for a rendezvous problem $\mathbf{R}(t) = [0,0,0]^T$) one can apply the CW equations to determine $\mathbf{V}_0$, the velocity at t_0, to complete a maneuver in a specified time, t. The velocity, $\mathbf{V}_0$, is measured in the CW reference frame. The CW equations in matrix form, solved for $\mathbf{V}_0$, are (Ref. 2)

(2) $$\vec{V}_0 = \Phi_{rv}^{-1}\left[\vec{R}(t) - \Phi_{rr}\vec{R}(t_0)\right]$$

where the matrices $_{rr}$ and $_{rv}$ are given by

(3) $$\Phi_{rr} = \begin{bmatrix} 4-3\cos(\psi) & 0 & 0 \\ 6(\sin(\Psi)-\Psi) & 1 & 0 \\ 0 & 0 & \cos(\Psi) \end{bmatrix} \qquad \Phi_{rv} = \begin{bmatrix} \frac{\sin(\Psi)}{n} & \frac{2}{n}(1-\cos(\Psi)) & 0 \\ -\frac{2}{n}(1-\cos(\Psi)) & \frac{4}{n}\sin(\Psi)-\frac{3}{n}\Psi & 0 \\ 0 & 0 & \frac{\sin(\Psi)}{n} \end{bmatrix}$$

where

(4) $\Psi = nt$

The equations work well if an Earth point mass gravity model is used to propagate the vehicles, and if the reference vehicle is in a perfectly circular orbit, and if the transfer time is small compared to the reference orbital period. However, if these conditions are not met then the performance of the CW equations will suffer. Figure 2 shows how the performance degrades when a 21 x 21 gravity model is used to propagate the vehicles. The vertical axis is the magnitude of the relative range error after a CW maneuver has been conducted in the high fidelity Earth gravity field. The horizontal axis is the transfer time, measured in units of the reference vehicle's orbital period. The range errors shown in Figure 2 are large compared to the requirements of modern rendezvous missions. Notice that when the transfer time is a multiple of ½ period there is a singularity in the Φ_{rv} matrix. For presentation purposes, the CW equations were evaluated at 0.48P and 0.98P instead of 0.5P and P (where P is the period of the reference satellite).

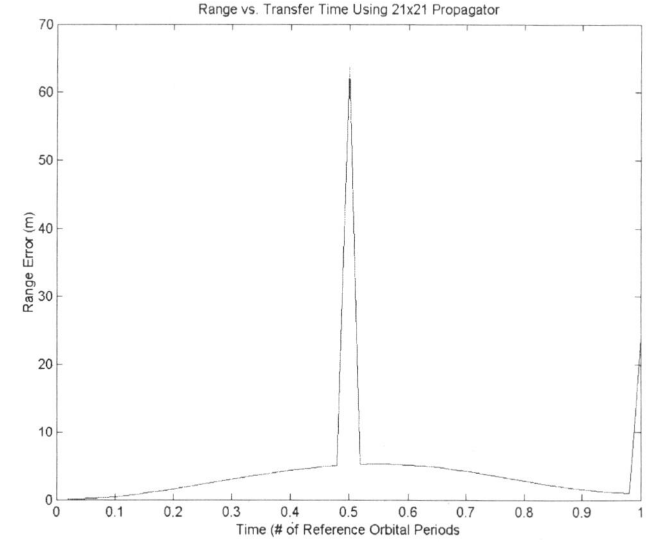

Figure 2. Range Error When Using CW Equations in the Presence of a 21 x 21 Earth0 Gravity Model

Formal Problem Definition

In this problem, the satellites are phased such that the distance between the two is about 25 m. The initial state vector of the reference satellite, chosen by propagating the reference vehicle from a perfectly circular orbit at 500 km altitude and 45 deg inclination for several hundred seconds under the influence of the perturbing forces, is

$$\begin{bmatrix} X \\ Y \\ Z \\ \dot{X} \\ \dot{Y} \\ \dot{Z} \end{bmatrix} = \begin{bmatrix} 6001.3435\,\text{km} \\ -2393.5073\,\text{km} \\ -2355.4969\,\text{km} \\ 3.7216\,\text{km/sec} \\ 4.6853\,\text{km/sec} \\ 4.7063\,\text{km/sec} \end{bmatrix}$$

The epoch time for the simulation is midnight 1 Jan 2001. Furthermore, the two vehicles have ballistic coefficients of 45.455 kg/m^2 and 170.455 kg/m^2. The ballistic coefficient is defined in this paper as

(5)
$$B^* = \frac{m}{C_D A}$$

where m is the vehicle mass, A is the cross-sectional area and C_d is the coefficient of drag (assumed to be 2.2 in this study). The ballistic coefficients are chosen such that the differential drag between the two vehicles yields a considerable range difference when propagated for the transfer time.

Solution of Eqs. (2-4) with a transfer time of 1500 s yields

$$\vec{V}_0 = \begin{bmatrix} -0.007201816 \\ 0.028542409 \\ -0.000009609 \end{bmatrix} m/s$$

in the CW reference frame. The range error, in the CW reference frame, after applying $\mathbf{V}_0$ is

$$error = \begin{bmatrix} 0.3556652 \\ 0.0705224 \\ 0.0095388 \end{bmatrix} m$$

FUNDAMENTAL APPROACH

This section outlines the fundamental approach used for solving the Rendezvous Optimization Problem, including the need for a global search technique, the rationale used to select the fundamental approach, a short tutorial on genetic algorithms, and a discussion of problem representation.

Local vs. Global Search Techniques

Local search techniques explore only a subset of the solution space, and the solutions they consider are highly dependent upon the starting point. Given an initial guess, a local search technique such as gradient descent would examine points within a small neighborhood of the guess; as long as the next point considered has a smaller value than the previous point, the algorithm will continue moving downhill. When the next point has a larger value, the algorithm stops and returns the previous point as the minimum. This technique works well for simple functions with a single point of inflection, but in the case of a function with multiple points of inflection, the gradient descent algorithm may find only a local minimum. Hence, local search techniques have difficulty dealing with ill-behaved solution spaces containing local minima.

Global search techniques generally employ a stochastic search process to traverse a much larger subset of the solution space, and thereby escape being trapped in local extrema. Global search techniques are much more robust because they are largely independent of the solution space or on initial conditions. Commonly used global search techniques include simulated annealing, neural nets, and genetic algorithms.

Rationale for Selecting Optimization Approach

Genetic algorithms were deemed to be an appropriate optimization approach to this problem because of the inherent difficulty of the problem. The Rendezvous Optimization Problem may be considered to be a variant of the classical Traveling Salesman Problem (TSP). In the classical TSP, a salesman must make a complete tour of n cities, visiting each city only once, in the most efficient manner possible (i.e., traveling the least Euclidean distance). No general method of solution is known, and the problem is classified as NP-hard. NP-hard problems cannot be solved in polynomial time, and are among the most difficult problems that can be conceived. Genetic algorithms have been used with great success against the classical TSP and related problems, and are appropriate for the Rendezvous Optimization Problem as well.

Genetic Algorithm Fundamentals

Genetic Algorithms (GAs) are a class of global search algorithms based on the principles of natural selection and genetic evolution. They differ from traditional optimization algorithms in several ways:

1) GAs work on a coding of problem parameters, not the parameters themselves
2) GAs search a population of points in the solution space, not a single point
3) GAs use objective function metrics to guide the search process, not derivatives or other auxiliary knowledge of the solution space
4) GAs use stochastic transition rules to traverse the solution space, not deterministic rules

In a genetic algorithm, a population comprised of some number of individuals is allowed to evolve over time under specified selection rules in order to produce more fit individuals. The operation of genetic algorithms depends greatly on the particular parameter coding used. In this analogy, an individual is represented by a chromosome (a numerical string), which corresponds to a candidate problem solution. Each numerical component of the chromosome (called a gene) is an encoding of some type of problem parameter. There are two fundamental types of parameter coding representations: discrete and continuous. Discrete representations model each individual as a binary string chromosome in which each gene is a bit. Continuous (also called real-valued) representations model each individual as a string of real numbers corresponding to specific traits.

A genetic algorithm is only as good as its fitness metric. The fitness metric is a quantitative measure that gauges a candidate solution's merit, according to some predetermined criteria. Without the fitness metric, there would be no way for the algorithm to differentiate good solutions from poor ones. As the algorithm proceeds, less fit solutions are discarded in favor of more fit solutions. The overall population then becomes dominated by solutions that rate favorably according to the fitness metric. These dominant individuals are then able to reproduce and pass on their genetic material to the next generation for further evolution and refinement. When the algorithm selects two parent individuals for mating, these individuals exchange genetic material to produce an offspring, which will share certain similarities with each parent. This process is known as crossover. Due to crossover, the offspring are generally sufficiently different from their parents that new regions of the solution space are explored. An additional mutation operator randomly permutes small sections of the offspring's chromosome to guarantee that at least some of the offspring are distinct from their parents. This mutation operation keeps the algorithm from converging prematurely. After the new generation is created, selected members of the parent generation are removed from the population. This process is known as replacement. The selection, crossover, and mutation operations are implemented by independent stochastic operators that determine the performance properties of the algorithm. Algorithm convergence is controlled by judicious selection of the stochastic operator parameters, population size, and replacement strategy.

Problem Representation

The solution space in this problem is explored by a population of 200 individuals, each of which represents a candidate maneuver vector $\mathbf{V}_0 = [V_x \ V_y \ V_z]^T$. The individual components V_x, V_y, and V_z are real-valued quantities. The initial population is formed by first using the Clohessy-Wiltshire equations to generate an initial solution. While this initial solution will not be optimal, it should be within a reasonable neighborhood of the optimal solution[1]. This initial solution serves as a

[1] Where the CW equations do not yield a solution that is within a reasonable neighborhood of the optimal solution, the GA may be applied with a different initial estimate. E.g., when the reference vehicle has an eccentricity greater than zero, a

seed which is randomly perturbed 200 times to generate an initial population whose individuals are all within a small neighborhood of the Clohessy-Wiltshire solution.

Input Information— Each candidate solution in the population is passed to Satellite Tool Kit (STK) via the Connect interface socket. The STK application then propagates the orbits of the maneuvering and rendezvous satellites using a 21 x 21 Earth gravity model, and returns the inertial position vector of the reference satellites, $\mathbf{R}_{ref}$, and the maneuvering satellite, $\mathbf{R}_{man}$, after the transfer time. STK also provides the third body, drag, and solar radiation accelerations. The relative range vector in the inertial coordinate system is $\mathbf{R}_{rel} = \mathbf{R}_{man} - \mathbf{R}_{ref}$, and the relative range magnitude is the root-sum-square of $\mathbf{R}_{rel}$.

Output Information— The output of the optimization algorithm is as follows:

- Proposed maneuver vector
- Final range error
- Graphical display of solution cost for each generation

OPTIMIZER ALGORITHM DESIGN

The algorithm used to solve the Rendezvous Optimization Problem uses the stochastic optimization technique of genetic algorithms. This section describes the overall design of the optimizer algorithm.

Solution Fitness Metric

The only solution fitness metric currently used is range error of the maneuvering satellite relative to the rendezvous satellite, R_{rel}. In more advanced applications, it would be possible to include other simultaneous metrics, such as V magnitude.

Selection

At the start of each generation, the entire population is rank ordered by solution fitness. A tournament pool of 40 individuals is randomly selected from the top half of the population. The individuals within this tournament pool then compete for the right to reproduce. The most fit individual of the 40 competitors wins the tournament, and is selected as a parent. Another random tournament pool is chosen, and the entire process is repeated until 50 parents have been selected.

Crossover

The 50 selected parents reproduce with each other to produce 100 offspring. Each pair of parents (denoted as mother and father) produce two offspring (denoted as son and daughter). These offspring are generated according to:

Lambert solution could be applied to determine the initial $\mathbf{V}_0$. Of course, the Lambert solution also uses simplifying assumptions, so the Lambert range will not be zero when a high fidelity gravitational model is used. Rather, the Lambert solution will provide a reasonable estimate of the optimal $\mathbf{V}_0$ and the GA can be applied as described in this study.

(6) $$p_{son} = p_{m_i} - \beta(p_{m_i} - p_{f_i})$$

(7) $$p_{daughter} = p_{f_i} - \beta(p_{f_i} - p_{m_i})$$

where p_{mi} are the corresponding genes (i.e., the components of V_0) from the mother, p_{fi} are the corresponding genes from the father, and β is a random number between 0 and 1. The crossover process ensures that the offspring have some genetic resemblance to both parents, and is the primary mechanism for traversing the solution space.

Mutation

Mutation is a random perturbation of genes that is performed at a relatively low frequency (typically 5 to 10%) in order to prevent premature algorithm convergence. M is the total number of gene mutations per generation, and is given by

(8) $$M = 3N\alpha$$

where N is the number of individuals in the population, and α is the mutation rate. The individual gene mutations themselves are produced by adding a random perturbation to the selected gene of the selected individual:

(9) $$p_i = p_i + random(-K, K)$$

The random range specified by K is given by

(10) $$K = index * A_{\max}$$

where A_{max} is the maximum perturbation amplitude and index is defined as follows:

(11) $$index = \begin{cases} 1 - \dfrac{gen\#}{m} & gen\# < m \\ e^{-r(gen\#-1)} & gen\# \geq m \end{cases}$$

where r and m are user-defined parameters (r = 0.03 and m = 70 in this study) and gen# is the current generation number.

The presence of the index forces the amplitude of the perturbation to decrease as the algorithm progresses. This dynamic perturbation technique prevents the mutation process from disrupting convergence near the end of the algorithm.

Replacement

After the 50 parents have mated and produced 100 new offspring, these offspring are inserted into the population, and the 100 worst individuals are removed. Therefore, after each generation, the least fit half of the population is replaced by the newly generated offspring. Note that there is no guarantee that every offspring will be more fit than every member of the half of the population being replaced; replacing the worst half of the population every generation irregardless of the relative fitness of the offspring is essential to maintaining population diversity. Figure 3 is a graphical overview of the selection and replacement process.

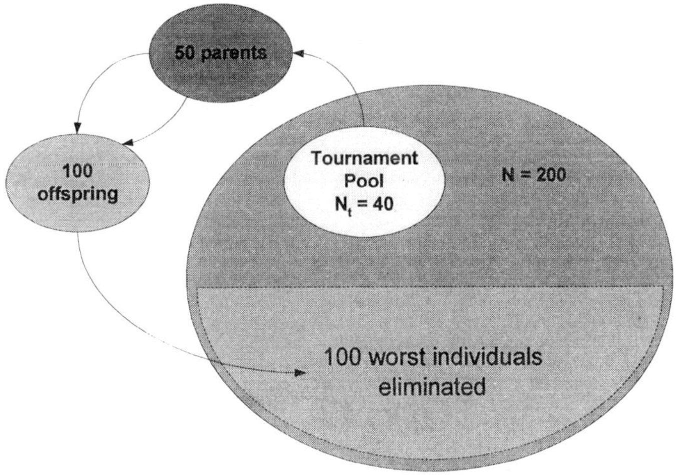

Figure 3. Population Selection and Replacement Process

Braking Criteria

Genetic algorithms traditionally suffer from the problem that they have no convergence criteria. In other words, the algorithm doesn't know when to stop. For this reason, algorithm braking criteria are used to halt the algorithm when certain conditions are met. These conditions are:

1) The range of the best individual in the population is less than or equal to 5e-6 meters.
2) The GA has not updated the best individual for at least 40 generations.

The algorithm must meet both conditions to stop.

ALGORITHM PERFORMANCE

This section discusses the performance of the optimization algorithm, and presents some representative results.

General Results

The test case outlined previously was run with a GA 100 times. The GA was created in a Matlab m-file. The m-file used STK to propagate both vehicles in a 21 x 21 Earth gravity model. Atmospheric drag was introduced via the different ballistic coefficients of the vehicles. Accelerations from the sun and moon were included in the experiment. Solar radiation was also included, however its contribution was very small compared to the other perturbing forces.

For each of the 100 cases evaluated the data gathered includes the final range error and the $\mathbf{V}_0$ vector chosen by the GA to minimize the range error. For some cases, chosen at random, the $\mathbf{V}_0$ vector as a function of GA generation was also returned. Figure 4 shows 100 final range errors as selected by the GA. Every iteration of the

GA returned a result that is better than the CW solution by a large margin. Figure 5 shows the magnitude of $\mathbf{V}_0$ for each iteration of the GA. The magnitude of the CW solution is 0.029437 m/s and the magnitude of the mean of the velocities chosen by the GA is 0.029159 m/s.

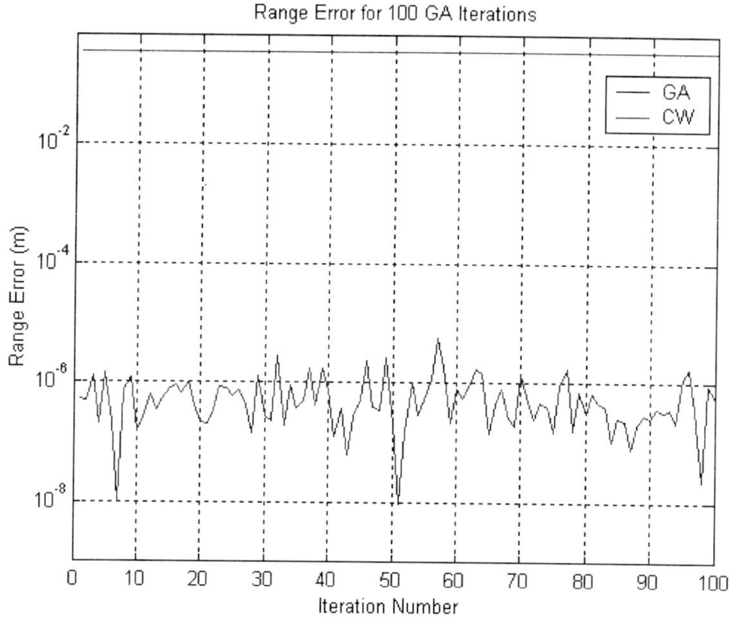

Figure 4. Range Error for 100 Iterations of the GA

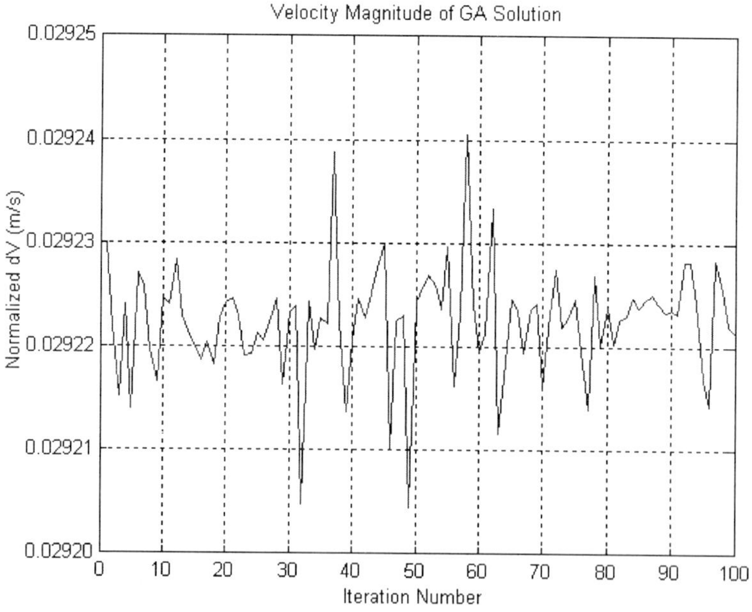

Figure 5. Normalized velocity for 100 iterations of the GA. The normalized velocity for the CW solution is 0.029437 m/s.

Figure 6 shows the range magnitude of one complete iteration of the GA. The horizontal axis is the generation number of the GA and the vertical axis is the range error of the individual with the best (i.e., the smallest) range error.

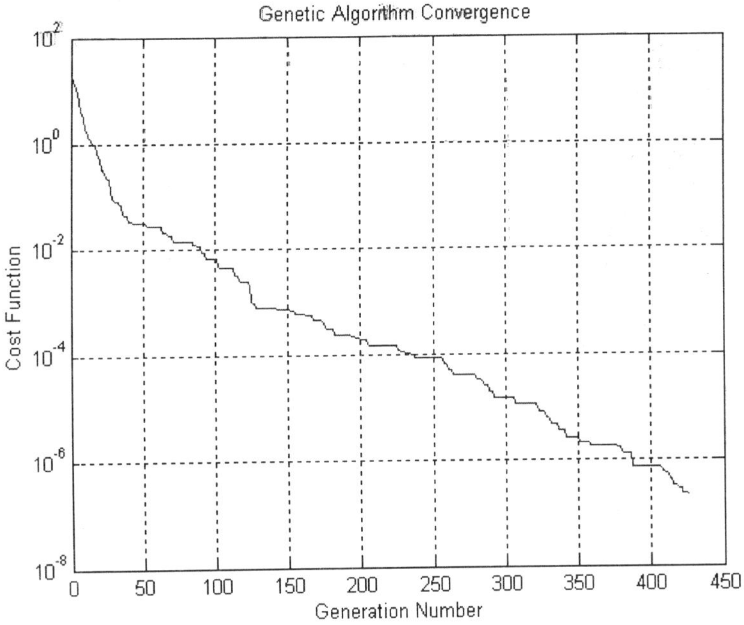

Figure 6. Range Error As A Function Of Generation Number For One Typical Iteration Of The GA

Statistical Performance

Figure 7 is a histogram of the range data. The data is shown in ten bins, each of width 2.832640e-7 m. The data appears to be normally distributed on the range $(0,\infty)$, which is indicative of the "half-normal" distribution function. The expected value and variance of the half-normal distribution are given by[2]

(12)
$$\hat{x} = \frac{\sum_i x_i}{N} = 6.362061\text{e-}007$$
$$\hat{\sigma} = \frac{\sum_i (x_i - \hat{x})^2}{N} = 8.537045\text{e-}007$$

where N is the number of data points in the set (100 in this case). Figure 7 also shows the half normal probability density function for this data set. The reduced χ^2 for the data shown is less than 1, indicating a reasonable fit of the half-normal probability distribution function (PDF) to the observed data. It is expected, based on prior research (Ref. 3), that more runs of the GA will produce more evenly distributed data that will approach a half normal fit.

[2] The expected value and standard deviation are calculated via the maximum likelihood estimator.

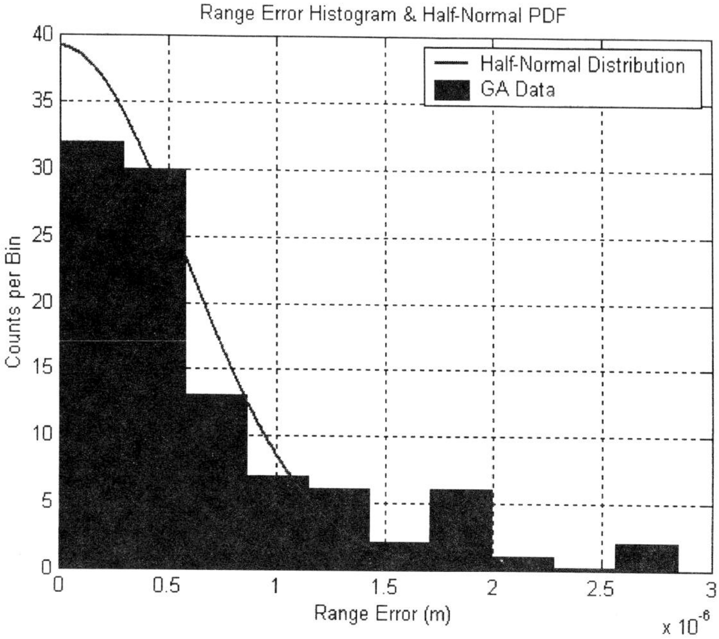

Figure 7. Range Error Histogram and Half-Normal Probability Density Function

Figure 8 shows a histogram of the magnitude of the velocity chosen by the GA. The distribution appears to be normal. The mean and standard deviation[3] of the magnitude of the range error is 6.36206E-07 m and 5.72154E-07 m. The magnitude of the range error after applying the velocity predicted from the CW equations is 0.362715 m. Thus, the GA range error is only about 1.8e-4% of the CW range error. Figure 9 shows the distribution of the Radial, In-track, and Cross-track velocity components. The mean and standard deviations are summarized in Table 1.

Table 1. Mean and Standard Deviation Summary

Component	Units	Mean of GA Data	Std Dev	CW Result	Ratio GA/CW
Radial Velocity	m/s	-7.05185E-03	4.70915E-10	-7.20182E-03	0.9792
In-Track Velocity	m/s	2.82932E-02	4.45047E-10	2.85424E-02	0.9913
Cross-Track Velocity	m/s	-2.01036E-05	1.69485E-10	-9.60873E-06	2.0922
RSS Velocity	m/s	2.91588E-02	5.41890E-10	2.94370E-02	0.9905
Range Error	m	6.36206E-07	5.72154E-07	3.62715E-01	1.754E-06

[3] The standard deviation is defined as the square root of the variance, which is the same as the standard deviation for a normal distribution.

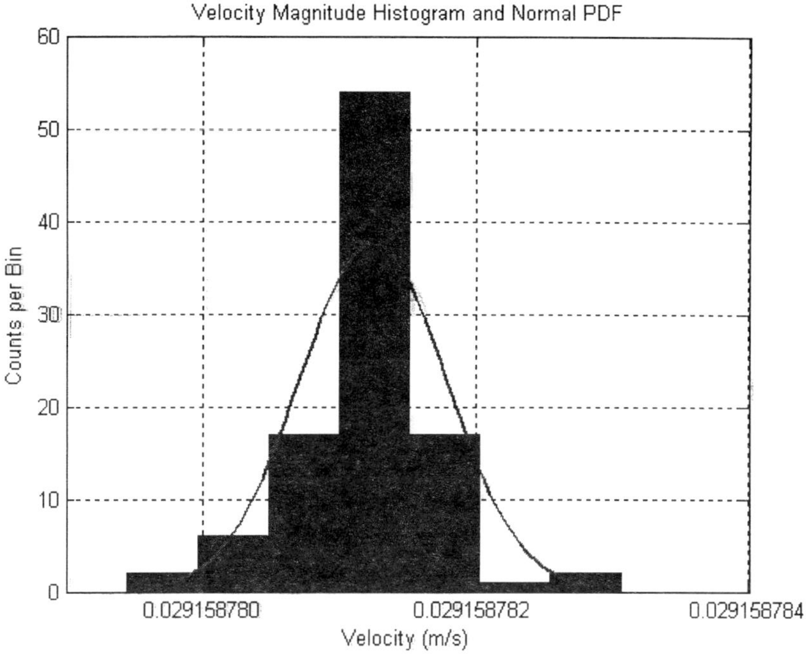

Figure 8. Velocity Magnitude Histogram with Bin Size equal to 5.16840e-010 m/s.

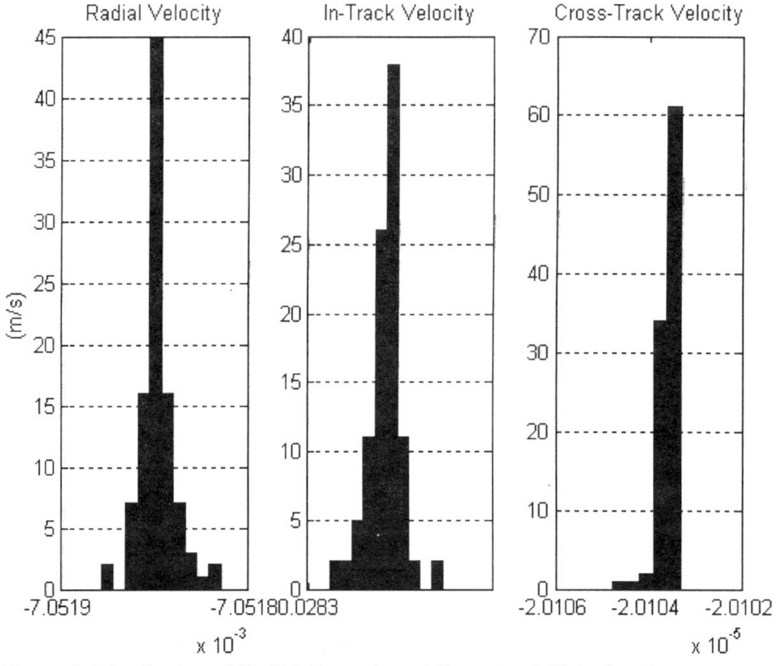

Figure 9. Distribution of Radial, In-track, and Cross-track Velocity Components

CONCLUSIONS

This paper has presented a methodology for optimizing a satellite rendezvous trajectory based on input from the closed-form Clohessy–Wiltshire equations. A Genetic Algorithm was used to minimize the range error after an impulsive maneuver is performed. After running 100 iterations of the GA optimizer, the resultant range errors were always significantly less than that for the Clohessy-Wiltshire approach, and the magnitude of the required V was about the same (however, the direction of the V vector was slightly modified). Furthermore, a statistical analysis has been presented which shows that the GA tends to choose radial, in-track, and cross-track V components which are normally distributed about a mean value with very tight standard deviations. It is expected that further refinement of this type of tool would be a valuable resource for precision rendezvous maneuvers in the presence of high fidelity gravity models and other high fidelity accelerations, including drag, third body, and solar radiation acceleration.

REFERENCES

1. W. H. Clohessy and R. S. Wiltshire, "Terminal Guidance System for Satellite Rendezvous," *1959 IAS National Summer Meeting,* June 16-19, 1959.
2. W. Weisel, *Spaceflight Dynamics*, : McGraw-Hill, Boston, MA: McGraw-Hill, 1997.
3. B. Carpenter and W. Jackson, "Stochastic Optimization of Spacecraft Rendezvous Trajectories," in Proc. IEEE Aerospace Conference, 2003.
4. M.M. Flood, "The Traveling Salesman Problem," Oper. Res. 4, pp. 61-75, 1956.
5. T.C. Fogarty, "Varying the Probability of Mutation in the Genetic Algorithm," in Proc. 3rd Int. Conf. On Genetic Algorithms, 1989, pp. 104-109.
6. B. Fu and H. Li, "Closeness of NP-Hard Sets to Other Complexity Classes," SIAM J. Comput., vol. 23, #2, pp. 255-260, 1994.
7. M. R. Garey and D.S. Johnson, *Computers and Intractability - A Guide to the Theory of NP-Completeness.* San Francisco: W.H. Freeman & Co., 1979.
8. D. Goldberg, *Genetic Algorithms in Search, Optimization, and Machine Learning.* New York: Addison-Wesley, 1989.
9. J. Grefenstette, R. Gopal, B. Rosmaita, and D. Van Gucht, "Genetic Algorithms for the Traveling Salesman Problem," in Proc. 1st Int. Conf. On Genetic Algorithms, 1985, pp. 160-168.
10. J. J. Grefenstette, "Optimization of Control Parameters for Genetic Algorithms," IEEE Trans. Systems, Man, and Cybernetics, vol. 16, #1, pp. 122-128, 1986.
11. R. Haupt and S. Haupt, *Practical Genetic Algorithms.* New York: John Wiley & Sons, 1998.
12. W. C. Jackson and M.E. McDowell, "Simulated Annealing with Dynamic Perturbations: A Methodology for Optimization," in Proc. IEEE Aerospace Applications Conference, 1990, pp. 181-191.
13. C. Papadimitriou, K. Steiglitz, *Combinatorial Optimization: Algorithms and Complexity.* Englewood Cliffs, NJ: Prentice-Hall, 1982.
14. J. Potvin, "Genetic Algorithms for the Traveling Salesman Problem," Annals of Operations Research, vol. 63, pp. 339-370, 1996.

AAS 03-018

MODIFICATIONS TO CONTROL, FAULT PROTECTION, AND OPERATIONS OF CONVENTIONAL SPACECRAFT TO SUPPORT AUTONOMOUS RENDEZVOUS BY A SERVICING SPACECRAFT

Kenny Epstein, Jeff Bladt,
Paul Alles, Jay Bloom, David Kaufman and Brad Porter*

INTRODUCTION

Ball Aerospace is part of the DARPA Orbital Express program. The mission is funded through the Defense Advanced Research Projects Agency, the central research and development organization for the Department of Defense. It is scheduled to launch in 2006.

The Orbital Express Advanced Technology Demonstration Program is intended to develop and demonstrate autonomous techniques for on-orbit refueling and reconfiguration of satellites – See Figure 1. During Phase II (Demonstration Phase), the team will perform an on-orbit servicing demonstration using two satellites: the unmanned service vehicle, Autonomous Space Transport Robotic Operations Spacecraft (ASTRO) - built by Boeing Phantom Works, and a surrogate next generation serviceable spacecraft that emulates both a next generation serviceable spacecraft and commodities depot, the Next Generation Commodity Spacecraft (NEXTSat/CSC) - built by Ball Aerospace – See Figures 2 & 3. A key aspect of this project is the development and demonstration of industry standard servicing interfaces and protocols that could be utilized by future spacecraft developers.

On-Orbit Servicing of spacecraft places requirements both on the Servicing vehicle (ASTRO) and the Next Generation Spacecraft (NEXTSat). While the bulk of these requirements fall onto the ASTRO spacecraft, conventional spacecraft that require servicing will require some modifications. The NEXTSat/CSC spacecraft represents future spacecraft that will utilize servicing capabilities. These servicing modifications are not extensive, however there are some specific requirements that will need to be implemented to support servicing. These minimal modifications will allow future spacecraft to focus on their primary mission, while adding features that will allow autonomous servicing by spacecraft.

* Ball Aerospace & Technologies Corporation (BATC), P.O. Box 1062, Boulder, Colorado 80306-1062.

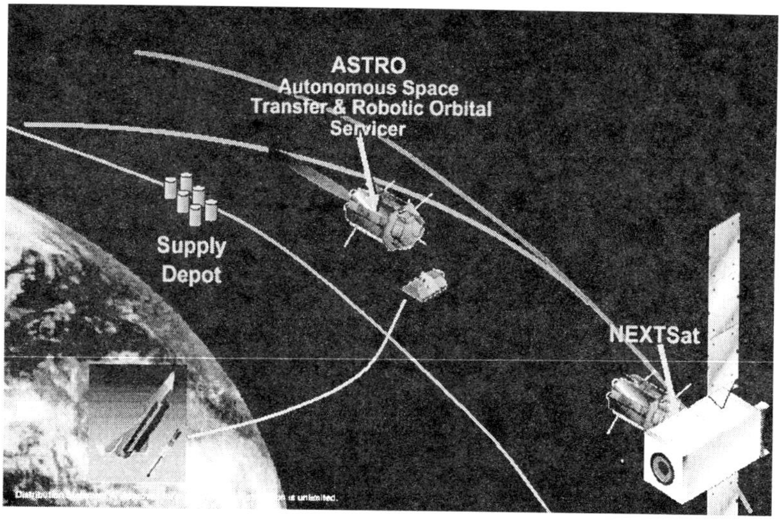

Figure 1 – Orbital Express On-Orbit Infrastructure

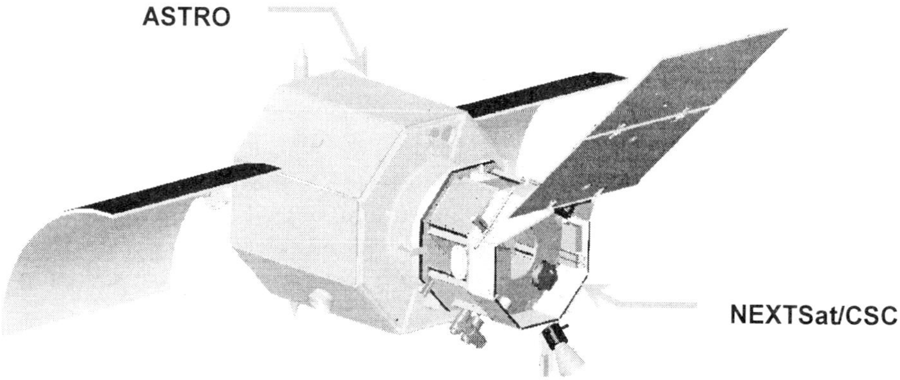

Figure 2 – Phase II Autonomous Space Transport Robotic Operations Spacecraft (ASTRO) & Next Generation Commodity Spacecraft (NEXTSat/CSC)

This paper discusses some of the particular modifications to the demonstration NEXTSat to support rendezvous and proximity operations. The demonstration NEXTSat spacecraft is based on an existing LEO spacecraft, called RS300. Modifications to the RS300 control system, flight software, and operational use will be outlined. Control interactions between NEXTSat and ASTRO, will also be discussed including modifications to NEXTSat required to mitigate interaction effects. Robust servicing solutions also require some Fault Protection features to keep both spacecraft safe during proximity operations. Finally servicing requires some specific flight operation planning to support these rendezvous and proximity operations.

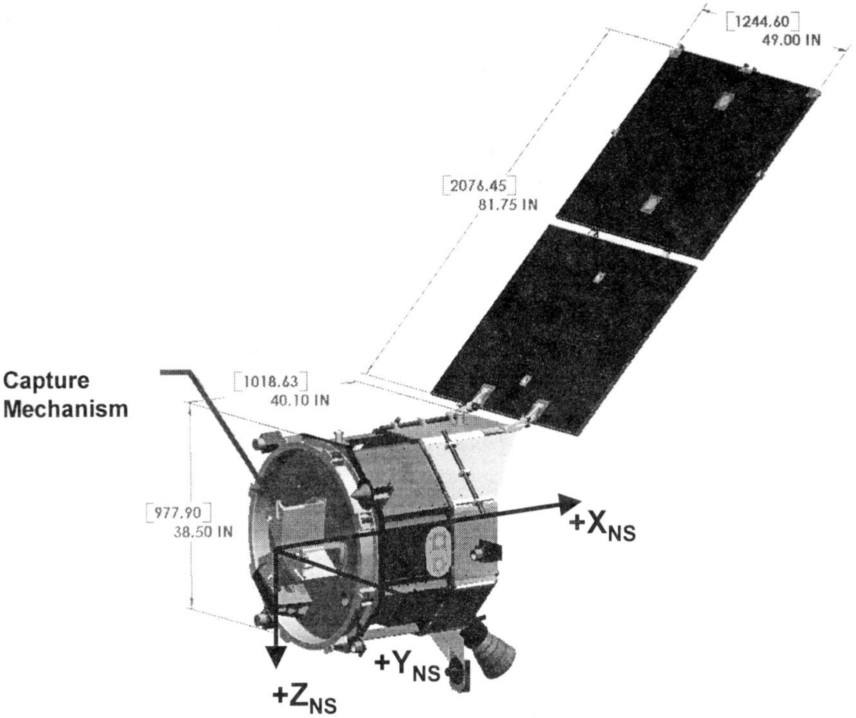

Figure 3 – Phase II NEXTSat/CSC Spacecraft

ATTITUDE CONTROL CONSIDERATIONS

NEXTSat attitude control considerations focus on preserving the capability of the client vehicle to counter disturbance torque sources without adversely impacting the ability of the servicing spacecraft to control the mated configuration. Another important consideration is preventing collisions from attitude slew maneuvers (e.g., sun acquisition slew) from occurring when the two vehicles are operating in close proximity.

In addition to the typical disturbance torque sources (i.e., aerodynamics drag, solar radiation pressure, gravity gradient, and magnetic field interaction), a rendezvous and docking mission introduces spacecraft-to-spacecraft interaction disturbances. Thruster plume impingement, separation-induced angular rates (analogous to launch vehicle tip-off rates), and contact forces imparted during a failed capture (docking) attempt will impact the ability of the NEXTSat vehicle to maintain steady attitude and manage angular momentum accumulation.

Attitude control actuators are sized to accommodate (with adequate design margin) the aforementioned disturbance sources during any given phase of the mission. The angular momentum accumulation budgets in Tables 1, 2, and 3 indicate the actuator storage capacity required for Far Field Operations, Capture Operations, and De-Mate and Separation Operations, respectively.

Angular Momentum Accumulation (Newton-meter-seconds) Far Field Operations, 420-km Orbital Altitude, Flat-Panel Solar Array			
Disturbance Source	Attitude Trajectory		Comments
	Solar-Inertial	LVLH	
Environmental Torque	0.53	0.84	Mean + 2 atmosph. density w/ diurnal bulge; TR30CFR TORQRO
Slew Maneuver	0.50	0.50	180-degree Solar-Inertial-to-LVLH slew time less than 12 minutes
Total	1.03	1.34	Arithmetic sum
Requirement	4.00	4.00	TW4A12 RWA angular momentum capacity
Margin	290%	198%	

Table 1. NEXTSat Angular Momentum Accumulation During Far Field Operations

Angular Momentum Accumulation (Newton-meter-seconds) Capture Operations, 420-km Orbital Altitude, Flat-Panel Solar Array			
Disturbance Source	Attitude Trajectory		Comments
	Solar-Inertial	LVLH	
Environmental Torque	0.53	0.84	Mean + 2 atmosph. density w/ diurnal bulge; TR30CFR TORQRO
Failed Capture Contact	0.25	0.25	0.3 centimeters per second relative velocity at contact
Plume Impingement	0.90	0.90	0.59 N-s net thruster impulse, 0.5% to 4% orbit-avg. duty cycle
Total	1.68	2.00	Arithmetic sum
Requirement	4.00	4.00	TW4A12 RWA angular momentum capacity
Margin	139%	100%	

Table 2. NEXTSat Angular Momentum Accumulation During Capture Operations

Angular Momentum Accumulation (Newton-meter-seconds) De-Mate & Separation Operations, 420-km Orbital Altitude, Flat-Panel Solar Array			
Disturbance Source	Attitude Trajectory		Comments
	Solar-Inertial	LVLH	
Environmental Torque	0.53	0.84	Mean + 2 ? atmosph. density w/ diurnal bulge; TR30CFR TORQROD
Sep.-Induced Ang. Rates	0.56	0.17	0.32 degree per second "tip-off" rate from ASTRO separation
Plume Impingement	0.90	0.90	0.59 N-s net thruster impulse, 0.5% to 4% orbit-avg. duty cycle
Total	1.98	1.92	Arithmetic sum
Requirement	4.00	4.00	TW4A12 RWA angular momentum capacity
Margin	102%	109%	

Table 3. NEXTSat Angular Momentum Accumulation During Separation Operations

Once the servicing spacecraft (ASTRO) has captured (docked) with the client spacecraft (NEXTSat), the former performs attitude control functions for the mated configuration and the latter vehicle's attitude control system enters a passive mode in which zero control torque is commanded. During Far Field Operations the client spacecraft retains full attitude control capabilities.

The transitional phase between separated and mated configurations, where the two vehicles are in close proximity to each other or have made contact but have not retracted the docking mechanisms to achieve a mated configuration, requires that the two vehicles' attitude control systems do not interact adversely. One approach is to place the NEXTSat attitude control subsystem into an operational state in which reaction wheel torque commands are limited to very small magnitudes. This approach precludes the introduction of large disturbance torque values to the ASTRO vehicle from NEXTSat, while allowing NEXTSat to counter the disturbance torque sources described above. This ADCS operational state is termed "Point Limited" Operational State, and it differs from the normal

"Point" Operational State (in which full attitude control is enabled) in that the torque commands are limited to small magnitudes and large angle attitude slew maneuvers are disabled. Disabling attitude slew maneuvers is implemented to prevent collisions during close proximity operations.

Figure 4 illustrates the NEXTSat attitude control subsystem operational state transition sequences. During normal Far Field operations, the NEXTSat attitude control subsystem is in "Point" Operational State. Shortly before the capture (docking) sequence is initiated, NEXTSat is placed in "Point Limited" Operational State. Once successful capture has been achieved the ADCS transitions to "Disengaged" Operational State, where all attitude control torque commands are zero, allowing the ASTRO vehicle to perform attitude control for the mated configuration. During Mated Operations, while ASTRO is controlling attitude, the ground controllers may elect to turn off power to the NEXTSat ADCS sensors and actuators (called "Off" Operational State).

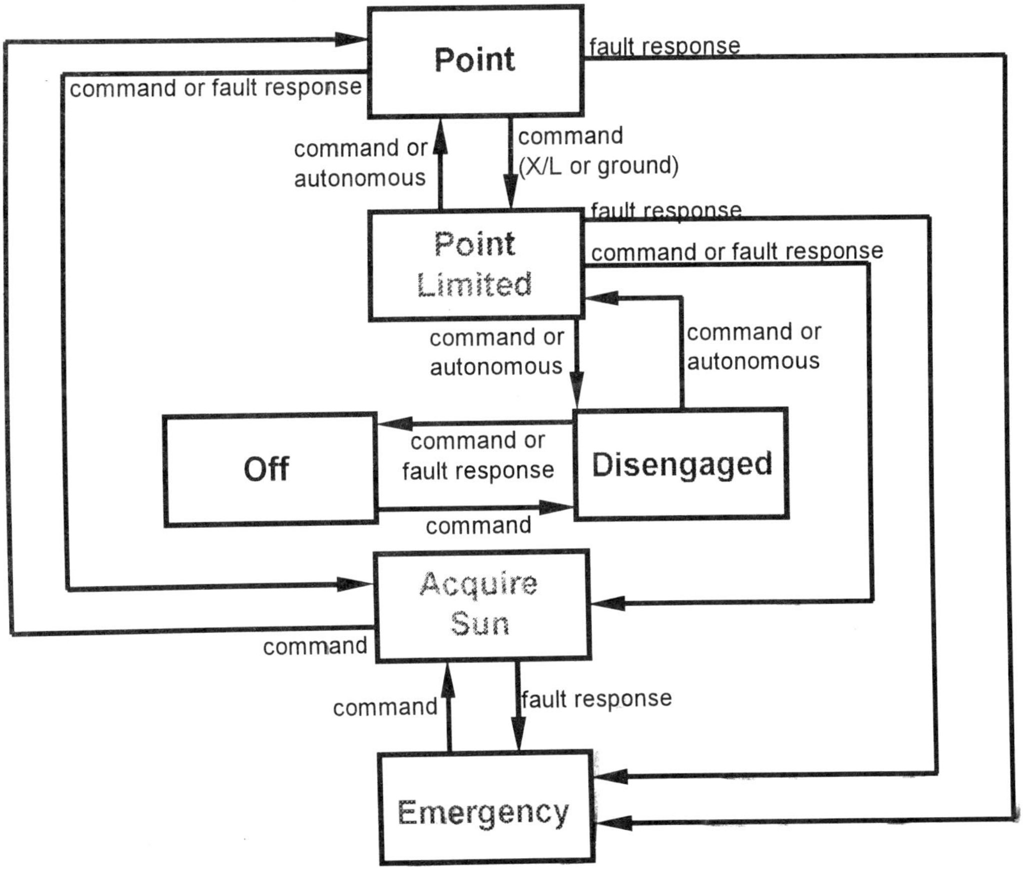

Figure 4 – NEXTSat Attitude Control Subsystem Operational State Transition Diagram

FAULT PROTECTION CONSIDERATIONS

Near Proximity Operations as well as the mating and de-mating events have particular requirements associated with protecting both the NEXTSat and ASTRO from damage, during faults on either NEXTSat or ASTRO. During rendezvous and proximity operations, the burden of fault protection falls on the ASTRO spacecraft. However, the NEXTSat will need some tailored fault responses to protect NEXTSat during proximity and mated operations. These responses need to be non-interfering with potential ASTRO fault responses during the proximity and mated operations.

The NEXTSat provides an automated rendezvous and mating health & fault status telemetry set at 1 Hz during rendezvous and docking operations. This telemetry set constitutes a heartbeat from NEXTSat/CSC Space Vehicle. For the Orbital Express demonstration the NEXTSat automated rendezvous and mating health & Fault status telemetry set shall be broadcast via the optional space-to-space link and space-to-ground links during rendezvous and docking operations. Within this heartbeat is pre-defined telemetry that assists ASTRO in determining the health of the NEXTSat. Depending on the heartbeat telemetry ASTRO may choose to hold until NEXTSat reaches the required state, proceed to the next phase of servicing if NEXTSat is ready, or perform a breakout (collision avoidance) maneuver if NEXTSat indicates a dangerous state. Note that for the demonstration we have chosen to provide this heartbeat via an RF link. Operational systems may choose other implementation methods (IR, optical, etc.) as best suits their Concept of Operations (CONOPS).

In addition to the heartbeat NEXTSat provides an internal hardware latch that persists across processor resets. This latch provides a queue for NEXTSat fault protection system during mated and non-mated operations. Typically during safeing or emergency states the NEXTSat will point the solar arrays towards the Sun and shed non-essential power loads. This fault response during mated operations would interfere with the control system of ASTRO. During mated operations attitude control responses are disabled. When this latch is set to the mated configuration, attitude control actuation remains disabled until ground intervention. These modes are summarized in table 4.

Orbital Express Operations	NEXTSat Modes	Available Attitude Control States
Safe Operations	Mated Safe	Off
	Unmated Safe	Acquire Sun
Emergency Operations	Mated Emergency	Off
	Unmated Emergency	Emergency

Table 4 NEXTSat Operational Modes during Safeing or Emergencies

During near proximity (or unmated) operations the NEXTSat delay's sun pointing fault responses by 60 seconds. This allows ASTRO the time to perform a breakout maneuver. As an example NEXTSat (during near proximity operations) may have severe processor reset or multiple resets. This would invoke an emergency mode response on

NEXTSat. During this fault the heartbeat would be interrupted, which gives ASTRO an indication of a NEXTSat fault. ASTRO would be responsible for ensuring safe separation, knowing that NEXTSat may start a sun-pointing (acquire sun) maneuver within 60 seconds.

OPERATIONAL CONSIDERATIONS

The Orbital Express program will demonstrate rendezvous and proximity operations that include commanding of NEXTSat via communications directly from the ground or indirectly via an optional ASTRO to NEXTSat space-to-space link. In addition mating operations of NEXTSat will be achieved by two different methods.

1. Direct mating by ASTRO to NEXTSat at the capture mechanism. The ASTRO rendezvous directly with NEXTSat with no intermediate actions

2. The ASTRO captures the NEXTSat using the ASTRO Manipulator Arm Assembly, and then berths (or mates) the NEXTSat to ASTRO at the capture mechanism.

The modifications to the NEXTSat support all of these capture operations. The future serviceable spacecraft may choose to only a subset of these operations, but the Orbital Express program will demonstrate all of these options. As discussed previously NEXTSat will demonstrate the use of Disengaged and Point limited Attitude Control states. The availability of these states are summarized in table 5

Orbital Express Operations	NEXTSat Modes	Available Attitude Control States
Servicing Operations	Approach (or departure)	Point Limited
	Separation	Disengaged or Point Limited
	Capture	Disengaged or Point Limited
	Grapple	Disengaged or Point Limited

Table 5 NEXTSat Operational Modes during Servicing

During the approach or departure phase of ASTRO the NEXTSat spacecraft will be in point limited state. If the entire sequence is done in Point limited no intermediate transitions are required. In this scenario the NEXTSat would be transitioned to the disengaged state after successful capture is completed, as illustrated in figure 4. This simplifies ground tasking or contact time, as well as increasing autonomy, if the servicing is being performed without a space-to-space link. If transitions to a disengaged state are desired before final capture, then this can be accomplished through additional tasking from the ground or via the optional space-to-space link. In a similar fashion the NEXTSat can remain in a disengaged state until fully separated from ASTRO. Transition to point limited (or other pointing states), can be achieved by time sequences, ground command, or by ASTRO after safe distances are achieved between ASTRO and NEXTSat.

Depending on the operational desires of future serviceable spacecraft all or some portion of these operational scenarios can be employed.

CONCLUSION

The modifications to the RS300 LEO spacecraft, outlined in this paper, show a flexible and robust approach to supporting autonomous on-orbit servicing. The modifications, while supporting specific servicing requirements, are kept at a minimal level. This approach provides additional design considerations for future spacecraft, which also desire autonomous servicing.

A successful Orbital Express demonstration will show how on-orbit servicing can reduce space program costs and improve mission utility through spacecraft life extension and/or improved maneuverability, which could have a significant positive impact on future space systems.

ACKNOWLEDGMENT

Ball Aerospace & Technologies Corp. is member of the Boeing Phantom Works lead Orbital Express Team, which is under contract from DARPA. The Ball team participates actively in supporting our direct customer Boeing as well as Boeing's customer DARPA. Ball Aerospace & Technologies Corp. works closely with the other members of the Boeing Phantom Works Orbital Express Team, which includes Northrop Grumman (formerly TRW Spacc and Tcchnology), McDonald Dettwiler Robotics, Draper Laboratory Inc., and Starsys Research Corp.

Section III
INTERPLANETARY MISSIONS

SESSION III

Joint Session Chairperson:	Luisella Giulicchi ESA/ESTEC
Joint Session Chairperson:	James E. Randolph Jet Propulsion Laboratory
Local Session Chairperson:	Dave Sonnabend Analytical Engineering

The following papers were not available for publication:

AAS 03-024 "Multiple Asteroid Survey Missions Using Solar Sail Propulsion," by G.W. Hughes, C.R. McInnes, University of Glascow (Paper Withdrawn)

AAS 03-025 "Mars Odyssey Interplanetary Navigation Strategy," by R.A. Mase, P. Antreasian, J. Bell, NASA/JPL (Paper Withdrawn)

AAS 03-027 "Genesis Early Flight Experience," by K.E. Williams, R.S. Wilson, NASA/JPL (Paper Withdrawn)

The following paper numbers were not assigned:
AAS 03-028 to -030

AAS 03-021

THE ROSETTA MISSION: RENDEZVOUS WITH AND LANDING ON A COMET

S. P. Airey , J. van Casteren, G. Coupé, T. van Overbeek and J. Rouquet*

The European Space Agency will launch the Rosetta spacecraft in January 2003† on a more than 10 years voyage to comet Wirtanen. The Rosetta orbiter will detach a scientific package which will land on the comet surface to perform in situ measurements of the comets composition. The orbiter will escort the comet for about one year and its extensive array of scientific instruments will study the comet in the using optical imaging, remote sensing, chemical composition measurements and magnetic field measurements.

The goal of this paper is to provide a detailed account of both the Rosetta mission and the spacecraft designed to cope with the principal challenges of the mission. The long mission duration is driven by the need to perform three planetary flybys in order to increase the spacecraft velocity to synchronise its orbit with the Wirtanen orbit. Included is a flyby of Mars. Passing at not more than 200 km altitude above the planets surface, this is considered a very critical manoeuvre for the spacecraft. In addition to the planetary flybys, two close asteroid flybys are planned, further increasing the scientific return. Even exploiting these multiple flybys, the spacecraft still must carry about 55% of its launch mass in propellant in order to adjust its orbit and finally to rendezvous with the comet.

As well as the long duration and complex flybys, another real challenge of the mission comes from the very large solar distances at which the spacecraft will have to travel and operate. Flying beyond the orbit of Jupiter, Rosetta will be the most distant spacecraft ever to be powered using only solar cells. Special solar cells have been developed to be optimised for the low temperatures and low intensity operation, even so, the solar arrays need to be huge. At nearly 33 meters long, they produce about 10KW at earth distance reducing to about 400 W past Jupiter.

The thermal control over distances varying between less than 1 AU and more than 5 AU is also a tough challenge requiring both super-insulation and heating in deep space and considerable heat rejection close to the earth orbit.

* European Space Agency, ESTEC, Keplerlaan 1, 2200 AG Noordwijk ZH, The Netherlands.

† Due to problems with the launcher, the Rosetta mission has been delayed. The mission will be re-scheduled with a different comet but the general mission outline is expected to remain similar. This paper discusses the original target comet, Wirtanen.

The large Earth to spacecraft distances, coupled with the long duration mission and the presence of 'key' phases needing rapid responses also presents problems. Communication turn-around times will be up to 1.5 hours in deep space and real-time intervention from the ground is clearly impossible. Moreover the spacecraft will be inaccessible or invisible for some considerable periods due to sun-earth-spacecraft conjunctions. Also ground operations have to be minimised to keep costs within limits. As a result of these considerations, the spacecraft must to be equipped with considerable redundancy and on-board, real time, autonomy to cope with hardware and software failures. Due to the nature of the mission, the spacecraft must rapidly, safely and autonomously recover from such failures and continue its nominal operations.

A final challenge of this mission is that, after enduring a 10 years journey to the comet, the spacecraft will have to perform it's critical lander delivery manoeuvres and then perform exacting scientific measurements while flying inside the comet gas and dust cloud. This imposes great difficulties, in particular, on the operation of the attitude control system. The problems and their solutions will be presented but a detailed discussion is outside of the scope of this paper

INTRODUCTION and Mission Objectives

Rosetta is the third Cornerstone mission in the scientific programme from the European Space Agency. It comprises a large orbiter, which is designed to operate for a decade at large distances from the Sun, and a small lander. Each of these carries a large complement of scientific experiments designed to complete the most detailed study of a comet ever attempted. After entering orbit around Comet Wirtanen in 2011, Rosetta Spacecraft will release a small lander for a soft landing on the icy nucleus, then spend the next two years orbiting the comet as it heads towards the Sun. On the way to Comet Wirtanen, Rosetta will receive gravity assists from the Earth and Mars, and fly past two main belt asteroids Otawara (2006) and Siwa (2008).

Figure 1: The Comet Wirtenan

The large Earth to spacecraft distances, coupled with the long duration mission and the presence of 'key' phases needing rapid responses drive much of the spacecraft design philosophy. Communication turn-around times will be up to 1.5 hours in deep space and real-time intervention from the ground is clearly impossible. Moreover the spacecraft will be inaccessible or invisible for some considerable periods due to sun-earth-spacecraft conjunctions. For cost reasons, ground operations during such a long mission must also be minimised to meet budget constraints. As a result of these considerations, the spacecraft must be equipped with considerable redundancy and on-board, real time, autonomy to cope with hardware and software failures. Due to the nature of the mission, the spacecraft must rapidly, safely and autonomously recover from such failures and continue its nominal operations.

The Flybys of Otawara and Siwa will provide, for the first time, detailed images and scientific data on these asteroids that previously have only been seen as dots of light. Very little is known about these asteroids other than their mass, orbits and assumed dimensions. Siwa is the largest asteroid that will ever have been studied by a spacecraft with an estimated size of 110 km. Otawara on the other hand will be one of the smallest a mere 3 km. The respective sizes of these asteroids compared to other famous asteroids is shown in Figure 2.

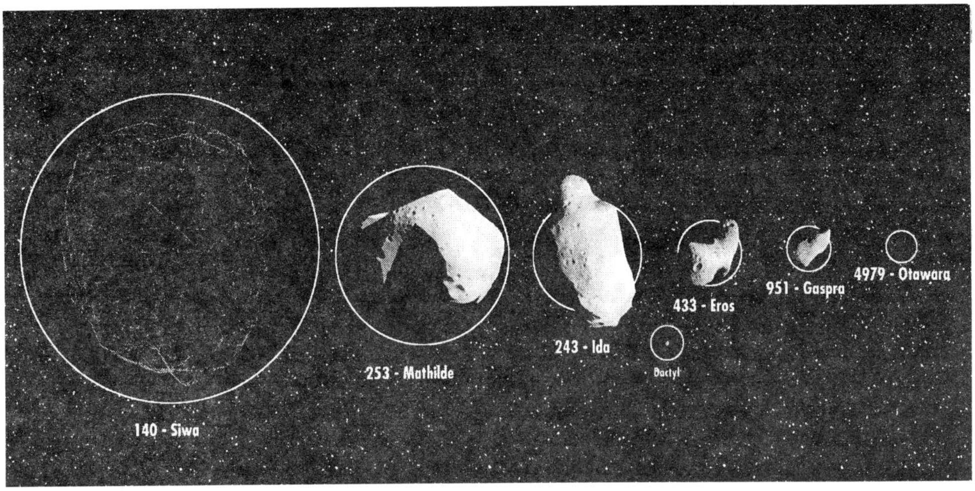

Figure 2: Siwa and Otawara

With less than 4 years from the beginning of the development phase to the launch, the tight programme required tight schedule control combined with highly efficient teamwork. Some 60 European companies formed the industrial team, led by Astrium GmbH of Friedrichshafen (Germany). Astrium SAS of Toulouse (France) were given the responsibility for the design and development of the Avionics which includes the Data Management System (DMS) and the Attitude and Orbit Control and Measurement Subsystem (AOCMS).

This paper serves as an introduction to the mission and concentrates on providing a overview of the key parts and concepts of the design from an AOCMS perspective. Due to the very complex nature of the mission and the large amount of on-board autonomy, certain simplifications and generalisations are made in the descriptions. This is done to ease the understanding of the concepts and high level design which are the focus of the paper, preventing the reader becoming swamped with details and exceptions.

Mission Overview

Rendezvousing with a comet in deep space is not an easy task and requires a large amount of energy. In order to rendezvous with Wirtanen in 2011, at a distance of 4.2 AU, Rosetta had to be launched during a very small launch window extending from 13th to 31st January 2003. Due to the rotation of the earth the launch opportunities were further restricted to only 20 minutes per day.

After launch, Rosetta still requires three planetary swingbys in order to gain sufficient energy to reach its target. These gravity assists come from a swingby of Mars in 2005, closely followed by a first swingby of the Earth in the same year. Rosetta then performs the Otawara asteroid flyby in 2006 before heading back for a second Earth swingby in 2007 that will set the spacecraft on its final trajectory. In 2008 Rosetta will perform the Siwa flyby on its way out to it maximum distance of 5.2 AU from the sun. During the time around apihelion the energy received from the sun is only 4% of that at earth meaning the solar arrays generate only 400 watts of power. In order to conserve power and survive this period the spacecraft enters a spin stabilised hibernation mode during which almost all units are powered down. After around 2 years of hibernation the spacecraft autonomously 'wakes up' and preparations for the comet rendezvous begin. Rosetta eventually rendezvous with Wirtenan in 2011.

The spacecraft will first enter a global mapping phase, orbiting the comet at a distance of 8 km during which time preparations for the lander separation will be made. Once a landing site is selected, the lander will be dropped for a soft landing on the surface and the orbiter will continue it's detailed study of the comet for the following 12 months. An overview of the mission is shown schematically in Figure 3 and summarised in Table 1.

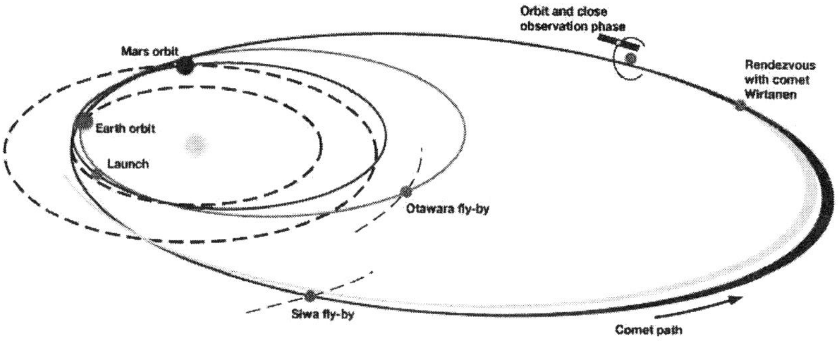

Figure 3: Mission Overview

Launch	13 January 2003 (Central European Time)
Mars gravity assist	26 August 2005
First Earth gravity assist	28 November 2005
Asteroid Otawara flyby	11 July 2006
Second Earth gravity assist	28 November 2007
Asteroid Siwa flyby	24 July 2008
Comet Wirtanen rendezvous	29 November 2011
Comet mapping	May-June 2012
Landing on the comet	July 2012
Escorting the comet	July 2012 – July 2013
Nominal end of mission	10 July 2013

Table 1: Mission Summary

Configuration

The design and layout of the Rosetta spacecraft is dominated by the large solar arrays and the central box structure. The structure consists of a central cylinder of aluminium honeycomb with shear panels connecting the side panels on which the spacecraft and payload equipments are mounted. In order to maintain communication and power while performing the flybys and the cometary mapping, the 2m diameter HGA and the solar arrays must both be steerable. Despite using three gravity assists, Rosetta still needs to make some large delta-v manoeuvres. These, together with the attitude control fuel needs, require that more than half of the spacecraft launch mass be made up of propellants. This large mass of propellant is stored in two large identical tanks, stacked one on top of the other at the centre of the spacecraft. A summary of the spacecraft characteristics is given in Table 2.

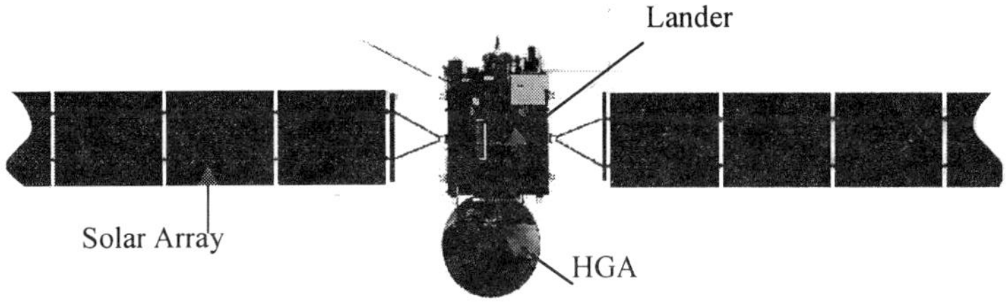

Figure 4: Rosetta External View

Size	
main structure	2.8 x 2.1 x 2.0 meters
wingspan of solar arrays	32 meters
Launch mass	
total	2,950 kg
propellant	1,645 kg
science payload	165 kg
lander	100 kg
Solar array output	850 W at 3.4 AU, 395 W at 5.25 AU
Propulsion subsystem	24 bi-propellant 10N thrusters
Operational mission	10,5 years

Table 2: Spacecraft characteristics

Avionics Architecture and Interfaces

The AOCMS provides the spacecraft with a largely autonomous capability to encounter two asteroids and a comet, to make extensive asteroid and comet observations and to place a scientific package on the cometary surface. The Avionics Major Subsystem embodies all central intelligence on board the spacecraft. In order to achieve a high success probability for this long lasting deep space mission the design must be, and is, very flexible and highly autonomous. The Avionics, through Data Management Subsystem (DMS), provides data management of all payload and spacecraft subsystem data, controlling the overall autonomy of the system, the command acquisition and distribution, data storage, data compression and formatting.

The Avionics also includes the Attitude and Orbit Control and Measurement Subsystem (AOCMS) which handles all of the attitude measurement and control needs of the Spacecraft. The DMS and AOCMS are both hosted within a single computer, the Central Data Management Unit (CDMU). This consists of two physically separated mechanical units that together house 4 processor modules, the TC-decoders, Transfer Frame Generators (TFG), reconfiguration modules and communication buses drivers. This unit is the heart of the avionics and is heavily internally cross-strapped to provide multiple failure tolerance. Each of the four processor modules are capable of being configured, on boot up, to host either the AOCMS or DMS software. The one selected for the DMS function acts as the bus master and is also in charge of platform subsystem management (TTC, Power etc). The AOCMS processor module is in charge of all sensors, actuators and the High Gain Antenna (HGA) and Solar Array (SA) drive mechanisms. The CDMU also interfaces to the Solid State Mass Memory (SSMM) for data storage. The four reconfiguration modules also house the clock generators and are used, with majority voting, to monitor and control the CDMU operation and reconfiguration.

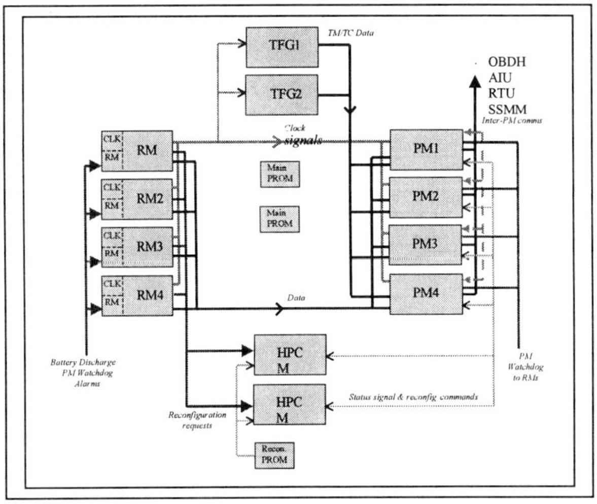

Figure 5: CDMU schematic architecture

Data Management Subsystem (DMS)

The Data Management Subsystem (DMS) is in charge of telecommand distribution to other spacecraft subsystems and payload, of telemetry data collection from spacecraft subsystems and payload and formatting, and of overall supervision of spacecraft and payload functions and health. The DMS is based on a standard OBDH bus architecture enhanced by high rate IEEE 1355 serial data link between the different Avionics processors, the SSMM and the Navigation Cameras. The OBDH bus is the data route for data acquisition and commands distribution via the RTUs. Payload Instruments are accessed via a dedicated Payload RTU. Subsystems are accessed via a dedicated Subsystem RTU.

Attitude and Orbit Control Measurement Subsystem (AOCMS)

The AOCMS is in charge of all the attitude and orbit measurement and control functions. It is in command of all the sensors and actuators required for autonomous attitude determination and control and performs all the pre-programmed manoeuvring. AOCMS subsystem hinges around the AOCMS Interface Unit (AIU), which is used by the AOCMS software to exchange functional data with a complete set of sensors and actuators via a variety of different electrical interfaces (IEEE1355, RS422, MACS and analogue). The sensors and actuators accessed through the AIU are; three Inertial Measurement Packages (IMP), two Star Trackers (STR) and Navigation Cameras (CAM), four Sun Acquisition Sensors (SAS) and four Reaction Wheels (RWA).

The AOCMS AIU also interfaces with and has control over the Solar Array Drive Mechanism, the HGA Drive Mechanism and the Reaction Control Subsystem.

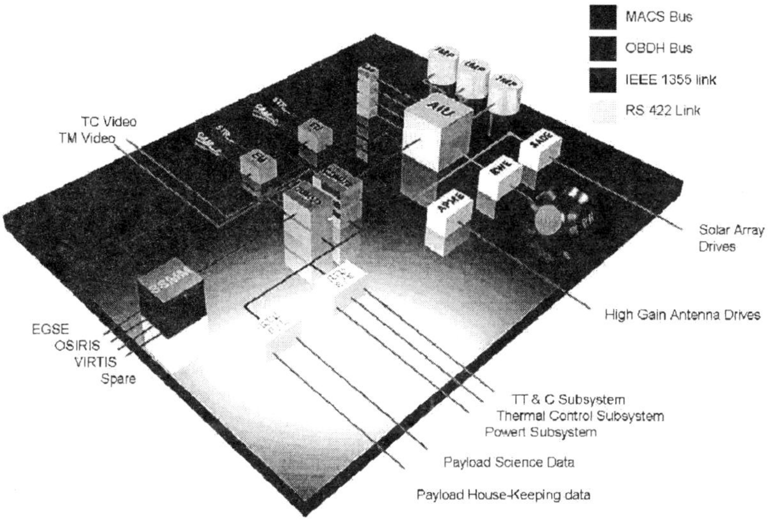

Figure 6: Schematic of Avionics Architecture

The on-board Software for Rosetta is built on a multitasking operating system. The data transfer within the spacecraft is asynchronous, which offers great flexibility. A high level of autonomy has had to be incorporated into the design. This is implemented almost entirely in the software and enables the spacecraft even to cope with failure cases and recovery autonomously.

AOCMS Modes

In order to perform the large number of varied operational tasks demanded by the mission, the AOCMS for Rosetta has been split into 10 software-operating modes of three different types: survival, operational and hibernation. Most of these modes are further divided into several subphases which are in change of well-defined specific activities. The mode diagram, which also shows the allowed mode transitions, is shown in Figure 7. The mode subphases are not discussed further in this paper.

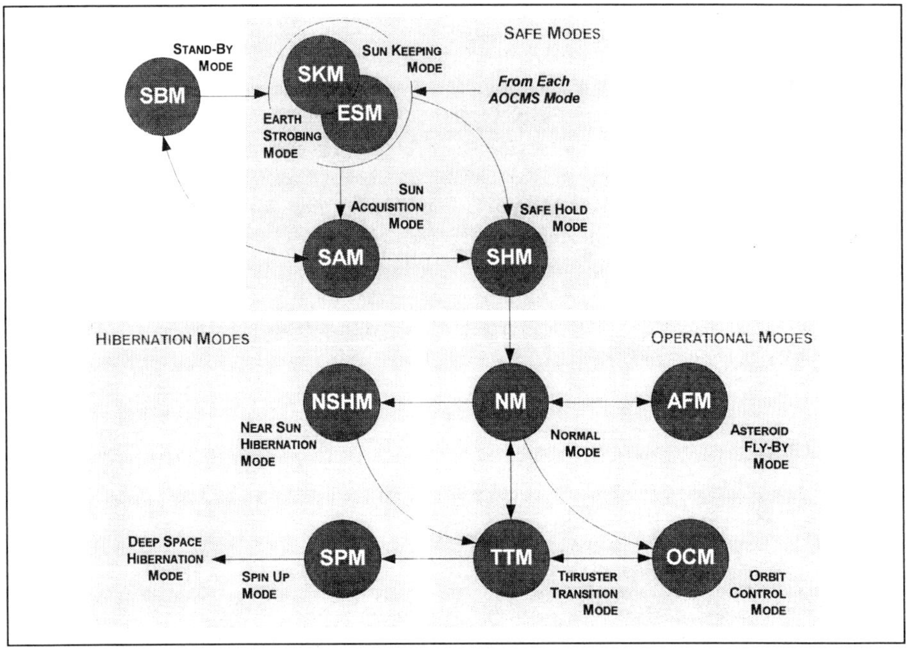

Figure 7: AOCMS Mode Architecture

The **Stand-By Mode (SBM)** is used in Pre-launch and Launch Modes for general check supervision. Only the DMS functions are activated. However, It is possible to command thrusters through AIU for RCS Priming.

Safe Modes

The **Sun Acquisition Mode (SAM)** performs the first attitude acquisition after launch and then again after solar array deployment to enable the solar arrays to be pointed towards the Sun. The mode is also used as a second level backup mode. SAM uses first the gyros and thrusters to reduce the potential spacecraft rate to less than 0.25 deg/s on each axis. It then ensures a permanent 2-axes stabilisation of the Spacecraft using the body mounted SAS and maintains a slow spin around the sun line. The transition to a 3 axis attitude control state is started by commanding the STR (which is still outside the control loop at this stage) into Autonomous Attitude. Once a valid 3-axis attitude is provided by the star tracker, its consistency with the other attitude information available on board (i.e. Sun direction measured in spacecraft frame and on-board ephemerides) is tested. This enables the autonomous transition to Safe Hold Mode (SHM) which complete the transistion to three axis attitude control.

The **Sun Keeping Mode (SKM)** is the first level backup mode when a major attitude failure is detected and continuation of the current operational mode is no longer feasible. It is equivalent to the SAM, except that the control is based on 2 SAS mounted on the solar arrays, instead of the body mounted SAS. This mode is used to perform, in case of failure, a transition to SHM without the loss of the solar array to sun pointing (i.e. reacquisition without reliance on the battery). Once SKM has converged, the autonomous transition to SHM is completed in the same manner as for SAM.

If the automatic SKM to SHM transition is not possible for some reason, then **Earth Strobing sub-Mode (ESM)** is entered. Its aim is to allow the ground to recover the communications. The general principle is to implement a rotation of the Spacecraft around the sun direction, while keeping the solar arrays Sun pointed, and to ensure that the Medium Gain Antenna (MGA), whose boresight points towards the spacecraft +X axis, describes a cone defined by the Sun-Spacecraft-Earth angle, providing a periodic link with the Earth (carrier signal only), once per revolution. In order to achieve the (along mission variable) adequate angle between the Sun direction and the MGA, the sun sensors implemented on the Solar Arrays are used, and the Solar Arrays are rotated. The Spacecraft attitude is then controlled in Solar Array axes, as in SKM. The duration between two consecutive links with the ground is very stable, enabling ground to compute the right instant to send a "stop strobbing" TC. This TC leads the spacecraft to a 3-axis stabilised attitude with the +X axis (and hence the MGA) Earth pointed and the SA perpendicular to the Sun.

The **Safe Hold Mode (SHM)** completes the sequence initiated by the Sun Acquisition Mode or the Sun Keeping Mode with the achievement of full 3 axis attitude control. The X axis (and also therefore the MGA) is permanently oriented toward the earth and the SA to the sun. This means the spacecraft is able to remain safe in this mode for an indefinitely. On reaching this holding attitude, the HGA control is enabled allowing high bandwidth communication. The reaction wheels are then spun up to a predefined rate ready for use in the control loop. Once the commanded wheel rates have been reached, actuation capability is autonomously transferred from thrusters to reaction wheels.

Operational Modes

The **Normal Mode (NM)** is designed to be used for most of the mission operational phases. Attitude control in Normal Mode is based on reaction wheels, gyros and star trackers. The gyroless estimation is reserved to Earth/Sun pointing quiet cruise phases. The spacecraft commanded attitude is processed autonomously on board for sun or Earth pointing and for slew maneouvres. Alternatively, say for comet observation, the attitude can be commanded to follow an on ground calculated and uplinked quaternion profile. However, whatever spacecraft pointing is commanded, the Solar arrays are kept oriented towards the Sun and the High Gain Antenna pointed towards the Earth. The attitude is estimated on board using a Kalman filter and can operate with or without gyro inputs.

The reaction wheels are subject to autonomous momentum management meaning that off-loading is performed autonomously with thrusters as soon as the on-board measured wheel rate is higher than an upper threshold, or lower than a lower threshold. In this way both wheel saturation and zero crossings are avoided. As soon as the wheel rate reaches a warning threshold (which is lower than the autonomous offload threshold), a warning message is downlinked to ground to enable ground to take control of the off-loading should they wish. This could be done to save fuel for instance through the proper orientation of the solar arrays thereby using solar pressure torque as the off-loading control torque.

It is also possible to use NM for performing small orbit manoeuvres without the need to entering TTM or OCM. In order to do this, Ground directly commands short thruster impulses while in Wheel Controlled Damping Phase on NM. The disturbance torques induced by any thrusters firings are then backed off by the wheels.

The **Orbit Control Mode (OCM)** is specifically intended to perform orbit manoeuvres and corrections. For large manoeuvres a 'turn and burn' method is used. The Spacecraft Z axis is first orientated to the proper direction and then a thrust is generated along Z-axis. X and Y axis attitude control is performed by thrusters off-modulation, and the Z-axis attitude by on-modulation of complementary thrusters. During some time critical mission phases (asteroid final approach, comet approach and comet orbit phases), a 3 axis "vectored thrust" can be also be generated. This allows a delta-v to be imparted in any direction without turning the spacecraft, the 3-axis attitude control being realised by thruster on-modulation. The exact length of any delta-v is autonomously controlled on board using either integrated measured accelerations provided by accelerometers contained in the IMPs, or integrated on-times commanded to the thrusters. The attitude estimation is performed in the same way as for normal mode.

The **Thruster Transition Mode (TTM)** ensures a smooth two-way transition between the Normal Mode, designed with reaction wheels control, and the orbit control mode, the spin-up mode and the near sun hibernation mode which are designed with thrusters control.

A dedicated **Asteroid Fly-by Mode (AFM)** is used to point the payload Z axis to the asteroid optical center throughout the flyby. This will enable the scientific instruments to track the visible part of the asteroid for the maximum length of time. Guidance and control for this mode is based on closed-loop tracking of the asteroid body using the navigation camera to determine the angular measurements with respect to the asteroid optical center. An on board estimator is used to make optimal use of all sensor information and feedforward torques minimise the pointing error. This phase lasts a few hours and is particularly critical during the few tens of minutes around closest approach when Rosetta has to rotate very fast (0.3 deg/s) and still guarantee accurate and stable asteroid observations. Due to the short duration of this phase, the redundant STR, CAM and IMPs are powered on to allow for rapid reconfigurations in the event of a failure.

Hibernation Modes

Rosetta has two hibernation modes. These have been introduced in order to preserve resources and minimise ground use during the long cruise phases between critical events. The objective of the **Near Sun Hibernation Mode (NSHM)** is to maintain the Solar Arrays pointed towards the Sun (and optionally an antenna pointed towards the Earth to enable communications) while minimising the fuel consumption and the number of thruster cycles. It uses an adaptive thrusters pulse width control algorithm in order to achieve, whenever possible, a single-sided limit cycle which uses the external disturbance torque in place of actuators for the control in one direction. The thruster control must keep the spacecraft Earth/Sun pointed attitude within a programmable attitude limit cycle in the range ± 15 °.

The **Spin-Up Mode (SPM)** is used to enter the Deep Space Hibernation phase of the mission. Its goal is to spin-stabilise the Spacecraft around 4°/s about a given inertial direction, such that the principal axis of inertia remains inertially fixed during the whole hibernation phase. When entering this mode, the Spacecraft is three-axis stabilised with its estimated principal axis towards a commanded inertial direction, which is typically 20 ° away from the Sun direction. The final spacecraft dynamic state shall be unambiguously checked on-board and on ground for authorising the shut down of AOCMS and transition to Hibernation. The Spacecraft will remain in this state for about 2 years around the 5.2AU aphelion where the power generation from the solar arrays falls to 4% of that at 1AU. The little power that is generated is used mainly for heating during this phase.

AOCMS Equipments

The Spacecraft carries four coarse **Sun Acquisition Sensors (SAS).** Two of these sensors are situated on the main body and together generate more than hemispheric coverage providing measurements to enable initial sun acquisition. A further unit is situated on each solar array allowing for fast recovery of the sun on thc solar arrays in case a serious sun pointing failure has occurred. These units are very robust, consisting only of a structure and solar cells, and totally passive.

There are two redundant autonomous **Star TRackers (STR)**, which provide the spacecraft inertial attitude without any ground intervention. They are each capable of tracking up to nine stars between magnitudes 2 and 5.5 and can determine autonomously from those coordinates the star tracker quaternion in inertial space. Each STR consists of two separate units, an optical head and a processing unit that is shared with Navigation Camera. These STR have been designed to operate even while the spacecraft is within the cometary dust cloud. This is significant achievement is described in a separate paper (Ref 2). The spacecraft is also equipped with two redundant **Navigation CAMeras (CAM)** which share their processing electronics with the STR. These will detect the asteroids at long range and will be used to control the spacecraft pointing during the very fast flybys by real time determination of the position of the asteroid optical centre. They will also be used to detect the comet some months before the encounter to allow accurate determination of its orbit for the precise planning of the various rendezvous manoeuvres.

The avionics sensor suite finally includes three **Inertial Measurement Packages (IMP)** composed of a set of three mutually orthogonal high performance Ring Laser Gyros. Each package is also equipped with three accelerometers for monitoring and controlling the orbit correction manoeuvres. The 2 first packages are mounted in a skewed configuration, which allows the AOCMS to use any combination of 3 gyros among the 6, to calculate a 3 axes attitude. The third unit is an ultimate alternative used by the AOCMS, when it is not able to identify 3 safe gyro axes among the 6 first ones.

For actuation, the AOCMS has both reaction wheels and thrusters. There are four **Reaction Wheel Units (RWU)** of which three are normally used at any one time, the fourth being kept in cold redundancy. Each reaction wheel has a 40Nms momentum capacity, can deliver 0.2Nm torque and are equipped with a high-resolution wheel speed monitor. During the asteroid flybys all four reaction wheels may need to be used simultaneously in order to meet the pointing requirements during the fast observation slew.

The coarse attitude control and orbit manoeuvres are performed with the **Reaction Control System (RCS)**, which is composed of two redundant sets of 12 bi-propellant thrusters of 10N. Two sets of four parallel thrusters are available for all the major orbital manoeuvres and the two sets of eight remaining thrusters are configured to generate force free attitude control torques during orbit manoeuvres coarse pointing modes and wheel-off loading. Each thruster features individual flow control valves and latch valves so that it can be isolated in case of failure.

The AOCMS is also in charge of monitoring and steering the HGA towards Earth and the solar array towards the Sun using respectively the **HGA Pointing Mechanism (HGAPM)** and the **Solar Array Drive Mechanism (SADM)**. For this purpose Sun and Earth ephemerids are generated on-board using Chebyshev polynomials for which coefficients are regularly updated from the ground to cover the different sequence of the mission phases.

The equipment use in the various AOCMS modes is summarized in table 3, below. Note that many of the modes consist of sub-phases which are not explained in detail in this paper.

	AIU	IMP	SAS	STR	CAM	RWU	RCS	SADM	HGAPM
Sun Acquisition Mode	1	2	(1+2)				1	HOLD/ OFF	
Sun Keeping Mode	1	2	1/2	(1)			1	HOLD	
Safe Hold Mode	1	2	(2)	1		(3)	1	2	OFF/1
Normal Mode	1	(2/1)	(2)	1		3	(1)	2	1
Thruster Transition Mode	1	2/1	(2)	1		OFF/3	1	2	1
Orbit Control Mode	1	2	(2)	1		OFF/3	1	2	1
Asteroid Fly-by Mode	1	2		1+(1)	1+(1)	4		2	1
Near Sun Hibernation Mode	1		(2)	1			1		
Spin-up Mode	1	2	(2)				1		1

Table 3: AOCMS Equipment use per mode.

Brackets indicate equipment is used out of the loop or only in a restricted sub-phase of the mode.

AOCMS Performances

Rosetta has to meet it's most demanding pointing requirements towards the very end of it's life. For the initial 9 year journey to the comet the pointing requirements, with the exception of AFM, are driven almost entirely by maintaining the spacecraft safety and minimizing fuel use. During the brief asteroid flyby manoeuvres, the payload has to be kept stably pointed to a rapidly changing target, likewise the pointing requirements around the comet are also driven by the demanding payload. The requirements and achieved performances of Rosetta for NM and AFM are given in Table 4. As can be seen, these requirements are very demanding and require very good gyro and ASTR performance in order to be met. Even so, the requirements are made even more difficult to achieve by the configuration of the spacecraft as the payload and ASTR must be mounted almost perpendicular to each other. As a result, the worst performing axes of the ASTR (around boresight) are coupled directly to the most demanding payload axes. This essential but non-optimal configuration leads to large margins around the Z axis and gives difficulties around the X and Y axes. From the achieved performances shown in Table 4 it can be seen that the requirements for RPE appear not to be met. In addition to the ASTR configuration problems mentioned earlier, another effect is responsible for this. The power requirements during payload operation require that the solar arrays be maintained pointing accurately towards the sun. The solar arrays must therefore be constantly rotated. However, the speed of rotation needed is not constant and the solar arrays must therefore occasionally jump between different discrete speed levels. This speed jump dominates the reported RPE figures but only has an effect around 4% of the time. For the remaining 96% of the time the RPE requirements are met.

	Achieved (arcsec)	Required (arcsec)
Comet Observation (NM)		
APE (Z axis halfcone)	80.2	108
RPE (around X and Y, 1 sec)	2.7*	1.1*
RPE (around Z. 1 sec)	2.1	22
Asteroid Flyby (AFM)		
APE	508	540
RPE (around X and Y)	119.6*	10.8
RPE (around Z)	35.1	72

***Requirement is 2 sigma for 95% of the time, achieved performance is given 2 sigma for 100% of the time. Requirement is met when the additional time constraint is taken into account.**

Table 4: Main AOCMS pointing requirements (2 sigma)

ON-Board Autonomy

Owing to the complex mission requirements and the high communications return trip time the spacecraft needs a high level of autonomy, both for operations and for contingency cases. The ground control centre cannot exert any real-time control because of the long signal roundtrip time, occultations and hibernation periods, when the control centre is not manned for cost reasons. Rosetta has therefore to resolve any critical problems autonomously and an elaborate FDIR service has been implemented for this purpose. The on-board autonomy capabilities of Rosetta are achieved through three main tools. These are the Mission Timeline (MTL), the use of On Board Control Procedures (OBCPs) and an extensive rule based FDIR system.

OBCPs are essentially short 'macros' for the spacecraft operation. They are stored on board and run by a simple telecommand. The OBCPs are capable of simple logical descisions and of generating telecommands and initiating checks and calculations. The combination of these simple facilities leads to OBCPs being capable of performing rather complex tasks such as conditional reconfigurations.

The MTL is stored on board and controls the timing of events. It consists of a series of time-tagged telecommands that are automatically run at the indicated time. The MTL is regularly uplinked from ground and makes full use of the OBCPs to minimise the amount of commanding needed.

A detailed description of the OBCPs and the MTL is not covered in this paper but the operation of the AOCMS part of the FDIR is given below.

AOCMS FDIR

While most spacecraft cannot perform automatic recovery for more than one failure, Rosetta can keep the solar arrays pointed towards the sun and its HGA to Earth in the event of multiple independent failures. It is even able to maintain the mission, un-interrupted, in the case of less severe failures. The complexity of such a FIDR design is one of the major drivers in the Rosetta spacecraft development.

The required autonomy in the FDIR is implemented in many ways ranging from very simple checks to highly complex filters and monitors. Almost all of the FDIR is performed in software. This software monitors and detects hardware and software failures using a tiered system of monitors, checks and recovery actions. The AOCMS implements three main levels of failure detection and several possibilities for isolating any given failure, these are especially critical as Rosetta does not have a separate independent set of hardware to transfer control to in the case of a major problem. As a result great care must be taken that the reconfigurations ensure that the problem is removed from the loop. The failure detection monitors are constructed in such a way that a triggering of a monitor occurs at the lowest level possible and wherever possible unambiguously leads to the suspicion of a single or small group of units in which the problem must lie. In this way the amount of reconfiguration needed can be minimised.

The first level of failure detection, the Local Monitoring (LM) performs checks on individual units, including their communication path and their data. The thresholds on the checks and filters that make up the LMs are set in such a way that a LM should always trigger before either of the next two levels of monitor in the case of simple failures. There are a total of eleven (11) LMs. Each of which are made up of a number of sub tests that typically include voltage monitors, temperatures, telemetry counters, BIT results and data range checks. Each of the tests are also subjected to filtering to ensure tolerance against SEUs and spikes. A good example of these LMs is the STR Health Status (L5) monitor which monitors the results of the STR built in tests, it's temperature at various points, it's operating mode and the age of the received data. If any of these sub-tests are out of limits for a given time then the unit is suspected failed and its data ignored or it may be reconfigured to the redundant unit. LMs are enabled once a unit is declared available and after a pre-defined time after unit switch on.

A second level of failure monitoring is provided by the Functional Monitors (FM). These eight monitors are also built up of several lower level tests. They aim to mainly check the consistency and validity of the information coming from one unit against that coming from another in order to catch functional and operational failures that are not detectable by unit level Local Monitors. As a result, the triggering of a FM results in a wider number of units being labeled as suspected failed. A good example of a functional monitor is the STR – Gyro consistency monitor (F5) which compares the rates measured from the gyro against the attitude changes measured by the STR. The triggering of this monitor would lead to the immediate suspicion and perhaps reconfiguration of the gyros, the STR and the AIU. Even if an FM is used in a mode it is only enabled once the units it uses are declared valid. A valid unit is one that is available, not declared failed and has no related LM surveillance triggering.

The final level of failure detection operational in the AOCMS are the Global Monitors (GM). These ten monitors are intended to catch all failures not detected at a lower level. These monitors check the global status of the spacecraft against it's expected status and known limits. A triggering of a Global Monitor indicates a significant, complex or possibly even multiple failures has occurred and almost always results in a mode change. Specific examples of these monitors are timeouts on specific actions such as slews (G10), sun presence monitors (G6) and thruster activity monitors (G9). The GMs very dependent on the operating mode are automatically enabled on mode entry and disabled on mode exit. A number of the GMs are only enabled after a 'relax timer' has elapsed. This ensures that transient mode entry effects do not falsely trigger these high level monitors.

The failure detection monitors can additionally be individually forced active or deactivated in each AOCMS operating mode by TC. Furthermore, the action to be performed should a monitor trigger is also dependent on the surveillance type associated with the monitor. This is also settable on a per mode basis. The four types of surveillance are Blocking Surveillance (BS), Emergency Surveillance (ES), Non Emergency Surveillance (NES) and Off Line Surveillance (OLS). All surveillance types, with the exception of ES, trigger an attempt to isolate the failure and recover the mission without exiting from the current operating mode. An ES will cause an immediate transition to a back-up mode with subsequent full reconfiguration should a further monitor trigger. The differences between the other three surveillance types are mainly due to the use of the equipment being monitored.

Emergency Surveillances usually cause an immediate reconfiguration to a safe mode with a reboot but without any reconfiguration of the CDMU but with the surveillances remaining active. Any further emergency surveillance trigger then causes entry into the Earth Strobing submode of SKM but with a reconfiguration of the CDMU which is the only unit common not reconfigured in the safe mode transition. The CDMU then reconfigures the processor modules to use the redundant modules and reboots again. On re-entry to safe mode the surveillances are disabled and this finally guarantees the removal of any fault.

Non Emergency Surveillances are triggered when a non-critical failure has occurred and redundancy for that unit exists. This will then cause the unit to be hot swapped for the redundant unit, usually without any need to change the mode or interrupt the mission.

Blocking and Off Line Surveillances are used to monitor equipment not directly in the loop in order to prevent failure propagation. The usual course of action on the triggering of such a surveillance is to simply switch off the unit.

As the active monitors and their associated surveillance types are dependent on the mode, the available redundancy, and the criticality fo the unit, it is obvious that there are a very large number of combinations of monitors, surveillances and threshold settings exist. Therefore a detailed description and mode by mode justification is outside of the scope of this paper. The practical implementation and operation of the AOCMS FDIR is therefore best illustrated by an example.

The example chosen starts in Normal Mode. This mode uses the equipment stated in Table 3. Local monitors are active covering each of these units (with the exception of the CDMU). Several FMs are enabled such as the Gyro consistence check (F1) and the STR/SAS/Ephemerides consistency check (F2). There are global monitors active against, amongst others, over rate, sun pointing/presence and thruster torques and total impulse. Additionally some timeouts are active. Most of these are active on mode entry. Firstly we can imagine a failure in the Reaction Wheels caused by a damaged bearing. This will be detected by the RWA friction torque monitor (L9) causing an immediate reconfiguration of the suspected units (i.e. the AIU and the Reaction wheel, the fourth reaction wheel being used in its place) without leaving the mode. The reconfigured reaction wheel will now be marked as failed and any further attempts to reconfigure the reaction wheels would lead to safe mode entry. A second failure can be imagined in the IMP. If detected by the gyro status monitor (an ES surveillance in this mode) a transition to SKM (first safe mode) will occur and a counter incremented. The spacecraft will autonomously recover and attempt to enter SHM. Here we can imagine a third failure (in either SAS, STR or on-board ephemeredes) that would trigger the related monitor. A further transition to SKM will occur and the counter incremented once more.

Now, no further attempts will be made to enter SHM autonomously, with the ephemerides suspected the spacecraft will now enter Earth strobing and attempt to recover communication, after which further recovery must be by ground. Should yet another failure trigger before communication is re-established, the spacecraft would once again enter Earth strobing but only after having reconfigured the CDMU and disabled all FDIR monitors.

With this example, the great robustness of Rosetta to autonomously handle on-board failures can be seen. The failures selected in the example are somewhat arbitrary but a similar logic and result would occur for any set of un-related failures until no more redundancy exists in the system.

CONCLUSION

This paper has given an introduction to and overview of the Rosetta mission and spacecraft with a focus on the AOCMS and AOCMS FDIR aspects. The requirements of this long duration mission are very demanding and require large amounts of autonomy. These needs have been answered with a complex but highly competent design that was developed, tested and delivered in time for launch under a very tight schedule.

ACKNOWLEDGMENT

Astrium GmbH and Astrium SAS, together with all of the AOCMS unit suppliers are thanked for their hard work during the programme. ESOC's large role in flight dynamics and operations preparation should also be acknowledged.

REFERENCES

1. Rosetta AOCMS Studies Team, AOCMS FDIR Definition and Justification, RO-MMT-TN-2008 Issue 4

2. S. P.Airey, G. Berrighi, D. Procopio, "Seeing through dust – tracking stars from within a cometary dust cloud", Breckenridge 2003.

AAS 03-022

SEEING THROUGH DUST – TRACKING STARS FROM WITHIN A COMETARY DUST CLOUD

S. P. Airey,* G. Berrighi† and D. Procopio†

In January 2003, the European Space Agency was scheduled to launch the Rosetta spacecraft on its 10 year journey to comet Wirtanen. On arrival at the comet, some 4.5 AU distant from the sun, Rosetta was to rendezvous and deliver its cometary lander.‡ Thereafter it will commence detailed close up scientific observations of the comet and accompany it throughout a 1 year long hurtle towards the sun. During it's time with the comet, the increasing solar intensity will transform the dormant ice 'rock', vaporizing and ablating the ice, producing a dense and extensive cloud of dust and ice crystals that will be seen from the earth as a fiery tail. During this entire time Rosetta must continue to operate flawlessly and entirely autonomously and be able to maintain the highly precise and stable pointing required by the scientific payload even while it's cometary orbit is within this cloud.

The primary sensors used on the Rosetta spacecraft are the Autonomous Star Trackers (ASTR) designed and manufactured by Galileo Avionica. These are optical sensors that detect, identify and then track stars within their Field of View (FoV). These types of sensors are by their nature, highly susceptible to 'false' stars such as those that are likely to be generated by a cloud of dust reflecting the light from the sun. This paper aims to explain the problems that can be caused to an ASTR in such an environment, to characterize the expected environment around the comet - as seen from an ASTR, to detail the ways in which these problems were overcome for the Rosetta mission and finally to give an overview of the achieved performances of the sensor.

INTRODUCTION

Rosetta is an ESA mission that was scheduled for launch in January 2003. It is a long duration cometary exploration mission continuing cometary research from the highly successful ESA Giotto mission. After many years of travel, Rosetta will rendezvous with comet Wirtanen at a solar distance of 4.5 AU. After deploying a lander, the spacecraft will continue to orbit the comet at altitudes as low as 1km, during it's one year fall towards the sun. The Rosetta mission shall study in detail and from unprecedented proximity, the awakening and subsequent behaviour of the comet.

---

* Mr. Airey is affiliated to: European Space Agency, ESTEC, Keplerlaan 1, 2200 AG Noordwijk ZH, The Netherlands.

† Mr. Berrighi and Mr. Procopio are affiliated to: Galileo Avionica Space Equipment B.U., Via A. Einstein 35, 50013 Campi Bisenzio (Firenze), Italy.

‡ Due to problems with the launcher, the Rosetta mission has been delayed. The mission will be re-scheduled with a different comet but the general mission outline is expected to remain similar. This paper discusses the original target comet, Wirtanen.

The two main sensors onboard Rosetta are the two redundant Autonomous Star Trackers (ASTR) and the high accuracy Inertial Measurement Packages (IMP). Both of these sensors are required to work together via an on board gyro-stellar attitude estimator, to ensure the successful highly accurate pointing control of the spacecraft during scientific observations. Further details of the Rosetta mission and avionics can be found in Ref (4).

Star trackers are highly accurate and reliable attitude sensors. Autonomous Star Trackers operate by taking successive images of the sky, usually using CCD or APS detectors. Stars appearing in theses images are located by the ASTR software, uniquely identified with respect to an on-board catalogue and thereafter tracked. The successful identification of multiple stars in one image allows the unique determination of the sensor attitude with respect to the stellar background, typically to a few arcseconds.

The ASTRs on board Rosetta have been produced by Galileo Avionica (Italy). They are required to provide their inertial attitude fully autonomously, within 20 seconds of being switched on. Their performance requirements are derived directly from the required scientific pointing requirements imposed on the spacecraft during it's cometary observation phase which occurs towards the end of the 11 year mission. The main requirements are those of the Noise Equivalent Angle (NEA): 5 arcseconds (2σ), the random bias: 4 arcseconds (2σ) and the pointing stability of 3 arcseconds.

Very late in the development program it was realised that the comet dust may have a serious impact on the ASTR. A study was initiated by ESOC to characterise the likely environment. A difficult task considering the dearth of information available on the subject. The outcome of the study showed that there was indeed likely to be a serious problem with any optical pattern recognition sensor. Hence, with the ASTR hardware already almost complete a redesign of the ASTR software was initiated to attempt the seemingly impossible – to successfully determine and track the inertial attitude in the presence of not hundreds, but thousands, of false stars and to do so without impacting the hardware which would have incurred a catastrophic schedule impact.

THE ENVIRONMENT AROUND THE COMET

As the comet moves closer to the sun, the increasing intensity of the solar radiation causes gases and particles to be emitted by the comet nucleus. This debris forms a cloud around the nucleus and is blown by the solar wind into the familiar cometary tail.

The debris ejected from the comet may vary greatly in size and shape and is assumed to have a density distribution depending mainly on the distance from the comet. This debris can have three main disturbing effects on star tracker measurements:

1. The obscuration of stars
2. An increase in background light levels (due both to diffuse scattering and increased lens contamination which produces a corresponding degradation in the real star signal)
3. The direct reflection of sunlight into the ASTR lens giving rise to 'false stars'

The relative importance of these effects for any given debris distribution varies depending on the distance from the comet and the ASTR – debris – sun angle. For Rosetta, two key scenarios corresponding to different cometary orbit phases, and several distributions were considered. The scenarios were those of the orbit required for delivery of the comet lander and that required for global mapping of the comet. These orbits, which have a period between 3 and 6 days, are illustrated in Figure 1. For the Star Tracker, the worst case condition occurs shortly after the time of closest approach when the sun angle, together with the high particle density, conspire to give rise to the maximum number of 'false stars'.

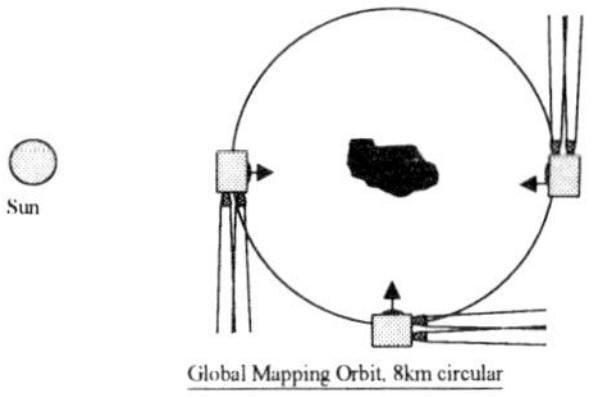

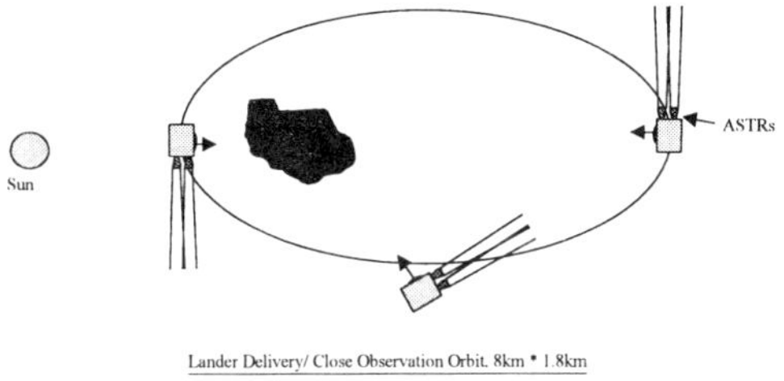

Figure 1: Comet orbit scenarios

Three distributions of debris make up were also considered, 'nominal', 'more large particles' and 'less large particles'. The results for each of these were based on mean conditions and hence the worst case of these from the ASTR point of view was taken as the development goal. The main results, detailing the number of directly illuminated dust particles ('false stars') in the ASTR FoV, are recalled below:

Table 1
FALSE STARS IN FOV

	Min	Max
Lander delivery	41	8987
Global mapping	20	229

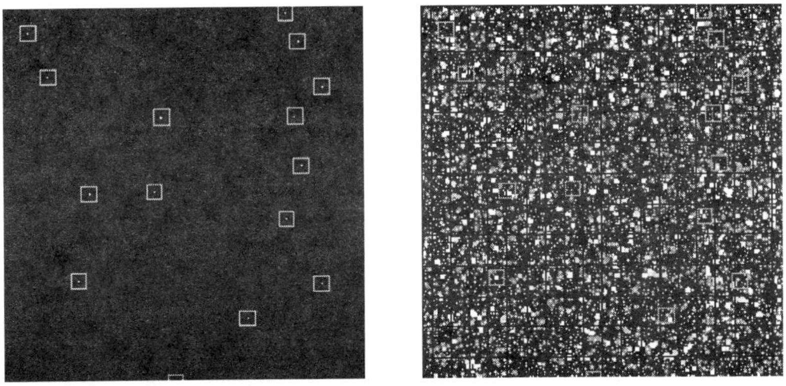

Figure 2: Simulated CCD images, without any false stars (left) and with 10,000 false stars (right), illustrate the difficulty of autonomous operation in the predicted conditions

It was seen that, even with very high densities of particles, the risks of obscuration of stars by un-illuminated particles was very low. Further studies also showed that the increase in straylight and the reduction in lens transmissivity (around 10.5% lens area contamination) were also only minor effects when compared to the problems caused by the direct sunlight reflection from dust particles.

This direct reflection gives rise to point sources of light that are essentially indistinguishable from stars. That is, their movement relative to the celestial background may be very small (and the direction vectors co-aligned), their apparent magnitude range is dependent on particle size and distance but covers the magnitude range of the ASTR and the spectral content of the reflected light is nearly identical to a typical star (originating from reflected sunlight). Hence comet dust, when viewed from within the dust cloud, gives rise to true 'False Stars' by the thousand.

Following on from the work done in characterising the expected environment, and taking full account of the large uncertainties in the results, the following requirements were set for the ASTR 'Dust Software' upgrade:

- The ASTR must be able to perform attitude determination with up to 1000 false stars in the FoV
- The ASTR must be able to continue tracking the attitude with 10,000 false stars in the FoV.
- The solution must have no impact on hardware in order to protect the schedule.

Further, the assumption was made that these false stars had magnitudes evenly distributed within the operating band of the ASTR. Together, these requirements and assumptions ensure that, for any orbit, there are always times during which the attitude can be determined and that, once determined, the attitude should never be lost due to the presence of the dust.

FALSE STAR EFFECTS ON STAR TRACKER OPERATIONS

The presence of False Stars in the FoV of the ASTR has several effects on the nominal operation of an ASTR that is designed only for the nominal environment;

- Data storage issues become critical
- Lost in Space capability may fail or, worse, give a false attitude
- Attitude tracking performances may degrade and tracking may even be lost

Data Storage

For most ASTR using fully space qualified parts, memory – due to the high cost of the chips – is an issue. In order to reduce the amount of memory required in nominal conditions a certain amount of real time pre-processing is performed and only the pixels related to an area of interest around a potential star are stored for later processing. In this way only a small fraction of the number of pixels (< 1%) need to be stored and processed further and hence much less memory is required. The memory for the Rosetta ASTR was sized in this manner resulting in an allocation of only 28Kwords for the storage of a full frame. Each potential star object can take up to 35 words to store. If unchanged, this would limit the number of objects (real and false stars) that can be handled to 800 which is insufficient to cover all cases. The additional 7Kwords of storage was found by re-structuring and optimising the remainder of the software's memory requirements. Loss-less compression was also studied to alleviate the memory problems. This would allow the storage of 1750 objects. It was decided not to implement the compression due to the risk involved in achieving the real time performances (@ 3.6 μsec/pixel) and the additional memory already freed up provided sufficient capability to meet the requirements.

With the ability to store 1000 objects, the requirement to be able to perform attitude determination in the presence of 1000 false stars is not limited by the data storage. However, as the previous section has shown, higher numbers of false stars – up to around 9000 – may be expected in the FoV. With the readout and storage routine implemented in the Rosetta ASTR, this scenario is not disastrous. As the readout and storage is performed in real time, once the storage area is full the remaining pixels are simply ignored. This effectively results in a STR with a reduced FoV, as illustrated in Figure 3, on which the processing algorithms may still operate.

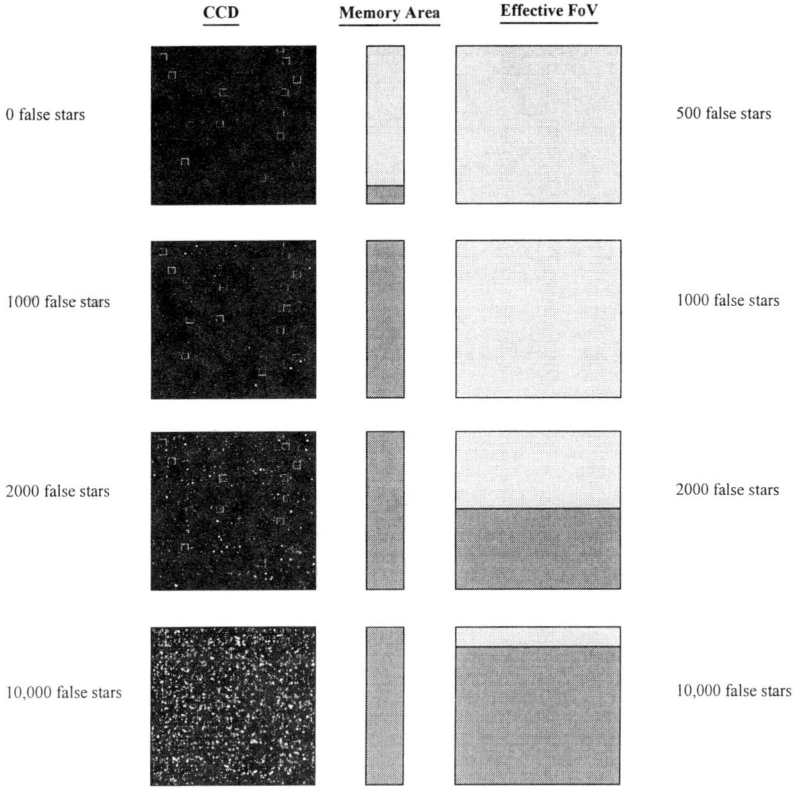

Figure 3: Illustration of the effect of pixel storage strategy

Lost in Space

For most cases where none stellar objects appear in the FoV, a study of the phenomena producing the false targets allows for implementing image processing filtering based on congruence checks such as image shape, size and, during tracking, also on the magnitude with respect to the expected star signal. However, in the case of Rosetta, such checks are useless against the dust environment as the dust is expected to produce images with the same shape, size and magnitude as real stars and another method to filter out the False Stars must be used.

Attitude determination is generally achieved by acquiring an image and applying a pattern recognition algorithm to recognise the stars, then correlating them against an onboard star catalogue. The presence of great number of false objects in the acquired image induces a reduction in the probability that the correct pattern recognition will be achieved. In the 'nominal' Rosetta ASTR (i.e. prior to having the requirements imposed by the dust environment) the attitude determination was performed using triads. Each triad was constructed using a target 'star' and its two closest neighbours. With such an algorithm it is sufficient that a false star is closer than one of the two neighbouring stars to induce failure of the algorithm. A quick study of the original algorithm showed that the probability of successful attitude determination decreased rapidly with the number of false stars and that it would be clearly impossible to reach the requirements of 1000 false stars without significant modification.

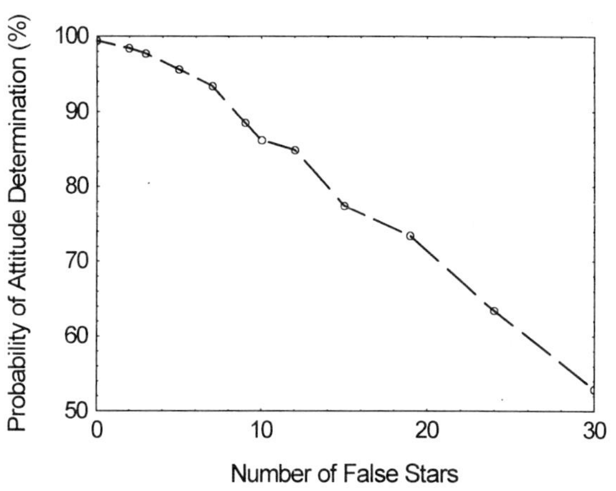

Figure 4: Attitude Determination Probability vs. False Stars for Original Triad Method

The original triad algorithm was adapted to increase its robustness to False Stars. In the modified version a target 'star' is chosen and the closed K stars in small annulus R1 to R2 around the star are found. Triads are made with all combinations of the target star and the K found stars. This improves the robustness in two ways. Firstly it provides filtering by allowing a search for stars only in a limited area, secondly it performs many more combinations, hence increasing the chance that the correct combination is found. This modified algorithm was also studied and showed a marked improvement over the original, though still far short of the requirements.

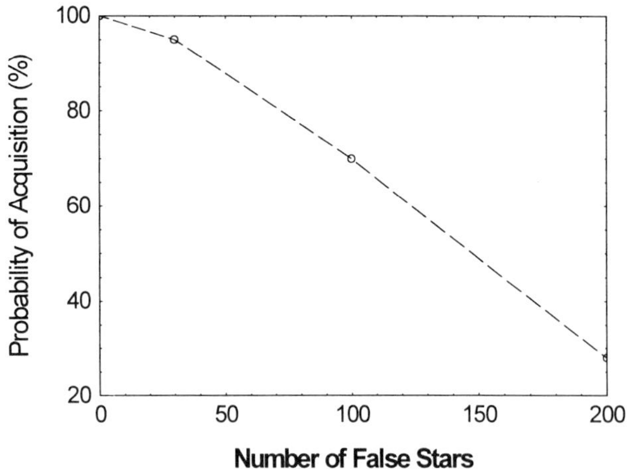

Figure 5: Attitude Determination Probability vs False Stars for Modified Triad Method

Additionally, the amount of processing time needed increases dramatically as the number of triads to be processed per star is given by:

$$Number\ of\ Triads = \frac{K!}{(K-2)! \times 2!}$$

As a result a different approach was found to be needed and this is detailed later in this paper.

Attitude Tracking

When the attitude is recognised the ASTR enters tracking mode. In this phase the position of the expected stars in the next image are predicted and only small windows of the image detector around these expected targets are considered. The ASTR tracks 9 stars and opens a 21 by 21 pixel window around each, storage problems are therefore not an issue during tracking. The use of windows also acts as an area filter on the CCD with respect to the false stars and results in an obvious, but dramatic reduction of the false objects to true stars ratio. This greatly aids the robustness of the sensor. In fact, to obtain a sensor truly robust against false stars with currently available technology it is felt that such windowed tracking must be implemented.

Even though the number of False stars to stars has been reduced, when trying to cope with 10,000 false stars there will still be, on average, 4 false stars per window. It is essential that these false stars are correctly identified and removed from further processing. If they are not then at best the information from that window will become corrupted, as the False objects can cause an incorrect baricentre position to be computed, and the ASTR output will become noisy or, at worst, tracking could even be lost.

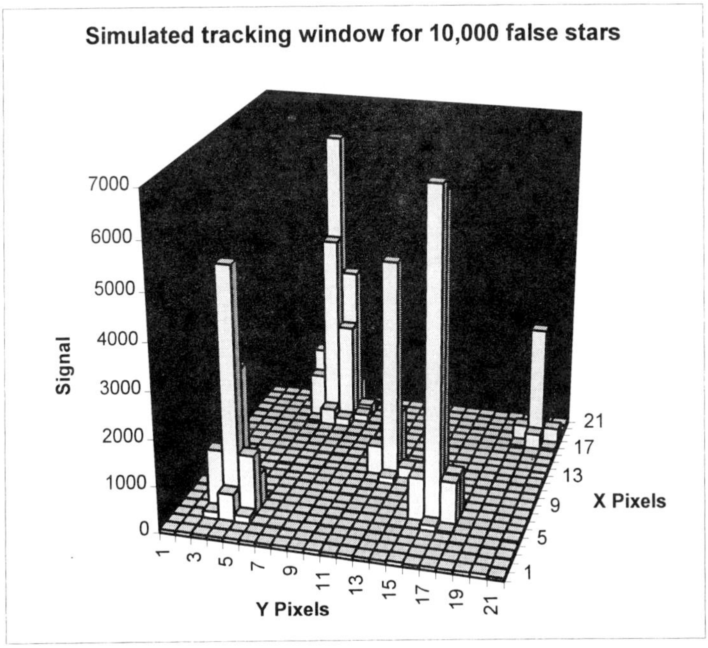

Figure 6: Simulated tracking window with 4 false stars. The real star is the one closest to the centre.

THE EFFECTS ON THE SPACECRAFT

The Rosetta Spacecraft AOCS will also be affected by the dust environment, but these effects are almost entirely linked to the problems of the ASTR as the other sensors (Gyros and Coarse Sun Sensors) are not effected. The effects on the spacecraft come therefore mainly from the mode behaviour of the ASTR in the presence of dust. Details of the effects on the spacecraft are outside of the scope of this paper but an overview will be given to set in context other changes required to the ASTR.

Initially the ASTR had three main modes;

- Standby
- Attitude Determination
- Attitude Tracking

The ASTR could be commanded to Attitude Determination, after which it would autonomously enter Attitude Tracking. Should the Tracking be lost for whatever reason (e.g. dust) the ASTR would autonomously revert to Attitude Determination. If this failed then it would revert to Standby and await a command from the AOCS.

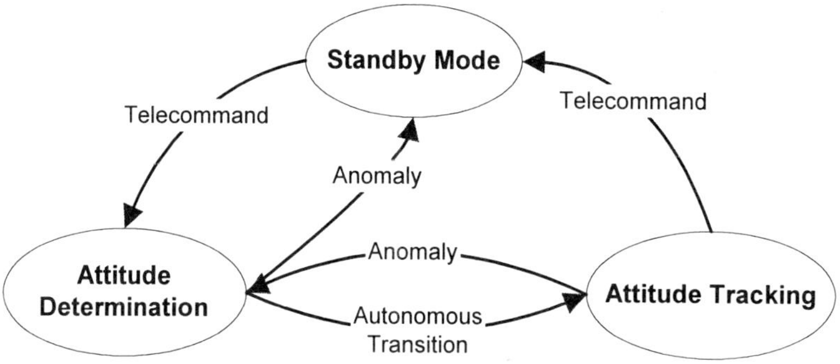

Figure 7: Original ASTR Main Mode Transitions

When close to the comet, the AOCS is nominally in Normal Mode (NM). The initially implemented logic in this mode would force an autonomous switch from ASTR A to ASTR B (i.e. switch to the redundant ASTR) on loss of tracking, in case of a further problem in establishing or maintaining tracking, the AOCS would then switch to a sun pointing mode, without using the ASTRs, and the mission would be interrupted. A return to the Normal Mode is then not possible without absolute measurements from the ASTR. In the comet dust environment, should tracking be lost due to the dust then this strategy would almost certainly result in the spacecraft entering sun keeping mode and eventually requiring both the spacecraft to exit the dust environment and the use of the ground operators for recovery. This would happen even though a very good knowledge of the absolute attitude is maintained by the AOCS throughout NM by using the high precision gyros and a Kalman filter based attitude estimator. Loss of tracking (either through dust or due to a STR hardware failure) would therefore be a major event for the mission. In order to resolve this and ease the risks to the mission a new strategy was determined which affects both the AOCS modes and the ASTR modes. In order to take full advantage of all the attitude information available to the AOCS, the ASTR Mode Structure was required to be changed to allow the use of a priori information in addition to its nominal fully autonomous operation. It was decided that when the AOCS commands Attitude Acquistion it should have the capability to also pass attitude and rate information, together with information on the expected uncertainty in that information.

Depending on the uncertainty level, the ASTR uses and reacts to the provided information differently. In the best case, should the attitude information be provided with an uncertainty off less than ±0.1 degrees, the ASTR is able to start immediately in Tracking mode, hence totally bypassing the most dust sensitive operations. These changes in the ASTR mode strategy enabled an AOCS, robust to the dust environment, to be designed.

OPERATING WITH FALSE STARS

The techniques needed to cope with high numbers of false stars are different for attitude determination and attitude tracking, hence these are discussed separately.

Attitude Determination

As shown earlier in the paper, the original attitude determination algorithms, though highly suitable for nominal use, were unable to cope with the demands of the dust environment in a satisfactory manner. It was therefore decided to change the method of performing attitude determination to increase robustness. The selected concept was proposed by ESOC (Ref. 8) with the development and detailed design, including all hardware constraints, performed by Galileo Avionica.

Attitude determination is essentially a pattern recognition problem and the method implemented on the Rosetta star tracker performs in the opposite way to the more usual ASTR pattern recognition algorithms. Normally a pattern is constructed using the measured stars and this is then compared with a set of reference patterns stored in the on board memory. In the case of the Rosetta algorithm each of the stored reference patterns are considered in turn and an attempt is made to fit this to the acquired image. When there are large numbers of false stars this approach minimises the searching operation and thus the processing time.

Such a change alone would not bring about the required increase in tolerance as, if a triad pattern were maintained, there would be a large number of false recognitions. In order to remove the false recognitions and successfully filter out the false stars it is necessary to increase the amount of information, or fitting criteria, for each star. Once the star magnitudes have been considered, the only way to do this is to increase the complexity of the pattern to be matched by incorporating more stars in each pattern.

The more information used for each star pattern description the higher the confidence will be in any matched pattern, the drawback is that a more complex comparison is required and the difficulties in assembling and checking the pattern catalogue also increase.

A trade off between the reliability of the information and the memory occupation led to define a search patterns consisting of a total of 6 stars. For each real star, named the 'central star', a pattern is built using the 5 closest stars. The pattern is then made up of the spherical angles and the separation angles of each star with respect to that central star.

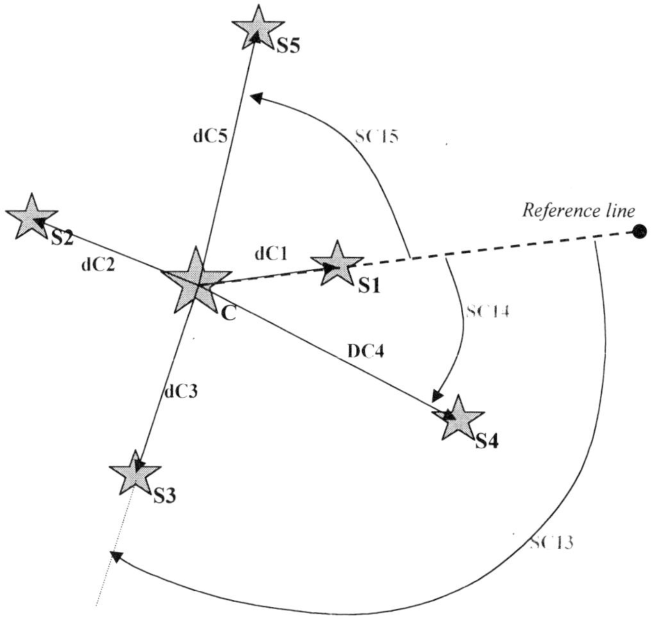

Figure 8: Pattern parameters definition

Defining "C" as the central star and S1, S2...S5 as its five closest neighbours, the stored data for the pattern are the five angular separation distances (dC1, dC2, dC3, dC4, dC5) between the central star and each of its neighbours, the four rotation angles (RC12, RC13, RC14, RC15) of stars S2 to S5 with respect to the reference direction of S1 and, finally, the index of each of the stars in the onboard star catalogue.

During the attitude determination, for each object in the FOV – starting at the centre - a 'measured' pattern is constructed using with the N closest objects. In the dust environment the majority of these objects will be false stars so N has to be selected high enough to ensure that the real stars will also be included. For the each of these N objects the separation angles and rotation angles are calculated and the resulting single 'measured pattern' is associated with the central object.

The pattern recognition then consists in comparing each reference pattern in the catalogue with the measured pattern until the catalogue pattern that matches a part of the measured pattern geometrical form. When dealing with large numbers of false stars there is a chance that one of the stars in the real pattern is corrupted or occulted by a false star, therefore a match is deemed to have been found when 5 out of the 6 stars in the pattern match to within a given tolerance (in both angles and magnitude). Once a match is found an attitude estimation is computed.

In figure 9, a measured star pattern is shown. This pattern contains also the corresponding catalogue pattern (shown in figure 10) that is hidden among the false objects.

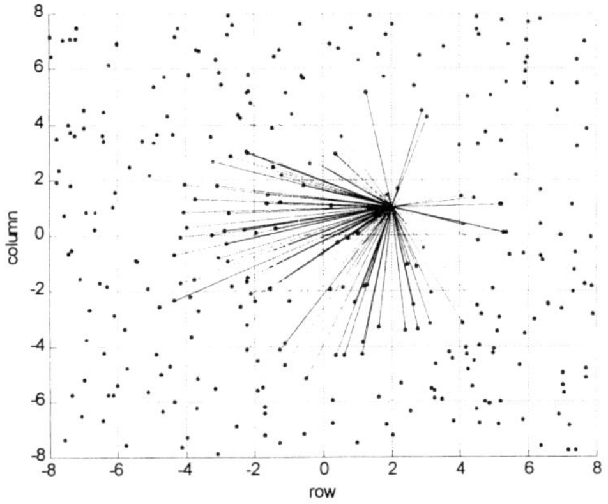

Figure 9: Measured Pattern representation with N=100 in presence of 300 false objects

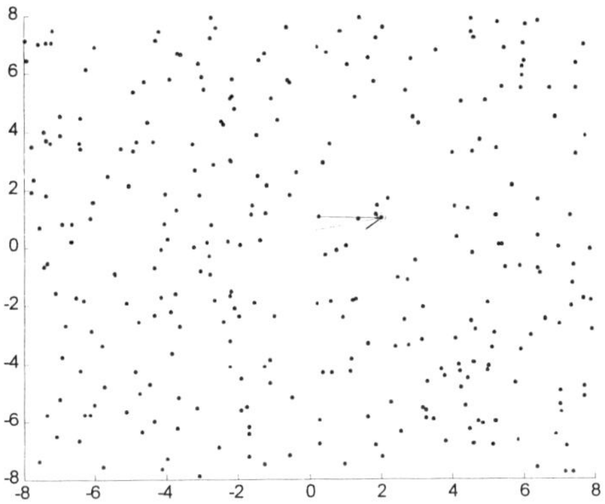

Figure 10: True reference pattern corresponding to the measured pattern in figure 6.

Once a pattern matching has been found, a further validation check is performed in order to confirm the attitude found is correct. Using the computed attitude, the on board star catalogue is projected onto the field of view. This gives the expected positions of all stars within the entire FoV, these expected stars are then searched for in the measured objects. If the position and magnitude match (again within a certain tolerance), the star is validated. If greater than a set percentage of the stars expected to be in the FoV are validated then the determined attitude is deemed valid. If the attitude validation step fails then the pattern search is resumed, continuing the comparison of the current measured pattern with the other stored reference patterns. When the whole catalogue has been compared with the current measured pattern without success, the whole process is re-started using the next object until all objects in the FoV have been checked.

As can be seen from the above description, there are a large number of parameters to be set and tuned. The setting of these parameters was done by trading off processing time, memory availability and performance (i.e. robustness) requirements.

Without time or memory constraints, this algorithm is able to reliably recognise a pattern in presence up to 2000 persistent false objects. Beyond this number, the false object density is so high that the number of real stars totally or partially overlapped by the disturbing target and hence rejected on the basis of geometry or magnitude filtering criteria becomes so high that reliable and robust attitude determination is not possible.

When the problems of dust on the ASTR were identified, the ASTR development programme was well advanced and it was too late to make hardware changes. This resulted in limited resources being available to the software. When considering these limited resources, such as memory and CPU time, one of the most important parameters to tune in this application was the number of objects (N) taken form the measured local pattern. Conceptually, a high value of this parameter will decrease the probability of loosing a valid measured pattern but increase the elaboration time during the matching phase. Where the elaboration time is limited by a time out counter, this increased elaboration time gives rise to an increased probability of triggering the time-out.

The size of the stored reference patterns significantly affects the performance of the pattern recognition algorithm. A reference pattern built with a large number of close stars reduces the probability of mismatching. On the other hand, large patterns require large storage areas and also reduce the probability that a pattern is completely contained in the acquired image which decreases the probability of attitude determination.

Finally, memory and CPU time limitations reduce the number of manageable objects and, for Rosetta, this is the limiting factor which constrains the attitude determination to 1600 objects. When there are greater than this number, the attitude determination mode will still attempt to perform determination but will effectively operate on a reduced FoV. This will further decrease the probability of a successful attitude determination. When the effective FoV falls below around 1.5 times the pattern size, the probability for successful determination will become very small.

The above described algorithm has been implemented and was extensively tested. When implemented on the DSP21020 processor of the Rosetta star tracker and the parameters are correctly set, this algorithm allows for a probability of successful attitude determination greater than 95%, when 1600 objects are processed within 800 sec. In the remaining 5% of cases a time-out usually occurs.

The very long processing duration of some hundreds of seconds means that, in a dust environment, the ASTR must be operated in a significantly different manner than in other, more typical, applications.

In most cases, when the number of false stars is not so high, the star tracker is able to operate in a fully autonomous mode as described previously: Two independent attitude determination cycles are performed within a few seconds: the slew rate is computed using the two measured quaternions, a check of rate consistency is performed and a tracking phase is then autonomously started.

In the case of the Rosetta dust enhanced algorithm, when starting from a lost in space situation, a direct rate measurement based on two quaternions acquired at hundreds second of distance is meaningless. This implies that a transition from attitude initialisation to tracking phase cannot be performed with an autonomous rate measurement. The angular rate information must therefore be provided by the AOCS as input to the star tracker.

When considering an attitude acquisition cycle duration of 800 seconds it is not only the s/c angular rate that can be problematic, the stability of that angular is also an issue. A rate stability of ±0.01 deg/sec (which is the rate guaranteed in by the Rosetta AOCS safe mode), would leads to an error in the attitude prediction of ±8 deg (0.01 deg./s × 800 sec). This accuracy is clearly not sufficient to initialise the tracking phase, where small windows placed around the expected star positions within the 16.4 * 16.4 degree FoV. Typically entry to tracking requires a maximum error in the positioning of those windows of around 0.2 deg. This would appear to make the problem unfeasible, however, a rough knowledge of the attitude allows to perform a second attitude acquisition cycle with a much reduced duration by severely reducing the extent of the catalogue to be searched.

A cycle of attitude determination, initiated with 8 degrees uncertainty, can be executed within about 120 s, in the case of the Rosetta Star tracker H/W. The spacecraft angular rate stability of 0.01 deg/s is then translated into only 1.2 deg of uncertainty in the attitude propagation. The process can be then repeated until the required accuracy to enter in tracking mode (better than 0.2°) is obtained. As the uncertainty in the a priori knowledge decreases, so further effects help to further speed up the attitude determination. Not only does the fraction of the catalogue needed continue to decrease but also the expected positions of the stars can be used to filter out false star objects not within the uncertainty area.

The design for Rosetta was tailored for 0.01 deg/s rate stability. In this case, the accuracy in the attitude propagation is compatible with the allowed attitude uncertainty of the following cycle and allows a 4 step converging process.

Table 2
TIME DURATION, STAR CATALOGUE AND FIELD OF VIEW REDUCTION FACTOR FOR THE ROSETTA STAR TRACKER

Cycle	Allowed Attitude Uncertainty (Deg)	Catalogue Fraction Needed	Fraction Of Fov To Be Processed	Allocated Cycle Duration (Secs)
1	>8°	>7% <100%	1	800
2	8	7%	1	120
3	1.2	2%	0.6	50
4	0.5	1.7%	0.3	10

In view of the new method of attitude determination and the impacts this has, together with the knowledge that the AOCS nominally has an internal and very good estimate of the attitude available, it was decided to re-structure the ASTR modes. This re-structuring allows the existing attitude knowledge in the AOCS to be used to the best advantage, enabling entry directly in one of the attitude initialisation cycles, or even attitude tracking, skipping the previous longer cycles and significantly speeding up the process.

As example, if no attitude information is known by the AOCS (only rate measurement is available), the star tracker will execute all the four cycles reported in Table 1, prior to enter in tracking mode 980 seconds later. On the contrary, if the attitude is already known with accuracy better than 1.2 deg, the star tracker will autonomously skip the first two cycles and will enter in tracking after 60 seconds (only cycles 3 and 4 are performed) and if the attitude is known to better than 0.2 degrees, the ASTR enters directly into attitude tracking mode after only a 0.1 second initialisation cycle.

In order to further speed the recognition and recovery from attitude tracking loss due to dust, some of the anomaly transitions were also changed. The revised mode structure is shown in Figure 11.

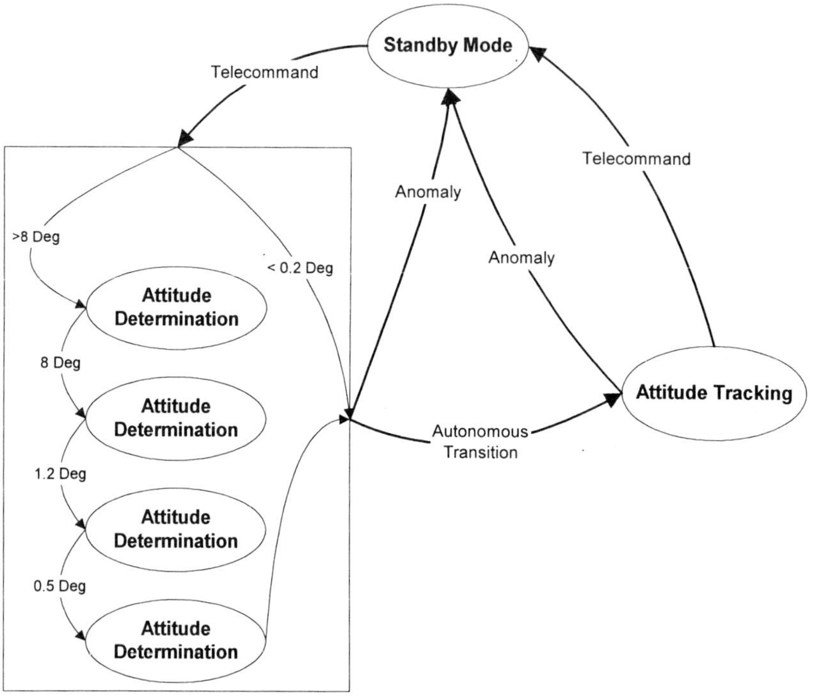

Figure 11: Revised ASTR Mode Diagram

Attitude Tracking

As previously mentioned, the attitude tracking mode of the Rosetta ASTR operates in a windowed mode where 9 21*21 windows are readout and processed. Such windowing acts as a very strong area filter and greatly improves the star to false star ratio. However, it can be expected, in high false star densities, that several false stars exist per window. These need to be filtered out by processing. Several checks are performed in order to ensure that all the false stars are removed and the maximum number of real stars recovered and used in the attitude update. Firstly, each of the separate objects in the window are explicitly identified and their shape and magnitude calculated. Any objects that do not have the expected star shapes are rejected. Now, from the on board catalogue, the ASTR has a priori knowledge of which star it expects to find in the window so the remaining objects are filtered with respect to the magnitude (with some applied uncertainty) of the expected star. Then, should there still be more than one potential object, the object nearest to the expected position is taken. This object is then compared with the expected position and if it is within a further tolerance of the expected position it is deemed valid. A high resolution centroiding is then performed that ensures that only pixels unaffected by false stars (or other disturbing effects) are used.

This approach ensures that only an object (i.e. a false star) of the same magnitude and that lies closer to the expected position than the real star can be used and corrupt the attitude measurement. The effect of this is a very slight increase in noise. Tracking loss is never caused by tracking only false stars as 9 stars are tracked and the probability of this happening simultaneously for 5 or more stars is negligible. Tracking loss can therefore only be caused by the simultaneous occultation of more than 7. Any loss of tracking (less than 2 stars validated and tracked in a given cycle) causes the ASTR to immediately enter standby mode. The introduction of continued attitude propagation inside the ASTR was considered as this would have allowed a further filtering on the attitude loss criteria. However, this was rejected for Rosetta due to the availability of a more reliable propagation within the ACMS, hence the transition to Standby enables a faster recovery using ACMS provided information.

Simulations with up to 10,000 false stars (See Figure TBD) show very good robustness using this approach with the average number of stars tracked only falling slightly and the minimum number never falling below the criteria for attitude loss.

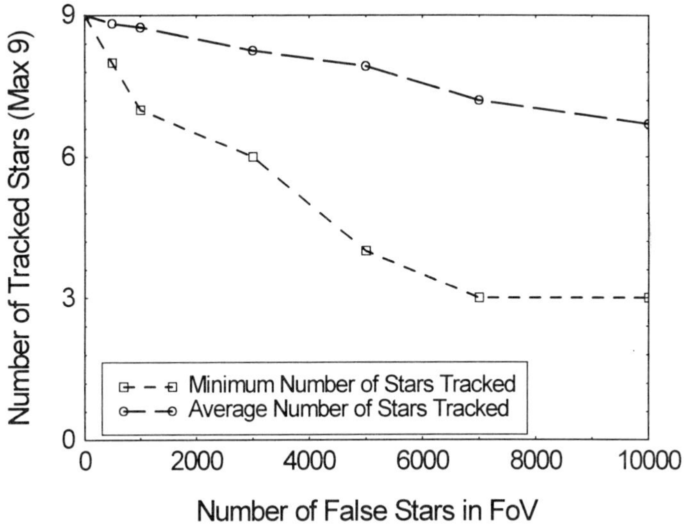

Figure 12: Tracking performance in presence of false stars

Obviously, the reduction and changes in the number of stars tracked has an impact on the noise performance of the sensor. The noise performance, under differing false star conditions, was found by simulation and shows that, even in the extreme case, a good performance is achieved. The original performance requirements continue to be met in the presence of 3000 false stars and only degrade slightly thereafter up to 10,000.

TESTING

The thorough testing of an autonomous star tracker is difficult, mainly due to the lack of accurate optical star pattern simulators. For Rosetta, the testing is made more complicated due to the requirements for robustness against false stars. A common approach to testing, that works well is as follows;

- Verification of BOL radiometric accuracy is performed using an high accuracy single star simulator (optical stimulation)
- Verification of BOL star position measurement accuracy, is performed using the above mentioned star simulator and placing the unit under test on high accuracy rotating tables.
- Verification of the nominal BOL functional operation and end to end testing is performed using a Night Sky test
- Verification of robustness and full sky coverage (BOL and EOL) via electrical stimulation using EGSE
- EOL radiometric and star position measurement accuracy is verified by analysis only.

As sufficiently representative optical star pattern simulators do not exist, the functional, robustness and performance testing of the ASTR, and in particular the pattern recognition and attitude measurement, must be verified using other means. Dedicated Electrical Ground Support Equipment (EGSE) was developed for this purpose. The EGSE allows for the direct injection into the ASTR of a digital image. This is performed via a dedicated test port and bypasses only the optics, CCD and ADC.

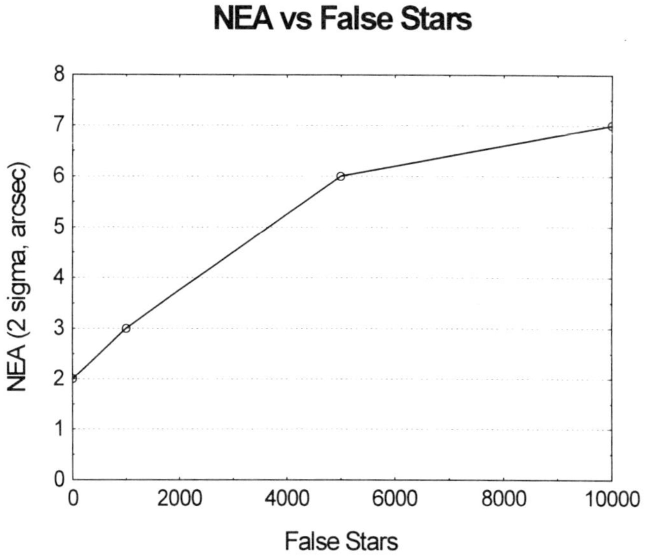

Figure 13: Noise performance in presence of false stars

This approach allows for a large flexibility in the generation of the digital image while still maintaining the correct environment for the software. In this manner a wide variety and large number of test cases can be produced that would otherwise be impossible. It is possible to verify by test the full sky coverage requirements and to simulate the large numbers of false stars, hence verifying the robustness to the dust environment. Large obscuring objects may also be simulated. During the development of this every care was taken to ensure that the simulated images are as realistic as possible hence all optical and CCD characteristics are simulated.

CONCLUSION

The experience of Rosetta has shown that, through co-operative and interactive system, subsystem and unit design, it is possible to design an AOCS that is extremely robust to the harshest of external environments. The Rosetta ASTR, together with the AOCS modes, were, in a very short timescale, re-designed to robustly handle the extreme environment of 10,000 true false stars in the FoV thereby significantly reducing the risk of the mission.

The engineering effort and investigations performed during this work have further shown that it is possible, with existing technology, to design ASTRs that are capable of robustly and reliably operating in harsh environments and that suitable algorithms can be found to cope with not only false stars but also single event upsets and objects in the FoV. Hopefully it will not be long before these advances are seen also in commercially offered products.

ACKNOWLEDGMENT

The authors would like to extend their thanks to ESOC for their assistance in the development of the Rosetta dust environment STR algorithms and the Rosetta industrial consortium (Astrium GmbH, Astrium SAS and Galileo Avionica) for their co-operative system wide approach to solving the problems.

REFERENCES

1. M. Buemi, A. Landi, D. Procopio, "Autonomous Star tracker for Rosetta", 4th Spacecraft Guidance, Navigation and Control System, Noordwijk, 1999.

2. G. Berrighi, A. Landi, E. Monnini, D. Porcopio, "Space attitude sensors: new technologies using APS", OPTRO 2002 Symposium, Paris, 2002

3. G. Berrighi, D. Procopio, "Star Sensors and Harsh Environments", ESA GNC Conference, Frascati 2002

4. S. P.Airey, J. van Casteren, G. Coupe, J. Rouquets "The Rosetta Mission: Rendezvous with and Landing on a Comet – a mission overview", Breckenridge 2003.

5. M. Muller, E. Grun, "An Engineering Model of the Dust Environment of the Inner Coma of Comet P/Wirtanen" Parts 1&2, RO-ESC-TA-5501

6. P. Lelong, P. Riant, "Improvement of the robustness of the Rosetta Avionics to the comet dust, Feasibility Phase Analysis Report", RO-MMT-TN-2105 Issue 2, Nov 2000

7. G. Berrighi, D. Procopio, "Improvement of the Robustness of the STR to the Dust Environment – Requirement Specification", RO-GAL-RS-2009 Issue 2, Dec 2001

8. J. Fainberg, : An Algorithm for Star Pattern Recognition in Presence of False Stars", RO-ESC-TN-5517, Mar 2000

AAS 03-023

AUTONOMOUS ON-BOARD NAVIGATION FOR INTERPLANETARY MISSIONS

B. Polle,* B. Frapard,* O. Saint-Pé,* V. Crombez*
S. Mancuso,† F. Ankersen,†
J. Fertig‡

Autonomous navigation is one of the key technologies for future space exploration. Autonomous navigation during a close encounter with an asteroid was one of the Deep Space One mission experiment. This paper presents the first results of the ESA Technological Research Program "Autonomous Navigation for Interplanetary Missions". The objective of this study is to take a major step towards a European capability for autonomous onboard navigation. The study involves design, prototype implementation and test of the developed autonomous onboard navigation system covering all the phases of the mission: cruise, low thrust deep space manoeuvre, planetary or small bodies (asteroid or comet) encounters. Navigation is based on the measurements of a navigation camera (NavCam) supplemented by the platform star sensor when required. The main phases have been analysed to derive navigation requirements and specify the navigation camera. The On board Navigation (ObNav) system two main functions (GNC and Image Processing) are being designed to cover all the mission phases with the minimum set of functions. The system will be validated using simulated images and ground images and flight images from two space missions SMART-1 and Rosetta.

* Astrium SAS, 31, avenue des Cosmonautes, F-31402 Toulouse Cedex 4, France.

† European Space Agency, ESTEC, Keplerlaan 1, NL-2200 AG Noordwijk ZH, The Netherlands.

‡ European Space Agency, ESOC, Robert-Bosch Str. 5, D-64293 Darmstadt, Germany.

INTRODUCTION

Autonomous navigation is one of the key technologies for future space exploration. Autonomous navigation during a close encounter with an asteroid was one of the Deep Space One mission experiment. This paper presents the first results of the ESA Technological Research Program "Autonomous Navigation for Interplanetary Missions". The objective of this study is to take a major step towards a European capability for autonomous onboard navigation. The study involves design, prototype implementation and test of the developed autonomous onboard navigation system (ObNav) covering all the phases of the mission: cruise, low thrust deep space manoeuvre, planetary or small bodies (asteroid, comet or planet) encounters.

REFERENCE MISSIONS

Two future European Space Agency (ESA) interplanetary space missions SMART-1 and ROSETTA have been used as reference missions to derive navigation requirements and specify the navigation camera. These two missions are also potential candidates to provide flight images for ObNav validation.

SMART-1 is the first of the SMART (Small Missions for Advanced Research and Technology) programmes. SMART-1 will test a range of new technologies for spacecraft and instruments and will be the first European spacecraft to travel to and orbit around the Moon. It will be the first ESA spacecraft making use of electric propulsion as primary propulsion system. After injection into a conventional Geostationnary Transfert Orbit, a first phase of continuous SEP thrusting aims at escaping from the radiation belts as quickly as possible. When the perigee radius of 20000 km is reached, limited thrust arcs around pericentre will raise the apogee towards the Moon. When the gravitational effect of the Moon is sufficiently strong, it is utilized during several revolutions to achieve considerable changes in inclination and increases in perigee radius with minimal fuel expenditure. This enables capture into an orbit around the Moon followed by a breaking manoeuvres to place the satellite on an eccentric polar observation orbit around the Moon.

AMIE (Asteroid and Moon micro-Imager Experiment) is a miniature camera imaging with 4 different spectral bands (450, 750, 847 and 950 nm) by means of a thin film filter deposited on the CCD surface. It has a square field-of-view of 5.3° and weighs about 1.8 kg. The camera will be used by several experiments (See Ref 1) and especially by this study to test (off line) autonomous navigation using images taken by the camera and Star Tracker measurement.

Figure 1: SMART-1 Artist view

ROSETTA primary mission goal is a rendezvous with comet 46 P/Wirtanen. On its eight-year journey to the comet, the spacecraft will pass close to two asteroids, (Otawara and Siwa). The trip requires three gravity assist maneuvers, one with Mars and two with the Earth (See Figure 2). Rosetta launch has been postponed after Ariane 5 launch failure in December 2002. This means that Rosetta mission has to be redesigned. However, the primary mission, rich of events for an autonomous navigation study is used as the reference scenario for planet and asteroid encounter and two Rosetta encounters will be used for ObNav performance validation. Rosetta's flyby of Otawara takes place on 11 July 2006, when the asteroid is 1.89 AU (about 283 million km) from the Sun. Travelling at a relative velocity of 10.63 km/s, the spacecraft passes by Otawara's sunlit side at a distance of about 2200 km. Rosetta flies to within 3500 km of Siwa on 24 July 2008. The spacecraft flies past at a velocity of 17.04 km/s, approaching the sunlit side and looking at a crescent phase as it moves away. At this time, Siwa is at 2.75 AU (411 million km) from the Sun and 3.11 AU (465 million km) from the Earth.

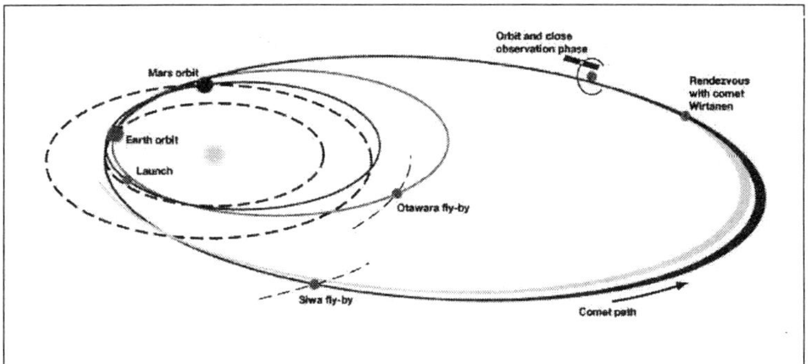

Figure 2: Rosetta primary mission overview

Optical Navigation Principle

Between two encounters, the space vehicle is either in ballistic cruise phase, or performing low thrust trajectory correction maneuver. During these phases, optical navigation relies on Line Of Sight (LOS) measurements of distant visible objects. These can be planets or asteroids, which are used as beacon. Measurement of the line of sight of a known object locates the vehicle in a cone starting from the beacon with a half cone angle equal to the camera LOS measurement accuracy. Measurement of two beacons LOS enables positioning of the vehicle in 3D as illustrated in Figure 3. A dynamical filter is used then to estimate the vehicle velocity from the observations of positions at (at least) two different times and enables to cope with non-simultaneous measurements.

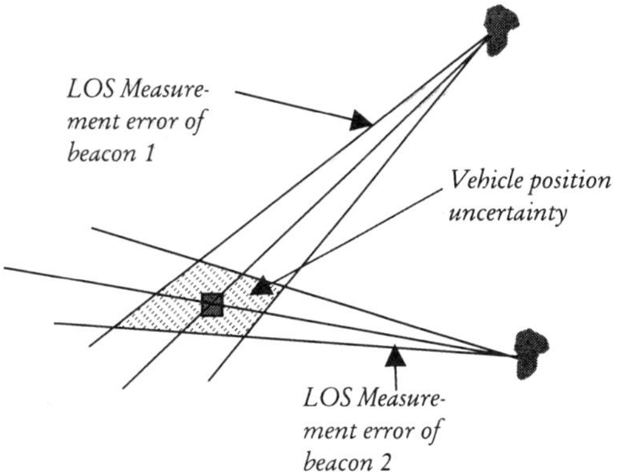

Figure 3: Principle of 3D localization using 2 beacons LOS measurement

During encounter phase, the target becomes closer than any other objects and target observation provides the highest accuracy positioning in the direction perpendicular to the LOS vector. The LOS vector being very close to the velocity vector, the encounter accuracy is mainly improved in the cross track directions. At large distances, along track (or range) position estimation is still based on cruise accuracy (based on distant object observation). When the target becomes resolved, along track position measurement can be based on the object size observation and comparison to the a priori knowledge. Moons orbiting around the target can also be used for this purpose. The drawback of this approach is that it requires a good knowledge of the target size or the presence of Moons, and thus lacks genericity. Observability analysis shows that, assuming the vehicle velocity modules is known and the target not aligned with the velocity, the along track error can be observed without any a priori knowledge or presence of Moons. The accuracy is proportional to the distance between the velocity vector and the target. This concept is selected as the baseline of the ObNav along track position estimation during close encounters. For resolved objects, the generic observable retained for navigation is therefore the reconstructed LOS to the center of the body. This generic approach permits to derive achievable performance figures in all phases of a generic interplanetary mission. Target figures for navigation performance are derived in the next §.

Navigation Requirements

The first navigation objectives are to satisfy mission requirements during encounter phase. Small bodies encounter and planet encounter requirement shall be distinguished as they have quite different requirements. The main objective of a small body encounter is scientific observation, while most demanding planetary encounters are gravity assist manoeuvres or aerocapture.

Small bodies encounters are divided in four phases: cruise phase, early encounter, late encounter and target tracking. Early encounter phase starts when the positioning accuracy provided by the target observation becomes better than the accuracy provided by far beacons, late encounter starts when the remaining time before encounter do not allow to perform anymore Trajectory Correction Manoeuvres (TCM). Target tracking phase starts when the mission requires pointing the vehicle scientific instruments to the target. Small bodies encounter navigation requirements concerns mainly cross track errors at the encounter time and Time Of Flight knowledge. Cross track accuracy at the encounter time is measured by the B-plane delivery accuracy and shall be as small part of the encounter nominal flyby distance. On Deep Space One the flyby distance was about 10km and the targeted accuracy less than 1km. On Rosetta Otawara and Siwa flyby distance is about 2000 km and is far more less demanding. In order to demonstrate the capability of ObNav, the objective has been set to the challenging 1km cross track accuracy for small bodies encounters. Time Of Flight knowledge accuracy is mainly driven by the necessity to schedule on board operations. Without any knowledge of the TOF, it is not possible to program TCM or to schedule scientific instruments operations. A TOF error of max (3%, 0.6km) has been allocated, ensuring good margins for operation scheduling up to the close observation of the target.

From navigation point of view, planetary encounters for gravity assist trajectory corrections are similar to small bodies encounters except that there is no need to improve knowledge accuracy after the last TCM nor to point instruments to the target. Cross track accuracy shall be such that appropriate safety margins are guarantied and that the consumption of trajectory correction maneuvers required to control trajectory before encounter and compensate for flyby errors after encounter be low. Allocations have been set to 10km cross track and 5 m/s consumption.

Cruise phase navigation accuracy requirements are less straightforward to derive. The most demanding requirement comes from the encounter phase TOF prediction (3%). TOF prediction based on target observation is accurate enough only during the late encounter phase. Before, TOF prediction is based on far body observation and drives cruise navigation accuracy requirement. The requirement is function of the flyby distance. For close flyby, like DS1 encounter with Braille, 500 km cruise accuracy is required to maintain the 3% TOF relative error over the whole trajectory. Close to encounter, the TOF estimation accuracy improves up to 0.6km at the end of the target-tracking phase. For a more distant encounter (or a planet encounter), several thousand of kilometers cruise accuracy is sufficient to meet the 3% TOF relative error requirement.

For the case where low thrust trajectory manoeuvres are permanently performed during the deep space phase, navigation requirements come mainly from the desire to control the thrust errors using the tail of the maneuver rather than using specific TCM. Using chemical propulsion, the maneuver duration is too short to enable close loop control. Thus, the typical 1-2% realization errors shall be controlled later using dedicated TCM. With electric propulsion, these errors can be controlled in closed loop, limiting the over consumption due to the realization error to a figure significantly lower than 1%.

Phase	Requirement	Justification
Asteroid Encounter	• 1 km cross track • 3 % along track (Up to encounter)	• To allow close encounter. • To enable proper instruments operation scheduling.
Planet Encounter (Gravity Assist Maneuver)	• 10 km cross track • 3% along track (Up to -3 hours)	• To minimize gravity assist manœuvre error correction consumption.
Cruise (For Asteroid Encounter)	• 500 km	• To enable detection of faint targets. • To meet along track error requirement.
Cruise (For Planet Encounter)	• 10 000 km	• To meet along track error requirement.
Low Thrust Deep Space Maneuver	• 500-1000 km	• To allow manœuvre realisation error correction at almost no cost.

Table 1: Navigation Requirements Synthesis

ASTEROID DATABASES ANALYSIS

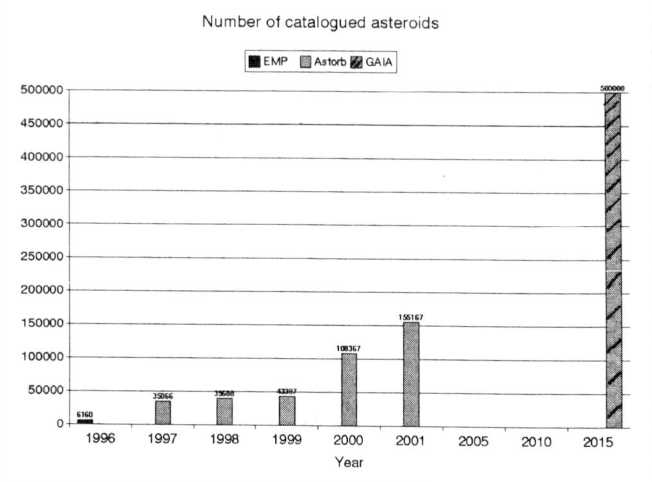

Figure 4: expected evolution of the number of detected asteroids up to 2015

The determination of the amount of detected asteroids is considered as a first basic element to be studied in order to determine the capability for this type of solar system objects to represent efficient beacons for autonomous navigation. Thanks to the progress of sky survey instruments, the total amount of detected asteroids has considerably increased during the past 10 years, growing up from few thousands to more than 150 000 at the end of 2001. As the large majority of the known asteroids have been observed from the Earth, the closest asteroids have been detected first. It is therefore not expected that the number of asteroids which are at the same time potential beacons (i.e. bright enough) and close from the Earth will increase a lot as nearly all have been detected today. On the other hand, a large amount of asteroids with diameter > 1 km is expected to be discoverer from new observations, mainly located far from the Earth. These objects are generally called TNO, for Trans Neptunian Objects. Therefore, the number of detected asteroids will continue to grow up during the next 10 years. One of the major new discoverers should be the ESA GAIA astrometric space mission. Even if study of solar system objects is not its main objective, at least 500 000 new asteroids will be observed (see figure 4). The plotted value ($5\ 10^5$) represents the minimum number of asteroids that GAIA should observe. It includes all the TNOs (Main Belt Objects (MBOs), Scattered Disk Objects (SDOs), Kuiper Belt Objects (KBOs) and Plutinos up to the 20.0 magnitude).

Figure 5 presents the expected Cumulative TNO+SDO Surface Density (CSD, in deg^{-2}) as a function of the limiting R magnitude (see RD2). It allows estimating, at the present time, the number of TNOs observable by degree square on the sky brighter than the apparent R magnitude m. In Jewitt et al (RD5), it is suggested that one can estimate the population of TNOs with diameters in excess of 100 km to be about 10^5. They orbit the sun between 30 AU (orbit of Neptune) and 50 AU (the practical limit of existing surveys). These objects obey a power law size distribution with index $q \sim -4$ which extends smoothly to Pluto (diameter about 2300 km) and allows the possibility that other Pluto-sized objects await discovery.

As a conclusion about the total number of asteroids, it is not expected that the number of potential beacons close to the Earth will increase a lot in the future, as such objects which will be discovered will feature faint magnitudes and/or a high relative speed (i.e. worse case for accurate orbit determination). On the contrary, a considerable amount of large objects far from the Earth will be discovered during the 10 next years, representing potential interesting beacons for autonomous navigation mission far in the outer solar system.

The achievable navigation performance is function of many parameters: the LOS accuracy, the asteroid spatial distribution, the position of the vehicle... In order to specify the camera limit magnitude, a statistical analysis has been performed to evaluate the RSS error as a function of these parameters. Given the asteroid distribution, the key parameters are expected to be the distance of the vehicle from the Sun and the camera limit magnitude. For each distance and limit magnitude the position of the vehicle and the date are randomly selected. For each case, the database is sorted by visual magnitude and only asteroids brighter than the limit magnitude are retained. Then the resulting list is sorted by distance to retain the closer visible asteroids. Finally, the navigation error ellipsoid is computed by successively including asteroids by distance order. The error ellipsoid size is proportional to the distance and to the LOS error. Normalized to a LOS error of one, the result is homogeneous to a distance. The semi major axis represents the worst-case navigation error. Figure 6 shows the result of this analysis for the observation of 20 asteroids. To obtain the navigation accuracy for a given camera, the value has just to be multiplied by the camera LOS accuracy. It can be seen that performance is between 0.02 and 0.05 AU in the range [2-3.5] AU, which correspond to the main asteroids belt. Before and after performance degrades very quickly. From this analysis we can conclude that the minimum limit magnitude for asteroid based navigation is 12 and that 13 allows navigation up to 3 AU from the Sun. Above 3AU a much more sensitive camera is required to detect a less dense population of asteroid receiving less light from the Sun. Assuming 8μrad LOS accuracy, the performance at 2.5 AU from the Sun is 0.03AU*8μrad=40km and navigation performance be dominated by Asteroid ephemeris errors. At 1AU from the Sun, the navigation accuracy is growing up to 800 km.

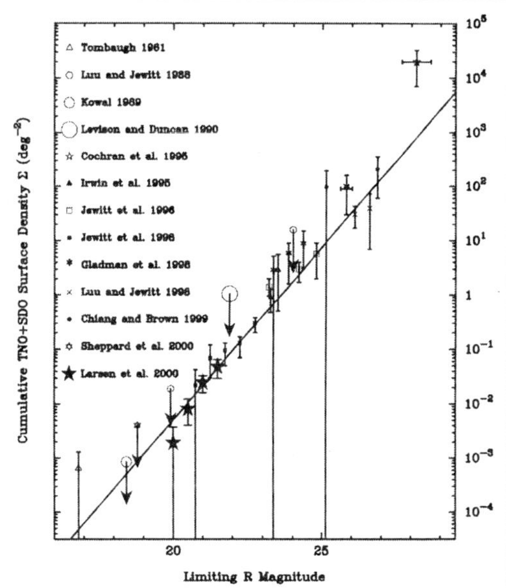

Figure 5: cumulative TNO+SDO Surface Density

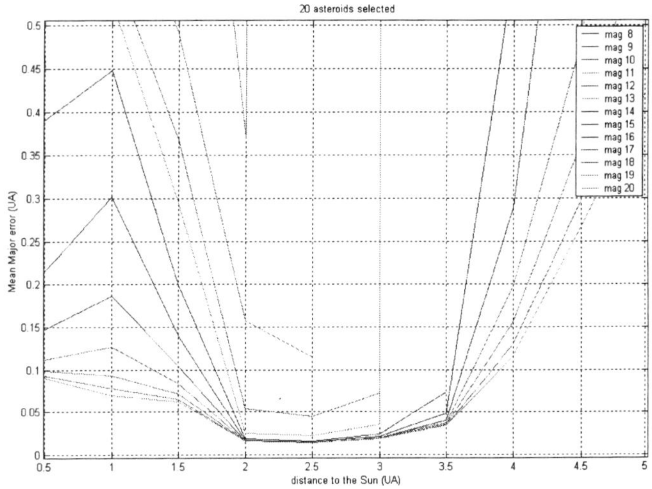

Figure 6: Statistical analysis: mean error on Major axis

NAVCAM CAMERA SPECIFICATION

Starting from mission level requirements ObNav subsystems requirements have been derived. For camera design they summarize to 2 major design goals: a 8 µrad LOS resolution, and a 13 magnitude for detection limit. Figure 8 shows the allocation logic that permits to derive these specifications. The cruise accuracy requirement of 500 km is not a mission level requirement but derived from the 3% Time Of Flight (TOF) knowledge requirement. The planetary encounter allocation logic is similar to the small body encounter logic except that LOS requirement allocation can be relaxed to 50 µrad and cruise control accuracy to about 5000 km. Low thrust deep space manoeuvre navigation accuracy required to allow the control of thrust realization errors in closed loop is relatively constraining (typically 1000 km) but can be easily relaxed with a small fuel consumption penalty. Figure 8 presents a synthesis of the derived allocations where critical performances are indicated in bold. Asteroid ephemeris accuracy requirement is 500 km. This is typical of the state of the art achieved for asteroids within 3 AU from the Sun (Ref 2). Accuracy can be improved by dedicated observations if required by the mission.

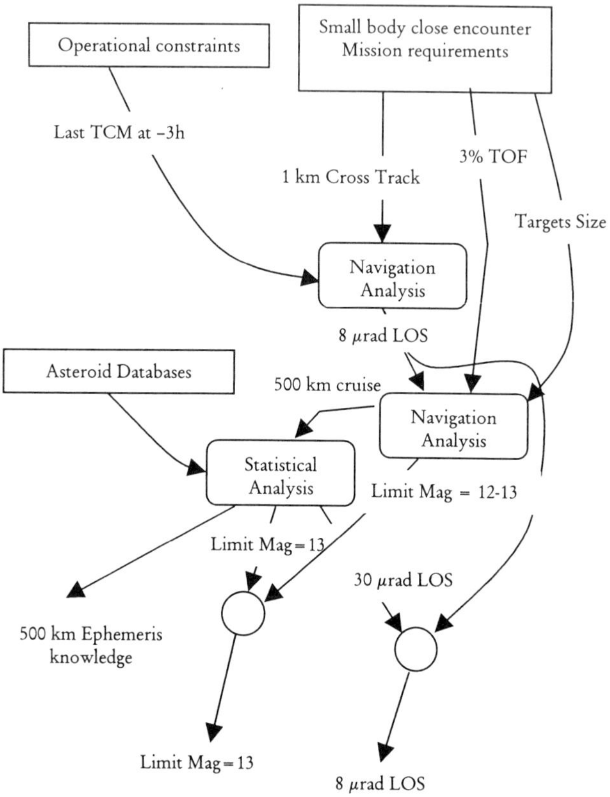

Figure 7: Small encounter derived requirements allocation logic

Small body close encounter derived requirements	
Camera LOS measurement error	**8 µrad**
Camera Limit magnitude	**13**
Asteroid ephemeris errors	500 km
Planet encounter derived requirements	
Camera LOS measurement error	**50 µrad**
Camera Limit magnitude	**12**
Low thrust deep space manoeuvre derived requirements	
Camera LOS measurement error	20 µrad
Asteroid ephemeris errors	1000 km

Figure 8: Autonomous navigation derived requirements

NAVIGATION SENSORS ARCHITECTURE

To achieve the performance required to support the small body encounter, the selected ObNav baseline is to use a small FOV camera imaging both asteroids and stars in the background. For planet encounter, this solution cannot be used anymore since a small FOV camera cannot simultaneously measure the planet and faint stars. One solution could be to increase the field of view to decrease the magnitude of the brightest visible stars, allowing simultaneous measurement of the planet and stars. The main drawback of this solution is that increasing the FOV increases also the probability to have the Sun in the FOV. Indeed, planetary swing by analysis shows that the separation angle between the Sun and the planet is quite low and tends to decrease as the S/C gets closer; In the case of Rosetta, value of between 7° and 50° have been observed. In order to cope with these situations and not to induce too strong constraints at mission design level, the choice has been made to use also a small FOV sensor for planetary encounters. The camera is used to make planet limb/space transition observations, the direction being referred to the inertial frame thanks to the platform star tracker. The main limitation of this design becomes the NavCam / Star tracker alignment stability, which shall be better than 50μrad. Although challenging, this stability can be achieved by an appropriate interface mechanical design. Depending on the magnitude of the beacon under observation, three modes of measurement are foreseen: beacon with stars in the background in a single exposure, beacon and stars in the background with adaptation of the exposure time in a sequence of three images (stars/beacon/stars) to compensate for platform motion between the measurements and finally beacon measurement with the NavCam and star measurement with the star tracker.

OBNAV NAVIGATION CAMERA

The field of view of the camera for the cruise phase is driven by the density of reference stars necessary for the asteroid position measurement in the camera reference frame. The reference stars to be taken into account shall provide signal to noise ratio better than the one of the beacon. As the asteroid minimum magnitude is 13, it is proposed to fix the reference star limit equal to 11. The star density changes a lot with galactic latitude. Based on data available about star density per square degree a 0.5°x0.5° FOV is selected in order to have a minimum of two reference stars even in the mean worst-case (90° galactic latitude). In this case, less than 50 stars with magnitude lower than or equal to 13+/-0.5 are expected for the mean best case (0° galactic latitude). This choice ensures that there is enough stars for measurement but not too many, reducing the probability of potential confusion between the asteroid and faint stars.

Thanks to its maturity, its availability and its close to perfect electro-optics performances, CCD technology has been selected as detection technology by nearly all the space missions observing in the visible and NIR bands these last 20 years. The emerging CMOS detector technology (also called "APS" for "Active Pixel Sensor") is considered as a promising alternative or successor. CMOS Imaging System (CIS) offer interesting features for the ObNav application: 1) Extension of domain when bright objects and stars can be imaged simultaneously. 2) Reduction of smearing with bright objects. In addition, CIS will allow to design a less expensive camera, with reduced power and mass budgets when compared to a camera using a CCD. CCD offers better electro-optics performances (and particularly a low read-out noise coupled to a high quantum efficiency), and has a long space heritage, which makes its use well predictable.

The camera associated with image processing shall enable unambiguous detection of Mag 13 beacons. One of the driving parameter is the LOS motion due to the vehicle attitude motion during signal integration by the detector. Platforms using reaction wheels have a significant advantage over platform using thrusters with a limited Minimum Impulse Bit. The later have been considered for the camera design assuming typical attitude rate for such platforms is 20 μrad/s. Another driving parameter is the detector degradation with radiation. CCD. Radiation causes degradation of the Dark Signal Non Uniformity (DSNU) but also of the Charge Transfer Efficiency (CTE) (mainly due to protons). To mitigate CTE degradation, the background level is maintained at a certain level (typ 1000 electrons) increasing the detection noise. To limit the camera size and mass, a 70 mm aperture diameter has been selected. This choice does not allow detection of Mv 13 beacons in a single frame, requiring to use advanced image processing algorithms.

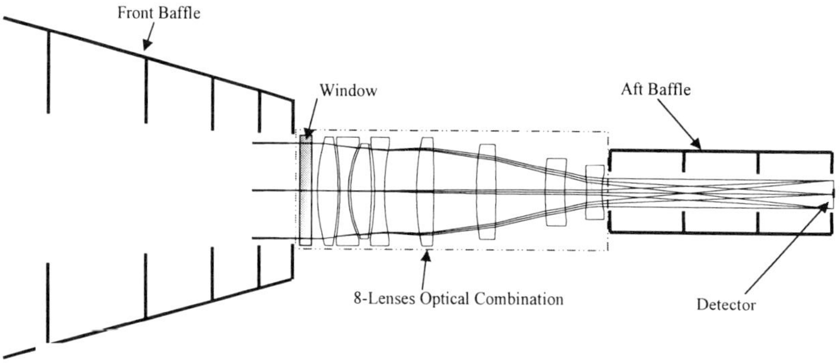

Figure 9: NavCam preliminary design

In nearly all the existing star trackers, the image of the sky is defocussed in the detector plane in order to achieve sub-pixel resolution. This is mainly because the ratio between the required field of view -say 20°- and the required LOS accuracy -say 5 arcsec- is larger -7500 in this case, considering a +/- 1/2 pixel accuracy- than practical detector size. Achieving a 1/20 sub pixel accuracy thanks to centroiding processing allows to select 750 pixels for detector size which is acceptable for most star trackers. For autonomous navigation, considering a detector pixel pitch of 13 μm, a 512x512 detector size and 800 mm focal length have been selected. In focused operating mode with +/-1/2 pixel accuracy, the corresponding 8 μrad LOS accuracy is met. After image processing, a significant improvement of this raw estimation is expected and will be the subject of intensive testing, allowing further optimization of the design to reduce mass and size. The current camera mass budget is 7.5 kg.

NAVIGATION FILTER

Filter Sizing

Covariance simulations have been performed to assess the filter behavior during the encounter phase and to link camera performance requirements to the navigation requirements. Variation of the approach trajectory shape and flyby distance enables to identify key parameters and simplified performance evaluation formulas. As an example, cross track estimation accuracy convergence is illustrated in Figure 10 on a case chosen close to the DS1 Braille approach scenario. This simulation assumes observation of the target body LOS only (no other beacons).

Trajectory Correction Maneuver separation interval is selected proportional to the remaining distance. An interval increasing factor of 2 has been chosen for this simulation while the study analysis concludes that the optimal ratio between two successive orbit control maneuvers is close to 2.7 (to minimize consumption). The target LOS observation is assumed performed just before the correction maneuver and improved the estimation by steps. The divergence between updating is mainly due to velocity estimation errors. This simulation confirms that the cross track accuracy is well approximated by $\sigma_y = D\sigma_{LOS}$, where σ_y is the cross track estimation accuracy, σ_{LOS} is the camera LOS measurement error and D the remaining distance to encounter.

Covariance simulations show that the shape of the approach trajectory is of prime importance for the along track position estimation. The vehicle/target trajectory curvature improving the observability of this low observability state during the encounter phase.

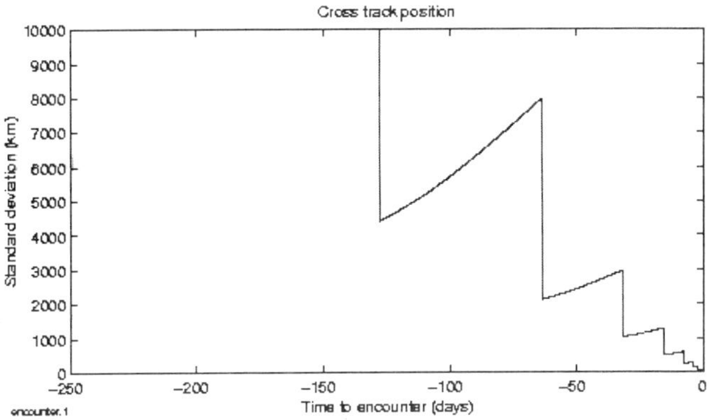

Figure 10: Cross track estimation covariance simulation (close encounter with asteroid)

Filter Implementation

The Orbit Determination filter is implemented as a sequential filter using UD decomposition of the covariance matrix for numerical stability. This formulation has been preferred to the least square batch SRIF batch filter for its flexibility and its ability to cover all the phases (cruise and encounter) using the same formalism. The filter state vector includes the vehicle position, the vehicle velocity, the non-gravitational acceleration bias and the thrust scale factor. One of the main advantages of the sequential formulation is that it more easily copes with situations where some of the states are observable while others are not. This is the case during early encounter phase, when cross track error is easily observable through target LOS measurement but along track error is not observable. The sequential filter allows to limit the observations to the target during the last days, the along track position being propagated from the cruise estimation up to the time where the target observation makes the along track position observable. In this phase, a batch filter would require to image other asteroids to make all the states observable. With a sequential filter, the imaging requests number (and associated attitude maneuvers) is reduced to the minimum in this critical phase of the encounter.

IMAGE PROCESSING

Image Processing (IP) is a central function of autonomous navigation. The core of the IP functions relies on Multiple Time Integration (MTI) and Multiple Cross Correlation (MCC) algorithms. MCC has been used on DS-1. The underlying assumption of the algorithm is that long exposure will be necessary to image dim objects and that due to the spacecraft attitude motion the images will be smeared in patterns called glyphs (See Figure 11). As the sensor FOV is very small, all glyphs are almost identical. The principle of MCC algorithm is to extract the glyph for each object and to perform correlation between the extracted glyph and all the other objects. A least square filter is then used to estimate the relative position of each object (Ref 3). MCC is a powerful algorithm but suffers of two drawbacks: first the resulting accuracy is highly dependent of the glyph shape. If the attitude rates does not change during the integration, the glyphs will resume to straight lines and the cross correlation accuracy will be much more accurate on the transverse direction than along the line. Next, the objects spend only a small fraction of time in each pixel, and the rest of the time pixels are integrating background noise instead of useful signal. To improve MCC performance, ObNav uses MTI as a front end to MCC. MTI consist in taking successive small exposure images and to integrate these images numerically. MTI increases the signal to noise ratio, reduces dispersion of correlation maxima during MCC and increases separation between glyphs (See Figure 12). The drawback of MTI is that many frames have to be processed while only one frame is processed in the MCC case. It is planned to accumulate about 30 successive frames, increasing the signal to noise ratio by a factor of 5 and reducing the camera entrance pupil diameter by more than a factor of two. In order to accumulate the images it is required to know the camera rotation between the successive frames. To limit interface between IP and Attitude control, attitude information will be reconstructed using guide stars with sufficient magnitude to be detected in a single frame (Mv >11). MTI/MCC enables to achieve two important goals: reduce the camera aperture requirement and achieve sub-pixel accuracy LOS measurement. Achievable performance evaluation will be one of the main output of Image Processing validation task, allowing to optimize the camera sizing.

Two other IP algorithms are of prime importance for ObNav: background cleaning and contour detection of extended objects. Background cleaning is required to remove most of the dark current or stray light signal, improving the performance of MCC and MTI. Measurement of extended objects position is achieved by limb observation. Limb observation has been selected for its genericity and simplicity. Interface between IP algorithms and navigation is the orientation of the S/C to body center of mass vector. This information is reconstructed by IP algorithms based on the object model. When only fractions of the limb is observable in the FOV, IP will provide an estimation of the measurement covariance. Fusion of successive observations being achieved by the navigation filter.

In what concerns structured noise (stray light) removal, JPL has published interesting and easy to implement methods to ensure flat field correction, namely the MICAS algorithm described in [Ref 4], which we intend also to use. A composite image is build from multiple images of the same field taking advantage of the fact that jitter actually shifts the images of the star relative to the CCD reference system between images, allowing easy identification of the star field vs. background photons. This composite image includes the stars of interest and also the background; the most relevant component of the former is surrounding stray light. The composite background image is then subtracted from each image in order to generate a much cleaner image of the star field.

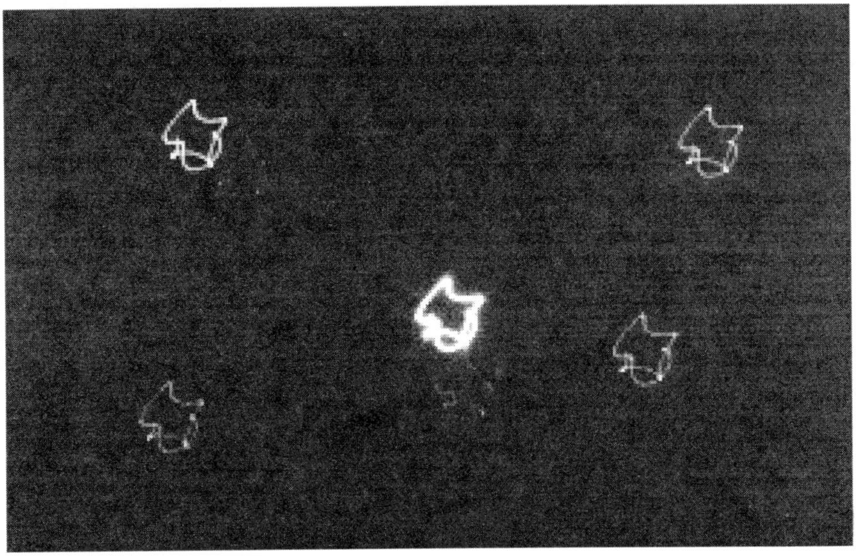

Figure 11: Long exposure image

Figure 12: Multiple Time Integration image

OBNAV ARCHITECTURE

ObNav top level architecture is illustrated in Figure 13. ObNav includes two types of function: real time blocks and executive experts.

Real time blocks: *ObNav Executive* is the real time task in charge of managing execution of activities related to autonomous navigation by sending appropriate requests to the subsystems involved in the navigation and monitoring their states. *Navigation Real Time* includes all the ObNav functions with RT interfaces with ACS. *ObNav image processing Real Time* includes the image processing functions working in real time. This can include for example processing of successive frames of the MTI mode. *AOCS Real Time* handle the ObNav executive requests and interfaces (S/C manoeuvring requests, non gravitational forces evaluation. *Camera management* handles the ObNav requests, image processing interfaces and is in charge of detailed programming of the navigation camera, retrieving the images and feeding the images to the image processing.

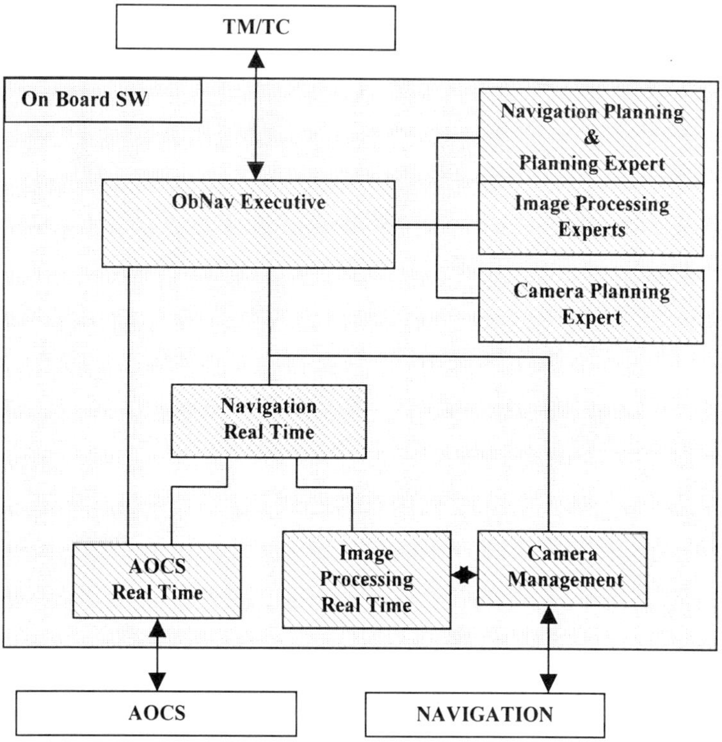

Figure 13: ObNav software architecture

Planning Experts: Planning experts function are functions which are triggered by the executive at some time and required to program further operations. They are not real time blocks in the sense that they are not called on a cyclic basis but they have to provide their outputs within a known deterministic duration. *Navigation Planning and planning experts* is the function in interface with the Executive, which is in charge of computation of parameters or sub-plans. This function will typically include the functions related to guidance (TCM manoeuvre amplitude and direction computation or planned DSM manoeuvres parameters adjustment) and to planning/scheduling (computation of effective date of programmed sequences of operations). *Camera planning Expert:* This block is in charge of the camera on board programmation, based on camera model eventually updated by dedicated calibrations. *Image processing Expert* groups IP functions that do not need to be performed in real time like MCC.

OBNAV DEVELOPMENT AND VALIDATION

Filter Development

ObNav development will use extensively automatic code generation. Real time blocks will be developed under Matlab/Simulink development environment. Some algorithms, mainly image processing algorithms will be developed on top of Matlab Image Processing Toolbox and code will be generated using Matlab Compiler. The produced code being integrated into Simulink S-functions. Two benches will be used to validate ObNav performances. A first bench is dedicated to functional validation and is only functionally representative of the On Board software. It is based on Matlab/Simulink with no specifying embedded software coding constraints. The second bench is derived from the previous one using automatic code generation. The on board software is generated using both Matlab Real Time Workshop and Matlab Compiler. The generated code is encapsulated in a single multirate S-function, which includes the on-board part of the software. The S-function is instrumented to continuously monitor computing time of cyclic tasks and expert function calls by the executive. The bench will run on a workstation using the same processors architecture as the on board computer (SPARC architecture) and on board processing time requirement will be derived by extrapolation using dedicated benchmarks.

Validation Approach

A first set of images is used to support design and validation of the image processing concepts and timings: robustness against starry background, for more or less crowdy conditions (See Figure 16 and 17). Then dedicated images acquisition campaigns are planned for both ground based and flying telescope for a complete validation.

Three type of images will be used to validate ObNav. Close loop validation of image processing, and GNC will be achieved using simulated images. Simulated images are essential to validate closed loop behavior, but representativity of simulated images is always questionable. This is why validation has been complemented by tests using real images. Two sources of real images will be used: flight images taken by the AMIE camera of SMART-1 or the navigation camera of ROSETTA and images taken by a ground observatory.

The Moon Fly By period of SMART-1 mission has been selected to be the most favorable phase for the experiment (figure 15). During about three months, the spacecraft trajectory semi major axis is maximum, going from the earth at perigee up to the Moon to acquire the gravity impulses, which will enable capture by the Moon. The trajectory in this phase is very like an encounter trajectory with the Moon or with the Earth. Images of Moon limb, Earth limb and stars (for alignment calibration) have been requested and will enable validation of the autonomous navigation applied to encounter with large bodies with and without atmosphere.

The ground images are planned to be taken by a small astronomical instrument at the “Pic du midi” observatory. The instrument field of view is identical to the ObNav camera FOV (0.5°). Figure 16 shows an image of the Psyche asteroid taken by the ground instrument selected to provide images to this study. The instrument aperture will be reduced to be representative of the onboard camera. As no simple solution has been found to be representative of the on board camera noise, noise will be added by post processing. Satellite attitude motion will be simulated by modification of the telescope earth rotation compensation rate. The most demonstrative test envisaged consist to feed the images taken in the navigation filter, and to let the filter estimate the observatory location and velocity. The dynamic model will be initialized to be representative of the observatory trajectory: A virtual planet will be placed on the Earth spin axis at the observatory longitude with a mass such that the orbital period at the “pic du midi” position be 24 hours (See Figure 14). Real images tests cannot be used to validate guidance and control, but is an essential tool to validate image processing performance and robustness.

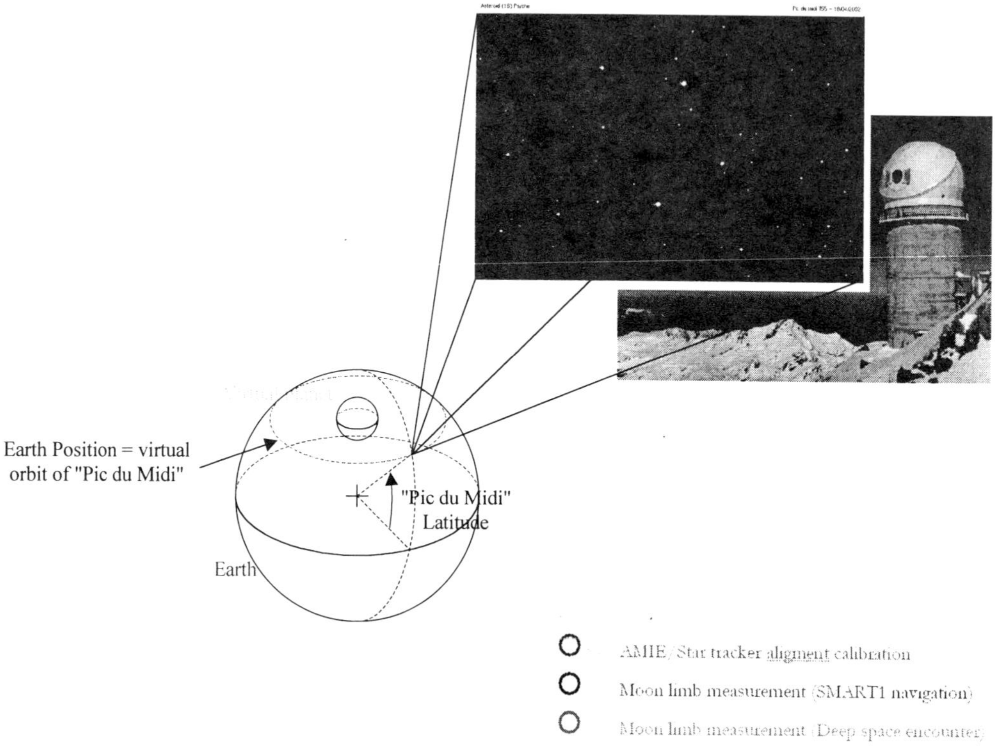

Figure 14: Experiment using ground images principle

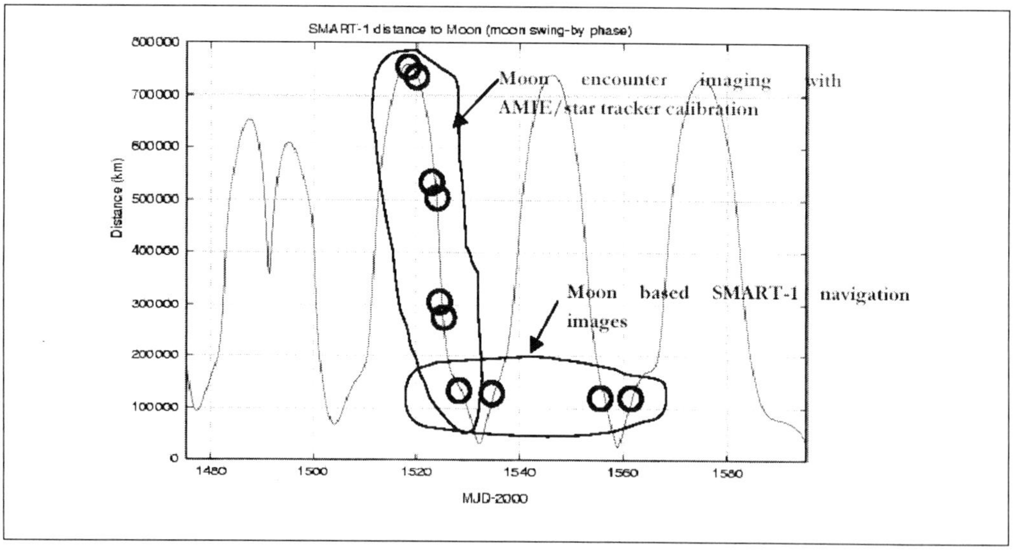

Figure 15: Experiment using Smart-1 Flight Images

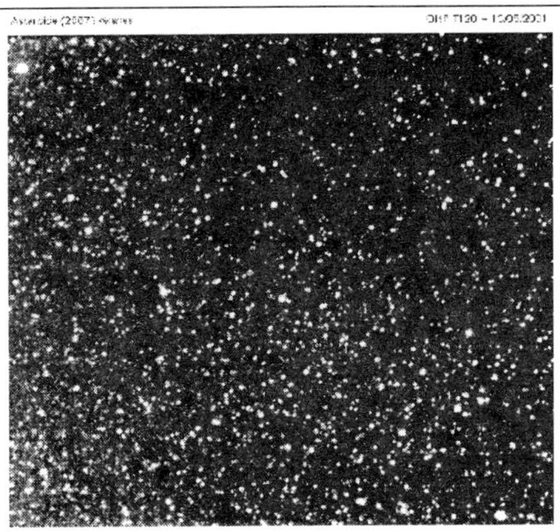

Figure 16: Aknes asteroid observed with the 120cm telescope (Illustrates a dense star background).

Figure 17: Psyche asteroid observed with the Pic du Midi 55 cm telescope on 18/04/02

CONCLUSION

Starting from the lesson learned on Deep Space 1, a generic optical navigation system has been designed and is currently being implemented. A dedicated camera capable to handle the various phases of the mission has been specified and a preliminary design proposed. The ObNav system validation campaign starts 3rd quarter of 2003 for ground based telescopes images acquisition and processing. Flight experiment with AMIE on SMART-1 starts first part of 2004.

ACKNOWLEDGMENTS

The authors wish to acknowledge J.Berthier form the "Institut De Mécanique Céleste et de Calcul des Éphémerides" who provides expertise on the Asteroid databases and provides the ground based images to this study.

REFERENCES

1. Andrea MARINI, Giuseppe RACCA, Bernard FOING, "SMART-1 TECHNOLOGY EXPERIMENTS IN PREPARATION TO FUTURE ESA PLANETARY MISSIONS" European Space Agency - ESA/ESTEC Noordwijk - The Netherlands
2. J. Berthier, Définitions relatives aux éphémérides de position des corps célestes, Notes Scientifiques Et Techniques Du Bureau Des Longitudes
3. J. E Riedel, S. Bhaskara Autonomous Optical Navigation DS1 Technology Validation Report, Jet Propulsion laboratory
4. Composite Image Subtraction: A Technique For Calibrating MICAS Images, Justin Maki, JPL, Section 388, 21, December 1998, in http://www-mipl.jpl.nasa.gov/~jnm/ds1/micas.htm.

AAS 03-026

DEEP IMPACT FLYBY SPACECRAFT INSTRUMENT POINTING IN THE PRESENCE OF AN INACTIVE IMPACTOR SPACECRAFT

Daniel G. Kubitschek, Nick Mastrodemos and Stephen P. Synnott*

The engineering goal of the Deep Impact mission is to impact comet Tempel 1 on July 4, 2005, with a 350 kg active Impactor spacecraft (s/c). The relative velocity will be just over 10 km/s. The impact is expected to excavate a crater of approximately 20 m deep and 100 m wide. The science objective is that of exposing the interior material and understanding the properties of the nucleus. In order to achieve the engineering goal and science objective, Deep Impact will use the autonomous optical navigation (AutoNav) software system to guide the Impactor s/c to Tempel 1 intercept near the center of brightness (CB), while a second s/c, the Flyby s/c, uses identical software to determine its comet-relative trajectory providing the attitude determination and control system (ADCS) with the relative position information necessary to point the High Resolution Instrument (HRI) and Medium Resolution Instrument (MRI) at the expected impact site during encounter.

If the Impactor s/c is determined to be functioning improperly prior to release, the issue of predicting the impact location to correctly point the instruments at key science epochs (TOI: Time of Impact; and TOFI: Time of Final Imaging), becomes important and therefore must be studied. This relies, fundamentally, on the ability to determine the trajectory of the Impactor s/c relative to the Flyby s/c by treating the Impactor s/c as an optical beacon, relative to which the Flyby s/c's trajectory is estimated using images of the Impactor s/c. Simulation results show that for an inactive Impactor s/c, the pointing error is improved from 519 μrad (1 μrad = 10^{-6} radians) to 72.3 μrad (3σ) at TOI, and from 3.96 mrad (1 mrad = 10^{-3} radians) to 441 μrad (3σ) at TOFI. Then compared to the baseline CB targeting/tracking approach, results show the pointing error contribution due to knowledge of the impact location changes from 60 μrad to 72.3 μrad (3σ) at TOI, and from 495 μrad to 441 μrad (3σ) at TOFI. This paper deals only with the pointing error contribution due to errors in predicting the impact location and describes the acquisition of optical data of the Impactor s/c and associated errors using the Flyby instruments; the expected uncertainty in predicting the impact location and the resulting pointing errors; and the algorithm for autonomously computing a pointing correction during encounter.

INTRODUCTION

The engineering goal of the Deep Impact mission is to impact comet Tempel 1 on July 4, 2005, with a 350 kg active Impactor spacecraft (s/c). The impact velocity will be just over 10 km/s. The impact is expected to excavate a crater of approximately 20 m deep and 100 m wide. A second spacecraft, the Flyby s/c, is responsible for delivering the Impactor s/c and will perform a

* Jet Propulsion Laboratory, California Institute of Technology, 4800 Oak Grove Drive, Pasadena, California 91109.

slowing maneuver (deflection maneuver), following Impactor release, to observe the impact event, ejecta plume expansion, and crater formation, which will take place over a period of approximately 800 seconds. Figure 1 shows the flight system configuration with the Impactor s/c stowed in the lower portion of the Flyby structure. Figure 2 shows the encounter geometry for the Deep Impact mission. The science objective is that of exposing the interior material and understanding the properties of the nucleus.

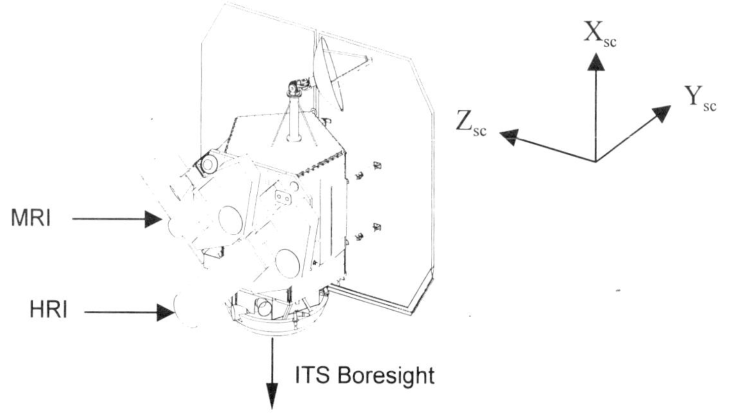

Figure 1 Deep Impact flight system configuration showing the instrument platform, High Gain Antenna, ITS boresight and solar array[2]

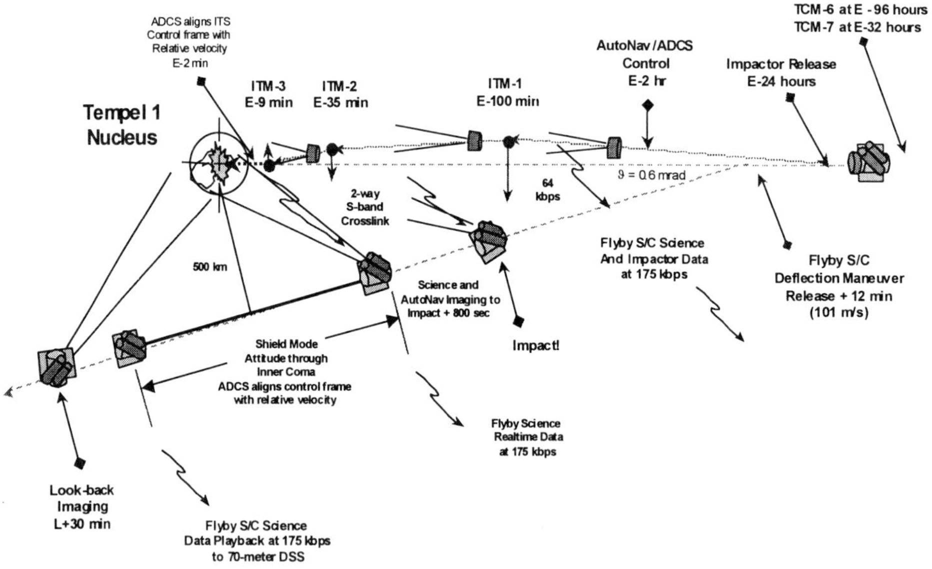

Figure 2 Tempel 1 encounter geometry for the Deep Impact mission

Deep Impact will use the on-board autonomous optical navigation (AutoNav) software system[1] to guide the Impactor s/c to Tempel 1 intercept near the center of brightness (CB). The Flyby s/c uses identical software to determine its comet-relative trajectory in order to provide the attitude determination and control system (ADCS) with the relative position information necessary to point the High Resolution Instrument (HRI) and Medium Resolution Instrument (MRI) at the nucleus CB during encounter. The Impactor s/c and the Flyby s/c operate in an independent fashion with the Flyby s/c pointing the MRI/HRI instruments at the impact site in an indirect way by assuming that the Impactor s/c will impact at or near the nucleus CB. Figures 3 and 4 show simulated images of the comet nucleus using the Impactor ITS camera and the Flyby MRI/HRI cameras along their respective trajectories.

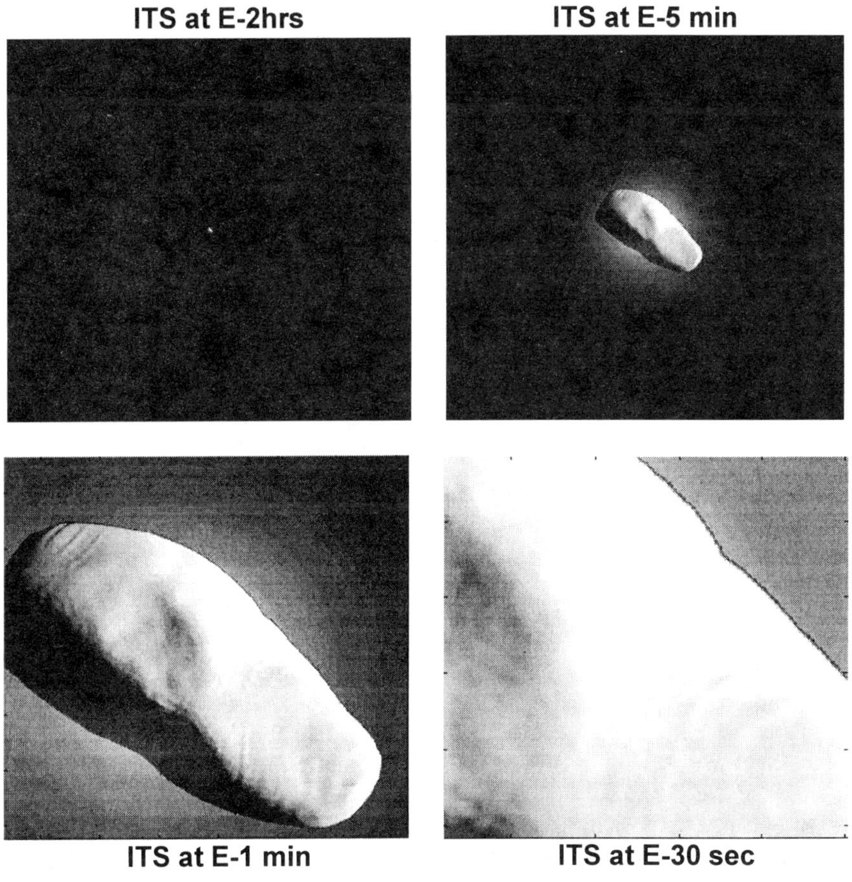

Figure 3 Simulated ITS images of the comet nucleus (based on Halley-Stooke data) during encounter where E- designates time to impact

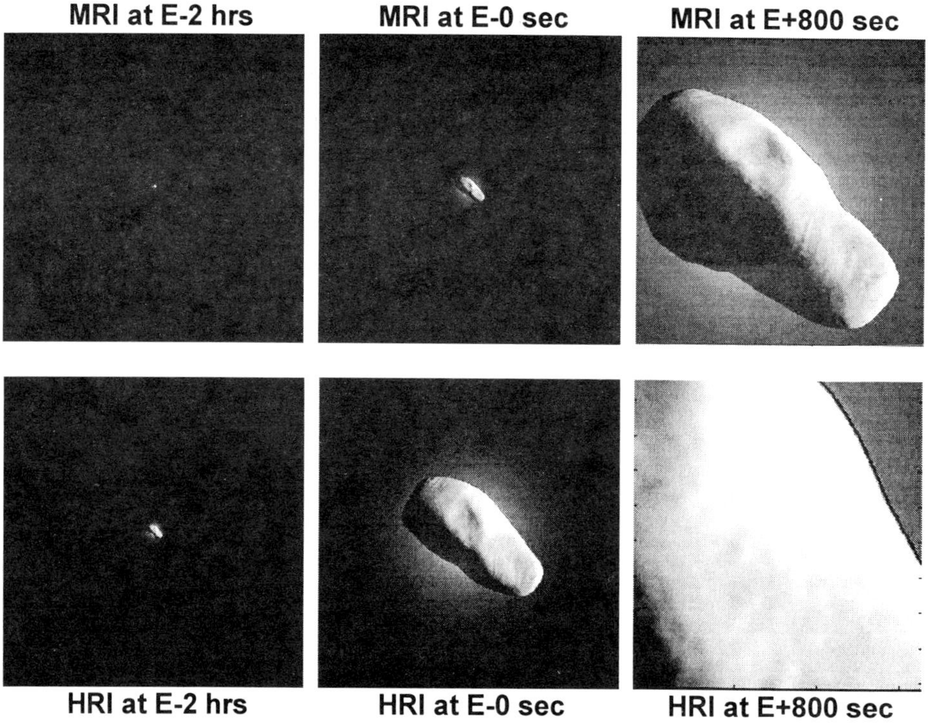

Figure 4 Simulated MRI and HRI images of the comet nucleus (based on Halley-Stooke data) during encounter where E- designates time to impact. Note the apparent rotation of the nucleus, seen in the MRI images, as the Flyby s/c passes underneath

Two key science epochs drive system performance during the Tempel 1 encounter: Time of impact (TOI), and Time of final crater imaging (TOFI). To obtain the highest possible temporal resolution imaging at the TOI, the HRI will be operated in a 128x128 pixel sub-frame mode (see figure 5) which allows images to be taken more rapidly than if they were full-frame exposures. The HRI instrument has a 1008x1008 active pixel charged-couple device (CCD) detector with a pixel scale of 2 μrad/pixel (1 μrad = 10^{-6} radians), giving it a 2 mrad (1 mrad = 10^{-3} radians) field-of-view (FOV). Therefore, overall pointing error must not exceed 128 μrad at TOI in order to capture the impact event in the HRI 128x128 pixel sub-frame. At the time of final imaging, the Flyby s/c will be at a range of 700 – 1000 km from the surface of the nucleus. At this range, the HRI FOV covers only a small portion of the nucleus (mean nucleus radius is estimated to be 2.6 km with an approximate axial ratio of no larger than 2:1). Pointing at both of these science epochs requires good knowledge of the impact site relative to the observed CB. Analysis shows that for an active Impactor maneuvering to intercept the CB, pointing at TOI, to capture the impact site in the 128x128 pixel subframe, is achievable and less than 100 μrad (3σ) of which 60 μrad (3σ) are due to uncertainties in knowledge of the actual impact site location. Studies have shown that the probability of capturing a high-resolution image of the fully developed crater is 97% for the expected HRI instrument performance[3].

Due to the short lifetime requirement (7 days) of the non-redundant Impactor s/c, the inactive Impactor failure scenario has warranted attention. There are a number of considerations that must be addressed under this contingency, broadly divided into two aspects: 1) achieving impact in an illuminated area with an inactive Impactor s/c; and 2) autonomously predicting the impact location of the inactive Impactor s/c to compute and apply a pointing correction relative to the CB in an effort to minimize Flyby pointing performance degradation at the key science epochs. The first problem is addressed by fine-tuning the trajectory of the flight system to intercept the nucleus of Tempel 1 prior to the release of the Impactor s/c. If it is determined that the Impactor s/c is not in an operational condition prior to release, then the maneuver to fine-tune the trajectory will be delayed to allow for additional optical navigation data collection and the release of the Impactor will be postponed to E-12 hrs to increase the probability of an illuminated impact. Here we discuss the method for solving the second problem.

If we consider only the contribution of errors in knowledge of the impact site location, the baseline targeting and tracking approach provides TOI and TOFI pointing performance as shown in table 1. In addition, table 1 shows that for an inactive Impactor s/c our ability to point the Flyby s/c instruments is substantially degraded due to uncertainties in knowledge of where the impact will occur relative to the nucleus center of brightness as observed from the Flyby s/c during the last 2 hrs of the encounter. The results shown in table 1 for a maneuvering Impactor s/c are only part of a larger pointing error budget and must be combined, in a root-sum-square (RSS) sense with errors such as knowledge of the Flyby s/c's position relative to the observed CB, motion of the CB due to nucleus rotation, ADCS alignment errors, ADCS alignment drifts, and ADCS control errors to arrive at the total pointing error at TOI and TOFI.

Table 1

Expected Flyby s/c pointing errors due to impact site location uncertainties at TOI and TOFI for the baseline targeting/tracking approach and for the current inactive Impactor s/c scenario

Approach	**TOI Pointing 3σ Error (μrad)**	**TOFI Pointing 3σ Error (μrad)**
Baseline (Maneuvering Impactor)	60	495
Inactive Impactor	519	3960

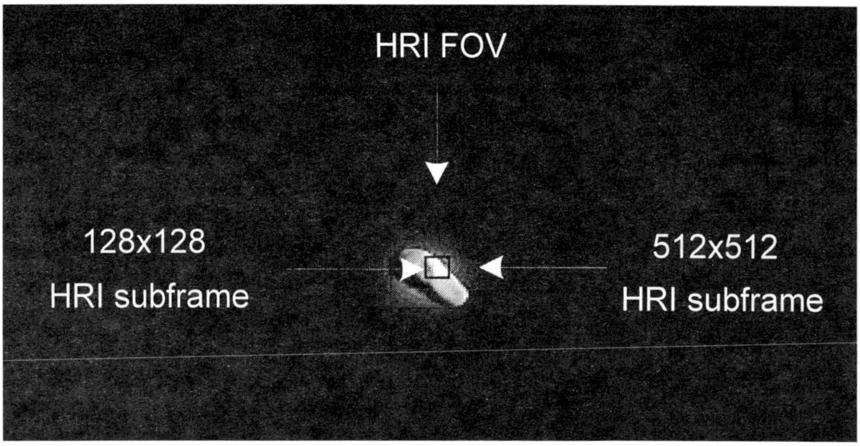

Figure 5 MRI image at TOI showing the HRI FOV (blue), the 512x512 pixel HRI subframe (green) and the 128x128 pixel HRI subframe (red) within which the impact site must reside at TOI

IMPACT SITE PREDICTION FOR AN INACTIVE IMPACTOR

The basic problem with an inactive Impactor s/c is that it cannot maneuver itself to impact at a location that is expected, independently, by Flyby s/c. Although the Impactor s/c has a high likelihood of impacting somewhere on the surface (studies show ~ 95% probability of delivering an inactive Impactor on an impact trajectory[3]), pointing the narrow FOV HRI instrument with the Flyby s/c is degraded: the Flyby s/c will point at the CB which may be as much as 4.5 km (3σ) from the actual impact site; an approach that may only capture crater images with the wider FOV MRI instrument (7 m resolution) instead of capturing images with the desired spatial resolution (3.4 m) obtained using the HRI instrument.

The basic idea of impact site prediction relies on the fact that following release, the inactive Impactor's trajectory remains unperturbed until impact. The Flyby s/c will initiate a deflection maneuver which is designed to control the Flyby miss-distance to 500 ± 50 km and slow the Flyby spacecraft to provide 800 ± 20 seconds of science imaging from the time of impact to the time of shield mode entry prior to passage through the inner coma dust environment (shield mode occurs approximately 50 sec before the Flyby s/c reaches it's closest approach point). If the Impactor s/c is healthy, then the deflection maneuver will nominally take place at E-23:48 hrs (12 min after Impactor release), where E- designates time of impact, and will be ~ 102 m/s in magnitude. This maneuver results in execution errors that map to a 32 km B-plane position error (3σ) at encounter and an 8 sec (3σ) time-of-flight (TOF) error.

The orthogonal triad that represents the orientation of the B-plane coordinate system relative to the International Celestial Reference Frame (ICRF), as seen in figure 6, is determined as follows:

$$\hat{s} = \frac{\overline{V}_\infty}{\left|\overline{V}_\infty\right|}$$

Here, $\overline{V}_\infty$ is the comet-relative velocity in the inertial frame of reference. The unit vector, $\hat{t}$, is derived by rotating $\hat{s}$ by and angle of -90° about the ICRF $\hat{z}$ axis, projecting the result onto the equatorial plane, and normalizing (The unit vector $\hat{t}$ is always in the ICRF *x-y* plane).

$$\hat{t}' = \begin{bmatrix} 0 & 1 & 0 \\ -1 & 0 & 0 \\ 0 & 0 & 1 \end{bmatrix} \hat{s} = \begin{pmatrix} s_y \\ -s_x \\ s_z \end{pmatrix}$$

$$\hat{t} = \frac{1}{\sqrt{s_y^2 + s_x^2}} \begin{pmatrix} s_y \\ -s_x \\ 0 \end{pmatrix}$$

And finally, $\hat{r}$ is the cross product of $\hat{s}$ and $\hat{t}$.

$$\hat{r} = \hat{s} \times \hat{t} .$$

The transformation from ICRF to B-plane coordinates is given by:

$$R_{ICRF \to Bplane} = \begin{bmatrix} \hat{t} & \hat{r} & \hat{s} \end{bmatrix}^T = \begin{bmatrix} t_x & t_y & t_z \\ r_x & r_y & r_z \\ s_x & s_y & s_z \end{bmatrix} .$$

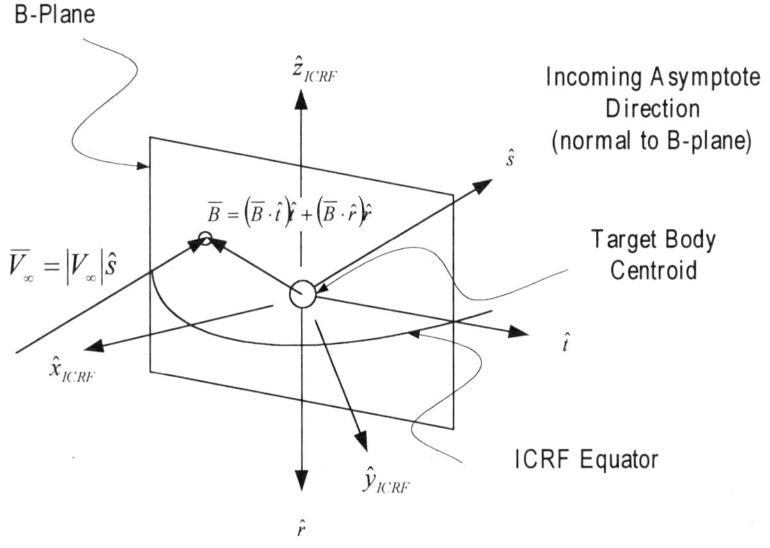

Figure 6 Definition and orientation of the B-plane relative to the International Celestial Reference Frame (ICRF)

If, following the deflection maneuver, the trajectory of the Flyby s/c can be estimated relative to that of the Impactor, then the Flyby's trajectory can be integrated and its position evaluated relative to the Impactor s/c at TOI. The key is to treat the Impactor as an optical beacon, relative to which the Flyby s/c's trajectory may be estimated using optical images of the Impactor. The expectation is that these images will provide very accurate determination of the Flyby's trajectory crosstrack to the line-of-sight relative to the Impactor. Alongtrack information will come primarily from radiometric tracking of the Flyby (range and Doppler measurements). This post-deflection radio tracking is part of the baseline approach and serves to evaluate the need for a contingent deflection trim maneuver, at E-12 hrs, as well as to reduce the TOF error introduced by the deflection burn. A summary of the post-deflection Flyby navigation plan is as follows:

1. Continue radio tracking of the Flyby s/c for 6 hrs after end of deflection burn
2. Image the Impactor from the Flyby, beginning 35 minutes after release
3. Estimate the Flyby's heliocentric trajectory with a radio-only solution
4. Estimate the Flyby's Impactor-relative trajectory with an optical solution using initial conditions obtained from the radio-only solution

OBSERVING THE IMPACTOR SPACECRAFT

Following separation, the Impactor remains on a trajectory, which is essentially the incoming asymptote relative to comet Tempel 1. The Flyby s/c slows by ~ 101 m/s with an ~ 5.7 m/s velocity component perpendicular to the incoming asymptote, giving a constant inertial view angle of the Impactor from the Flyby s/c of ~ 3° with respect to the Impactor's comet-relative trajectory. This angle does not change until impact. Images of the Impactor and background stars will be acquired, beginning 35 minutes after release, and processed on the ground. The accuracy of the orbit determination based on these optical data depends, among other things, on the Signal-to-Noise Ratio (SNR) of the background stars and the observability of the Impactor s/c using the Flyby instruments.

Impactor Spacecraft Surface Properties

The Impactor has an irregular shape and a mix of materials with varying reflective properties, e.g. metal, paint, thermal blanket etc. The ITS anti-boresight side of the Impactor s/c bus, the one that for most of the time will be facing the Flyby s/c, has a hexagonal shape (figure 7). The ITS boresight direction has a hemispherical shape made up of copper plates, with six panels connecting the two sides. An accurate determination of the optical signal strength obtained from the Impactor at a given orientation would require knowledge of the reflective properties, exact shape modeling and attitude history. The Impactor, during the time it is imaged from the Flyby, may be turning to point the ITS off-nucleus to obtain calibration images, or in the case that it is inactive it will be tumbling at an unknown rate and direction. The overall phase darkening of the Impactor at different attitudes will vary in a complicated way, requiring a modeling process beyond the scope of a contingency treatment. The approach is not to compute the exact signal from the Impactor, but rather to establish whether reasonable, as well as conservative assumptions, result in a sufficient SNR.

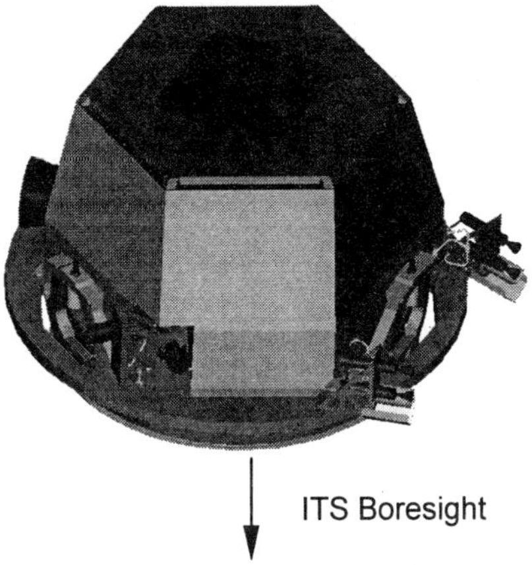

Figure 7 Impactor s/c flight system configuration[4]

Our simplified assumption is an irregular shape with an average diameter of 0.8 m and albedo of ~ 0.3 based on drawings and data from Ball Aerospace Technologies Corporation (BATC). Assuming that we view mostly a flat Impactor surface and that the single particles on the surface scatter isotropically, we may adopt a Lommel-Seeliger reflection law with an average incidence angle, i, equal to the phase angle ≈ 65°; and an average emission angle, e, equal to the view angle from the Flyby s/c ≈ 3°. The integrated intensity at that phase angle compared to intensity at a phase of 0° is: $\cos(i)/(\cos(i) + \cos(e)) = 0.29$ corresponding to a phase darkening of ≈ 1.3 magnitudes or a linear phase coefficient of 0.02 mag/degree. By parameterizing the phase darkening of the Impactor with a phase coefficient, we can consider a variety of more conservative phase functions: 0.04 mag/degree, a value typical for comets and asteroids, or 0.06 mag/degree corresponding to some of the most phase-darkened objects, such as comet Encke[5].

Impactor Spacecraft Optical Signal

Figures 8 and 9 show the expected SNR in the peak pixel for a 1 second and 3 second exposure, as a function of time for 3 different phase functions for the HRI and MRI, respectively. The SNR is computed with the current data for each instrument and with the rather conservative assumption of 4 DN (≈ 120 e-) of read noise. As shown in figure 8, HRI imaging guarantees a high SNR for many hours after separation. For the MRI, however, the SNR is significantly lower. To maintain a value of SNR ≥ 7, which will typically guarantee a center-finding accuracy of ~ 0.1 pixels (1σ), we may need to consider longer exposures. Figure 9 shows that a 3 second exposure can provide a SNR ≥ 7 for ~ 3 hours of imaging. In general, the choice of camera (MRI, HRI) depends on many factors, which include:

1. Sufficient signal from at least two background stars. The FOV of the MRI (10 mrad) is 5 times larger than that of the HRI (2 mrad) FOV, but the HRI can detect stars fainter by at least 4 magnitudes compared to the MRI. Depending on the approach trajectory, background star availability may favor one instrument over the other.

2. Ability to acquire the Impactor in a narrow HRI FOV. Because of errors in the deflection maneuver, which map to errors in the Flyby s/c position at time of Impactor acquisition, the Impactor will not necessarily be at the predicted location during imaging.

These considerations will be examined in detail after the approach asymptote of the flight system becomes known (i.e. after launch). Another issue, which is common to both instruments, is that of obtaining star-relative Impactor images in the presence of an extended diffuse coma, which is foreground to the stars and background to the Impactor. Based on the Deep Impact Science Team predictions for the coma brightness, a worst-case scenario whereby all of the observed out-gassing is the result of a narrow jet, gives the peak pixel brightness due to coma of up to 80,000 e-/s in the pixels adjacent to the nucleus. For a late release at E-12 hrs, imaging of the Impactor could extend to as late as E-9 hrs. At that time the Impactor will be projected, relative to the Flyby instruments, in a direction where the background coma is ~ 15,000 km from the nucleus. At that distance the coma is quite faint. Assuming the coma brightness decreases as r^{-1} with distance from the nucleus, the coma background will be ~ 3 - 3.5 DN for an exposure of 1 – 3 sec (i.e. comparable to the CCD system noise). This will result in a small decrease of the Impactor's optical signal relative to the background, and therefore, has no affect on center-finding accuracy. Figure 10 shows a simulated image of the Impactor s/c as seen in the HRI instrument at a range of approximately 180 km and with an exposure duration sufficient to bring the signal up near full-well (16,383 DN).

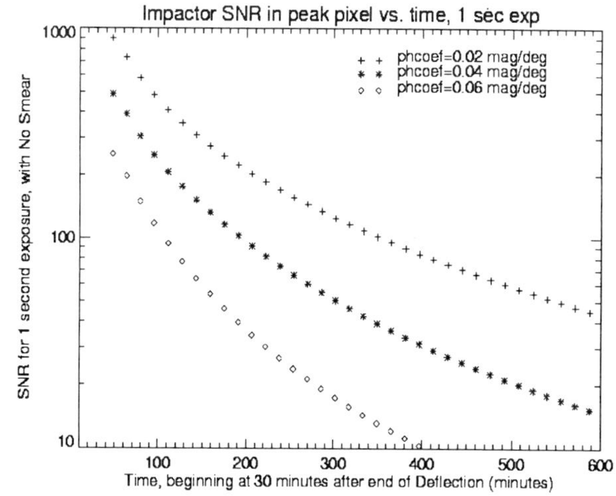

Figure 8 SNR of the Impactor s/c as observed with the HRI

SIMULATIONS AND MONTE CARLO BASED PERFORMANCE ANALYSIS

Given that sufficient optical signal from the Impactor s/c is expected, as shown in the previous section, a suite of MATLAB algorithms have been developed to:

1. Simulate images of the Impactor s/c using either the MRI or HRI instrument
2. Integrate the Impactor and Flyby s/c truth trajectories
3. Perform image processing and orbit determination[6]
4. Evaluate the Flyby position error (true position – estimated position) at the key science epochs (TOI and TOFI)
5. Compute the pointing error at TOI and TOFI
6. Perform a Monte Carlo analysis to assess the pointing error due to the uncertainty in the impact site location at the key science epochs

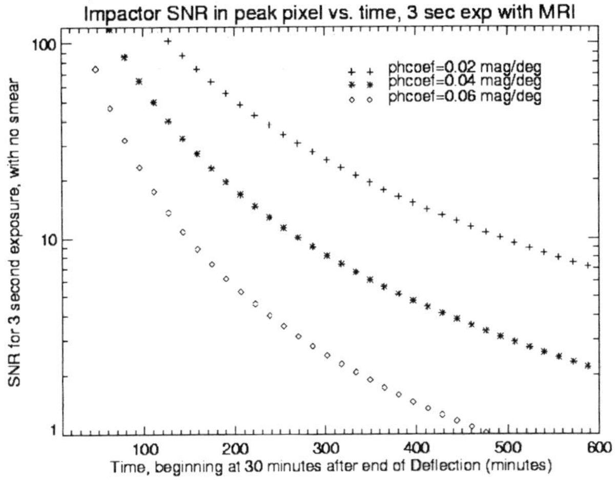

Figure 9 SNR of the Impactor s/c as observed with the MRI

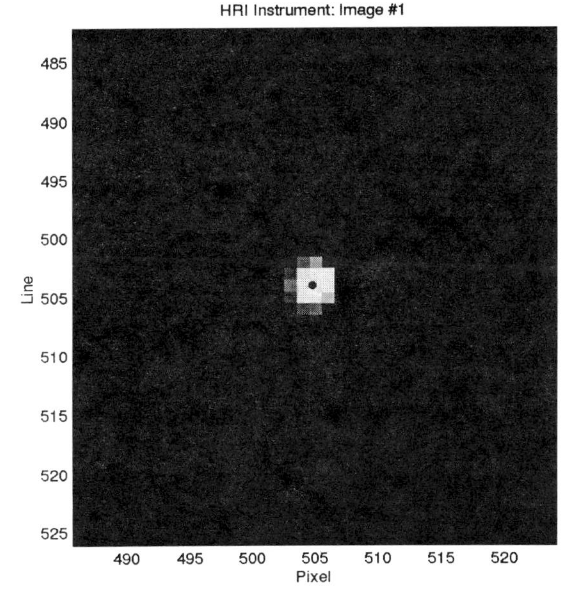

Figure 10 Simulated subframe image of the Impactor s/c in the HRI at a range of 179.9 km and phase of ~ 65°; blue dot represents location of CB

The random parameters in the Monte Carlo simulations are the Flyby trajectory deflection maneuver errors, and the center-finding errors described previously (0.1 – 0.2 pixels; 1 – 2 μrad for the MRI). The epoch time is taken to be Release plus 35 minutes (start OpNav imaging arc) which results in a range of ~ 180 km from the Flyby s/c to the Impactor s/c. The deflection burn is modeled as an impulsive maneuver occurring at release plus 5 minutes, with maneuver execution errors randomly sampled in the alongtrack direction (0.50 m/s, 1σ), relative to comet Tempel 1, and crosstrack directions (0.12 m/s, 1σ). These errors represent the expected flight system deflection maneuver execution capability and are used to integrate the Flyby s/c truth trajectory.

The beginning of Impactor image acquisition occurs at Release plus 35 minutes and ends at Release plus 1 hour and 35 minutes to give a batch orbit determination (OD) arc length of 1 hour. Images are sampled at a 15 min interval providing 5 observations that are used in the estimation process. At the beginning of image acquisition, differences between the Flyby s/c's true position and the Flyby s/c's expected position result in pointing errors on the order of 3.8 mrad (3σ). Here, the position error (3σ) results in a pointing error that is approximately 2 times larger than the HRI FOV. Since immediate detection of the Impactor is not guaranteed with the HRI, it will be preferable to use the MRI throughout. In 500 simulation runs, the uncertainty in the initial position of the Impactor never exceeded the MRI FOV. The maximum pointing error was found to be 6.5 mrad at initial image acquisition. The maximum dispersion from the MRI boresight was found to be 650.5 pixels or approximate 6.5 mrads in 500 simulation runs. Figure 11 shows the dispersion of locations of the Impactor in the MRI FOV at initial acquisition. The HRI FOV is superimposed (blue) and shows that the associated Flyby s/c maneuver execution errors require use of the MRI instrument for Impactor tracking to be assured of capturing the Impactor observations.

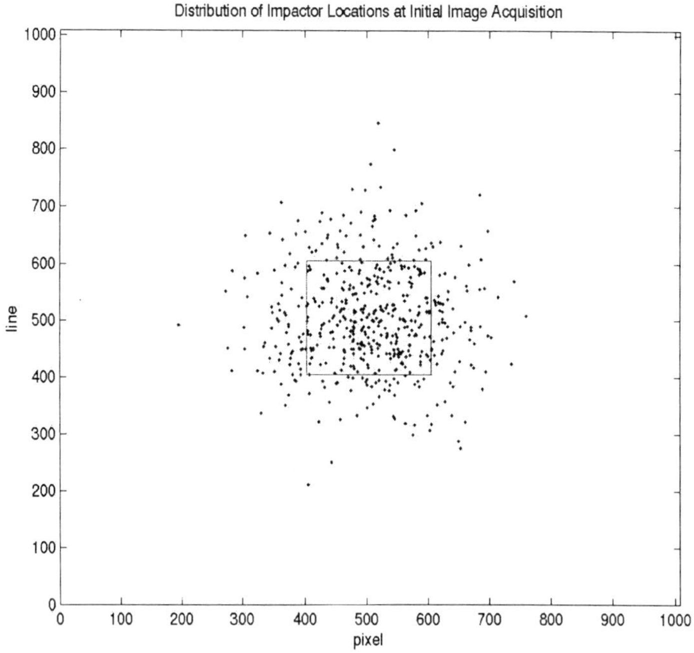

Figure 11 Dispersion of Impactor locations in the MRI FOV at initial image acquisition (Release + 35 min) for 500 simulation runs; the HRI FOV is superimposed in blue

After computing the orbit solution, the estimated Flyby position is evaluated at TOI (Release + 24 hrs). The angular difference between the unit direction vectors from the true Flyby s/c position and estimated Flyby position, relative to the Impactor s/c gives the pointing error at TOI.

$$\vartheta_{TOI\_Error} = \arccos(\hat{u}_{True} \cdot \hat{u}_{Est})$$

where

$$\hat{u}_{True} = \frac{\overline{X}(t_{TOI})_{Flyby \to Impactor\_True}}{\left|\overline{X}(t_{TOI})_{Flyby \to Impactor\_True}\right|}, \qquad \hat{u}_{Est} = \frac{\overline{X}(t_{TOI})_{Flyby \to Impactor\_Est}}{\left|\overline{X}(t_{TOI})_{Flyby \to Impactor\_Est}\right|}$$

Here, $\overline{X}(t_{TOI})_{Flyby \to Impactor\_True}$ and $\overline{X}(t_{TOI})_{Flyby \to Impactor\_Est}$ are the true and estimated position of the Flyby s/c, relative to the Impactor s/c at TOI, respectively.

At TOFI (TOI + 750 sec) the pointing error is computed differently, since there is a discontinuity in the Impactor's trajectory at that time. Following impact, the B-plane error vector, $\Delta\overline{x}$, remains constant and represents that contribution to the pointing error, which is due to the uncertainty in our knowledge of where the impact site is located. Here we compute the B-plane error vector by first transforming the vector difference between the Flyby s/c's estimated position and the Flyby s/c's true position relative to the Impactor s/c at TOI into the B-plane reference frame.

$$\Delta\overline{r} = \left[R_{ICRF \to Bplane}\right]\left(\overline{X}(t_{TOI})_{Flyby \to Impactor\_True} - \overline{X}(t_{TOI})_{Flyby \to Impactor\_Est}\right)$$

The error vector is then the projection of $\Delta\overline{r}$ onto the B-plane to give the direction. The projection is scaled by the range and pointing error at TOI to give the magnitude.

$$\Delta\overline{B} = \frac{\vartheta_{TOI\_err}\left|\overline{X}(t_{TOI})\right|}{\sqrt{\Delta r(1)^2 + \Delta r(2)^2}}\begin{pmatrix}\Delta r(1)\\ \Delta r(2)\\ 0\end{pmatrix}$$

Transforming the projection into the ICRF frame gives the inertial correction vector that is used to evaluate the pointing error at TOFI.

$$\Delta\overline{x} = [R]^T_{ICRF \to Bplane}\,\Delta\overline{B}$$

Using the error vector and the true Flyby s/c range to the comet nucleus at TOFI, the pointing error is computed from the law of cosines as

$$\vartheta_{TOFI\_Error} = \arccos\left(\frac{R^2_{Flyby \to Comet} + R^2_{Flyby \to Impact\_Site} - \left|\Delta\overline{x}\right|^2}{2R_{Flyby \to Comet}R_{Flyby \to Impact\_Site}}\right)$$

where

$$R_{Flyby \to Comet} = \left|\overline{X}(t_{TOFI})_{Flyby \to Comet\_True}\right|$$

and

$$R_{Flyby \to Impact\_Site} = \left| \overline{X}(t_{TOFI})_{Flyby \to Comet\_True} + \Delta \overline{x} \right|$$

Table 2 shows the error contribution of the uncertainty in the impact site location to the overall pointing error at TOI and TOFI. The OD arclength is 1 hr with observations taken every 15 min to give a total of 5 optical navigation (OpNav) observations. The following section will describe the autonomous algorithm that is necessary to point the MRI and HRI instruments at the key science epochs using the estimated Flyby s/c's trajectory.

Table 2

Monte Carlo results for Flyby s/c pointing error contribution due to impact site location uncertainty at TOI and TOFI based on 500 simulation runs

Key Science Epoch	Min (μrad)	Max (μrad)	3xRMS Error (μrad)
TOI Pointing	3.47	41.2	72.3
TOFI Pointing (TCA-100 sec)	21.3	328	441

POINTING ERROR SENSITIVITY TO OPNAV ORBIT DETERMINATION

For this approach, pointing errors depend directly upon the accuracy of the orbit determination process using optical images of the Impactor s/c. To fully understand the problem and to assess the best approach, we consider the sensitivity to OD arc length, center-finding errors, and range from the Flyby s/c to the Impactor s/c during the OpNav imaging arc.

To assess the sensitivity to the OD arc length, 3 different OD batch lengths were considered. Each batch has the same starting epoch corresponding to release (R) + 35 min and the same imaging interval of 1 image per 15 min. Table 3 shows the results for 1 hr, 2 hr, and 3 hr OD batch solutions.

Table 3

Sensitivity to OD Arc Length

OD Arc Length Number of Observations	RMS TOI Pointing Error (μrad)	RMS TOFI Pointing Error (μrad)
1 hr arc ⇒ 5 OpNavs 500 sim runs	24.1	147
2 hr arc ⇒ 9 OpNavs 100 sim runs	45.8	297
3 hr arc ⇒ 13 OpNavs 100 sim runs	37.5	231

This analysis assumes center-finding errors of 0.1-0.2 pixels, which represents a random angular measurement error of 1-2 μrad (1σ) in the MRI instrument. Using the HRI, this error would be 0.2-0.4 μrad. Here, we considered a single 1 hr OD batch arc length starting at R + 35 minutes with 1 observation every 15 min. Table 4 shows the results for 3 different center-finding uncertainties: 2 μrad (0.2 MRI pixels), 10 μrad (1 MRI pixel), and 20 μrad (2 MRI pixels).

Table 4
Sensitivity to center-finding errors

Center-finding Error (1σ)	RMS TOI Pointing Error (μrad)	RMS TOFI Pointing Error (μrad)
2 μrad (0.2 MRI pixels) 500 sim runs	24.1	147
10 μrad (1 MRI pixel) 100 sim runs	47.0	289
20 μrad (2 MRI pixels) 100 sim runs	88.2	547

Finally, the sensitivity to the Impactor range during OpNav imaging was considered. Three different 1 hr OD arcs were considered: the first batch started at R + 35 min and ended 1 hr later; the second batch started at R + 1 hr 35 min and ended 1 hr later (R + 2 hr 35 min); and the third batch started at R + 2 hr 35 min and ended at 3 hr 35 min. Table 5 shows the RMS pointing errors at TOI and TOFI for each of the 3 OD arcs. The degradation with range is not unexpected since the OD process relies on angular measurements whose spatial error increases as the range increases.

Table 5
Sensitivity to Impactor range during OpNav imaging

OD Arc Start	OD Arc End	RMS TOI Pointing Error (μrad)	RMS TOFI Pointing Error (μrad)
R + 35 min 500 sim runs	**R + 1 hr 35 min**	24.1	147
R + 1 hr 35 min 100 sim runs	**R + 2 hr 35 min**	28.7	176
R + 2 hr 35 min 100 sim runs	**R + 3 hr 35 min**	29.0	178

THE AUTONOMOUS POINTING CORRECTION ALGORITHM

Until now we have discussed the process of ground-based OpNav to determined the location of the Impactor s/c relative to the Flyby s/c and the uncertainty in the location of the impact site, which results in pointing errors at the key science epochs. The next step is to develop an algorithm for computing a pointing correction on-board the Flyby s/c and in real time so that the ADCS software can adjust the MRI/HRI pointing during encounter.

Once the Flyby s/c's inertial, heliocentric trajectory has been determined based on optical observations of the Impactor s/c, the Chebyshev coefficients that represent the Flyby s/c and the Chebyshev coefficients that represent the Impactor s/c trajectory as a function of time throughout the entire encounter period, are uploaded to the AutoNav flight software on the Flyby s/c. As shown in figure 2, AutoNav acquires and processes images of comet Tempel 1 starting 2 hrs prior to the expected TOI. Every minute, AutoNav updates its position and velocity based on observations of the CB of comet Tempel 1 and generates an ephemeris in the form of Chebyshev coefficients. These coefficients are passed to the ADCS software, which evaluates the position of the Flyby s/c and the position of Tempel 1 (also from a set of Chebyshev coefficients) and computes the desired pointing direction to center the comet CB in the HRI FOV. With every AutoNav OD update, the difference between the Flyby s/c's ground-based inertial position (based on optical observations of the Impactor s/c) and AutoNav's updated estimate of the Flyby s/c's inertial position (based on optical observations of Tempel 1) is computed and the resulting vector is applied directly to correct the Impactor's heliocentric position as shown in figure 12.

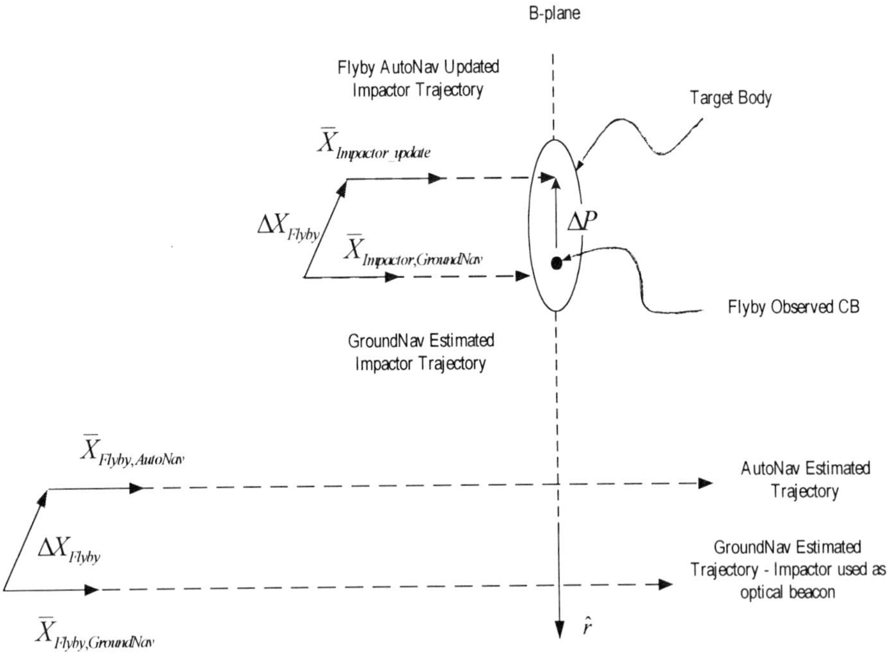

Figure 12 Graphical depiction of the inertial pointing correction, ΔP, which is computed by AutoNav. All state vectors correspond to the same AutoNav OD update epoch time (i.e. 5 min prior to estimated TOI); B-plane is normal to view

The difference between the Flyby s/c AutoNav solution (based on images of the comet) relative to the inertial frame and the Flyby s/c GroundNav solution (based on images of the Impactor s/c) relative to the inertial frame is computed to be

$$\Delta\overline{X}(t_{OD})_{Flyby} = \overline{X}(t_{OD})_{Flyby,AutoNav} - \overline{X}(t_{OD})_{Flyby,GroundNav}$$

This correction is applied directly to the Impactor s/c's inertial state obtained by evaluating the Impactor s/c ephemeris, which resides on-board the Flyby s/c in the form of Chebyshev coefficients, at t_{OD}

$$\overline{X}(t_{OD})_{Impactor\_update} = \overline{X}(t_{OD})_{Impactor,GroundNav} + \Delta\overline{X}(t_{OD})_{Flyby}$$

The updated Impactor s/c state is then mapped into B-plane coordinates, where $B \cdot r, B \cdot t$ give the 2-D pointing correction (the impact location), relative to the observed CB, in the B-plane frame. Since the orientation of the B-plane is known relative to the ICRF, the 2-D vector, ΔP, is transformed into an inertial 3-D pointing correction via the transpose of $R_{ICRF \rightarrow Bplane}$ and passed to ADCS for application.

$$\overline{X}_{Impactor\_update} = \begin{pmatrix} B \cdot r \\ B \cdot t \\ TOF \end{pmatrix} \Rightarrow \Delta P = \begin{pmatrix} \Delta P_x \\ \Delta P_y \\ \Delta P_z \end{pmatrix}_{ICRF}$$

The process of computing the pointing correction occurs with every OD update until TOI after which time the applied correction will remain constant.

SUMMARY

Predicting the impact location to point the Flyby s/c instruments in the presence of an inactive Impactor s/c can be achieved by treating the Impactor s/c as an optical beacon, estimating the Flyby s/c's inertial trajectory using images of the Impactor, uploading the estimated trajectory to the Flyby s/c's AutoNav software and autonomously computing a pointing correction for ADCS. This correction addresses only that component of the overall pointing error at key science epochs (TOI and TOFI), which is due to the uncertainty in the impact site location for an inactive Impactor s/c. Simulations show that uncertainties in the Flyby's position relative to the Impactor at the time of initial image acquisition require use of the MRI instrument for Impactor optical navigation purposes and acquisition of OpNav images as soon as possible following release of the Impactor s/c provide the best measurements. Results also show that for an inactive Impactor s/c, the pointing error due to impact site location uncertainty is driven back down to a level comparable with that of the baseline approach of a maneuvering Impactor s/c that will target the nucleus center of brightness. By treating the Impactor as an optical beacon pointing errors improve from 519 μrad to 72.3 μrad (3σ) at TOI, and from 3.96 mrad to 441 μrad (3σ) at TOFI; this is the difference between a mission with seriously degraded science return and a mission with flight system performance that equals the current baseline approach, provided that the inactive Impactor impacts in an illuminated area. The maneuvering Impactor design was selected to first, improve the probability of impacting in an illuminated area and second, impacting in a location, which is viewable from the Flyby s/c.

AKNOWLEDGEMENTS

The authors would like to acknowledge Bob Gaskell, NASA-Jet Propulsion Laboratory, for his work on modeling cometary nuclei for simulation analysis and for providing us with a sphere model of the Impactor s/c and Robert Werner, NASA-Jet Propulsion Laboratory for his comments and suggestions. The research described in this paper was carried out at the Jet Propulsion Laboratory, California Institute of Technology, under contract with the National Aeronautics and Space Administration.

REFERENCES

1. Bhaskaran, S., Riedel, J. E., Synnott, S. P. (1996) "Autonomous Optical Navigation for Interplanetary Missions." *Space Sciencecraft Control and Tracking in the New Mellennium*, Proc. SPIE, pp. 32-34.

2. Bank, T., Frazier, W., Blume, W., Kubitschek, D., Mastrodemos, N., Null, G., and Synnott, S. (2001) "Deep Impact: 19 Gigajoules Can Make Quite An Impression." AAS 01-022, in *Guidance and Control 2001*, R. D. Culp and C. N. Schira, eds., *Advances in the Astronautical Sciences*, Vol. 107, pp. 295-310 (paper presented at the 24th Annual AAS Guidance and Control Conference, Breckenridge, CO).

3. Mastrodemos, N. (2002) "Rationale for High Rate of Optical Data Acquisition During Approach Phase." *JPL Interoffice Memorandum DI/2002-2.*

4. Phinney, A. (2001) "Mass on the Impactor." *BATC Systems Engineering Report DI-IMP-STR-004.*

5. Fernandez, Y.A., et al. (2000) "Physical Properties of the Nucleus of Comet 2P/Encke." Icarus, **147,** pp. 145-160.

6. Bhaskaran, S., et al. (1998) "Orbit Determination Performance Evaluation of the Deep Space 1 Autonomous Navigation Software." AAS 98-193, in *Spaceflight Mechanics 1998*, J. W. Middour et al., eds., *Advances in the Astronautical Sciences*, Vol. 99, pp. 1295-1314 (paper presented at the 8th AAS/AIAA Space Flight Mechanics Meeting, Monterey, CA).

Section IV:
GUIDANCE AND CONTROL STORYBOARD DISPLAYS

SESSION IV

Joint Session Chairperson:	Rick Jackson CDA Co.
Joint Session Chairperson:	Joe Vellinga Lockheed Martin Astronautics
Local Session Chairperson:	Don Mackison University of Colorado

The following Storyboard papers were not available for publication:

AAS 03-032 "Non-Contacting Isolation System Resides in a 25mm Envelope," by Larry Germann, Verl Dunham, Joe Mahlberg, Helmut Soehn, Left Hand Design Corp.; D. Lawrence, University of Colorado (Not in Paper Format)

AAS 03-033 "GP-B, Last Lap," by D. DeBra, Stanford University (Not in Paper Format)

AAS 03-034 (Withdrawn)

AAS 03-036 "Radiation Hardened High Performance Space Processor," by R. Campbell, R. Elmhurst, K. Bogdan, Honeywell, Inc. (Not Available)

AAS 03-038 "Honeywell's Small Control Moment Gyro Development," by B. Abbott, Honeywell, Inc. (Not Available)

AAS 03-045 "High Performance Fiber Optic Gyro Inertial Attitude Reference Unit (HP-FARU) Development and Performance," by W. McArthur, R. DeMaio, Honeywell, Inc. (Not Available)

AAS 03-046 "Mission Enabling GN&C Technology," by Neil Dennehy, Joel Gambino, NASA/GSFC (Not Available)

AAS 03-047 (Withdrawn)

AAS 03-048 "Terma Star Trackers," by H. H. Bonde, Terma A/S, Space Division (Withdrawn)

AAS 03-051 "Genius I Have Known," by D. Sonnabend, Analytical Engineering (Not in Paper Format)

AAS 03-052 "A Distant Observer," by D. Sonnabend, Analytical Engineering (Not in Paper Format)

AAS 03-053 "Demonstration of Autonomous Rendezvous Technology (DART) Using the Advanced Video Guidance Sensor (AVGS)," by G. Dorsey, Orbital Sciences Corp. (Not in Paper Format)

AAS 03-054 "SimuLogix and the Aerospace Industry," by D. LaMont, L. Schlom, SimuLogix (Not in Paper Format)

The following paper numbers were not assigned:
AAS 03-055 to -060

AAS 03-031

ASSESSMENT OF PARASITIC TORQUES FROM MAGNETTORQUERS

Stephan Theil,[*] Peter Offterdinger[*] and Oliver Matthews[†]

Magnetic torquers are widely used as actuators for satellite attitude control. Mainly they are applied for the coarse and acquisition mode. For fine attitude control other actuators like reaction wheels are utilized. In this case the torquers are used for support only, or they are not used at all.

So far the effect of inactive magnetic torquers was simplified or neglected. But for recent and future scientific mission with high accuracy requirements for attitude determination and control this may become a significant factor. These parasitic magnetic torques are mainly produced by the ferromagnetic core. Three effects can be distinguished; the residual dipole, the magnetization by an external magnetic field and the damping due to hysteresis.

This paper evaluates all three effects for the ZARM-Technik/Microcosm magnetic torquers. The effects are assessed by evaluating FEM analysis results and measurements.

INTRODUCTION

Magnetic torquers are usually applied for coarse attitude control as well as for the acquisition phase. Depending on the mission they are used for various purposes in the nominal mission phases e.g. for momentum wheel discharging but also for active attitude control. In some cases especially in high precision pointing missions they are not used anymore.

Nevertheless they will produce a torque due to interaction of the magnetic core with the external magnetic field of the Earth. This torque can be a disturbance to the attitude control system. Especially for scientific satellites needing an ultra low disturbance environment these parasitic torques may become not negligible. For that reason this paper will evaluate the magnitude of these torques.

* ZARM, Center of Applied Space Technology and Microgravity, University of Bremen, Am Fallturm, D-28359 Bremen, Germany.

† ZARM-Technik GmbH, Am Fallturm, D-28359 Bremen, Germany.

The parasitic torques result mainly from three different effects on the magnetic core of the torquer. These effects are:

- the residual magnetic dipole (remanence),
- the magnetization due to the external magnetic field,
- the damping due to hysteresis of the core material.

In the following sections all three contributors to parasitic torques will be evaluated. This will be done by measurements, Finite-Elements-Method analysis and worst case estimation.

REFERENCE MODEL

For the assessment of the parasitic torques a reference model was defined. The ZARM-Technik torquer MT-30 was selected for this purpose. It has a nominal maximal linear dipole moment of 30 $A \cdot m^2$. The overall dimensions are $349mm \times 60mm \times 44mm$ where the cylindrical core has a size of $\varnothing 13.4mm \times 320mm$ (see figure 1).

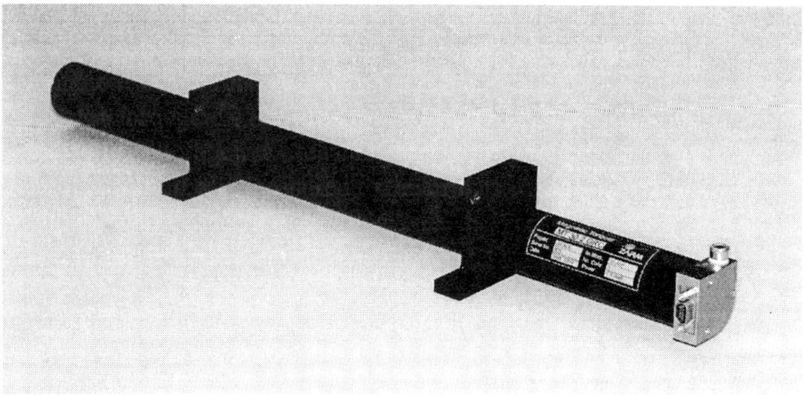

Figure 1: The ZARM-Technik MT-30 torquer used as a reference model for parasitic torque analysis.

In order to compare the different torques a external magnetic field of $B = 50,000\ nT$ is assumed.

MEASUREMENT OF RESIDUAL DIPOLE

The residual dipole respectively remanence was indirectly measured by determining the torque produced by the switched off magnetic torquer. For that purpose a torsion balance measurement setup (see figure 2) was used. It was assembled within a set of Helmholtz coils which are used to apply a homogeneous magnetic field (see figure 3).

Figure 2: Torsion balance for torque measurement with assembled MT-30 torquer

Figure 3: Helmholtz coil assembly for applying homogeneous magnetic fields

For measuring the torque resulting from the residual dipole moment of the unit the following procedure was applied:

- The magnetic torquer is operated to the maximum linear dipole value of $30\ A \cdot m^2$.

- The drive current through the torquer is switched off.
- An external homogeneous magnetic field of $B = 100,000\ nT$ is applied perpendicular to the torquer axis and the direction of the torsion spring.
- The deflection caused by the residual magnetic torque against the torsion spring is measured using a laser beam.

Using this method the residual dipole moment of the torquer could be determined to a value of:

$$m_{res} = 2.42 \cdot 10^{-2}\ A \cdot m^2 \tag{1}$$

In case of an external field of $B = 50,000\ nT$ the maximum torque from the remanence is:

$$T_{Rem} = 1.21 \cdot 10^{-6}\ N \cdot m \tag{2}$$

TORQUE DUE TO MAGNETIZATION FROM EXTERNAL FIELD

Finite Elements Analysis

The magnetic behavior of the magnetic torquers was analyzed using a finite elements analysis (FEM). The results of this method are compared to measurements of the magnetic field produced by the torquer. In this case an excitation coil is modeled which magnetizes the core of the magnetic torquer. At selected points outside the torquer the magnetic flux density could be computed and compared to the measured data. Using this approach the FEM model and computation method could be verified.

Computation of Parasitic Torque

For the analysis of the torque due to external magnetization only the core of the magnetic torquer is modeled (see figure 4). Using this model the magnetic field intensity and the flux density for each element can be computed. Figure 5 shows the magnetic flux for each element as a contour plot and a vector plot. The results displayed in this figure are for an external magnetic field vector with an angle $\alpha = 45°$ to the torquer axis.

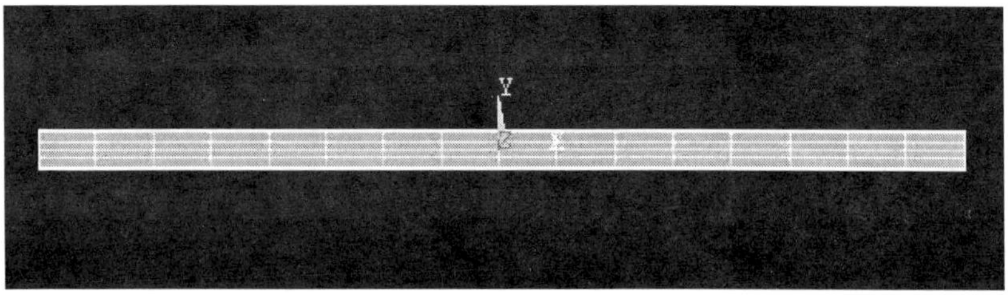

Figure 4: Finite elements of magnetic torquer core

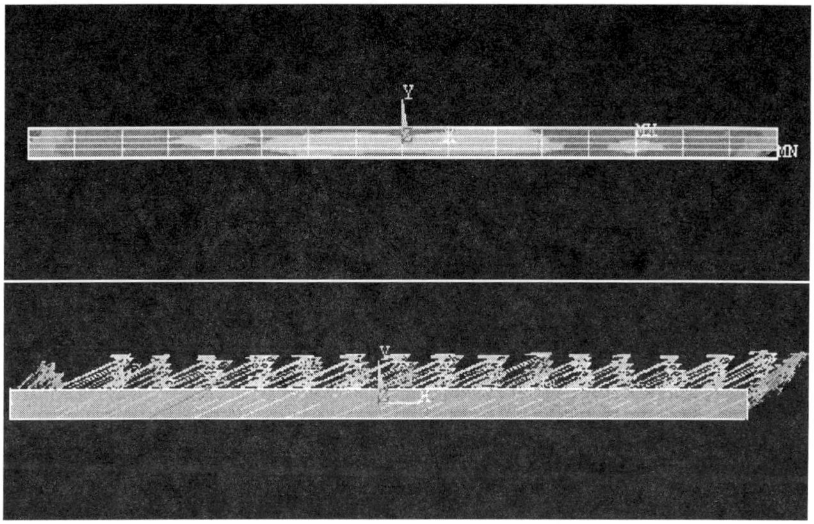

Figure 5: FEM solution of magnetic flux for magnetic torquer core

For the same case in figure 6 the magnetic flux lines are shown. It can be seen that the magnetic core distorts the field. From this interaction a torque can be expected which can be computed as follows. The computation of the FEM model delivers for each element of the core the volume V_i, the magnetic field intensity H_i and the magnetic flux density B_i. From these data for each element the magnetic dipole as well as the total dipole can be calculated by using the following equations:

- The magnetization $\underline{M}_i$ of each element is calculated from magnetic field intensity $\underline{H}_i$ and magnetic flux density $\underline{B}_i$ inside the element:

$$\underline{M}_i = \frac{\underline{B}_i}{\mu_0} - \underline{H}_i \qquad (3)$$

- The magnetic dipole of each element is calculated from the magnetization and the volume:

$$\underline{m}_i = V_i \cdot \underline{M}_i \qquad (4)$$

- The total dipole moment of the core cylinder is the sum over all elements:

$$\underline{m}_{total} = \sum_i \underline{m}_i \qquad (5)$$

- The torque due to interaction with the external field is derived using:

$$\underline{T} = \underline{m}_{total} \times \underline{B}_{ext} \qquad (6)$$

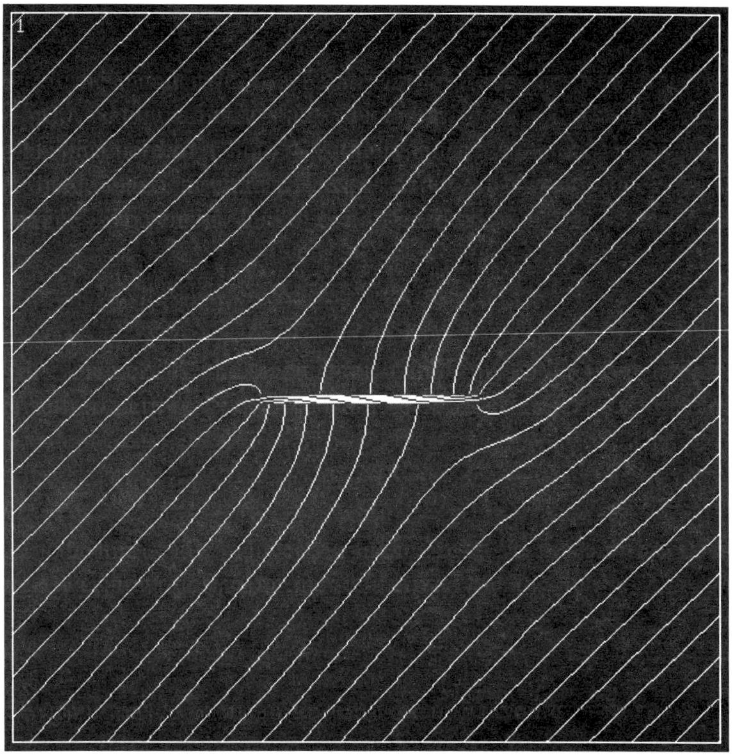

Figure 6: Magnetic flux lines around magnetic torquer core.

In order to get the dependency of the parasitic torque from the direction and magnitude of the external field the external homogeneous field is simulated at different angles between the torquer axis and the flux lines. The magnetic flux density is varied too.

The variation of the angle and the strength of the external magnetic field renders the dependencies shown in figures 7 and 8. It can be seen that the dependency of the torque T from the angle α between torquer axis and flux vector is:

$$\begin{aligned} T &\sim \sin\alpha \cdot \cos\alpha \\ T &\sim \sin 2\alpha \end{aligned} \tag{7}$$

The function of the torque from the flux density of the external field B is quadratic. It can be described as:

$$T \sim B^2 \tag{8}$$

So the maximum value for the parasitic torque due to magnetization is reached at an angle $\alpha = 45°$. For an external field of $B = 50,000\ nT$ we get a torque of:

$$T_{Mag} = 8.51 \cdot 10^{-8}\ N \cdot m \tag{9}$$

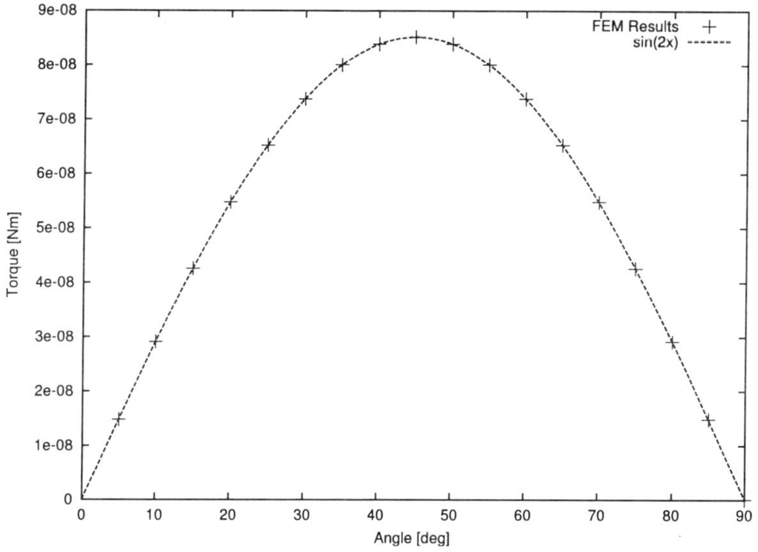

Figure 7: Torque from magnetic field interaction vs. angle between symmetry axis and magnetic field for a magnetic torquer core (3D FEM analysis)

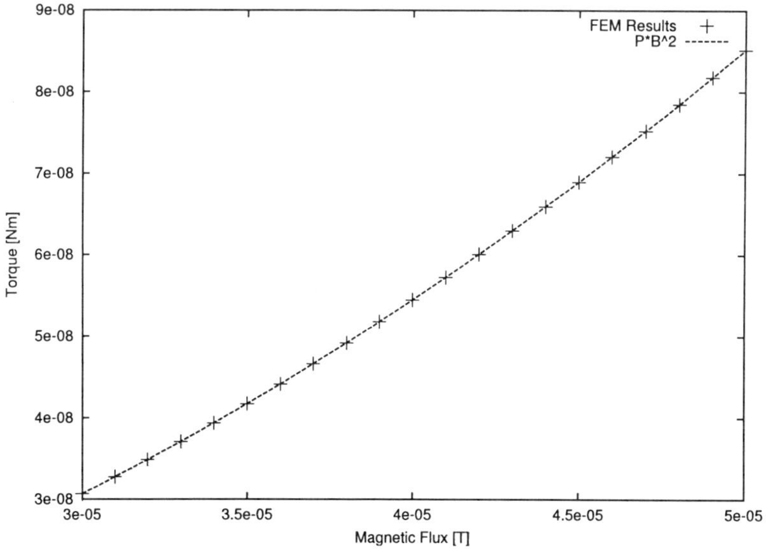

Figure 8: Torque from magnetic field interaction vs. magnetic flux density of the external magnetic field for a magnetic torquer core (3D FEM analysis)

TORQUE DUE TO HYSTERESIS DAMPING

The rotation of the magnetic torquer core within a magnetic field causes an oscillation of the field w.r.t. the symmetry axis. During a single cycle of the oscillation the core material will be magnetized in both directions. Since the material has a magnetic hysteresis the rotation of the magnetic field causes an energy dissipation due to the friction of the magnetic domains. The energy loss over one period of the rotation is described by:

$$E_{mag} = V \cdot \oint H \mathrm{d}B \tag{10}$$

where V is the volume, H is the magnetic field intensity and B is the magnetic flux density. The magnetization energy is converted from kinetic mechanical energy to heat. The kinetic energy loss is defined as:

$$E_{kin} = \int T_{Hyst} \cdot d\varphi \tag{11}$$

where T_{Hyst} is the torque due to hysteresis acting over the angle φ. If we assume the torque to be constant we can write for the energy loss over one revolution:

$$E_{kin,2\pi} = T_{Hyst} \cdot 2\pi \tag{12}$$

From this we can derive the mean torque over one revolution from the hysteresis of the magnetic torquer core:

$$T_{Hyst} = \frac{E_{kin,2\pi}}{2\pi} = \frac{E_{mag}}{2\pi} = \frac{V}{2\pi} \oint H \mathrm{d}B \tag{13}$$

It can be seen that the torque is independent from the rotation rate because the energy dissipated over one period is also independent from the frequency.

In order to assess the torque from the hysteresis effects the term $\oint H\mathrm{d}B$ must be evaluated. For that purpose the hysteresis curve of the core material is used which is shown in figure 9(a). It can be seen that the saturation is reached at a magnetic flux density level of $B \approx 1.5\ T$. The finite element analysis has shown that the maximum flux density inside the core material from an external field of $B_{ext} = 50,000\ nT$ is at an level of $B_{max,i} \approx 250,000nT$. So the field can be assumed to be a weak field which allows the analytical description of the hysteresis loop (see [3]). It can be written as:

$$B = (\mu_A + \nu H_m) H \pm \frac{\nu}{2} \left(H^2 - H_m^2\right) \tag{14}$$

where the $\pm$ sign denotes the two branches of the hysteresis loop. The parameters μ_A, ν and H_m have to be determined either by measurement or must be estimated. For the worst case estimation the last option was chosen. The parameters are chosen in a way that the hysteresis loop fits the rectangle $[-H_c, +H_c; -B_{max,i}, +B_{max,i}]$ in figure 9(b). The resulting curves of the hysteresis loop are shown there too.

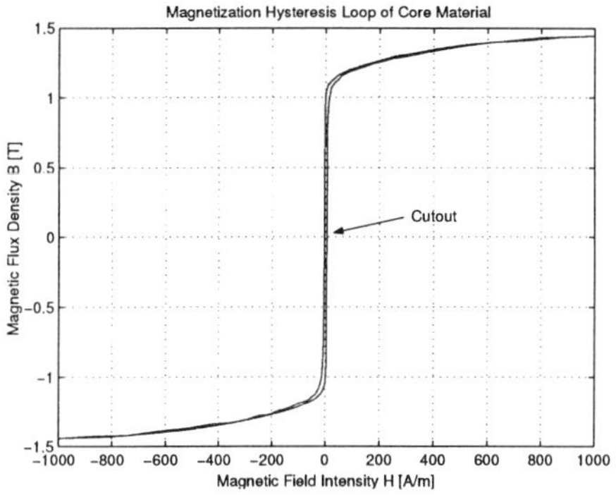

(a) Saturation hysteresis loop

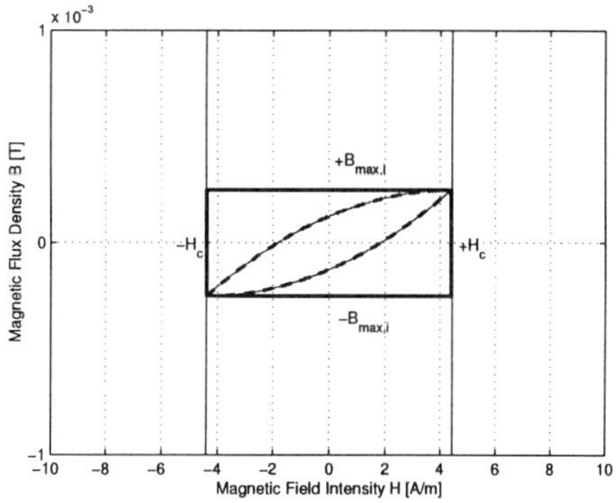

(b) Estimated worst case hysteresis loop

Figure 9: Derivation of worst case hysteresis loop from saturation hysteresis loop

The dissipated energy can be computed using the analytical description. The result is:

$$E_{Mag} = V \cdot \oint H \mathrm{d}B = 6.62 \cdot 10^{-8} J \tag{15}$$

From that the mean torque due to hysteresis renders a value of:

$$T_{Hyst} = \frac{E_{mag}}{2\pi} = 1.05 \cdot 10^{-8} N \cdot m \tag{16}$$

SUMMARY AND CONCLUSIONS

The parasitic torques produced by a magnetic torquer due to interaction with an external magnetic field were derived by measurement, FEM computation and worst case estimation. The results are obtained for the ZARM-Technik MT-30 torquer model which has a nominal linear dipole moment of $m = 30\ A{\cdot}m^2$. This dipole produces in interaction with an external magnetic field of $B_{ext} = 50,000\ nT$ a nominal torque of $T_{Nom} = 1.50 \cdot 10^{-3}\ N \cdot m$. The results for the parasitic torques are shown in table 1.

The largest of the parasitic torques is due to the remanence. The other two effects are one respectively two orders of magnitudes smaller. It can be seen that the parasitic torques are at least 3 orders of magnitude smaller than the nominal torque from the maximum linear dipole momentum.

Table 1 shows in addition to the torque values the dependencies of the different effects on the geometry of the magnetic core. They can be predicted if the same core material is used and if the ratio between diameter and length of the core is in the same range.

If we now take into account larger magnetic torquers the parasitic torques may reach values where they have to be considered as a major disturbance source for the attitude

Type	Value	Dependency
${T_{Rem}}^1$	$1.21 \cdot 10^{-6}\ N \cdot m$	$\sim$ core volume
${T_{Mag}}^2$	$8.51 \cdot 10^{-8}\ N \cdot m$	$\sim$ core length
${T_{Hyst}}^3$	$1.05 \cdot 10^{-8}\ N \cdot m$	$\sim$ core volume

[1] - At an angle of 45 degrees between flux lines and torquer axis.
[2] - At an angle of 90 degrees between flux lines and torquer axis.
[3] - Mean torque per revolution w.r.t. flux lines.

Table 1: Comparison of parasitic torques for an external field of $B_{ext} = 50,000\ nT$; Nominal torque $T_{Nom} = 1.50 \cdot 10^{-3}\ N \cdot m$

control system. Therefore the parasitic torques may not be negligible anymore especially for future high accuracy attitude control systems.

REFERENCES

[1] Daniel Damm. Mechanical Measurement of the Magnetic Dipole from Magnetic Coils for Satellite Attitude Control; German: Mechanische Vermessung des magnetischen Dipolmomentes von Magnetspulen zur Lageregelung von Satelliten. Master's thesis, Fachhochschule Wedel, August 2002.

[2] Hermann Ebert, editor. *Physikalisches Taschenbuch.* Vieweg & Sohn Verlagsgesellschaft, 1976.

[3] Eckart Kneller. *Ferromagnetismus.* Springer Verlag, 1962.

[4] Stephan Theil. *Satellite and Test Mass Dynamics Modeling and Observation for Drag-free Satellite Control of the STEP Mission.* PhD thesis, Department of Production Engineering, University of Bremen, December 2002.

[5] Roald K. Wangsness. *Electromagnetic Fields.* John Wiley & Sons, New York, 2nd edition, 1986.

AAS 03-035

EARTH RADIANCE MODELING

Todd J. Bednarek*

Goodrich Corporation, a long time producer of high quality Earth sensors, has developed new analytical tools to more accurately predict the attitude determination error associated with Earth radiance variation. Infrared Earth sensors provide spacecraft attitude knowledge by sensing the radiant energy from the Earth's horizon in contrast to space. Radial and tangential radiance gradients within the field of view of the Earth sensors create error signals that limit the accuracy of attitude determination.

Traditional Earth radiance models, such as LOWTRAN, MODTRAN, and CORPS, are seasonal in nature. They provide typical latitudinal radiance variation for different times of the year, but lack longitudinal variations and changes that can occur from one day to the next. Goodrich's new Earth radiance models use empirical Earth radiance databases created from remote IR sensing instruments such as HIRS/2 and VAS, which are onboard the NOAA and GOES satellites respectively.

Earth radiance error analyses using actual radiance maps provide more accurate calculation of the errors to be expected during flight. This paper provides examples of the various radiance models that have been used and of the predicted attitude errors using Goodrich's latest generation Earth sensor, the Multi-Mission-Sensor (MMS).

INTRODUCTION

Earth sensors are commonly used on spacecraft for 2-axis attitude determination relative to the local vertical or nadir. Earth sensors view the Earth horizon in the infrared spectrum and, with the use of a locator algorithm, locate the infrared horizon with respect to the sensor's reference axes. Given three or more horizon locations, or two horizon locations and spacecraft altitude, the direction to the center of the Earth disk as seen from space can be determined in two axes, pitch and roll.

The errors associated with spacecraft attitude determination can be separated into two categories, instrument error and Earth radiance error. Earth radiance error is defined as the change in calculated pitch and roll from Earth sensor outputs due solely to the change in Earth radiance. Instrument errors such as noise, end of life (EOL) drift, and thermal sensitivity are present even with uniform Earth radiance. It is convenient to separate instrument errors from

* Goodrich Corporation, Optical and Space Systems Division, Electro-Optical Systems, 100 Wooster Heights Road, Danbury, Connecticut 06810.

radiance errors to better compare the performance of Earth sensors of different configurations and of those produced by different manufacturers. Instrument errors can be accurately predicted by analysis and verified for the most part by test. Accurate predictions of Earth radiance errors, however, are limited by the fidelity of the Earth radiance model.

Earth radiance variations can be over simplified as linear gradients with latitude and/or longitude. Such models result in under estimation of Earth radiance errors. Historically, Goodrich has used a somewhat more complex model based on meteorological data. The **CO**mprehensive **R**adiance **P**rofile **S**ynthesizer (CORPS) uses atmospheric temperature and pressure measurements to calculate Earth radiance[1]. As will be discussed later, the CORPS model is limited to monthly radiance variations with latitude and assumes uniform radiance with longitude. More recently, Goodrich has developed an Earth radiance model using actual radiance measurements from space-based instruments. Two such instruments are the **V**isible infrared spin scan radiometer **A**tmospheric **S**ounder (VAS) on board the GOES satellites and the **H**igh-resolution **I**nfra**R**ed **S**ounder/2 (HIRS/2) on board NOAA satellites. Measurements from these instruments provide databases of daily Earth radiance that varies with both latitude and longitude.

ATMOSPHERE-BASED RADIANCE MODELS

CORPS was developed in the 1960's as part of a study to define the variations in the Earth horizon radiance in the 15μm CO_2 spectral band. The program calculates horizon radiance profiles from measurements of atmospheric temperature and pressure versus altitude[1]. A horizon radiance profile is a curve of radiance versus the tangent height of the line-of-sight (LOS) of the observer. Figure 1 depicts the definition of tangent height.

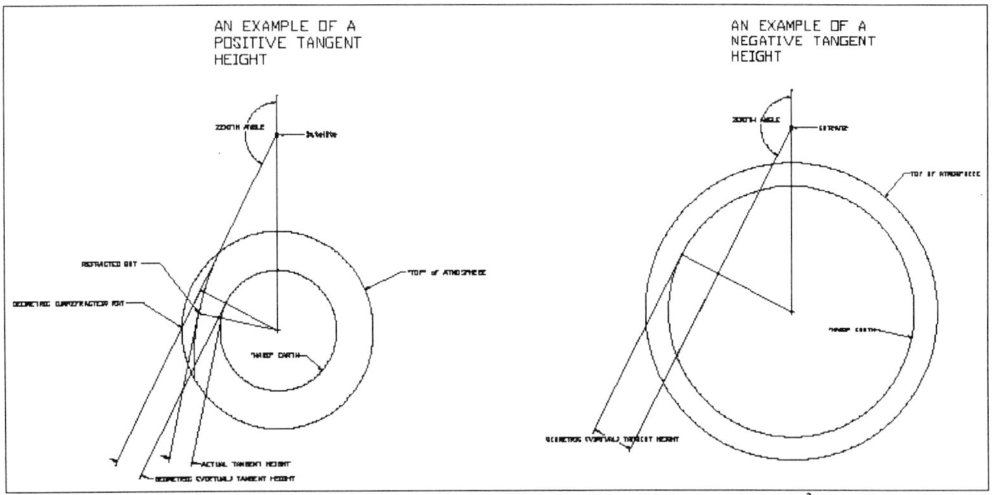

Figure 1: Illustration of Positive and Negative Tangent Heights[2]

The study was limited to the 14μm to 16μm spectral band because it had previously been shown to be a stable reference of the horizon location. The CO_2 band exhibits relatively lower sensitivity to atmospheric variations, such as clouds and water vapor, than do other IR spectra[1]. Goodrich has used CORPS to generate a set of radiance profiles from the temperature and pressure profiles of the 1978 U.S. Air Force Reference Atmospheres[3]. The radiance profiles represent monthly averages of radiance versus tangent height from the equator to the North Pole in 15° latitudinal increments. Profiles for southern latitudes are estimated by time shifting the northern profiles by six months. Figure 2 plots the horizon radiance profiles for January and July. Figure 3 is a color contour of the January and July profiles mapped onto the geosynchronous Earth. Note that the July profiles are mapped onto the southern hemisphere.

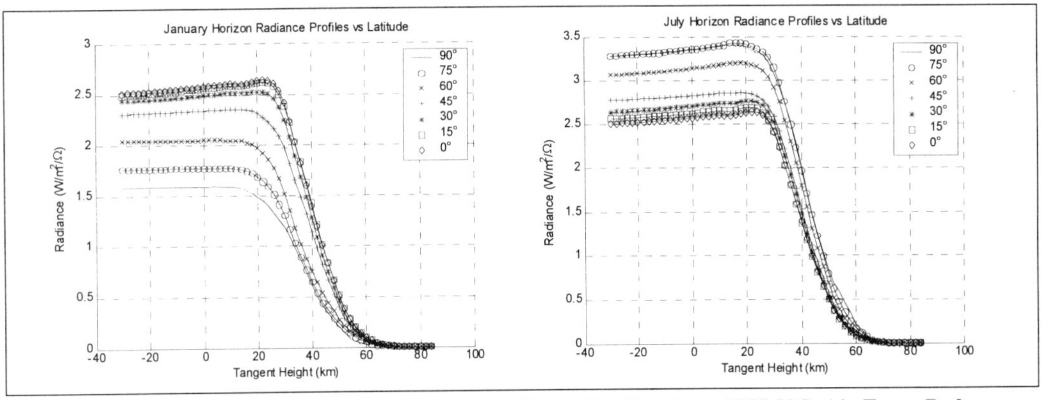

Figure 2: January and July Northern Horizon Radiance Profiles from 1978 U.S. Air Force Reference Atmospheres (Actual 14μm to 16μm Filter)

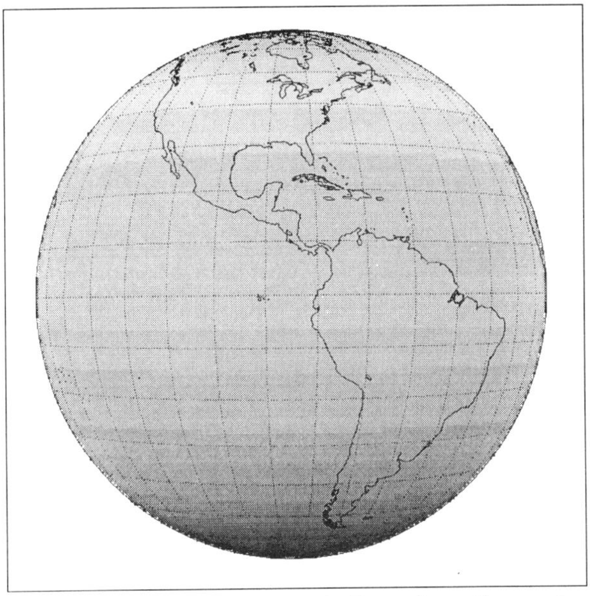

Figure 3: CORPS Model of January Earth Radiance from Geosynchronous Orbit.

LOWTRAN and MODTRAN are also atmosphere-based models that can be used to calculate Earth radiance as seen from space. These, unlike CORPS, are not limited to the 15μm CO_2 spectral band. MODTRAN is an enhanced moderate spectral resolution version of LOWTRAN and is more suitable to calculating Earth radiance at the horizon[4]. An analysis by Jay Shepherd for Goodrich shows that the shape of horizon radiance profiles calculated from built-in MODTRAN atmospheres are comparable to those calculated with CORPS[5]. Given the good comparison, Goodrich continued to use the CORPS radiance model.

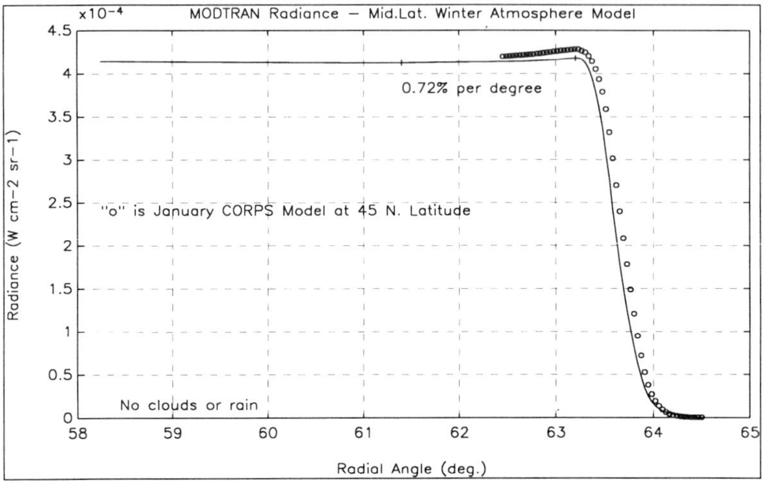

Figure 4: Sample Comparison of CORPS and MODTRAN Horizon Radiance Profiles (Uniform 14μm to 16μm Filter)[5]

SPACE-BASED RADIANCE MODELS

Although the CORPS model provides representative monthly averages of Earth radiance versus latitude and tangent height, it lacks longitudinal spatial variation and daily temporal variation. To overcome this limitation, Goodrich first considered using empirical radiance measurements from space borne instruments in the late 1980's[6]. A few days of VAS data from the geosynchronous GOES satellites were used to predict Earth radiance error for the "classic" GEO static sensor. Figure 5 is a representative map of the VAS measured Earth radiance[7].

As can be seen in figure 5, the coverage of the VAS data is limited, making it less than ideal for analyzing lower orbits, higher inclinations, or arbitrary yaw. In the late 1990's Goodrich acquired Earth radiance maps measured by the HIRS/2 instruments onboard NOAA satellites. Since the NOAA satellites are low-earth polar orbiting, the HIRS/2 instruments provide global Earth radiance data. The HIRS/2 is available from NOAA as the TOVS (TIROS Operational Vertical Sounder) Sounding Product-Global[8]. Figure 6 is a representative map of the HIRS/2 measured Earth radiance.

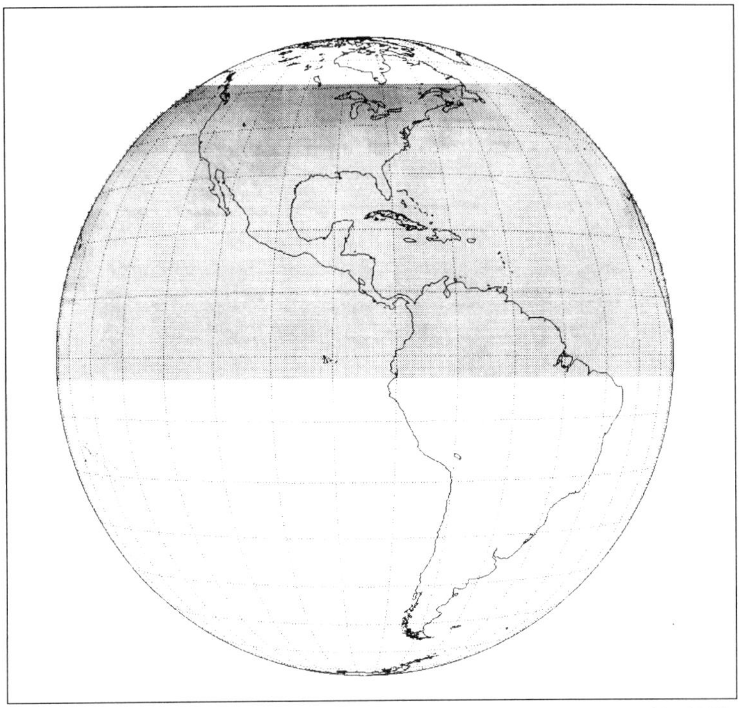

Figure 5: Map of VAS Measured Earth Radiance from January 29, 1989.

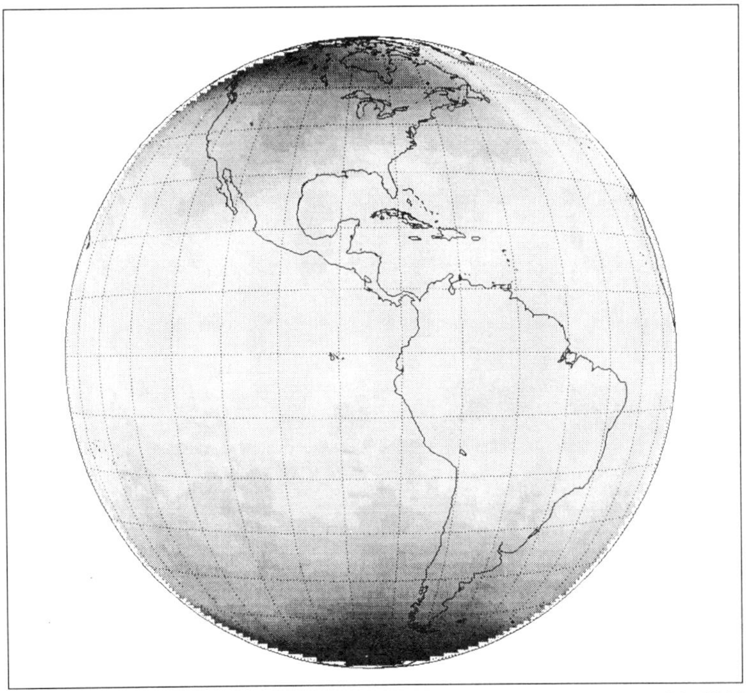

Figure 6: Map of HIRS/2 Measured Earth Radiance from January 29, 1989.

Note that the HIRS/2 instrument only scans ±49.5° from nadir and therefore does not measure the horizon radiance directly as does the VAS instrument[8]. Use of the HIRS/2 data alone would omit horizon effects such as the height of the CO2 horizon and its variation, limb brightening, and the gradual transition of radiance from Earth to space. To overcome this, Goodrich has developed a method of combining the HIRS/2 radiance maps with the horizon radiance profiles from CORPS. The radiance near the horizon for a particular day is initially calculated from the HIRS/2 radiance maps as a function of latitude and longitude. An appropriate CORPS horizon profile is then selected based on the month of the HIRS/2 data and on the latitude of the observed point on the Earth. The radiance is scaled as a function of tangent height using the CORPS profile. The resulting model of the horizon radiance has the same shape and height as predicted by CORPS, but the absolute radiance is consistent with the actual radiance measured by the HIRS/2 instrument. Figure 7 shows the result of modifying the horizon profiles of radiance map shown in figure 6. Note that this map extends to the CO2 horizon and has a smoother transition from Earth to space. This technique was validated against the VAS model in an analysis by Jay Shepherd[9]. Pitch and roll radiance errors for a specific sensor configuration were calculated using both the VAS and modified HIRS/2 radiance maps. Predicted errors from the same day were compared and sample of 118 days was found to have a good correlation of about 80%.

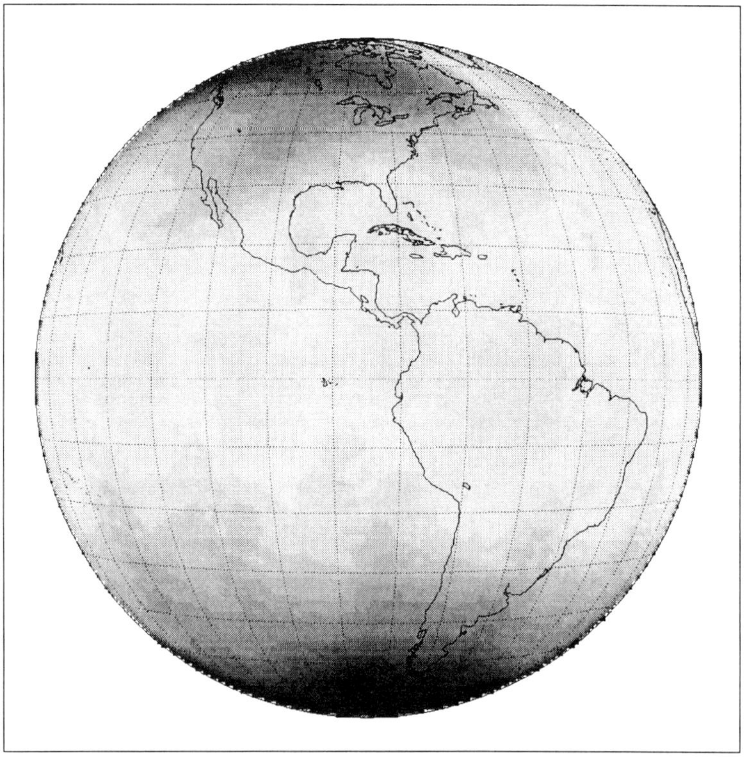

Figure 7: HIRS/2 Radiance Map Modified with CORPS Horizon Profiles.

Goodrich's standard Earth radiance model currently consists of over 270 daily HIRS/2 radiance maps from 1995, 1996, 1997, and 1999 and 12 sets of monthly CORPS horizon radiance profiles generated from the USAF reference atmospheres. The horizon radiance for each day is modified as described above using the CORPS horizon profiles.

EARTH RADIANCE ANALYSIS

Radiance error analyses are performed by simulating the projection of the Earth sensor's field of view (FOV) onto the Earth and calculating the detector's output signal level based on the observed radiance. Spacecraft pitch and roll attitude is ultimately calculated from the simulated sensor outputs. The results obtained when using a non-uniform radiance model are compared with those from a uniform Earth radiance simulation and the difference is defined as the radiance error.

The FOV is first segmented into a finite number of relatively small solid angles. A vector defining the center of the FOV element is projected out from the sensor in order to calculate its intersection with the Earth. The radiance at the point of intersection is multiplied by the solid angle of the element, the area of the entrance pupil of the optical system, the detector responsivity, and the effective transmission of the optical system in the direction of the element. The resulting detector voltages are summed to calculate the output of the detector. Note that the actual detector output is proportional to the net flow of energy between the detector, its environment, and the Earth target. For the purpose of analyzing errors due to non-uniformities in Earth radiance, the radiances of the detector and its environment are assumed to be a constant and are normalized to zero.

The detector signal is amplified and processed according to the type of Earth sensor. Goodrich Corporation has produced Earth sensors of various configurations including Conical Scanners and Static Earth Sensors for 3-axis-stabilized spacecraft and body-mounted Horizon Crossing Indicators (HCI) for spin-stabilized spacecraft. The processing and horizon locator algorithms for scanning type sensors are time dependent. A common approach is to differentiate the detector signal and use a percentage threshold to locate the transition of the FOV from space to Earth and from Earth to space[10]. Figure 8 depicts a FOV scanned across the Earth and the resulting detector signal and processed waveform. Earth radiance errors are caused by changes in the height of the IR horizon and radiance gradients observed by the FOV while scanning the Earth below the horizon. These gradients induce errors in setting the threshold level in the locator algorithm.

Historically, static Earth sensors produced by Goodrich use detectors with three or more fields of view for each horizon locator. Multiple horizons are measured by integrating multiple detectors within one sensor assembly, as done with the "classic" GEO and GPS sensors. Alternatively, Goodrich's mini-static Earth sensors sense only one horizon and multiple sensors are integrated onto a bracket designed to optimize the horizon locations for a particular mission. Figure 9 shows an example of an ideal static earth sensor FOV and the locator algorithm[10]. Earth radiance errors for these sensors are caused by changes in the height of the IR horizon and by differences in the mean radiance observed by the A and B detectors.

Goodrich's latest generation of Earth sensors, the Multi-Mission Sensor (MMS), uses a detector that is an array of over 19,000 pixels[11]. The pixels are grouped into four earth-sensing fields designated AB1, AB2, AB3, and AB4 and two space reference fields designated S1 and

S2. The space fields are used in conjunction with an offset radiation source to reference the Earth fields to space and thus make the detector outputs independent of the mean detector temperature.

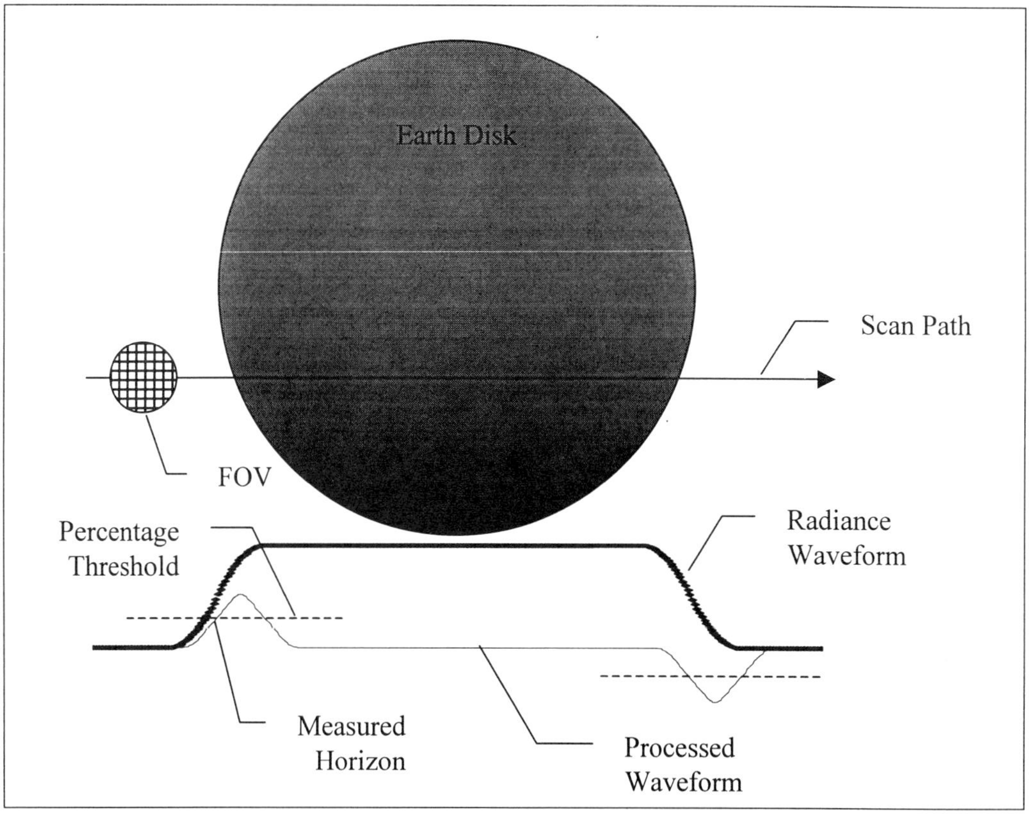

Figure 8: Example of Scanned and Processed Earth Radiance Waveforms.

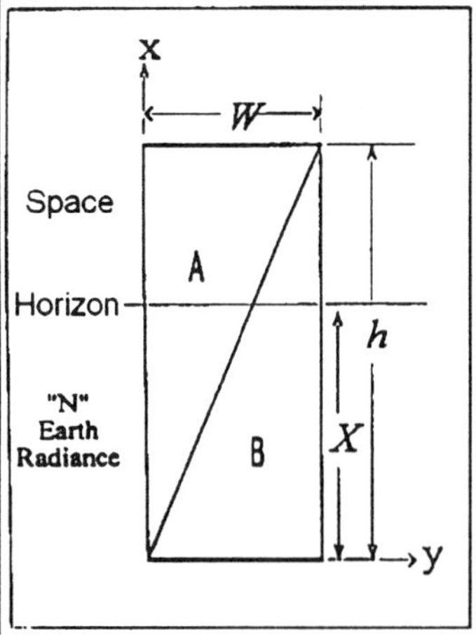

Signal on Detector A = $\int_0^x N\frac{Wx}{h}\,dx = \frac{NW}{2h}X^2$

Signal on Detector B = $\int_0^x N\left(W - \frac{Wx}{h}\right)dx = NWX - \frac{NW}{2h}X^2$

Ratio of Signals = $\frac{A}{B} = \frac{X}{2h - X}$

Solving for X Yields Basic Algorithm: $X = \frac{2hA}{A+B}$ (Eq. 1)

Sum of Signals = A + B = NWX $\Rightarrow X = \frac{A+B}{NW} = \frac{A+B}{N^1}$ (Eq. 2)

Eq. 1 and Eq. 2 are combined to find $N^1 = \left(\frac{A+B}{2hA}\right)^2$

Figure 9: Ideal Static Earth Sensor FOV and Locator Algorithm.

Each AB field consists of A and B sub-fields with the output of all pixels within a sub-field combined to produce a single output. The detector pixels are wired such that the segments of the A sub-field are interleaved with the segments of the B subfield. Figure 10 shows the total MMS FOV projected onto the GEO Earth. Different colors are used to distinguish the alternating segments of the A and B sub-fields that are viewing the Earth. The detector pixels are also wired such that the polarity of every other segment of an A or B sub-field is inverted. The result is two phase-shifted sinusoidal outputs versus horizon location as shown in Figure 11. The relative horizon location within one cycle is calculated as the arctangent of the ratio of the A and B signals. This technique of edge detection is categorized as a quadrature sensing. The absolute horizon location is calculated using a proprietary cycle-picking algorithm and other detector signals that are available but not discussed here.

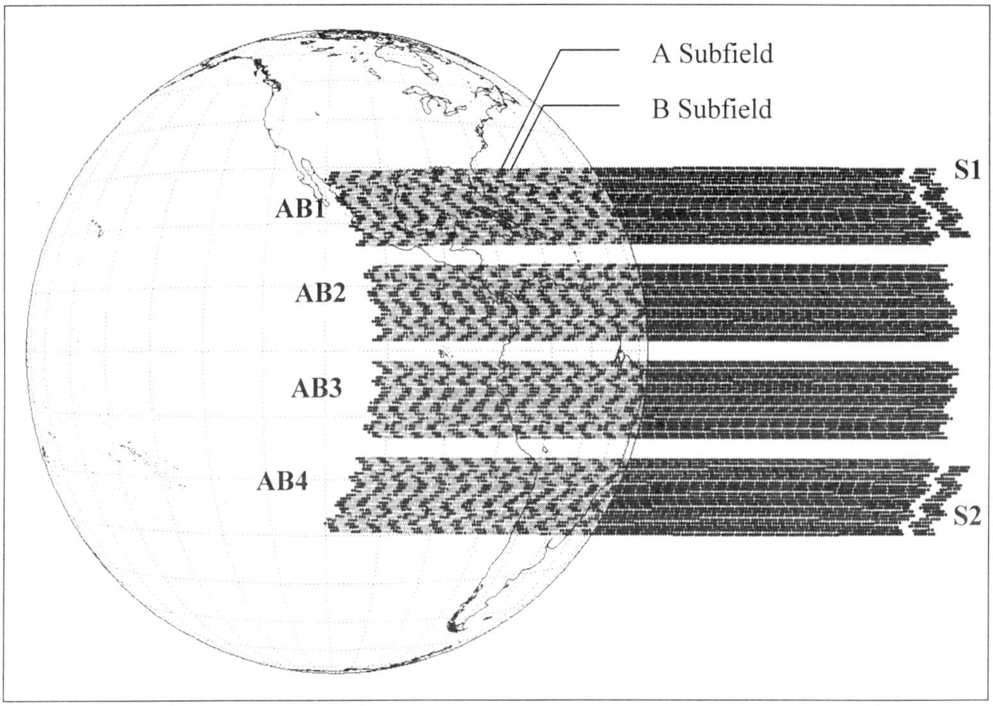

Figure 10: MMS Field of View on GEO Earth.

With four horizon locators, the MMS provides 2-axis attitude determination at GEO and MEO altitudes even with sun or moon intrusion. The MMS provides large operating ranges similar to scanning sensors but with higher reliability since there are no moving parts. The quadrature sensing technique provides lower radiance errors than would other static sensors if configured to provide an equivalent operating range.

The design described above makes the MMS highly insensitive to Earth radiance variations. The simplified radiance models used in the past would predict near negligible radiance errors as shown in Figure 12. Peak pitch and roll errors are predicted to be 0.001° and 0.005° respectively. Figure 13 provide histograms of pitch and roll radiance errors for an equatorial GEO orbit. The 3-sigma pitch and errors about the mean are 0.007° and 0.016°

respectively. Even with a more sophisticated radiance model based on actual measurements of Earth radiance, the MMS Earth sensor is predicted to provide excellent performance.

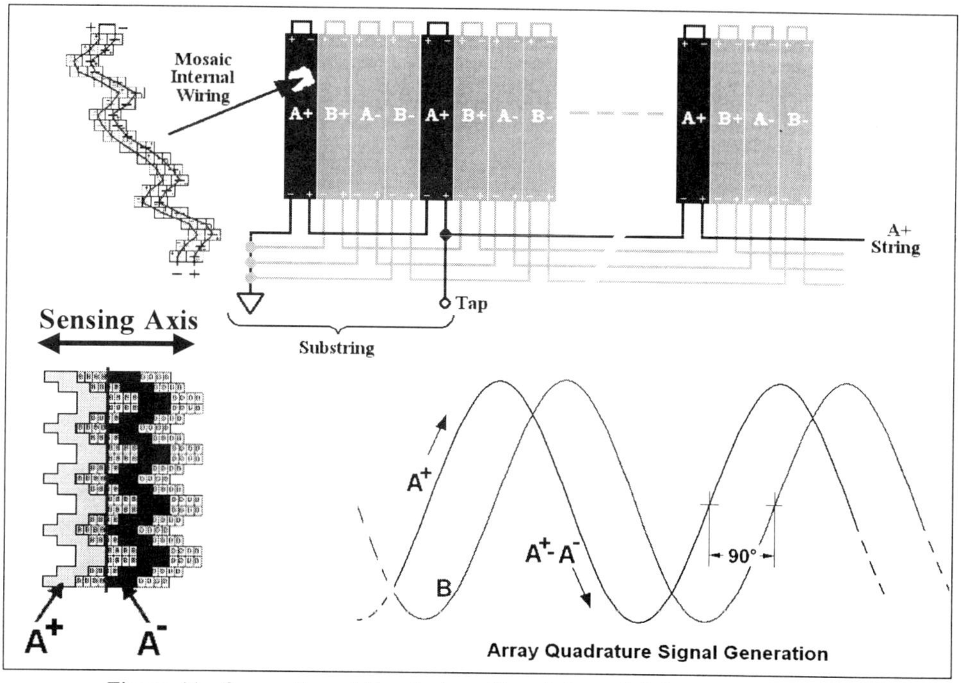

Figure 11: Generation of Quadrature Signals from MMS Detector Design.

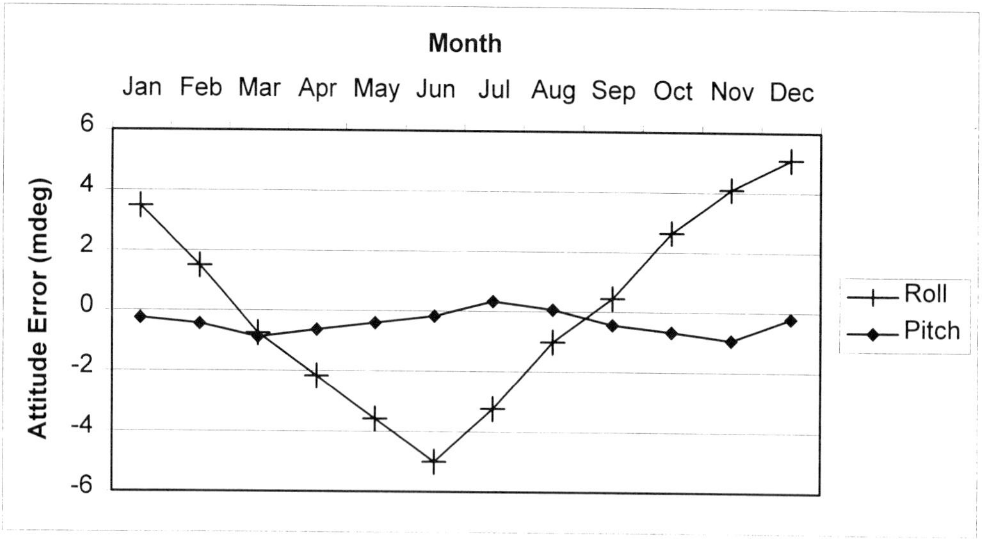

Figure 12: MMS Seasonal Radiance Error from CORPS Model.

CONCLUSION

Throughout its history of producing high quality Earth Sensors, Goodrich has strived to accurately predict flight performance with the use of sophisticated testing and analysis. A key aspect of sensor performance is error due to variations in the Earth's radiance. Although the sensitivity of a particular Earth sensor design to radiance variations can be readily analyzed, the ultimate performance prediction is based on the Earth radiance model used to provide input for the sensor simulation.

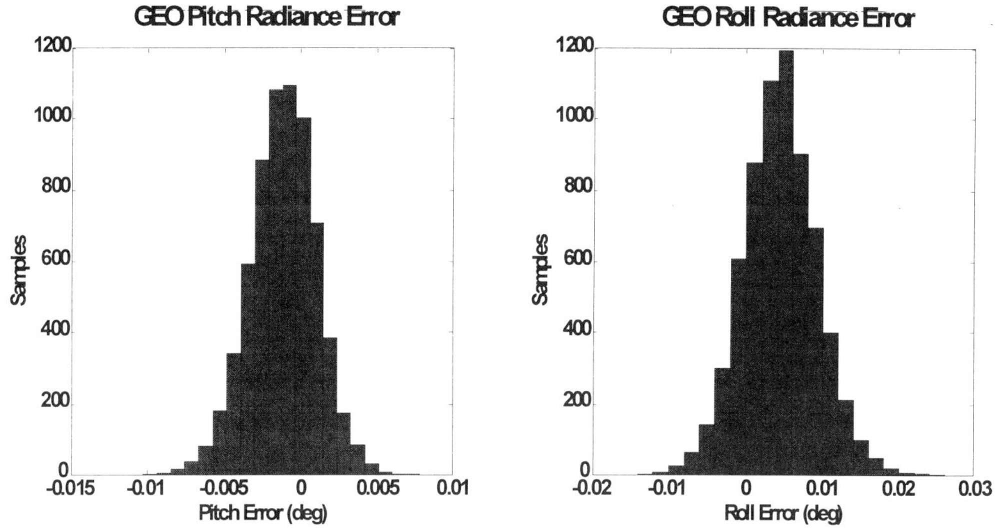

Figure 13: Histograms of MMS Radiance Error from HIRS/2-CORPS Model.

Goodrich has continued to update its radiance modeling tools to take advantage of the most recent information practical. Key to radiance error predictions is the inclusion of radiance non-uniformities as well as shifts in the height of the CO_2 atmosphere. Goodrich's latest Earth radiance model uses empirical radiance measurements from space to accurately capture both spatial and temporal radiance variations. The model predicts radiance errors more than 3 times greater than those predicted by models with latitudinal gradients alone. Use of this model has lead to the development of the latest generation Earth sensor, the MMS, to provide excellent performance over a generous operation range in the presence of realistic Earth radiance distributions.

REFERENCES

1. J.C. Bates, D.S. Hanson, F.B. House, R.O'B. Carpenter, and J.C. Gille, (1966) *The Synthesis of 15 μm Infrared Horizon Radiance Profiles from Meteorological Data Inputs*, NASA CR-724.

2. J. Shepherd, (2001) *A Brief Description of Radiance Modeling at Barnes*, Goodrich Document PA 5021.

3. A.E. Cole, and A.J. Kantor, (1978) *Air Force Reference Atmospheres*, AFGL-TR-78-0051.

4. A. Berk, L.S. Bernstein, and D.C. Robertson, (1989) *MODTRAN: A Moderate Resolution Model for LOWTRAN 7*, GL-TR-89-0122.

5. J. Shepherd, Internal Goodrich Memorandum to P. Cirino and D. Patrick dated July 28, 1993.

6. W.L. Smith, (1989) *VAS Geostationary Satellite Radiance's for the Study of Attitude Sensing Errors Due to Natural Variations in Atmosphere Radiance Near 15 Microns*, Goodrich Document PR C30-0529.

7. J.D. Clark, ed. (1983) *The GOES User's Guide*, National Oceanic and Atmospheric Administration (NOAA).

8. K.B. Kidwell, ed. (1995) *NOAA Polar Orbiter Data Users Guide.*

9. J. Shepherd, Internal Goodrich Memorandum to D. Patrick dated May 7, 1997.

10. R.A. Gontin and K.A. Ward, (1997) *Horizon Sensor Accuracy Improvement Using Earth Horizon Profile Phenomenology*, AIAA 87-2598.

11. T.J. Bednarek, (2002) *Performance Characteristics of the Multi-Mission Earth Sensor for Challenging, High-Radiation Environments*, AAS 02-036, in Guidance and Control 2002, R.D. Culp and S.D. Jolly, eds., Advances in the Astronautical Sciences, Vol. 111, pp. 239-248.

AAS 03-037

HONEYWELL SIGI (SPACE INTEGRATED GPS/INS) H-764G SYSTEM

Zygmunt Zubkow*

The Honeywell Space Integrated GPS/INS system is discussed providing information on applications for space. The SIGI is configurable for various missions and has been used with a number of different hardware and software missionizations. Present configurations are discussed as well as configurations that have been used in aircraft and shipboard applications that may be used for space applications.

INTRODUCTION

The Honeywell SIGI is a space version of the H-764G EGI, which is widely used for a variety of military applications, particularly for high dynamic situations such as fighter jets. The SIGI configuration uses the same H-764G EGI with minor changes, such as enhanced connector coatings, space application specific software, and higher environmental qualification testing. The SIGI has been used on a wide range of space platforms and can easily be missionized for specific requirements. The SIGI has had hardware missionizations such as a 4 channel receiver for providing a GPS derived attitude measurement system, and an embedded S-band RF telemetry card. Software missionizations are typically created for the SIGI to provide messages unique to a customer's application, as well as consideration for flight profiles or launch requirements. Honeywell also has the capability to add any of the following embedded functions to the military version of the H-764G: Configurable I/O (Analog, Discrete, Synchro), Radar Altimeter, ARINC 429 interface, Stability Augmentation, GPWS/TERPROM/TAWS, Multi-Mode Receiver, and Navigation Management.

The Honeywell H-764G systemhad its first production contract in 1994. Since then, the H-764G has been missionized for over 100 different platforms, in over 35 different countries around the world to date, over 9000 SIGI and EGI systems have been delivered.

* Honeywell Defense and Space Electronics, Advanced Space Systems, 13350 U.S. Highway 19 North, Clearwater, Florida 33764.

Figure 1. H-764G SIGI with Mount

SYSTEM OVERVIEW

H-764G SIGI Missionizations

The off-the-shelf SIGI design, is a member of Honeywell's family of stand-alone inertial and GPS/INS systems being used in land, aircraft, missiles, and space vehicles.

All of the SIGI chassis configurations are based upon electronic and software module building blocks common to H-764G products. By integrating digital laser gyro (DLG) sensors and accelerometers with a modularized circuit card packaging, computing, and software technologies, the SIGI provides accurate navigation units that can readily be adapted for specific space vehicles. Some examples of SIGI missionizations are as follows;

- Hardware
  - Radiation Tolerant SRAM's on system processor
  - 5 Channel GEM or 12 Channel Trimble Embedded GPS
  - S-Band RF Telemetry Card
  - Nickel plated external connectors
- GPS Space Software Functions
  - Differential GPS for autonomous landing
  - On-orbit PVT and attitude initialization and updates
  - Orbit propagation and extended dynamics

- Negative mask angle for orbital operations
- Continued output of 1PPS when TFOM>7

- Navigation Space Software functions
  - Modified gravity model with local level gravity error states
  - Sensor compensation parameter inputs
  - Custom sensor outputs for On-orbit maneuvering
  - Flight control delta V, delta theta outputs (200hz)

Baseline SIGI System

Honeywell's off-the-shelf Baseline H-764G SIGI design is a modular system which is integrated (tightly coupled) with either an embedded Collins or Trimble GPS receiver (either P(Y) and C/A Code, as required). Both 5 and 12 channel GPS systems have been used. Figure 2 is a simplified electrical interface diagram for the Honeywell Baseline SIGI.

The H-764G SIGI is a self-contained navigation system providing outputs of linear and angular acceleration, linear and angular velocity, Universal Transverse Mercator (UTM) coordinates of position, attitude (roll, pitch, and platform azimuth), magnetic and true heading, altitude, body angular rates, time tags, and Universal Time Coordinated (UTC) synchronized time.

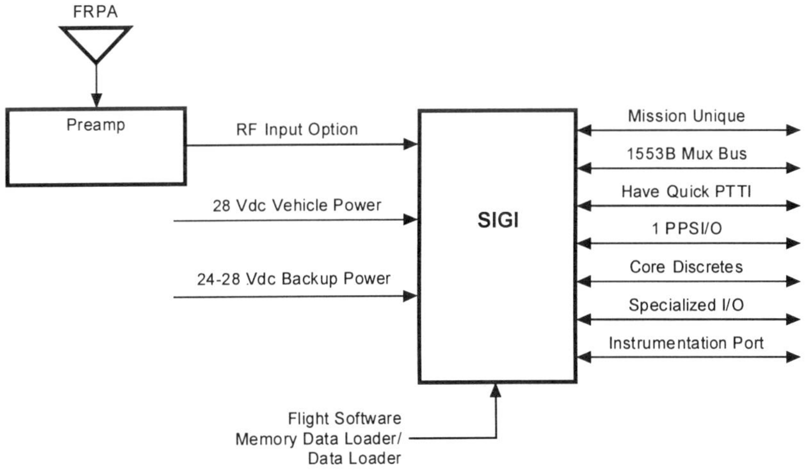

Figure 2. Baseline SIGI Interface Diagram.

The H-764G SIGI is specified to provide position accuracy to better than 0.8 nmi/hour @ 1 hour after a 4 minute ground gyrocompass (GC) alignment. For space launch vehicle applications where ground alignment time is not critical an extended alignment time is typically used. Velocity accuracy (V_{north} and V_{east}) is better than 2.5

ft/sec rms with Vertical Velocity less than 2.0 ft/sec rms. Attitude accuracy (pitch, roll and azimuth) is less than 0.05 deg rms and true heading is specified at 0.1° rms.

Table 1 provides the Free Inertial Performance of the SIGI system. Table 2 provides the blended GPS/INS performance of the SIGI system

Parameter	Performance[1]	Metric Approximation
Position (CEP) (3 min. GC)	1 nm/hr (1 hour)	~1.9 Km/hr (1 hour)
Position (CEP) (4 min. GC)	0.8 nm/hr (1 hour)	~1.5 Km/hr (1 hour)
Position (CEP) (4 min. GC)	0.8 nm/hr (2 hour)	~1.5 Km/hr (2 hour)
Altitude(RMS)[2]	50 ft	~15 m
Velocity X, Y (RMS)[3]	2.5 ft/sec	~0.76 m/sec
Velocity Z (RMS)[3]	2.0 ft/sec	~0.61 m/sec
Acceleration X, Y, Z (RMS)[3]	0.064 ft/s/s	~0.02 m/sec/sec
Pitch, Roll (RMS)	0.05 deg	
Platform Azimuth (RMS)[4]	0.05 deg	
True Heading (RMS)	0.1 deg	
Magnetic Heading (RMS)[5]	0.3 deg	
Pitch, Roll (synchro)	+/- 0.1 deg	
True Heading (synchro)	+/- 0.1 deg	
Magnetic Heading (Synchro)	+/- 0.5 deg	

Notes:
1. Proper GC align has been performed
2. Relative to Pressure Altitude Input
3. Navigation Frame as defined in SNU 84-1
4. Platform Azimuth as defined in SNU 84-1
5. Assuming periodic updates of Mag. Var. Tables.

Table 1. Free Inertial Performance.

Parameter	Performance
Position	
SPS	76 meters SEP
PPS	16 meters SEP
Velocity (RMS)	
SPS	N/A
PPS	0.1 m/sec
Time(RMS)	
SPS	<155 nsec
PPS	<100 nsec
Pitch, Roll (RMS)	0.05 deg
Platform Azimuth (RMS)	0.05 deg

Table 2. GPS/INS Performance

The H-764G SIGI provides for 3 alignment modes: Gyrocompass ground (GC) Align; Stored Heading Align (SHA); and In-Motion Alignment (IMA).

The H-764G SIGI provides three simultaneous navigation solutions: 1) GPS-only, 2) GPS/INS (blended), and 3) INS-only. The GPS navigation solution is available whenever the EGR is tracking at least four satellites. GPS/INS navigation combines line-of-sight data from the EGR with inertial navigation data to form a navigation solution, which is superior in accuracy and noise to either the GPS-only or INS-only navigation solution. The EGR communicates with the SIGI inertial navigation computer processor via a Dual-Port Random Access Memory (DPRAM) interface. The DPRAM is accessible by both the EGR and the INS; i.e., they both can read and write to/from this memory storage area.

Currently the Collins receiver allows for the tracking of up to 5 Space Vehicles (SVs), while the Trimble receiver allows for the tracking of up to 12 SVs simultaneously. Both the Collins and Trimble receivers provide independent GPS position and velocity data, which is available via a 1553B or RS-422 data bus.

The H-764G SIGI accepts and processes turn-on discrete, mode commands, and initialization data via the host vehicle 1553B Mux data bus or via RS-422 data bus, depending on system configuration due to customer requirements. Upon request from the data bus, the SIGI then transmits output data. The 1553B-message interface structure is very similar to the U.S. Government specification SNU 84-1 as well as messages defined in U.S. Government specification ICD-GPS-059. The RS-422 message structure is basically a subset of the MIL-STD-1553 message structure.

The Baseline H-764G SIGI operates with 28 Vdc host vehicle primary electrical power. If 28 Vdc primary power is lost, the H-764G SIGI will continue to operate on backup power when provided. A timer can be activated to initiate a software-controlled shutdown after the H-764G SIGI has operated on the backup power for 10 consecutive seconds. If primary power is restored to the H-764G SIGI within the first 10-second period, the unit shall revert to operation on primary power and the shutdown timer is then reset to zero. The 10-second period is a default value set in the SIGI's software and is a configurable value.

SYSTEM HARDWARE DESCRIPTION

Baseline H-764G SIGI Hardware Subassemblies. Figure 3 is an exploded view of the H-764G SIGI design. This view shows the five major electronic hardware subassemblies:

1) Inertial Sensor Assembly (ISA),
2) Inertial Electronics (IE),
3) Common Advanced System Processor (CASP)
4) Embedded GPS Receiver (EGR), and
5) DC Power Supply (DCPS).

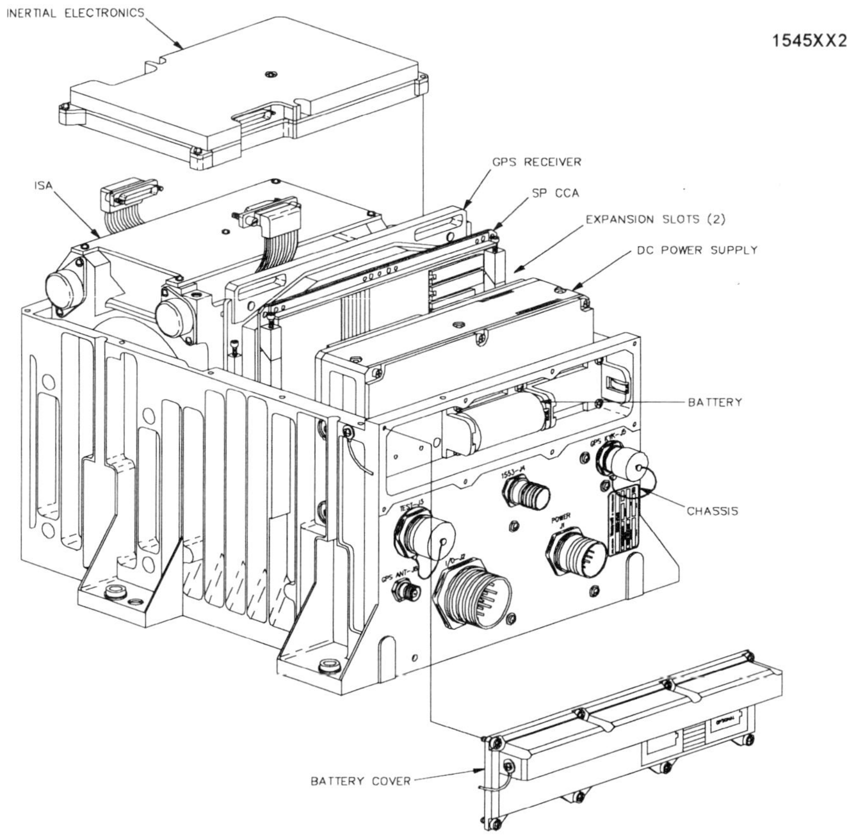

Figure 3. Exploded View of the 9.8-inch H-764G SIGI / EGI System.
The five major subassemblies shown are common to most of Honeywell's SIGI designs. Open slots provide for optional growth potential.

Also shown are the Master Interconnect Board (MIB), Battery, and Chassis with outboard mounting feet (Note: A rack mounted chassis configuration is also available). In addition, there are two expansion slots available for additional circuit card assemblies (CCAs) to provide increased functional capability.

The standard SIGI has two slots available for future growth or mission specific I/O. Figure 4 is a top-level H-764G SIGI functional block diagram that depicts the NDI Baseline H-764G SIGI common architecture.

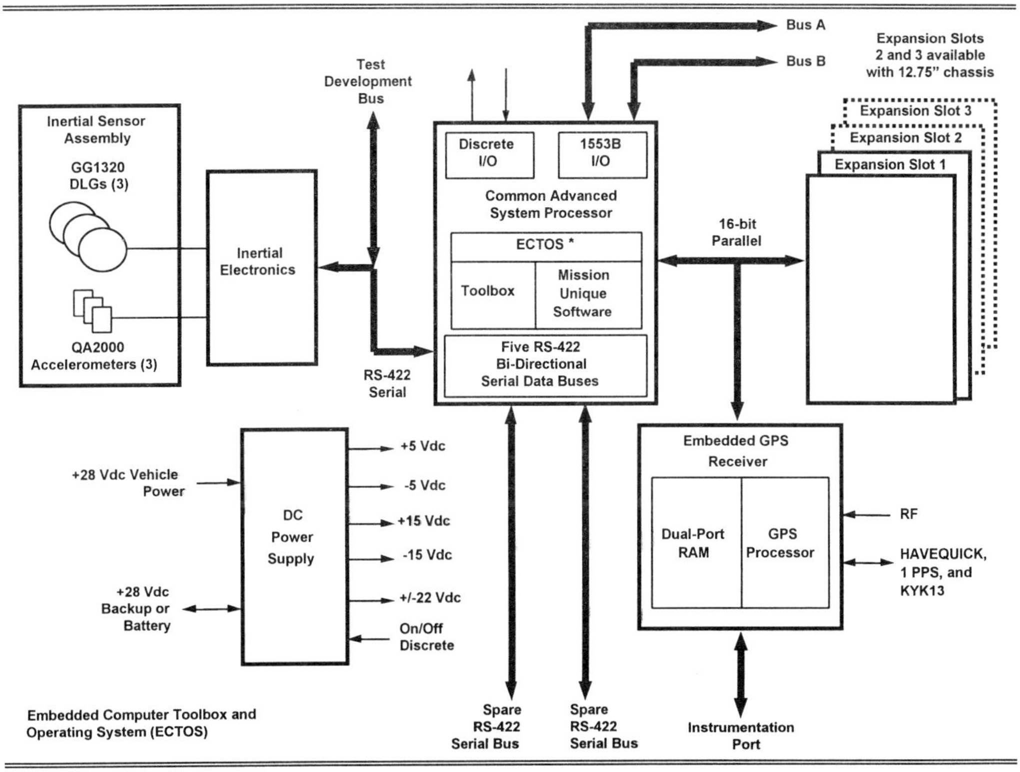

Figure 4. H-764G SIGI Functional Block Diagram.
The number of circuit cards in the design has been reduced because all of the gyro control circuitry is packaged within the case of the Honeywell DLG.

ISA and Inertial Electronic Functional Descriptions. The Inertial Sensor Assembly (ISA) and the Inertial Electronics (IE) are shown functionally in Figure 5.

Inertial Sensor Assembly. The ISA contains all the essential sensing elements to perform strapdown free inertial navigation. The ISA, shown in Figure 6, is composed of the following subassemblies mounted on a rigid aluminum block:

- Three mutually orthogonal Honeywell DLGs,
- Three mutually orthogonal Honeywell QA-2000 accelerometers, and
- The Temperature Calibration/Programmable Read-Only Memory (Temp/Cal PROM) CCA

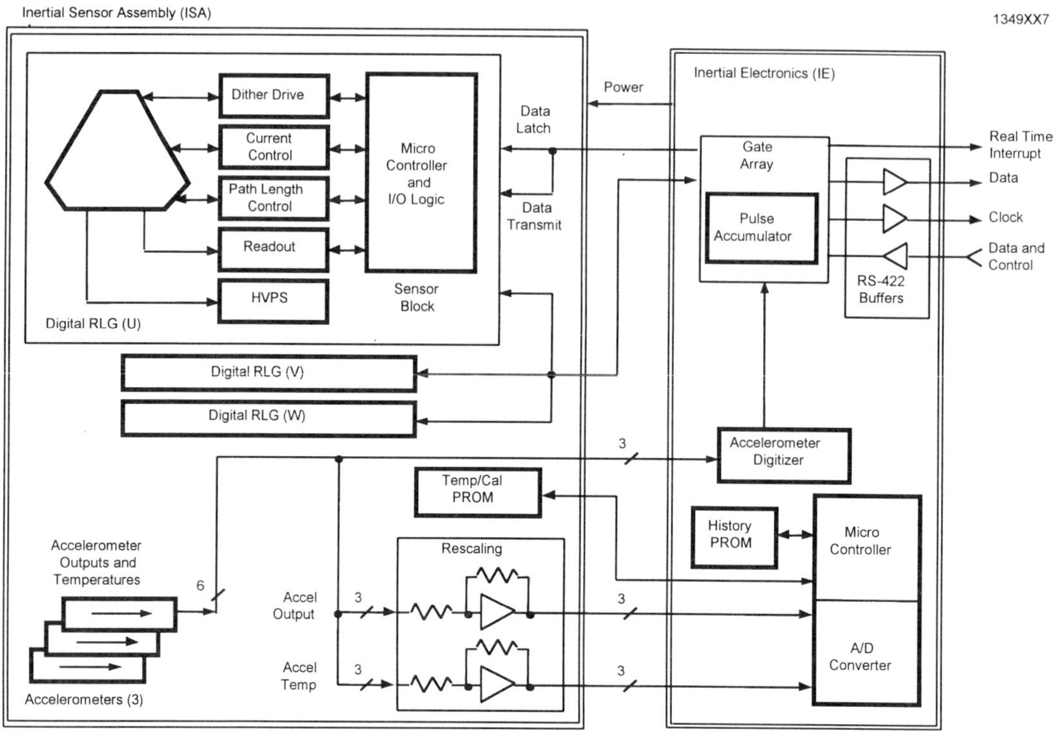

Figure 5. ISA and IE Assembly Functional Block Diagram.

Figure 6. Honeywell Inertial Sensor Assembly.
The ISA containing three DLGs, three accelerometers, and a Temp/Cal PROM CCA provides a compact solution to inertial sensing.

Honeywell DLG Functional Description. Honeywell's DLG, model GG1320AN, is the world's first production DLG requiring no external sensor control electronics. High voltages and sensor control electronics are contained completely within the gyro housing. Figure 7 is a photograph of the DLG.

1530XX5

Figure 7. The Honeywell DLG.
Our DLG is a self-contained device requiring only a low-voltage power input.

High-voltage and sensor-control electronics for the DLG are all contained within the gyro. This integration eliminates several circuit card assemblies from the system, and greatly simplifies internal interconnections, sensor assembly cabling, and fault isolation.

Communication with the DLG is via an RS-232 serial data bus. Input voltage requirements are +5.0 Vdc and +15 Vdc, with power consumption totaling a low 2.0 watts per gyro. The DLG sensor element, the DLG laser block, has 2.0-inch legs.
An embedded microcontroller controls the following electronic functions:

1. Startup and control of the gas discharge
2. Control of the current balance between the two laser discharge legs
3. Acquisition and control of the proper laser mode over temperature
4. Gyro dither and control. Pseudo-random noise is added to properly randomize the gyro output
5. Reading of the gyro output signals
6. Compensation of gyro output for several small errors of gyro bias (as a function of temperature), gyro angle (as a function of temperature), and dither motion and pick-off linearities

Honeywell's GG1320 DLG is the world's first production DLG requiring no external support electronics. DLG development capitalized on our experience producing more than 200,000 ring laser gyros and years of research and development into low-cost laser gyro concepts and advance technology areas. The end result of this effort is a DLG with performance matching the larger GG1342, but with a 70 % reduction in weight and 68 % reduction in volume. To date, Honeywell has produced more than 30,000 GG1320 gyros since placing it in full-scale production in 1990, equating to more than 30 million hours of operational service.

Accelerometer Functional Description. Honeywell's Baseline SIGI uses the Honeywell QA2000 Q-Flex™ accelerometer. These accelerometers are single-degree-of-freedom, pendulous, force-rebalance accelerometers. The QA2000 model is used for all U.S. military Tri-Service EGI applications.

Honeywell has produced the Q-Flex™ family of accelerometers for over two decades, and is delivering over 8,500 QA2000 accelerometers yearly. The demonstrated MTBF in some INS applications is over 500,000 hours. The QA2000 sensor design benefits from two decades of developmental refinements in materials and processes.

A temperature sensor is mounted inside the accelerometer housing and its output is used at the system level for temperature modeling. To obtain fast reaction and accurate operation over a -54°C to +71°C temperature range, thermal modeling of bias, scale factor, and misalignment are used. Using the temperature models, the accelerometer outputs are temperature compensated by the software executing in the System Processor CCA.

Temperature Calibration/PROM CCA. The Temp Cal/PROM CCA provides the following functions:

Serves as an electrical interconnection point for the gyros and accelerometers, passing strobes to the gyros and serial data from the gyros to the IE

Multiplexes and rescales analog temperature inputs from each accelerometer and the analog voltage inputs from each accelerometer and sends this data on to the IE microcontroller ADC

Non-volatile memory storage of system calibration data.

Accelerometer BIT circuitry.

Inertial Electronics Assembly. The IE assembly controls the interface between the gyro and accelerometer sensors and the SP CCA. This microcontroller also contains an Analog-to-Digital Converter (ADC) which is used to convert analog temperature and accelerometer data into digital quantities. The IE CCA also contains circuitry that provides the following functions:

1. A Programmable Gate Array (PGA) microcircuit, acting under the control of the microcontroller:

2. An accelerometer-digitizer microcircuit that converts analog accelerometer data into digital pulses
3. A separate non-volatile memory storage area is provided for storage of flight history and BIT data.

GPS Receiver Functional Description. The SIGI uses either a 5 or 12 channel Embedded GPS Receiver (EGR), which provides the resources necessary to select, acquire, track and navigate using the GPS satellite signals at the L1 and L2 frequencies using the C/A-codes (and P-codes if available) from a Radio Frequency (RF) source. Both SPS and PPS EGR receivers have been used in SIGI systems. The EGR provides the high speed interface and handshake signals necessary to communicate with the SP. The EGR also provides the resources necessary to communicate over the Precise Time and Time Interval (PTTI) interface and the instrumentation port.

DCPS Functional Description. The SIGI Direct Current Power Supply (DCPS) is a compact, self-contained modular design which contains filter assemblies, a DC/DC Converter CCA, and a BIT and Control CCA. The DCPS requires host vehicle 28 Vdc prime and, if available, 24 Vdc backup power as an input. The DCPS then provides all other voltage levels required by other SIGI circuit cards. The power supply is designed with more power capability than needed by the baseline SIGI design so that additional missionization circuit cards in expansion slots can be powered. Figure 8 is a functional block diagram of the DCPS. The DCPS features:

1. Switchable 12 Vdc output for re-programming of the SP flight software via the 1553B Bus
2. 22 Vdc output to power analog synchro outputs when added
3. Under- and over-voltage/current monitor
4. Input voltage monitor and transparent switchover to backup power if primary power is lost
5. "Diode-or'ing" of cryptographic keying batteries to prevent battery charging when two batteries are installed
6. Selection of one of three turn-on modes to enable use of one DCPS design for most SIGI missionizations.

The DCPS is compliant with MIL-STD-704C (Appendices A, B, C, D) and has been tested to worst case conditions.

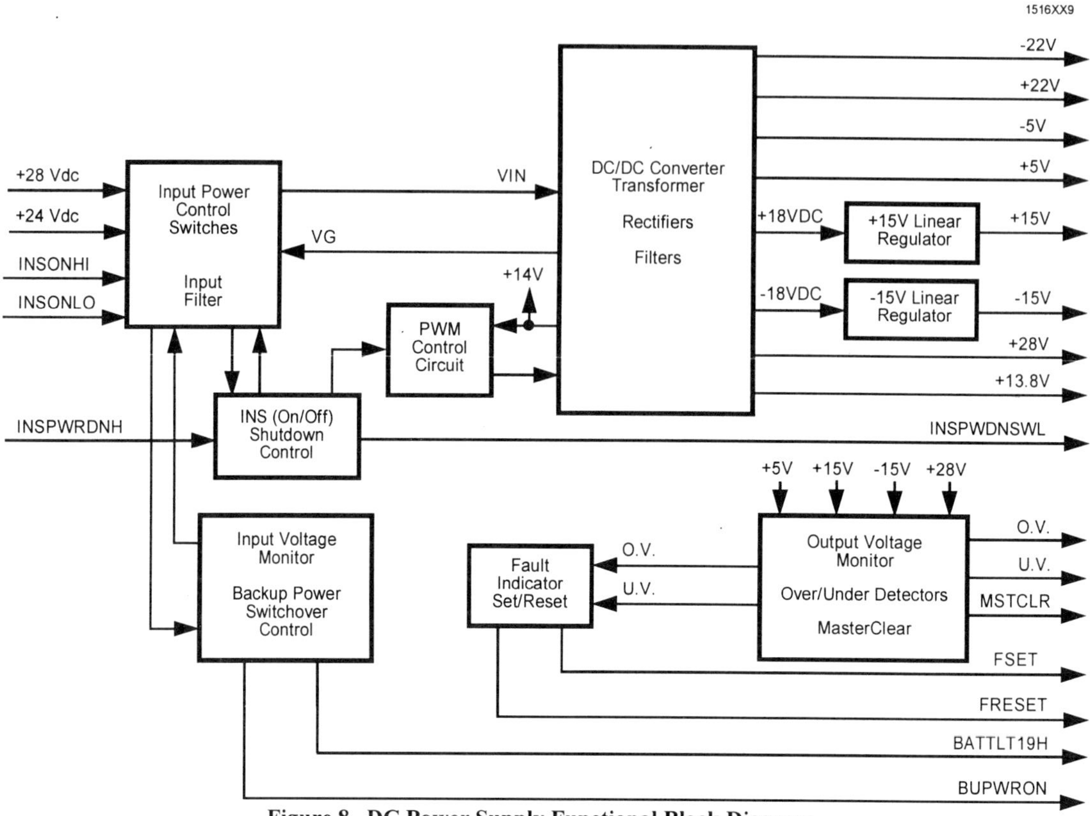

Figure 8. DC Power Supply Functional Block Diagram.

H-764 SIGI Chassis. The H-764G SIGI mechanical installation drawing is shown in Figure 9. The size of the H-764G SIGI is 7.0 inches wide by 7.0 inches high by 9.8 inches long and weighs ≈18.5 pounds.

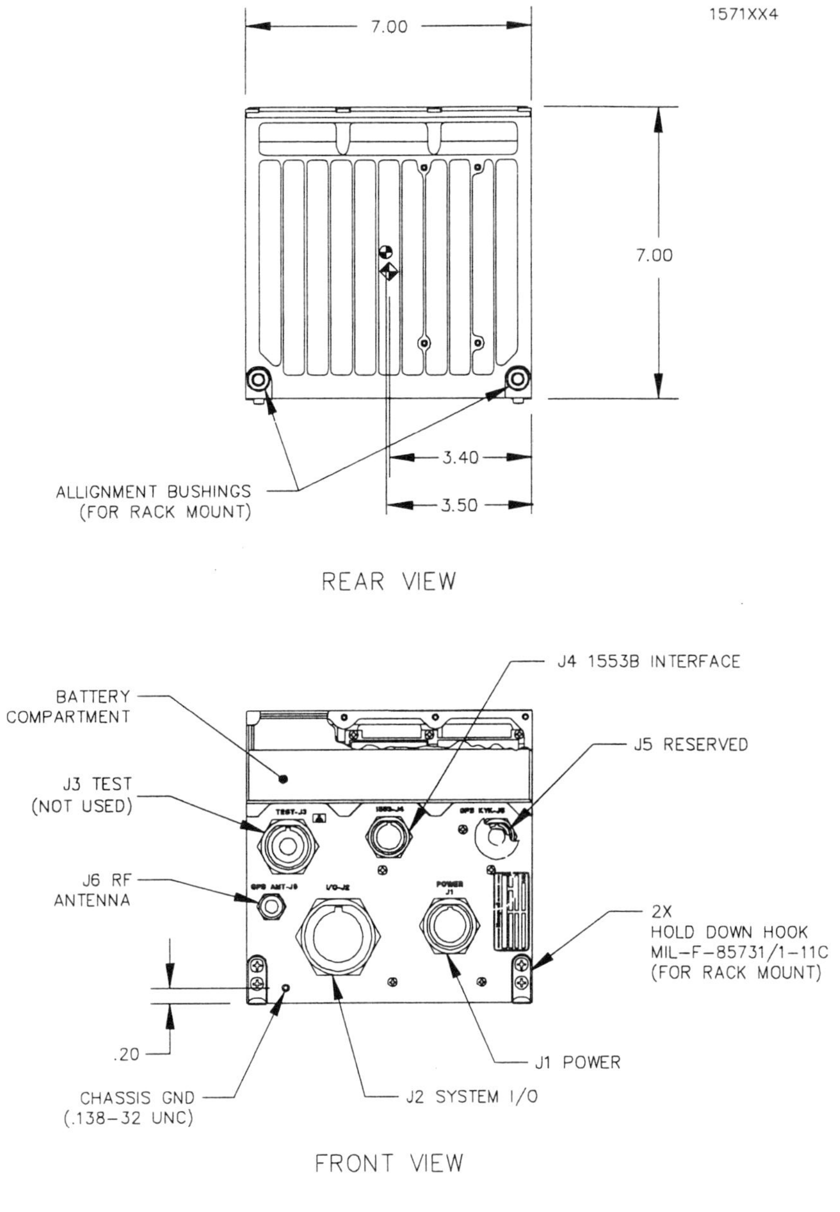

Figure 9. H-764G SIGI Mechanical Installation

CONCLUSION

The Honeywell SIGI has been a successful implementation of the adaptation of a military embedded GPS/INS system for various space vehicle uses. With relatively minor changes to the hardware and software it has been found that very good performance has been achieved for a variety of space applications. With the spare slots available in the system various options exist to allow mission specific I/O interfaces.

AAS 03-039

CLOSED LOOP TEST ENVIRONMENT FOR SPACE APPLICATIONS USING GPS SIGNALS

M. Mittnacht,* E. Gottzein,† M. Hartrampf* and A. Konrad*

Numerous space applications use GPS receivers today. They perform orbit determination for different kinds of missions from LEO (Low Earth Orbit) to GEO 6 (Geostationary Orbit). For LEO satellites the usage of GPS receivers is becoming a standard. Navigation satellite systems are not only used for real-time absolute orbit determination, but also for relative navigation between multiple vehicles, attitude determination, precise orbit determination and a growing number of other areas of interest. The so-called MosaicAODS[5] is a spaceborne Attitude & Orbit Determination System developed at Astrium GmbH. It is based on the MosaicGNSS receiver, extended by a star sensor and an optional INS. In order to validate the functionality of the MosaicAODS and to ensure its proper functionality in space a modular real-time test environment has been developed, where single AODS modules as well as the full system can easily be tested. Additional to these tests, the system provides the capability to test precise and relative navigation using GPS signals for rendezvous maneuver, formation flying or spacecraft co-positioning. This paper outlines the AODS test environment and the GPS precise and relative navigation test environment.

INTRODUCTION

In recent years numerous spaceborne GPS receivers have been developed all over the world. The area of applications is growing constantly. GPS signals have been explored in different types of orbits, i.e. in Low Earth Orbits (LEO), Highly Elliptical Orbits (HEO) and even in Geostationary Orbit (GEO). For LEO satellites the usage of GPS receivers is becoming a standard, examples are the GLOBALSTAR satellites or the TerraSAR-X satellite. Navigation satellite systems are not only used for real-time absolute orbit determination, but also for relative navigation between multiple vehicles, attitude determination, precise orbit determination and a growing number of other areas of interest. State-of-the-art and future scientific missions for Earth and space observation need high accuracy in knowing the position of the observing vehicle. This can be done by using dual frequency receivers and the correction data from the Space Based Augmentation Systems (SBAS) in combination with carrier phase smoothing of the pseudorange to reduce the noise of measurements which will be used for navigation. Formation flying of multiple spacecrafts requires an even higher accuracy on the relative position determination. Carrier Phase Differential GPS (CDGPS) can provide the necessary means to achieve this goal and to carry out such missions. The GPS/AODS test environment was set up as a hardware-in-the-loop test facility to evaluate the MosaicGNSS receiver and the multi-sensor MosaicAODS system. The MosaicAODS is based on the MosaicGNSS receiver, extended by a star sensor and an optional INS. The system provides the full state information

* Astrium GmbH, Ludwig Boelkow Allee, D-81663 Munich, Germany.

† Scientific Consultant, Astrium GmbH, Munich, Germany.

consisting of position, velocity, attitude and time allowing a large degree of on-board autonomy independent of the mission (LEO, MEO, HEO, and GEO).

A Spirent Ltd. STR4760 GPS signal simulator is used to provide simultaneously GPS signals for up to two different user satellites. The GPS signal simulator system is equipped with a real-time remote control option, which allows integrating the simulator in the closed loop test system. The remote system (spacecraft simulator) is synchronized with the GPS simulator for correct operation. In order to validate the GPS precise and relative navigation algorithms running on the MosaicGNSS receiver, the test environment was extended by the simultaneous simulation of two spacecrafts

UNIT UNDER TEST - SPACEBORNE GPS Rx/AODS

The MosaicGNSS[3] receiver (Figure 1, Table 1) is a radiation tolerant (up to 100 krad) GPS single board receiver for applications in LEO, MEO and GEO. It is a L1 C/A code receiver, which is capable of acquiring and independently tracking up to eight GPS satellites. A key feature of this receiver is the implementation of main parts of the signal processing in software[1]. The MosaicGNSS receiver will be flown on the German TerraSAR-X[8] satellite, due to be launched in 2005 and operational in 2006. The MosaicAODS is based on the MosaicGNSS receiver and complemented by star sensors and optional gyros. Therefore, the MosaicAODS makes the full state information continuously available. The MosaicGNSS receiver hardware serves as the central part of the MosaicAODS. A set of high performance processing, communication and storage modules form the common hardware platform. A Digital Signal Processor (DSP) TSC21020, clocked with 20 MHz, serves as the central processing unit. The image processing algorithms of the star sensor as well as the GPS algorithms are executed on the MosaicAODS board on the same DSP.

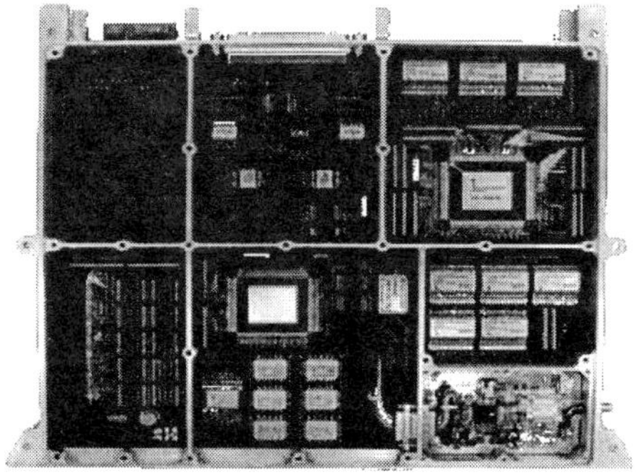

Figure 1: MosaicGNSS Rx/MosaicAODS Common Hardware Platform

Accuracy, 3D, no SA, RMS			
	Position	Velocity	Time
LEO	< 10 m	< 0.01 m/s	< 100 ns
MEO*	< 30 m	< 0.02 m/s	< 200 ns
GEO	< 150 m	< 0.02 m/s	< 750 ns
TTFF (LEO)			
Cold start	< 8 min	Warm start	< 4 min
* altitude 10000 km			

Table 1
MosaicGNSS Rx PERFORMANCE

TEST ENVIRONMENT

The closed loop test environment will be described in the following by means of the GPS/AODS test environment and the relative navigation test environment. The term closed loop is used here in the sense of a closed loop test system and not in the sense of a closed loop control loop.

GPS/AODS Test Environment

The real-time closed loop test environment (Figure 2, Figure 3) comprises the MosaicAODS, a simulation computer, which calculates the orbit and attitude of the user satellite, a 12 channel GPS RF signal simulator unit and a star sensor head simulation. All simulation equipment is timely synchronized with the GPS RF signal simulator. The orbit and attitude data of the user satellite is fed to the GPS RF signal simulator unit, which stimulates the GPS receiver module of the AODS and to the star sensor head simulation where the attitude information is used to simulate the output of an optical head and the pre-processing part of a real star sensor. Using data from the GPS RF signal simulator and the star tracker the MosaicAODS calculates the full state information and sends it back to the simulation computer for further analysis, visualization and storage.

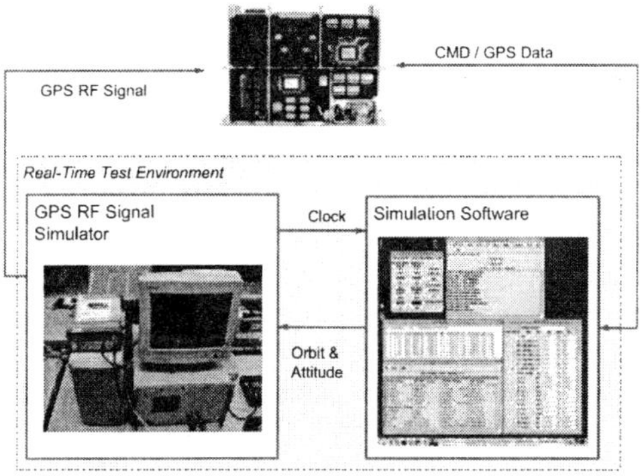

Figure 2: GPS Test Facility

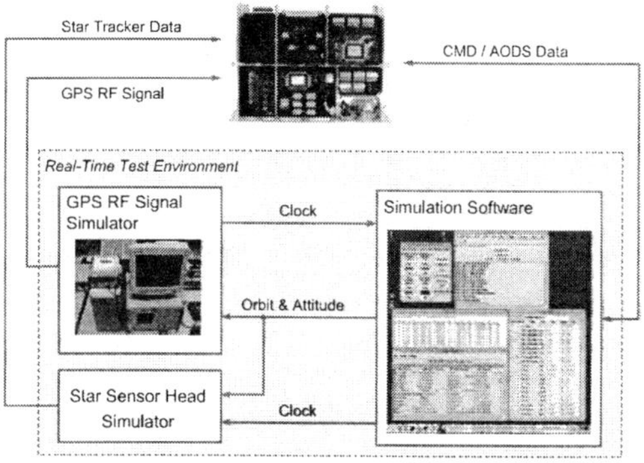

Figure 3: GPS/AODS Test Facility

The purpose of the AODS test environment is

- to provide an extensible user interface to send and log commands to the MosaicAODS board and to display and store messages coming from the board,
- to perform a simulation of the orbit and attitude motion of a spacecraft and to supply the GPS receiver via the GPS RF signal simulator with the corresponding stimuli inputs,
- to perform a simulation of the star sensor head and to supply the Image Processing (IP) module running on the MosaicAODS board with the corresponding inputs.

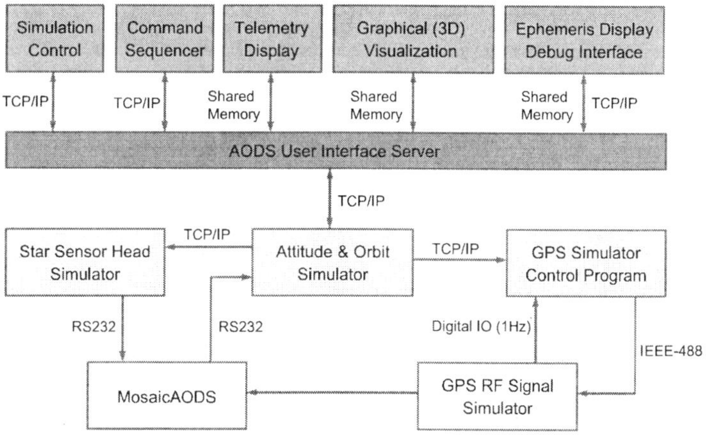

Figure 4: Functional Architecture & Interfaces of the AODS Test Bench

The goal of the new test bench was to develop a framework of applications, which can be used for various different purposes in the space checkout field. The system should be suitable from simple and closed unit tester to distributed checkout systems. To achieve this goal a networked and distributed data base system is used. The communication between single modules is based on the current space-standards for data handling defined by CCSDS. The functional architecture of the AODS test bench is shown in Figure 4. A central element of this new development is the exclusive use of free software. Long time availability, export licenses and costs are the most important reasons for their employment. In order to obtain a real-time test environment the core programs are written in C using GNU Lib C as standard library. The control of the test environment can be done manually by using the appropriate graphical user interface or alternatively from a script file. As script language the widely used Tcl/Tk is chosen.

Figure 5 shows the graphical user interface of the test environment, from where all parts of the test set-up are controlled.

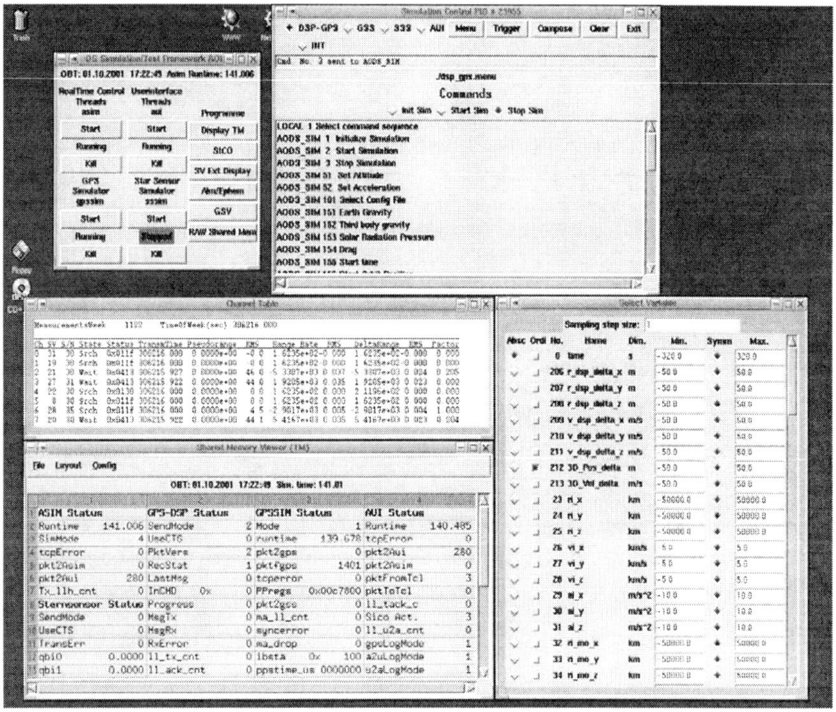

Figure 5: Graphical User Interface (GUI)

GPS Relative Navigation Test Environment

In order to validate the GPS precise and relative navigation algorithms running on the MosaicGNSS receiver, the test environment (Figure 6) was extended by the following features:

- simultaneous simulation of two spacecrafts,
- corresponding remote control of the GPS RF signal simulator (two spacecrafts),
- communication with two MosaicGNSS receivers,
- SBAS extensions:
  - New telemetry and telecommands from/to the MosaicGNSS receivers,
  - Display of navigation and correction data.

Due to the modularity of the test environment these changes didn't affect the base environment architecture and are pure extensions of the previous facility.

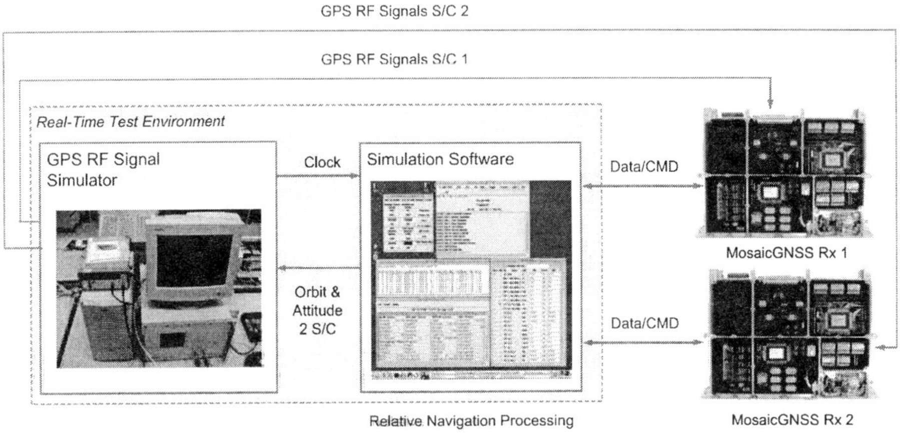

Figure 6: Relative Navigation Test Environment

TEST RESULTS

The feasibility of the MosaicAODS has been demonstrated using the real-time closed loop test environment described above. For the tests the ASTRO 15[7] star sensor from Jena Optronik GmbH was used to complement the MosaicAODS. The modular design of the sensor allows operating it as a star tracker, or equipped with an additional hardware module, the Image Processing Unit (IPU), as an autonomous star sensor. Used as a star tracker it picks up the visible stars and calculates their centers with sub-pixel accuracy (centroiding). The functionality of the IPU has been relocated to the MosaicAODS board. Therefore, the IPU software is executed on the MosaicAODS board and computes inertial 3-axis attitude based on star measurements and an on-board star catalogue. The star tracker and the MosaicAODS board communicate via a serial interface. Inertial 3-axis attitude information can be provided with spacecraft rates up to two deg/s.

Two orbits have exemplary been chosen here for the MosaicAODS tests:

- LEO (altitude: 1400 km, inclination: 64°, one zenith pointing antenna, no SA)
- MEO (altitude: 10350 km, inclination: 51.6°, one zenith pointing and one nadir pointing antenna, no SA)

LEO Application

It could be shown that a continuous operation of the integrated system with up to 6 tracked GPS satellites and without performance reduction of the attitude determination is possible. Starting with 8 GPS channel a time-to-first-fix of < 8 minutes (LEO) can be achieved. Using a filtered navigation solution the 3D accuracies (rms) were:

- position < 10 m,
- velocity < 0.01 m/s,
- time < 100 ns.

Attitude performance during a slew maneuver (0.5 °/s) in LEO:

- pitch/yaw error ≤ 3.6 arcsec (1σ),
- roll error, ≤ 40.6 arcsec (1σ).

The attitude performance tests have been performed assuming end-of-life (EOL) conditions for the star tracker.

MEO Application

In Figure 7 the GPS constellation and a MEO satellite (the parameters are based on the multimedia demonstration experiment – MEDIS[4]) are shown as well as the visible GPS satellites (green line between GPS and MEO satellite).

The visibility of the GPS satellites for the period of 24 hours is illustrated in Figure 8.

As seen from Figure 8 a continuous point solution is not possible. Sequential estimation algorithms have to be applied. Therefore, a model of the system (user) dynamics and statistical knowledge of the process and measurement noise have to be used.

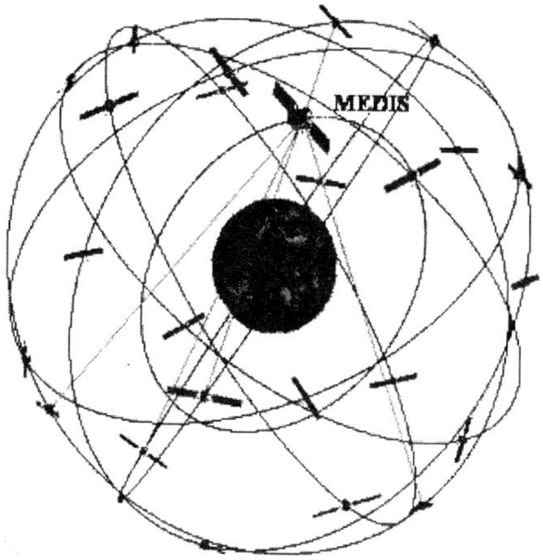

Figure 7: MEO User Satellite & GPS Constellation

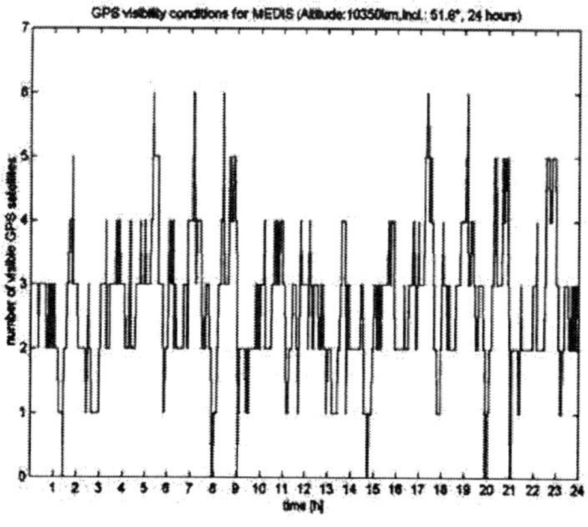

Figure 8: Visibility Conditions

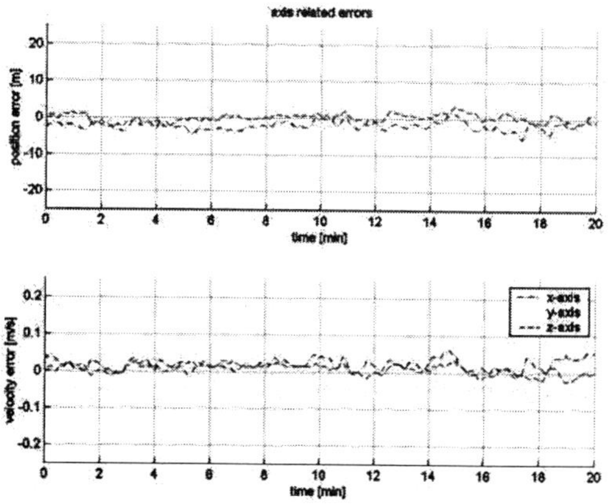

Figure 9: Position & Velocity Performance

The axis related position and velocity results are shown in Figure 9 (x-axis – flight direction, y-axis – radial, z-axis – out of plane). The 3D accuracies (rms) are:

- position < 30 m,
- velocity < 0.02 m/s,
- time < 200 ns.

Attitude performance in MEO:

- pitch/yaw error ≤ 2.3 arcsec (1σ),
- roll error, ≤ 24.6 arcsec (1σ).

As before, the attitude performance tests have been executed assuming end-of-life (EOL) conditions for the star tracker.

Relative Navigation

Within a constellation or formation of satellites the relative position between the single satellites is often needed with a higher accuracy than the absolute position of the satellites. Therefore more elaborated algorithms are needed for the relative positioning.

Relative navigation[2] of two or more spacecrafts is done by differencing the measured pseudoranges and carrier phase measurements of the GPS constellation. By this differencing, errors common to all user satellite, such as GPS ephemeris errors or ionospheric errors can be eliminated. When using the carrier phase measurement an ambiguity of the number of wavelengths has to be resolved in order to obtain utilizable data.

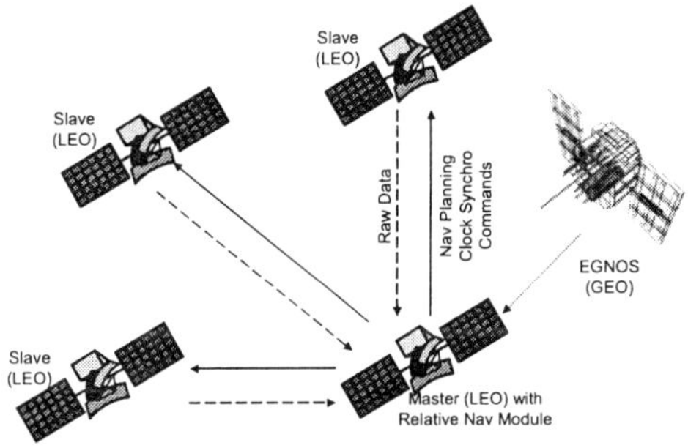

Figure 10: LEO relative navigation

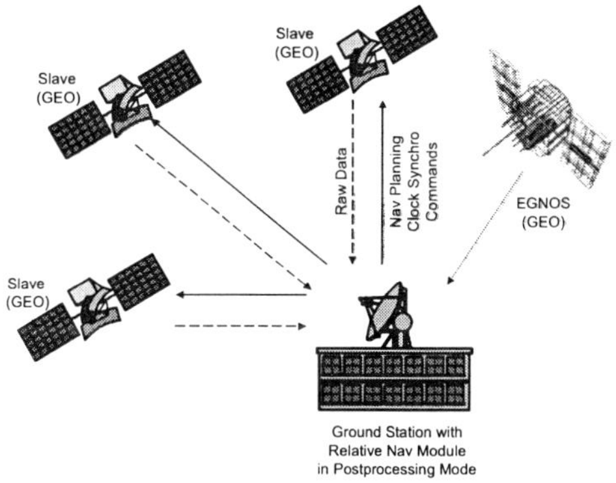

Figure 11: GEO relative navigation

In the above figures two options of the relative navigation processing are shown. The first option is the on-board processing of the relative navigation software. Therefore, the satellites of the constellation have to be connected via an inter-satellite link. One spacecraft of the constellation is chosen as the 'master' satellite (Figure 10). The collected data from the different GPS-receivers is transmitted regularly from the 'slave' satellites to the 'master', where the relative navigation module is executed. This execution is either done within the GPS receiver of the 'master' satellite or in the onboard computer unit (OBCU).

The second option is the ground station processing (Figure 11). The relative navigation is then executed on ground with the possibility to post-process the data and further increase the accuracy of the relative position solution. Due to the fact, that, with the exception of geostationary satellites, there is no continuous link between the user satellites and the ground station, the data are collected on the user satellites and transmitted to the ground only from time to time. This means, that the relative position of the user constellation is only known in post-processing.

For the tests of the relative navigation algorithms, a test set-up with two receivers has been used as shown in Figure 6. As example the following conditions have been chosen:

- two satellites in a circular low earth orbit,
- semi-major axis: 7800 km,
- inclination of 40°,
- distance between the two satellites during the formation flight was set to approx. 1 km.

Figure 12 shows the results of this test. All results are shown in the Earth-centered, Earth-fixed (ECEF) coordinate system. The first three graphs in this figure represent the errors in the different axis of the orbit. The enveloping curve represents the 3-sigma value of the error. The fourth graph shows the number of GPS satellites tracked by the GPS receiver (solid line) and the Position Dilution of Precision (PDOP, dashed line). The time to first fix was about 200 seconds, followed by the time required for the ambiguity resolution of about 800 seconds. After this, the receiver operates in normal operation mode for the rest of the run. Overall, the accuracy of the positioning in normal operation mode is 0.59 cm (1- σ) on the x-axis, 0.37 cm (1- σ) on the y-axis and 0.59 cm (1-σ) on the z-axis. The relatively bad accuracy on the y-axis at the end of the test run is due to the high PDOP (>8) at this time.

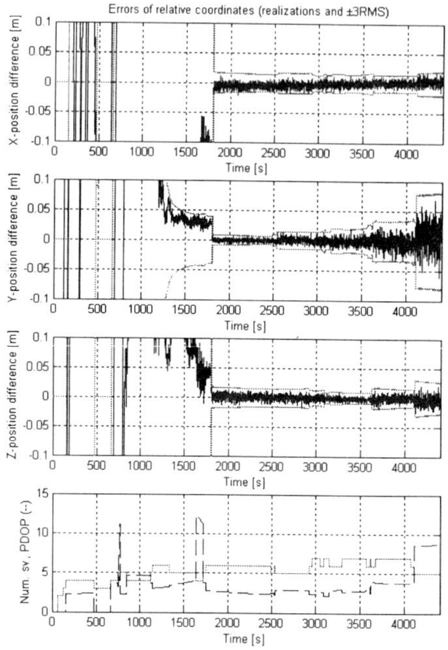

Figure 12: Relative Positioning Errors

CONCLUSIONS

The real-time test facility enables tests of spaceborne GPS receivers and integrated systems (e.g. AODS).

Features:

- real-time closed loop tests,
- modules distributable on computers,
- manual/script commanding modes,
- automatic test sequences,
- automatic data analysis,
- Spirent Ltd. GPS Simulator with 2 x 12 channels, capable of L1 and L2.

Tasks:

- spacecraft (precise orbit & attitude) simulation,
- simulation command & control,
- user/data interface,
- data storage,
- real-time comparison of data,
- visualization,
- star sensor simulation (optional).

Satellite missions range from LEO to GEO. Single or dual frequency receivers as well as SBAS capable receivers can be tested. Absolute navigation, relative navigation (e.g. formation flying, co-location of GEO satellites), attitude determination using GPS signals and tests of integrated systems are within the range of the test facility. A central element of this new development is the exclusively use of "free of charge software". Long time availability, export licenses and costs are the most important reasons for their employment. An advantage of the test bench is that the simulation control can be performed either manually by using the Graphical User Interface (GUI) or by employing script files. Usage of script files enables creation of reproducible simulation runs.

The above described test environment has successfully been used for test with the MosaicGNSS receiver, different other GPS/SBAS receivers and the MosaicAODS.

ACKNOWLEDGEMENTS

Basic research was performed with partial funding of the German Aerospace Center (DLR)/BMBF, contracts no. 50 NB 9808, no. 50 YC 9903, 50 NA 0114 and no. 50 NA 9807 as well as by INTAS projects 00-0407 and 00-0420. The authors wish to thank their colleagues (especially A. Rieger and T. Bückle) at Astrium GmbH and the team of Soft Nav Ltd. for their support. The authors are responsible for the contents of this paper.

REFERENCES

1. Botchkovski, A., V. Chistyakov, M. Golubev, N. Mikhailov, S. Pospelov, M. Vasilyev, P.A. Krauss, W. Fichter, M. Mittnacht, E. Gottzein, Softflex: An Advanced Approach to Design of GNSS Receiver with Software Correlator, Proceedings of the 12th International Technical Meeting of The Institute of Navigation, ION-GPS99, Nashville, Tennessee, 1999.
2. Hartrampf, M., Gottzein, E., Mittnacht, M.,Müller, C., Relative Navigation of Satellites Using GPS Signals, DGLR Annual Meeting, Stuttgart, Germany, September 2002.
3. Krauss, P. A., MosaicGNSS Receiver Flyer, Astrium GmbH, email: peter.a.krauss@astrium-space.com, Munich, Germany, 2002.
4. MEDIS information available at http://www.dlr.de/KN/KN-DN/projects/atm-sat/pub2/workshop 2002/vortrag-10.pdf.
5. Mittnacht, M., et al., Multi-Sensor Navigation and Attitude Determination System for Space Applications, Proceedings of the 15th International Technical Meeting of the Satellite Division of the Institute Of Navigation, ION GPS-02, Portland, Oregon, September 2002.
6. Mittnacht, M., Fichter, W., Real Time On-board Orbit Determination of GEO Satellites Using Software or Hardware Correlation, Proceedings of the 13th International Technical Meeting of the Satellite Division of the Institute Of Navigation, ION GPS-00, Salt Lake City, Utah, September 2000.
7. SETIS Star Sensor Technical Description, Jena-Optronik GmbH, Internal Report, Jena, Germany, 21.08.1997.
8. TerraSAR-X information available at http://www.astrium-space.com/en/programs/index_earth_ob. htm?/programs/part7/00000247.htm.

AAS 03-040

FIRST RESULTS FROM A HARDWARE-IN-THE-LOOP DEMONSTRATION OF CLOSED-LOOP AUTONOMOUS FORMATION FLYING

Eberhard Gill,[*] Bo Naasz[†] and Takuji Ebinuma[‡]

A closed-loop system for the demonstration of autonomous satellite formation flying technologies using hardware-in-the-loop has been developed. Making use of a GPS signal simulator with a dual radio frequency outlet, the system includes two GPS space receivers as well as a powerful onboard navigation processor dedicated to the GPS-based guidance, navigation, and control of a satellite formation in real-time. The closed-loop system allows realistic simulations of autonomous formation flying scenarios, enabling research in the fields of tracking and orbit control strategies for a wide range of applications.

The autonomous closed-loop formation acquisition and keeping strategy is based on Lyapunov's direct control method as applied to the standard set of Keplerian elements. This approach not only assures global and asymptotic stability of the control but also maintains valuable physical insight into the applied control vectors. Furthermore, the approach can account for system uncertainties and effectively avoids a computationally expensive solution of the two point boundary problem, which renders the concept particularly attractive for implementation in onboard processors.

A guidance law has been developed which strictly separates the relative from the absolute motion, thus avoiding the numerical integration of a target trajectory in the onboard processor. Moreover, upon using precise kinematic relative GPS solutions, a dynamical modeling or filtering is avoided which provides for an efficient implementation of the process on an onboard processor. A sample formation flying scenario has been created aiming at the autonomous transition of a Low Earth Orbit satellite formation from an initial along-track separation of 800 m to a target distance of 100 m. Assuming a low-thrust actuator which may be accommodated on a small satellite, a typical control accuracy of less than 5 m has been achieved which proves the applicability of autonomous formation flying techniques to formations of satellites as close as 50 m.

INTRODUCTION

As the objectives in earth observation, communication and navigation become increasingly complex, single satellite solutions may no longer be adequate. Instead, multi-satellite systems may be required which are implemented as formations, constellations, or swarms. The orbit control of near-term satellite formations, such as the presently operated US-German GRACE mission (Ries et al. 2002) or the planned American TechSat-21 project (Chien et al. 2001) may well be operated by a conventional ground control center. In contrast, missions planned for the next decade, like the ESA missions XEUS (Bleeker & Mendez 2002) and Darwin (Chapman & Gillett 2002), call for an onboard autonomous relative navigation control. Research and development in the field of autonomous relative navigation and control is thus an essential contribution to a forward-looking orientation in contemporary space flight technology.

[*] German Aerospace Center (DLR), German Space Operations Center, Münchener Str. 20, D-82230 Wessling, Germany. E-mail: eberhard.gill@dlr.de.

[†] NASA Goddard Space Flight Center, Greenbelt, Maryland 20771.

[‡] Center for Space Research, University of Texas at Austin, Austin, Texas 78712.

The described closed-loop system for formation acquisition and keeping using hardware-in-the-loop attempts to demonstrate the feasibility of autonomous formation control in a realistic environment. This comprises the use of real GPS signals with typical systematic errors such as ionosphere errors and GPS broadcast ephemeris errors, real GPS measurements with their associated statistical errors, as well as execution of the control task on a flight processor. Moreover, the demonstration accounts for the complex dynamics of satellite orbital motion which renders the demonstration as realistic as possible. In the next chapter, the architecture and the constituents of the closed-loop system are described, followed by an introduction to Lyapunov's stability analysis method and its specific adaptation to the formation flying problem. In the last chapter, the results of an autonomous formation flying demonstration are presented which prove the feasibility of the proposed concept. Finally, benefits and drawbacks associated with the herein adopted formulation of Lyapunov's stability method are discussed.

SIMULATION SYSTEM DESCRIPTION

System Architecture

The formation flight system (Fig. 1) has been established at the Formation Flying Test Bed (Leitner 2001) of NASA's Goddard Space Flight Center. It consists of a GPS signal generator, two GPS ORION receivers, a Power PC navigation processor, a remote control PC as well as a monitoring Laptop. Communications between the system components apply both serial and parallel interfaces as well as radio frequency links.

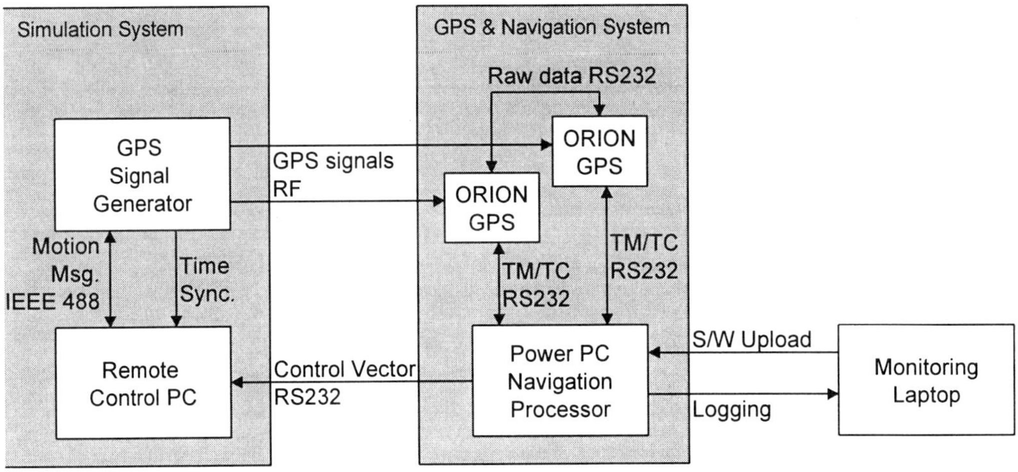

Fig. 1. Closed-loop autonomous formation control setup.

The system uses a Spirent STR4760 GPS signal simulator (GSS) capable of simulating L1 signals for 2 vehicles on up to 16 channels each. Upon running a closed-loop formation simulation with two satellites, the GSS is provided every 100 ms with two satellite motion messages from an external Remote Control PC, which comprise the current spacecraft position, velocity, acceleration, and jerk values.

Two GPS Orion receivers, described in more detail in a separate section below, are directly connected to the simulator R/F output of the GSS via co-axial cables and low noise amplifiers. This is considered an acceptable restriction and avoids the detrimental effects of a direct signal transmission inside the laboratory environment. In the described demonstrations, absolute GPS position and velocity fixes are used as well as their mutual relative state. The latter is based on the exchange of raw measurements via a dedicated serial data link (Montenbruck et al. 2002). Thus, no raw GPS measurements have to be processed on the flight computer. Therefore this study will focus on the autonomous formation control problem, considerably reducing the complexity of the onboard processor software.

Both GPS receivers were connected to two RS-232 ports of the Power PC navigation processor, which is described below in more detail. The navigation processor acts as flight computer on one of the two satellites and executes the formation flight guidance and control functions. As output, a control request message may be issued which specifies the time and velocity increment associated with a single thruster activation. This message is in turn transmitted via the third RS-232 port of the navigation processor for application within the remote control PC. The monitoring laptop allows a direct upload of the Real-time Formation Flight (RFF) application software to the processor via a dedicated parallel port of the processor board. Monitoring of the control process at run-time and logging on the monitoring laptop is conveniently supported making use of a dedicated serial interface.

Control requests issued from the navigation processor are received by the Remote Control PC. This PC numerically integrates the satellite state vectors and adds the velocity increments of the control requests, representing the thruster activities, to the orbital velocity of the satellites. The Remote Control PC also controls the GPS signal simulator through the provision of the current satellite states in specific motion messages at a rate of 10 Hz, thus closing the loop of the autonomous formation control system. To this end, an IEEE 488 interface is applied and time synchronization is established using a dedicated timer card.

GPS Receiver Hardware

The GPS Orion receiver, developed by DLR's German Space Operations Center, is based on the Mitel (now Zarlink) GP2000 chipset. It comprises a GP2015 R/F front end and DW9255 saw filter, a GP2021 correlator as well as an ARM60B 32-bit microprocessor (Montenbruck et al. 2001). It provides C/A code tracking on 12 channels at the L1 frequency and operates with an active antenna having a total gain of roughly 28 dB. The receiver consumes a total of 2.4 W when accounting for the main receiver board as well as the interface board. The latter provides line drivers for the two I/O ports as well as a rechargeable NiCd battery to maintain the real-time clock and non-volatile memory inside the receiver while disconnected from the main power supply.

The receiver provides navigation solutions, smoothed and unsmoothed C/A code pseudoranges and integrated L1 carrier phase measurements. The carrier phase measurements are obtained using a 3rd order (frequency lock loop (FLL) assisted) phase lock loop (PLL), which offers good signal acquisition performance as well as accurate tracking of the carrier phase under high dynamics. Upon connecting the two Orions, the receivers process single difference measurements to obtain their mutual relative state (Montenbruck et al. 2002). This differential processing scheme allows a high level of common error cancellation while minimizing the resulting noise by appropriate use of carrier phase measurements.

Flight Computer

The navigation processor features an industrial Power PC 823e processor operated at 48 MHz clock rate (without floating point support), which provides a performance of 66 MIPS. A total of 8 MB of DRAM memory are available as well as 8 MB of shadow mirror memory and 128 kB of ROM to save critical run-time parameters (Montenegro 2001). The DRAM is parity protected and duplicated which allows for error detection and correction. The application software can be directly uploaded to the processor via a parallel connection located at the processor board. Run-time monitoring and logging on a remote PC is conveniently supported making use of a dedicated RS-232 interface.

The navigation processor is running under the real-time operating system BOSS, which separates the kernel run-time system and a hardware dependant layer. BOSS is a preemptive multitasking operating system well suited for real-time and onboard applications. Processes are executed as separate threads, which are controlled by a central scheduler based on pre-assigned priorities and timers. In this way, short and high-priority activities (e.g. commanding) can be separated from computation intensive tasks with long duty cycles (e.g. orbit determination). BOSS is implemented in C++ and provides a C++ interface for the software application development (Leung et al. 2002). The Power PC board and the real-time operating system BOSS were developed for the German BIRD small satellite mission and have been successfully operating onboard BIRD since October 2001.

LYAPUNOV CONTROL THEORY

Theory Outline

Since the equations of satellite motion are non-linear, the relative motion of a satellite formation is governed by non-linear equations as well. Although linear control may in general be applied to non-linear systems and may even turn out to be highly efficient, non-linear control strategies can be beneficial in various ways. In general, they may improve the control robustness to parametric uncertainty, enable control for systems not controllable by linear approaches, and preserve and exploit physical insight.

Out of the many non-linear control theories, Lyapunov's direct method (second method) appears as a feasible approach to the problem of relative motion control. Here, the control drives the system dynamics along a direction opposite to the gradient of a specific Lyapunov function in such a way, that the current state approaches and finally reaches the terminal state conditions. Although Lyapunov's method is necessarily neither time-optimal nor fuel-optimal, it provides globally asymptotically stable solutions to linear and non-linear systems. As the computation of the feedback control requires only a moderate computational effort, it is especially suited for real-time and onboard applications. Furthermore, large matrices as in predictive control strategies (Rossi & Lovera 2002) are avoided since the involved matrix/vector dimensions are just of the size of the state of the dynamical system.

Lyapunov's theory analyzes the stability of dynamical systems described by n-dimensional first-order equations of the form

$$\dot{\boldsymbol{x}} = \boldsymbol{f}(\boldsymbol{x}) \qquad \boldsymbol{x}, \dot{\boldsymbol{x}}, \boldsymbol{f} \in \boldsymbol{R}^n \tag{1}$$

where dotted symbols denote derivatives with respect to time. If we select an m-dimensional feedback control vector $\boldsymbol{u}(x)$ which denotes the acceleration due to a thruster system, the closed-loop system becomes

$$\dot{\boldsymbol{x}} = \boldsymbol{f}(\boldsymbol{x}, \boldsymbol{u}(\boldsymbol{x})) \qquad \boldsymbol{u} \in \mathbf{R}^m . \tag{2}$$

Based on a particular guidance law, a reference (target) state $\boldsymbol{x}^*$ is given and a scalar Lyapunov function $V(\boldsymbol{x}-\boldsymbol{x}^*)$ may be introduced to analyze the system's stability. Similar to the discussion of dynamic systems in terms of the scalar energy potential, Lyapunov's method proves the state $\boldsymbol{x}(t) = \boldsymbol{x}^*$ to be globally asymptotically stable, if a function $V(\boldsymbol{x}-\boldsymbol{x}^*)$ is found, which is continuous in its first derivative and satisfies the following conditions (Ilgen 1993)

$$V(\boldsymbol{0}) = 0 \tag{3}$$

$$V\left(\boldsymbol{x}-\boldsymbol{x}^*\right) > 0 \quad \forall\left(\boldsymbol{x}-\boldsymbol{x}^*\right) \neq \boldsymbol{0} \tag{4}$$

$$\dot{V}\left(\boldsymbol{x}-\boldsymbol{x}^*\right) < 0 \quad \forall\left(\boldsymbol{x}-\boldsymbol{x}^*\right) \neq \boldsymbol{0} \tag{5}$$

$$\lim_{\|\boldsymbol{x}-\boldsymbol{x}^*\| \to \infty} V = \infty . \tag{6}$$

Applied to the problem of relative motion control, Lyapunov's method may in principle be based on the Cartesian state vector, as well as on Keplerian or non-singular elements. It turns out, however, that the region in state space that could be controlled using a Cartesian formulation is much smaller than using Kepler elements (Naasz et al. 2002). Thus, in the following the state vector $\boldsymbol{x} = (a,e,i,\Omega,\omega,M_0)^T$ is considered.

Although Lyapunov's direct method does not comprise a procedure to construct the function $V(\boldsymbol{x}-\boldsymbol{x}^*)$, a quadratic positive-definite function is often an effective starting point, so

$$V\left(\boldsymbol{x}-\boldsymbol{x}^*\right) = \frac{1}{2}\left(\boldsymbol{x}-\boldsymbol{x}^*\right)^T \boldsymbol{K}\left(\boldsymbol{x}^*\right)\left(\boldsymbol{x}-\boldsymbol{x}^*\right) \tag{7}$$

where $\boldsymbol{K}$ is a positive definite gain matrix. It is obvious that (7) satisfies the Lyapunov conditions (3), (4) and (6). If we consider, furthermore, that the state vector time derivative is independent on the system dynamics in the absence of any control (as is the case for the Keplerian element state), we may write

$$\dot{\boldsymbol{x}} = \boldsymbol{g}(\boldsymbol{x})\boldsymbol{u} \tag{8}$$

where $\boldsymbol{g}$ is a matrix function of the state. Applied to orbital dynamics, Gauss's formulation of the Lagrange Planetary Equations (LPE) takes the form of (8) with

$$\frac{\partial}{\partial t}\begin{bmatrix} a \\ e \\ i \\ \Omega \\ \omega \\ M_0 \end{bmatrix} = \begin{bmatrix} \frac{2a^2 e s_v}{h} & \frac{2a^2 p}{hr} & 0 \\ \frac{p s_v}{h} & \frac{(p+r)c_v + er}{h} & 0 \\ 0 & 0 & \frac{r c_{\omega+v}}{h} \\ 0 & 0 & \frac{r s_{\omega+v}}{h s_i} \\ -\frac{p c_v}{he} & \frac{(p+r)s_v}{he} & -\frac{r s_{\omega+v} c_i}{h s_i} \\ \frac{p c_v - 2re}{na^2 e} & -\frac{(p+r)s_v}{na^2 e} & 0 \end{bmatrix} \begin{bmatrix} u_H \\ u_L \\ u_C \end{bmatrix} \tag{9}$$

where the acceleration vector components (u_H, u_L, u_C) are given in Hill's frame (Height, Along-track, Cross-track). Here $p = a(1-e^2)$, h denotes the orbital angular momentum and s_α and c_α denote the sine and cosine of the argument α. Based on (8) the derivative of Lyapunov's function can be written as

$$\dot{V}(\boldsymbol{x}-\boldsymbol{x}^*) = \frac{\partial V}{\partial \boldsymbol{x}}\dot{\boldsymbol{x}} = \frac{\partial V}{\partial \boldsymbol{x}}\boldsymbol{g}(\boldsymbol{x})\boldsymbol{u}. \tag{10}$$

If we furthermore set the control vector to be

$$\boldsymbol{u} = -\boldsymbol{g}^T(\boldsymbol{x})\left(\frac{\partial V}{\partial \boldsymbol{x}}\right)^T \tag{11}$$

the time derivative of Lyapunov's function becomes

$$\dot{V}(\boldsymbol{x}-\boldsymbol{x}^*) = -\left(\frac{\partial V}{\partial \boldsymbol{x}}\right)\boldsymbol{g}(\boldsymbol{x})\,\boldsymbol{g}^T(\boldsymbol{x})\left(\frac{\partial V}{\partial \boldsymbol{x}}\right)^T \tag{12}$$

which satisfies Lyapunov's condition (5). Based on a diagonal gain matrix, the Lyapunov function as given in (7) and applied to the control of the Keplerian elements is thus

$$V = \frac{1}{2}\left[K_1\delta a^2 + K_2\delta e^2 + K_3\delta i^2 + K_4\delta\Omega^2 + K_5\delta\omega^2 + K_6\delta M_0^2\right]. \tag{13}$$

This Lyapunov function satisfies the Lyapunov stability criteria for any system of the form as given in (8). Substituting the LPE into (11) leads to

$$\boldsymbol{u} = -\boldsymbol{g}^T(\boldsymbol{x})\left(\frac{\partial V}{\partial \boldsymbol{x}}\right)^T = -\begin{bmatrix} \frac{2a^2 e s_\nu}{h} & \frac{2a^2 p}{hr} & 0 \\ \frac{p s_\nu}{h} & \frac{(p+r)c_\nu + er}{h} & 0 \\ 0 & 0 & \frac{r c_{\omega+\nu}}{h} \\ 0 & 0 & \frac{r s_{\omega+\nu}}{h s_i} \\ -\frac{p c_\nu}{he} & \frac{(p+r)s_\nu}{he} & -\frac{r s_{\omega+\nu} c_i}{h s_i} \\ \frac{p c_\nu - 2re}{na^2 e} & -\frac{(p+r)s_\nu}{na^2 e} & 0 \end{bmatrix}^T \begin{bmatrix} K_1\delta a \\ K_2\delta e \\ K_3\delta i \\ K_4\delta\Omega \\ K_5\delta\omega \\ K_6\delta M_0 \end{bmatrix} \tag{14}$$

which exhibits the characteristics of Lyapunov's method to maintain a transparent relation between Keplerian element errors and the resulting control vector. However, it turns out that a proper selection of the six free control gain parameters K_i (i=1...6) is not a trivial task and still has to be solved in the sequel. Furthermore, to fully exploit the laws of orbital motion it turns out that mean anomaly errors are most efficiently removed by a change of the semi-major axis, which calls for a specific mean motion control. Both problems are addressed in the subsequent sections.

Gain Selection

While the derivation of the control law is relatively straightforward, the main difficulty in an appropriate implementation of the algorithm is the selection of the gain vector and the specification of the control on/off logic. This requires on one hand an educated guess of appropriate gain values and on the other hand extensive testing of the chosen values, similar to a Kalman filter tuning within orbit determination. One way is to relate the LPE in (9) with the achievable maximum changes in the orbital elements and the available control acceleration (Naasz 2002). This approach assures, that

1. Control is enforced only in such parts that effectively modify the orbital element
2. Control requests for different orbital elements are reasonably scaled w.r.t. each other
3. Control is enforced only when the orbital element error is above a certain threshold.

Although this approach is basically empirical, it provides a reasonable approach to arrive at suitable values for the gain vector. As a result, the following gain matrix has been derived to lowest order in eccentricity (Naasz 2002)

$$\boldsymbol{K}' = \frac{(n^* a^*)^2}{4\Delta t} \operatorname{diag}\left(\frac{1}{a^{*2}}, 1, 4, 4\sin^2 i^*, e^{*2}, e^{*2}\right) \tag{15}$$

where Δt is the thrust duration and the star refers to the reference orbit. Basically, the gains in (15) are not varying with time. However, use of time varying gains can be made to encourage certain orbit element corrections to occur during particular phases of the orbit. From the LPE it is concluded, that the response to thrusts for the argument of perigee is shifted by 90° as compared to changes in the semi-major axis and eccentricity, which are encouraged at apogee and perigee. Thus the gains for the in-plane elements may be varied depending on the true anomaly ν^* of the reference orbit according to

$$\begin{aligned} K_1 &= K_1' \left|\cos(\nu^*)\right| \\ K_2 &= K_2' \left|\cos(\nu^*)\right| \\ K_5 &= K_5' \left|\sin(\nu^*)\right|. \end{aligned} \tag{16}$$

To prevent thrusting when the orbital element deviations are too small to require activation of the thruster, a velocity increment of

$$\Delta \boldsymbol{v} = \frac{T}{m} \frac{\boldsymbol{u}}{|\boldsymbol{u}|} \Delta t \tag{17}$$

is applied only when $|\boldsymbol{u}|$ exceeds the threshold T/m, where T is the available thrust level from the actuator and m is the spacecraft mass. No specific spacecraft thrust model is assumed, where a thruster on/off logic would apply for each thrust direction individually.

Mean Motion Control

Lyapunov's method, as given in (14), provides a working control law for satellite formations. However, a modification for enhanced efficiency may be obtained, when we consider that mean anomaly errors δM could be corrected for by a change of the semi-major axis, affecting the mean motion. This approach, also termed as mean motion control, fully exploits the laws of orbital motion to reduce in-plane errors. Furthermore, it reduces the control problem from a six-dimensional control state by one dimension at the expense of complicating the semi-major axis control. The dynamics of relative mean anomaly are, using $dM/dt = n$, given as

$$\frac{d}{dt}(\delta M) = n - n^* = \sqrt{\frac{GM}{a^3}} - \sqrt{\frac{GM}{a^{*3}}} \tag{18}$$

where GM denotes the Earth's gravitational coefficient. A suitable approach to minimize the mean anomaly error is given by

$$\frac{d}{dt}(\delta M) = -K\delta M \tag{19}$$

with some positive gain K. Using (18) and (19) we can solve for the semi-major axis a, which removes the mean anomaly error by a suitable chosen modified target semi-major axis a^{**} according to

$$a^{**} = \left(a^{*-3/2} - \frac{K\delta M}{\sqrt{GM}} \right)^{-2/3} \tag{20}$$

which reduces to a^* for vanishing δM. Since the required modification of the semi-major axis due to mean anomaly errors is small as compared to the semi-major axis itself, a first-order Taylor series expansion leads to

$$a^{**} \approx a^* + \frac{2}{3}\frac{K\delta M a^*}{n^*}. \tag{21}$$

Hence, if we replace $\delta a = a - a^*$ by $\delta a^* = a - a^{**}$ in (14), we obtain the along-track control law due to semi-major axis and mean anomaly errors as

$$u_L \approx \frac{K\delta M\, a^*}{3\Delta t} - \frac{\delta a\, n^*}{2\Delta t}. \tag{22}$$

Thus, the along-track control acceleration may conveniently be written as the sum of a component due to mean anomaly errors (first term) and due to semi-major axis errors (second term). Considering a scenario just with an initial along-track position difference of δL_0, the total initial along-track control is due to the mean anomaly error. As consequence, a semi-major axis offset δa is built up until the two contributions balance. In the sequel, the semi-major axis error has to be removed again to stop the drift and arrive at the targeted along-track position.

An appropriate selection of the mean motion gain K is particularly simple, if we consider impulsive velocity increments from a thruster system, not constrained in magnitude. To this end, an along-track offset δL is considered which may be removed by a semi-major axis offset of δa accumulated over a time t_c, according to

$$\delta L = \delta Ma^* = 3\pi\delta a \frac{t_c}{T^*} \tag{23}$$

with the orbital period T^*. When (23) is inserted into (22) an expression for K is obtained

$$K = \frac{n^* T^*}{2\pi \, t_c} = \frac{1}{t_c}. \tag{24}$$

Thus, the mean motion gain for unconstrained thrusters is simply given by the inverse of the characteristic time t_c, over which the mean motion control should accumulate along-track offsets. The shorter the time for along-track offset compensation should be, the higher we have to choose the mean motion gain and the more aggressive is the mean motion control. For a Low-Earth Orbit satellite, a mean motion gain of $K = 1.4 \cdot 10^{-4}$/s would remove along-track errors in a period of about 120 min.

AUTONOMOUS FORMATION FLYING DEMONSTRATION

Simulation Setup

The formation consists of two spacecraft, flying in a polar Low Earth Orbit of 87° inclination and 450 km altitude, which involves a strongly varying GPS satellite visibility and notable signal dynamics. A close formation with an initial along-track separation of 800 m is considered.

A pulsed plasma thruster (PPT) system has been selected which can be accommodated on a small satellite (Cassady et al. 2000). In a doubled configuration the thrusters provide an impulse-bit of 0.112 mNs and fire at a rate of 1 Hz. In the framework of the simulations, an effective thrust level of 0.1 mN and a spacecraft mass of 20 kg is assumed. Thrust accelerations are accumulated over 10 s intervals to arrive at velocity increments to be applied instantaneously to the current satellite velocity. Although specific satellite designs may significantly constrain available thrust directions, no such constraints were considered in the present setup.

As part of the simulation system the satellite orbits are propagated by a precision software propagator running on the Remote Control PC. It applies the numerical integration algorithm DE of variable order and stepsize (Shampine & Gordon 1975) to integrate the equations of motion in an Earth-Centered Earth-fixed (ECEF) frame. As the DE method provides interpolation for dense output it is ideally suited to support the required 100 ms period of motion messages to the GPS signal simulator. Furthermore, use of the ECEF system for integration avoids frequent coordinate transformations between an inertial system and the ECEF system, which is the system to be applied for motion messages.

The adopted force model accounts for the Earth's complex gravity field by using a 20x20 subset of the Joint Gravity Model 3 (JGM-3). Furthermore, it accounts for perturbations due to atmospheric drag, solar radiation pressure, as well as sun and moon third body perturbations. In addition to the natural perturbations, orbit maneuvers are accounted for by adding velocity increments in the ECEF frame to the ECEF velocity components of the respective state vector. A summary of the simulation setup is given in Table 1.

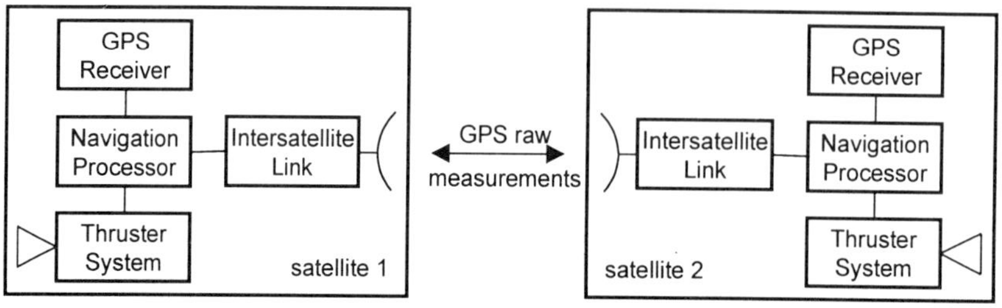

Fig. 2. Space segment of a formation with two thrusting spacecraft.

Table 1. CONFIGURATION OF THE FORMATION FLYING SYSTEM.

Subsystem	Description
Remote Control PC	
Earth gravity field	20x20 subset of JGM-3 model
Atmospheric drag	modified Harris Priester model
Solar radiation pressure	Spherical model
Sun third body force	Analytic sun ephemeris
Moon third body force	Analytic moon ephemeris
Numerical integration algorithm	DE variable order and stepsize
Reference system for integration	WGS-84
GPS Signal Simulator	
Ionospheric errors	Based on constant TEC of 20 TECU
GPS broadcast ephemeris errors	Maximum of 12 m per component
Navigation Processor	
Applied GPS measurements	Absolute & relative navigation solutions
Number of thrusting spacecraft	2
Control cycle period	10 s

To demonstrate the autonomous control concept in the presence of realistic GPS signals, systematic ionospheric errors and GPS broadcast ephemeris errors were modeled in the GPS signal simulator and applied for the generation of GPS signals. Ionospheric errors were based on a total electron content of $2{\cdot}10^{-17}$ electrons/m^2 (= 20 TECU). Intentional GPS broadcast ephemeris errors have been added with maximum values of 12 m per component, as derived from observed offsets between GPS broadcast ephemerides and precise IGS ephemerides for the simulation date.

A symmetric space segment is assumed with two thrusting spacecraft as indicated in Fig. 2. The intersatellite link has been replaced in the hardware setup by the direct connection of the two GPS receivers through a dedicated serial link. While a generic concept would perform the relative navigation solution within the navigation processors, the Orion feature to provide the relative navigation solutions has been utilized instead. In order to emulate two independent navigation processors on a single flight computer, two separate control threads have been implemented on the Power PC. It is noted, that no control vectors have to be exchanged between the two spacecraft according to the guidance concept described in the following section.

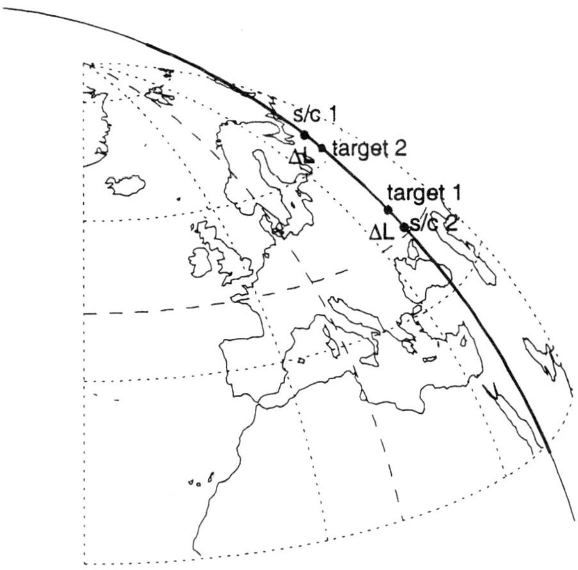

Fig. 3. Guidance strategy for formations separated in along-track direction by ΔL. The target states for spacecraft 1 and 2 (black) are denoted by 1_t and 2_t (gray).

Guidance Concept

The adopted formation control algorithm requires the absolute current and target state of the formation as input. While the absolute current state vector of the spacecraft is readily available from its GPS position and velocity fixes, the derivation of an appropriate target state is part of the guidance problem of formation flying. Although this problem might be solved by numerically integrating a reference trajectory, this approach is inefficient since it requires a high computational effort on the navigation processor. Furthermore it is not flexible, since the guidance is defined with respect to a fixed trajectory.

In general, guidance of a formation may well be separated into guidance with respect to the absolute orbit of one of the spacecraft and guidance with respect to the relative formation geometry. Scientific or technological objectives of a formation flying mission are in fact in many cases only related to the relative formation geometry.

Thus, for a formation separated in along-track direction with spacecraft 1 preceding spacecraft 2 and a positive target separation of ΔL, the target elements for the first spacecraft are the current elements of the second spacecraft, except for the mean anomaly, which is

$$M_1^{\text{target}} = M_2 + \frac{\Delta L}{a_2} \tag{25}$$

where the indices refer to the two spacecraft. It is important to note that a control of the osculating motion of the formation in the presence of a realistic force model is effectively avoided by replacing the osculating current and target states in the control by a suitable set of mean elements, e.g. using the algorithm of Brouwer (1957) modified by Lyddane (1963). This guidance approach based on the relative formation geometry has the following characteristics:

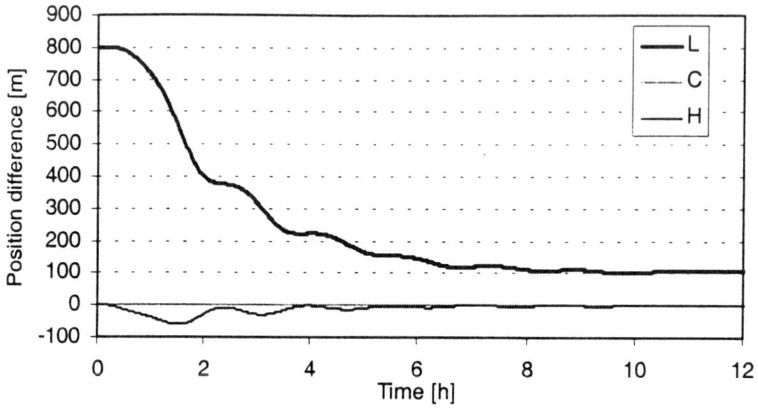

Fig. 4. Demonstration of closed-loop autonomous formation control. The initial along-track separation is decreased to the target separation of 100 m.

1. Guidance depends on absolute current states and relative target formation geometry
2. Dynamic guidance law separates the absolute and relative guidance problem
3. Numerical integration of the target trajectory is avoided
4. Measurement-driven absolute state errors do not affect the control to first order.

In the conducted demonstration, this approach has been extended to allow both spacecraft to be active and control the formation in parallel as indicated in Fig. 3. Although not fuel optimal, this concept allows for a completely autonomous and dynamic acquisition of the final formation geometry. Even if one spacecraft's thruster system would fail, no ground intervention would be required to achieve the target relative formation geometry in this case.

Results from the Hardware-in-the-Loop Demonstration

A hardware-in-the-loop (HWIL) demonstration of closed-loop autonomous formation control has been performed based on the above described setup. In the course of the demonstration which covers a period of 12 hours, the initial along-track separation of 800 m is effectively removed by changing the relative mean semi-major axis of the two orbits, as indicated by the height component in Fig. 4.

Following a coarse acquisition of the target formation within 8 hours into the demonstration, the formation keeping starts at about 9.5 hours after start of the simulation. The performance of the control algorithm for formation keeping may be evaluated by the maximum, mean and standard deviation figures of the current spacecraft state with respect to the target state. A summary of the formation keeping results in the orbital frame components (L, C, H) and relative position (R) covering a period of 2.5 hours is given in Table 2.

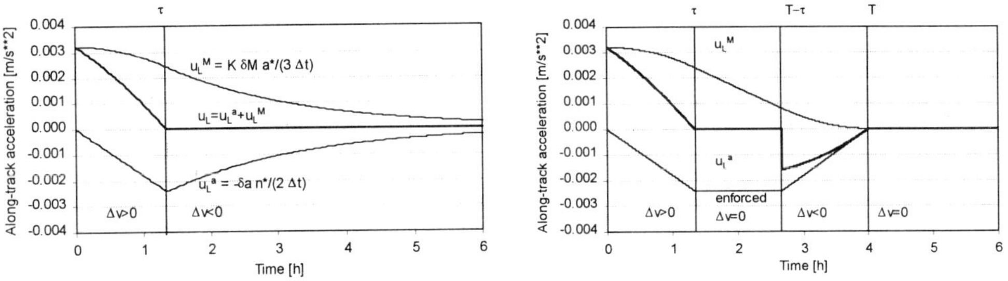

Fig. 5. Along-track control requests in the formation acquisition phase. Gain scheduling is absent in the left figure and present in the right figure.

Table 2. CHARACTERISTICS OF AUTONOMOUS FORMATION KEEPING CONTROL.

	L [m]	*C* [m]	*H* [m]	*R* [m]
Minimum	100.3	-0.1	-2.7	100.3
Maximum	111.1	0.1	1.5	111.1
Mean	106.2	0.0	-0.3	106.2
RMS	2.8	0.1	1.1	2.8

Thus, the targeted along-track separation of 100 m was kept with a mean position error of about 6 m and a standard deviation of 3 m.

Performance Analysis

The above results demonstrate that the closed-loop system and the applied control algorithm are working well for autonomous formation control. In the following section, the performance of the system is addressed in terms of the achieved GPS measurement accuracy, the computational load on the navigation processor, and the fuel consumption.

GPS Performance

The Orion absolute navigation solutions derived from pseudorange measurements exhibit a typical data noise of 1 m and 0.5 m/s in the absence of systematic errors. With broadcast ephemeris errors, the resulting position errors are directly proportional to the assumed ephemeris errors and amount to 5–10 m for an S/A free scenario. The high level of common error cancellation reduces the errors of the relative navigation solutions to $\sigma_p^r = 0.6$ m and $\sigma_v^r = 0.005$ m/s even in the presence of systematic errors.

Computational Load on the Flight Computer

The computational load on the navigation processor is governed by the computation of the mean elements and the control vectors. The computation of mean elements based on the Brouwer/Lyddane algorithm requires about 50 ms per element set, comparable to the computation of mean elements using the SGP4 algorithm. Thus, the computational load for mean element computation of two satellites is 100 ms. The control algorithm including the computation of the control gains takes 50 ms, hence a total of 150 ms is required per execution of the control thread, which is evoked every 10 s. As the absolute and relative GPS navigation solutions are applied, no dynamical smoothing or numerical trajectory integration is required. Thus the implemented concept proves to be very efficient, requiring only about 1.5% of the navigation processor resources.

Fuel Consumption during Formation Acquisition

An almost continuous thruster firing is observed in the formation acquisition phase. This is caused by a lacking drift phase, present in the classical burn-coast-burn scenario, and a thruster on/off logic which is very susceptible to GPS measurement noise. This is illustrated in the left diagram of Fig. 5. Here, the along-track acceleration and its components due to mean anomaly and semi-major axis errors are displayed according to (22). Continuous thrust in the first phase is requested almost entirely due to the mean anomaly error. As a consequence, an intentional semi-major axis error is built up until the two contributions almost cancel each other. Subsequently, a delicate balancing of the two decreasing components is observed which nevertheless results in a residual net acceleration, sufficient to pass the thruster on/off logic. Thus, the drift is reduced almost immediately after it has been built up and reduces gradually until the formation keeping phase is entered.

Instead of a permanent thrusting, a dedicated drift phase can be enforced by an appropriate gain scheduling. To this end, the gain is set to zero in the drift interval $[\tau, T - \tau]$ where T is the acquisition period and τ is thruster activity period according to

$$\tau = \frac{T}{2} - \sqrt{\frac{T^2}{4} - \frac{\Delta L_0}{3u_{s/c}}} \qquad (26)$$

where $u_{s/c}$ is the thrust acceleration and ΔL_0 is the initial along-track offset from the target. As a consequence, the semi-major axis stays constant during the coast phase as seen in the right diagram of Fig. 5. At the end of the coast phase, the semi-major axis offset causes a strong request for drift stop, which dominates the measurement noise and, thus, renders the formation acquisition phase essentially independent from measurement noise. Instead of choosing an arbitrary acquisition period T, the elapsed time between the center of the two burns, $T - \tau$, may be chosen as an integer number of orbits, thus avoiding relative eccentricity errors at the end of the acquisition phase. In the above example with $T = 4$ h, the required velocity increment for the acquisition phase with gain scheduling is 0.048 m/s, matching that of the conventional burn-coast-burn strategy, as opposed to 0.144 m/s without gain scheduling.

Fuel Consumption during Formation Keeping

The GPS measurement noise of the relative navigation solution dominates in particular the formation keeping phase. This is obvious when we compare the acceleration threshold T/m, being $5 \cdot 10^{-6}$ m/s^2 for the considered case, with the quotient of the relative velocity noise and the thrust accumulation interval $\sigma_v^r/\Delta t$, which is $5 \cdot 10^{-4}$ m/s^2. As a consequence, a permanent thrusting is observed which renders the control ineffective. Two approaches may basically solve the problem: a decrease of the data noise and an increase of the thrust accumulation interval. Since the data noise may, additionally, be reduced by an averaging algorithm over the interval Δt, the noise scales with $\Delta t^{-1/2}$. In order to improve the control noise of acceleration by a factor of 100, Δt should be changed by a factor of $100^{2/3} \approx 22$ from 10 s to about 3.5 min. Verification of noise reduction techniques are therefore among the next steps to be performed.

CONCLUSIONS

A hardware-in-the-loop demonstration of real-time autonomously controlled closed-loop formation flying has been performed. To that end, two interconnected GPS receivers delivered their absolute and relative navigation solutions to a flight computer, which derived the necessary control vectors for the autonomous formation acquisition and keeping. The guidance law has been formulated completely in relative navigation terms, thus avoiding any numerical integration of the reference trajectory on the flight computer and rigorously separating absolute and relative control of the formation. The control vectors were computed based on Lyapunov's direct method modified for mean motion control to fully exploit the laws of orbital mechanics for along-track separated satellite formations.

A sample scenario was considered consisting of two micro-satellites in a near-polar Low Earth Orbit, separated initially in along-track direction by 800 m. Making use of a pulsed plasma thruster system on each satellite, the autonomous acquisition and subsequent keeping of the formation with two active spacecraft controlling dynamically their relative geometry has been demonstrated. The formation acquisition phase ended at about 8 h into the simulation, when the satellites achieved their targeted along-track separation of 100 m. Making use of a realistic force model and accounting for typical systematic ionospheric and broadcast ephemeris errors, a formation keeping accuracy of about 3 m (3D rms) has been obtained. According to a generic requirement, the control accuracy of a formation should be at least one tenth of the minimum spacecraft separation. Thus, the demonstration proved the feasibility of the autonomous control of a satellite formation down to a minimum separation of at least 50 m. The achieved results are considered of relevance in the field of autonomous navigation and enhance potential mission scenarios for upcoming formation flying missions.

ACKNOWLEDGEMENT

The described system has been established at the Formation Flying Test Bed at NASA's Goddard Space Flight Center (GSFC). Work was carried out from June to September 2002 in the framework of a DLR sabbatical. T. Stengle's support and organization of the sabbatical at GSFC is greatly acknowledged. R. Burns is thanked for helpful discussions on the subject and L. Jackson for his support to establish the system interfaces.

REFERENCES

Bleeker J., Mendez M.; *The XEUS Mission*; Proc. Symposium *New Visions of the X-ray Universe in the XMM-Newton and Chandra Era* 2, ESTEC, The Netherlands, ESA SP-488 (2002).

Brouwer D.; *Solution of the Problem of Artificial Satellite Theory Without Drag*; The Astronautical Journal **64**, 378-397 (1957).

Cassady R.J., Hoskins A.W., Campbell M., Rayburn C.; *A Micro-Pulsed Plasma Thruster for the Dawgstar Spacecraft*; Proceedings of the 2000 IEEE Aerospace Conference, Big Sky, Montana (2000).

Chapman P., Gillett L.; *Formation Control for Darwin Demonstration in the SMART-2 mission;* International Symposium Formation Flying Mission and Technologies, Toulouse, France (2002).

Chien S., Sherwood R., Burl M., Knight R., Rabideau G., Engelhardt B., Davies A., Zetocha P., Wainwright R., Klupar P., Cappelaere P., Surka D., Williams B., Greeley R., Baker V.; *The Techsat-21 Autonomous Sciencecraft Constellation*; ESA Workshop on On-board Autonomy, WPP-191, Noordwijk, The Netherlands (2001).

Ilgen M.R.; *Low Thrust OTV Guidance Using Lyapunov Optimal Feedback Control Techniques*; AAS 93-680, in Astrodynamics 1993, A.K. Misra, et al., eds., Advances in the Astronautical Sciences, Vol. 85, pp. 1527-1545 (1994); (paper presented at the AAS/AIAA Astrodynamics Specialist Conference, Victoria, British Columbia, Canada, Aug. 16-19, 1993).

Leitner J.; *A Hardware-in-the-loop Testbed for Spacecraft Formation Flying Applications*; 2.0908, IEEE Aerospace Conference, Big Sky, MT (2001).

Leung S., Gill E., Montenbruck O., Montenegro S.; *A Navigation Processor for Flexible Real-Time Formation Flying Applications*; International Symposium Formation Flying Mission & Technologies, 29-31 October 2002, Toulouse, France (2002).

Lyddane R.H.; *Small Eccentricities or Inclinations in the Brouwer Theory of the Artificial Satellite*; The Astronomical Journal **68**, 555-558 (1963).

Montenbruck O., Markgraf M., Leung S., Gill E.; *A GPS Receiver for Space Applications*; B1-Ax; ION GPS 2001 Conference, Salt Lake City, 12-14 Sept. 2001 (2001).

Montenbruck O., Ebinuma T., Lightsey E.G., Leung S.; *A Real-Time Kinematic GPS Sensor for Spacecraft Relative Navigation*; Aerospace Science and Technology **6**, 435-449 (2002).

Montenegro S.; *Spacecraft Bus Computer*; BIRD Spacecraft Description **3**, TN-BIRD-5800-GMD/073 DLR (2001).

Naasz B.J., *Classical Element Feedback Control for Spacecraft Orbital Maneuvers*, Virginia Polytechnic Institute and State University, Master Thesis (2002).

Naasz B.J., Karlgaard C.D., Hall C.D.; *Application of Several Control Techniques for the Ionospheric Observation Nanosatellite Formation*, AAS 02-188, in Spaceflight Mechanics 2002, K.T. Alfriend, et al., eds., Advances in the Astronautical Sciences, Vol. 112, pp. 1063-1079 (2002); (paper presented at the AAS/AIAA Space Flight Mechanics Meeting, San Antonio, Texas, Jan. 27-30, 2002).

Ries J.C., Tapley B.D., Bettadpur S.; *The GRACE Mission: Status and Plans*; DDA 33rd Meeting, Mt. Hood, OR, (2002).

Rossi M., Lovera M.; *A Predictive Approach to Formation Keeping for Constallations of Small Spacecraft in Elliptical Orbits*; 5th International ESA Conference on Guidance Navigation and Control Systems, October 22-25, Frascati Italy (2002).

Shampine L.F., Gordon M. K.; *Computer Solution of Ordinary Differential Equations*; Freeman and Comp., San Francisco (1975).

AAS 03-041

USE OF VARIABLE SPEED CONTROL MOMENT GYROS FOR A FINE POINTING DUAL LINE-OF-SIGHT SPACECRAFT

Marcello Romano* and Brij N. Agrawal*

This article introduces the researches on course at the Spacecraft Research and Design Center on the dynamics and control of a dual line-of-sight spacecraft for relay of optical signals. This proposed spacecraft, called Bifocal Relay Mirror, consists of two large telescopes, that are optically and mechanically coupled. A new analytical model of the dynamics of the Bifocal Relay Mirror spacecraft with four Variable Speed single gimbal Control Moment Gyros in a pyramid configuration has been developed. Moreover, a new complex Three Axes Simulator test-bed, with the same set of actuators as the analytical model, has been designed and is now in the development/integration phase.

INTRODUCTION

The mission task of the Bifocal Relay Mirror spacecraft is to redirect a laser beam from ground-based sources to distant points on the earth or space. Possible applications would be for example laser communication, environmental polluting or chemical agents detection.

The Main challenges of the spacecraft control design are double line-of-sight Pointing and Tracking capability and the multibody configuration.

In the recent past we have studied the dynamics and control of the Bifocal Relay Mirror spacecraft, considering the use of Reaction Wheels as actuators. Both numerical simulations and on the ground experiments on a Three Axis Simulator have been carried out, see ref. [1] and [2].

We are currently focusing on studying the possible use of Variable Speed Control Moment Gyros (often called also Gimbaled Momentum Wheels) instead of Reaction Wheels for the Bifocal Relay Mirror. The main advantages would be an increased slewing agility, especially useful during the retargeting phase, and an increased efficiency. Moreover, the wheel speed variation can be used to avoid the singularity configurations problem typical of the Control Moment Gyros.

A new analytical model of the dynamics of the Bifocal Relay Mirror spacecraft has been developed. Moreover a new complex Three Axes Simulator test-bed, with four Variable Speed Control Moment Gyros in a pyramid configuration, has been designed and is now in the development/integration phase. It is planned to be soon operational at the Joint Optical Relay Spacecraft Laboratory of Air Force Research Lab & Naval Postgraduate School, located at the Spacecraft Research and Design Center.

---

* Spacecraft Research and Design Center, Naval Postgraduate School, Monterey, California 93943.

ANALYTICAL MODEL & NUMERICAL SIMULATIONS

A complete analytical model of the dynamics of the Bifocal Relay Mirror spacecraft is being developed. In the model, the BRM spacecraft is considered a rigid multi-body dynamics system, consisting of:

- Two main bodies, transmitter telescope and receiver telescope, mechanically coupled through a rotational joint (7 dof)

- Four single gimbal Variable Speed Control Moment Gyros mounted in a pyramid configuration. The model takes in account also the dynamics of the gimbal.

For the derivation of the equations of motions the high-level block diagrams of the system, reported in fig. 1, are considered. Moreover, an approach is taken, that is similar to the ones used in ref. [3] and [4].

Numerical simulations of the attitude estimation and control are going to be carried out during a sample relay mission (see fig. 2), with laser source in Monterey (CA) and target in Albuquerque (NM). During that sample mission the attitude and telescope joint angle is commanded to slew along a reference maneuver, which guarantees, in the ideal case of no disturbances, the laser connectivity between source and target. The reference attitude and joint motion is computed by a guidance algorithms, which has been coded by considering the constraints given by the geometry of the mission, the orbital dynamics of the spacecraft, the location of the source and target and the rotation of the earth (see ref.[1] for the details).

We plan to study and propose solutions for the following main issues:

1. Attitude and pointing control law for the optical relay spacecraft with VS Control Moment Gyros.

2. Steering law: in particular a comparison between the usual local optimization of the steering law with a global optimization along the reference attitude motion will be carried out

3. Singularity avoidance.

The solutions proposed will be first tested by numerical simulations, integrating the equation of motions in matlab/simulink, then by experimental tests on the ground.

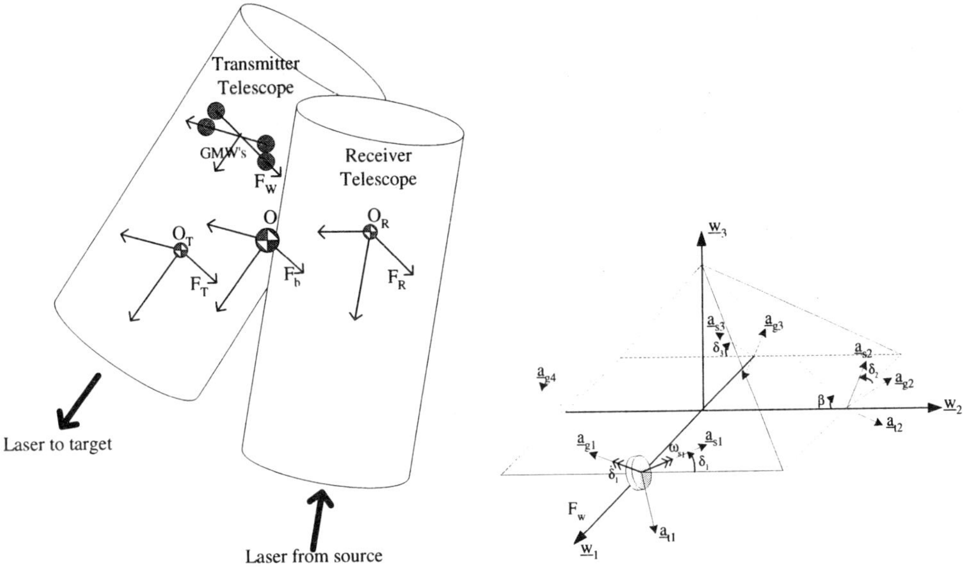

Overall Optical Relay Spacecraft

Detail of the Control Moment Gyros

FIGURE 1. High-level block diagram of the system model used for the simulations.

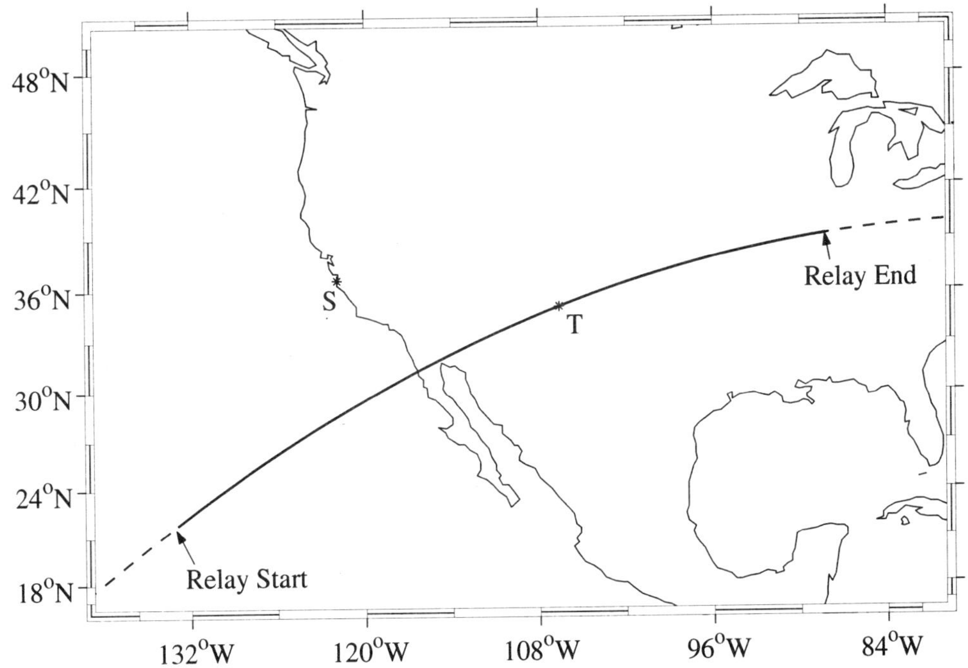

FIGURE 2. Sample relay mission of the Bifocal Relay Mirror spacecraft, used in the simulations.

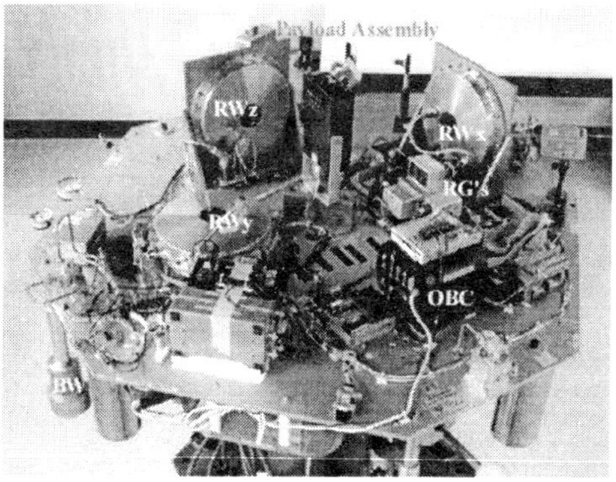

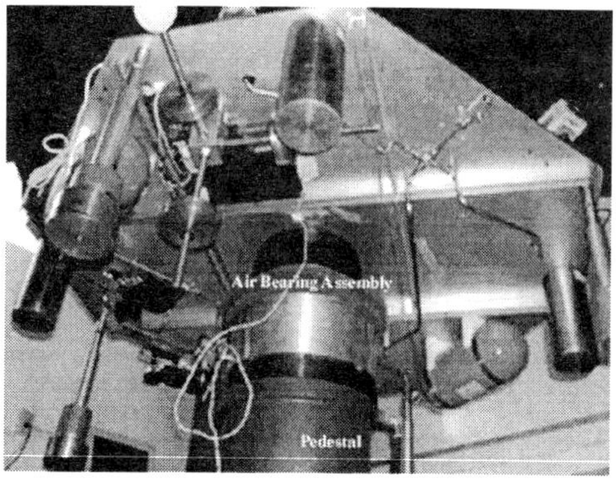

Top view Bottom view

FIGURE 3. The first Three Axes Spacecraft Simulator at the NPS-AFRL Optical Relay Spacecraft Laboratory of the Spacecraft Research and Design Center.

EXPERIMENTAL TESTS ON THE GROUND

The basic concept exploited to simulate on the ground the attitude motion of a spacecraft in microgravity conditions, is to free float a hardware spacecraft model over a hemispherical air bearing. Indeed, in principle, a torque-free environment is achieved if the center of mass of the spacecraft model coincides with the center of rotation of the air bearing.

THE FIRST THREE AXES SIMULATOR TEST-BED

This Three Axes Simulator test-bed, located at Spacecraft Research and Design Center, is actuated by Reaction Wheels and represents a simplified version of the Bifocal Relay Mirror spacecraft and it has been used to preliminary validated possible attitude stabilization and laser jitter suppression solutions (see ref. [2]).

The core of the TASS spacecraft model consists of a 1.3*m* wide octagonal metallic structure, which is mounted on top of a 1.4*m* high pedestal, trough a hemispherical air bearing interface. The main bus elements of the TASS, used for the presented research, are listed here below and shown in figures 3:

- Three orthogonally mounted reaction wheels, with max angular momentum of 20.3*Nms @2500 rpm* and max torque of 0.16*Nm*, manufactured by Ball Aerospace (RW in figure 3).
- Three orthogonally mounted mechanical rate gyroscopes, by Humprey (RG) in figure 3).
- Two analog inclinometers, by Seika, used as attitude sensors around the pitch (x) and roll (z) axes.
- One infrared sensor used for the attitude around the yaw axis (y).

- One on board embedded computer and input/output cards (OBC in figure 3). Commands are sent from and telemetry data are retrieved to a desktop computer via wireless Ethernet connection.

- Balancing and ballast weights for the precise positioning of the center of mass (BW in figure 3).

THE NEW THREE AXES SIMULATOR TEST-BED

A New Three Axes Simulator test-bed with four Variable Speed Control Moment Gyros in a pyramid configuration has been completely designed and is currently in the development and integration phase, by two contractors (Automated Control Environment and Guidance Dynamics Corporation).

The experience accumulated through the use of the previous Three Axes Simulator has been exploited in the design of the new simulator. In particular, the following improved features have been included in the new simulator:

1. Fiber optics rate gyroscopes
2. Automatic auto-balancing system
3. Laser based attitude sensors
4. A standard patterned optical equipment deck

All the bus elements of the test-bed have been located on the bottom of the platform. The top of the platform will be hosting a new optical payload. Each Variable Speed Control Moment Gyro will provide a Maximum Torque of 12*Nm* (1:*1Nm* in reaction wheels mode).

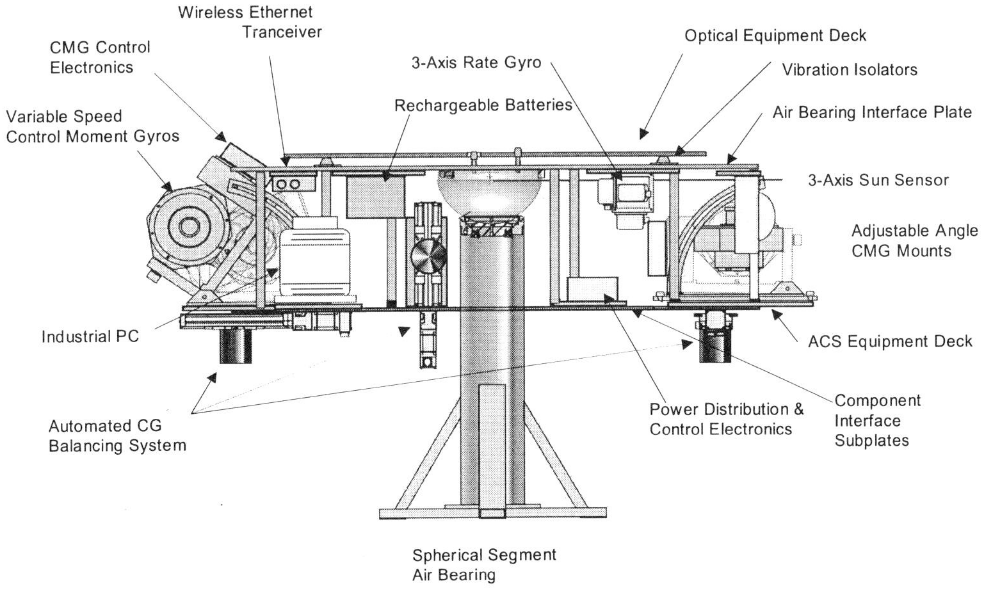

FIGURE 4. The final design of the New Three Axes Simulator, front view (Picture courtesy of E. Rasmussen, GDC)

The new optical payload is being designed and developed at the Air Force Research Laboratory of Albuquerque (NM). The payload will consist of two telescopes optically and mechanically coupled through a rotational joint. The new test bed in its final configuration will then allow the simulation on the ground of the laser relay capability of the Bifocal Relay Mirror spacecraft. In particular it will be a valuable tool, in addition to the numerical simulations, for the preliminary validation of the concepts currently under study for the attitude and pointing control of the Bifocal Relay Mirror spacecraft.

The first Three Axes Simulator will continue to be operational in our laboratory and may be used, in the future, for integrated experiments with the new Simulator test-bed.

REFERENCES

1. M. Romano and B. Agrawal, "Acquisition, tracking and pointing control of the bifocal relay mirror spacecraft," in *Proceedings of the 53 rd International Astronautical Congress*, (Houston, TX), October 2002.

2. M. Romano and B. Agrawal, "Tracking and pointing of target by a bifocal relay mirror spacecraft, using attitude control and fast steering mirrors tilting," in *Proceedings of the AIAA Guidance, Navigation and Control Conference*, (Monterey, CA), August 2002.

3. K. Ford and C. Hall, "Flexible spacecraft reorientations using gimbaled momentum wheels," *The Journal of the Astronautical sciences*, vol. 49, pp. 421–441, July-September 2001.

4. H. Schaub, S. Vadali, and J. Junkins, "Feedback control law for variable speed control moment gyros," AAS 98-140, in AAS *Advances in the Astronautical Sciences: Spaceflight Mechanics 1998*, Vol. 99, pp. 581–600, 1998.

AAS 03-042

EVALUATION OF ATTITUDE CONTROL SYSTEM PERFORMANCES USING OPTICAL STIMULATION OF A STELLAR SENSOR

Jean Mignot and Jean-Marc Julio*

With regard to targeted accuracy, stability and autonomy performances, the new AOCS are based on the use of stellar sensor. The new generation equipment provides three-axis attitude measurement which can be used directly by an on-board computer. Unfortunately, simulation studies are delicate, as it is very difficult to obtain a mathematical model representing real flight sensor behavior . Therefore it is a great advantage to be able to include real equipment for "hardware in the loop" simulation.

Developed at Cnes for our own needs, **BRAHMS** (**B**anc de **R**estitution d'**Att**itude **H**ybride par **M**esure **S**tellaire - hybrid attitude restitution test bench using stellar measurements), can be adapted to different stellar sensors.

The stimulation device is paired with a satellite simulation environment to evaluate attitude control law performances in a closed loop using a stellar sensor. For the AOCS Myriade microsatellite the environment is provided by the Satellite System Validation Test Bench (BVSS), as it is computed by simulation environment called **BACH** (**B**anc de test d'**A**ttitude en **C**onfiguration **H**ybride - hybrid configuration attitude test bench) for generic AOCS activities.

1 ATTITUDE CONTROL AND STAR TRACKER

Satellite attitude control requires on-going improvements to take into account changes to satellites requirements. This is clearly shown by increasing demand for pointing accuracy, mission availability, autonomy while targeting at controlled cost of the system.

These trends explain the increasing use of stellar sensors during the last few years. In fact, taking into account the generalization of CCD detectors (Charged Coupled Device) and now APS (Active Pixel Sensor), the performances of these equipments have considerably progressed. The on-board attitude determination software has also made great progress and the sensors are therefore capable of providing reliable and quick measurements. Today, they occupy the core of the attitude detection for the new attitude and orbit control systems (AOCS). Their performance is appreciated both in initial acquisition mode when the satellite is "lost in space" as well as during tracking phase when the stellar sensor (SST) is used to follow identified stars in his field of view.

* Centre National D'Etudes Spatiales, 18, Avenue Edouard Belin, F-31401 Toulouse Cedex 4, France. E-mail: jean.mignot@cnes.fr; jean-marc.julio@cnes.fr.

SSTs may be associated with other sensors like GPS, gyros, magnetometers, etc depending on the mission involved. However, in many cases in order to reduce costs, the AOCS architecture is often based on the sole use of stellar measurement. DEMETER, (Figure 1) the first CNES satellite in the Myriade family, takes up the challenge for normal attitude control law.

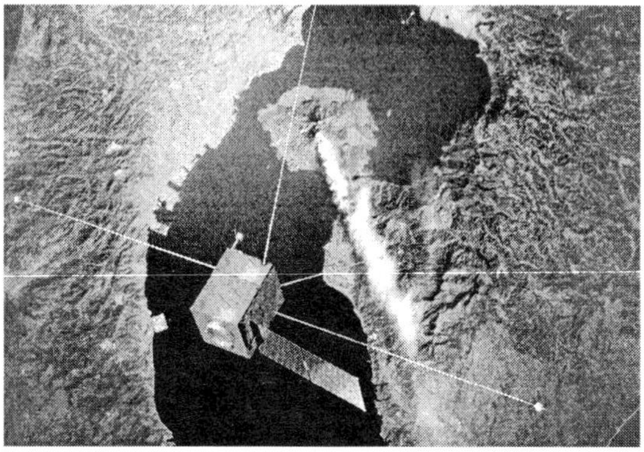

Figure 1 Artist view of DEMETER satellite.

The demonstration of the good operation of this sensor is primordial. The assessment of the performance of the attitude control loop on ground therefore requires a stimulation facility making it possible to use the effective stellar sensor in a representative simulated satellite environment.

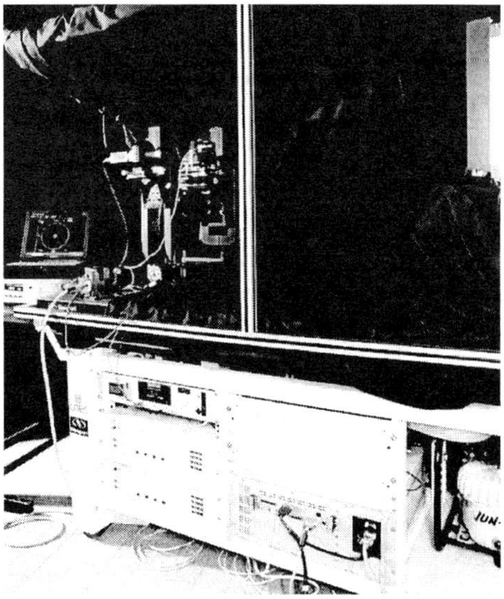

Figure 2 BRAHMS test Bench.

The **BRAHMS** stimulation bench (Figure 2)meets these requirements by in particular making it possible to assess the robustness and measurement availability of a stellar sensor for a given satellite mission. The originality of this bench is to enable tests in a laboratory environment while minimizing real sky tests and, at the same time, maintaining a large field of view of the sensor. The intensive use of a video display The effect due to disruptive elements in the field of view may also be assessed by means of realistic scenarios (Figure 3). Optical stimulation even though initially more complicated than electrical stimulation makes it possible to perform an independent realization of the sensor used given the field of view characteristics similar to current SSTs.

2 AN EVALUATION STELLAR ENVIRONMENT

BRAHMS comprises an "optical" component incorporating the sky display system (high-resolution screen) and the screen acquisition system (stellar sensor optical head). This equipment is located in a protected environment, from seismic excitation by means of an optical suspended table, and from optical disturbances as it is confined in a dark enclosure limiting the entry of extraneous light, (eliminating the need of a specific room to work). The overall dimensions of this unit are 1.9m x 1m x 1.7m.

The sky is generated by lighting the pixels of a faceplate type technology high-resolution color monitor (1600 * 1200 pixels) diagonal 0.553 m, the technology minimizes edge-effect due to the curvature of the tube. So as not to alter SST measurement (up to 10 Hz) and to ensure continuity in the movement of stars, the screen refresh frequency is 85 Hz. The size of the display area at 1m covers a 22° by 30° field. The size of each pixel (around 250µm) provides a minimum star diameter of 51 arc-seconds (1 pixel).

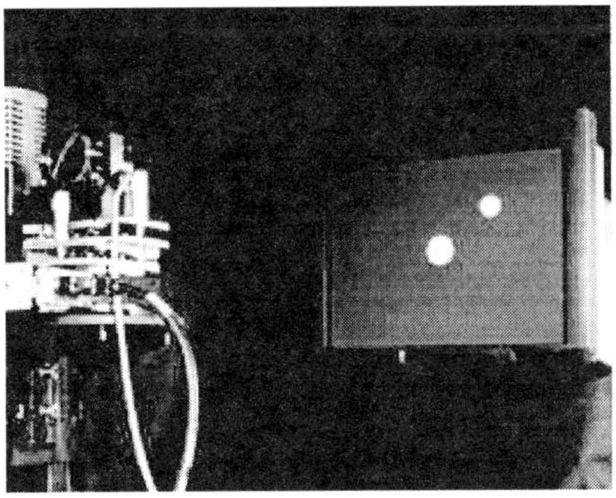

Figure 3 Moon and disruptive element in the sky.

The implementation of the stellar sensor in this restricted environment requires the installation on the optics of a collimator cap consisting of achromatic lenses, with a 1-meter focus, to reduce displayed pixels to infinity.

The two units making up the "optical" assembly are mechanically independent. Therefore, the reciprocal position of the units affects the overall attitude restoration performance of the BRAHMS bench. Due to its overall dimensions and weight, the monitor is stationary. Final positioning of the stellar sensor is through motorized mechanical movements with 6 degrees of freedom. Control of the sensor and monitor optical head alignment is provided by an auto-collimating sight.

3 A SKY

The BRAHMS bench is controlled by a processor fitted with a real-time commercial executive. The processing unit is constructed using a SPARC processor board associated with a high-resolution graphic board. The stimulation software receives the screen background primitive commands to display to the stellar sensorʌ, after calculating the corresponding imageʎ, by means of an associated graphic board, it controls displayʏ on the high-resolution screen (Figure 4).

Input data is read according to 2 formats, the satellite environmental data (viewing direction of the celestial canopy and perturbators) or a calibration panel.

A test management PC is used to implement, calibrate and control the BRAHMS bench. In addition, it displays on line the sky backgrounds requested and viewed by the sensor. It is used also to back up data exchanged during testing.

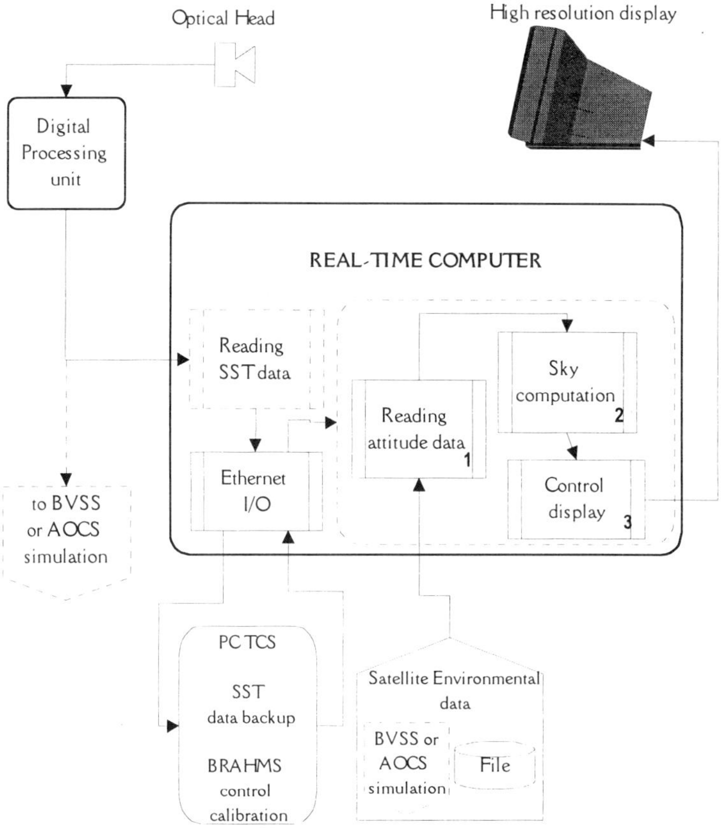

Figure 4 Synoptic.

4 OPERATING MODES

BRAHMS can operate coupled with the Satellite System Validation Bench (BVSS) of the µSat Myriade family or with the attitude hybrid configuration generic test bench (BACH) developed by the AOCS Technical department. In addition, an operating mode called "open loop", which was the first development phase, can be used to work with an SST coupled to its own Electrical Ground Support Equipment.

The functional validation of the Attitude Control System of the first µSat – DEMETER – with the DTU (Denmark Technical University) stellar sensor in the loop will be done using BRAHMS connected to the BVSS. A specific interface is used for the communication between the two benches in order to display the sky corresponding to a viewing direction.

In the Open loop mode, only the optical assembly and the "real-time" processor are involved This BRAHMS bench configuration makes it possible to validate the static and dynamic stimulation function of a stellar sensor by means of a controlled optical facility. It is also necessary for positioning and calibration of the optical assembly. The stimulation aspect has been processed by comparing angular values (ascension, declination, rotation) of a sky viewing direction transmitted by the test tool (installed on the SUN workstation) to the values returned by the SST.

The optical assembly (video display + SST) are calibrated by comparing target pattern images controlled by the Test Management (CDT) PC and images recovered by the SST.

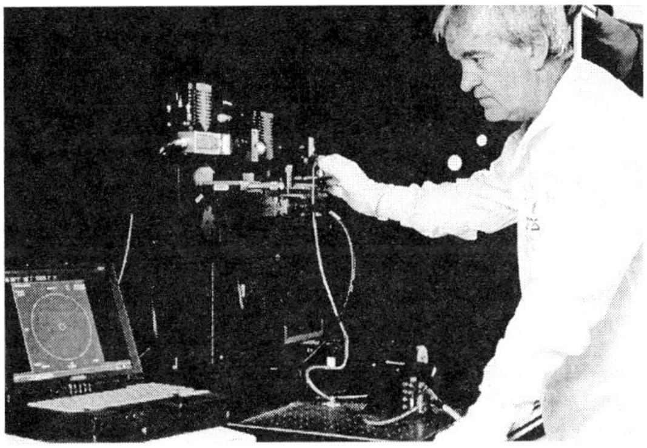

Figure 5 SST Alignment.

The last mode is part of the generic assessment studies of AOCSs using a Stellar Sensor. The satellite dynamics and environment simulator was developed by AOCS department for his own studies. This software, runs on a SUN workstation, takes into account the effective stellar sensor in the loop. This environment is functionaly similar to the operational one used for DEMETER AOCS validation but is devoted to performance evaluation during preliminary phases of AOCS design.

BRAHMS were also used for fine tuning and validation of the Control-Monitoring chain (ground and on-board segment) for implementation of the Denmark Technical University (DTU) stellar sensor on the aimed balloon gondola.

REFERENCES

"DLR-TUBSAT : a microsatellite with high precision attitude control", M. STECKLING, U. RENNER, H-P. RÖSER, Technical University of Berlin / German Aerospace Research Establishment, juin 1996.

"The star field simulator for the spacelab instrument pointing system fixed head star trackers", F.C. WESSLING (McDonnell Douglas), M.A. VANDER DOES (TRW), Hunstville, Alabama, SPIE, vol.2221, pp.116-127.

"Techniques et technologies des véhicules spatiaux", Cours de technologie spatiale, volume 3, module XII, Cepaduès Ed., 2002.

AAS 03-043

NIGHT SKY TESTS AND IN-FLIGHT RESULTS OF SED16 AUTONOMOUS STAR SENSOR

Ludovic Blarre, Sylvain Thomas, Philippe Jacob, Thierry Foisneau, Didier Vilaire and Marc Pochard*

The SED16 is an autonomous multi-missions star tracker, which delivers three-axis satellite attitude in an inertial reference frame and the satellite angular velocity with no prior information. The first flight of this star sensor took place in May 2002 on board the SPOT5 satellite. A lot of in-flight data of the sensor has been made available to SODERN by the CNES and an analysis of in-flight performances has been achieved and is presented. This data is completed by information from the second SED16 flight, which started end of May on a classified program and presently by a third mission, SORCE, launch by the end of January 2003. In flight results are in accordance with results issue from additional ground night sky tests achieved at SODERN throughout Q3 and Q4 2003. Complete picture allows an almost complete validation of major SED16 requirements: Accuracy, robustness, availability and reliability.

SED16 AUTONOMOUS STAR TRACKER MAIN FEATURES

SED 16 is an autonomous, multi-missions and cost-effective star tracker. SED 16 delivers three-axis satellite attitude and the satellite angular velocity to the Attitude Control Sub-system.

The main functions of the sensor are:

- An optronic sensor comprised of the lens and a 1024 x 1024 pixel CCD detector with a cooling device,
- Electronics combining functions for implementing the detector, digital processing and interfaces as well as for shaping the video signal,
- A mechanical structure to maintain the required stability of the lens and the detector,
- A sun shade to protect the star tracker against stray-light,
- A software enabling the 3-axis attitude restitution; this software incorporates star catalogues as well as recognition and tracking algorithms.

In-house software routines were developed by SODERN to improve tracking accuracy, high angular rate operation, tracking even with a high level of stray-light and to operate with the Moon in the field of view, either in tracking or in acquisition.

* SODERN, 20 avenue Descartes, F-94451 Limeil-Brévannes Cédex, France. E-mail: marc_pochard@sodern.fr.

The star tracker's design involves a small number of components. This makes the SED 16 easy to integrate and leads to greater cost-effectiveness.

This new family of star sensors is a breakthrough for attitude measurement due to the performance achieved their easy integration on the satellite bus, their answer to the satellite's mission requirements as well as their low recurring cost.

The Fig. 1 represents the SED16 with a line of sight parallel to the mounting plane (SED16-B) associated with a 25° baffle and the SED16 with a line of sight perpendicular to the mounting plane (SED16-A) associated with a 40° baffle.

Functional features	
Attitude	3-axis attitude determination
Acquisition	Autonomous (lost in space function), 99.9 % in less than 3 s.
Tracking	Autonomous, 10 stars tracked simultaneously
Polluting objects	Automatic exclusion (debris, satellites, ...)
Moon	Able to operate with the Moon in FOV
Standard Sunshades : • D=160 mm / L=160 mm / M=0.48 Kg • D=200 mm / L=240 mm / M=0.67 Kg	Sun / Earth rejection angle : 40 degrees / 32 degrees 25 degrees / 25 degrees
Output data rate	1- 10 Hz
Accuracy at 20°C, 10 HzEOL : • Bias (X, Y and Z axes) • Noise (X and Y axes) • Noise (Z axis)	 10 arcsec 3σ 11 arcsec 3σ 70 arcsec 3σ
Angular velocity	Up to 20 degrees/s at 10 Hz in tracking
Environmental features	
Operating temperature range	-30°C to +60°C
Vibration	25 g rms
Mechanical interfaces	
Size	170 x 120 x160 mm without sun shade
Mass	2.5 kg without sun shade
Mounting plane	SED-B parallel or SED16-A perpendicular to line of sight
Electrical interfaces	
Typical power consumption	8 W
Power supply	20 - 55 V, 50-122 V
Data I/O	RS 422 or 1553 B
Lifetime	15 years in GEO
Reliability	1000 to 2700 fit according to part quality grade and interfaces

Table 1 : SED16 main features

Figure 2: SED16-B and SED16-A

IN-FLIGHT RESULTS

The STR in-flight results are based on three missions, SPOT5, a second mission, which is classified and a NASA mission named SORCE. They are both LEO orbit but with different characteristics:

- SPOT5 uses a single Star Tracker with a 40° exclusion sunshade. It works with a frequency of 1 Hz and uses an ASCS16 dialog interface. The optical axis is perpendicular to the base plate. The angular rate is low and constant.
- The second mission uses 2 Star Trackers with a 40° exclusion sunshade. The operating frequency is 10 Hz and they use ASCS16 dialog interface. The optical axis is parallel to the base plate. The satellite is more agile than SPOT5 and higher angular rates and accelerations have been experienced.
- SORCE mission (Orbital spacecraft) uses 2 Star Trackers with a 25° exclusion sunshade. The operating frequency is 5 Hz and they use 1553B dialog interface. The optical axis is parallel to the base plate. The attitude control loop does not have any gyros and the attitude measurement relies on star trackers.

The SPOT5 satellite launch date was beginning of May 2002 and the second mission less than one month later, end of May 2002. SORCE was launched by end of January 2003.

Global Behavior and Calibration

Acquisition Mode and Transition Between Acquisition and Tracking Mode

The Autonomous Acquisition mode works nominally in the three missions. On SPOT5, the Star tracker has recognized stars pattern *in less than 4 seconds*. The In-flight Star Tracker use two different propagation options in acquisition mode. The Star Tracker propagates only according to the angular rate in the SPOT5 case (no angular acceleration) or both according to the angular rate and to the acceleration in the other mission. The transition between the acquisition mode and the tracking mode was always successful. After the loss of tracking due to the sun or the earth blinding, the star tracker récovers automatically the tracking mode without any problems. The "Lost in space" functionality gives fully satisfaction.

Tracking Mode

The STR works nominally in this mode. The number of coherent stars is always between 9 and 10 when there are no polluting objects in the Field Of View. During One orbit, the average number of coherent stars measured on SPOT5 was 9,92. The STR delivers a valid attitude at each telemetry.

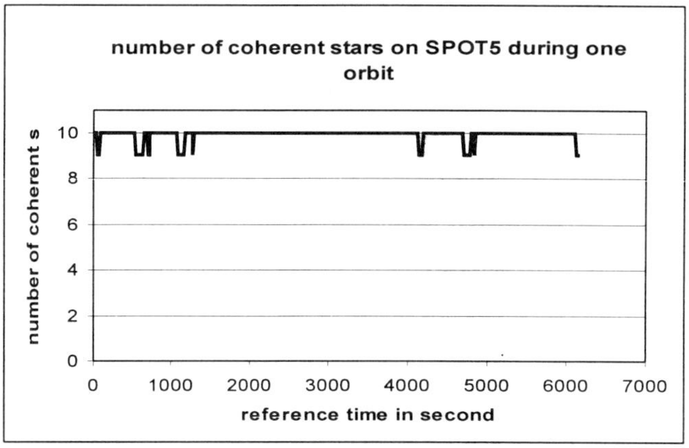

Figure 3 : Number of coherent stars during one orbit (SPOT5)

Star catalogue and calibration

The star catalogue gives fully satisfaction. The star sensor has rejected only three stars of the onboard catalog. They are all due to a difference between star catalog expected magnitude and measured magnitude. The stars are systematically rejected due to the coherence magnitude criteria and there is no degradation of the performances thanks to the high number of tracked stars (x10). The following figure shows the measured magnitude in one window during 500s. Several stars are measured during this time. The gaps are due to new selected star in the window. On this figure, the star catalogue error is easily detectable.

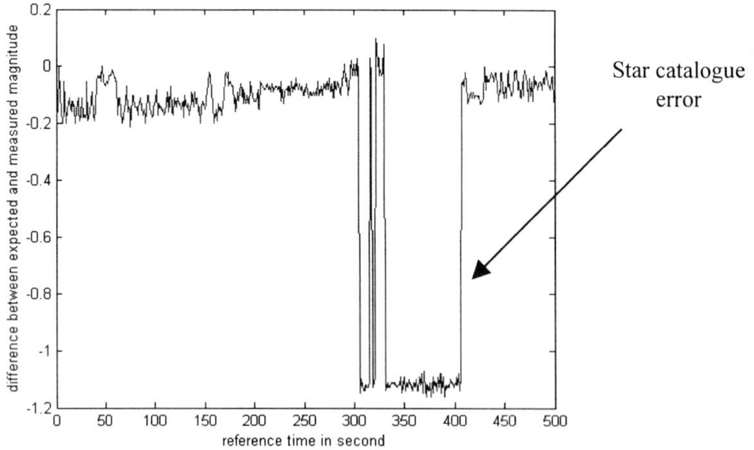

Figure 4 : Star magnitude measurement in one window with one star catalogue error

As we can see on the previous figure, the photometric calibration is good. The measured star magnitude is closed to the expected one. The gap is less than 0.2 magnitude and is largely covered by the Star Tracker magnitude tolerance.

Accuracy Performances

The Accuracy budget is mainly based on SPOT5 data. All the Spot5 specifications are respected by the SED16. Comparison between simulated performances and real one in Beginning of Life conditions (low dark current, no stray light...) shows that SODERN simulation are coherent and slightly pessimistic with respect to reality. According to these results, we can assume that real performances in EOL conditions will be much better than the specification.

Noise Equivalent Angle:

The specification is given in EOL conditions. The values given in the following table correspond to one typical position on the celestial vault. They are not global performances. This value is compare to night sky test measurement on ground and in both cases to calculated value

Axis	*Measured NEA On ground in arcsec 1σ*	*Calculated NEA On ground in arcsec 1σ*	*Measured NEA SPOT5 in arcsec 1σ*	*Calculated NEA In-flight in arcsec 1σ*	*Specification NEA in arcsec 1σ*
X & Y	1.5	1.9	0.87	1.17	8.3
Z	8.9	12.8	4	7.3	24

Table 5: In-flight and night sky test results

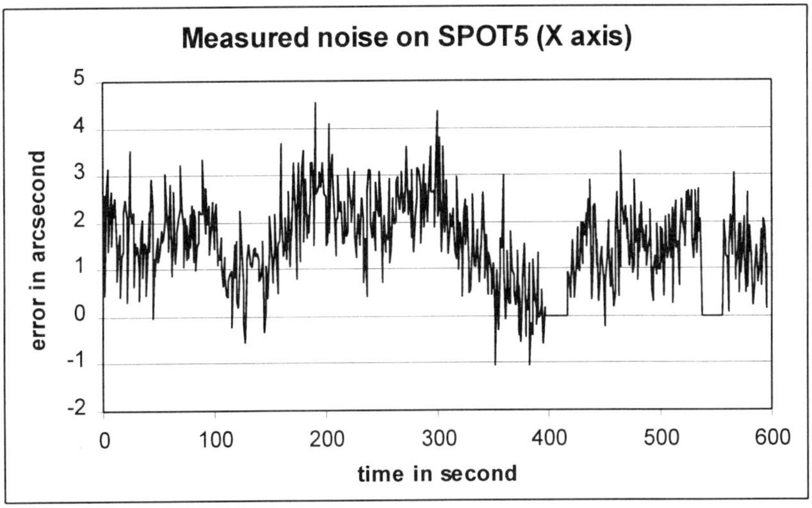

Figure 6 : Measured noise on SPOT5 (X axis)

The effect of angular rate was checked during night sky test and the SED16 star tracker demonstrates a outstanding robustness with respect to this condition. Results are indicated in table 7 with a velocity of 2 deg./s.

Axis Angular rate 2deg./s	Measured NEA On ground in arcsec 1σ	Calculated NEA On ground in arcsec 1σ
X & Y	3.5	3.3
Z	18.8	22.5

Table 7: night sky test results with 2deg./s angular rate

The same performances are given with the Moon crossing the Field of View. They are given for another point on the celestial vault. The SPOT5 specification is a degradation of the performances of 100% with an objective of 20% of degradation. During the Moon crossing, the number of coherent stars varied from 10 to 7.

Axis	Measured NEA SPOT5 in arcsec 1σ	Specification NEA in arcsec 1σ	Goal NEA in arcsec 1σ
X & Y	2.6	16.7	10
Z	10	48	28.7

Table 8: In-flight results with Moon in the FOV

The following figures show the impact of the moon crossing on the delivered attitude and number of coherent stars. The impact of the Moon is visible on the attitude but not critical. The Star tracker delivers good performance even with 7 or 8 stars.

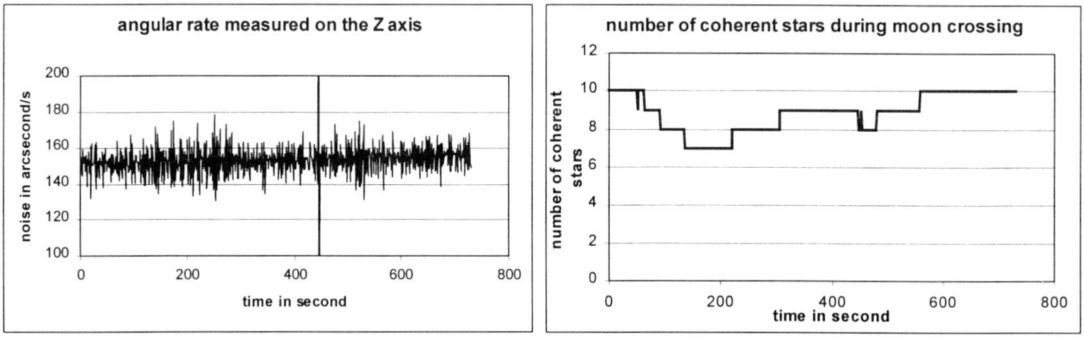

Figure 9 : Angular rate and coherent stars during Moon crossing (SPOT5 - Z axis)

Robustness

The SED16 measures 10 stars simultaneously in tracking mode, which offers a great robustness to specific events like small objects in the Field Of View. The SED16 also uses many coherence criterions. They remove automatically wrong or false star measurements and allow the Star Tracker to deliver an accurate attitude.

Kinematics Robustness

Many night sky tests have been realized with high angular rate and acceleration. The Star Tracker passed tests up to 20°/s of angular rate and acceleration up to $5°/s^2$ (at a frequency of 10Hz). This robustness has been tested in flight to an angular rate up to 4.5°/s and acceleration up to $4°/s^2$ (with low angular rate).

Moon in the Field Of View

SPOT5 data have demonstrated that the Star Tracker is robust in tracking with the Moon in the Field Of view. Nevertheless, some loss of tracking due to the Moon has been noticed (less than 10 since the satellite launch). This has revealed a software problem with the processing of the Moon in acquisition. This Moon processing function has been de-clutched at the moment by a configuration parameter change and correction of anomaly was continued on ground.

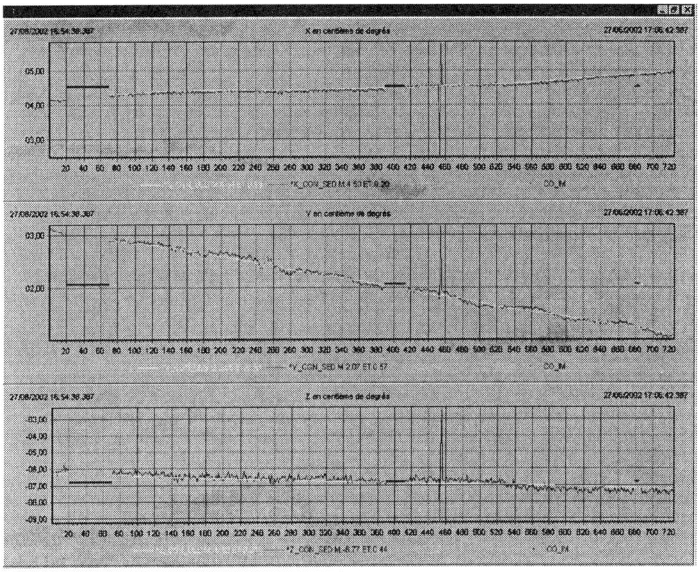

In yellow AOCS data and in green STR data

Figure 10 : Star tracker attitude measurements during Moon crossing on SPOT5.

A new Software version has been validated since the problem occurred. Night sky tests have been performed with the Moon. 256 successful autonomous pointing (acquisition followed by tracking) have been achieved during these tests. Now the Star tracker is able to give an attitude in acquisition mode with the moon in the FOV. This modification was in-flight tested by the end of 2002.

Sun and Earth effects

For two missions the star trackers use in flight a 40° baffle. They have in fact a very good behavior with the Sun in their exclusion angle. The STR keeps the tracking mode with the Sun at 33° of the optical axis and sometimes until 27° (instead of 40°). When it is in acquisition mode, it recovers the tracking mode with the Sun at 33°. Performances with Earth are also very good. During night, The STR keeps the tracking with the Earth until +/- 12.5° (which is the edge of the Field Of View).

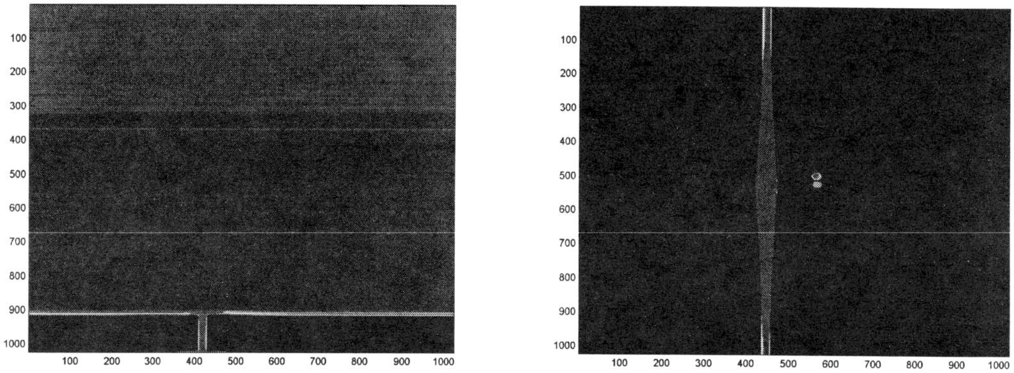

Figure a : No Moon processing

Figure b : Moon processing

Figure 11 : CCD images in acquisition with the Moon in the Field of View

Stray light source :	**θ < 12.5°**	**12.5° < θ < 27°**	**27° < θ < 29°**	**29° < θ < 33°**	**33° < θ < 40°**	**θ > 40°**
SUN	No tracking	No tracking	No tracking	Possible loss of tracking Degraded performances	Degraded performances	Nominal performances
MOON	Degraded performances	Nominal performances	Nominal performances	Nominal performances	Nominal performances	Nominal performances
EARTH (night)	Possible loss of tracking Degraded performances	Nominal performances	Nominal performances	Nominal performances	Nominal performances	Nominal performances
EARTH (day)	No tracking	Possible loss of tracking Degraded performances	Nominal performances	Nominal performances	Nominal performances	Nominal performances

Figure 12 : Impact on the STR performances for different stray light sources

Protons effects

SPOT5 data have shown that the Star Tracker was robust to proton effects. SED16 is able to maintain the Tracking mode up to 5 000 p+/cm²/s. We have no information with greater values. Nevertheless, we have noticed some peaks in attitude during South Atlantic Anomaly crossing. They are due to confusion between stars and protons in the tracking windows. Some peaks have also been detected on the star magnitude and positions measurement.

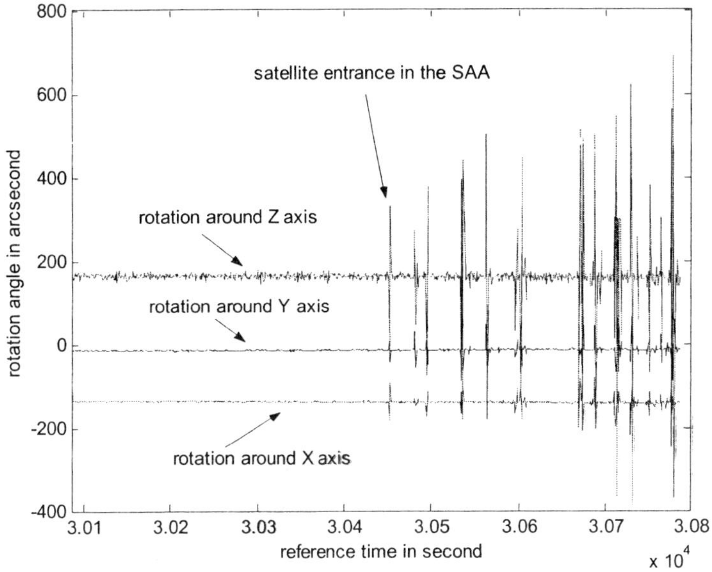

Figure 13 : Angular rate measurements during SAA crossing

These peaks are not critical for the SPOT5 mission since filters and gyros are used. SODERN has reproduced the South Atlantic Anomaly behavior on ground with simulated image sequences. The correlation with in flight results is excellent.

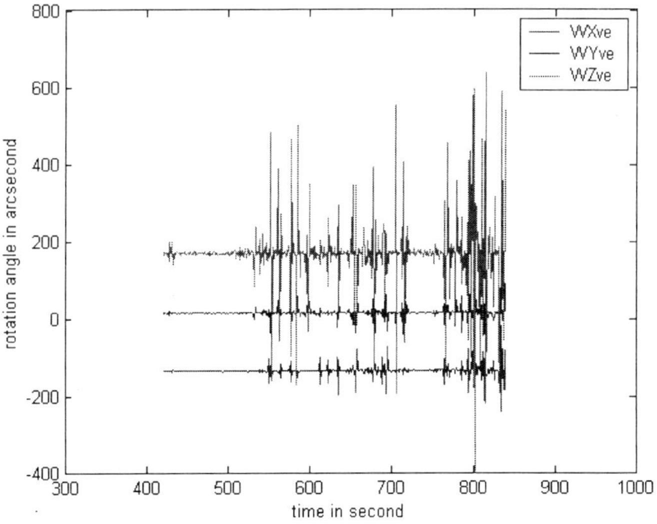

Figure 14 : Star tracker measured angular rate during simulated SAA crossing (SPOT5 configuration with simulated digitized image sequence)

New parameter was implemented in Star Trackers in order to reduce the peak. They allow a reduction by more than a factor 2 of the noise.

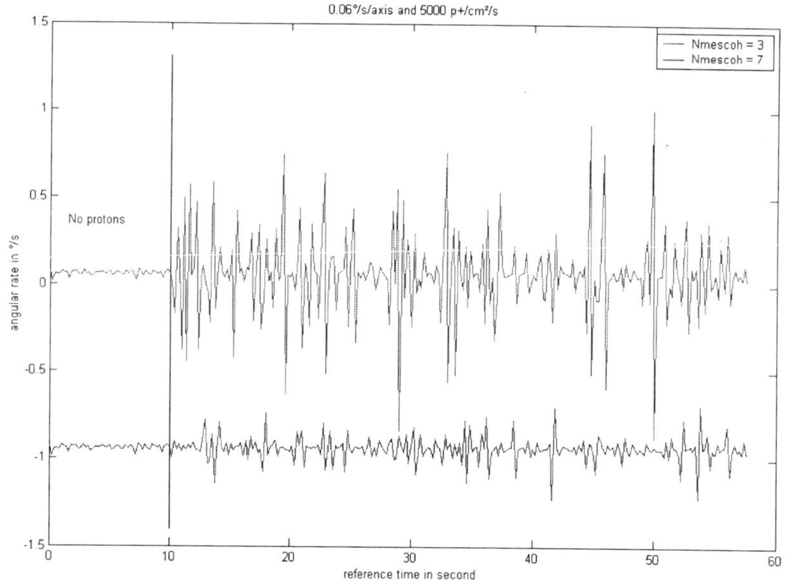

Figure 15: Impact of the new parameters on the peaks during simulated SAA crossing

CONCLUSION

The trend of Avionics aboard satellite towards high performance, low-cost, longer lifetime and better robustness led Sodern to developing a new family of star trackers. Extensive on ground qualification, numerous robustness-tests and night sky tests demonstrated the suitability of the SED16 star tracker with respect to Customers specifications. The SED16 in-flight performances have met the Customer expectations for this new generation of autonomous and multi-mission star trackers. Accuracy performance appears to be slightly better than expected. Critical issues like protons in the South Atlantic Anomaly and moon crossing through the field of view were successfully tested. Sodern is committed to pursue its improvement effort on the SED16 product line in order to propose to the market an extremely reliable and robust product. This is a key point to firmly rooted star tracker as prime attitude measurement device. The SORCE mission with a gyro-less attitude control sub-systems and SED16 star trackers aboard is demonstrating the operability of this concept.

AAS 03-044

VIDEOMETER: A NEW RENDEZVOUS SENSOR

Agnès Mestreau-Garreau, Carole Moussu, Ludovic Blarre, Nicolas Perrimon, Philippe Jacob, Marc Pochard,* Sabine Ludwig,† Paul Da Cunha‡ and Stein Strandmoe**

This paper presents the design of a new rendezvous sensor named Videometer. The aim of the Videometer Assembly (VDMA) is to support the correct docking of the Automated Transfer Vehicle (ATV) to the International Space Station (ISS) providing relevant Range, Line Of Sight and relative attitude information.

The Videometer Assembly operates from a distance of 300 m between the ATV and the ISS to docking. The VDMA is composed of the Videometer Device (VDMD) implemented on the ATV and the rendezvous Target (RVT) mounted on the ISS. The operating principle is based on VDM laser diode beams illuminating the RVT. Reflected light towards the VDM is detected and processed to compute target data.

This Rendezvous Sensor could be implemented onboard any chasing vehicle to ease an automatic rendezvous and docking.

At short distance, the accuracy on range is a few centimeters, the accuracy on Line Of Sight and attitude is some tenth of degrees. The robustness of this rendezvous sensor is a key point for this development and the performance of the sensor is secured even with the sun in the smearing area of the detector and during proton hits when the spacecraft is crossing through the South Atlantic Anomaly.

INTRODUCTION

One of Europe most ambitious contributions to the International Space Station is the Automated Transfer Vehicle (ATV), which will deliver up to 9 tons of supplies, payload and propellant and will provide during 6 months re-boost and attitude control for the ISS.

ATV will be the first automated rendezvous with a space station using optical sensors and the first ever where the target's attitude motion is compensated for by the chaser. For this mission Sodern will deliver two key contributions with the delivery of star trackers for the attitude control of ATV and the Videometer primary sensor, which measurement are used in the active guidance, navigation & control (GNC) loops to control ATV's motion.

* SODERN, 20 avenue Descartes, F-94451 Limeil-Brévannes Cédex, France. E-mail: marc_pochard@sodern.fr.

† JENA OPTRONIK, Iena, Germany.

‡ EADS LV, Les Mureaux, France.

** ESA, Les Mureaux, France.

To ensure proper capture and acceptable docking loads, ATV's probe has to meet ISS docking cone within a radius of 10 cm (4 inches) and a lateral velocity of less than 2cm/s (0.8 inch/s).

The Videometer delivers range and line-of-sight angles to the GNC system, and, from within 30 m, relative attitude angles with a great accuracy.

OPERATING PRINCIPLE

The Videometer Assembly aims at measuring **Range, Line Of Sight (LOS) and attitude** (named target data) of the Videometer on board the chasing vehicle with respect to the Rendezvous targets (RVT) located on the ISS SM. Range and LOS are measured if Range is lower than 300m, attitude if Range is lower than 30m.

The measurements are achieved from the observation through a detection system (CCD and lens) of Laser reflections on Laser Retro-Reflectors. The illumination is performed by Laser diodes of the Videometer Illuminator Head. The target illumination system is made of two kinds of Laser diodes; 6 Long Ranges and 2 Short Ranges (one nominal and one redundant), differing in the size of their illumination cone.

They are located very close to the Videometer lens aperture pupil. A lens set is used for each diode to match the beam divergence to the requested shape. Each group of laser diodes can be selectively powered during a variable duration selected by the Videometer software and synchronized with detector integration.

Two Videometers are used: one nominal, and one in hot redundancy. Their functioning is synchronous.

The Rendezvous target is composed of:

- The outer target; 3 clusters of 7 Laser Retro-Reflectors (LRR) forming a triangle.
- The inner target base-plate composed of 4 LRR's (square shape with 1 LRR at each corner) and 1 forward LRR (on a tower).

During mission, two main basic functions are used with this sensor. The first function determines the target data (and eventually motion rates) with no previous information. This function is based on a pattern recognition algorithm. This function is performed by the **Target Acquisition (TA)** operating mode.

The second function determines recurrently the precise target data (and eventually motion rates) from rough target data. The sensor processes detection windows containing the expected target positions. The pixels in these windows are processed to get the best accuracy on each single target position and signal. The single target data are then processed to deliver the best target data. This function is performed by two different operating modes: **Target Tracking (TT)** and **Hybrid Target Tracking (HTT)**. The chaining-up of TA, TT and HTT is called **Target Pointing (TP)**. In Target Pointing, target data are delivered in TM at **1Hz**.

According to the range and operating mode sequencing, the target pattern is seen as illustrated on figure 1

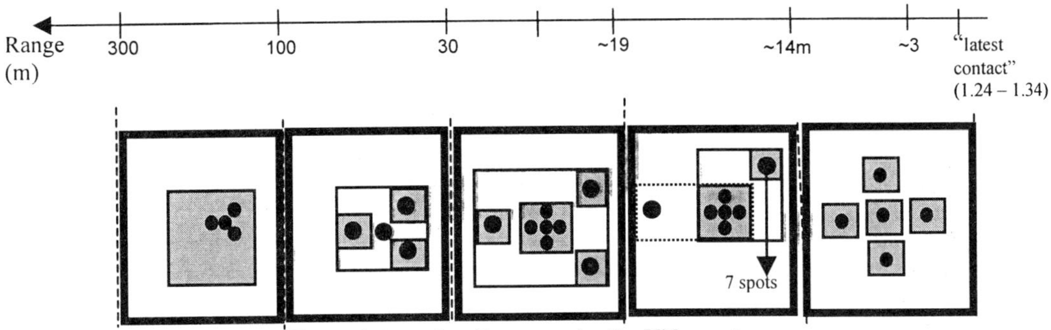

Figure 1:_target pattern seen by the Videometer

VIDEOMETER DESCRIPTION

The VDM Device includes in two units aboard the chasing vehicle the following items (see figure 2):

➢ The " Videometer" comprised of :

- A lens imaging the FOV on the CCD,
- The Illuminator Head (IH), including the laser diodes, the radiator, the heaters,
- The Detection Unit (DU), including :
  - The CCD detector and its window,
  - The thermo-electric cooler,
  - The electronics necessary to operate the detector, to process the video signal, to process the digital signal, to manage the data exchanged with the on board computer (GNC/FCM) and the DC-DC converter which generates the secondary voltages.
- An on board software necessary to determine the target data (Range, LOS and attitude),
- A mechanical structure which maintains the overall stability of the different sub-assemblies.

➢ The Electronic Box (**EB**) which :

- Receives the power supply coming from ATV (primary voltage, redundant and nominal "survival" heaters),
- Distributes a part of it to the VDM for heater power supply and power supply of the detection unit,
- Power supplies the laser diodes mounted of the VDM with appropriate current pulses.

Aboard the ISS the VDM includes the Rendezvous target, which consists of Inner and Outer targets.

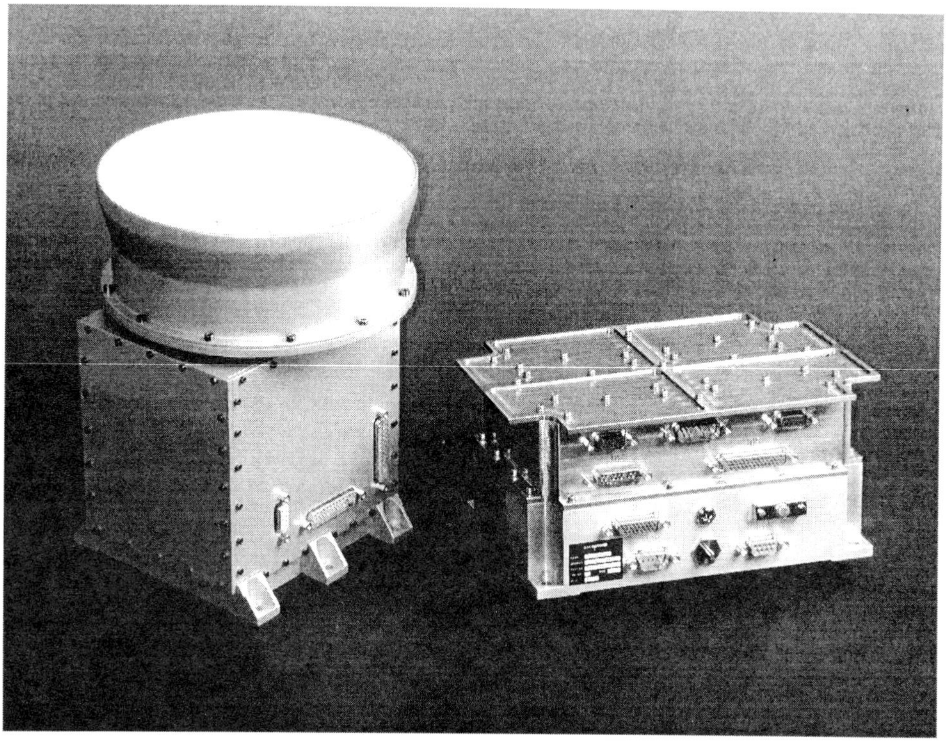

Figure 2: ***Videometer Device***

VDM Detailed Technical Description

Illuminator Head

The illuminator (see figure 3) is composed of a structure on which are mounted :

- 8 laser diode sets (6 for long range and 2 for short range illumination : 1 nominal and 1 redundant). The laser diode set includes a lens calculated to adapt the divergence of the laser diode to the needed diverging light beam.
- a radiator, a thermistor and heaters to stabilize laser diode temperature between 20°C and 30°C when operating.

The characteristics of the laser diode sets are the following:

- Wavelength of the laser diode : 810 nm
- Long range laser diode sets :
  - Minimal optical power larger than 2 W/sr in a 13.5 ° half angle cone
  - Timing: 1 Hz with a 30 ms maximum pulse width or 5 Hz with a maximum pulse width of 6 ms
- Short range laser diode sets :
  - Minimal optical power larger than 2 W/sr in 15° half angle cone
  - Timing: 10 Hz with a maximum pulse with of 10 ms ; the minimum pulse duration is 20 μs

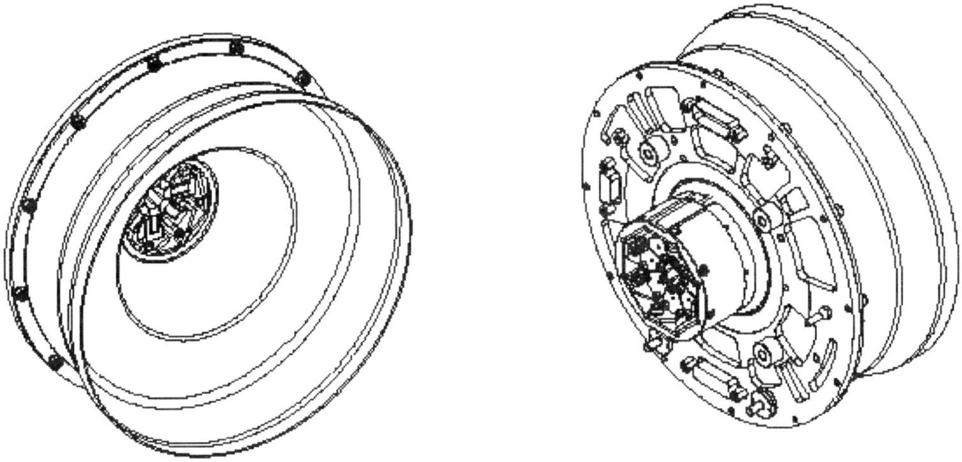

Figure 3: Illuminator Head

Lens

The lens is a telecentric system made of 4 lenses realized with radiation-hardened glasses with an anti-reflection coating. The lens includes:

- An optical stop aperture located in front of the lens.
- A band pass filter centered on the laser diode wavelength and with a bandwidth of 11 nm. This filter is necessary to ensure a good rejection of light outside the spectral bandwidth of the laser diodes.
- A neutral filter to adjust the transmission of the lens such as the typical Sun smearing equals 20% of typical pixel full well capacity.
- Field of view 24°×24° (square)

Detection module

The Detection module behind the lens includes a CCD detector with a protective window, a thermoelectric cooler and the focal plane support for connection with the housing and a the proximity board. (see figure 4),

The detector is the CCD 47-20 with shielded anti-blooming from E2V. Its main features are:

- Front Face illuminated,
- Frame Transfer detector,
- Image area of 1024x1024 pixels,
- Square pixels of 13 x 13 μm^2,
- Anti-blooming.

Furthermore, a DUMP command is implemented near the readout that allows quick access to the windows of interest by dumping all unnecessary lines allowing a fast transfer of the memory zone to the register.

In order to guarantee sighting performances over the possible operational temperature range and at the end of life (deterioration due to the radiation environment), the detector is assembled on a thermoelectric cooler, which reduces the temperature. The standard operating temperature is nominally set to 0°C.

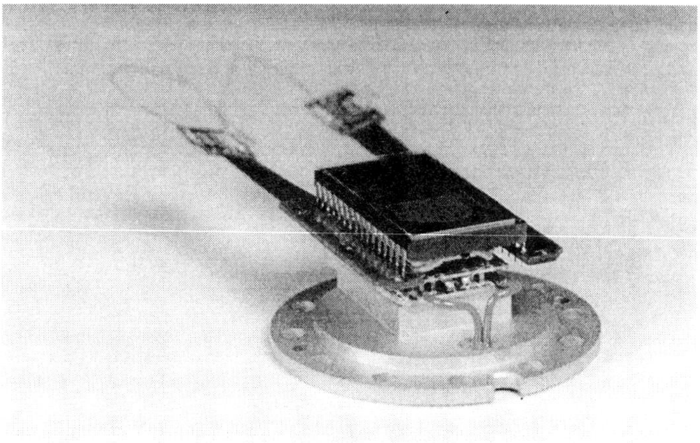

Figure 4: Detection module

Detection unit electronics

The electronic architecture is made up of the following functions:

- Detection and control of the CCD,
- Command of the laser diodes, laser diodes temperature, CCD temperature,
- Digitizing of the video signal,
- Video pre-processing,
- Analogue acquisitions,
- Computer function,
- Power supply of the detection unit

This architecture is built around an ASIC circuit, which is involved in most of the VDM electronic functions including:

- Generation of the CCD and laser diodes clocks,
- Alignment of the video signal,
- Video pre-processing,
- Control of analogue acquisition DAC,
- Control of the Thermo-Electronic Cooler power supply current,
- Processor monitoring watch-dog,
- Built in self-tests.

This component assumes a real time processing of the pixels, which could not be assumed by the processor.

Figure 5: ***Detection Unit Electronics***

Electronic box

The Electronic Box (**EB**) :

- Receives the power supply coming from ATV (primary voltage, redundant and nominal "survival" heaters),
- Distributes a part of it to the VDM for heater power supply and power supply of the detection unit,
- Power supplies the laser diodes mounted of the VDM with appropriate current pulses.

The electronic box is physically separated from the VDME.

Rendezvous Target

The RendezVous target aboard the International Space Station, close to the Zvezda' port, comprises:

- The inner target comprises of five laser retro reflectors. The central LRR is not in the same plane as the others such as to process the attitude at Short-Range. These retro reflectors are mounted on a base-plate which has to be stable versus thermal constraints to limit measurement errors.
- The 3 outer target assemblies are made up of seven retro reflectors each mounted on a base plate. The central retro reflector is uncoated, the six others are coated to be the less reflective as possible at 920 nm which is the operating wavelength of an other optical sensor the Telegoniometer.

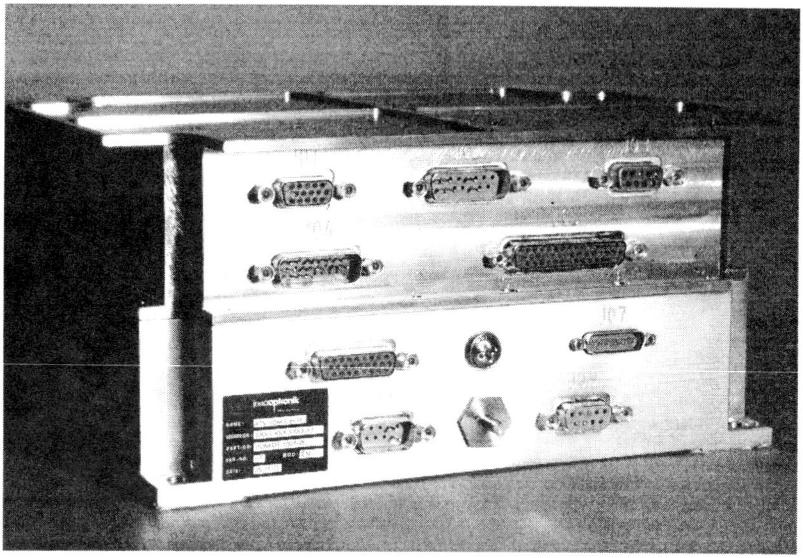

Figure 6: Electronic Box

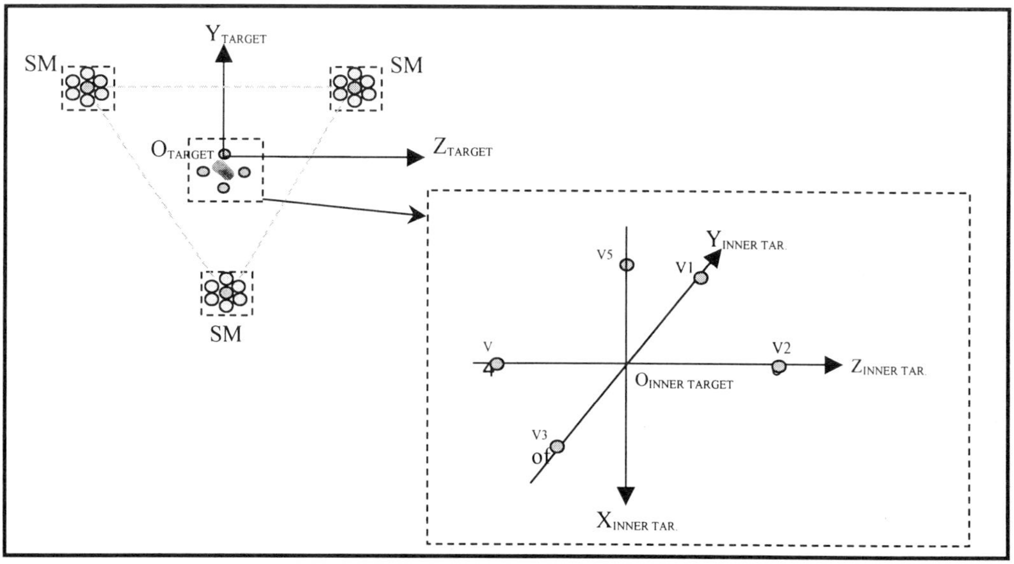

Figure 7: Rendezvous Target

Software

There are two types of software processing tasks:

- Sequencing tasks for all software tasks and the control of electronic functions,
- Operating modes tasks.

The inspection tasks controls the electronic functions of initialization and permanent monitoring. The permanent monitoring tasks performs:

- The acquisition of physical enlargements for remote measurements and internal settings,

- The inspection of the various settings (Cooler, etc...),
- The periodical reset of the Watch-dog (monitoring loop for the processor function),
- The control of software and calibration parameters checksums.

These tasks are run automatically upon power-up of the VDM.

The operating modes themselves apply specific algorithms that provide the functions required ensuring the autonomy of the VDM. The task-sequencing monitor ensures the sequencing in real time of the VDM tasks

The VDM is provided with 11 operating modes (sequencing described in figure 8). Only four operating modes are for standard use during nominal operation of the VDM.

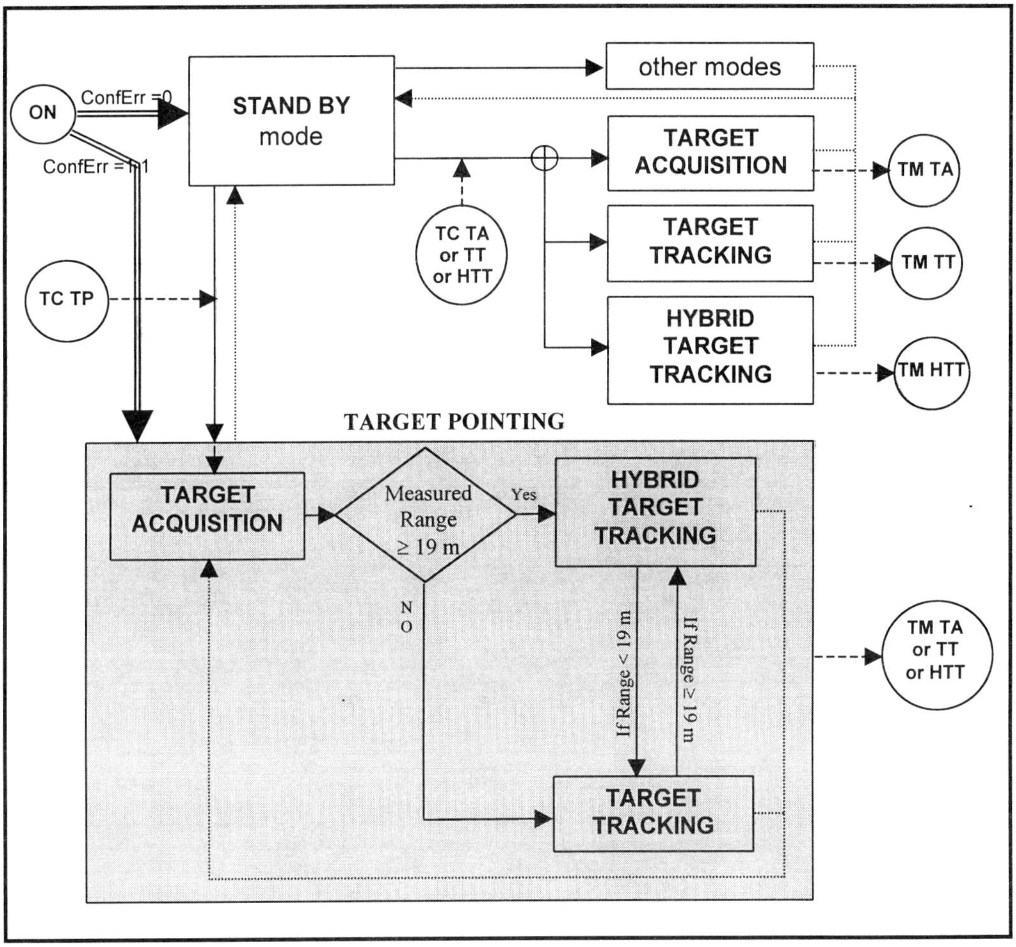

Figure 8: Operating modes sequencing

Videometer Main Features

The videometer main characteristics are indicated in the table 9.

Autonomy	Fully autonomous
Operating range	Range from 300 m up to 1.24 m Operate with the sun in the field of view
Outputs	- Range - Line Of Sight - Relative attitude, if Range is lower than 30m.
Field of view	24° (square)
Measurement frequency	1 Hz
Dialogue interface	1553
Power supply	UPB 20V-50V
VDMD mass	VDM = 4 kg ± 20% EB = 2.6 kg ± 5% Inner target ≤ 2.8 kg Outer targets ≤ 3 x 0.8 kg
VDMD typical power consumption	26W

Table 9: Videometer features

Robustness

To demonstrate the robustness of the Videometer we simulated a worst case scene, which is shown on figure 10:

- Sun in the FOV (14.3 times the saturation level).
- Range of 150m.
- Targets located in the Sun smearing zone.
- Extended permanent objects.

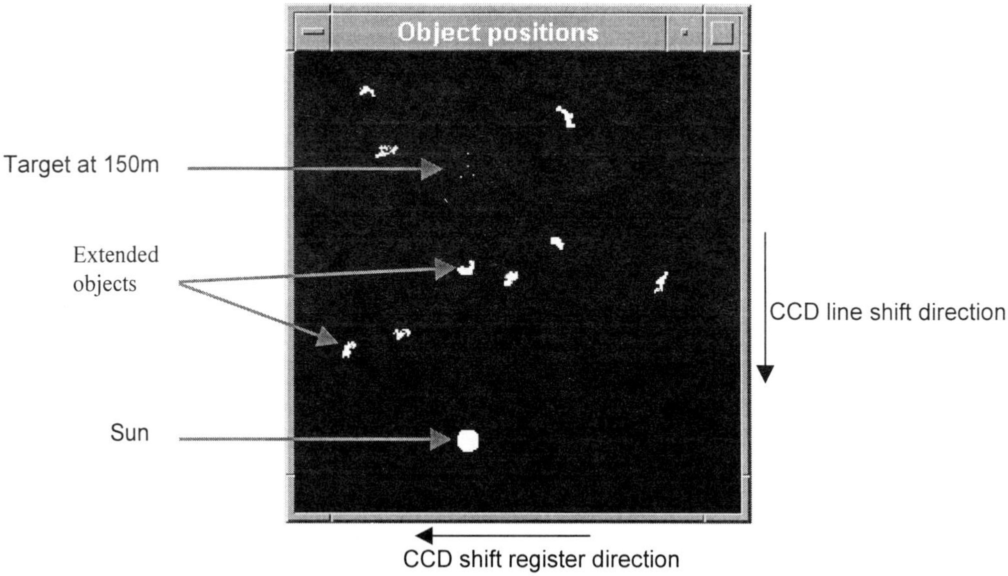

Figure 10: Worst case image

In order to simulate the VDM optics, the previous image is convoluted with the diffraction spot of the VDM optics (gaussian with a diameter at 80% of energy equal to 30μm).

Figure 11 illustrates the double integration, the reading-out of the CCD and the digitalization. It shows the smearing of Sun inthe place where targets are located and the dark signal slope. All the permanent objects (Sun, extended permanent objects and targets) are doubled. Moreover, the protons transient effects are simulated: on this image of 256×256 2×2 binned pixels, 500 false objects due to protons are simulated, which corresponds to a worst case of transit above the South Atlantic Anomaly.

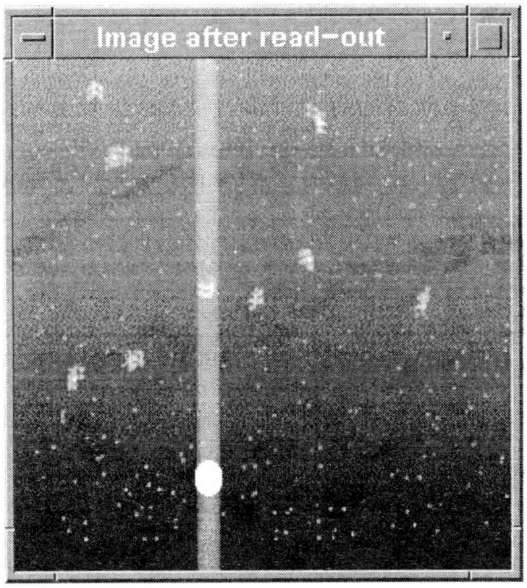

Figure 11: Worst case image seen by the CCD

Figure 12 image illustrates the ASIC processing:

- Smearing is suppressed,
- Extended objects are suppressed,
- Most of objects due to protons are suppressed,
- Dark current slope is corrected.

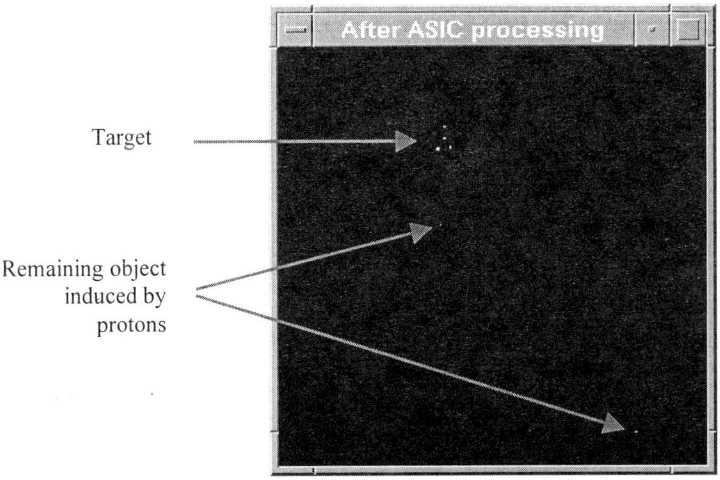

Figure 12: Worst case image after ASIC Pre-processing

Accuracy budget

The performance budget is indicated for several ranges.

a: RVT out of Sun smearing area.

b: RVT in Sun smearing area.

Range :

Range (m)	Bias (max) Performance (m)	Bias (max) Specification (m)	Noise (at 3σ) Performance (m)		Noise (at 3σ) Specification (m)
	rss		(a)	(b)	
1.23	0.006	0.01	0.0005	0.001	0.005
10	0.04	0.09	0.01	0.04	0.09
30	0.3	0.3	0.03	0.1	0.4
300	**3.4**	3	4.6	**17.7**	15

- ***LOS :***

Range (m)	Bias (max) Performance (°)	Bias (max) Specification (°)	Noise (at 3σ) Performance (°)		Noise (at 3σ) Specification (°)
	rss		(a)	(b)	
1.23	0.38		0.002	0.005	
10	0.18	0.5	0.001	0.002	0.2
30	0.18		0.001	0.004	
300	0.17	0.3	0.002	0.008	

- ***Roll :***

Range (m)	Bias (max) Performance (°)	Bias (max) Specification (°)	Noise (at 3σ) Performance (°)		Noise (at 3σ) Specification (°)
	rss		(a)	(b)	
1.23	**2.10^{2}**		0.02	**0.07**	
10	**2.10^{2}**		**0.06**	**0.22**	0.05
30	0.38	1.5	0.04	**0.17**	

- ***Pitch and Yaw :***

Range (m)	Bias (max) Performance (°)	Bias (max) Specification (°)	Noise (at 3σ) Performance (°)		Noise (at 3σ) Specification (°)
	rss		(a)	(b)	
1.23	**0.61**	0.55	**0.06**	**0.17**	0.05
10	0.64	0.7	0.17	**0.59**	0.5
30	0.89	1.4	0.80	**3.33**	1.4

Development Plan

The European Space Agency with EADS Launch Vehicle as prime Contactor develops the Automated Transfer Vehicle. Sodern is in charge of the Videometer with Jena Optronik as the main Sub-contractor. This design was started in mid-2001 and the preliminary design review was held in mid-2002. This review concluded that the overall design was sound and give authorization to proceed to the next phase. The Critical design review will be held in Mach 2003.

The manufacturing of an Engineering and Qualification Model has already started and this model will start Qualification Test by May 2003. These tests will be completed by September 2003 and the first flight set of Videometer will be delivered before the end of 2003.

A key point to qualify this sensor is the development of test benches. Three test benches are currently developed.

- The first one is dedicated to calibrate the sensor and to check its short range performance (1.4 to 4 m),
- The second one is to check the functionality in dynamics and the performance in static in long range, up to 300 m,
- Finally the Videometer will run tests on a facility (EPOS) developed specifically for the ATV by the European Space Agency. This facility will allow testing the sensor in static and dynamics, with the Sun in the field of view, in the range from 1.2 m up to 20 m.

To check the robustness of this sensor and to verify its properly operating with the ATV's Avionics, two Electrical Ground Support Equipment units are developed. One will allows running close loop tests and the other one to simulate worst case conditions.

CONCLUSION

A fully optical sensor is designed as the primary sensor for the automatic rendezvous and docking of the Automated transfer vehicle to the International Space Station. This sensor will provide to the active control loop range, Line of sight and attitude with centimeter and tenth of degree accuracy. The robustness was a key design driver for this sensor and simulations demonstrate its capacity to operate with the sun in the field of view and with proton hits. In mid-2003, qualification test will be run on the qualification and delivery of flight models is scheduled for the end of 2003.

These new optical sensor are a breakthrough for automatic rendezvous between a chasing vehicle and a target due to their robustness, low mass and low power consumption.

AAS 03-049

PROPERTIES OF BEI'S NEW PROXIMITY SENSOR, THE LINEAR GAP DISPLACEMENT TRANSDUCER (LGDT)

Andrew Chouinard*

INTRODUCTION

In today's high-technology marketplace the automation of processes is growing, component sizes are shrinking and the corresponding demands placed on sensors are ever increasing. Included in this trend are proximity and displacement sensors. As the applications for proximity and displacement sensors expand, so does the demand for sensors that can operate on almost any material, in diverse and harsh environments, over a variety of ranges, and with ever increasing resolution. This article presents test results for a BEI patented (U.S. Patent # 5,017,772) optical proximity sensor designed to meet some of those demands.

The Linear Gap Displacement Transducer (LGDT) measures the distance from the probe tip to a flat surface. The LGDT uses a fiber optic probe consisting of concentric fiber rings to illuminate the target surface, measure the reflected light, and derive a measurement of the distance to the target. The LGDT is designed to measure target surface movement over a small range very precisely. The output of the LGDT is an analog voltage, which varies with the distance from the target to the tip of the fiber optic probe minus the standoff distance. The linear range, sensitivity, and resolution of an LGDT probe depend primarily on the probe geometry, thus probes can be designed to meet application needs.

PRINCIPLE OF OPERATION

The LGDT consists of a fiber-optic probe and an electronic module. A complete unit is shown in Figure 1. The fiber-optic probe contains three fiber bundles. At the sensor end of the probe, the bundles are arranged in concentric rings with separation bands between them, as shown in Figure 2. At the opposite end of the probe, the bundles are separated into independent fiber cables for coupling to the electronic module.

The target surface is illuminated by a long-life, infrared (880nm) LED via the central fiber bundle. Divergent light emitted from the central fiber bundle is reflected from the target surface back to the probe end and is collected by the two outer fiber bundle rings. As the target moves further away from the probe end, the outer fiber ring receives progressively more reflected light, while the inner ring receives less. Each of

* Precision Systems & Space Division, BEI Technologies, Inc., 1100 Murphy Drive, Maumelle, Arizona 72113.

the receiver rings transmits the collected light to a detector in the electronic module. An analog divider provides the ratio of the detector signals, which is then multiplied by a gain factor. Finally, an offset voltage is subtracted to yield the final analog LGDT output.

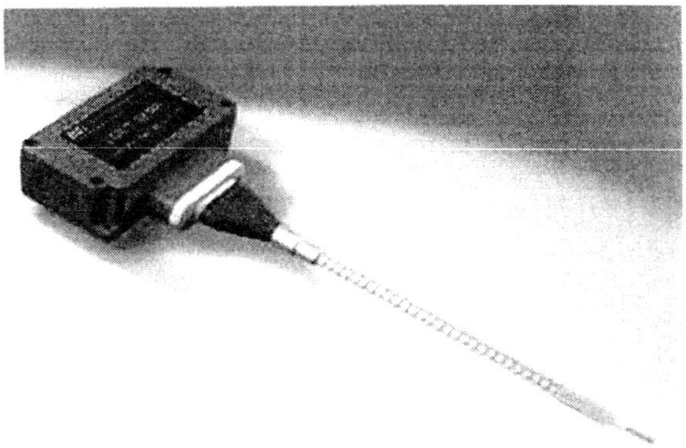

Figure 1. SERB model Linear Gap Displacement Transducer (LGDT) unit.

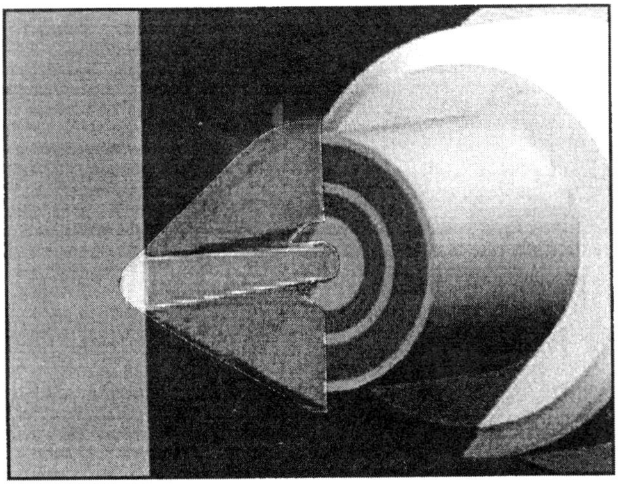

Figure 2. A close-up view of the probe end showing emitted light in blue and light reflected from the target and received by the LGDT in red.

OUTPUT CHARACTERISTICS

Figure 3 illustrates optional settings of the sensor gain and offset. The blue plot illustrates the dynamic range response of an LGDT unit with a 0.375" diameter probe. In this plot, the gain is five and the offset is zero volts. This low-gain configuration provides the greatest dynamic range from a given probe. An alternative configuration uses increased gain and an offset voltage to provide output of zero to ten volts over just the linear range of the probe, as shown in Figure 3's red plot. This high-gain configuration provides the highest sensitivity for this probe, 2 V/mm. With an output noise of just 0.3 mV RMS, an LGDT unit with a 0.375" diameter probe (LERR model LGDT) provides resolution to 150 nm over a 5mm range. The dynamic and linear measurement ranges of a probe depend on its opto-mechanical geometry; smaller probes have shorter ranges and higher sensitivities, yielding correspondingly higher resolution.

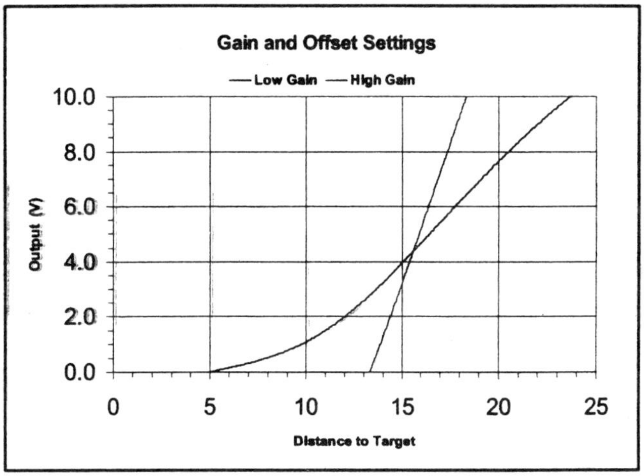

Figure 3. Adjustments to gain and offset provide output flexibility. Low gain utilizes the probes entire dynamic range. High gain provides the greatest sensitivity over the probe's linear range.

The repeatability of the LGDT was measured by acquiring four scans of its response over a 7mm region centered on the probe's linear region (linear region for the probe is 5mm long). For each scan point, the difference between the maximum and minimum of the four output values was calculated. The average difference over the 7mm range was 2.7 mV. This value is within the noise floor of the data acquisition system used to acquire the scans; therefore, the true repeatability of the LGDT is expected to be less than the 2.7 mV obtained in this test.

LGDT has other characteristics that qualify it for a large variety of applications. The ratiometric optical technique utilized by the LGDT enables its use with bare and finished metals, paint, plastics, glass, paper products – anything that reflects light, including diffuse surfaces. Because the probe is optical, it can be placed in strong electric, magnetic, and RF fields without performance degradation. Additionally, the length of the fiber bundle has no effect on the measurement, thus allowing remote placement of the electronics.

TARGET TESTING

Tests were conducted with an LERR model LGDT unit to determine the response of the LGDT to various target conditions. The conditions tested were variances in target lighting, reflectivity, diffusivity, and angular misalignment. In each of these trials, the condition under test was altered and a scan of the LGDT response to the new condition was acquired via LabVIEW control of an Instranet data acquisition system. The test results are presented below.

Lighting

In this test, the target was illuminated with differing amounts of external light, which included darkness, ambient sunlight within the room, fluorescent room lights, low intensity spotlight, and high intensity spotlight. The target irradiance under each condition is given in Table 1, with the resulting LGDT responses shown in Figure 4.

Table 1
LIGHT CONDITIONS

Lighting	Irradiance (W/m^2)
Dark	0
Ambient	3.9e-3
Fluorescent	0.39
Low Spot	6.3
High Spot	10.4

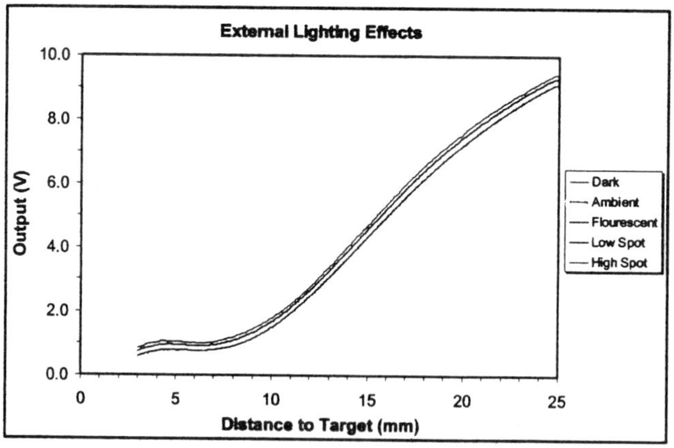

Figure 4. Without optical bandpass filters, the LGDT exhibits a slight offset sensitivity to high intensity external illumination, but not to normal room lighting. Optical filtration eliminates this sensitivity.

Normal lighting conditions had no measurable effect on the LGDT response, while high intensity target illumination caused an offset voltage. This is expected, because the outer receiver ring, which is larger in cross-sectional area than the inner receiver ring, picks up more external light than the inner ring. This causes a greater proportional increase in signal from the outer ring than from the inner ring, the net result of which is a positive shift in the LGDT output voltage.

Ambient lighting conditions have almost no effect on the LGDT sensitivity, but do cause a signal offset. For applications in which only target movements are measured, the LGDT can be used without calibration for external lighting, as long as the amount of external light remains constant during measurements. For absolute distance measurements in high light environments, the LGDT requires calibration for lighting conditions.

NOTE: These results apply to an LGDT unit without optical filtration. For applications in which ambient light effects must be eliminated, optical bandpass filters can be inserted between the receiver bundles and detectors.

Reflectivity

The LGDT response was tested with target reflectivities of 98%, 75%, 42%, and 3% at the wavelength of the LED. The results, shown in Figure 5, clearly indicate that the LGDT performs well with targets of almost any reflectivity. While the LGDT utilizes a ratiometric measurement technique, and thus has an output independent of illumination intensity, the automatic LED power control increases LED output as target reflectivity decreases This increases light levels, and hence, signal-to-noise ratios for targets of low reflectivity. Below a certain reflectivity the LED power limit is reached and signal-to-noise ratio begins to degrade. The exact reflectivity at which this occurs depends on the optical probe used. As can be seen in Figure 6, the effect does not occur with a target of 42% reflectivity, but is beginning to appear at 3% reflectivity.

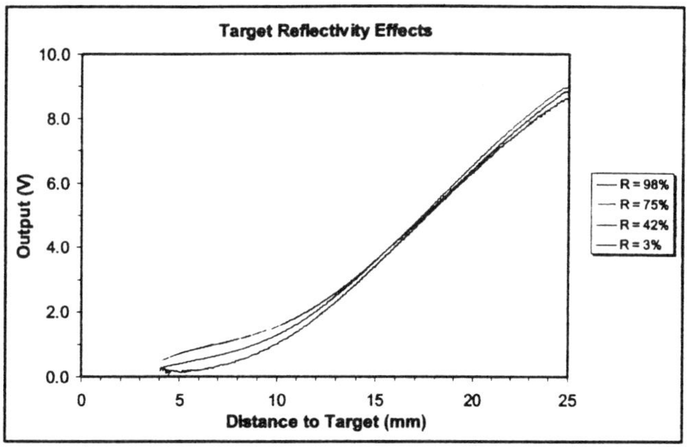

Figure 5. Testing of targets with reflectivity from 3% to 98% shows excellent performance over an extremely broad reflectivity range.

Diffusivity

The target characteristic with the greatest effect on LGDT response is diffusivity. In order of increasing diffusivity, the four targets tested were: an aluminum front-surface mirror, bare aluminum with surface oxidation, aluminum scoured with 120 grit sandpaper, and plain white paper. The LGDT response to each of these targets is shown in Figure 6.

Increased target diffusivity results in a positive voltage offset and anincrease in sensitivity. The explanation for this effect is found in the way in which light leaves the target. Almost all light reflected from a specular target is reflected at an angle that is equal to, and opposite in direction from, the angle of incidence. The pattern of light reflected from the target is constrained by the divergence angle of the light emitted from the probe illumination bundle. Highly diffuse targets, in contrast, scatter rays in random directions about the theoretical reflection angle, causing a spread of the reflected light. The more diffuse the target, the greater the spread of the reflected light. This results in more light striking the outer receiver ring at closer target distances, causing an increase in the LGDT output. The effect also increases with increasing target distance, resulting in higher sensitivity for diffuse targets than for specular targets.

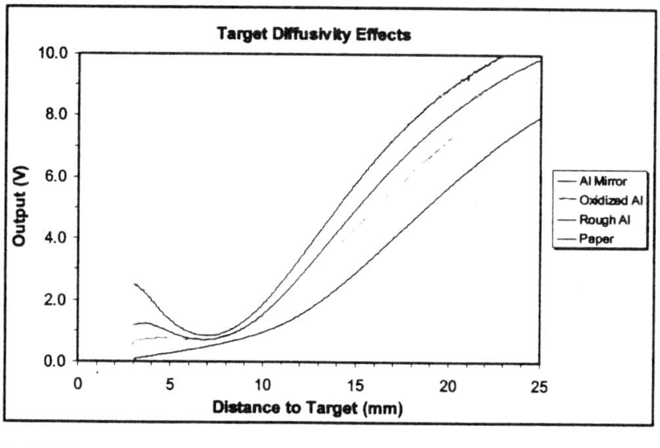

Figure 6. The LGDT works with either specular or diffuse targets. Diffuse target surfaces produce a positive offset voltage and increased sensitivity as compared to specular targets.

Angular Misalignment

The degree to which angular misalignment affects LGDT response depends on the diffusivity of the target. Each of the four diffusivity test targets above was used to test angular misalignment. Starting with the targets parallel to the probe end face, the targets were rotated through nine degrees, with a response scan acquired at one-degree intervals. Figure 7 shows the target angle effects for the front-surface aluminum mirror, oxidized aluminum, aluminum scoured with 120 grit sandpaper, and plain white paper, respectively.

The effect of target angle is much greater for highly specular targets than it is for diffuse targets. Because highly specular targets reflect light in a specific direction that is dependent on the incident light angle, rotation of the target causes a twice as great rotation of the light rays leaving the target, shifting the reflected light pattern across the probe. The pattern of light reflected from a diffuse target is broadened by random fluctuations in the reflected light angle. If the magnitude of the random scattering angles is greater than the rotation angle of the target, there will be little change in the light pattern incident on the probe receiver rings. This effect is seen quite clearly in the data.

Target diffusivity can be used to an advantage in applications in which the target angle may be changing or unknown. The use of a diffuse target reduces, or even eliminates, target angle effects. Care must be taken, however, to balance target angle immunity with signal-to-noise ratio. Extremely diffuse targets can cause low signal levels, and hence, low signal-to-noise ratios.

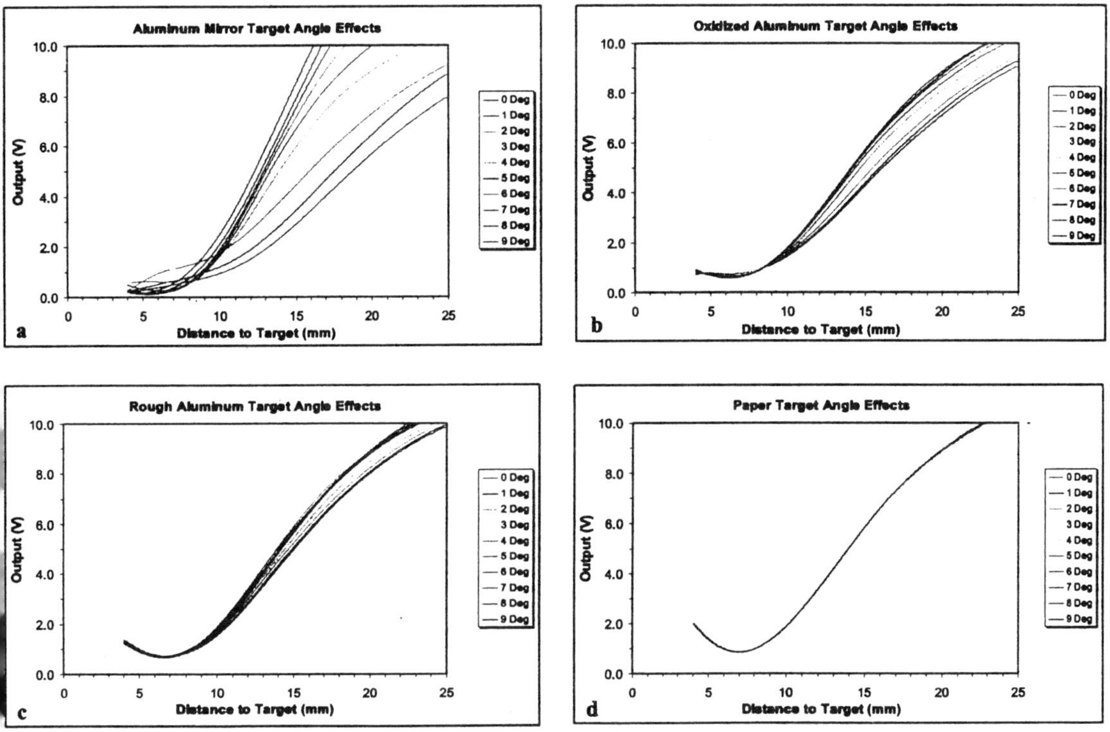

Figure 7. Target surface angular misalignment can have a significant effect on the LGDT output with highly specular targets such as a) an Al front-surface mirror. The effect decreases with targets of increasing diffusivity such as b) oxidized Al and c) rough Al, and almost disappears completely with highly diffuse targets such as d) paper.

APPLICATIONS

The ability of the LGDT to work with a myriad of target materials makes it an attractive solution to an equally diverse spectrum of proximity sensing needs. The following list presents just a few of the many applications of LGDT:

- Fast-steering mirror displacement sensing.
- Voice coil actuator closed-loop control.
- Control of shaft levitation in magnetic bearings.
- Web position sensing in web processing.
- Real-time concentricity measurements.
- Bearing runout measurements.
- Sensing mechanical displacements under dynamic load.
- Vibration measurements.
- Pressure diaphragm sensing.
- Precision liquid level monitoring.
- Audio speaker response testing.
- Flow valve closed-loop control.
- IC package and PC board inspection.

CONCLUSION

The Linear Gap Displacement Transducer presented offers excellent features for a broad range of applicability. These features include: high resolution, outstanding repeatability, broad linear range, diverse operational environments, and adaptability to numerous target types. BEI's LGDT, employing a patented fiber optic probe, has the ability to perform in harsh environments, while delivering high performance and reliability.

AAS 03-050

EXTREME ACCURACY STAR TRACKER IN SUPPORT OF HYPER PRECISION COLD ATOM INTERFEROMETRY

S. P. Airey,* G. Bagnasco,*
M. Barilli,† S. Becucci,† G. Cherubini† and A. Romoli†

HYPER stands for Hyper-precision cold atom interferometry in space. HYPER is a potential Future Flexi-mission of the European Space Agency, ESA, in the fundamental physics domain.

Following a preliminary feasibility study (Ref 1) conducted in the Concurrent Design Facility of the European Space Technology Center, ESTEC, which was supported by Professor W. Ertmer of the "Institut für Quantenoptik, Hanover" and his team of scientists, ESA awarded Astrium Germany and its industrial team with a "follow-on" Feasibility Study Contract. As part of this team, Galileo Avionica was responsible for the design and performance requirements simulation of the optical bench and precision star tracker, two of the components at the heart of the HYPER payload.

This paper provides first of all an essential but quick introduction to cold atom-wave interferometry which is at the core of the high-performance HYPER payload and provides the capability of measuring accelerations and rotations with extremely high accuracy.

A summary of the scientific mission objectives, which entail the repeated measurement of extremely small rotations and accelerations over at least one-year period with respect to an inertial reference frame, is described. Thereafter, a short resume of the main mission requirements is given. Among these requirements are two "critical" requirements, which dominate the overall spacecraft design:

1. The superior thermo-mechanical stability between the HYPER instruments and a Precision Star Tracker, PST, which provides the inertial attitude reference in science mode (Spectral Density=2.75×10^{-10} rad/Hz$^{-0.5}$, in the frequency range between 3.2×10^{-5} Hz and 0.15 Hz);
2. The PST Absolute Measurement Accuracy (Spectral Density=1.75×10^{-9} rad/Hz$^{-0.5}$, in the frequency range between 3.2×10^{-5} Hz and 5 Hz).

Finally, this paper highlights the challenges the Industrial Team faced in meeting those requirements and describes the optical design of this extremely accurate Precision Star Tracker. The preliminary results of the performance requirements simulation are also presented.

* Affiliated with the European Space Agency (ESA).

† Affiliated with Galileo Avionica.

ATOM-WAVE INTERFEROMETRY

In the last two decades matter-wave interferometry has been pursued mainly in order to demonstrate and test the wave character of matter, one of the most puzzling effects in physics, predicted by quantum mechanics. Only recently has the technique of matter-wave interferometry also been used to realise very high-performance instruments for the measurement of accelerations, rotations and frequency shifts (Ref. 2). The technological advances that make atom-wave interferometry useful as a high accuracy sensor are those related to developments in the following two main fields:

1. Laser-controlled high precision Raman spectroscopy for the manipulation and control of cold atoms

2. Cooling techniques applied to matter-wave atom clouds to increase their coherent interaction time.

These two techniques are brought together in an Atomic Sagnac Interferometer (ASI), where the Raman laser spectroscopy allows, as in a classical optical interferometer, the coherent splitting, redirection, recombination and spectroscopic detection of atoms by modifying their internal electronic state. At the same time, it modifies their macroscopic state, by absorbing photon momentum, leading to the formation of a so-called Sagnac loop, as shown in Figure 1, conceptually identical to a standard optical Sagnac interferometer used in Fiber Optic Gyros.

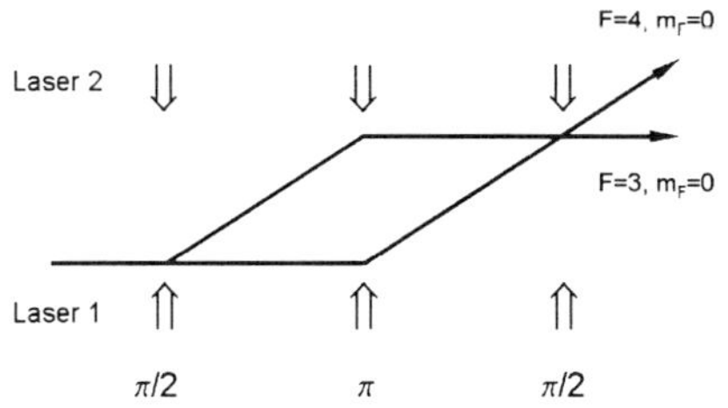

Figure 1: Split, redirection and recombination of atoms by lasers in the two interferometer arms, leading to the formation of a Sagnac Loop.

In an interferometer experiencing rotations and/or accelerations the symmetry between the two matter-wave beams, which travel in opposite directions along the interferometer arms, is broken. This symmetry breakdown causes a measurable interference pattern, formed when the arms re-join. It is this measurement that is used to sense the accelerations and rotations to which the loop has been subjected.

In order to discriminate between accelerations and rotations two ASIs, propagating in opposite directions, are combined into a single instrument, called the Atom Sagnac Unit (ASU). The effects of the rotations will affect these ASIs differently whereas the accelerations will cause the same effect. The signals from the two ASIs can therefore be combined to measure, independently, either the pure rotation or the pure acceleration without cross talk from the other effect.

The conceptual configuration of an ASU is shown in Figure 2.

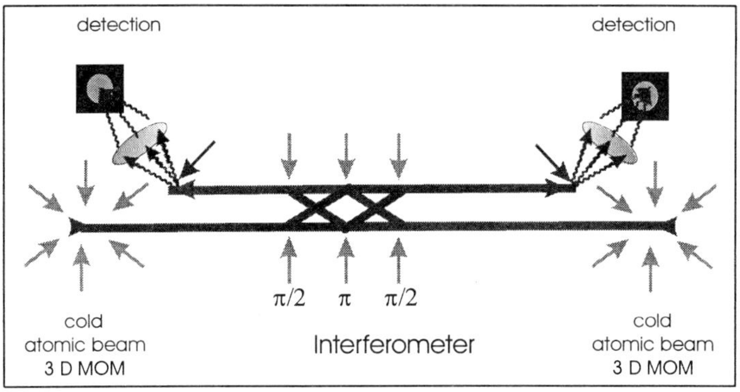

Figure 2: Conceptual configuration of an ASU.

In each ASI, as in a classical optical Sagnac interferometer, a phase shift ΔM is created.

$$\text{While } \Delta M_{Sagnac\text{-}Optical} = 4\, B\, A_{opt}\, S / c\, \lambda \quad (1)$$

It can be shows that the equivalent phase shift in a matter –wave interferometer is

$$\Delta M_{Sagnac\text{-}Atoms} = 4\, B\, m_{atom}\, A_{atom}\, S / h \quad (2)$$

This phase shift between the two waves, travelling along the two arms of the interferometer, is caused by the Doppler effect between the atomic waves and the manipulating lasers, which are essentially fixed. Comparison of equation (1) and (2) suggests a theoretical superiority in sensitivity of the matter-wave interferometer, as:

$$\Delta M_{Sagnac\text{-}Atoms} / \Delta M_{Sagnac\text{-}Optical} = m_{atom}\, c\, \lambda / h \quad (3)$$

For typical experimental parameters (λ = 633 nm, m_{atom} = Caesium mass) the increase in sensitivity would result in a factor 10^{10}, if the areas A were equal. In reality A_{atom} is << A_{optics}

It can be shown that A_{atom} in equation (2) is equal to

$$A_{atom} = L^2\, A_{atom}\, V_T / V_L \quad (4)$$

where L is the length of the interferometer, $V_T = hk/ 2B\, m_{atom}$ is the transverse velocity due to the two-photon momentum transferred to the atomic beam during the Raman transition, $V_L = (2k_BT/ m_{atom})^{1/2}$ is the drift velocity of the atomic beam in the interferometer. From equation (2) and (4) it can be seen that the highest potential for further improvements of the sensitivity in the detection of $\Delta M_{Sagnac\text{-}Atoms}$ in an atom-wave interferometer resides in the lowering of the drift velocity V_L of the atoms by means of cooling techniques.

Since 1975 several techniques have been developed to cool a large amount of atoms (10^9 - 10^{11}) hence slowing them down. Table 1 identifies some of these techniques together with the temperatures and drift velocities they can typically achieve.

Laser cooling does not only reduce the mean drift velocity, V_L, of the atoms but also increases their coherence. Indeed, if all the atoms could be controlled to exactly the same velocity they would then behave as a single coherent wave whose wavelength would be given by the de-Broglie relation:

$$\lambda_{d\text{-}B} = h / m_{atom} V_L \qquad (5)$$

Thermal Atomic beam	300 K	300-2000 m/s
Magneto-optical trap	0.1 – 10 mK	0.1 – 10 m/s
Sub-Doppler Molasses	1 – 100 μK	1 – 10 cm/s
Bose-Einstein Condensation	10 – 100 nK	1 mm/s

Table 1: Atom Cooling Techniques and Performances.

At present, ASU rotation measurement accuracy of the order of 5.0×10^{-12} rad/sec or better with respect to an inertial reference frame, at a sampling frequency of 0.3 Hz are considered feasible.

MISSION OBJECTIVES

The primary scientific objectives of the HYPER mission is are:

- To map, for the first time, the latitudinal structure (magnitude and sign) of the Lense-Thirring effect (LTE) induced by the Earth's rotation and predicted by the General Relativity with a measurement accuracy of 10%
- To determine, independently from Quantum Electro-Dynamics theories, the "fine structure" constant, $(=2Be^2 / hc)$, with a measurement accuracy improvement of one order of magnitude with respect to the present experimental results
- To investigate various distinct sources of matter-wave de-coherence effects, predicted by quantum gravity theories

Furthermore, as a secondary objective the successful completion of this mission would demonstrate for the first time the superior performance of cold atom matter-wave interferometer instruments. HYPER might therefore lead the way to a novel generation of inertial sensors and gyroscopes that would be needed to enable other future missions. The scientific objectives given above are only feasible in space, where the potential of cold-atom matter-wave interferometry sensors can be fully exploited thanks to the provision, via drag free satellite control, of an environment almost totally free of disturbing accelerations, such as those due to gravity on Earth.

The present Industrial Feasibility Study concentrated only on demonstrating the feasibility of the Lense-Thirring (Ref 3) effect measurement. This effect causes the Gravitational Field surrounding a rotating mass to differ from that surrounding a non-rotating mass. A simple analogy can be drawn from Electromagnetic Field theory: a rotating charged sphere will create both an electric and a magnetic field, while a non-rotating charged sphere will produce only an electric field. In general, slowly rotating bodies like the Earth create only a tiny modification to the gravitational field with the consequence that their effects on the motion of a body in its immediate surroundings are almost totally insignificant. However, these tiny modifications to the local gravitational field will cause a gyroscope, placed in an orbit around the Earth, to process according to the following formula:

$$\Sigma_{LT} = G/r^3 \, [3 \, r \, / \, r^2 \, (S_e \, \theta_r) - S_e] \qquad (6)$$

(S_e is the spin angular momentum of the Earth and r is the vector identifying the position of the gyroscope).

Although sign and strength of the induced rotation, Σ_{LT}, depends on its latitudinal position along the orbit, the average precessional rate of the gyroscope over 1 orbit (in a 1000 km polar orbit) would be of the order of only 10^{-14} radians/sec.

This implies that with the previously quoted ASU rotation measurement accuracy (5.0×10^{-12} rad/sec, at a sampling frequency of 0.3 Hz) HYPER will need to gather the order of 10^7 measurements for about one year, in order to measure such a tiny effect with a 10% accuracy.

PAYLOAD REQUIREMENTS AND DESIGN SOLUTIONS

The performance requirements on the HYPER payload are driven by the need to measure the Lense-Thirring effect. The payload must therefore be capable of measuring the two perpendicular Σ_{LT} components and of distinguishing the rate measurements from other effects such as accelerations. One ASU, described before, allows the unambiguous measurement of rates around a single axis. Therefore, two independent and orthogonal ASUs are needed for HYPER. This baseline configuration is presented in Figure 3.

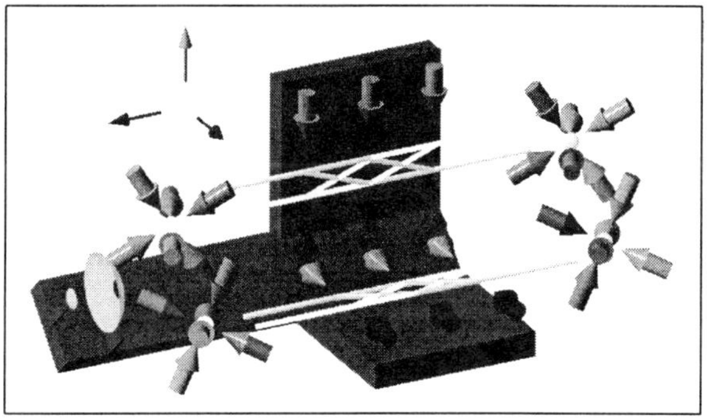

Figure 3: Conceptual Payload configuration showing two ASUs, each consisting of two counter propagating ASIs, mounted on a single optical bench together with the Precision StarTracker.

However, each ASU measures the relative rotation of the atomic waves with respect to the manipulating lasers. As the lasers and the s/c are contained within the same local inertial frame they too will experience the Lense-Thirring effect. It is therefore fundamental that the manipulating lasers are maintained fixed with respect to a 'non-local' inertial frame. A guide star provides the reference for this 'non-local' inertial frame, tracked by a Precision Star Tracker. The two main payload requirements can thus be summarised as follows, in terms of spectral density:

1. The ASU-PST alignment stability: 2.75×10^{-10} rad/$Hz^{-0.5}$, in the frequency range between 3.2×10^{-5} Hz and 0.15 Hz

2. The PST Absolute Measurement Accuracy: $1.75x10^{-9}$ rad/$Hz^{-0.5}$, in the frequency range between $3.2x10^{-5}$ Hz and 5 Hz).

The design concept adopted for HYPER foresees the two ASUs (made up of 2 ASIs each) and the PST to be integrated together in a single compact module, which acts as a common optical bench. This design concept allows minimising the misalignment error sources between the PST and the two ASUs co-ordinate frames. In order to meet the challenging relative stability requirement over one orbit (about 100 minutes) and respect at the same time the mass limitations imposed by the selected Rocket launcher (< 1000 kg) ULETM (glass ceramic Titanium-silicate) has been selected as the baseline material for the Optical bench. The overall arrangement is shown in Figure 4.

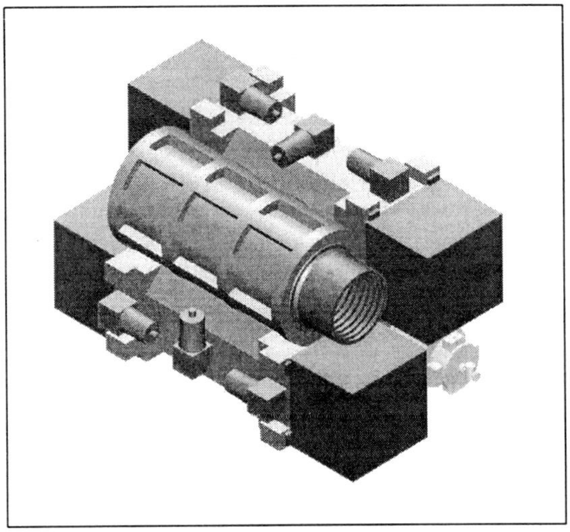

Figure 4: Payload Design configuration showing respectively the four ASIs (in red), the ULETM optical bench with the integrated Precision Star-Tracker (in blue). Three laser fibre injectors (in green) for each ASU determine the paths of the manipulating lasers, which in turn determine the sensitive directions of each ASU. The main ASU reference frame is therefore defined by the line through the centre of these reflectors and is nominally perpendicular with the Precision Star-Tracker boresight.

The task of fitting the PST telescope into a 700 mm optical bench and of ensuring its manufacturability and testability was another major design driver. Two kinds of optical concepts were considered initially for the PST. One foresaw a focusing telescope, based on a Cassegrain or Ritchey-Chretien configuration; while the other consisted of an a-focal Galileian telescope using a negative lens as collimating element after a Cassegrain telescope. The first concept has the advantage of simplicity and it is less critical for what concerns surface defects and cleanliness of the secondary mirror. This mirror is the effective optical surface, which is crossed by the smallest section of the beam. The smaller its area, the greater is the undesired effect of a surface defect on the final image.

The second concept of the a-focal telescope, having one additional element, is more sensitive to misalignments and its residual surfaces defects and particle contamination could result be more critical than the first option, changing the shape of Point Spread Function, PSF, and the background illumination depending from the star position. Both phenomena can be considered highly undesirable for the very high precision required to the PST. Indeed the background can decrease the sensitivity of the PST, and the change of PSF shape can affect ground, without taking into account in flight contamination. As a result of these configurations, the first concept was initially selected.

The following trade-offs led to refinement of the design concept on the basis of the system error analysis, which led to consider an optical system whose characteristics are resumed in table 2.

Effective Focal Length	36.000 mm
Useful diameter	190 mm
Central Obstruction diameter	90 mm
FOV	±25 sec
Resulting F/number	190

Table 2

Particularly important was the assumption of the effective focal length of 36 m. Such a large focal length could cause strong criticality when realized with a simple Ritchey-Chretien. Indeed, since the overall optical bench length is constrained to 700 mm, assuming that the CCD relief with respect to the primary mirror effective surface is (for instance) 40 mm, the first order optics layout can be set up as shown in Figure 5.

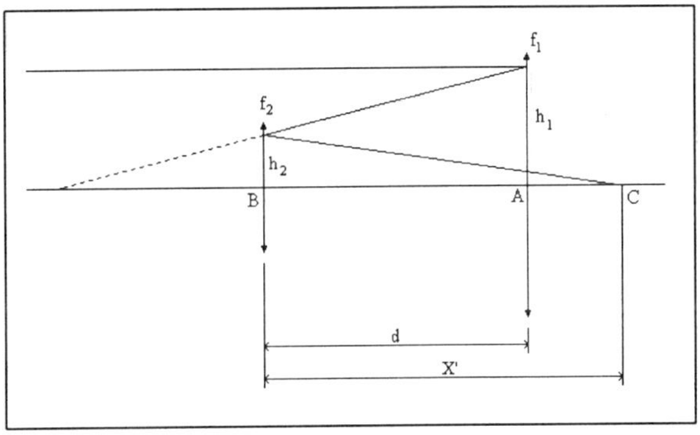

Figure 5: Ritchey-Chretien first order layout.

The layout of Figure 6 can be considered as a collimated telescope made by two elements, one positive with EFL f1 and the other negative with focal f2', followed by a positive element f2".

The telescope has magnification M given by:

$$M = -\frac{f_1}{f_2'} = \frac{h_1}{h_2} \qquad (7)$$

The second element shall have an EFL given by:

$$f_2' = -\frac{f_1}{M} \qquad (8)$$

And the distance d is given by:

$$d = f_1 + f_2' = f_1 \frac{M-1}{M} \tag{9}$$

And the back focal length X' will result:

$$X' = f_2'' \tag{10}$$

Therefore the focal length f2 is given by:

$$f_2 = \frac{1}{\varphi_2' + \varphi_2''} = \frac{f_2' f_2''}{f_2' + f_2''} = \frac{f_1 \cdot X'}{f_1 - F} \tag{11}$$

Then by putting, for instance:

$$X' = f_1 = 700\,\text{mm} \tag{12}$$

You obtain:

$$f_2 = -13.88\,\text{mm} \tag{13}$$

Which means a magnification M of about 50 x.

This magnification is critical if we consider that, for instance, a tolerance of 1 micron on the distance d causes a defocus just inside the depth of focus. Then, in order to reduce this and some other criticality, it is necessary to reduce M.

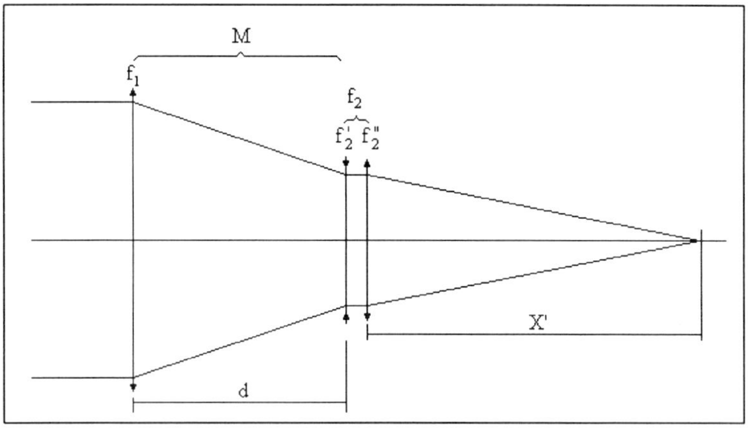

Figure 6 – First order layout with thin lenses.

The improved optical system, schematised in Figure 6, can be considered as a Galileian of magnification M, followed by a lens whose focal length is F/M. The basic idea is to increase the focal length f2" and to fold it twice by mean of two flat folding mirrors. The sizes of the elements were computed as follows:

By indicating with L = 700 mm the overall length of the PST telescope and with R=40 mm the relief of the sensitive area of the CCD detector with respect to the effective optical surface of the primary mirror, then d = L – R = 700 – 40 = 660 mm. A first flat folding mirror M3 is located in the centre of the hole of the primary mirror M1, and it is slightly tilted, in order to reflect the beam outside of the secondary mirror M2. The distance of M3 from M2 is d1 = d = 660 mm.

A second flat folding mirror M4 orthogonal to the optical axis is set very close to the secondary mirror M2. Therefore the distance d2 of M4 from M3 is given by d2 ≈ d1 = d = 660 mm.

M4 reflects the beam coming from M3 toward the CCD, which is mounted at distance R= 40 mm from the front surface of M1. The distance d3 from M4 up to the CCD is then given by d3 ≈ d + R = L = 700 mm. Therefore the Back Focal Length X' of the system will be given by:

$$X' = d_1 + d_2 + d_3 = 660 \cdot 3 + 40 = 2020 \text{ mm}$$

This means that the magnification of the Galileian is given by:

$$M = F/X' = 36000/2020 = 17.82 \text{ x}$$

reducing the magnification ratio of a factor 2.8. Figure 7 shows a perspective view of the optical layout.

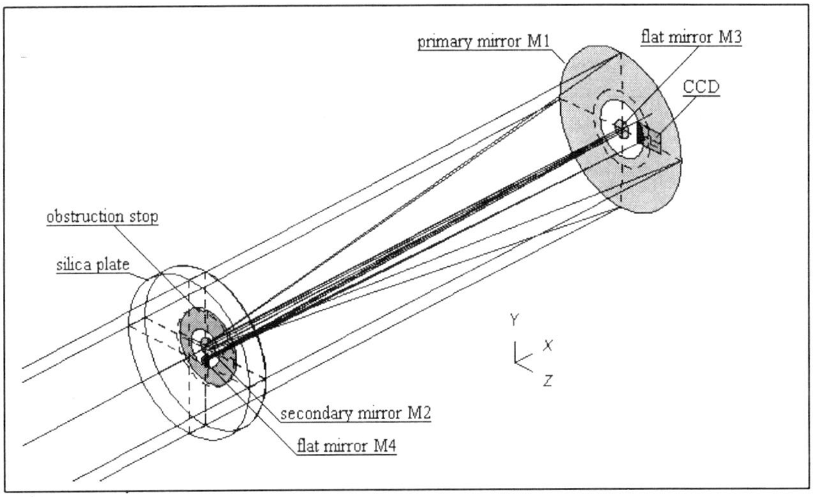

Figure 7: Perspective view of the PST optics with two flat folding mirror.

The beam coming from left crosses a flat parallel plate of silica, and it is reflected by the primary parabolic converging mirror M1 toward the secondary diverging parabolic mirror M2. The beam is folded a first time by M3, slightly tilted (0.6°) with respect to the optical axis of the telescope and a second time by M4, which is obtained on the inner surface of the silica plate. The sensitivity of the system to the distance between M1 and M2 is now decreased in such a way that a change of 5 microns is not critical. This renders easier the assembly of the system. Figure 8 shows another view of the optical layout.

The performances of such PST optics are well inside the diffraction limit and from the point of view of the geometrical aberrations, it can be considered perfect, showing a Strehl ratio very close to 100%

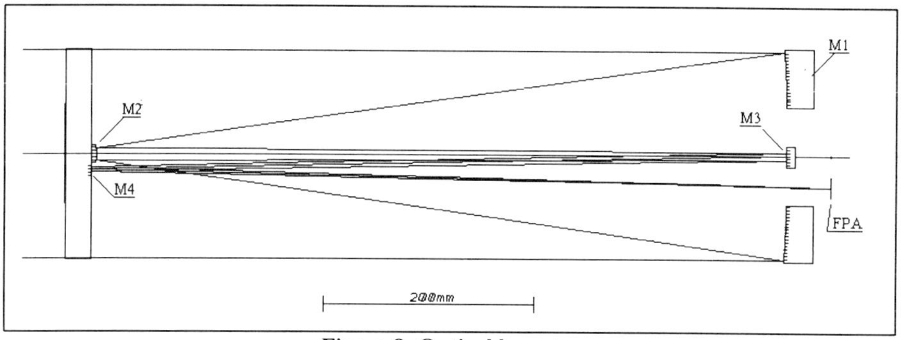

Figure 8: Optical layout.

A sensitivity analysis was performed in order to understand the alignment stability criticality of each optical system element. Table 3 provides the alignment sensitivity analysis preliminary results:

		M1	M2	M3	M4
Tilt	0.001°=3.6"	2.488"	0.405"	0.272"	0.140"
	1"	0.691"	0.113"	0.076"	0.039"
Shift 0.001 mm		0.296"	0.289"	-	-

Table 3

The linear shifts of M3 and M4 have no effects, since they are flat mirrors. Since the two flat mirrors are small and supported respectively by the primary mirror, M1, and secondary mirror, M2, these two mirror groups have been considered, in first approximation, as rigid bodies. For this reason no movement of M1 and M3 with respect of the CCD, supported as well by the M1, is expected and at the same time M2 movements should be equal to those experienced by M4. These considerations lead to table 4 with new, simplified alignment sensitivity values.

		M2-M4 Group
Tilt	0.001°=3.6"	0.545"
	1"	0.152"
Shift 0.001 mm		0.289"

Table 4

The combination of a large photon collecting area (useful diameter= 190 mm), of this compact Ritchey-Chretien optical design, modified with two flat folding mirrors and of a complex centroiding and error correction algorithms, developed for the selected 1024 x 1024 pixel CCD detector, allows the PST to get the best performances related to the limited volume allocated.

BAFFLING AND STRAY LIGHT ANALYSIS

A very important aspect of the PST baffle is that of the internal baffle design. Indeed the external baffle can be considered as a continuation of the internal envelope of the telescope, which is short (about 30 cm) compared with the telescope itself.

The preliminary internal baffle design, which was developed in a previous step taking into account symmetry around the optical axis, has been revised in order to accommodate the two folding mirror. The most part of the baffle is modelled in order to have rotational symmetry, as it results from Figure 9.

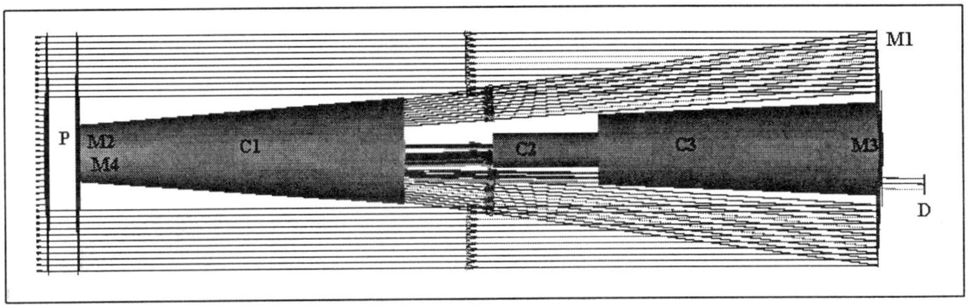

Figure 9 – Lateral view of the PST internal baffle.

On the left there are two obstructing stops, placed before and after the silica flat and parallel plate P. A first cone C1, on the left, based on the plate P, surrounds the secondary mirror M2 and the flat folding mirror M4. It acts a first baffle of the beams. After reflection on the secondary mirror M2 the beam passes through a cylinder C2, which is based on obstructed part of the primary mirror M1 and it surrounds the flat folding mirror M3. The cylinder C2 is better visible in the perspective view of Figure 10.

The beam reflected by M3, which is tilted of 0.6°, comes back towards the flat folding mirror M4, close to the secondary mirror M2.

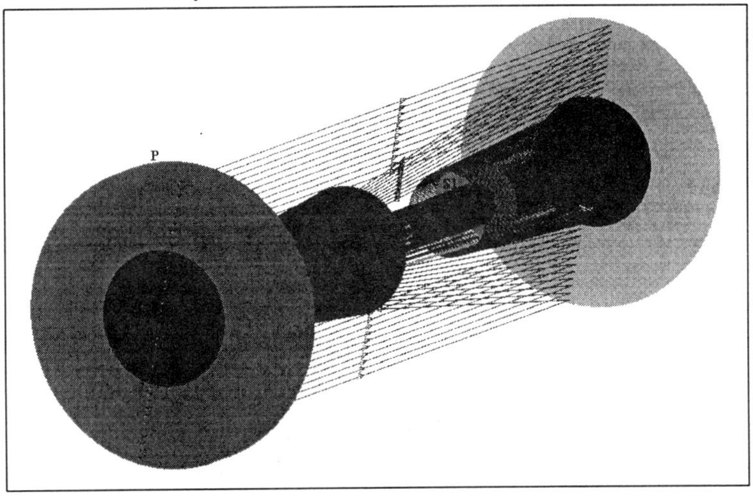

Figure 10 – Perspective view of the PST internal baffle, seen from object side.

After reflection on M4, the beam comes on the detector D, passing between the cylinder C2 and the cone C3, coaxial to C2. A flat stop S1, visible in Figure 10, with an hole allowing the beam reaches the focal plane, is located between C2 and C3. A last stop S2, close to the effective surface of the primary mirror M1, with a small lateral hole allows the primary beam impinging the detector.

Figure 11 shows another view of the internal baffle, showing more in details the cone C1 and the beams inside.

This configuration has a good efficiency, being based on simple geometrical concepts. A preliminary analysis of the efficiency was done with ASAP®™ of Breault Research Organisation, by considering reflective properties of the mechanical walls of the baffle.The result is that no reflected path arrives on the detector. Further analyses have been done. Figure 12 shows the efficiency of the baffle when painted with Chemglaze Z302.

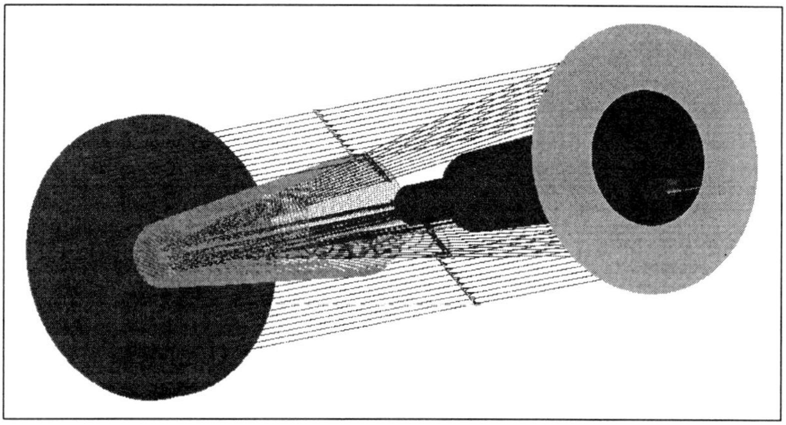

Figure 11 – Perspective view of the PST internal baffle, seen from image side.

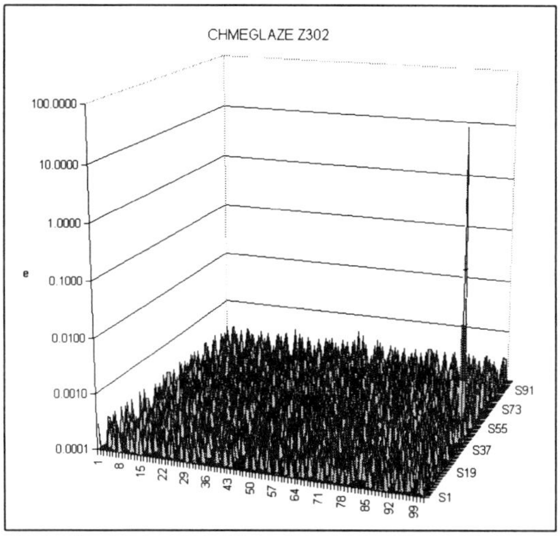

Figure 12 – Baffle efficiency with Chemglaze Z302.

Figure 13 is shows the efficiency of the baffle coated as a Lambertian diffuser with 5% of reflectivity. In the first case the attenuation of stray light is better than 10^{-4} and in the second it is well under the value of 10^{-5}.

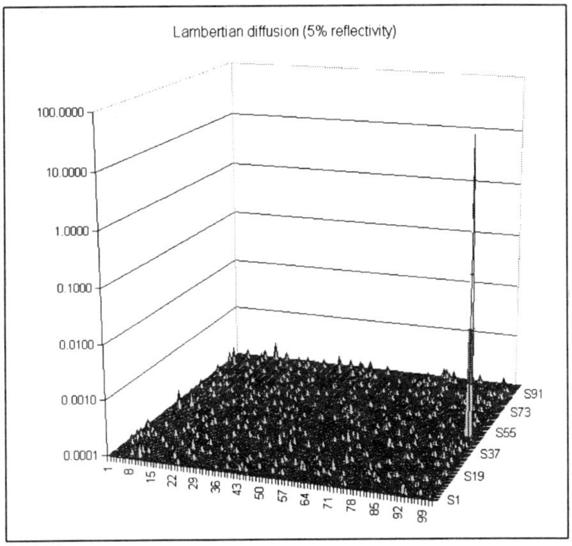

Figure 13 – Lambertian diffusion (5% reflectivity).

HYPER PRECISION STAR TRACKER

The characteristics of the PST baseline have been obtained as result of a trade-off performed with the following parameters:

- IFOV REDUCTION, in order to reduce the contribution of all errors that can be characterised in terms of fraction of pixels such as Centroid and Noise Equivalent Angle (NEA): such reduction can be obtained by a longer focal length.

- INCREASE OF INTEGRATION TIME in order to increase the star signal on the CCD

- INCREASE OF CCD Full Well Capacity (Pixel size), in order to avoid CCD saturation

- MAINTAIN THE OPTICS DIAMETER, in order to keep the PST assembly mass

- TRACKING MATRIXES, in order to match larger PSF produced by high F # and reduce Centroid error in terms of fraction of pixel

The limiting magnitude has been driven by the need to have a sufficient number of suitable guide stars always available as the orientation of the orbit changes over one year and the need to guarantee a good signal to noise ratio. In order to evaluate in detail all the optical configurations and centroid algorithms, the GA single star simulator has been used. The simulator is able to take into account the use of various methods of centroiding, including the required very large tracking matrixes and different ways to weigh pixel values. All the pointing errors have been considered, showing that the main error contributors from the PST are the Centroid error and NEA. These simulator runs have allowed the best compromise between noise and signal sensitivity, with full consideration of the centroiding algorithm method and CCD selection, to be found.

The best performance was seen to be obtained by a 13 mm pixel size CCD, an IFOV of 0.074 arcsec and a centroid algorithm based on 17x17 pixel tracking matrix.

Considering Hipparcos star catalogue data, a guide star list containing 48 stars was constructed. The selected stars have the following main characteristics:

+2.81	≤ V (Visual Magnitude) ≤	+3.99
+0.15	≤ B-V Colour index ≤	+1.67

Additionally there are no variable, binary or double stars and the proper motion of each of the stars is known with an accuracy of better than $8*10^{-3}$ arcsec/year and their right ascension, declination and parallax errors are known with an accuracy better than $3*10^{-3}$ arcsec

THERMO-MECHANICAL ASPECTS

As can be deduced from table 3, showing the telescope sensitivity analysis, a shift of the M2-M4 group of 1 nm (with reference to M1-M3 group) produces a tilt on the optical axis of the same order of magnitude of the required stability ($2.75x10^{-10}$ rad/Hz$^{-0.5}$, in the frequency range between $3.2x10^{-5}$ Hz and 0.15 Hz). This consideration gives the feeling about the sensitivity of the system and on which level of detail this study has to reach to "assess" the foreseen compliance of the instrument to the science requirements.

The best way to reach this stability is to minimize thermal and mechanical disturbancies of the OB environment. Causes of main thermal disturbances are:

- Earth albedo, producing a rotating heat load on the S/C around the instrument boresight, occuring at almost the same frequency of the LTE,
- Constant heat leackage through the optical entrance aperture to CS
- The payload elements heat dissipation

These disturbances have all been minimized by an active thermal control, that takes care of the functional elements dissipation by maintaining their I/F temperature very stably. MLI shielding has permitted to reduce the radiative effects of Earth Albedo on the bench and the leakage to space has been cut out by inserting a temperature controlled quartz window in front of telescope. The remaining thermal disturbances come from a residual temperature change on the structural I/F to S/C, which occurs at the orbit frequency, and by the heat losses within the Fibre injectors and the retro reflecting mirrors. To evaluate their impact a FE model of the optical bench structure was constructed and its behaviour verified under these thermal loads. Figure 14 shows the Finite Element model of the Optical Bench.

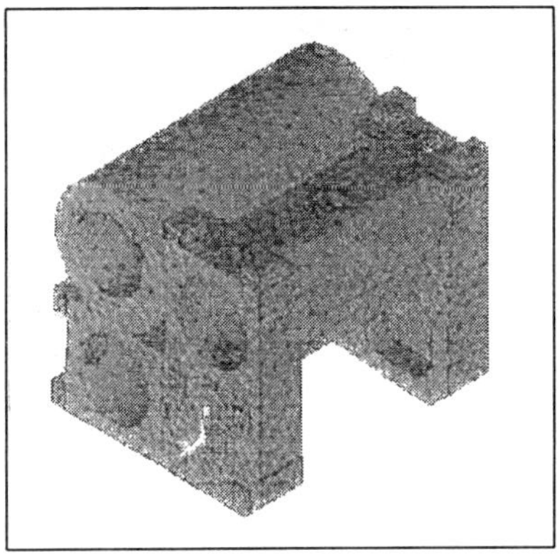

Figure 14: 3D View of the Optical Bench Finite Element Model.

This model has been used to track the movement (generated by the thermal environment changes) of all the 63 points of interest for the requirements verification. To track the Telescope optical axis movement during one orbit, maintaining the number of data within acceptable limits, the rotating orbital loads have been divided in 8 positions, as shown in Figure 15. Loads relevant to positions F to B (through A) are referring to both OB I/F to S/C temperature changes, of the order of 0.1°C, and to the heat injection from thermally uncontrolled functional elements. Loads from C to D show only the latter, being the Earth Albedo influence on the OB I/F shielded by the S/C structure. To each one of these positions the relevant thermal loads have been applied for three orbits, to reach a steady state.

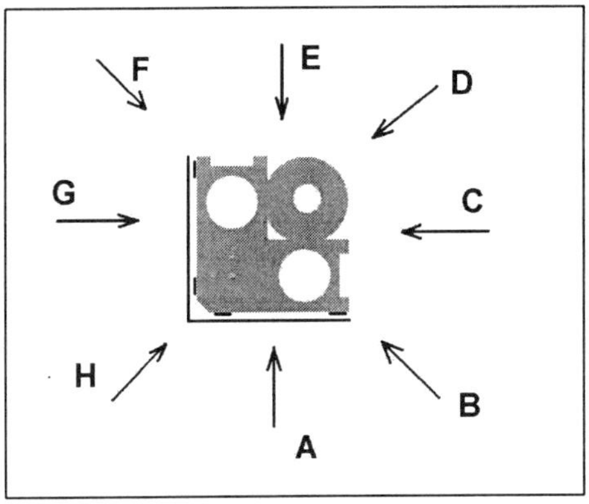

Figure 15: Orbit load steps: PST Boresight normal to figure plane.

Figure 16 shows the thermal surface behaviour of the OB under step H loads (temperatures in Kelvin), the maximum temperature gradient within the bench is 0.2 K. Figure 17 gives the relevant displacement profile (dimensions in mm) showing a maximum displacement of about 16 nm.

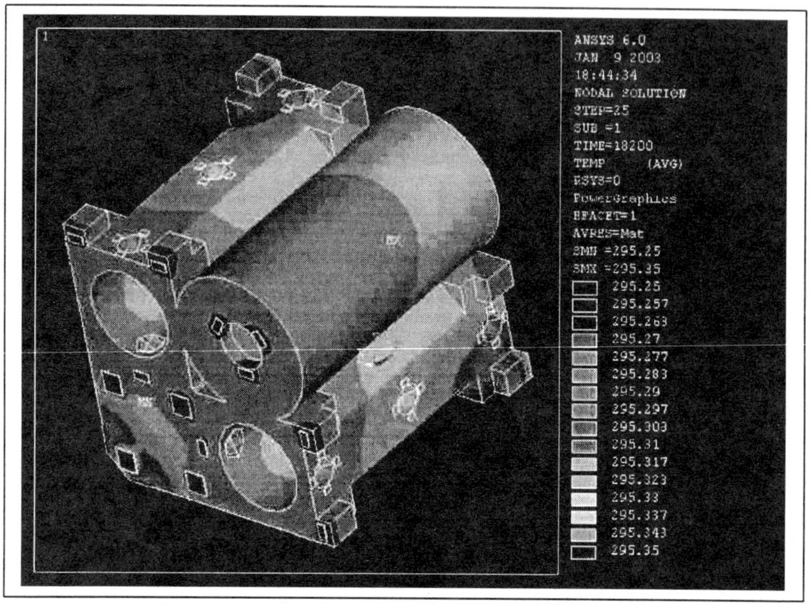

Figure 16: OB FEM thermal profiles for step H.

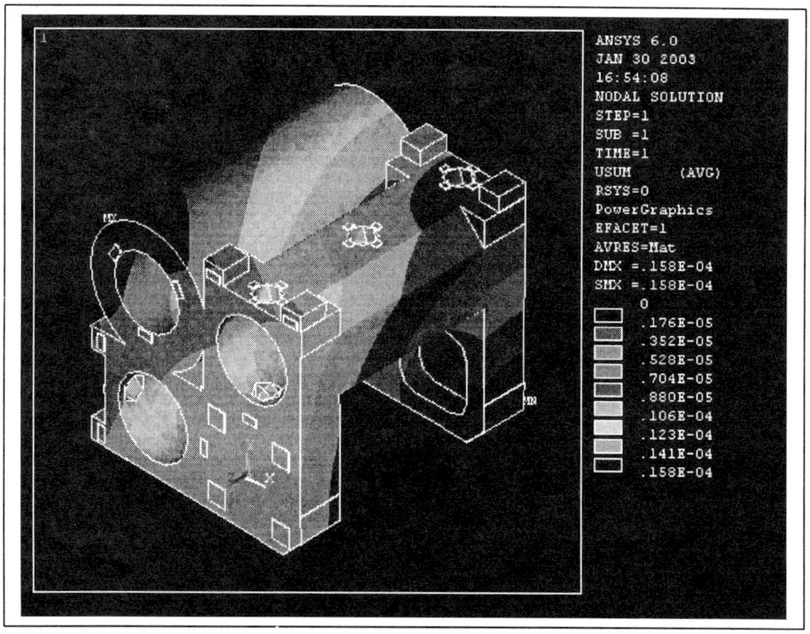

Figure 17: OB FEM displacements for step H

As shown in Figure 17, the maximum displacement (red / top left corner) is found at the opposite side of the fixed I/F point (blue / bottom right corner). These results are currently under evaluation. The maximum displacement dimension allows to be optimistic on the stability requirement compliance.

CONCLUSIONS

The HYPER feasibility study, with its extremely stringent requirements in terms of pointing accuracy and stability, allows to push the limits of Star Tracking technology and further expand the large precision sensing knowledge base in Galileo Avionica. The first rcsults from this feasibility study, which is still to be completed, show great promise that the extremely stringent requirements demanded by the HYPER mission can be met.

ACKNOWLEDGMENT

The authors would like to thank Astrium GmbH, who are running the HYPER study and were responsible for the mission requirement breakdown to the level of the Optical Bench and PST. The authors would also like to acknowledge the essential and longstanding contributions of Dr. Ernst Maria Rasel, Dr. Philippe Bouyer and Dr. Arnaud Landragin, who formed the core of the advisory science team. The key contributions to the study of the other members of the HYPER Scientific core team and the industrial feasibility study team are also gratefully acknowledged.

REFERENCES

1. "HYPER: Hyper Precision Atom Interferometry in Space" ESA CDF-09, September 2000.

2. Ch. Borde': "Atom interferometry and laser spectroscopy" in Laser Spectroscopy X, World Scientific, 239 (1991).

3. J. Lense, H. Thirring, Phys Z. 19, 156 (1918).

Section V:
GUIDANCE AND CONTROL DESIGN APPROACHES FOR PRECISION POINTING APPLICATIONS

SESSION V

Joint Session Chairperson: Brian Kawauchi
The Aerospace Corporation

Joint Session Chairperson: Steven Lee
Jet Propulsion Laboratory

Local Session Chairperson: Jim D. Chapel
Lockheed Martin Astronautics

The following paper was not available for publication:

AAS 03-062 "Optimal Maneuver Control of a Precision Imaging Satellite," by W. Martin, P. Heffernan, Lockheed Martin Space Systems (Paper Withdrawn)

The following paper numbers were not assigned:
AAS 03-068 to -070

HUBBLE SPACE TELESCOPE RELATIVE POINTING STABILITY POST-SERVICING MISSION 3B

A. Bradley,[*] J. L. Kelley,[†] J. H. Wirzburger,[‡] E. Nelan,[**] D. Smith,[††] O.G. Franz,[‡‡] L.H. Wasserman,[‡‡] M. Wenz[††] and D. Crain[‡]

We present an overview of the most recent Shuttle Servicing Mission of Hubble Space Telescope, conducted in March, 2002, and present the latest relative pointing stability measurement for Hubble Space Telescope. The measurement was obtained from observations with the Astrometric science instrument, a two-axis shear interferometer, which also functions as a Fine Guidance Sensor. Position and Transfer Function mode observations were performed on the 8.59 m_V star HD225146. We find on average $\sigma = 5.5$ mas.

INTRODUCTION

The Servicing Mission 3B (SM3B) EVA's were conducted from March 4 thru 8 of 2002, it was the fourth Shuttle Servicing mission since it's launch in April 1990. This mission accomplished the installation of shorter-rigid Solar Array's, Reaction Wheel #1 and a NICMOS CryoCooling System. Each of which could have impacted the spacecraft relative pointing stability (jitter).

Following the deployment of HST in 1990, it became clear that the flexible Solar Array's were the dominant source of jitter. MSFC initiated a pointing control system design improvement study in 1993, whose results[1-6] were presented at the Keystone conference. While the studies were of interest, the HST Project decided to continue with the existing control law redesign[7], known as SAGA-GA. These early studies addressed the vibrations from the Solar Array I flexible panels. On December 6, 1993, the Solar Array II flexible panels were installed during SM1. While the Solar Array II panels were an improvement, control law modifications were still necessary. In this most recent Servicing Mission (SM3B), rigid Solar Array panels were installed. One requirement was that the new Solar Array's be compatible with the existing HST interface requirements of mechanical, thermal, electrical, commanding, slews and settling time. The on-orbit results demonstrate that is no longer necessary to use the SAGA control law. Thus, we have switched back to the original as-launched PID control law.

* Spacecraft System Engineering Services, P.O. Box 91, Annapolis Junction, Maryland 20701.

† Lockheed Martin Missiles & Space, 7474 Greenway Center Drive, Suite 200, Greenbelt, Maryland 20770.

‡ Honeywell Technology Solutions Inc., 7474 Greenway Center Drive, Suite 200, Greenbelt, Maryland 20770.

** Space Telescope Science Institute, 3700 San Martin Drive, Baltimore, Maryland 21218.

†† Lockheed Martin Technical Operations, 7474 Greenway Center Drive, Suite 200, Greenbelt, Maryland 20770.

‡‡ Lowell Observatory, 1400 West Mars Hill Road, Flagstaff, Arizona 86001.

SPACECRAFT JITTER

Qualitative Jitter Indicator

The frequency of FGS disturbance related loss-of-lock events are a good qualitative indicator of jitter. When a vehicle disturbance is excessive in magnitude (usually > 80 mas), an FGS can lose Fine Lock on a guide star. Figure 1 reflects the fact that since SM3B, there have been no loss-of-lock events. As an aside, a seasonal influence on vehicle disturbances is also evident.

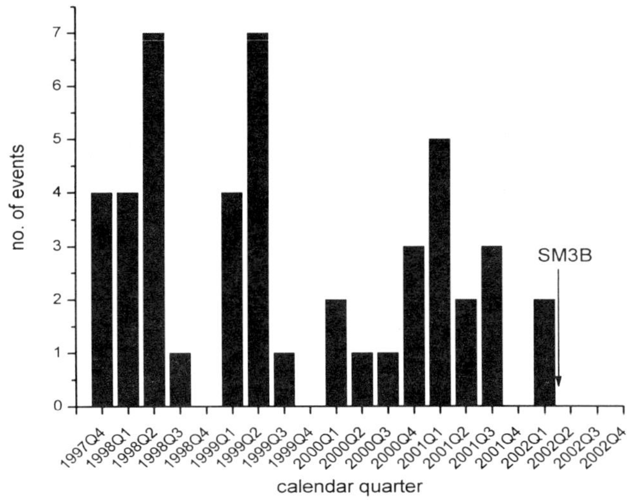

FIGURE 1 Frequency of loss-of-lock events due to vehicle disturbances.

FGS Jitter Test

The Fine Guidance Sensor (FGS) Jitter Test was executed on day 127, following SM3B. The Jitter Test consisted of four consecutive Position (POS) mode observations of a single star, followed by eleven consecutive Transfer-Function (TRANS) mode observations (105 transfer function scans) of the same star. The objective was to determine the vehicle jitter as perceived by an independent observer (an Astrometric instrument). Bradley et al.[8] provide an overview of the FGS instrument and describe the science modes. Table 1 summarizes the Jitter Test Conditions. FGS #2 was the dominant (DOM) FGS, which updates the PID control law with V_2-V_3-axis attitude error corrections. FGS #3 was the non-dominant FGS, which is used to determine the V_1-axis attitude error correction. HST has a complement of six gyros, of which three are required to conduct science observations under stable 3-axis control. The jitter will vary with the geometric projection of the active gyro-input-axes selected for control. During this jitter test the active control gyros were 2, 3 and 4, which have an expected jitter contribution of 3.26 asrcsec.

Table 1
FGS Jitter Test Conditions

Day of run	Ast FGS	Dom FGS	Sun Angle	Control Gyros	NICMOS CryoCooler
127	1	2	53.9°	3,4,2	Stabilized at cooling temp.

FGS #3 is the only FGS which has an active OCS control heater, which effects Fold Flat #3 in the FGS optical train and induces pointing changes. The OCS Control Heater can cycle from approximately 0.005 Hz to 0.02 Hz. Peak-to-peak pointing changes, as measured in ground test, were 1.2 mas in the X-axis and approximately 4.0 to 6.0 mas in the Y-axis. The OCS control heater modes are in the low frequency vicinity of the rigid body modes. Since FGS #3 was defined as non-dominant, these vibrations have no effect on the jitter.

POS Mode Results

The POS mode portion of each test consisted of four consecutive segments, with each segment containing about 18m30s of Fine Lock data. The Orbit Day length was 1h04m36s and the Orbit Night length was 31m39s. The start and stop times of the individual segments, relative to the orbital terminators, are shown in Table 2. The test had good coverage through both of the orbital terminators. The target star information is shown in Table 3, the target magnitude m_v is a Johnson magnitude.

Table 2
POS mode observations start and stop times

Segment	Start time	End time
1	EON + 24m21s	EOD + 11m59s
2	EOD + 12m37s	EOD + 31m09s
3	EOD + 31m47s	EOD + 50m15s
4	EOD + 50m54s	EON + 04m45s

Table 3
The target star for the POS and TRANS mode observation's

Star HD#	m_v	B_T	V_T	B-V	V-I	Spectral type	FGS 1 magnitude Obs. 1	Obs. 2	Obs. 3	Obs. 4
225146	8.59	9.0±.013	8.69±.014	.360±.014	.42±.01	BOIbp	9.412	9.414	9.415	9.414

When determining the jitter, the FGS signals were processed in the following manner. The Astrometer and Dominant FGS Star Selector Servo angles and PMT counts are converted into Fine Errors R_x and R_y, relative to the FGS X-Y frame. The mean and slope are removed, these signals are denoted as de-trended. The Dominant FGS errors and the Astrometer fine error signals are combined to remove the sensed vehicle disturbance. The resultant signals are then converted to the vehicle frame as translations, these signals are denoted as de-jittered (*post facto* jitter).

Table 4
POS mode jitter measurements from Ast and Dom FGS's

	AST de-trended (mas)			AST de-jittered (mas)			DOM de-trended (mas)		
Segment	V_3	V_2	rss	V_3	V_2	rss	V_3	V_2	rss
1	3.5	3.8	5.2	1.7	2.0	2.6	3.6	3.3	4.9
2	3.8	3.7	5.3	1.7	2.0	2.6	3.5	3.5	4.9
3	4.1	4.0	5.7	1.8	2.0	2.7	3.8	3.8	5.4
4	3.4	4.8	5.9	1.7	2.0	2.6	4.7	3.2	5.7
avg			5.5			2.6			5.2

The jitter measurements in Table 4 were obtained from the four segments of the test for the Astrometer (AST) and the Dominant (DOM) FGS, the data from each segment was processed in whole. The de-trended and de-jittered jitter measurements, as determined by the Astrometer, are shown in milliarcsec increments in the Tables 5 and 6. The final two entries of the De-Trended Table, denoted as POS and TRANS, are measurements from the DOM FGS, over the entire POS and TRANS part of the test. They are given in this Table since the processed DOM FGS signals are only de-trended.

Table 5
Percent of de-trended data contained within defined milliarcsec intervaLS

	Milliarcsecond intervals										
Seg.	±0.5	±1.5	±2.5	±3.5	±4.5	±5.5	±6.5	±7.5	±8.5	±9.5	Axis
1	21.8	42.2	59.4	72.8	81.5	88.7	93.2	96.3	98.2	99.2	X(V3)
	21.6	42.2	60.0	73.8	84.0	91.3	95.6	98.0	99.2	99.7	Y(V2)
2	21.0	40.8	58.2	72.4	82.8	89.8	94.3	97.0	98.4	99.3	X(V3)
	21.2	39.0	56.7	69.8	81.3	88.7	94.0	96.6	98.3	99.1	Y(V2)
3	19.6	37.6	52.6	67.4	78.7	87.0	92.4	95.8	97.7	98.8	X(V3)
	19.0	39.4	53.5	69.5	78.6	87.1	91.3	94.9	96.7	97.9	Y(V2)
4	22.6	42.3	58.3	72.0	81.8	88.1	92.8	95.8	97.5	98.6	X(V3)
	25.3	48.6	65.0	79.5	87.8	93.6	96.7	98.3	99.2	99.6	Y(V2)
POS	21.3	41.4	58.3	71.8	82.3	89.4	94.0	96.8	98.3	99.1	V3
	23.4	45.0	62.7	76.4	86.0	92.2	95.9	97.9	98.9	99.4	V2
TRANS	21.9	42.4	60.1	74.2	84.5	91.1	95.1	97.3	98.6	99.2	V3
	32.2	59.3	78.6	90.2	96.1	98.6	99.5	99.8	99.8	99.9	V2

Table 6
Percent of de-jittered data contained within defined milliarcsec intervals

	Milliarcsecond intervals										
Seg.	±0.5	±1.5	±2.5	±3.5	±4.5	±5.5	±6.5	±7.5	±8.5	±9.5	Axis
1	37.1	67.0	86.1	95.4	98.9	99.8	99.9	100			X(V3)
	43.7	75.2	91.8	98.0	99.6	99.9	99.9	100			Y(V2)
2	36.5	66.8	86.1	95.5	98.8	99.8	99.9	100			X(V3)
	42.9	74.4	91.3	97.8	99.6	99.9	99.9	100			Y(V2)
3	36.3	66.1	85.1	95.1	98.7	99.7	99.9	99.9	100		X(V3)
	43.1	74.1	90.9	97.7	99.6	99.9	99.9	99.9	100		Y(V2)
4	37.0	67.3	86.3	95.6	98.9	99.8	99.9	99.9	99.9	99.9	X(V3)
	43.6	75.5	92.1	98.1	99.7	99.9	99.9	99.9	99.9	99.9	Y(V2)

The de-trended and de-jittered Astrometer attitude error measurements are shown in Figure 2, which cover the EOD and EON terminators. There are definite EON disturbances in the V_3-axis, for the Sun Angle of 54 degrees (compare Figure 2A to Figure 2C). Nearly the entire POS mode and TRANS mode orbits can be examined with the Dominant FGS. Figure's 3 and 4 show the V_2-V_3-axis time traces over the entire POS and TRANS mode observations. The most frequent disturbances occur in the vicinity of the EON terminator. The spectral content indicates that the primary vibration mode is at 1.2 Hz.

TRANS Mode Results

The TRANS mode portion of the test consisted of eleven segments. The first 10 segments have 6m53s of TRANS mode data (consisting of 10 scans per segment), the last segment has 3m18s of data (consisting of 5 scans). The jitter removal approach from the Astrometer is similar to that performed in the POS mode. Since the Astrometer was scanning the target, the standard deviations shown in Table 7 were determined from the dominant FGS, after trend removal.

Table 7
TRANS mode jitter measurements from the Dominant FGS

	DOM de-trended (milliarcsec)		
Segment	V_3	V_2	rss
1	3.3	2.2	4.0
2	3.3	2.4	4.1
3	3.4	2.4	4.2
4	3.4	2.4	4.2
5	3.5	2.5	4.3
6	3.3	2.4	4.1
7	3.5	3.5	4.9
8	4.4	2.5	5.1
9	6.2	2.4	6.6
10	4.6	2.4	5.2
11	3.8	2.3	4.4
average			4.6

In TRANS mode, the Astrometer will repetitively scan across the target, which yields the Transfer Function. The 10 scans from a segment are then combined and fitted to a calibrated Transfer Function from the nearest location in the FGS field-of-view where the TRANS mode observation was located. Figure 5 shows the fitted results for segments 1 and 8, segment 1 was near the EOD terminator and segment 8 passed through the EON terminator. The residuals represent the difference in the fit between the TRANS mode scans and the calibrated Transfer Function.

Excessive jitter results in magnitude and phase deviations in the combined TRANS mode scans, which are reflected in the residuals. None of the individual scans had to be rejected because of distortions due to jitter. For both segments, the FGS X-axis (V_3-axis translation) is a bit high, as expected from the measurements in Table 7. The residual for segment 8 (which passed through the EON terminator) is higher than for segment 1.

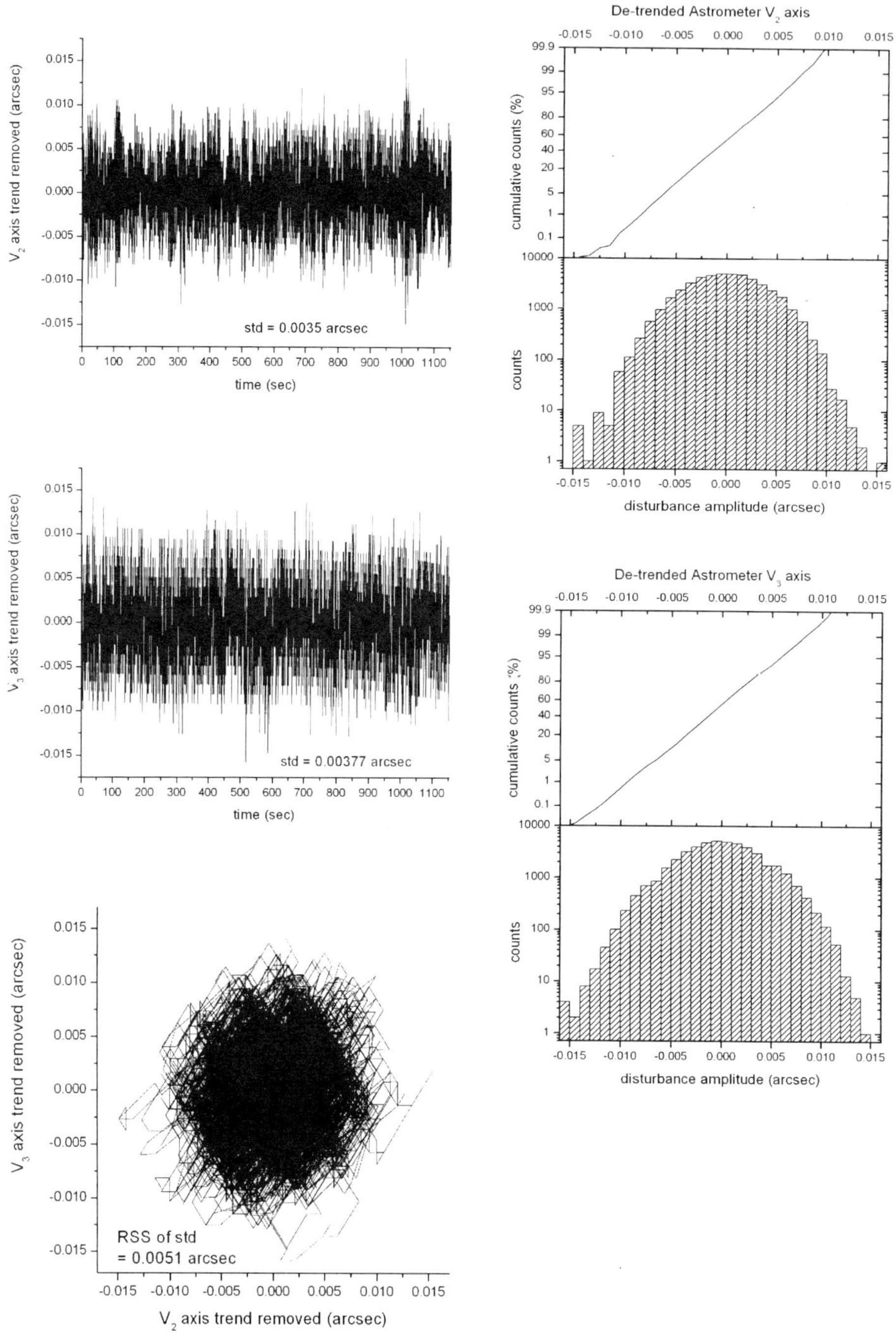

Figure 2A Astrometer FGS #1 de-trended attitude error. POS mode observation #1. Conducted from EON + 24m21s to EOD + 11m59s.

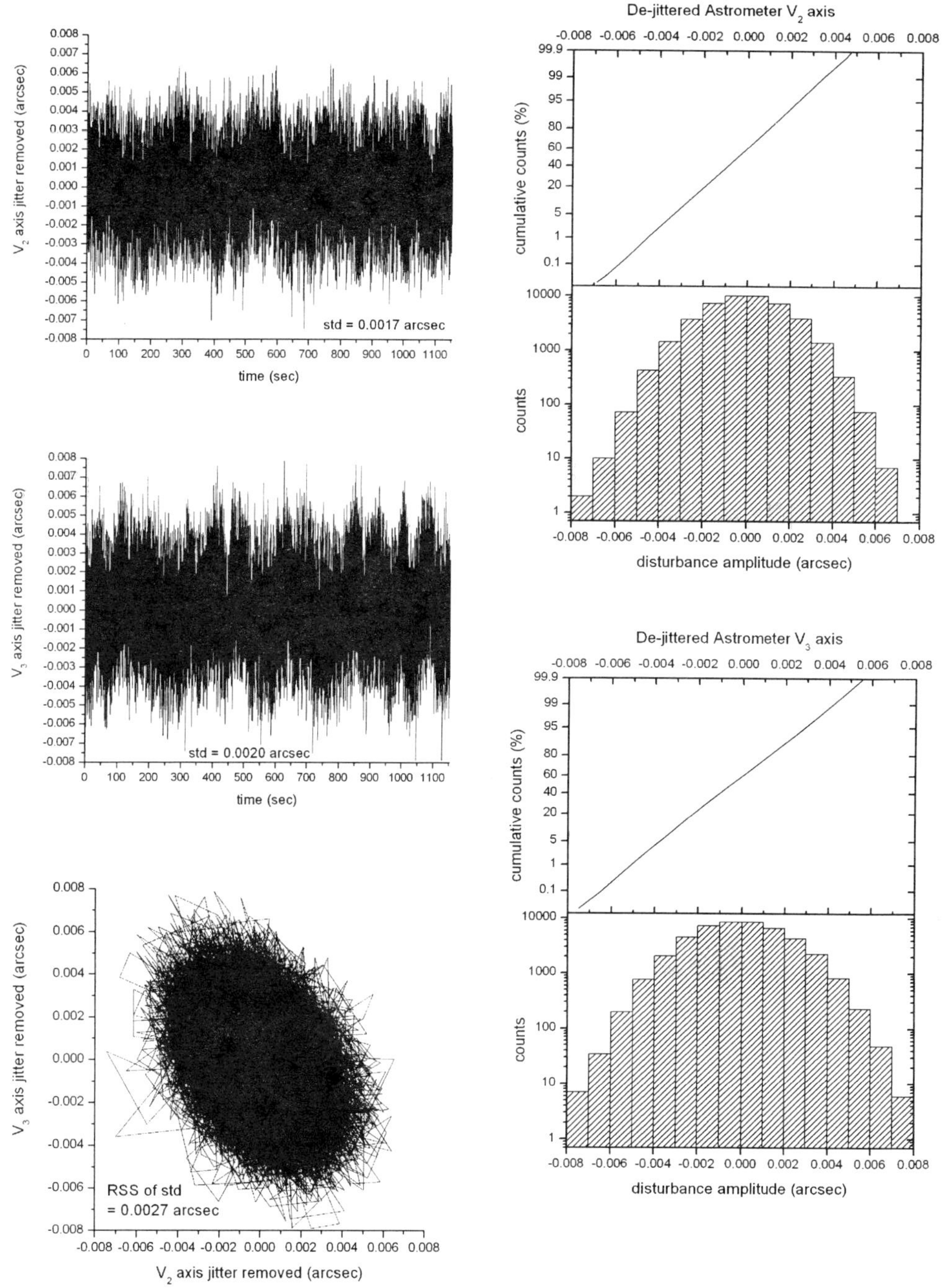

Figure 2B Astrometer FGS #1 de-jittered attitude error. POS mode observation #1. Conducted from EON + 24m21s to EOD + 11m59s.

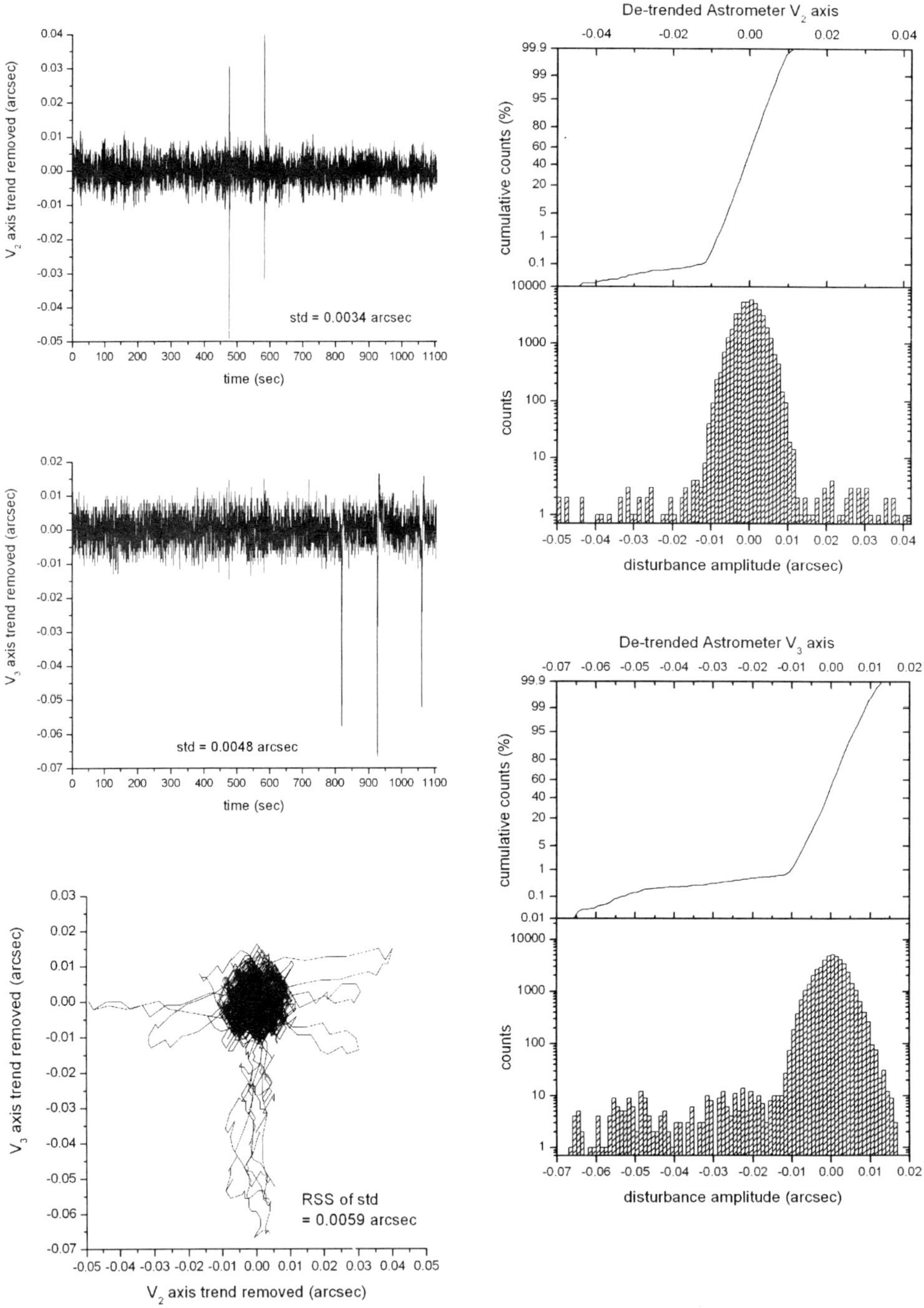

Figure 2C Astrometer FGS #1 de-trended attitude error. POS mode observation #4. Conducted from EOD + 50m54s to EON + 04m45s.

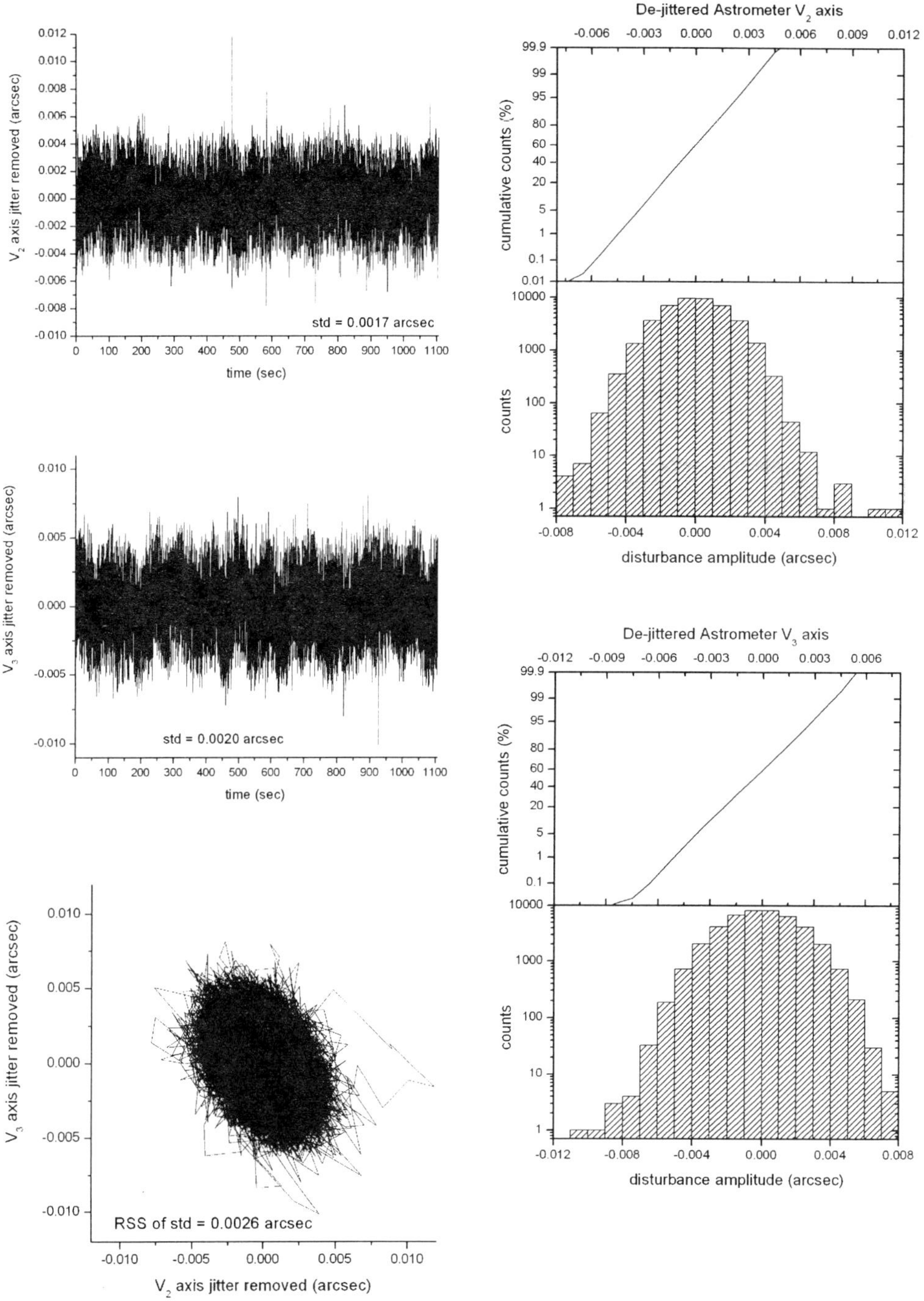

Figure 2D Astrometer FGS #1 de-jittered attitude error. POS mode observation #4. Conducted from EOD + 50m54s to EON + 04m45s.

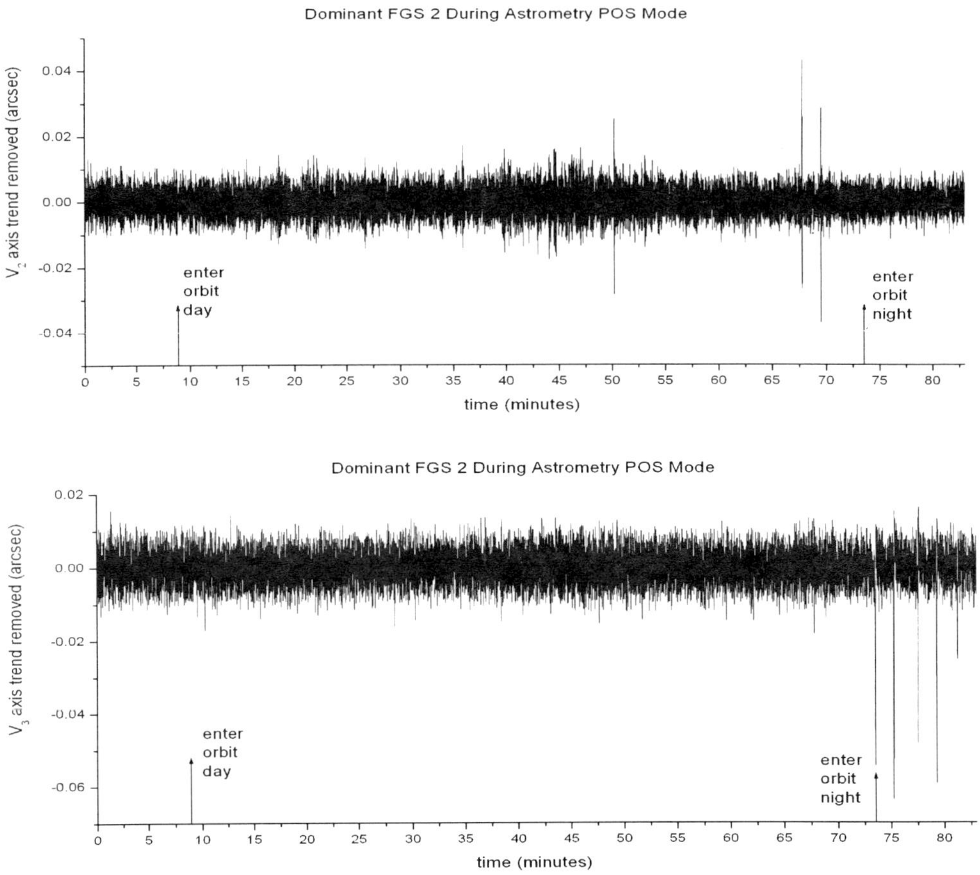

Figure 3 Dominant FGS #2 de-trended attitude error, over the entire POS mode observation set. Conducted from EON + 24m21s to EON + 04m45s, of the next EON terminator.

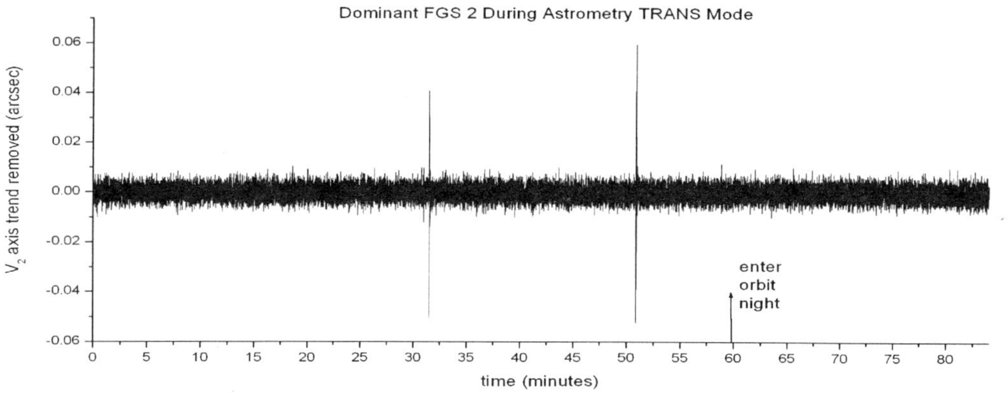

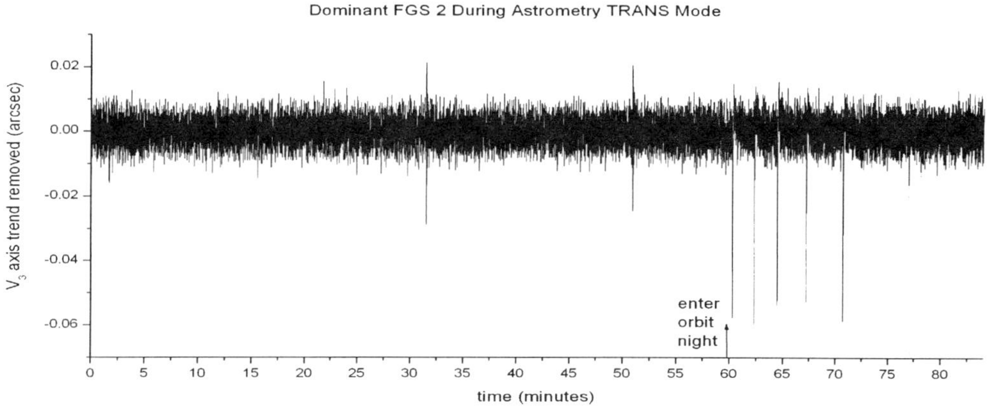

Figure 4 Dominant FGS #2 de-trended attitude error, over the entire TRANS mode observation set. Conducted from EOD + 07m06s to EON + 19m07s.

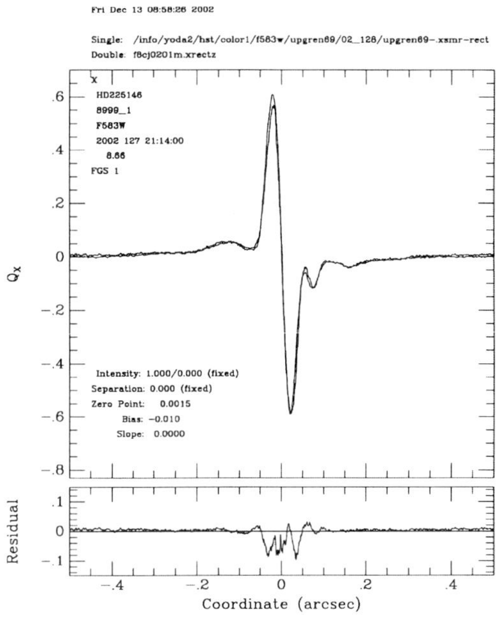

Figure 5A Segment #1 X-axis (V_3 translation).

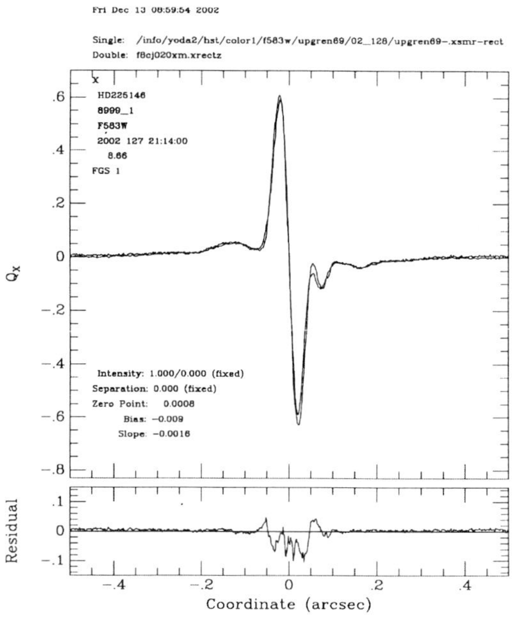

Figure 5C Segment #8 X-axis (V_3 translation).

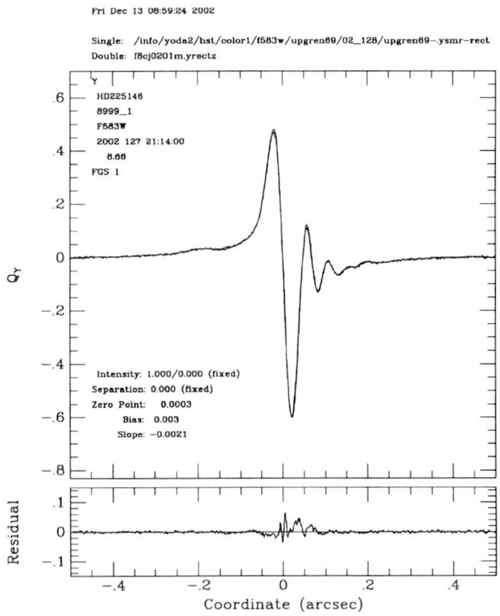

Figure 5B Segment #1 Y-axis (V_2 translation).

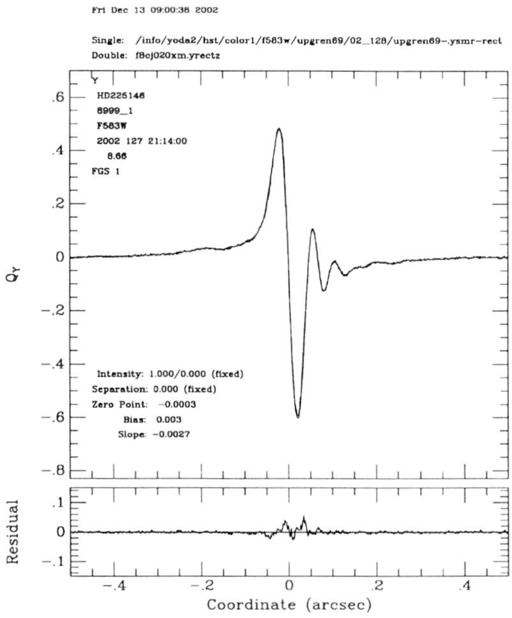

Figure 5D Segment #8 Y-axis (V_2 translation).

Figure 5 The TRANS mode results for segment #1 (just after EOD) and for segment #8 (passing through EON).

CONCLUSION

The spacecraft jitter is noticeably improved over the pre-SM3B jitter environment; both in qualitative and quantitative metrics, considering only three control gyros are used. The jitter will vary with the geometric projection of the active gyro-input-axes selected for control. During this jitter test the active control gyros were 2, 3 and 4, which have an expected jitter contribution of 3.26 asrcsec. The average two-axis jitter, as determined by the Astrometric Instrument, during the four POS mode observations is 5.5 milliarcsec, at a Sun Angle of 54 degrees from the V_1-axis. The Astrometer is independent of the control law and the Astrometer target star is brighter than the guiding Dominant star by approximately 2 m_v, it is a robust estimate of vehicle jitter. The *post facto* estimate of spacecraft jitter, as determined by the Astrometer, is 2.6 milliarcsec. The reduced jitter magnitude has permitted the adoption of the as-launched PID control law.

REFERENCES

1. A.P. Bukley, "Hubble Space Telescope Pointing Control System Design Improvement Study," AAS 93-001, in *Guidance and Control 1993*, R.D. Culp and G. Bickley, eds., *Advances in the Astronautical Sciences*, Vol. 81, pp. 461-466, 1993 (paper presented at the 16th Annual AAS Guidance and Control Conference, Keystone, Colorado, February, 1993).

2. R.D. Irwin, D.A. Lawrence, R.D. Glenn, and W.G. Frazier, "H-Infinity Control Design Specifications Applied to the Hubble Space Telescope," AAS 93-003, in *Guidance and Control 1993*, R.D. Culp and G. Bickley, eds., *Advances in the Astronautical Sciences*, Vol. 81, pp. 493-518, 1993 (paper presented at the 16th Annual AAS Guidance and Control Conference, Keystone, Colorado, February, 1993).

3. G. Zhu, K.M. Grigoriadis and R.E. Skelton, "Covariance Constraint Control Design for the HST," AAS 93-004, in *Guidance and Control 1993*, R.D. Culp and G. Bickley, eds., *Advances in the Astronautical Sciences*, Vol. 81, pp. 519-538, 1993 (paper presented at the 16th Annual AAS Guidance and Control Conference, Keystone, Colorado, February, 1993).

4. E.G. Collins, Jr. and S. Richter, "LQG and Maximum Entropy Control Design for the Hubble Space Telescope," AAS 93-005, in *Guidance and Control 1993*, R.D. Culp and G. Bickley, eds., *Advances in the Astronautical Sciences*, Vol. 81, pp. 539-551, 1993 (paper presented at the 16th Annual AAS Guidance and Control Conference, Keystone, Colorado, February, 1993).

5. M. Balas, L.R. Robertson, Y.J. Lee, J. Mohl and R. Garzella, "Hubble, Bubble, Toil and Trouble: A Reduced-Order Model Based Saga of the Hubble Space Telescope," AAS 93-006, in *Guidance and Control 1993*, R.D. Culp and G. Bickley, eds., *Advances in the Astronautical Sciences*, Vol. 81, pp. 553-574, 1993 (paper presented at the 16th Annual AAS Guidance and Control Conference, Keystone, Colorado, February, 1993).

6. S. Addington and C.D. Johnson, "A Dual-Mode Disturbance-Accommodating Controller for the Hubble Space Telescope," AAS 93-007, in *Guidance and Control 1993*, R.D. Culp and G. Bickley, eds., *Advances in the Astronautical Sciences*, Vol. 81, pp. 575-605, 1993 (paper presented at the 16th Annual AAS Guidance and Control Conference, Keystone, Colorado, February, 1993).

7. G.S. Nurre, J.D. Nelson and A.J. Bradley, "Current NASA HST Controller Design and Performance," AAS 93-002, in *Guidance and Control 1993*, R.D. Culp and G. Bickley, eds., *Advances in the Astronautical Sciences*, Vol. 81, pp. 467-492, 1993 (paper presented at the 16th Annual AAS Guidance and Control Conference, Keystone, Colorado, February, 1993).

8. Bradley, L. Abramowicz-Reed, D. Story, G. Benedict and W. Jefferys, "The Flight Hardware and Ground System for HST Astrometry," 1991, PASP, 103, 317.

AAS 03-063

PRECISION POINTING OF THE THURAYA SATELLITE

Richard A. Fowell and Hanching G. Wang*

Thuraya began cellular phone service for Europe, Africa and Southwest Asia in July 2001 from geosynchronous orbit at 44° E. The satellite attitude control system is a beacon-referenced, 3-axis stabilized, zero-net-momentum system. We present flight telemetry from the payload antenna beacon. The steady-state jitter is < 4 arcseconds, 3 sigma, and the daily autonomous thruster momentum dumping transients are less than 60 arcseconds. The beacon precision is due to the high gain, inertia and stiffness of the 12.25m L-band antenna and the dual high power spread-spectrum ground beacons. The payload pointing precision is enhanced by continual self-calibration, and by independent electronic steering of the 246 L-band spot beams to remove the parallax effects of Thuraya's 6 degree inclined orbit. The magnitude and update frequency of the electronic steering is minimized by the near-optimal spacecraft bus steering strategy of pointing at a subterranean target. Thruster firing transients are minimized using anticipatory preemphasis torque. We present block diagrams of the control system, and discuss performance of the hemispherical resonating gyros, digital sun sensors, and reaction wheel pyramid.

Figure 1: The Thuraya D1 satellite on-station at 44° E.

* Boeing Satellite Systems, El Segundo, California 90245.

INTRODUCTION

Thuraya (Figure 1) requires more precise pointing than traditional geosynchronous communications satellites. The L-band payload pattern consists of 246 circular spot beams with 0.7° beam diameter. The beam ground targets are a hexagonal grid, with coverage areas as small as 450 km diameter. To simplify system operations, every beam is to be held to its assigned ground target to better than 0.070 radial degrees plus an allocation of 0.025° (90 arcsec) for stationkeeping and momentum dumping transients. This task is complicated by Thuraya's 6 degree orbit inclination, which creates variations in subtended beam target separation of up to 0.600 degrees, and multiple firings per day of 22N (5 lbf) bipropellant thrusters to unload the momentum induced by solar torques on the large, highly asymmetric antenna. Eclipse thermal snap of the 34.5 m (131 ft.) span solar arrays, and bipropellant East-West stationkeeping maneuvers add to the challenge.

In-flight performance has well exceeded requirements, with steady-state line-of-sight jitter of < 4 arcseconds, 3 sigma. Steady-state yaw pointing errors seem to be under 100 arcseconds. Transients from stationkeeping, momentum dumping and eclipse thermal snap are under 70, 60 and 40 arcseconds, respectively.

The following key design features contribute to the pointing accuracy:

- The digitally formed, high gain, spread spectrum payload beacon sensor is more intimately tied to payload pointing than conventional beacons. This largely removes the effects of thermal deformation on payload line-of-sight (LOS) pointing.
- Two beacon sites near the center of the service region minimize the pointing effects of rotational errors about the beacon LOS, average errors, and provide hot redundancy.
- The fixed subterranean pointing target ~1200 km below the heart of the service region. This nearly minimizes the apparent motion of the ground target set in the satellite frame, and is significantly simpler than the truly minimizing pointing.
- Electronic beam steering to remove ground pattern parallax and gain/phase shifts.
- A rigid, high moment of inertia antenna and high-inertia satellite attenuate any disturbances above the control system bandwidth.
- Rotational control about the beacon line of sight using a a continuously running gyro calibrated by a wide field of view precision digital sun sensor.
- Anticipatory preemphasis torque for thruster momentum dumping pulses reduces momentum dumping transients.
- Dedicated, low thrust bipropellant stationkeeping thrusters are targeted through the satellite center of mass to reduce disturbance torque, and fired in a deadbeat pulse train with feedforward to reduce flexible mode disturbances of stationkeeping.

Attitude Control Subsystem Overview

Figure 2 is a simplified block diagram of the Thuraya attitude control subsystem during payload operation. The architecture is similar to that of other precision attitude control systems - compare, for instance the stellar-inertial system block diagrams in [1].

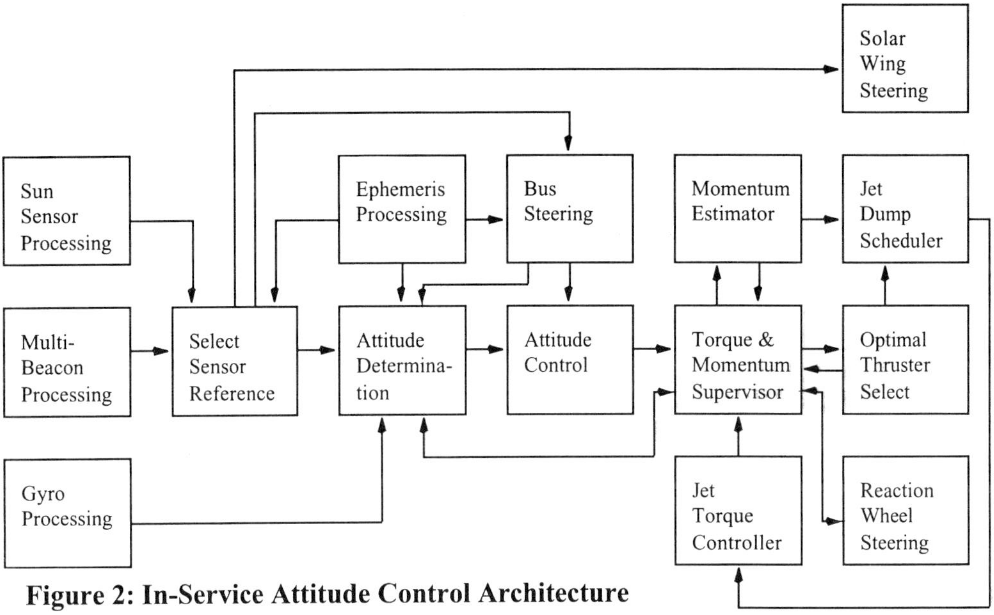

Figure 2: In-Service Attitude Control Architecture

On the left are the sensors. Normal operation uses two of the three Boeing digitally formed multibeacon sensors, a 49° x 49° field of view (FOV) Adcole digital sun sensor[2], and a 3-axis Litton hemispherical resonating gyro (HRG)[3]. Each sensor is processed by a dedicated sensor processing software module to produce measurement vectors. A Sensor Selection software module selects two measurement vectors and their two inertial reference vectors computed by the ephemeris processing module. The Attitude Determination module uses the TRIAD algorithm to produce an attitude update from these. This module also receives the gyro angular increments, the dynamics model predicted angular acceleration, and the commanded rate. The Attitude Determination module outputs an estimated quaternion and rate. The Bus Steering module generates the desired quaternion and rate based on the pointing target coordinates and ephemeris. The attitude control module compares the commanded and estimated quaternions and rates, generating a commanded acceleration. The "Torque and Momentum Supervisor" (TMS) sums the commanded acceleration with the gyroscopic coupling torques from a fixed-inertia dynamics model, converts the result to a commanded torque and sends it to the selected combination of wheel array or thruster array. Commanded wheel torques go through wheel array management to be distributed to the (normally four) reaction wheel control modules. Similarly, commanded thruster torques go to optimal thruster selection logic (OTS) which uses a Glandorf algorithm ([4], section 7.6.2). The resultant commanded wheel torques and jet on-times are processed by actuator specific processing, and sent to the array of 4 Honeywell 45.6 N-m-s wheels and 8 ARC 22N (5 lbf) bipropellant thrusters. The momentum control system consists of a momentum estimator, a Jet Dump Scheduler strategy module and a Jet Torque Controller momentum dump tactical module. The estimator estimates the current momentum and its projected growth over the next day. The strategy module creates a schedule of momentum dump events, subject to a ground selected thruster subset and on-time limit. Each is coded as a thruster torque impulse corresponding to a single

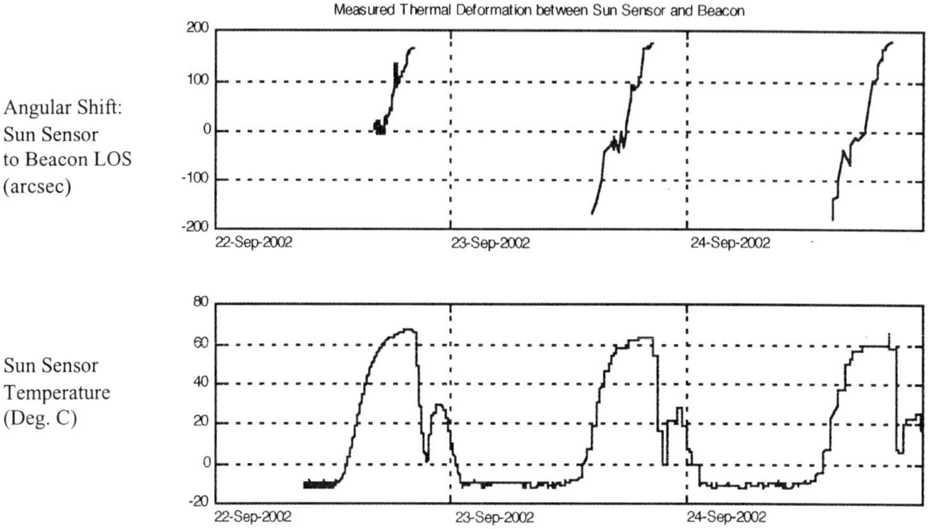

Figure 3: In-flight measured thermal distortion from body sun sensor to beacon

22N thruster firing. The tactical module executes the dump as an open-loop overlapping wheel torque command and thruster torque command into TMS. The Stationkeeping sequencer executes a ground-commanded sequence of open-loop zero-torque thruster events into OTS with the selected 9N stationkeeping thruster, alternating with high gain closed loop attitude control.

Digitally Formed Beacon Sensors

Precision pointing systems require tight coalignment of the payload frame with the attitude determination frame despite thermal deformation. One effective method for doing so is to use sensors that share the same transmission path as the payload, such as fine guidance sensors for optical payloads, or antenna beacon sensors for radio frequency payloads like Thuraya's.

For Thuraya, the time-varying azimuthal thermal deformation between the payload LOS and the digital sun sensor alone is at least +/- 180 arcsec in amplitude. This is demonstrated by Figure 3, which compares the known angular separation between the ground beacon transmitter and the sun with that reported by the beacon sensor and sun sensor in flight telemetry. Even the best stellar-inertial attitude determination system mounted to the satellite bus would, in itself, be powerless to eliminate this.

Conventional satellite beacons use separate receive horns, measurement chains and frequencies from the payload beams. The Thuraya beacon processing shares these resources with the payload, eliminating error sources and saving cost, weight and power.

The Thuraya beacon sensor processing supports three four-lobe monopulse beacon nulls simultaneously. The lobes can be reprogrammed by ground command to any point in

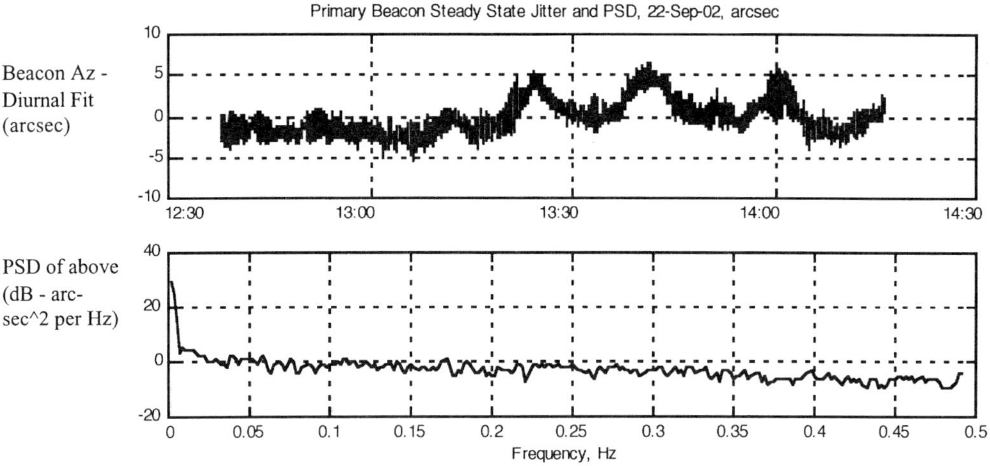

Figure 4: Thuraya D1 on-station pointing steady-state jitter for two hours on 9/22/02

the L-band payload pattern. This eliminates many of the drawbacks of conventional hard-wired beacon systems - the Thuraya beacon system easily supports large-angle offset pointing, satellite relocation, or ground beacon relocation.

The beacon signals emitted by the ground beacons are collected by the same 12.25 m aperture L-band reflector and 128 feeds used by the communications payload. The signals at the feeds are then processed by the payload digital signal processor processing depicted in Figure 2 of [5]. This processing digitizes the 128 channels, and weights them by the spread spectrum codes and the beam forming coefficients to compute each of the 12 beacon lobe measurements. These measurements are integrated to form each of the beacon beam outputs sent to the spacecraft attitude control processor (SCP). The SCP computes an azimuth and an elevation measurement for each of the three four-beam groups.

The result is a very low noise sensor. Figure 4 shows on-station steady-state jitter in the time domain and the frequency domain. The orbit-induced profile has been removed from the data to make the jitter visible.The frequency plot shows that torque noise sources are not significant enough to stimulate the 0.07 Hz and 0.3 Hz structural modes.

Since the beacon is a small part of the payload pattern, distortions over the aperture due to effects such as thermal distortion in the antenna and structure are not corrected by the beacon. Prior to flight, these effects were measured in ground testing. One flight measure of pattern stability, however, is the difference between the known angular separation between two beacon sites and that measured by the beacon sensor on-orbit. Figure 5 shows this measured separation for two days with transients removed. The amplitude of +/ - 40 arcsec, while significant, is much lower than the thermal distortion between the spacecraft bus and the payload LOS.

Beacon Sites

Rotational errors about the beacon LOS couple into payload beam radial pointing error by the separation angle between the beacon LOS and the payload beam in question. Such rotational errors have very little impact on radial pointing of beams close to the beacon LOS. On the Thuraya service region map in Figure 6, the Thuraya beacon sites at Sharjah in the UAE and Al Maadi in Egypt are indicated by crosses.

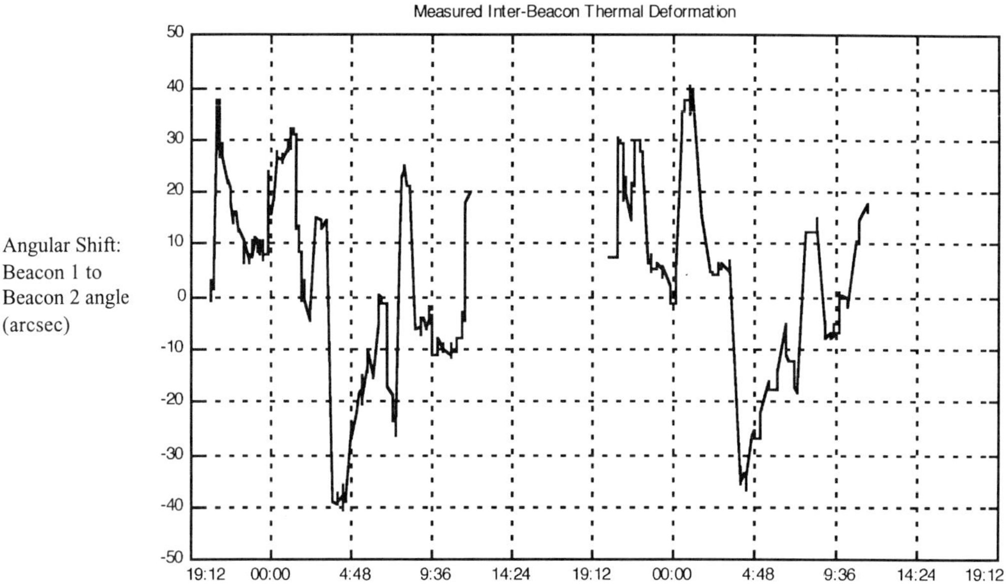

Figure 5: Error in beacon pattern over two days

Ephemeris error coupling to LOS transverse pointing error is also reduced. For geosynchronous stellar-inertial systems, an ephemeris error of 1,000 meters creates an LOS pointing error of 5.8 arcseconds at the subsatellite point. For Thuraya, the effect is zero at the beacon site, and less than 0.5 arcsecond at the farthest point in the pattern.

Using multiple beacon sites has several advantages. It provides redundancy in the case of any difficulties with one of the beacon sites, reduces the impact of independent random beacon errors into LOS error in the direction connecting the two sites by roughly 40%, and provides another way to measure rotation about the beacon LOS.

Subterranean Target Steering

Thuraya's inclined orbit means that the apparent shape of the array of beam ground targets changes as the satellite undergoes its daily motion from 6 degrees above the equator to 6 degrees below the equator. For example, the equator is a straight line when viewed from a geostationary point over the equator. When the satellite is north of the equator, the equator curves up. When the satellite is south of the equator, the equator curves down. For an

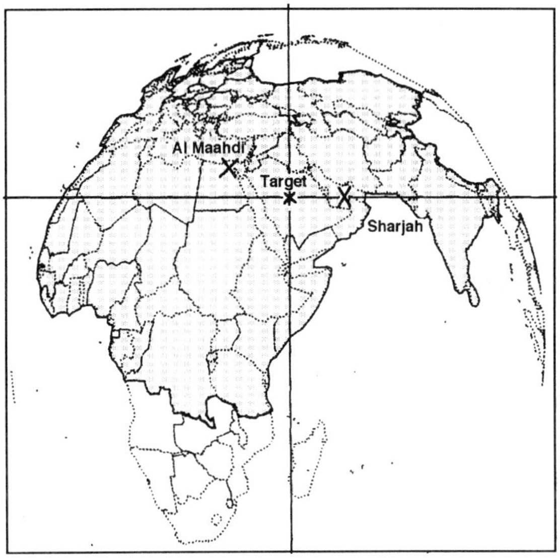

Figure 6: Thuraya service region with beacon sites and subterranean target marked

antenna beam pattern fixed in the satellite, there is no attitude which will hold all the beams fixed to the array of ground targets over the orbit. This is a long-standing problem of current interest, e.g. [6] from last year's conference.

As we stated over a decade ago [7]: "Our first step then is to define the desired payload orientation in the inclined orbit 14. In general, the best results will be obtained by taking the position of the satellite 10 in the inclined orbit 14, at relatively close time intervals, and determining what orientation the payload should have at each time interval to get a ground coverage pattern which best meets the payload requirements. This optimization, in general, will have different criteria for every satellite, depending on the customer's desires." For this application, since the residual errors will be corrected by electronic steering of the beams, the metric chosen was to minimize the maximum amplitude of required electronic beam steering over the pattern.

We computed this metric for three pointing strategies: unconstrained pointing, surface target pointing, and subterranean target pointing. In the first strategy, the three satellite rotational degrees of freedom at each time point were independent free variables. The second strategy was that flown on ATS-6 in 1974 [8]. There the design degrees of freedom were a "desired ground aim point (in latitude-longitude coordinated[sic])". Then "the control system is required to maintain the spacecraft yaw angle about its earth-pointing Z-axis so that the projection of its roll axis on the Earth points East." with the aid of uploaded "orbit ephemeris data for the purposes of holding a fixed ground aim point (by compensating for non-geostationary orbit effects)" since "it is necessary for the attitude control system to automatically vary its N-S (roll) and E-W (pitch) pointing angles so as to compensate for non-geostationary orbit effects (inclination, eccentricity, etc.)". The third strategy generalized the ATS-6 strategy by adding a third design degree of freedom -

radius [9]. For the third strategy, the optimal solution was: 39.8236 E, 29.0087 deg N, and 5160.5 km radius (roughly 1,200 km underground), marked in Figure 6 with an asterisk.

The required payload electronic offsets for the subterranean point were insignificantly more than for the unconstrained steering, and significantly less than the surface target strategy. The subterranean steering was selected because it was much simpler to implement and communicate than unconstrained steering. The on-board ephemeris and its maintenance were already required to use the digital sun sensor processing for earth-oriented pointing, and the incremental effort to make the target radius reprogrammable rather than fixed was negligible.

Rotational Control About the Beacon LOS

Thuraya supports three methods of sensing rotations about the beacon LOS. These are: a 49° x 49° field of view Adcole digital sun sensor mounted to the satellite bus, a Litton hemispherical resonating gyro (HRG) mounted to the bus for gyrocompassing, and the multiple beacon sensors.

The primary method of rotational control is to use the digital sun sensor for the few hours a day when the sun sensor has data, while using the sun sensor and beacon to calibrate the gyro. When the sun sensor lacks data, the gyro is used for gyrocompassing. to control this rotation. When the sun re-enters the sun sensor, the comparison between the gyrocompassing based bus attitude and the sun-sensor based attitude allows us to get an idea of the level of error in these methods. We examined these transients for four consecutive days: 2/9/01-2/12/01. The differences were 50, 43, 86, 76 arcseconds.

Digital Beam Steering

Payload beam weighting coefficients are sent up every hour, to update them for gain and phase shifts measured by the on-board calibration system. Since the coefficients are being updated for this in any event, the correction for beam parallax due to orbit motion is included at the same time. Figure 7 shows the apparent motion of the Thuraya beam targets in the satellite frame for the subterranean target steering and a six degree inclined orbit at 44° E, as the satellite travels from the equator to its southernmost point.

The beam coefficient updates shift the beams. Three reasons are:

- The latitude/longitude update granularity
- The shift in the estimated gain/phase of the payload channels
- The quantization of the beam coefficients

The beam shifts cause step changes in the beacon pointing at each update. These currently result in pointing transients of up to 70 arcseconds, shown in Figure 8.

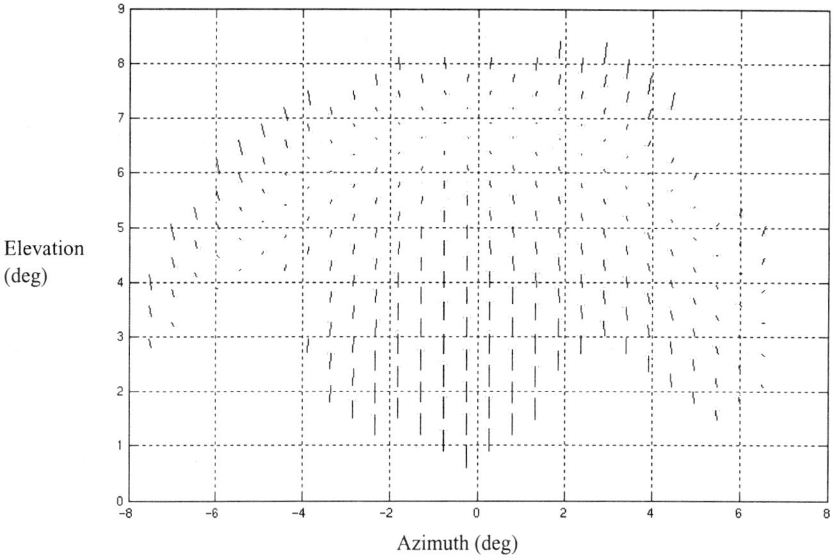

Figure 7: Apparent beam target motion over 1/4 orbit in 6 deg inclined orbit

Structural Contributions to Reduced Jitter

Key features in reducing jitter on Thuraya are the large inertias of the antenna and satellite, and the rigidity of the antenna. Despite all the clever tricks one can play with closed-loop control, performance always rests on the characteristics of the open-loop dynamics. The Thuraya structure provides a solid foundation.

The Thuraya antenna is a TRW Astromesh™ ring truss antenna [10,11]. The ring truss is an elliptical cylinder with principal axes of 12.25 m and 16 m. The depth of the truss greatly aids the stiffness - the first mode of the antenna clamped at the reflector - boom interface is 0.8 Hz. Since the ring truss places the mass of the truss elements as far from the antenna center of mass as possible, the inertia of the antenna is high as well. The minimum moment of inertia of the antenna about its center of mass is about 1000 kg-m^2. Figure 9 shows the Thuraya D1 antenna in the factory.

The overall deployed satellite inertias are [32000, 16000, 32000] kg-m^2 in roll, pitch and yaw. The first two free-free modes are low frequency modes caused by the relatively compliant solar arrays at 0.07 Hz and 0.3 Hz. The next three free-free modes are principally flexure of the antenna boom, at 0.7, 0.9 and 1.2 Hz.

Momentum Dumping Transients

Momentum unloading can be a significant source of transient disturbance. In many precision pointing applications, there are housekeeping periods set aside where this opera-

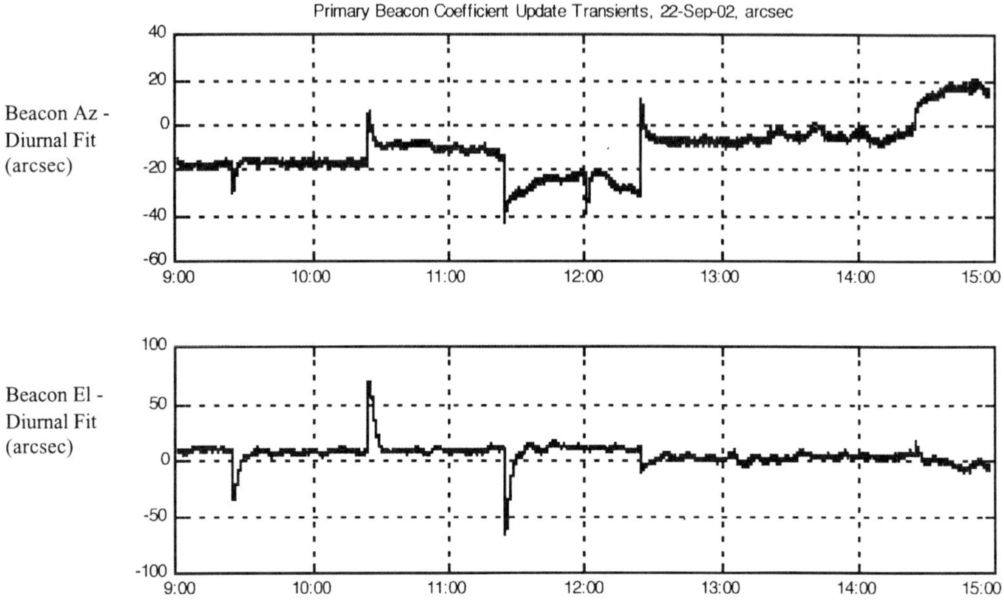

Figure 8: Beacon Coefficient update transients (orbit-period motion removed)

tion can be done without corrupting payload performance. In others the torque of the momentum dumping actuators (e.g., magnetic torquers or ion thrusters) is well within the torque capability of the momentum storage device. Neither case applies to Thuraya, although a separate 90 arcsecond allowance was given for infrequent transient events.

Thuraya is a good example of designing for low pointing disturbance while desaturating wheels with high-torque actuators. Momentum dumping on Thuraya uses the 22N thrusters sized to counterbalance imbalance torques of the 436 N orbit-raising thruster. Each 22N thruster generates roughly 40 N-m of torque.

Thuraya's large, asymmetric antenna gives rise to significant solar torques, and bus-antenna shadowing and eclipse result in large secular pitch momentum buildup in addition to the roll and yaw momentum buildup that are typically dominant in body-stabilized geosynchronous communications satellites. A typical day's buildup seen on-station is (8, 3, 0) N-m-s/day in the roll, pitch, yaw frame at midnight. The pitch momentum buildup is due to the East-West asymmetry of the satellite.

Diurnal momentum is stored in the wheel array. Figure 10 shows the wheel speeds over a week. The wheels are arranged in a conventional pyramid configuration. The magnitude of the angular momentum contributed by each wheel at 6000 rpm is [23, 32, 23] N-m-s in [roll, pitch, yaw]. The peak wheel torque is 0.2 N-m.

The Thuraya autonomous thruster momentum dumping system has a high degree of redundancy and fault protection. It needs only one of the four positive pitch thrusters, and one of the four negative pitch thrusters to provide full three-axis momentum dumping

Figure 9: Thuraya D1 antenna in the factory

capability. The daily momentum dumping plan is posted to telemetry for ground inspection at least 10 minutes before the first momentum dump. The software opens the latch valves only for the necessary window to support the dump sequence, and independently coded software fault protection will reject any thruster firing commands unless positive telemetry verification is received that both oxidizer and fuel latch valves have opened.

The size of the momentum dumping transient is kept low by anticipatory preemphasis torquing. The design philosophy is to maximize momentum dump fuel efficiency subject to a maximum allowable momentum dump attitude transient.

The dump is initiated by an open-loop, full torque wheel pulse. This is begun in advance of the thruster pulse in order to reduce the pointing transient induced by the thruster pulse. After roughly half the wheel unload is complete, the thruster pulse is fired. Since all of these torques have been fedforward to the estimator, the estimator is fully up to date on events at this point, and control is switched to high gain feedback control to damp the remains of the transient. The result is shown in Figure 11.

This anticipatory torquing, while effective, is not new. It has been called "preemphasis" compensation. For example ([12], p. 8-12) says: "The [1.5 Nm - RAF] thruster disturbance can be preemphasized by applying 0.15 Nm of opposing wheel torque for a period of ten times the thruster firing interval; the thruster pulse is centered within this period."

Stationkeeping Transients

Stationkeeping is like momentum dumping - frequently it can be done outside of payload operations, or using low disturbance actuators like electric thrusters. When these

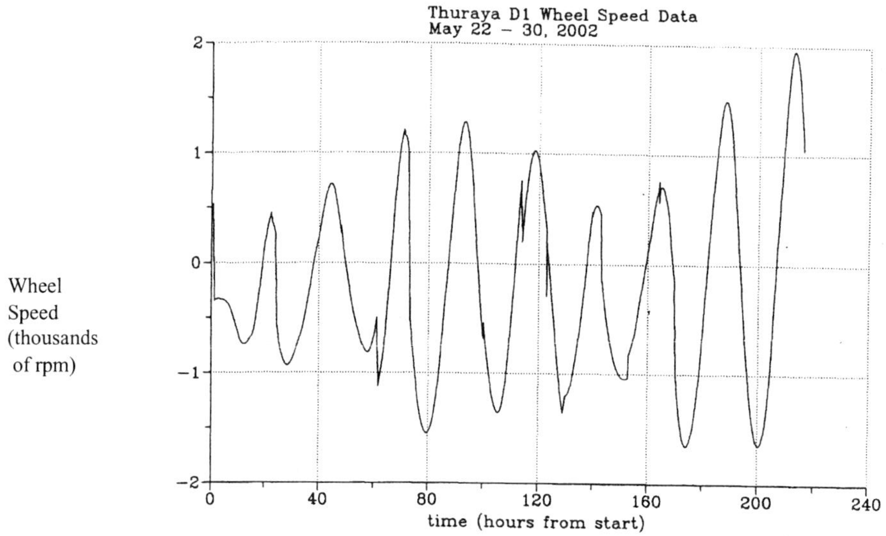

Figure 10: Thuraya D1 daily wheel speed variation for reaction wheel #1

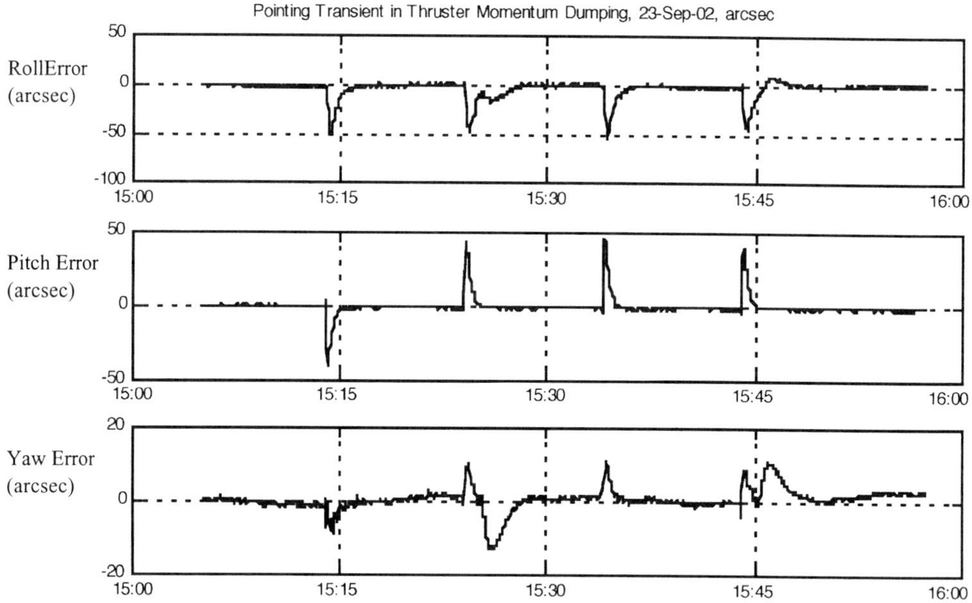

Figure 11: Pointing error transients during momentum dumping.

are not options, as on Thuraya, careful planning of the open loop maneuver plan combined with feedforward can be used to support the feedback control system in keeping the transients small.Thuraya D1 does not need north-south stationkeeping, and east-west maneuvers are accomplished using only thruster firing from one side. Maneuvers use one of the two redundant 9N (2 lbf) thrusters mounted to point through the predicted middle of life center of mass. This reduces the force disturbance level by the use of the lower force

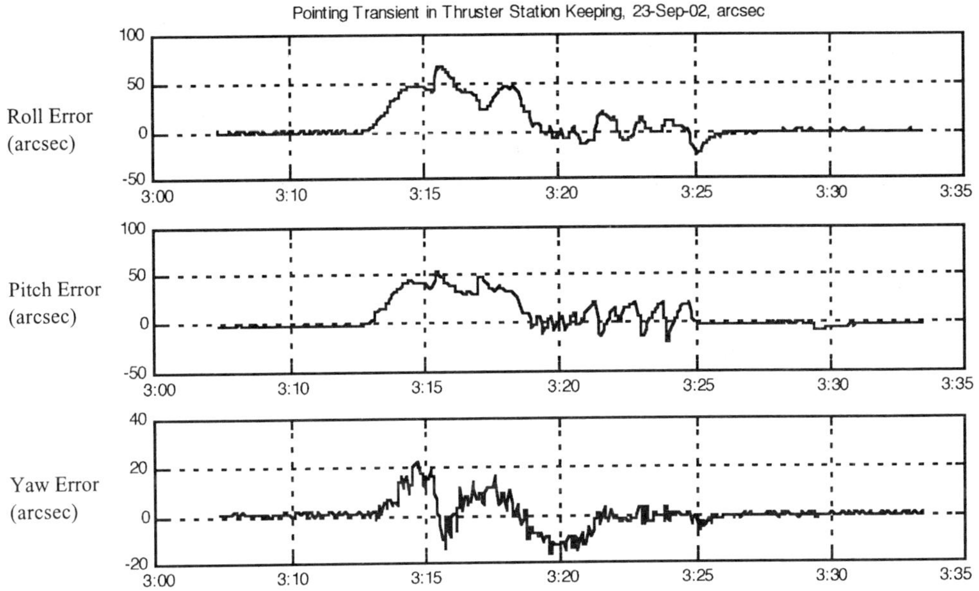

Figure 12: Pointing error transient during stationkeeping.

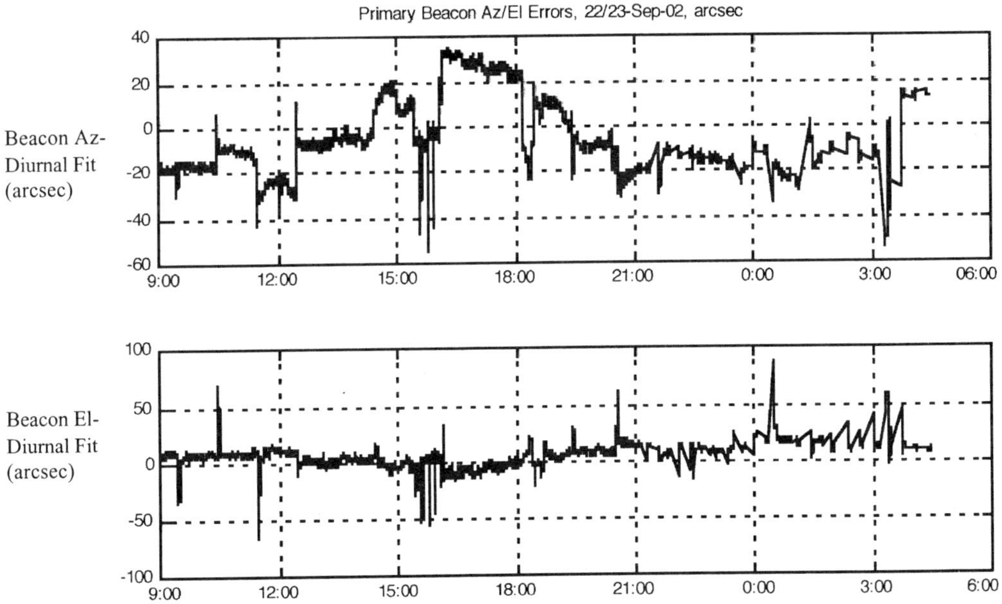

Figure 13: Beacon azimuth and elevation errors vs. predict over 19 hours

thruster (9N vs. 22N) and the torque disturbance level by both the reduced force and reduced moment arm. The open-loop stationkeeping pulse magnitude is sized to keep the size of the transient acceptably small. The open-loop pulse torque is kept near zero by commanding the stationkeeping thruster on to OTS with a commanded torque of zero.

Since the observed torque of the stationkeeping thruster has been uploaded to OTS, OTS selects up to three of the 22N thrusters to fire short pulses simultaneously with the 9N thruster. Between the stationkeeping pulses, the wheels are used under closed loop control to maintain the satellite attitude. The successive pulses are timed so that the solar array excitation from successive pulses are 180 deg out of phase in the solar array panel frequency - simple deadbeat control[13]. Flight telemetry is shown in Figure 12. The pointing transient is 70 arcsec - under the 90 arcsec requirement.

CONCLUSION

Figure 13 brings all of the transient pointing performance data together. It shows 19 hours of azimuth and elevation beacon flight telemetry with the ground-predicted values subtracted. Figures 8, 11 and 12 are subsets of this data, focusing on segments dominated by coefficient update, momentum dump, and stationkeeping transients. The eclipse exit transient at 2140 GMT on this day is much smaller than the other transients. Despite all these types of transients, the transient errors remain below 90 arcseconds, and the short-term steady state jitter below about 4 arcseconds.

ACKNOWLEDGEMENT

The Thuraya design was the result of the skills and ideas of many teams of people, and the authors wish to thank those who designed the Thuraya antenna and beacon systems in particular. These new technologies were key to the satellite's success. Thanks also to John Haskell, Jim Mc Ennan and Andy MacTavish for providing flight telemetry.

REFERENCES

1. K. Gwin, Attitude Determination and Control for the Kepler Spacecraft, Proceedings of the 25th Rocky Mountain Guidance and Control Conference, Feb. 6-10, 2002, Breckenridge, Colorado, pp. 297-317.

2. Lerner, Gerald M., "Digital Sensors", Section 6.1.3 of *Spacecraft Attitude Determination and Control*, James R. Wertz, ed., D. Reidel Publishing Company, 1978.

3. Wright, D., Bunke, D., "The HRG as Applied to a Satellite Attitude Reference System," Paper No. AAS 94-004, AAS Guidance and Control Conference, Keystone, CO, February 1994.

4. B. Wie, Space Vehicle Dynamics and Control, AIAA, 1998, pp. 448-451.

5. D.A. Sunderland et. al., "Megagate ASICs for the Thuraya Satellite Digital Signal Processor," International Symposium on Quality Electronic Design, March 18 - 21, 2002 San Jose, California http://microsys6.engr.utk.edu/~bouldin/darpa/satellite-asics.pdf

6. L Barker, et. al., Improved Pointing Techniques for Modern Geosynchronous Space-craft, Proceedings of the 25th Rocky Mountain Guidance and Control Conference, Feb. 6-10, 2002, Breckenridge, Colorado, pp. 79-91.

7. Fowell, R.A., Two-axis attitude correction for orbit inclination, U.S. Patent 5,184,790, February 9, 1993 (applied for July 22, 1991) http://patft.uspto.gov/netacgi/nph-Parser?patentnumber=5184790

8. W.N. Redisch & R.L. Hall, *ATS-6 Spacecraft Design/Performance,* EASCON `74, pp 44-44G.

9. Fowell, R.A., Subterranean Target Steering Strategy, U.S. Patent 6,135,389, October 24, 2000. http://patft.uspto.gov/netacgi/nph-Parser?patentnumber=6135389

10. Thomson, M.W., *The AstroMesh deployable reflector*, Proc. Fifth International Mobile Satellite Conference (IMSC'97), 16-18 June 1997, Pasadena, CA, 1997, pp. 393-398. JPL Publication 97-11.

11. M.W. Thomson, "AstroMesh Deployable Reflectors for Ku- and Ka-band Commercial Satellites", proceedings of the 20th AIAA International Communications Satellite Systems Conference (ICSSC 20), presented May 12-15, 2002, Montreal, Canada. AIAA2002-2032. http://www.astro-aerospace.com/datasheets_papers/aiaa-2032.pdf

12. NASA CR-152509, *Severe Storms Observing Satellite Study - Final Technical Report*, TRW, 30 June 1976.

13. Fowell, R.A., Robust resonance reduction using staggered posicast filters, U.S. Patent 5,610,848 (applied for May 13, 1994).

AAS 03-064

TRANSIENT POINTING ERROR REDUCTION IN SPACECRAFT PERFORMING STATIONKEEPING OPERATIONS WITH MOMENTUM WHEEL CONTROL

Lee Barker and Xen Price*

Short-term periodic disturbances to spacecraft attitude due to thruster firings are challenges to meeting pointing budget requirements. For commercial communications satellites, where pointing requirements apply continuously, even short-term infrequent out-of-spec pointing errors can mean customer dissatisfaction and loss of revenue. Spacecraft station keeping can generally be performed under thruster attitude control or under momentum wheel attitude control. While thruster control maneuvers remove the concern for momentum build up during the maneuver, attitude is dead band controlled, maximum pointing error is limited by thruster size and performance uncertainty, and considerable propellant is expended for attitude control. Under wheel control, pointing can be more precisely controlled, and much less propellant is required since only momentum unloading is required for attitude control. Conventional wisdom in wheel control maneuvers is to tune station keeping thrusters to impart the minimum change in momentum to the system (fire through the wet center of mass) in order to reduce the required momentum unloads and resulting disturbance torques. However, unattached liquids (fuel, oxidizer) contained within the spacecraft tanks are not instantaneously accelerated like the spacecraft body is. The transient resulting from a station keeping impulse is a function of the offset between the wet center of mass and the "dry" center of mass, spacecraft properties, and controller design. Rigid body modeling of the disturbance shows conservation of angular momentum, yet telemetry does not support this since angular momentum is stored the motion of the propellants within the tank and other flexible appendages. This paper will discuss a multi-body dynamics model used to predict the attitude behavior of a spacecraft during wheel controlled station keeping with liquid slosh or other flexible appendages, show how the model matches telemetry for spacecraft currently on orbit, and examine a number of tactics, strategies, and control designs that reduce the transient pointing errors through a better understanding of the dynamics involved.

INTRODUCTION

Short-term periodic disturbances to spacecraft attitude due to thruster firings are challenges to meeting pointing budget requirements. Under wheel control, pointing can be precisely controlled, and propellant is conserved since only momentum unloading is required for

* Space Systems/Loral, 3825 Fabian Way, Palo Alto, California 94303-4604.

attitude control. External torques, however, such as those resulting from station keeping thruster firings will cause pointing disturbances that the control system must handle. Conventional wisdom in wheel control maneuvers is to tune station keeping thrusters to impart the minimum change in momentum to the system, or fire through the wet center of mass (cm), in order to reduce the required number of momentum unloads and their associated disturbance torques. Unattached liquids (fuel, oxidizer) contained within the spacecraft tanks are not instantaneously accelerated with the spacecraft body, though eventually the motion of the fluid and the constraints imposed by the tanks walls transfers the maneuver acceleration to the fluid. The angular momentum of the sloshing fluids imparts torques on the spacecraft body. The transient resulting from a station keeping impulse is a function of the offset between the wet cm and the "dry" cm, spacecraft properties, the impulse magnitude, and controller design.

By modeling spacecraft slosh effects, the attitude transients for given thruster disturbances can be predicted and pointing error mitigation techniques studied. The goal is to identify spacecraft configurations that produce wheel mode maneuver pointing issues and gain improvements in pointing during wheel mode station keeping.

TELEMETRY OBSERVATIONS

The three basic telemetry points required to investigate pitch pointing disturbances to the spacecraft during a wheel mode station keeping maneuver are:

1) The pitch error,
2) Pitch rate, and
3) Pitch stored momentum.

For one spacecraft experiencing a single pulse of a wheel mode east-west maneuver, Figure 1 contains these three basic telemetry points. From top to bottom, the presented data are:

1) The Earth sensor pitch error,
2) Pitch rate measurement, and
3) Pitch momentum stored in the wheels as a function of wheel speeds and geometry. Figure 2 contains, for a second spacecraft experiencing three pulses of a wheel mode east-west maneuver, the same three primary telemetry points previously discussed.

Both spacecraft are of similar design, though they have different mass properties. Both spacecraft have roughly the same propellant fill state (beginning of life) at the time of the maneuvers shown, and both spacecraft have similar thruster strategies and on-times. Yet the pitch transient on the spacecraft in Figure 1 is significant while the transients on the spacecraft in Figure 2 are barely noticeable (they are circled to identify them). This difference motivates an understanding of the data in order to learn how to reliably improve pointing through maneuvers by design requirements and appropriate tactics in maneuvering.

Examination of the data in Figure 1 provides the following insight:

1) Angular momentum as defined by rigid body rotational dynamics is <u>not</u> conserved in the transient. Substantial angular momentum is stored in the propellant motion, inducing pitch-pointing errors.

2) The angular momentum imparted on the system by the thrusters, ΔH_{thrs}, can be measured from the pitch angular momentum telemetry by differencing the initial angular momentum and the final angular momentum after all slosh motion has subsided. Predicted changes in pitch angular momentum due to solar torques must be accounted for.

3) The initial pitch rate resulting from the maneuver, and the torque imparted on the spacecraft by the maneuver derived from ΔH_{thrs}, provides a direct measure of the offset between the wet center of mass (cm) and "dry" cm of the spacecraft. The "dry" cm is an effective dry cm that includes fluid in the tanks that acts as if it is rigidly attached to the tank wall. This concept is further explained later.

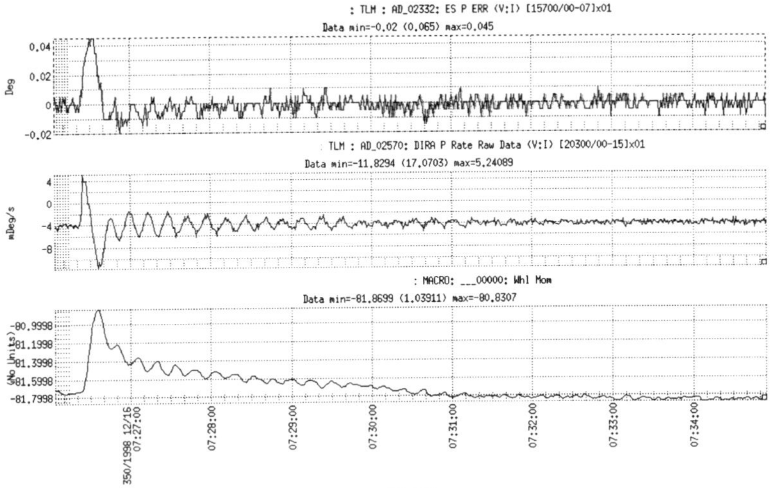

Figure 1 – Spacecraft 1 basic telemetry.

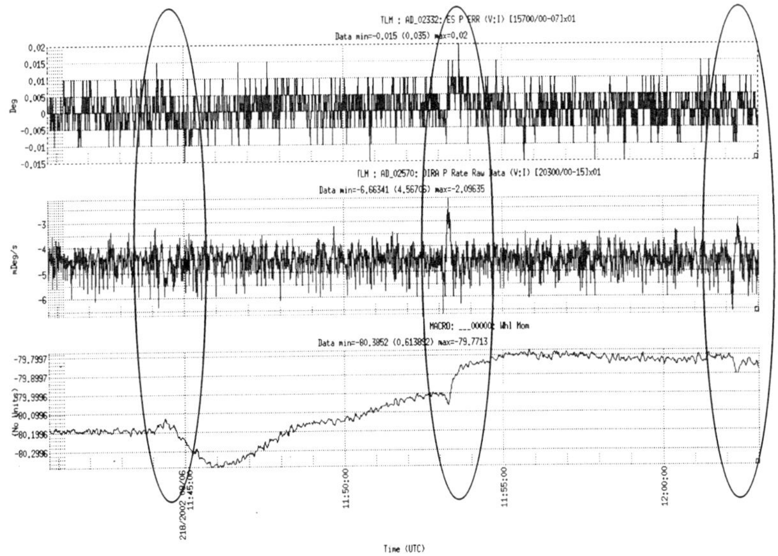

Figure 2 – Spacecraft 2 basic telemetry with highlighted maneuvers.

It is important to note that the difference between the wet cm and "dry" cm in the spacecraft in Figure 2 at beginning of life is, by design, roughly 25% of the difference in wet cm and effective dry cm in the spacecraft in Figure 1.

By modeling the propellant slosh motion during maneuvers, it becomes possible to predict the attitude disturbances resulting from maneuvers. The basic telemetry described above is used to develop, or 'tune', model parameters for given fill states. The calibrated model can then be used in new programs to add confidence that pointing requirements are being met, and provide a tool for investigating pointing improvement strategies.

DYNAMICS/SLOSH MODEL DESCRIPTION

The dynamics modeling employed does not attempt to predict the actual fluid motion, but rather attempts to predict the net effect of fluid motion on the spacecraft motion. The model applies in the low or zero 'g' environment, with low thrust, such as the environment found during wheel mode maneuvers on-orbit. The model assumes a portion of the fluid, modeled as a point mass, moves along the surface of an ellipsoid whose dimensions are a defined by tank geometry and fill percentage. The remainder of the fluid remains 'attached' to tank walls, tank structural ribbing, and the propellant management device (PMD). The fluid remaining 'attached' results in a new effective "dry" cm.

Fluid cm dynamics are constrained to motion on the ellipsoid. The cm derivative, or cm velocity, is parallel to the ellipsoid surface. A 'constraint' force perpendicular to the ellipsoid surface corrects deviations from that surface due to integration error. This 'constraint' force is sized to be the minimum force required to maintain the cm motion close to the constraint surface.

There is damping between the fluid particle and the tank. This is modeled as a combination of linear and quadratic damping. In addition, there is a PMD 'restoring' force that draws the fluid towards the PMD.

Figure 3 illustrates the constraint ellipsoid definition for a spherical tank with a cylindrical mid-section. The upper left tank in the figure shows the general fluid location and fluid cm in the tank during zero 'g' on-orbit flight. The propellant is generally attracted about the PMD and held there by surface tension. The upper right tank in Figure 3 illustrates the fluid cm in steady state acceleration from roughly 60 degrees off the central axis of the tank. (The integral for the solution to this cm location in the case of tanks with cylindrical midsections is not trivial, but can be solved. Of course the solution for purely spherical tanks is straightforward.) The lower tank in the figure illustrates how the constraint ellipsoid is fit through the two cm locations to provide a surface on which the fluid cm migrates during maneuvers.

The slosh excitation mechanism is illustrated in Figures 4 and 5. (Note - A spacecraft with only one tank is illustrated to simplify discussion.) If the external force is applied through the "dry" cm of the spacecraft, as in Figure 4, there is no initial torque on the "dry" box and no initial pitch rate disturbance. However, the propellant, which is not initially accelerated along with the box, is soon constrained to roll along the tank wall, resulting in a transient disturbance torque from the slosh. Momentum management usually precludes a strategy where thrust is applied through the "dry" cm as large momentum unloads would be required during or following maneuvers.

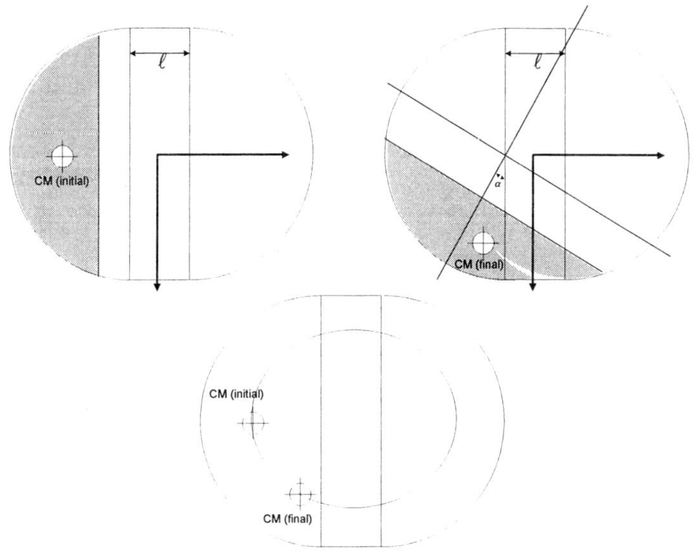

Figure 3 – Slosh cm motion constraint ellipsoid.

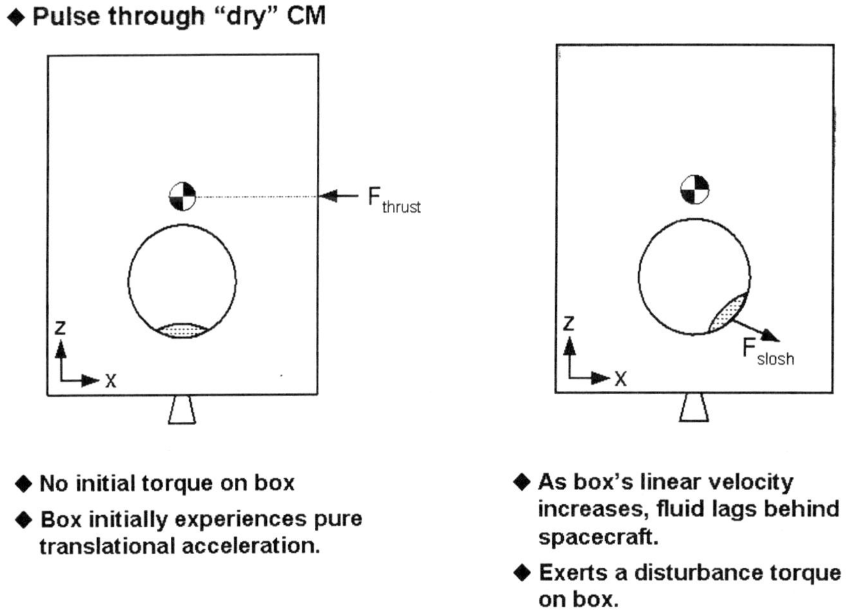

Figure 4 – Pulse through "dry" cm.

If the external force is applied through the wet cm of the spacecraft, as in Figure 5, the result is an initial torque and angular acceleration on the "dry" spacecraft as observed in the pitch rate data in Figure 1. The propellant, which lags behind the spacecraft, exerts a disturbance torque in the opposite sign of the initial thruster disturbance.

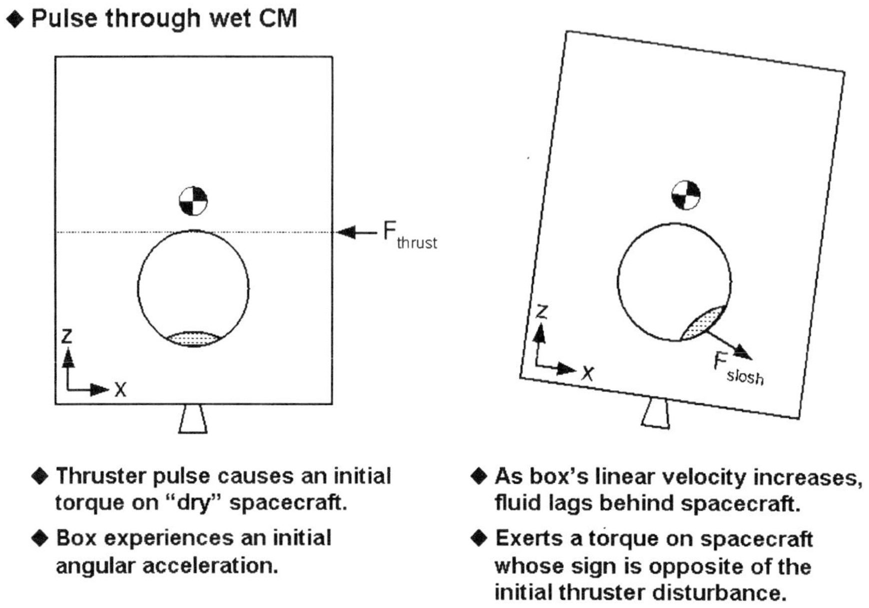

Figure 5 – Pulse through wet cm.

The typical bi-propellant spacecraft contains two tanks, one each for fuel and oxidizer. This configuration is shown in Figure 6. (Spherical tanks are illustrated.) The spacecraft dry cm is indicated, as are the two liquid cm's and the resulting wet cm. Typically the oxidizer tank is located below the fuel tank to meet launch vehicle cm constraints. The longer moment arm of the oxidizer cm, combined with the fact that the oxidizer is denser, and thus heavier, than the fuel, makes this configuration result in a wet cm aft of the dry cm.

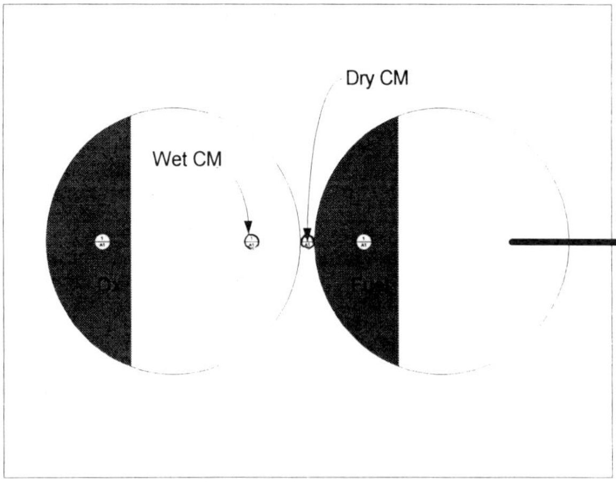

Figure 6 – Two tank spacecraft model.

As previously discussed, and illustrated in Figure 7, a percentage of the fluids behave as if they were attached to the spacecraft. This is modeled as fixed masses located on the tank PMD's and used to derive the effective "dry" cm.

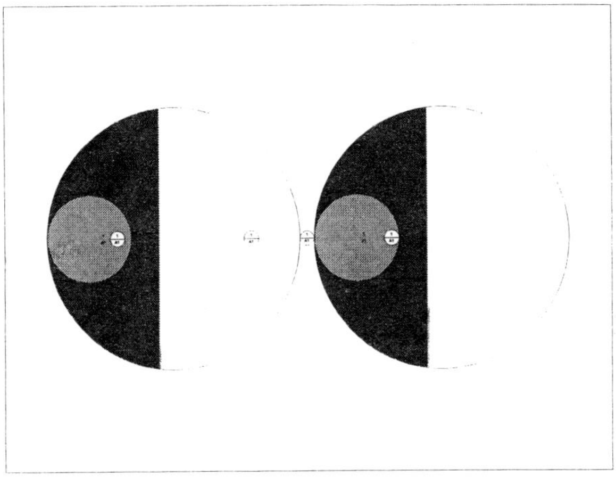

Figure 7 – Attached fluid vs. free fluid.

All other dynamics parameters being defined by physical properties of the fuel and oxidizer (viscosity, density, binding coefficient, etc), the validation and tuning process is limited to adjustment of the percentage of free fluid vs. attached fluid for a given fill state. In Figure 8, three different but similar spacecraft were observed for different fill states and the model free fuel and oxidizer percentage calibrated for each case. Fill states varied from ~40% fill down to ~8% fill. The trend line is indicated. The data shows that roughly 50% of the fluid in the tank behaves as if it was attached. While more data is desirable to add confidence to the model results, a reasonable value of the tuning parameters can now be estimated for any spacecraft of this configuration between zero and 40% fill state.

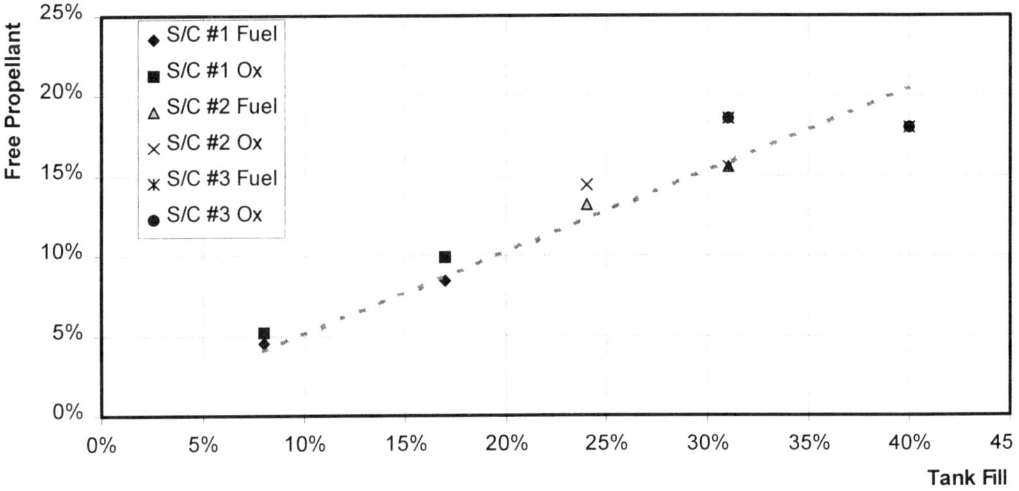

Figure 8 – Model Calibration.

Figure 9 illustrates a sample run of the model with appropriate output to show that linear and angular momentum is conserved in the slosh dynamics. In addition the cm motion of the fuel, oxidizer, and dry box relative to the wet cm are useful in interpreting telemetry.

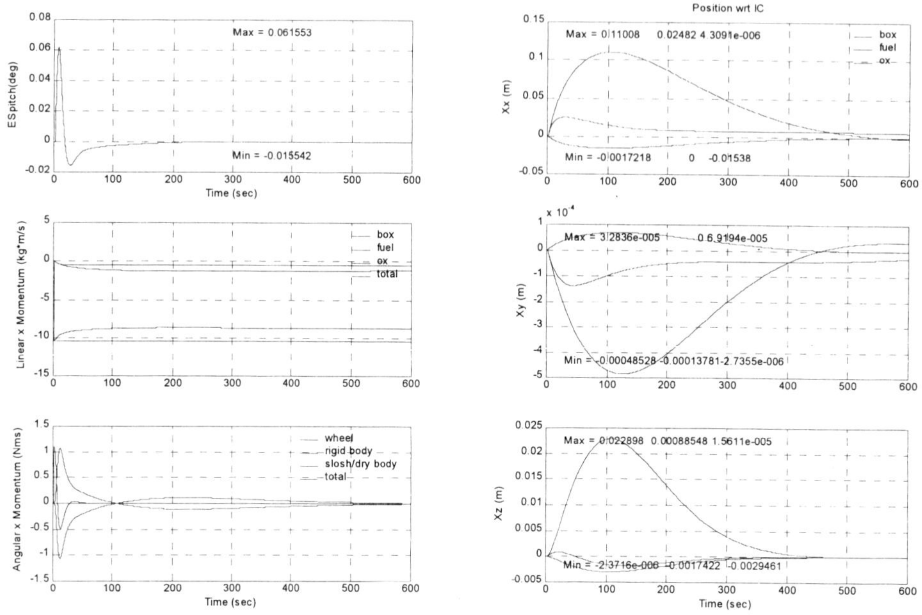

Figure 9 – Conservation of momentum and cm motion.

MODEL COMPARISON WITH FLIGHT DATA

Examples of model comparison with actual flight data are provided in Figures 10 and 11.

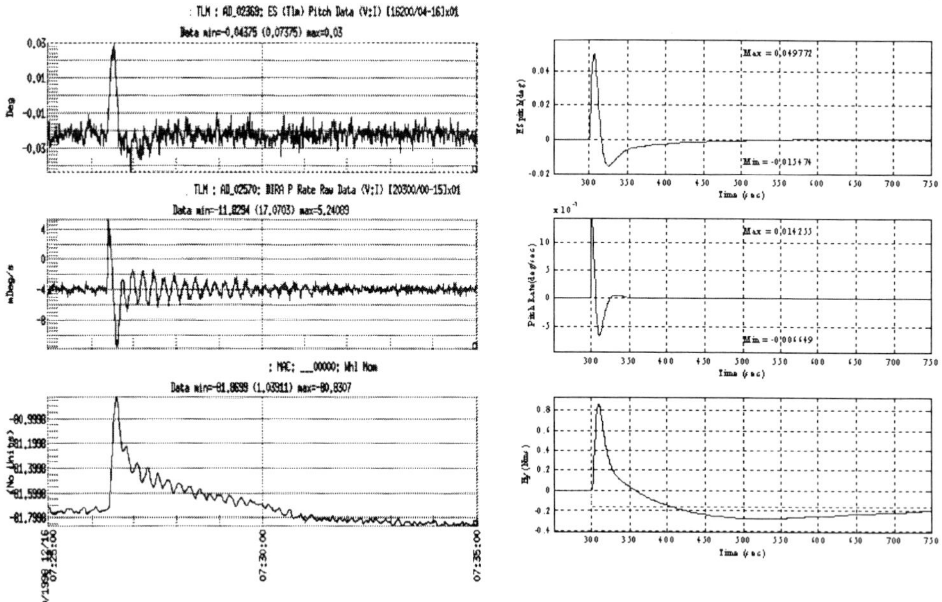

Figure 10 – Flight data vs. Model Prediction.

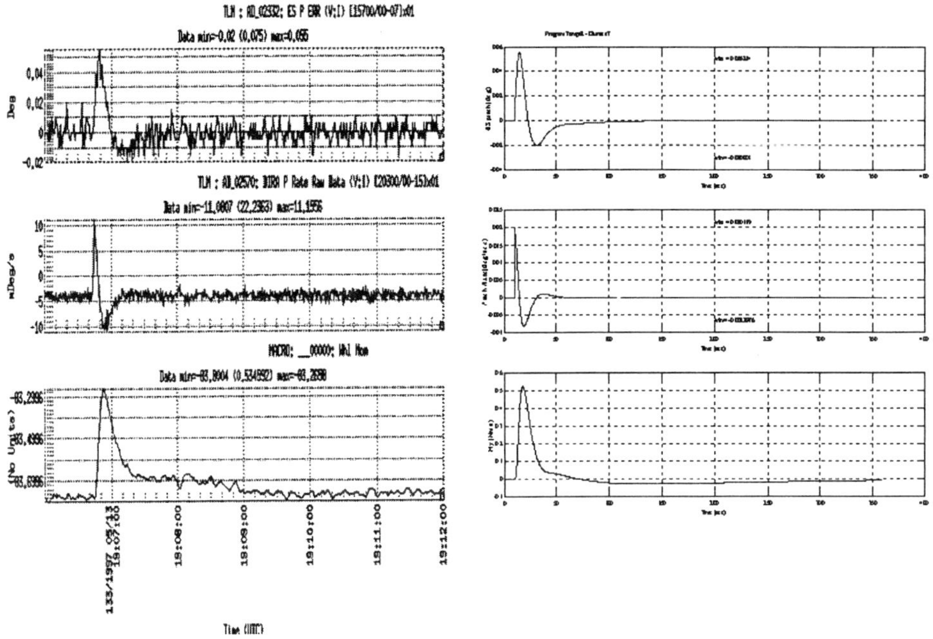

Figure 11 – Flight data vs. Model Prediction.

POINTING IMPROVEMENT STRATEGIES

Having established a representative model, it is desirable to study what induces transients and find effective tactics for reducing transients. A representative spacecraft and controller are used. Single parameter trades were conducted, varying wet and dry cm separation, pulse duration, thruster aiming, and wheel feed forward torque durations for two wheel feed forward torque tactics.

The first trade, shown in Figure 12, assumes the thrust acts through the wet cm and varies the separation between wet and dry cm. This separation is the basic source of pointing errors. Smaller separation leads to smaller pointing error and a large separation between wet and dry cm indicates pointing challenges. This type of trade can be used to levy requirements on S/C design (mass properties) based on pointing requirements.

In Figure 13, the thrust acts through the wet cm and the maneuver pulse duration is varied. Pulse duration reduction is a common method of meeting pointing requirements once on orbit. Smaller pulses lead to smaller pointing error. There is a mass consumption impact, however, as smaller pulses are less efficient. In addition, smaller pulses may mean more time required to complete a given maneuver, limiting maneuver size in the planning cycle.

By biasing the thrusters aim towards the dry cm, it is possible to reduce the initial pitch rate imparted on the spacecraft, thus reducing the overall transient. Biasing also alters the angular momentum imparted on the system by the thrusters, $\Delta H_{thrs.}$ This tactic is limited by momentum accumulation capacity, but may be allowable up to some amount. Figure 14 illustrates varying the thrust between the wet cm and the dry cm. Note that this trade covers the range of pitch response between the two cases described in Figures 4 and 5, firing through the wet cm and firing through the "dry" cm, respectively.

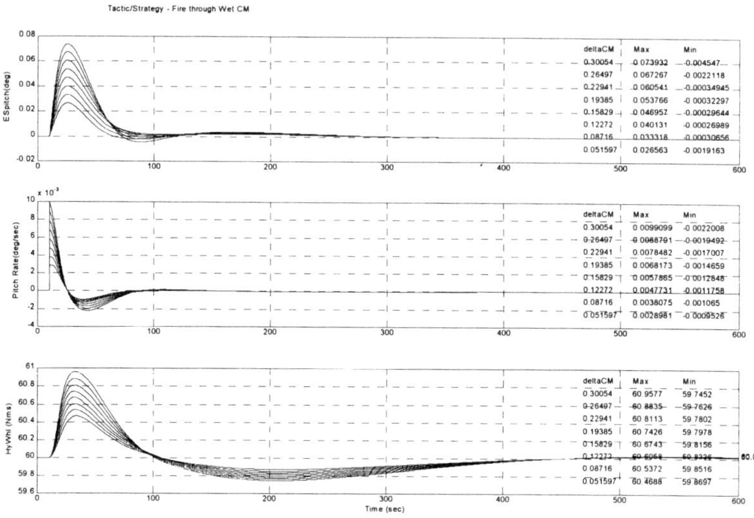

Figure 12 – Pitch response for varying wet and dry cm separation.

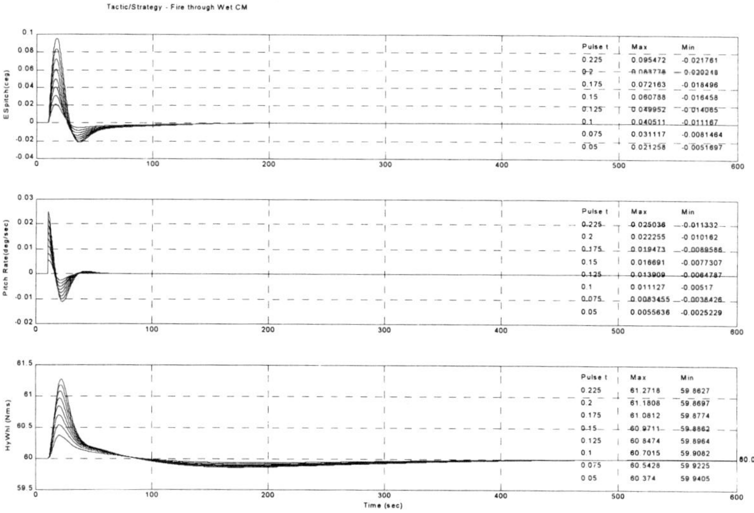

Figure 13 – Pitch response for varying maneuver pulse duration.

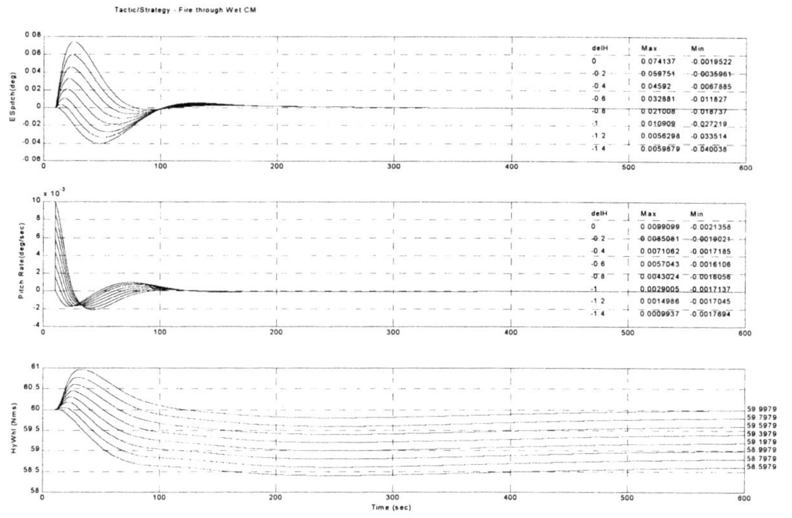

Figure 14 – Pitch response for varying $\Delta H_{thrs.}$

Given an understanding of the dynamics, and using the model as a prediction analysis tool, many feed forward torque techniques could be devised to attack the transient problem. Two are illustrated in Figures 15 and 16. Figure 15 varies a realizable feed forward torque to the wheel system at the time of the thruster disturbance. Figure 16 varies a realizable feed forward torque to the wheel system prior to and centered at the time of the thruster disturbance. Both assume the thrust acts through the wet cm. significant reductions in pointing error are achieved. This tactic is limited by wheel torque authority.

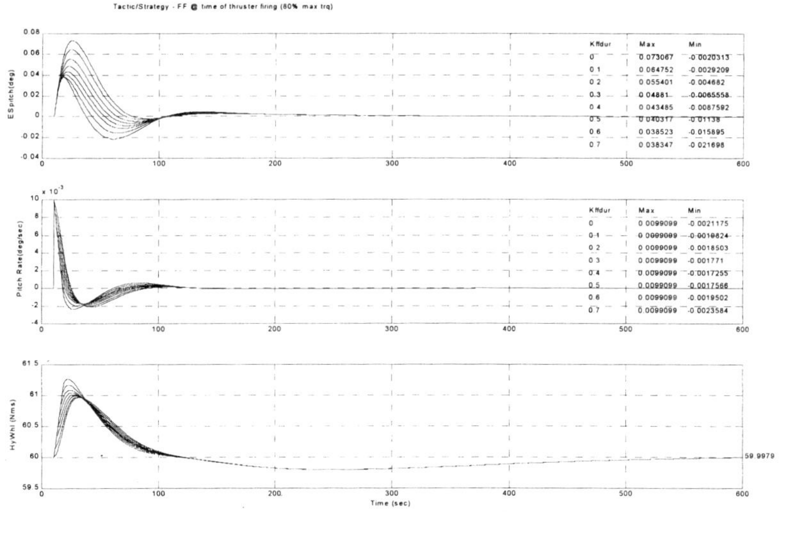

Figure 15 – Pitch response for varying feed forward torque at time of disturbance.

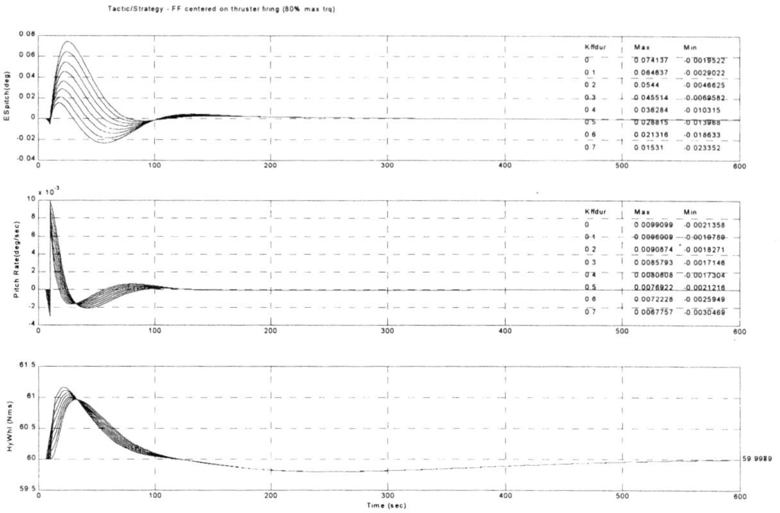

Figure 16 – Pitch response for varying feed forward torque prior to time of disturbance.

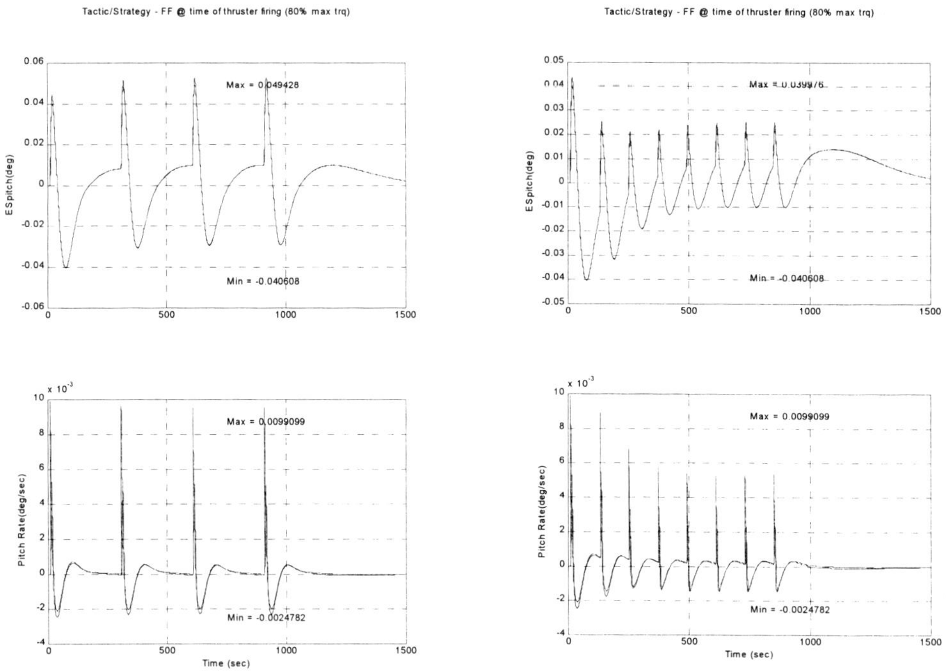

Figure 17 – Varying pulse interval.

The settling time for the disturbed fluid during any given maneuver pulse is between six and ten minutes for a single pulse. By decreasing the interval between maneuver pulses, it is possible to reduce the pointing error for individual pulses by "pumping up" the propellant. In other words, rather than allowing the propellant cm's to settle at the points of largest difference between wet and "dry" cm, continue to pulse such that the difference in wet and "dry" cm in the pitch axis is minimized. In this displaced location, the cm motion constraint now causes the fluid cm to be more rapidly accelerated by the subsequent maneuver pulses, in effect shifting the "dry" cm still closer toward the wet cm. These effects reduce the initial pitch rate acquired by each pulse. Figure 17 illustrates pitch and pitch rate response to a series of maneuver pulses performed at five-minute intervals and two minute intervals. The individual pulse size is the same in both cases. From the two minute interval case it can be seen that, once the propellant has been significantly displaced, the pointing error is roughly half that seen in the initial pulse, and half that seen in the five minute interval case. Note this simulation assumes the maneuver thrust is sufficiently small, or the propellant slosh adequately damped, to prevent exciting slosh modes.

Obviously more complex feed forward schemes can be investigated. In addition, combinations of various tactics described above can lead to significant reductions in pointing error during maneuvers due to slosh. Many of these strategies have been shown to reduce the slosh cm motion induced errors below the level of error due to thruster uncertainty. Use of these tactics and strategies provides very robust pointing to the level of thruster uncertainty. For large geosynchronous spacecraft used in this study, transient pointing error reductions greater than 50% peak to peak were attainable even with thruster uncertainty. Figure 18 illustrates predicted performance of one system with a large offset between the wet cm and "dry" cm. A combined strategy, including feed forward torque at the time of maneuver pulse and targeting the thrust 0.2 Nms towards the "dry" cm from the wet cm, is employed to meet a 20mdeg pitch pointing specification under thruster variability of up to ±0.2 Nms. Without these techniques, pitch pointing error would exceed 50mdeg for the pulse size chosen.

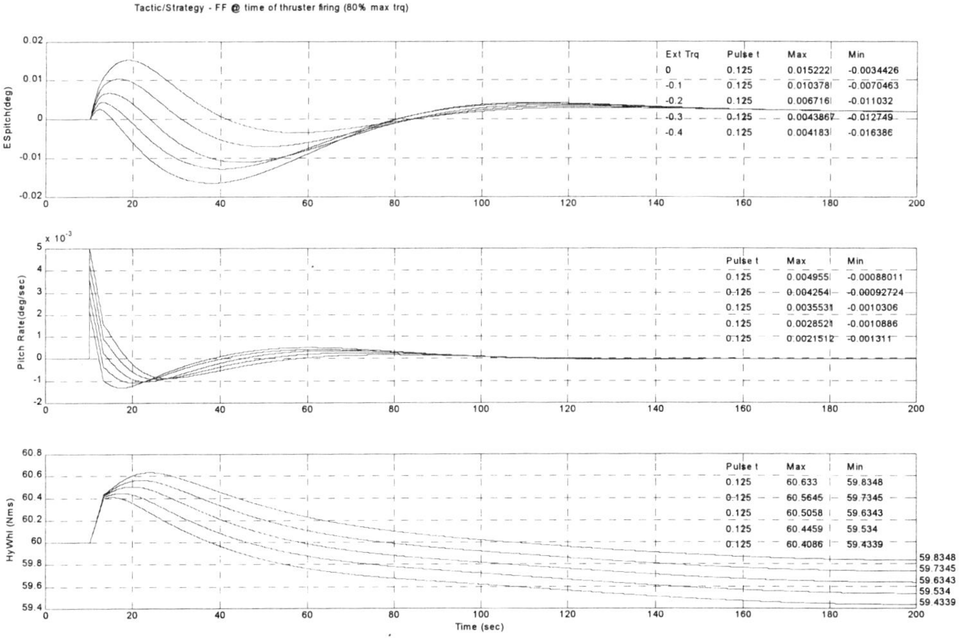

Figure 18 – Sample-pointing strategy under thruster variability.

CONCLUSION

Spacecraft in-flight data showing attitude disturbances during wheel mode operations was shown to illustrate the effects of propellant slosh on spacecraft pointing. The different responses seen in similar spacecraft designs with different propellant loads or mass properties prompted an investigation into the cause of the disturbances. This led to the development and validation of a slosh model for attitude disturbance prediction. The model was shown to reasonably represent fluid cm motion and spacecraft attitude response to thruster disturbances. The model was then used to investigate several tactics and strategies to reduce the disturbance transient. The tactics and strategies range from early design requirements on spacecraft mass properties to operational tactics that can be applied on-orbit. Finally, application of these tactics has been shown to produce significant pointing error reductions during maneuvers.

AAS 03-065

DESIGN AND ANALYSIS OF THE ST7 DISTURBANCE REDUCTION SYSTEM (DRS) SPACECRAFT CONTROLLER

P. G. Maghami,[*] F. L. Markley,[*] M. B. Houghton[*] and C. J. Dennehy[*]

The Space Technology 7 experiment will perform an on-orbit system-level validation of two specific Disturbance Reduction System technologies: a gravitational reference sensor employing a free-floating test mass and a set of micronewton colloidal thrusters. The Disturbance Reduction System is designed to maintain a spacecraft's position with respect to the free-floating test mass to less than 10 nm/√Hz, over the frequency range 10^{-3} Hz to 10^{-2} Hz. This paper presents the design and analysis of the coupled drag-free and attitude control system that closes the loop between the gravitational reference sensor and the micronewton thrusters while incorporating star tracker data at low frequencies. The effects of actuation and measurement noise and disturbances on the spacecraft and test masses are evaluated in a seven-degree-of-freedom planar model incorporating two translational and one rotational degrees of freedom for the spacecraft and two translational degrees of freedom for each test mass.

INTRODUCTION

NASA's New Millennium Program (NMP) recently selected the Disturbance Reduction System (DRS) flight validation experiment, managed by the Jet Propulsion Laboratory (JPL), for the Space Technology 7 (ST7) mission.[1] NMP missions are intended to validate advanced technologies that have not flown in space in order to reduce the risk of their infusion in future NASA Space Science missions. The ST7 DRS incorporates two specific technologies: a highly sensitive Gravitational Reference Sensor (GRS), provided by Stanford University, to measure the position and attitude of a spacecraft with respect to an internal free-floating test mass, and a set of micronewton colloidal thrusters, provided by the Busek Company. The ST7 DRS, scheduled to fly on the European Space Agency's SMART-II spacecraft in 2006, is designed to maintain the spacecraft's position, with respect to the GRS free-floating test mass, to less than 10 nm/√Hz, over ST7's science measurement frequency range from 1 to 10 mHz. The DRS instrument package consists of two gravitational reference sensors, two sets of four microNewton thrusters each for position and attitude control, and an interferometer to measure the distance between the two test masses, as shown in Figure 1.

This paper presents the overall design and analysis process of the spacecraft controller being developed at NASA's Goddard Space Flight Center to close the loop between the GRS and the micronewton colloidal thrusters. A two-dimensional (planar) model has been developed to capture the essential dynamics of the ST7-DRS package. It includes seven rigid-body dynamic degrees of freedom: two translations and a rotation for the spacecraft, and two translations for each test mass, ignoring the inessential rotational dynamics of the test masses. Actuation and measurement noise and disturbance sources acting on the spacecraft and test masses are modeled. The ST7 DRS comprises three control systems: the attitude control system (ACS) to maintain a sun-pointing attitude; the drag free control (DFC) to center the spacecraft about the test masses; and the test mass suspension control. This paper summarizes the control design and analysis of the ST7-DRS 7-DOF model.

---

[*] Guidance, Navigation and Control Division, NASA Goddard Space Flight Center, Greenbelt, Maryland 20771.

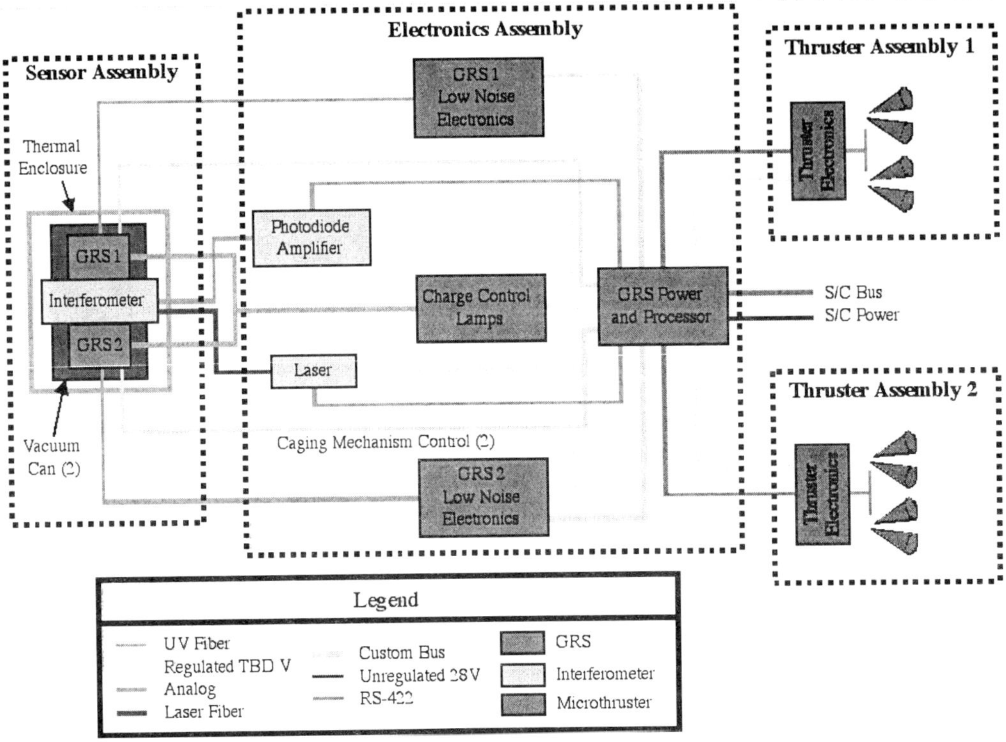

Figure 1. DRS Instrument Schematic

MODEL DESCRIPTION

The analysis presented in this paper uses the highly simplified 7-DOF model shown in Figure 2, which incorporates the essential dynamics of the system. Here the location and the orientation of the two test masses can be arbitrarily assigned. The nominal position vectors for the two test masses and their respective housings are chosen as $\mathbf{R}_1 = [0.1 \quad 0]$ m and $\mathbf{R}_2 = [0.3 \quad 0]$ m, which means that the sensitive axis is along the X-axis of the spacecraft. The mass and moment of inertia of the spacecraft are assumed to be 200 kg and 175 kg-m^2, respectively, and the test masses are assumed to be 1.3 kg each.

Two clusters, each containing two thrusters, are located on the $\pm$X faces to provide thrust capability for attitude and drag-free control. It should be pointed out that in the current analysis the thruster configurations are mainly used for thrust noise characterization. Thrust and torque commands for attitude control and drag-free control are computed in the spacecraft body frame, and are not resolved to the individual thruster commands.

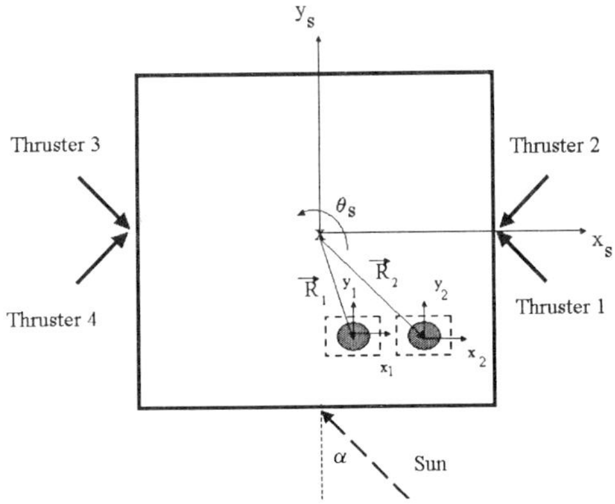

Figure 2 7-DOF Model Configuration

Two disturbances are included in this model. The first is the nominal solar radiation pressure and its variation. The Sun exposed face of the spacecraft corresponds to the –Y direction. The angle of the incident rays of the sun to the surface normal, α, may be arbitrarily assigned. However, it is assumed to be zero for the current analysis. The frequency spectrum used for solar radiation flux variations given in Figure 3 represents a conservative assessment of measured variations.[2,3] This plot indicates a constant spectrum at the frequencies below 0.1 mHz, followed by a $1/f$ roll off. This spectrum also includes the so-called 5-minute acoustic oscillation (at 3.5 mHz), and levels off at frequencies above 10 mHz. The second disturbance source modeled was the acceleration noise on the test mass. A number of sources contribute to this acceleration noise, including magnetic and Lorentz forces, thermal disturbances, cosmic ray impacts, etc.[2] The spectral density function for the test mass acceleration noise is assumed to have the following characteristics:

- $1/f^2$ rolloff at frequency range of 0.01-0.1 mHz,
- $1/f$ rolloff at frequency range of 0.1-1 mHz,
- constant spectral density $3x10^{-14}$ $m/s^2/Hz^{0.5}$ at frequencies above 1 mHz.

In order to incorporate this power spectrum within the frequency and time domain analyses, the linear filter approximation shown in Figure 4 was developed. Note that all requirements are either met or exceeded by this approximation. This acceleration noise was applied to both test masses in all directions.

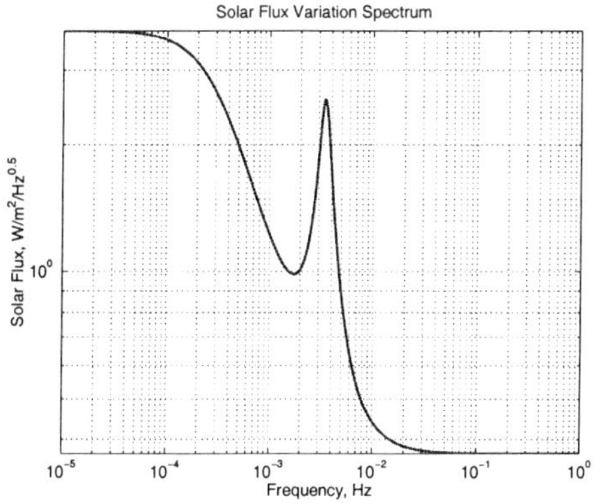

Figure 3 Root Power Spectrum of the Solar Radiation Flux Variations

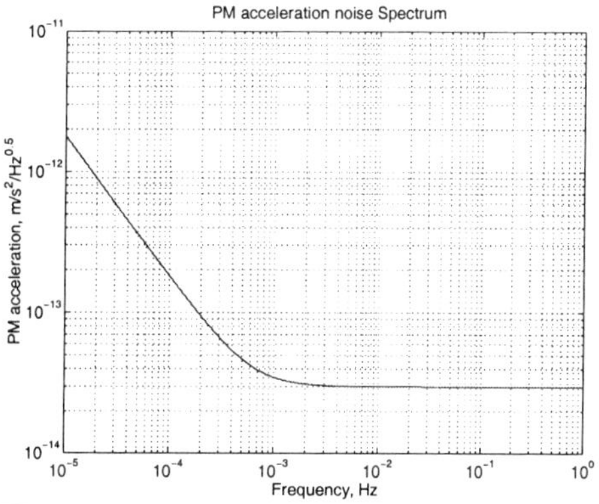

Figure 4 Root Power Spectrum for the Test Mass Acceleration Noise

White-noise models were used to capture thruster noise, electrostatic suspension force noise, star tracker noise, and the capacitive sensing noise (used to measure the positions of the test masses relative to the spacecraft). The intensity levels are captured in Table 1.

CONTROLLER DESIGN

A top-level block diagram of the system dynamics is shown in Figure 5. The control system uses five output measurements: the relative positions of test mass 1 in X and Y; the relative positions of test mass 2 in X and Y; and the spacecraft attitude error from the star tracker. The five control inputs are the thruster force commands in X and Y, the thruster torque command, and the suspension control force commands on test mass 2 in X and Y. It is important to note that no suspension forces are applied to test mass 1. A 1% parasitic cross-talk between the X and Y axis suspension forces on test mass 2 is assumed in the analysis. There are two main control loops required for the spacecraft control. First is the drag-free controller, which controls the position of the spacecraft (in X and Y) to establish the drag-free motion of test mass 1. The second controller is the spacecraft attitude control, which is primarily designed to orient the spacecraft in the low frequency band (DC and near DC) using the star tracker data. However, it is also designed to center the spacecraft about test mass 2 in the transverse (Y) direction in the ST7 science measurement band from 1 to 10 mHz. Both test masses are effectively freely falling in the measurement band.

Table 1
ACTUATION AND SENSING NOISE INTENSITIES

Noise Source	Intensity
Thrusters	0.1 μN/Hz$^{0.5}$
Suspension Force	2e-14 N/Hz$^{0.5}$
Star Tracker	20 arcsec/ Hz$^{0.5}$
Capacitive Sensing	3 nm/ Hz$^{0.5}$

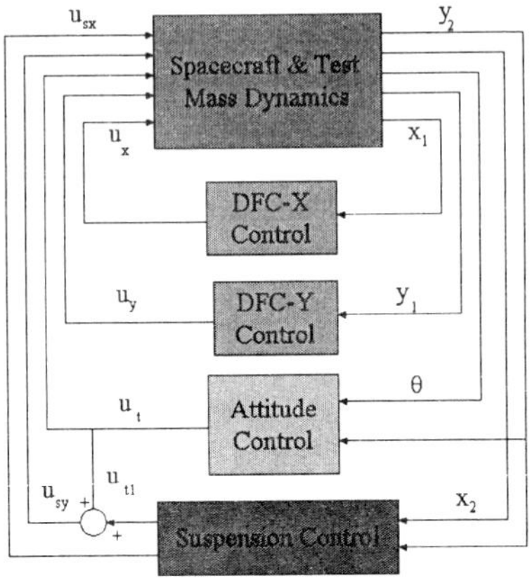

Figure 5 Block Diagram of the 7-DOF System

The drag-free controller, which includes two loops (one for X-axis control and one for Y-axis control), was designed using a classical control approach. However, in this model, the relative test mass positions, which are used by the drag free controller, are not only a function of the spacecraft and test mass translations, but also are affected by the attitude of the spacecraft. The drag-free controller was a Proportional-Integral-Derivative (PID) controller with a roll-off filter.

$$u_x(s) = K_d(s)x_1(s)$$
$$u_y(s) = K_d(s)y_1(s) \tag{1}$$

$$K_d(s) = [G_{pd} + sG_{dd} + \frac{G_{id}}{s}]F_d(s) \tag{2}$$

where $x_1(s)$ and $y_1(s)$ denote the measured relative position of test mass 1 from its caging in X and Y, respectively. The terms $u_x(s)$ and $u_y(s)$ denote the required control thrust force in X and Y. G_{pd}, G_{dd}, and G_{id} denote the proportional, derivative, and integral gains. The term $F_d(s)$ denotes the third-order attenuation filter. The current drag-free controller is defined as

$$K_d(s) = \frac{2811\, s^2 + 634.9\, s + 13.36}{s^4 + 5.969\, s^3 + 17.81\, s^2 + 26.58\, s} \tag{3}$$

This controller provides a cut-off frequency of 0.055 Hz. It should be noted that all sensor measurements are updated at 10 Hz sampling with the exception of the star tracker signal, which is updated at 1 Hz. Moreover, approximations to zero-order hold and expected computational/transport delays, in the form of a Pade approximation, were included in the plant dynamics.

The attitude controller is a two-input/single-output controller, which uses star tracker measurements and the relative position of test mass 2 in the Y-direction as inputs. This controller was designed following a classical design approach and uses blending filters to accommodate its two tasks. The structure of the attitude control may be summarized as follows.

$$u_t(s) = K_{al}(s)\theta(s) + K_{ah}(s)y_2(s) \tag{4}$$

Here $\theta(s)$ denotes the attitude error measurement obtained from the star tracker data, $y_2(s)$ represents the measured relative position of test mass 2 from its caging in the Y-direction, and $u_t(s)$ denotes the required thrust control torque. The term $K_{al}(s)$ denotes the low-bandwidth part of the attitude control designed to maintain the attitude of the spacecraft. Its design is a classical PID loop with an appropriate roll-off filter, given as

$$K_{al}(s) = \frac{b_2 s^2 + b_1 s + b_0}{s^5 + a_4 s^4 + a_3 s^3 + a_2 s^2 + a_1 s} \tag{5}$$

$$\begin{aligned} a_4 &= 6.568e-3 & b_2 &= 2.481e-12 \\ a_3 &= 2.157e-5 & b_1 &= 4.410e-16 \\ a_2 &= 4.148e-8 & b_0 &= 3.308e-20 \\ a_1 &= 3.990e-11 & & \end{aligned}$$

This controller provides a cut-off frequency of about 0.00005 Hz. The controller term $K_{ah}(s)$ represents the part of the attitude controller that centers the spacecraft about test mass 2 in the transverse direction (Y-direction) in the ST7-DRS measurement band. This controller is also designed based on the classical approach, and is a series combination of lead-lag filter, PD filter, and a roll-off filter, resulting in a sixth-order controller with the following transfer function.

$$K_{ah}(s) = \frac{b_5 s^5 + b_4 s^4 + b_3 s^3 + b_2 s^2 + b_1 s + b_0}{s^6 + a_5 s^5 + a_4 s^4 + a_3 s^3 + a_2 s^2 + a_1 s + a_0} \tag{6}$$

$$\begin{aligned} a_5 &= 3.579 & b_5 &= 392.8 \\ a_4 &= 6.406 & b_4 &= 70.81 \\ a_3 &= 0.1596 & b_3 &= 0.1764 \\ a_2 &= 0.0015 & b_2 &= 1.658e-4 \\ a_1 &= 6.273e-06 & b_1 &= 6.934e-08 \\ a_0 &= 9.845e-09 & b_0 &= 1.088e-11 \end{aligned}$$

The relative X and Y position of test mass 2 is controlled by the electrostatic suspension control internal to the GRS. The relative position in X is controlled via a low bandwidth PID controller to provide disturbance rejection at DC and near DC. The Y-position control comprises two compensators. One is a low bandwidth PID controller to provide disturbance rejection at DC and near DC. The other is a feedforward compensator that nulls out the compensation effects of the low bandwidth attitude controller. This compensation removes a significant part of the coupling between the attitude loop and the 2nd test mass position control in Y, and therefore makes the system more amenable to decentralized control. It also naturally corrects for any disturbances caused in the test mass position control loop by attitude control commands. The structure of this controller is given by

$$u_{sy}(s) = K_{sy}(s) y_2(s) + h * K_{al}(s)\theta(s) \tag{7}$$

Here $u_{sy}(s)$ denotes the suspension control force for test mass 2 in Y. The nominal controller, $K_{sy}(s)$, is a low-bandwidth PID loop, and the scaling constant h depends on the moment arm from spacecraft center of mass to the test mass, the spacecraft inertia, and the mass of the test mass.

STABILITY MARGINS

Each of the controllers was designed to have sufficient stability margins. However the 7-DOF system represented in Figure 5 is a MIMO system, by virtue of the cross coupling between the relative test mass positions and the attitude of the spacecraft. Hence, the loop gains at each input and output channel (while the remaining channels are closed) must be analyzed to obtain proper stability margins. Figures 6–11 show the loop gains for six of the ten channels. The loop gains of the other four channels (the X and Y components of the relative positions of the two test masses) are virtually identical to four of the plots shown. Figure 9 shows that the Y-axis suspension force on test mass 2 is the loop exhibiting the minimum margins: 7.6 dB of gain margin and 35° of phase margin. These margins are amply sufficient considering that the effects of zero-order hold and computational and transport delays are already included in the analysis. The margins for the attitude pointing error shown in Figure 11 are similar to those for the Y-axis suspension force, which is driven by spacecraft attitude accelerations.

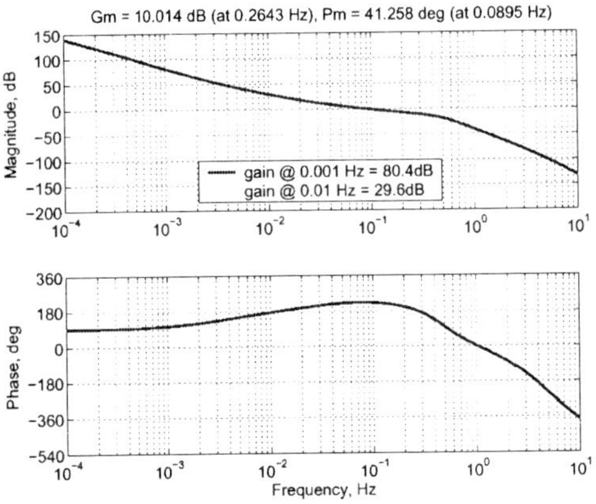

Figure 6 Loop Gain at the Thrust Force: X-Direction

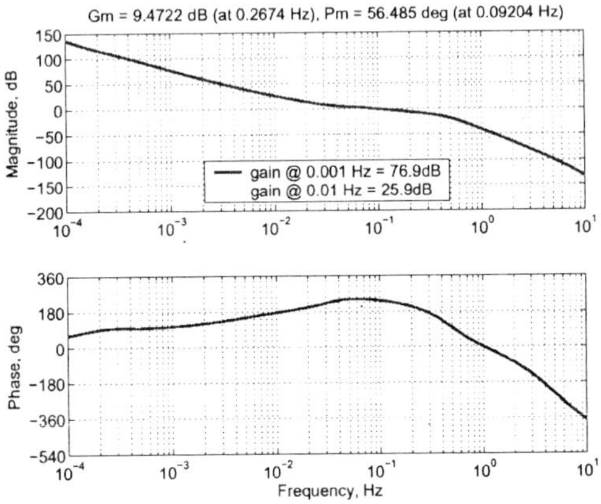

Figure 7 Loop Gain at the Thrust Force: Y-Direction

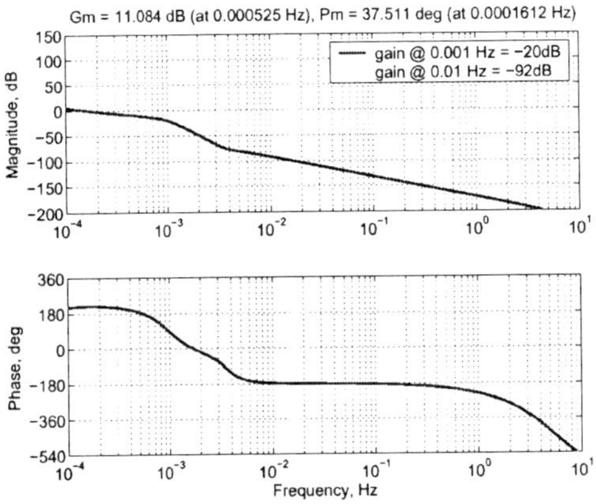

Figure 8 Loop Gain at the X-Axis Suspension Force on Test Mass 2

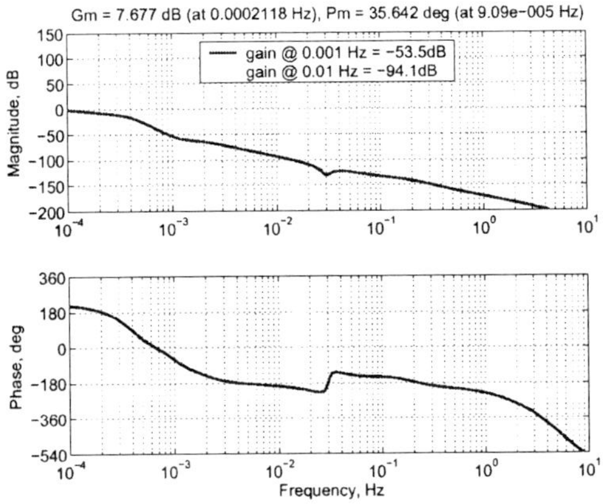

Figure 9 Loop Gain at the Y-Axis Suspension Force on Test Mass 2

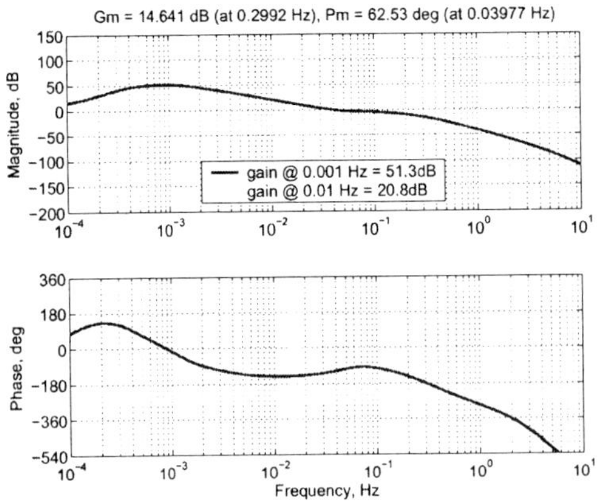

Figure 10 Loop Gain at the Spacecraft Torque Command

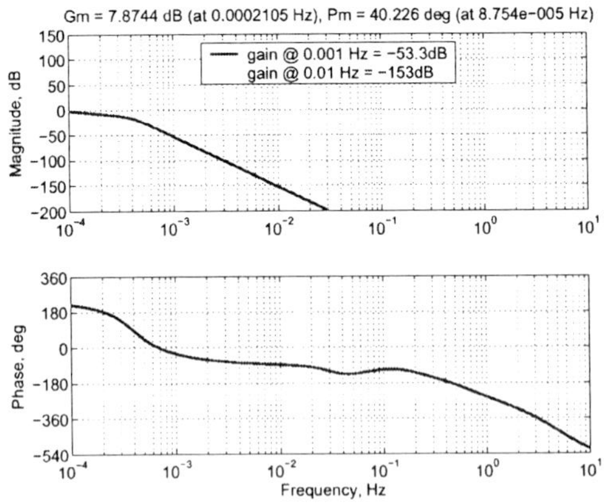

Figure 11 Loop Gain at Spacecraft Attitude Pointing Error

ANALYSIS RESULTS

The attitude and drag-free controllers were implemented in a MATLAB-based model of the system that serves as the design and analysis tool for the 7-DOF Model. The results of both time-domain and frequency-domain analyses are shown in Figures 12–21. The root power spectral density plots show the contributions of the various disturbance sources. The contribution of each disturbance category represents the root sum squared (RSS) values for that category; for example, the thruster noise plot is the RSS contribution of the noise from all four thrusters.

Figures 12a–15a illustrate the root power spectral densities of the relative positions of both test masses, in both X and Y directions. These are dominated by the thruster noise and capacitive sensing measurement noise in the measurement bandwidth. The cross coupling that exists in this MIMO system may be observed in the variations in the contributions from different thruster noise sources, as well as the contribution of solar radiation pressure noise in the X direction. Both test masses satisfy the positioning requirement of 10 nm/√(Hz) in both directions in the science measurement band (the top of the plots for test mass 1 and indicated by horizontal lines for test mass 2), although the spacecraft only follows test mass 1. The necessity of controlling test mass 2 to follow the spacecraft over long time scales is reflected in the higher root power spectral densities of its relative position components at frequencies below the science measurement band. Figures 14a and 15a show that the star tracker measurement noise contributes significantly to the test mass 2 relative position errors below the science measurement band. Figures 12b–15b show 42-hour time histories of the relative positions of the two test masses, with the lower plot being an expansion of the last 1000 seconds. It can be seen that the displacement of test mass 1 never exceeds 4 nm, and only very occasionally exceeds 3 nm, in agreement with the frequency domain results. The displacement of test mass 2 shows larger low-frequency motions, also agreeing with the frequency domain results.

Figure 16 shows the root spectrum for the spacecraft pointing error, which indicates that the error is well within its requirement of 1°/√(Hz). This error is also dominated by the thruster noise and capacitive measurement noise in the science measurement band and by star tracker measurement noise below the measurement band.

Figures 17–19 present the spectra and time histories of the thruster force commands in the X, Y and torque directions. The RSS levels are well within the 20 μN capability of the colloidal thrusters. The major contributor to the spectra in and below the measurement band is thrust noise, while capacitive sensing noise becomes dominant well above the measurement band.

Figures 20 and 21 illustrate the spectra and time histories of the suspension control forces on test mass 2 in the X and Y directions, respectively; these are well within the 10 nN/√(Hz) capability of the electrostatic suspension system of the GRS. Thrust noise, capacitive sensing noise, and star tracker noise all have major contributions to the suspension force spectral densities, within and below the measurement bandwidth.

DISCUSSION

The time-domain and frequency-domain analyses presented in this paper show that all the requirements for the ST7-DRS control system are met in a planar seven-degree-of-freedom model. These requirements include establishing drag-free motion of the test masses in the science band as well as spacecraft attitude control. The spacecraft position relative to the primary test mass will be maintained within the required precise limits. Successful spacecraft attitude control is accomplished by combining low frequency data from a star tracker and high frequency data from the transverse position of the second test mass. It was also proved that rapidly rolling off the electrostatic suspension forces on the second test mass between DC and the measurement frequency band provides adequate suspension while maintaining its drag-free state within the frequency range of interest. All these conclusions must be confirmed in a three-dimensional eighteen-degree-of-freedom dynamic model of the spacecraft and test masses. No essential complications are expected to arise at this level, however.

REFERENCES

[1] Keiser, G. M., Buchman, S., Byer, R. L., Folkner, W. M., Hruby, V., and Gamero-Castaño, M., "Disturbance Reduction System for Testing Technology for Drag-Free Operation," SPIE Paper 4856-02, Astronomical Telescopes and Instrumentation Conference, Waikoloa, Hawaii, USA, August 2002.

[2] Final Technical Report of the (Phase A) Study of the Laser Interferometer Space Antenna (Dornier Satellintensysteme GmbH – Matra Marconi Space – Alenia Aerospzio) ESTEC Contract no. 13631/99/NL/MS, Report No. LI-RP-DS-009, April 2000

[3] J. Pap, et. al., "Variation in Total Solar and Spectral Irradiance as Measured by the VIRGO Experiment on SOHO", *Adv. Space Res.*, 24:215-224, 1999.

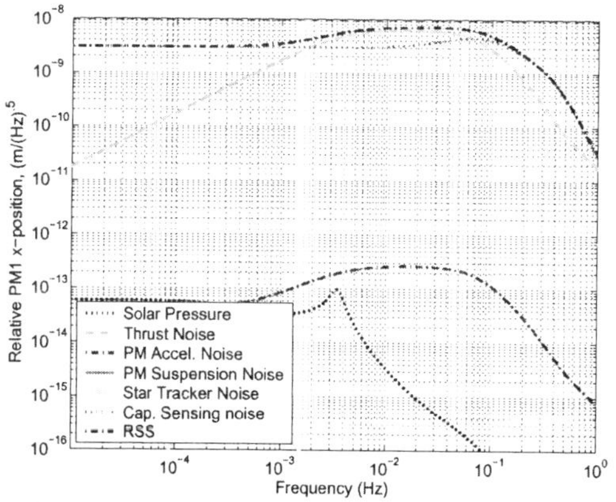

Figure 12a Root Power Spectrum of the Relative Position of Test Mass 1: X-direction

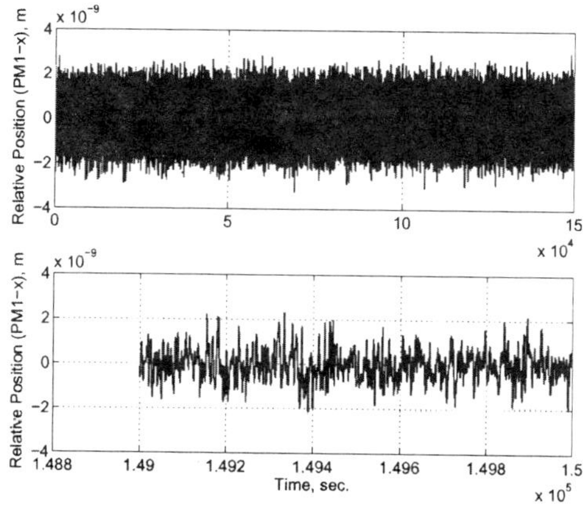

Figure 12b Time History for the Relative Position of Test Mass 1: X-Direction

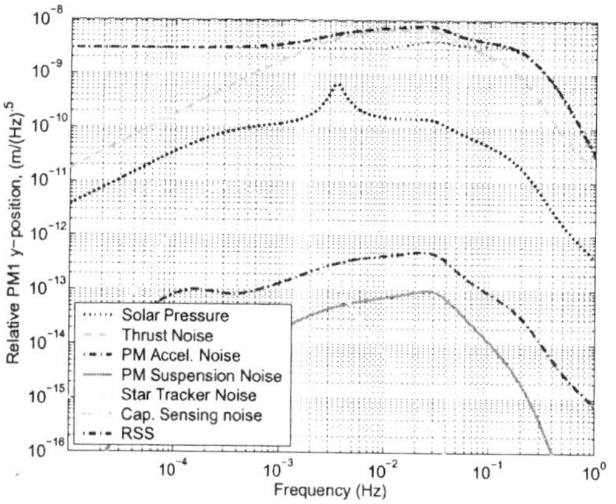

Figure 13a Root Power Spectrum of the Relative Position of Test Mass 1: Y-direction

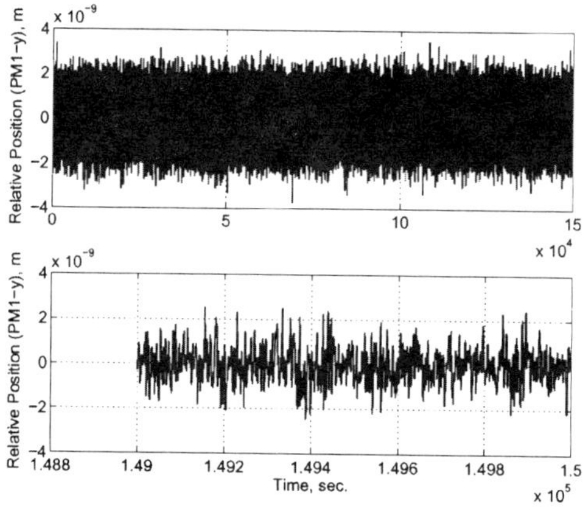

Figure 13b Time History for the Relative Position of Test Mass 1: Y-Direction

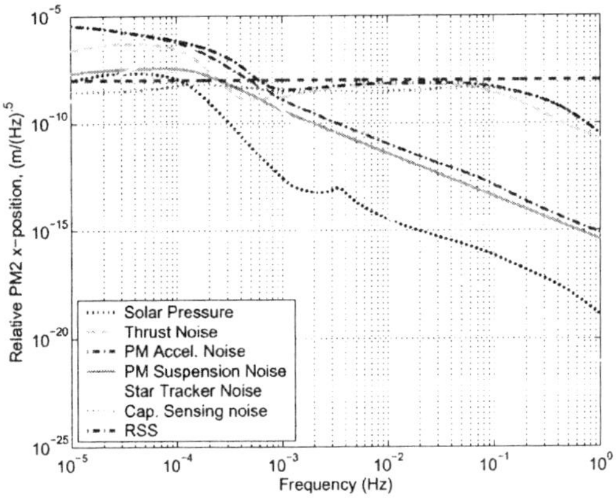

Figure 14a Root Power Spectrum of the Relative Position of Test Mass 2: X-Direction

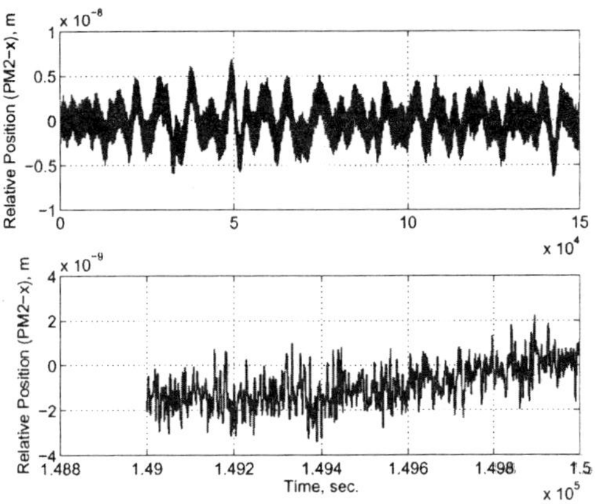

Figure 14b Time History for the Relative Position of Test Mass 2: X-Direction

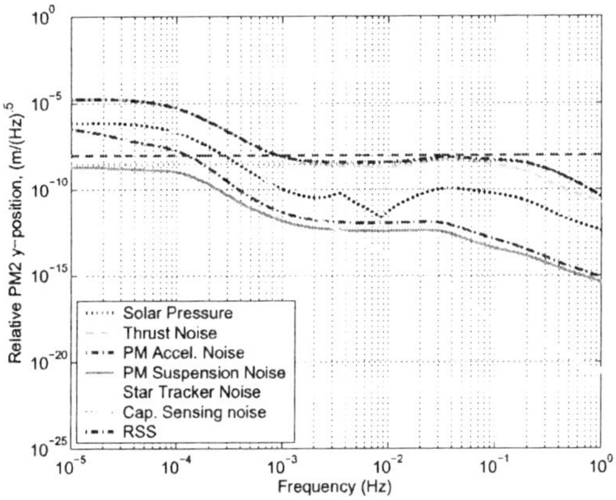

Figure 15a Root Power Spectrum of the Relative Position of Test Mass 2: Y-Direction

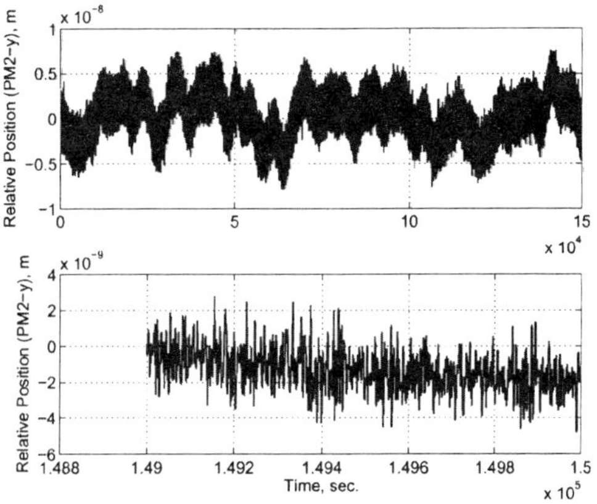

Figure 15b Time History for the Relative Position of Test Mass 2: Y-Direction

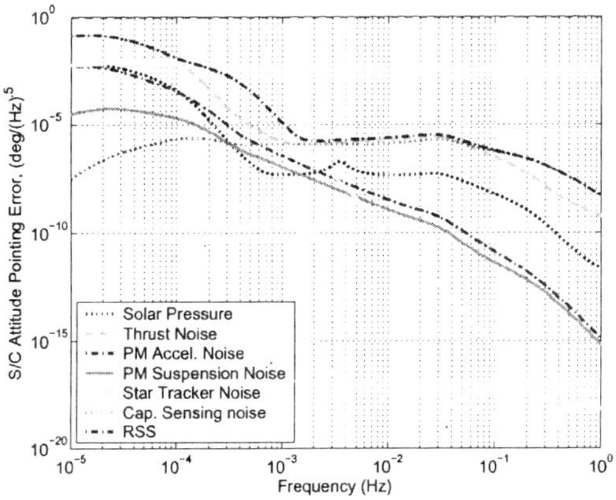

Figure 16a Root Power Spectrum of the Spacecraft Attitude Pointing Error

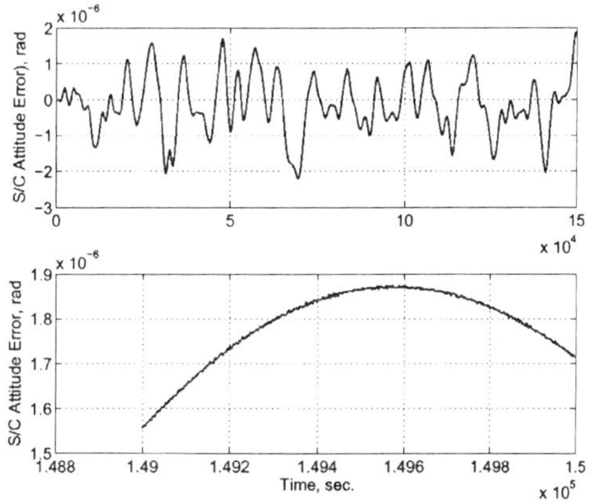

Figure 16b Time History for the Spacecraft Attitude Pointing Error

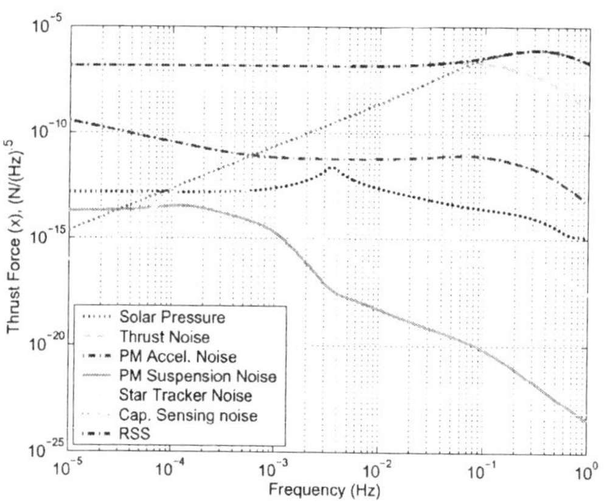

Figure 17a Root Power Spectrum of the Thrust Force: X-Direction

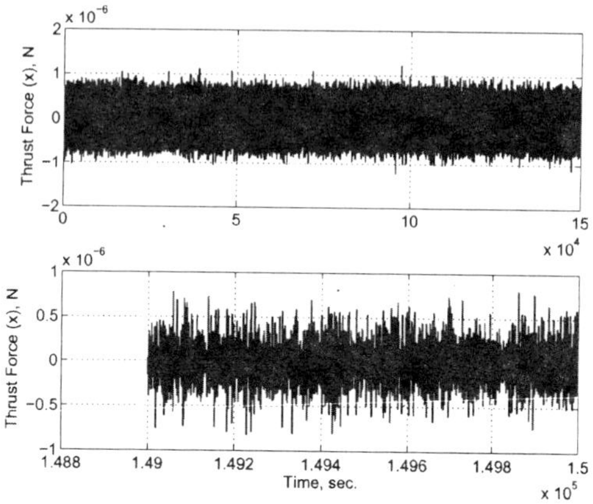

Figure 17b Time History for the Thrust Force: X-Direction

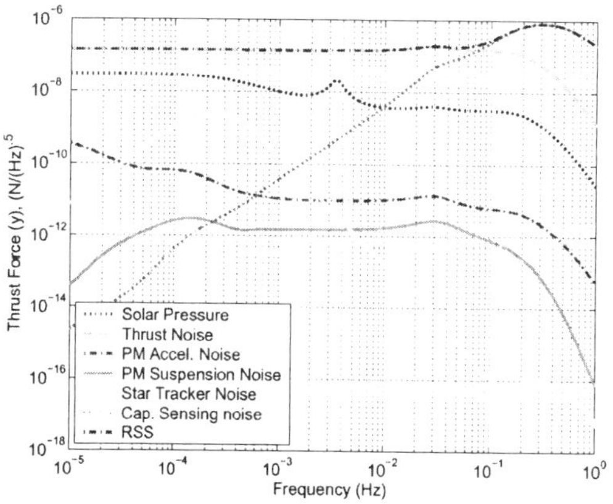

Figure 18a Root Power Spectrum of the Thrust Force: Y-Direction

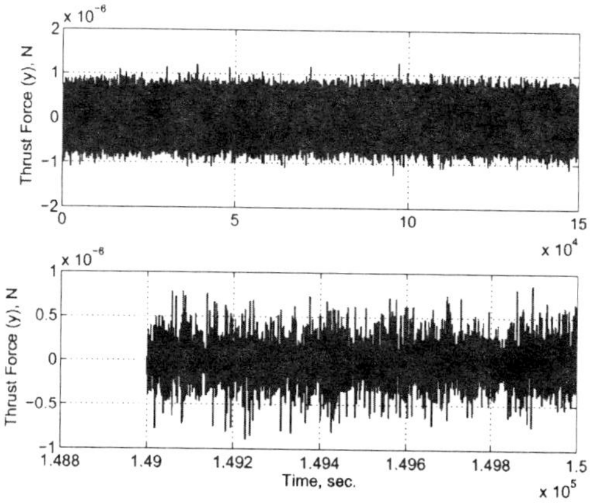

Figure 18b Time History for the Thrust Force: Y-Direction

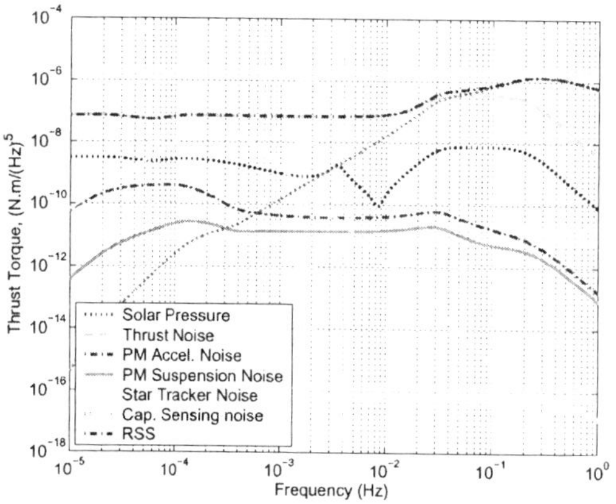

Figure 19a Root Power Spectrum of the Thrust Torque

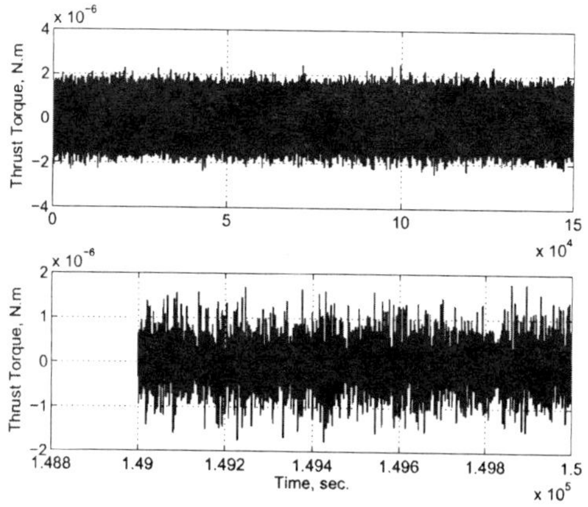

Figure 19b Time History for the Thrust Torque

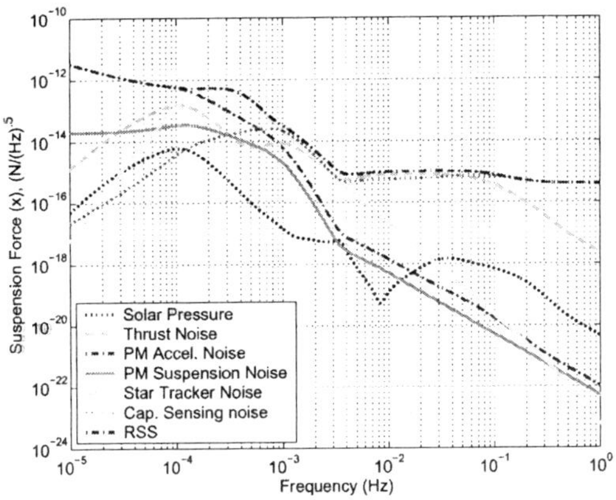

Figure 20a Root Power Spectrum of the X-axis Suspension Force on Test Mass 2

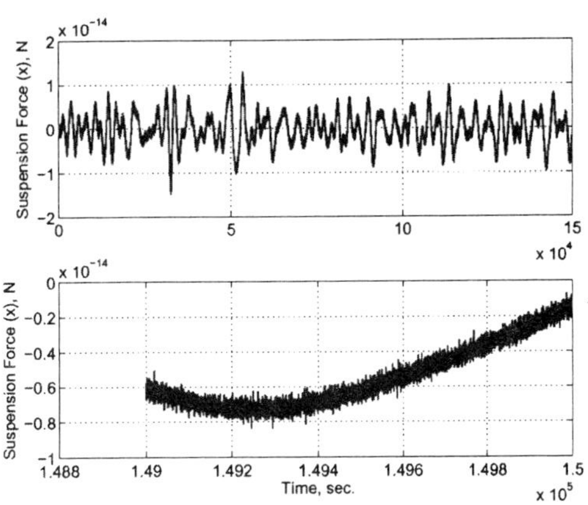

Figure 20b Time History for the X-axis Suspension Force on Test Mass 2

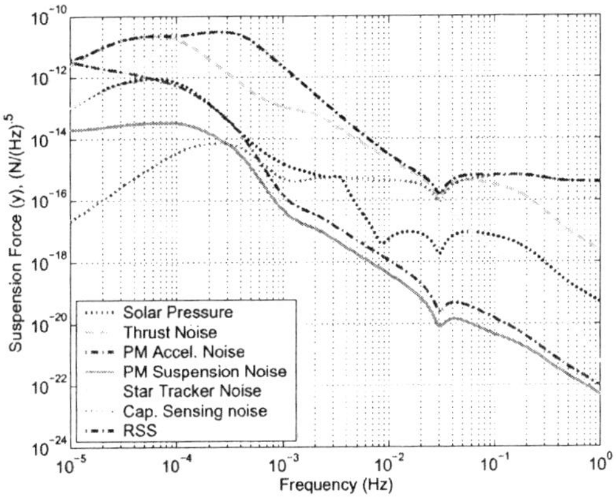

Figure 21a Root Power Spectrum of the Y-axis Suspension Force on Test Mass 2

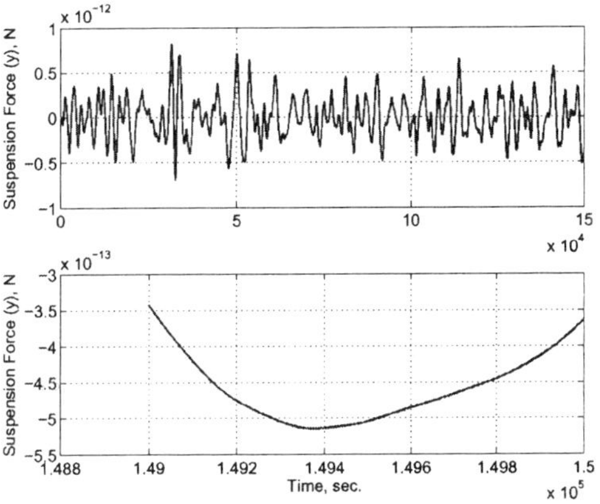

Figure 21b Time History for the Y-axis Suspension Force on Test Mass 2

AAS 03-066

PRECISION POINTING FOR THE LASER INTERFEROMETRY SPACE ANTENNA MISSION

T. T. Hyde* and P. G. Maghami†

The Laser Interferometer Space Antenna (LISA) mission is a planned NASA-ESA gravitational wave detector consisting of three spacecraft in heliocentric orbit. Lasers are used to measure distance fluctuations between proof masses aboard each spacecraft to the picometer level over a 5 million kilometer separation. Each spacecraft and its two laser transmit/receive telescopes must be held stable in pointing to less than 8 nanoradians per root Hertz in the frequency band 0.1-100 mHz. The pointing error is sensed in the received beam and the spacecraft attitude is controlled with a set of micro-Newton thrusters. Requirements, sensors, actuators, control design, and simulations are described.

THE LISA MISSION

The detection of gravitational waves will open a new window of observation on the universe. Unlike typical observatories, which detect electromagnetic waves traveling through space-time, The Laser Interferometer Space Antenna (LISA) will detect ripples in space-time itself [1]. Science targets include galactic binaries, merging supermassive black holes, intermediate-mass/seed black holes, and cosmological backgrounds. Gravitational waves are detected by measuring strain (change in length over length). In practical terms, this is done by measuring the change in distance between a set of freely falling masses separated by a great distance. Ground based detection of gravitational waves by Laser Interferometer Gravity-wave Observatory (LIGO) and other observatories is possible with laser interferometry, but the relatively short arm length and seismic noise limit the measurement band to above 10 Hz on the Earth. LISA also uses laser interferometeric measurement of the change in distance between test masses, but does it in space. Space allows very long arm lengths (5 million kilometers for LISA) and a very quiet acceleration environment (3 $\times 10^{-15}$ $m/s^2/Hz^{0.5}$ for LISA), which allows detection of gravitational wave strains to a best sensitivity of 10^{-23}strain (S/N=5) and over the measurement band of 10^{-4} to 1 Hertz for a one-year observation. Figure 1 shows a drawing of the LISA spacecraft with its Y-tube payload of two telescopes.

* Aerospace Engineer, GN&C Systems Engineering Branch, Code 591, NASA Goddard Space Flight Center, Greenbelt, Maryland 20771. E-mail: tupper.hyde@nasa.gov.

† Aerospace Engineer, Flight Dynamics Analysis Branch, Code 595, NASA Goddard Space Flight Center, Greenbelt, Maryland 20771. E-mail: peiman.g.maghami@nasa.gov.

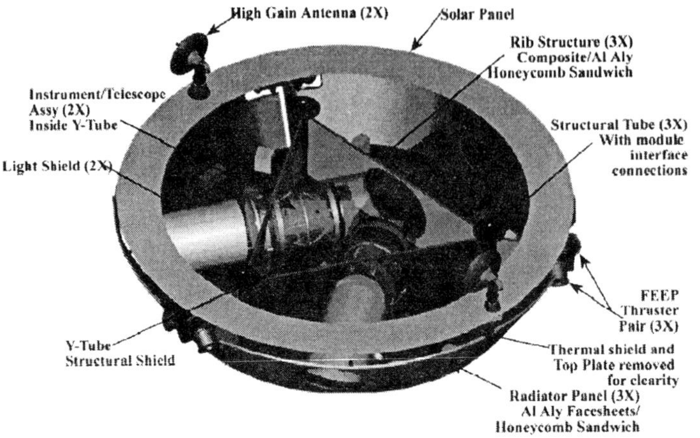

Figure 1 The LISA Spacecraft

The LISA mission consists of three spacecraft in heliocentric orbit. The orbits are chosen so that the three spacecraft form a roughly equilateral triangle with its center located at a radius of 1 AU and 20 degrees behind the Earth. Figure 2 shows how the LISA constellation triangle makes a "pin-wheel" motion about its center with a period of one year. The laser metrology arms between the three spacecraft are created by 1 watt, 1064 nm lasers that transmit through 30 cm aperture telescopes. Telescope diffraction creates a beam pattern about 40 km in diameter (first Airy ring) at the receive telescope which captures 70 picowatts of the laser power. The telescopes act in transmit and receive mode simultaneously, phase locking the return beam to enable the metrology in a way not unlike transponder based ranging for radio. The two telescopes on each spacecraft are mounted through a small articulation actuator that allows the angle between them to be slowly varied to cover the 59 to 61 degree range of triangle "breathing" created by the natural orbit dynamics.

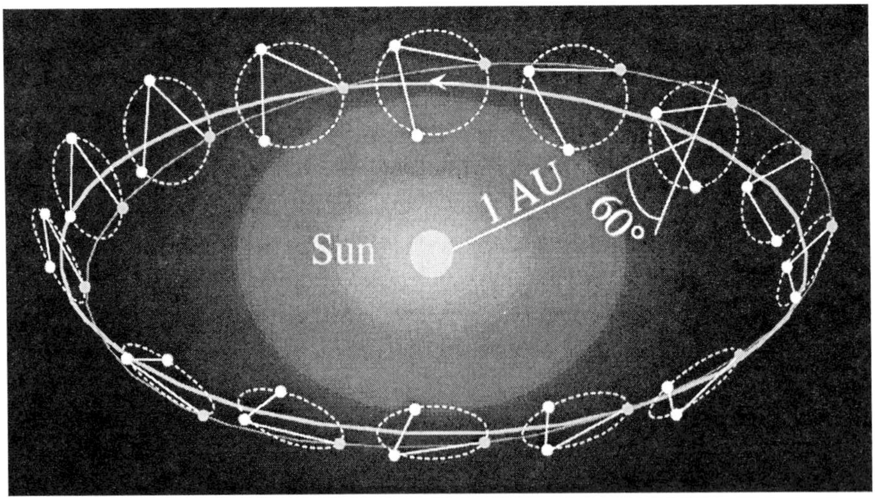

Figure 2 The LISA Constellation

The control system for LISA, called the Disturbance Reduction System (DRS), consists of five control functions:

1) Attitude control system (ACS): to orient the S/C to align the telescopes with incoming laser beams

2) Drag free control system (DFC): to maintain drag free motion of the proof masses in LISA measurement directions

3) Proof mass (PM) suspension control: to maintain relative attitude of the proof mass with respect to its housing and to maintain relative position of the proof mass with respect to its housing in the transverse directions

4) Telescope articulation (TA) control: to maintain the angle between the telescopes

5) Point ahead (PA) and acquisition control: to point the outgoing beam while sensing the incoming beam

LISA pointing is controlled through the closed loop ACS and open loop PA controls. The TA control may be open or closed loop depending on the mechanical impedance of the articulation actuator chosen. The proof mass displacement relative to its housing is measured in six degrees of freedom. This information is used to control the position of the spacecraft in translation to keep the proof mass centered in its housing and provide a drag free environment for the proof mass (DFC). A set of micro-Newton thrusters responds to the DFC force commands (in the three translation axes) and to the ACS torque commands (in the three rotation axes).

In order to realize the LISA design sensitivity very accurate laser metrology is required between the proof masses on the three LISA spacecraft. The change in distance measurement must be made with a noise level less than $4x10^{-11}$ m/$Hz^{0.5}$ in the band 1-100 mHz. One contribution to this noise budget is that caused by the imperfections in the wavefront coupled with pointing jitter in the outgoing laser beam. The current budget allocation to pointing jitter is 8 nanoradian per root Hertz. This requires that spacecraft attitude also achieve this level since the telescopes are fixed to the spacecraft. This challenging pointing stability requirement is achieved through a unique set of sensors and actuators and straightforward single-input single-output (SISO) control loops.

POINTING CONTROL

As in any control design task, the LISA pointing control is addressed with a good understanding of the plant, appropriate choice of sensors and actuators with the required range and resolution, and models of disturbance sources.

To understand the operation of LISA, one must "follow the light". The layout of the LISA optical bench is shown in Figure 3. The laser beam is launched from the fiber positioner and goes to the central polarizing beamsplitter (ps1). Most of the light (99%) is reflected at ps1 and goes out through the telescope into space. The received beam is collected by the telescope and comes onto the bench where a small amount (10%) is reflected by the beamsplitter (s1) and focused onto an acquisition CCD detector (marked qp2). The majority of the power (90%) goes through ps1 and to the proof mass. Upon reflection on the proof mass and return to ps1, the received beam is reflected, goes through the beam compressor (bc1), and onto the quadrant photodiode (qp1). Here it is mixed with the small amount (1%) of the outgoing laser in a heterodyne detection. The beat note is counted and the relative phase measured to better than 10^{-5} cycle/ $Hz^{0.5}$ to make the fundamental LISA measurement. The relative phase difference between the received and local wavefront on the four quadrant detectors is simultaneously used as the science mode pointing error sensor.

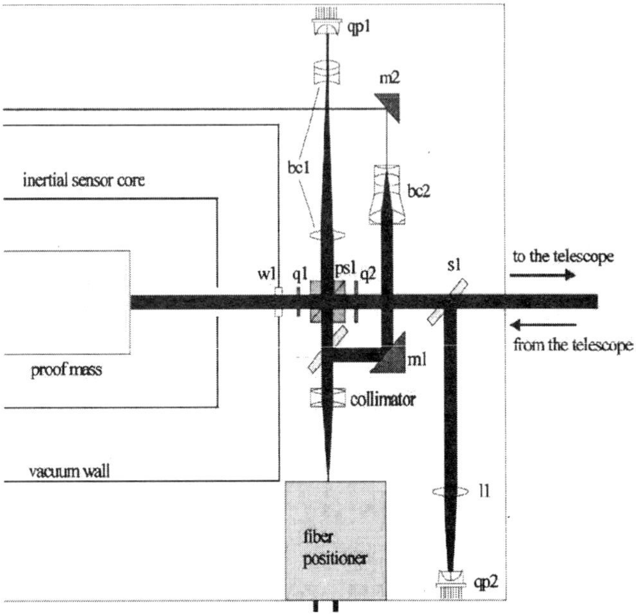

Figure 3 Layout of the LISA Optical Bench

A subtlety of the LISA transmit/receive configuration is the so-called "point ahead" angle required between the transmit and receive beam directions. The relative motion of the spacecraft and the finite speed of light requires that the beam be received from the location of the other spacecraft roughly 16 seconds ago and transmitted toward its location roughly 16 seconds in the future. Table 1 shows the magnitude of the fixed and time varying components of the point ahead angle. The point ahead compensation is handled by a commanded offset in the proof mass angle relative to the optics bench. Since the received beam bounces off the proof mass but the outgoing beam does not, the offset angle can be commanded open loop using orbit dynamics as the input.

Table 1
POINT AHEAD ANGLES

Component	Fixed	Time Varying
In-plane	3.3 microrad	55 nanorad
Out-of-plane	85 nanorad	5.8 microrad

The sensors for LISA are unique among satellite control applications. Precision pointing error in science mode is computed from the relative phase error between the received beam and the local laser on the four quadrant detectors of qp1. This measurement has a theoretical signal noise limit of:

$$NEA_{theoretical} = 0.56\ \lambda\ D^{-1}\ N^{-0.5}$$

where D is the telescope diameter of 30 cm, λ is the wavelength of roughly 1 micron, and N is the photon count. In science mode, the received beam of 70 picowatts provides about 260 million detected photons (0.85 transmission ratio and QE of 0.70) over a 1 second integration. The $NEA_{theoretical}$ then is about 0.12 nanoradians. To be conservative and to account for other detector, background, and local laser shot noise sources, a value of 3 nanoradians per root Hertz is assumed for this science mode sensor. The two axis pointing errors from the two telescopes on the spacecraft are used to create a three-axis attitude error quaternion. Other sensors include the displacement and attitude of proof mass relative to the housing measured with capacitive sensors, the acquisition CCD detectors (qp2), and two traditional star trackers. After acquisition (in science mode), none of these others sensors are used for fine pointing.

The actuators for LISA are also unique among satellite control applications. The spacecraft translation and orientation is controlled with a set of micro Newton thrusters. The baseline for these thrusters is a Field Emission Electric Propulsion (FEEP)[1] design that accelerates liquid metal ions with a voltage differential. The very high I_{sp} means that only a few grams of propellant are required for the entire mission. A minimum of six thrusters is required for six DOF control and twelve thrusters are provided for redundancy. Each thruster produces a linearly commandable force of up to 100 μN with a resolution of 0.1 μN. The other actuators are the telescope articulation actuators (1 degree stroke, 3 nanoradian resolution) and electrostatic forcers for six DOF proof mass control. Each telescope has an articulation actuator although only one is required (the other will be a cold spare). The electrostatic actuators use the same electrodes as the proof mass capacitive sensors.

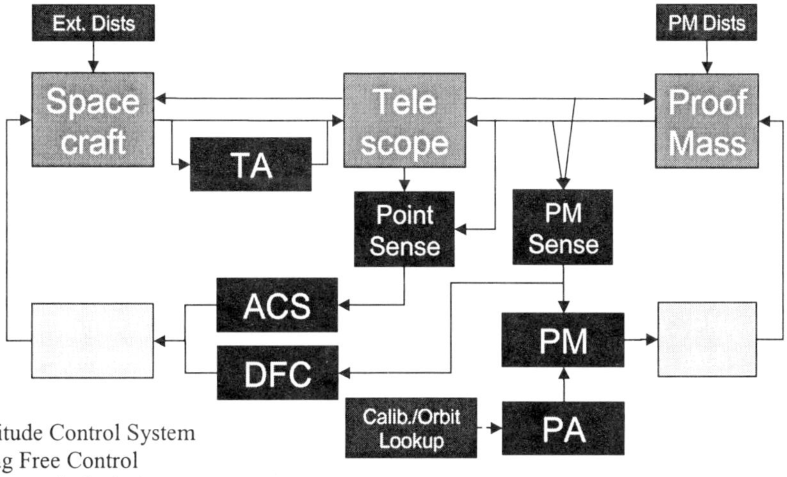

Legend:
ACS=Attitude Control System
DFC=Drag Free Control
TA=Telescope Articulation
PM=Proof Mass Suspension
PA=Point Ahead and Acquisition

Figure 4 LISA DRS Block Diagram

The LISA DRS control architecture is shown in the block diagram of Figure 4. During science mode, there are 16 sensor channels (2x2 pointing sensors, 2x6 proof mass sensors) and 19 actuator channels (6 thrusters, 2x6 proof mass actuators, 1 telescope articulator). Periodic calibration and orbit dynamics lookup provide an additional 4 point ahead values (2 axes for 2 telescopes). The multi input/multi output (MIMO) nature of the plant manifests in the coupling between the six axes within each proof mass sensor and actuator package, as well as between the proof mass and spacecraft coupled dynamics. However, straightforward SISO control of the individual pointing (ACS, 3 loops), drag free (DFC, 3 loops), and proof mass suspension (PM, 12 loops) feedback controls meets performance and robustness goals. This is due to the low coupling between controlled DOFs. The mass ratio between the proof mass (about 1.4 kg) and the spacecraft (over 300 kg) means that this coupling effect is relatively weak. The coupling in the proof mass electrodes is designed to be less than 1 percent between axes, meaning that this coupling effect is also relatively weak. Stability margins, acquisition operations, the 19 DOF model, and simulation results are described in the following sections.

POINTING ACQUISITION

In order for LISA to accomplish pointing acquisition, a series of steps are taken. This process is initiated from ground command and then accomplished autonomously. First the angle offset of each star tracker axis with respect to the associated CCD array center is measured by observing a bright star with both. Then the two spacecraft point at each other based on star tracker measurements, and set their proof mass angles according to the required point ahead angles based on orbit dynamics. The outgoing beam on one spacecraft is spoiled to provide a 10 times wider angle (but 100 times dimmer) beam that encompasses the accuracy (noise and mounting bias) of the star tracker. The other spacecraft looks for the received beam in the acquisition CCD. The CCD measurement is then used to center the received beam on the detector qp1 with the local laser turned temporarily off. The received beam centroid is measured by qp1 in a direct detection quad cell mode (Direct QC). The local laser is turned on, and the heterodyne differential wavefront tilt measurement at qp1 is used. The other spacecraft repeats the process and finally both spacecraft return their beam to the focused condition. This allows detection of received beam angle at the full science mode sensitivity. Biases are calibrated at each step allowing re-acquisition to be accomplished without the direct QC step. An alternative scheme involves scanning over the uncertainty cone instead of spoiling the beam.

Table 2

DYNAMIC RANGE ALLOCATIONS FOR ACQUISITION

Sensor	Resolution (1 Hz effective bandwidth)	Range	Alignment Bias to Next Sensor
Star Tracker	5 microrad	Full sky	20 microrad
CCD	0.3 microrad	360 microrad	0.9 microrad
Direct QC	30 nanorad	3.6 microrad	90 nanorad
Science Mode	3 nanorad	360 nanorad	N/A

The allocation of requirements to the resolution, range and alignment bias allowed for each sensor in the acquisition process was done according to the following rule:

$$\text{Range}_{i+1} > 2\,(\,3\,\sigma_i + \text{Bias}_i\,)$$

That is, the range of the next sensor must encompass the accuracy (worst case bias plus three times the resolution) of the proceeding stage with 100 percent margin. Note that the bias from the direct QC to science mode is due to the potential misalignment of the local laser. Table 2 and Figure 5 show the requirements that have been allocated to the resolution, range, and alignment bias of each of the sensors used in the pointing acquisition process.

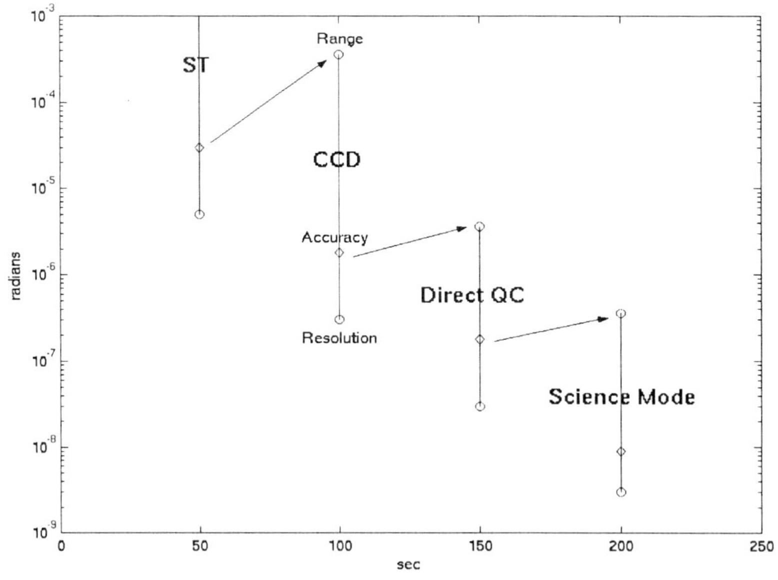

Figure 5 Graphical Depiction of Sensor Range, Resolution, and Bias for Acquisition

THE 19 DOF MODEL

The LISA 19 DOF model [3] was developed to capture the dynamics of a typical LISA spacecraft in formation with two others. Dynamically, it includes all rigid-body degrees of freedom, six for the S/C translation and rotation; six for each proof mass translation and rotation; and finally one rotational DOF for telescope articulation. A number of disturbance sources, both internal and external, are included. Measurement noise for quad detector and capacitive sensing are included. The five control systems that comprise the LISA DRS are designed and included in the model. The dynamics of the LISA spacecraft is modeled and simulated in SIMULINK, a dynamic simulation tool. The configuration used for the current 19-DOF Model is shown in Figure 6.

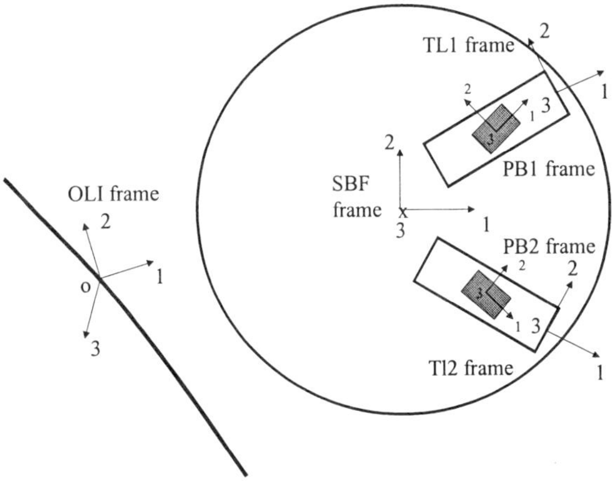

Figure 6 19 DOF Model Configuration

White-noise models were used to capture thruster force, electrostatic suspension force, capacitive sensor, star tracker, CCD, direct quad cell, and science mode pointing sensor. The intensity levels used are given in Table 3. Colored noise models were used for the solar radiation flux variations (Figure 7), and for the proof mass acceleration variations (Figure 8). This solar variation spectrum is based on refs. [1] and [2], and represents a conservative assessment of potential variations. This spectrum includes the so-called 5-minute acoustic oscillation. The proof mass acceleration noise includes a number of sources including magnetic and Lorentz forces, thermal disturbances, cosmic ray impacts, etc [1].

Table 3
ACTUATION AND SENSING NOISE INTENSITIES

Noise Source	Intensity
Thrusters	0.1 μN/Hz$^{0.5}$
Suspension Force	2e-14 N/Hz$^{0.5}$
Capacitive Sensing	3 nm/Hz$^{0.5}$
Star Tracker	5 microrad/Hz$^{0.5}$
CCD	0.3 microrad/Hz$^{0.5}$
Direct QC	30 nanorad/Hz$^{0.5}$
Science Mode	3 nrad/Hz$^{0.5}$

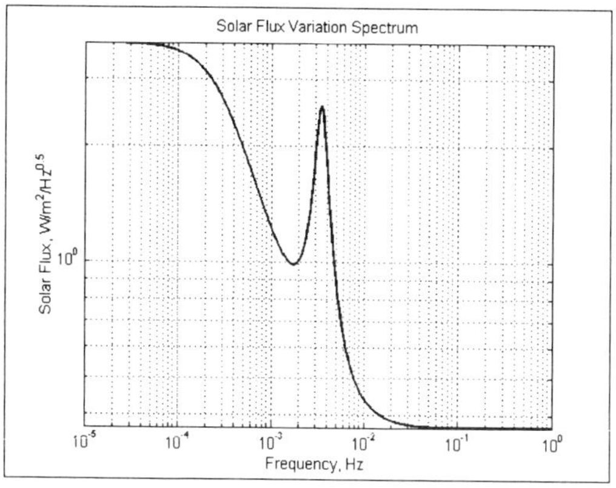

Figure 7 Root Power Spectrum of the Solar Radiation Flux Variations

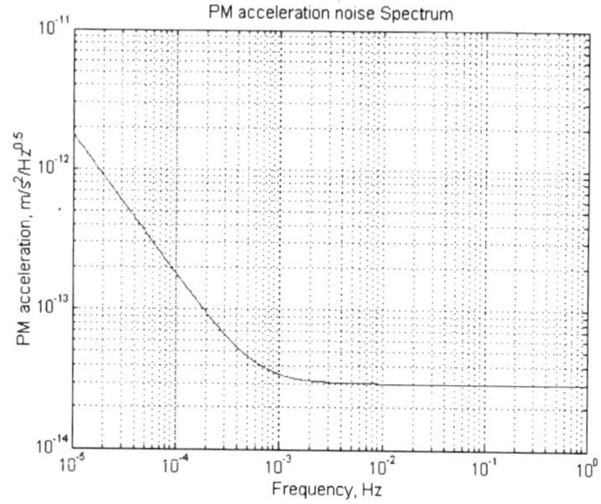

Figure 8 Root Power Spectrum for the Proof Mass Acceleration Variations

STABILITY MARGINS

Each of the controllers was designed to have sufficient stability margins. However the 19-DOF system represented in Figure 1 is a MIMO system, by virtue of the cross coupling between the relative test mass positions and the attitude of the spacecraft. Hence, the loop gains at each input and output channel (while the remaining channels are closed) must be analyzed to obtain proper stability margins. The stability analysis indicates that each input/output channel meets the requirements of at least 9 dB of gain margin and 35° of phase margin. These margins are amply sufficient considering that the effects of zero-order hold and computational and transport delays are already included in the analysis. Figure 9 illustrates the loop gain for one of the pointing axis loops.

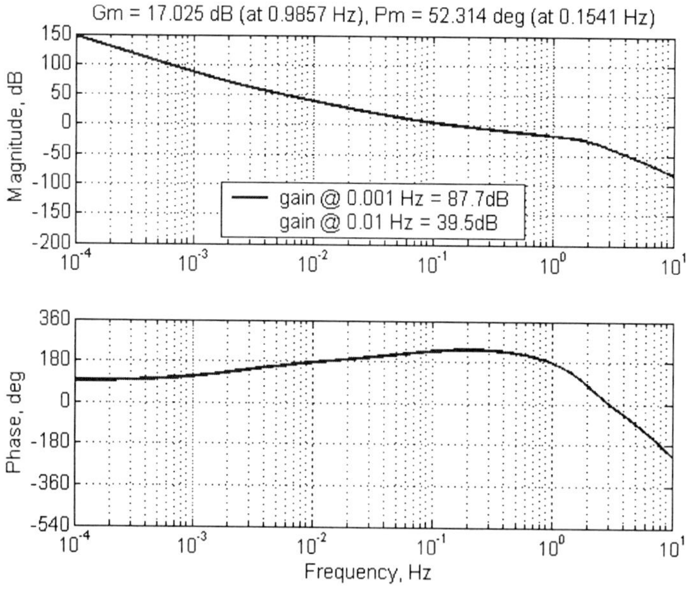

Figure 9 Gain and Phase Margins for one of the LISA Pointing Loops.

SIMULATION RESULTS

A time domain simulation was run for 1000 seconds. In order to see the effect of pointing acquisition operations, a sequence of sensors was "turned on". Table 4 lists these transitions. The bias error for each sensor was set according to its maximum allotment (see Table 2).

Table 4
SENSOR MODE TRANSITIONS DURING SIMULATION

Simulation time (seconds)	Pointing sensor mode
0	Star trackers only on
250	CCD turned on
500	Direct QC turned on
750	Science mode pointing turned on

Figure 10 shows the two axis pointing error for telescope 1 throughout the acquisition process. With star tracker only on (0-250 seconds), the pointing error is consistent with the alignment bias and noise of the star tracker. At t=250 seconds, the CCD is used, resulting in a transient to a new steady state condition (Figure 11). Here the pointing error is consistent with the bias and noise of the CCD sensor. At t=500 seconds, the quadrant detector is used in direct detection mode (Direct QC), resulting in another transient to a third steady state condition (Figure 12). Here the pointing error is consistent with the bias and noise of the direct QC sensor. Finally, the local laser is turned on, the received beam has been focused, and the science mode sensor therefore enabled. Figure 13 shows the final, post-acquisition, performance of just over 1 nanoradian rms.

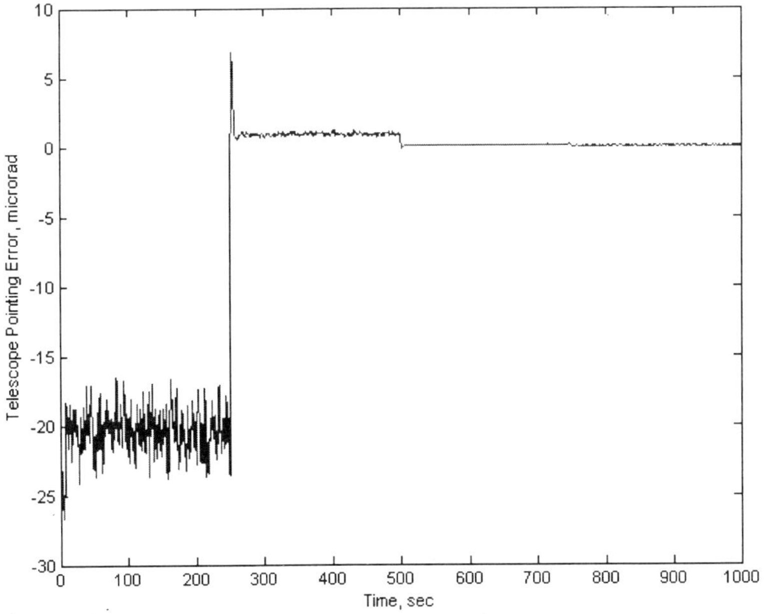

Figure 10 Beam Pointing Error throughout Acquisition

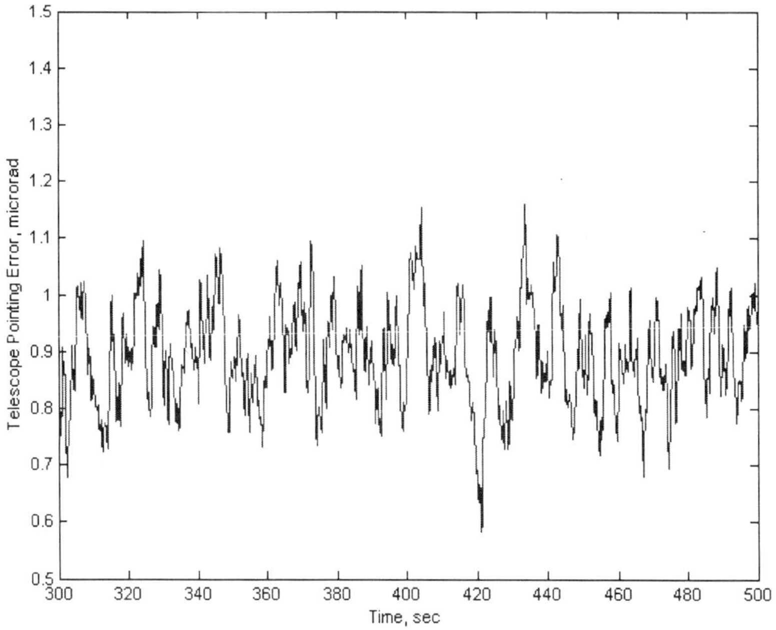

Figure 11 Beam Pointing Error in CCD Mode

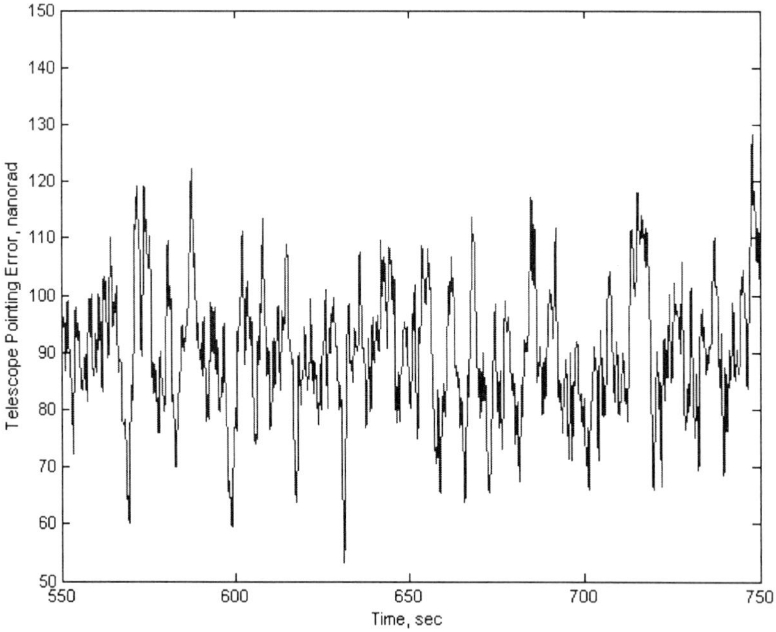

Figure 12 Beam Pointing Error in Direct Quad Cell Mode

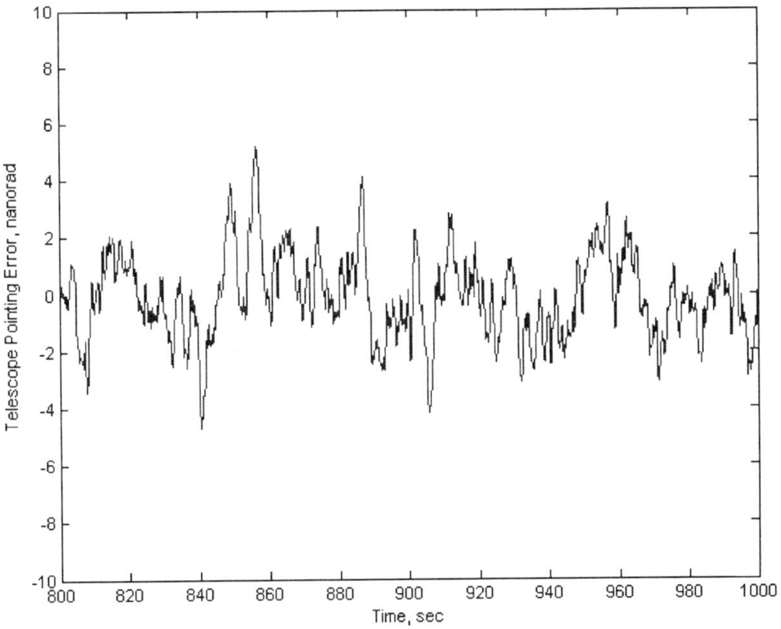

Figure 13 Beam Pointing Error in Science Mode.

The science mode pointing error shown in Figure 13 has an RMS of just over 1 nanoradian. This value (0.0002 arcsec) is more than an order of magnitude lower than any spacecraft pointing control system to date. The Hubble Space Telescope pointing performance is about 0.004 arcsec. It is both having a bright laser source to guide on and LISA's unique sensor based on the received laser wavefront that allows this level of pointing performance. The low force noise micro-Newton thrusters and the low disturbance spacecraft are also consistent with achieving this unprecedented spacecraft pointing performance.

The performance requirement for LISA pointing is specified in variation power over the frequency range of 1-100 mHz, so a longer simulation in science mode was run and the spectrum taken. Figure 14 shows the root power spectral density for the pointing error of telescope 1 in two axes. The peak level of about 3 nanoradians per root Hertz meets the requirement of 8 nanoradians per root Hertz with over 100 percent margin.

CONCLUSIONS

The precision pointing performance for LISA has been shown through detailed time domain simulation. Science mode pointing stability of a little over 1 nanoradian RMS is an order of magnitude better than any spacecraft known. This can be accomplished due to two factors. One is the bright laser source that is available to guide on. The second is LISA's unique pointing error sensor that uses the received beam wavefront tilt and precision phase measurement on four quadrants. The predicted level of less than 4 nanoradian/$Hz^{0.5}$ in the LISA measurement bandwidth proves significant performance margin to the 8 nanoradian/$Hz^{0.5}$ requirement. The control laws have shown to be robust through traditional frequency domain analysis. The acquisition scenario has been shown to work with successive layers of sensor precision. The dynamic range of each sensor encompasses the accuracy and bias in the preceding layer with appropriate margins. Time domain simulation of the acquisition process that shows no special control mode changes are required and that science mode lock can be achieved in the budgeted time allotment.

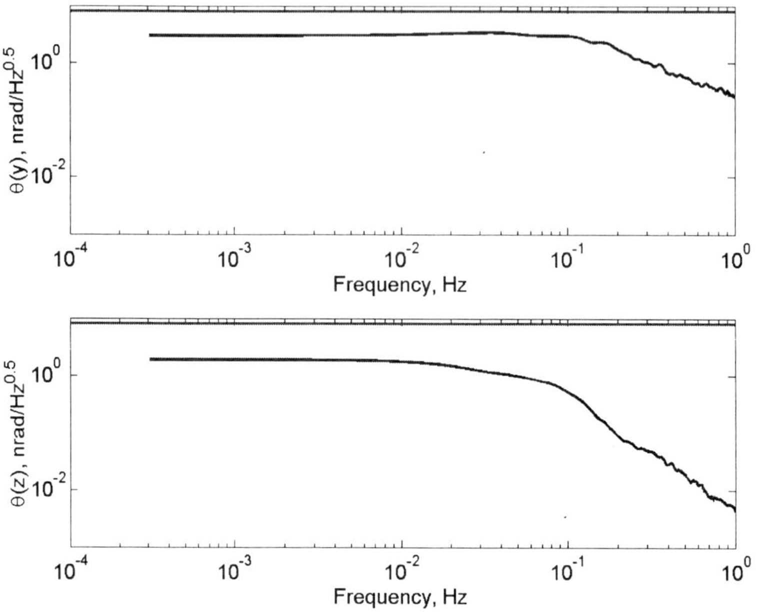

Figure 14 Root PSD of Pointing Error in Science Mode.

ACRONYM LIST

LISA=Laser Interferometer Space Antenna
LIGO=Laser Interferometer Gravity-wave Observatory
RMS=Root Mean Square
QC=Quad Cell
CCD=Charge Coupled Device
DRS= Disturbance Reduction System
ACS=Attitude Control System
DFC=Drag Free Control
TA=Telescope Articulation
PM=Proof Mass Suspension
PA=Point Ahead and Aquisition

REFERENCES

1. *Final Technical Report of the (Phase A) Study of the Laser Interferometer Space Antenna* (Dornier Satellitensysteme GmbH - Matra Marconi Space - Alenia Aerospazio), ESTEC Contract no. 13631/99/NL/MS, Report No. LI-RP-DS-009 (April 2000). Available via anonymous ftp at ftp.estec.esa.nl in directory pub/sci-spd/lisa/LISA Documents.

2. J. Pap, et. al., "Variation in Total Solar and Spectral Irradiance as Measured by the VIRGO Experiment on SOHO", *Adv. Space Res.*, 24:215-224, 1999.

3. P. Maghami, and T. Hyde, "Laser Interferometer Space Antenna 19-DOF Controls Model," to appear in Classical and Quantum Gravity 20.

AAS 03-067

MARS RECONNAISSANCE ORBITER DESIGN APPROACH FOR HIGH-RESOLUTION SURFACE IMAGING

S. W. Lee,* E. D. Skulsky,† J. Chapel,‡ D. Cwynar,‡
R. Gehling** and A. Delamere**

The Mars Reconnaissance Orbiter (MRO) will explore Mars equipped with a suite of six scientific instruments and two engineering experiments, and supporting two additional facility investigations. One of the objectives of the MRO mission is to use the High-Resolution Imaging Science Experiment (HiRISE) to provide 30 cm/pixel images of future Mars landing sites. To achieve such detail, MRO must meet some very challenging target-relative pointing and pointing stability requirements. A combination of analysis, operational constraints, and spacecraft design modifications were utilized to ensure that the necessary pointing requirements will be met.

INTRODUCTION

In this paper, we begin by briefly reviewing the ambitious objectives of the Mars Reconnaissance Orbiter (MRO) mission,[1] We then discuss the instruments and investigations supported by the mission, and provide a brief overview of the spacecraft. We follow with a more detailed discussion of the High-Resolution Imaging Experiment (HiRISE)—a camera which will provide detailed (30 cm/pixel) images of the Mars surface, and which drives the target-relative pointing and pointing stability requirements for the spacecraft. Next, we describe the MRO Guidance and Control system and present the driving HiRISE target-relative pointing and pointing stability requirements, along with a budgets and allocations for each requirement. In the final section, we identify and describe the disturbance sources and provide a detailed discussion of each along with the analysis approach for pointing stability requirements,

MISSION OVERVIEW

The Mars Reconnaissance Orbiter (MRO) is a key element in NASA's Mars Exploration Program. MRO will explore Mars for candidate sites where extinct or extant life might be found, contribute to our understanding of Mars' global climate, and provide detailed images of future landing sites. Echoing NASA's "follow the water" strategy,[2] the specific objectives of the MRO mission are to:

* All correspondence should be sent to Steven W. Lee, Jet Propulsion Laboratory, 4800 Oak Grove Drive, M/S 301-420, Pasadena, California 91109-8099. E-mail: steven.w.lee@jpl.nasa.gov.

† Jet Propulsion Laboratory, California Institute of Technology, 4800 Oak Grove Drive, Pasadena, California 91109.

‡ Ball Aerospace and Technologies Corp, P.O. Box 1062, Boulder, Colorado 80306.

** Lockheed Martin Space Systems, P.O. Box 179, Denver, Colorado 80201.

- Search for sites showing evidence of aqueous and/or hydrothermal activity.
- Map and characterize in detail the stratigraphy, geologic structure and composition of Mars surface features at many globally distributed targeted sites.
- Provide high-resolution imagery to support the selection of future landing sites.
- Recover the Mars Climate Orbiter (MCO) atmosphere and climate science objectives by characterizing Mars' seasonal cycles and diurnal variations of water, dust, and carbon dioxide.
- Detect on Mars the presence of liquid water and determine the distribution of ground ice in the upper surface, particularly within the near-surface regolith.
- Provide atmospheric observations in addition to the MCO capabilities to further define Mars' atmospheric structure and circulation.
- Characterize the Martian gravity field in greater detail to understand better Mars' geologic history and the structure of its crust and lithosphere.
- Provide telecom and/or navigation relay capability to support future Mars orbiter and lander missions.

MISSION SYNOPSIS

The Mars Reconnaissance Orbiter will launch in August 2005, and will arrive at Mars in March 2006. Similarly to two previous Mars spacecraft, MRO will use aerobraking[3,4] to attain the Primary Science Orbit (PSO), which will have a periapsis altitude near 255 km and an apoapsis altitude near 320 km. The PSO will be sun-synchronous with an ascending node orientation of approximately 3:00 pm local mean solar time (LMST). During the Primary Science Phase (PSP), repetitive (mapping) and targeted observations of the planet's surface and atmosphere will be conducted over a time span of one Martian year (687 Earth days), nominally beginning in November 2006. Throughout the PSP, the orbiter will be oriented with the primary science instruments pointing towards areodetic nadir. Using slew maneuvers, MRO will provide the capability to image targets up to 30 deg off-nadir. Science data acquisition will be planned such that data can be down linked to the DSN during at least two 8-hour tracking passes every day.

The MRO orbiter will also support the Mars Exploration Program by providing relay communications support to various Mars landers and orbiters through its telecommunications / navigation subsystem. During these support periods, relay and navigation functions will have priority over orbiter science observations (including site characterization), although such observations may continue if they do not interfere and are within the mission resources.

INSTRUMENT/INVESTIGATION OVERVIEW

A suite of six science instruments, two engineering payloads, and two facility investigations has been selected to support MRO's objectives:

- High-Resolution Imaging Science Experiment (HiRISE). A 0.5 m aperture, time delay integration (TDI) telescope with 12 m focal length. HiRISE provides very high resolution panchromatic and color images (25 cm/pixel at 255 km). At full resolution, a single HiRISE image is expected to be as large as 16 Gbit. HiRISE will investigate stratigraphy processes, and site morphology.
- Compact Reconnaissance Imaging Spectrometer for Mars (CRISM). A Ritchey-Chretien-based spectrometer providing high spectral (400 – 4050 nm) and spatial (15 m/pixel at 255 km) images. CRISM will investigate regional and local surface composition and atmospheric properties.

- Context Imager (CTX). A wide swath (> 30 km wide at 255 km orbit altitude) monochromatic telescope with 6 m resolution at 255 km. *telescope aperture and focal length?* CTX will investigate regional stratigraphy and morphology in addition to providing context for other, higher resolution investigations.
- Shallow Subsurface RADAR (SHARAD). A 15–25 MHz radar capable of providing 15 m vertical resolution to a depth of 1 km. SHARAD will investigate regional near-surface ground structure including detection of water ice.
- Mars Color Imager (MARCI). A multispectral (5 visible and 2 UV) imager providing 1 – 10 km/pixel nadir resolution, < 7 km limb resolution and a limb-to-limb field of view. MARCI recovers a portion of the Mars Climate Orbiter (MCO) science with its global weather and surface change observations.
- Mars Climate Sounder (MCS). A 4 cm aperture, f/1.7 telescopic filter radiometer in nine spectral channels between 0.3 – 30.0 μm. Like MARCI, MCS recovers MCO science by providing nadir and limb scans to investigate atmospheric fields, transport phenomenon, and polar processes, similar to the PMIRR (Pressure Modulated Infrared Radiometer) instrument aboard MCO.
- Optical Navigation Camera (ONC). An engineering/navigation experiment to validate optical navigation techniques. Upon Mars approach, ONC will image Mars, Phobos, Deimos, and background stars for navigation purposes.
- Electra. Electra provides a two-way UHF radio link to future Mars spacecraft for command and telemetry relay. Electra also supports one-way X-band Doppler for future incoming Mars spacecraft.
- Gravity Science facility investigation. Utilizing Doppler data, the gravity science team will develop an improved Mars gravity model and investigate transient mass changes.
- Atmospheric Structure facility investigation. Utilizing accelerometer data obtained during aerobraking drag passes, the atmospheric structure team will investigate upper atmospheric structure and variability.

SPACECRAFT OVERVIEW

The MRO spacecraft is a relatively large spacecraft with exceptional data collection and return capability. Figure 1 shows the MRO spacecraft configuration during the Primary Science Phase. The direction of flight is in the plane of the solar arrays with the instrument deck nominally oriented towards nadir. Total spacecraft wet mass is approximately 2000 kg at launch. Two solar panels provide 20 m^2 of total useable substrate area to generate up to 2880 W of power at Mars. A large (3 m diameter) High Gain Antenna (HGA) supports downlink rates as high as 6 Msps. Both solar arrays and the HGA are mounted on two-axis gimbals to maximize power and communications flexibility while maintaining a nadir body orientation. This structure is built around a large, 1187 kg capacity propellant tank, which provides up to 1578 m/s ΔV capability.

HiRISE SCIENCE INSTRUMENT

The primary functional requirement of the HiRISE imager is to allow identification of both predicted and unknown features on the surface of Mars to a much finer resolution and contrast than previously possible. As such, the HiRISE instrument is the driver for MRO pointing and stability requirement.

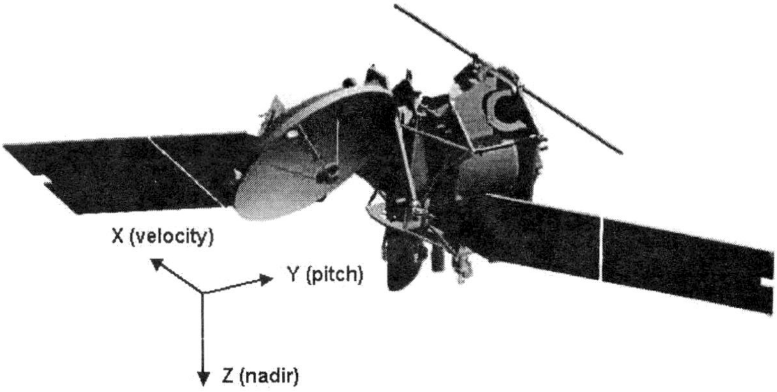

Figure 1. MRO Spacecraft Configuration in the Primary Science Orbit

HiRISE Design Features

The HiRISE instrument performance goals are listed in Table 1. The design features a 50 cm aperture and a detector with 128 lines of Time Delay and Integration (TDI).

Parameter	Performance	Comments
Ground Sample Distance (GSD)	30 cm/pixel	From 255 km altitude, 40 cm/pixel from 320 km
Telescope Aperture	0.5 m, f/24	For resolution and SNR
SNR – Blue-Green	Typically 100:1	Achieved with TDI, backside thinned CCDs, and 50 cm aperture
SNR – Red	Typically 200:1	
SNR – NIR	Typically 100:1	
Spectral range	400 to 600 nm 550 to 850 nm 800 to 1000 nm	Blue-Green (BG) Red NIR
Swath Width – Red	> 6 km	From 300 km altitude
Swath Width – Blue-Green & NIR	> 1.5 km	From 300 km altitude
Swath length	> 2× swath width	Along track
Dynamic range	3,750:1	Without binning, full well/system noise
Data Precision	14 bit A/D	12 to 13 bit usable
Data Compression	Real-time 14 to 8 bit	Look-up table
	Up to 16 × 16 binning	Increases areal coverage
	Felics compression at SSR	~ 2:1
Data storage	28 Gbits	All channels
Number of pixels across swath	20,264 4,072	From swath width and pixel scale
TDI line time	76 μsec to 184 μsec	To match ground track speed
CCD read noise	< 50 electrons rms at 10°C	Achieve SNR at low signal levels
FOV	1.14° × 0.1°	
IFOV	1 × 1 μrad	Detector angular subtense
Relative Radiometry	< 1 % pixel to pixel	Absolute 20%

Table 1. HiRISE Requirements and Performance Characteristics

The imager design is an all-reflective telescope with light-weighted Zerodur optics and a graphite-composite structure. The slightly off-axis Cassegrain objective with relay optics is optimized for diffraction-limited performance over the long, narrow field-of-view (FOV) required for "push-broom" scanning and imaging. Filters in front of the detectors provide images in the three wavelength bands: red (or panchromatic), blue-green (BG), and near infrared (NIR).

The detector-chip-assemblies (DCA) housing the charge-coupled devices (CCDs) are staggered to provide full swath coverage without gaps. Both the BG and NIR bands have two DCAs each to give a total swath width of 4072 pixels, and the red channel has ten DCAs to give a much larger swath width of 20,264 pixels.

HiRISE features a 50 cm diameter primary mirror. For compactness a relay optical section follows the F/8 Cassegrain telescope giving a system focal ratio of F/24 at the focal plane. The HiRISE optical system meets the science requirements by providing a 30 cm per pixel diffraction limited modulation transfer function (MTF) on 12 μm pixels for all 14 HiRISE detectors. The HiRISE color filters are located 40 mm from the detectors for all three channels. This distance avoids problems due to stray light and multiple reflections from the filters in the F/24 quasi-collimated beam. A Lyot stop, located between the last fold mirror and the tertiary mirror, ensures excellent stray light control.

Distortions in the large field of the panchromatic channel have been analyzed. As HiRISE points in the image of the Mars' surface track along each CCD column in TDI mode, a point in the image will remain in a single CCD column with no cross column smear for all HiRISE detectors.

The HiRISE telescope structure is made of graphite-fiber-reinforced composite. This produces a stiff, lightweight structure with low moisture absorption properties and low coefficient of moisture expansion. The negative coefficient of thermal expansion (CTE) of the composite elements, in conjunction with metallic, positive CTE fittings, is tailored to produce near-zero instrument CTE. Figure 2 shows the full-size model of HiRISE.

Figure 2. Full-size Model of HiRISE (Approximately 70 cm in Diameter by 1.4 m in Length)

HiRISE Focal Plane Subsystem (FPS)

The FPS consists of DCAs, a focal plane substrate of aluminum–graphite composite material, a spectral filter assembly, and CCD processing/memory modules (CPMMs). Each DCA holds a single line scan CCD, with 2048 12 × 12 um pixels in the cross-scan direction and 128 TDI elements (stages) in the along-track direction. The 14 staggered DCAs overlap by 40 pixels at each end of each CCD as shown in Figure 3. This provides an effective swath width of 20,000 pixels for the panchromatic images and 4,072 pixels for the blue-green and NIR images.

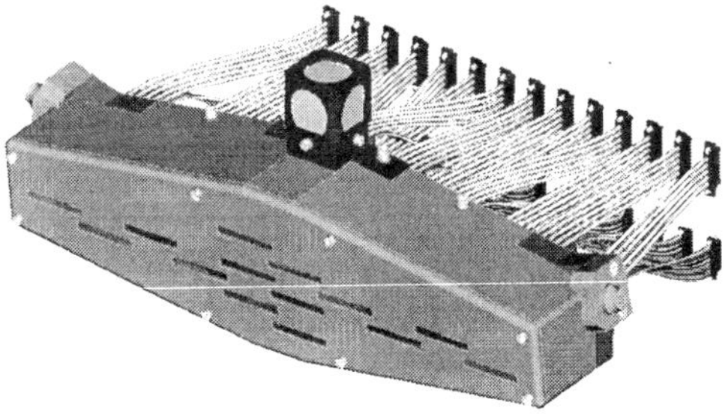

Figure 3. HiRISE Focal Plane Assembly

Using the TDI method increases the exposure time, allowing us to obtain both very high resolution and a high signal-to-noise ratio. As the spacecraft moves above the surface of Mars, TDI integrates the signal as it passes across the CCD detector by shifting the accumulated signal into the next row (line) of the CCD at the same rate as the image moves (see Figure 4). The line rate of 13,000 lines/sec corresponds to an integration time of 78 microsec for 300 km altitude. The pixel integration time is set to match the ground velocity so that charge from one image region is sequentially clocked into the next corresponding element in the along-track direction. The imager can use 8, 32, 64 or 128 TDI stages (detector elements in the along-track direction) to match scene radiance to the CCD full well capacity. Spacecraft orientation will compensate for image smearing during the integration period. A practical limit is reached when residual image smear and spacecraft pointing jitter seriously degrade the required resolution. The 128 lines is the largest number of lines that meets all requirements. Images with higher SNR and lower resolution images will be obtained by binning the signal from adjacent lines and pixels within the CCD, up to a maximum of 16 × 16 pixels.

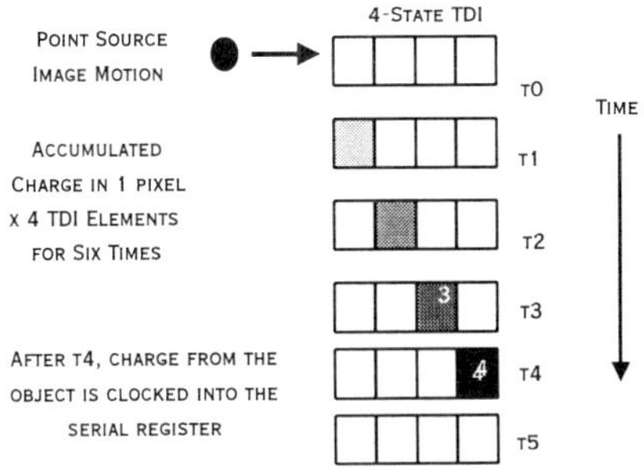

Figure 4. TDI Operation Using a 4-TDI Configuration to Illustrate Charge Accumulation

HiRISE Performance

The maximum signal is 76,000 electrons for the red channel at 300 km with no binning. Figures 5 and 6 show the expected un-binned SNR capability as a function of spacecraft latitude and regional albedo for the blue-green, red (pan) and NIR bands.

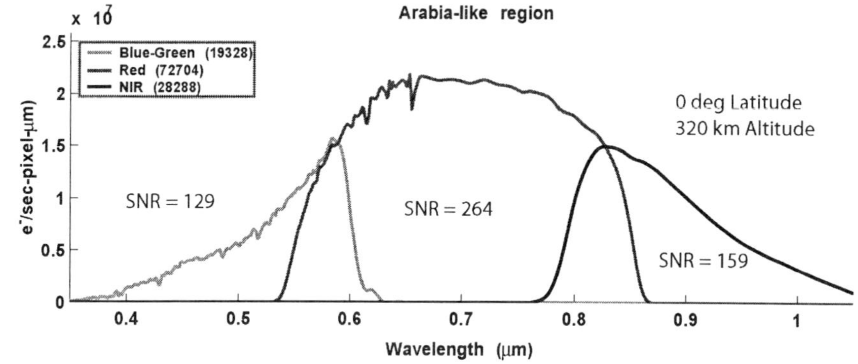

Figure 5. Signal in Electrons as a Function of Wavelength for a Bright Region of Mars

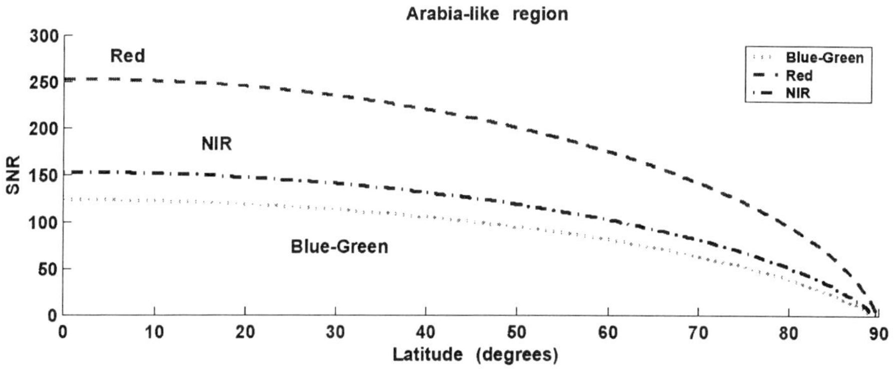

Figure 6. SNR for a Bright Region of Mars

MRO GUIDANCE AND CONTROL

Pointing Performance-Related Requirements

Two types of pointing performance requirements drive the design of the MRO Guidance and Control Subsystem. The first type, target relative pointing, specifies how well a body-fixed vector must point at a target on the surface of Mars. The second type, pointing stability or jitter, specifies how still the spacecraft must remain (inertially) over a specified period of time.

Target-relative pointing requirements were levied on MRO by each imaging instrument. However, we only describe the HiRISE requirements because these are, by far, the most challenging. The fundamental HiRISE target-relative pointing requirements are:

- The HiRISE mechanical boresight shall point to a target on Mars to within 2.6 mrad (3σ) in the HiRISE FOV cross-track direction and 6.6 mrad (3σ) in the HiRISE FOV down-track direction.

- and -

- The projection of the HiRISE instrument focal plane detector array onto the planet's surface is aligned perpendicular to the spacecraft direction of motion within 1.2 mrad (3) during HiRISE image acquisition.

The first target-relative requirement ensures that the target is captured in the instrument field of view. As described below, this requirement has implications for navigation accuracy, target location uncertainty, and spacecraft inertial pointing accuracy. The second requirement is necessary to minimize image smear.

Pointing stability requirements ensure that the image quality meets the science objective. As mentioned above, pointing stability refers to the allowable spacecraft rotational motion over a specified period of time. As with the target-relative pointing requirements, the HiRISE stability requirements are the most challenging of all the pointing stability requirements; however, in Table 2 we list each instrument's pointing stability requirements.

Instrument	Magnitude	Window
HiRISE	0.0032 mrad	12 msec
CTX	0.020 mrad	1 msec
CTX	0.020 mrad	2.5 msec
CRISM	0.032 mrad	200 msec
ONC	0.033 mrad	300 msec
CRISM	0.107 mrad	1 sec
MCS	1 mrad	2 sec
MCS	3 mrad	16 sec

Table 2. MRO Science Instrument Pointing Stability Requirements (magnitudes are 3σ)

GN&C Subsystem Description

The G&C hardware suite and the component locations on the spacecraft are shown in Figure 7. The subsystem consists of two Honeywell Miniature Inertial Measurement Units (MIMU), two Galileo Avionica A-STR star trackers, and four Honeywell Constellation Series HR16 reaction wheels. The A-STR star trackers provide attitude quaternion estimates at 10 Hz. The MIMUs utilize three orthogonal ring laser gyros to provide angular rate data at 200 Hz.

Attitude estimation is performed by a 6-state Extended Kalman Filter,[5] which combines attitude quaternion estimates from the star tracker with angular rate measurements from the inertial measurement unit to produce a 3-state attitude error estimate and 3-state gyro bias estimate. Ground-generated ephemeris and pointing commands are uploaded to the spacecraft, which uses them to construct an attitude command.

Pointing Strategy

Unlike most imaging spacecraft designed around a single high-resolution sensor,[6,7,8] the Mars Reconnaissance Orbiter will provide science observations from multiple science sensors. Therefore, many of the design optimization approaches applied to these Earth-based spacecraft simply cannot be applied to the MRO design. Specifically, the MRO spacecraft must be designed to support continuous nadir-oriented science operations at Mars, with periodic maneuvering to acquire high-resolution images of ground-specified targets. Although the MRO pointing accuracy and stability characteristics must support the HiRISE instrument requirements, they cannot be accomplished without regard to operations of the other instruments, or without regard to the power and telecommunications constraints of Mars-orbital operations. Unlike their Earth-based counterparts, MRO must provide precision-pointing capability while operating multiple instruments (some of which include gimbaled platforms), and while providing solar array and high-gain antenna tracking. Adding to the design challenge, the MRO spacecraft is a relatively large spacecraft bus (> 1000 kg for start of science operations), has large articulated appendages with low-frequency structural modes (< 1 Hz), and has a large fuel load with very low frequency slosh modes (< 0.01 Hz). These characteristics limit the agility of the MRO spacecraft needed to achieve targeted surface imaging.

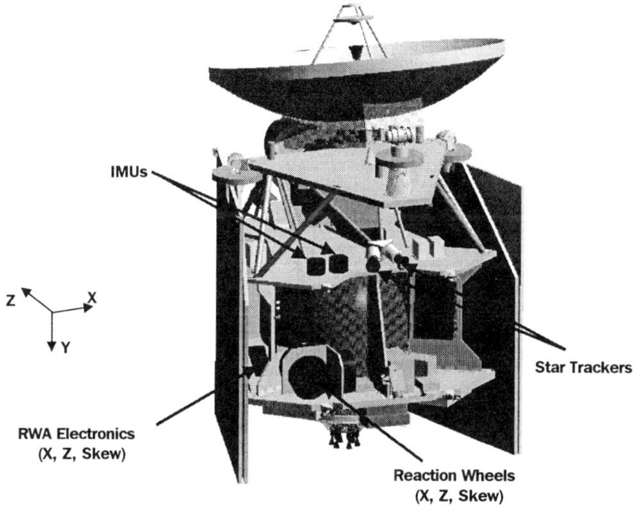

Figure 7. GN&C Subsystem Hardware Configuration

Because of its narrow field of view and extremely high resolution, the HiRISE target-relative pointing and jitter requirements are the most challenging of all the instruments. To this end, the spacecraft will point the calibrated HiRISE boresight at the imaging target during the primary science phase, with the other instrument teams accepting the misalignment between the HiRISE boresight and their instrument's boresight as part of their pointing error budget. One complication associated with TDI-type imagers is the image smear seen if the target apparent motion is not aligned with the image scan direction. For a nadir-pointed vehicle such as MRO, the target apparent motion is a combination of the spacecraft orbital motion and the rotational motion of the planet. Since the HiRISE boresight is nominally aligned with the spacecraft yaw axis, a small variable yaw offset can be added to the nominal nadir-pointing attitude to eliminate smear. For MRO, this yaw offset can be as large as 5 deg (at the equator).

Target-Relative Pointing Budget

As a TDI instrument, HiRISE requires control of the boresight direction to capture the target in the field of view as well as rotational control about the boresight for smear compensation. The primary HiRISE pointing error sources are listed in Table 3 followed by a brief description of each error.

Inertial attitude accuracy is the accuracy to which the spacecraft can orient a body-fixed reference frame in inertial space, and it includes control errors as well as inertial attitude determination uncertainty (for MRO, the body-fixed reference frame is assumed to be the HiRISE calibrated boresight). This allocation becomes a requirement levied on the spacecraft design team.

Error Source	Magnitude (all values are 3-sigma)
Inertial Attitude Accuracy (calibrated HiRISE boresight)	0.7 mrad about X, 1.0 mrad about Y and Z
Predicted Ephemeris Uncertainty	1.5 km in-track, 0.05 km cross-track, 0.04 km radial
Spacecraft Clock Error	10 msec
Target Horizontal Position Uncertainty	600 m per axis
Surface Altitude Uncertainty	500 m

Table 3. Target-Relative Pointing Error Sources and Allocations

MRO measures (via the star tracker) its inertial attitude; it has no Mars-relative attitude sensing capability. To point a body-fixed vector at a target on the surface of Mars, knowledge of the spacecraft position with respect to Mars is required. Updated spacecraft ephemeredes will be uplinked to the spacecraft every 48 to 72 hours, with maximum ephemeris uncertainty occurring just prior to uplink (and assumed in the pointing budget). Incidentally, the in-track ephemeris uncertainty is by far the largest contributor to the target-relative pointing budget. This allocation becomes a requirement levied on the Navigation team.

Errors in the onboard spacecraft clock have two effects on spacecraft pointing: for a given time they result in an error in the knowledge of the spacecraft in-track position and an error in the knowledge of rotational position of Mars. This allocation becomes a requirement on the operations team to calibrate spacecraft clock drift.

The final two error sources are simply caused by our inexact knowledge of Mars geography. Target horizontal position uncertainty is the uncertainty in the target location on the surface of Mars. For targets that are not directly below the spacecraft, surface altitude uncertainty results in a pointing error.

Pointing Stability Budget

The HiRISE pointing stability tree is shown in Figure 8, which illustrates and categorizes the disturbances and error sources that affect pointing stability. Figure 9 provides an illustration of how these errors affect image quality for the HiRISE TDI imager. The low-frequency disturbances from sources such as control loop rate error, uncompensated IMU bias, flexible body dynamics, propellant slosh, etc, affect image quality because they result in motion of the HiRISE boresight across the TDI columns. These *rate errors* are combined in a root-sum-square (RSS) sense because they are uncorrelated and all appear as roughly constant over the imaging window time period. *Jitter sources*, on the other hand, generally have distinct spectral components of at least 20 Hz and, as such, are added within the pointing stability budget. The largest jitter sources for MRO are due to reaction wheel imbalance, the CRISM cryocooler, and the HGA gimbal drives (the solar array gimbals, as explained in the following section, are paused during high-resolution imaging). Finally, *misalignments* between instruments and sensors also result in an error in the target direction of motion through the field of view, so these are also included in the pointing stability budget and are combined in an RSS sense.

The pointing stability budget for a 12 msec window is shown in Table 4 (12 msec corresponds to the maximum time window in the HiRISE pointing stability requirement). Among the rate error components, ACS control loop rate error is by far the largest contributor at 0.6 microrad, though this is relatively small compared to the overall budget of 3.2 microrad. Among jitter sources, reaction wheel imbalance, the CRISM cryocooler, and the HGA gimbal drive dominate the budget. It is interesting to note that, because of the low-frequency content of slosh disturbances relative to most of the stability requirement windows, slosh has been determined to be a insignificant contributor to stability performance. Strategies for managing the contributions of these disturbances are described in the following section.

Error Category/Term	Allocation (microrad)
Rate Error Contributors	**0.61**
IMU Uncompensated Bias Drift	0.02
IMU Scale Factor Error	0.01
S/C Flexible Body Dynamics (including SHARAD)	0.07
Propellant Slosh	0.05
Target Location Errors (including terrain)	0.01
Target Timing Errors	0.09
S/C Ephemeris Errors on Motion Compensation	0.02
S/C Pointing Errors on Motion Compensation	0.01
ACS Control Loop Rate Error	0.60
Jitter Contributors	**2.57**
RWA Imbalance	1.25
IMU Dither Disturbance	0.27
CRISM Gimbal Drive	0.10
RWA Motor Disturbances	0.13
CRISM Cryocooler	0.63
HGA Gimbal Drives	0.57
Thermal Snap	0.00
MCS Gimbal Drive (gimbal motion assumed paused)	N/A*
Solar Array Gimbal Drives (gimbal motion assumed paused)	N/A*
Alignment Contributors	**0.18**
Attitude Control Error w.r.t. Star Tracker	0.10
Attitude Knowledge Error w.r.t. Star Tracker	0.10
Calibrated HiRISE Alignment w.r.t. Star Tracker	0.08
Calibrated HiRISE Internal Alignment	0.08
HiRISE Mechanical and Thermal Stability	0.04
Total (3-sigma)	**3.20**

Table 4. HiRISE 12 msec Pointing Stability Allocation

MRO DISTURBANCE SOURCES

The following discussion presents the approach undertaken to address MRO-unique issues. The major error sources shown in the pointing stability allocation are discussed, along with their mitigations. Included are the gimbaled science platforms (MCS and CRISM), the CRISM cryogenic cooler, the solar array and high-gain antenna articulation, the reaction wheel imbalance and related desaturation strategy, and disturbances from the dither of the IMU's ring-laser gyro assembly.

Analysis Approach for Pointing Stability Requirements

MRO's many pointing accuracy and stability requirements reveal the HiRISE requirements to be the most stringent, and establish them as the principal design drivers. In this context, the approach taken to meeting MRO pointing accuracy and stability requirements has been to design to the HiRISE requirements, and then to verify that the other pointing stability requirements are met as well. This approach has generally proven successful, although some design tweaks have been needed even after meeting all the HiRISE requirements.

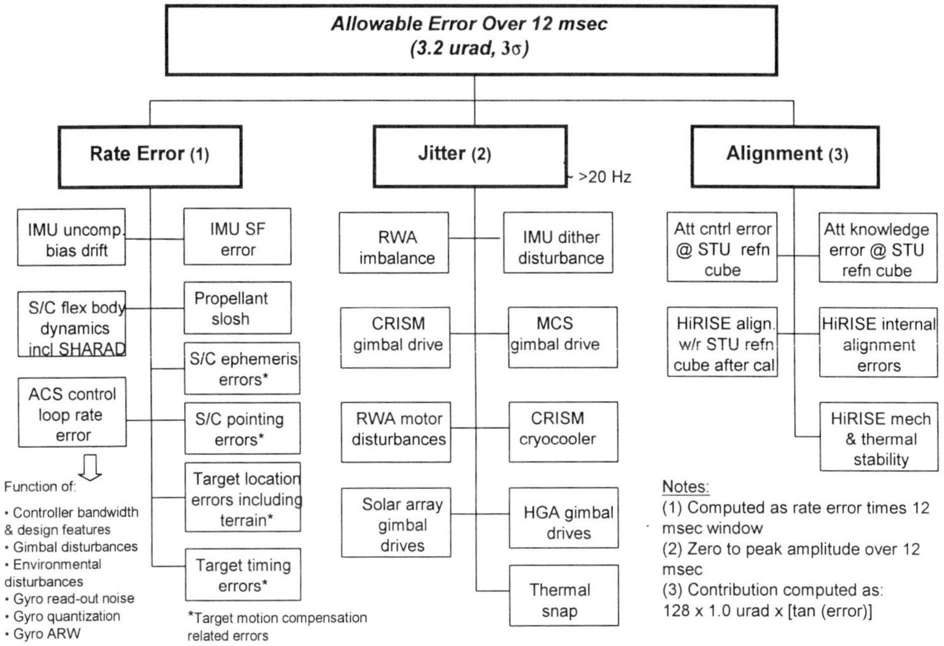

Figure 8. HiRISE Pointing Stability Tree

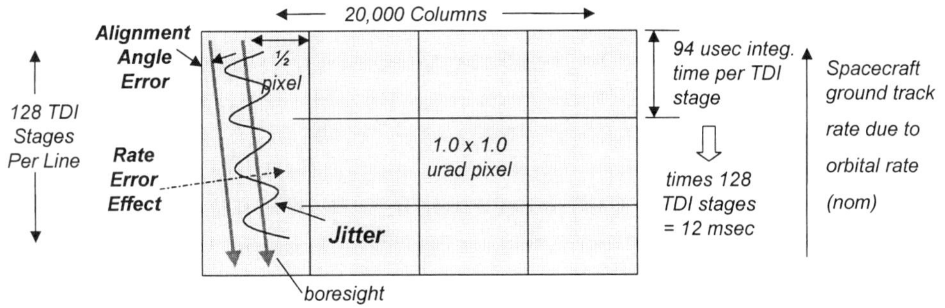

Figure 9. Effect of Jitter, Rate Error, and Misalignment on HiRISE TDI

The jitter and rate error contributors have been evaluated using various methods. Jitter was evaluated by first building a NASTRAN Finite Element Model (FEM), as shown in Figure 10, and then modeling the transmissibility of the various disturbances to the HiRISE mounting base. Typical and worst-case disturbance sources were analyzed and time histories created. The disturbance sources examined include stationary sources (reaction wheel imbalance, cogging, and ripple, IMU dither, CRISM cryocooler, gimbal DC motors, and MCS stepper motors) and transient sources (harmonic drive irregularity of the solar array and high-gain antenna gimbals, CRISM scan torque profile, and MCS scan torque profile). For the MRO analysis, jitter motion consists of dynamics higher than 20 Hz. Low-frequency motion, which shows up as essentially a rate-tracking error, has been analyzed with a combination of closed loop simulations, Monte-Carlo assessments and analytical tools. For the MRO analysis, low-frequency motion consists of dynamics less than 10 Hz.

The results of these assessments have been provided to the instrument providers, allowing them to assess their requirements and refine them as appropriate. For high-frequency jitter, the instrument providers were given time histories of crosstrack and downtrack motion. Data was provided for a 10 sec window with 10 microsec time steps. A single rate error value containing the combination of all of the low frequency components was also provided.

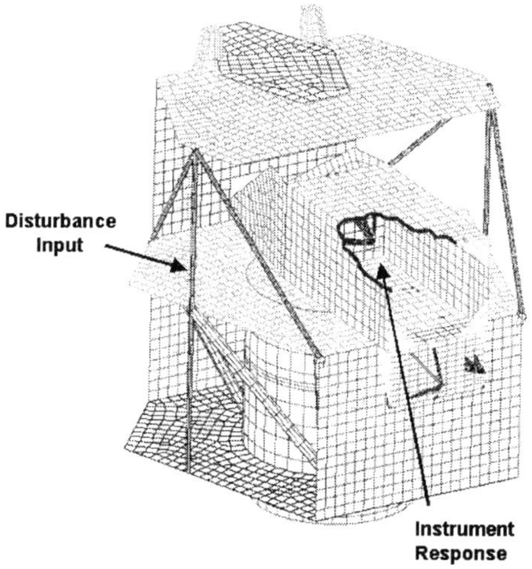

Figure 10. MRO Core Spacecraft FEM

The instrument providers reviewed and analyzed the data, and provided their evaluations. Ball examined the data in great detail, using their "TRADES" simulation tool to generate TDI image lines for relative evaluation of image quality. Based upon their assessment, several instrument stability requirements were revised and some spacecraft design changes were made to improve performance. These changes included stiffening all deck structure and redesigning the optical bench to accommodate HiRISE mounting and isolate it as much as possible from solar array and HGA gimbals. Several operational changes were also made, including "freezing" the solar array gimbals and the MCS scan platform during high-resolution imaging. The following discussion provides more detail on why these changes were needed.

Other Science Instruments (CRISM and MCS)

As discussed earlier, the MRO spacecraft must support simultaneous operations of multiple science instruments. Spacecraft disturbances from two instruments are of particular concern for high-resolution imaging, these instruments being CRISM and MCS. Both of these instruments have gimbaled platforms and the CRISM instrument design includes a cryocooler. The MRO science mission includes "coordinated observations" between HiRISE and CRISM, so CRISM must be able to operate during these high-resolution imaging periods. However, the MCS science collection is independent of the HiRISE instrument, and so could be interrupted for brief periods if necessary.

Representative disturbance profiles from the CRISM and the MCS instruments are shown in Figure 11. These profiles were constructed based upon typical scan motion profiles of both instruments and their respective dynamics models. As can be seen from theses profiles, the MCS torque disturbances are roughly an order of magnitude worse than the torque disturbances from the CRISM instrument.

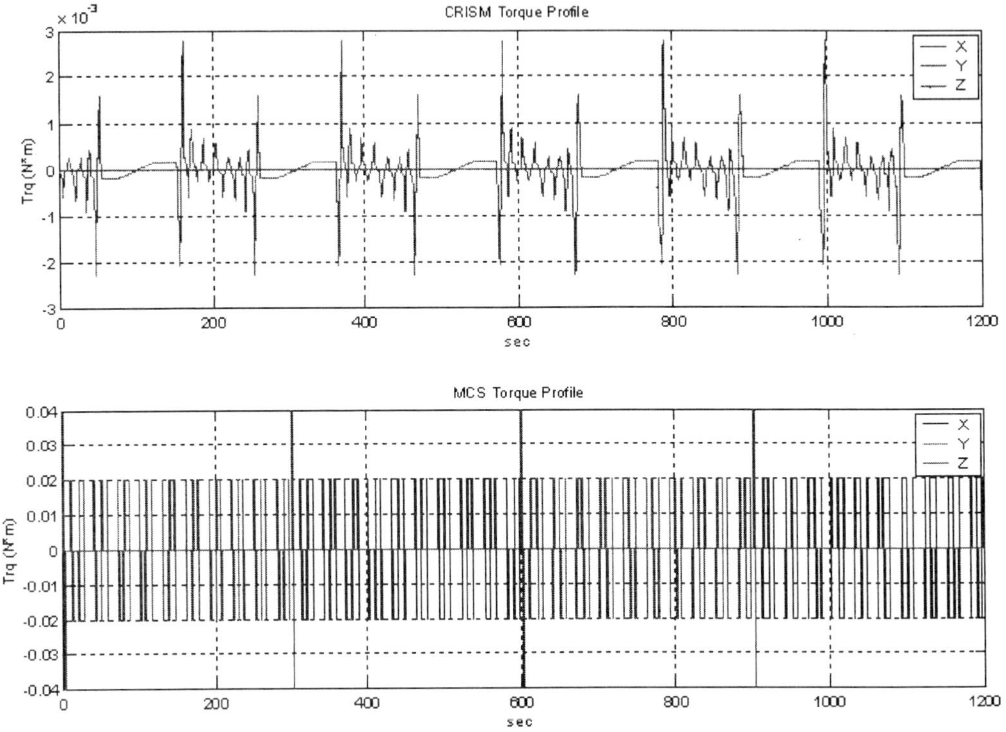

Figure 11. Representative CRISM and MCS Gimbal Torque Profile

An orbital simulation for MRO has been run with the above disturbance profiles to assess the pointing stability impacts. A preliminary design of the flight control laws was included in the simulation that has a bandwidth of approximately 0.1 Hz. To isolate the effects of the payload gimbal motion, no other gimbals were running (e.g., the solar arrays and the high-gain antenna were fixed in a nominal position). Additionally, the environmental disturbance models were disabled for this simulation. Representative structural models and slosh models were included, with the first structural mode at 1.08 Hz and the zero-g slosh mode at 0.007 Hz; the simulation results are shown in Figure 12. Because the payload disturbance torques have significant energy above the bandwidth of the controller, the simulation shows significant motion of the spacecraft (peak amplitude of approximately 35 microrad).

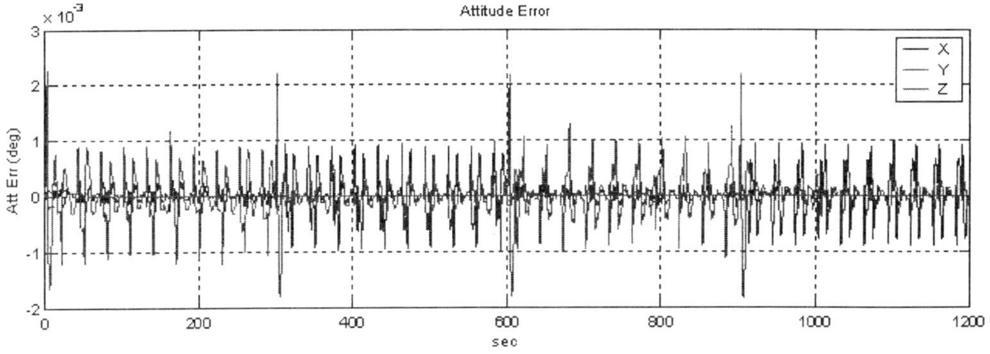

Figure 12. Spacecraft Attitude Error Response to CRISM & MCS Gimbal Motion

Detailed analysis of the attitude data shown in Figure 12 indicates that most of the motion is caused by the MCS gimbal motion and that the HiRISE pointing stability requirements cannot be achieved while the MCS gimbals are scanning. Therefore, an operational constraint has been imposed to stop MCS motion prior to any high-resolution HiRISE imaging. The analysis does indicate, however, that the CRISM gimbal motion is acceptable during high-resolution imaging and does not require any complex feed-forward or other control compensation schemes.

Another significant payload disturbance source is the CRISM instrument's Stirling cryocooler. Because of the high resolution and relatively long integration time of the HiRISE instrument, the high-frequency vibration of a cryocooler can easily violate HiRISE pointing stability requirements. The MRO science objectives require that HiRISE and CRISM perform coordinated observations of Mars surface targets, and therefore it would be desirable to have them co-located to achieve the best possible co-alignment. However, co-locating HiRISE and CRISM on the same deck would exacerbate the CRISM cryocooler microphonics effect upon HiRISE pointing stability. Analysis of cryocooler data provided by Johns Hopkins Applied Physics Lab (APL)[9] shows the need for significant structural attenuation of the cryocooler vibration. Therefore, the CRISM instrument has been located as far from HiRISE as possible to reduce cryocooler and gimbal disturbances on HiRISE boresight stability.

Solar Array and High-Gain Antenna

Because the MRO solar array and high-gain antenna are very large, they represent significant disturbance sources during tracking. The gimbals have been carefully designed to eliminate the motors as a significant vibration source. The gimbal utilizes a DC brushless motor, rather than a more traditional stepper motor, to eliminate the disturbances and vibration associated with stepper motors. Furthermore, the commutation of the DC brushless motor is implemented using synchronous resolvers instead of Hall-effect sensors. This allows a position-dependent current command, rather than a simple Hall-effect switching scheme, to reduce the motor cogging.

The gimbal design also includes a fairly large gear reduction (260:1) implemented with an harmonic drive. Although harmonic drives are efficient mechanisms from a mass perspective, they have many undesirable dynamics characteristics for a precision-pointing spacecraft. Chief among these is the nonlinear and positional dependent friction characteristics. The MRO design attempts to limit the friction effects by implementing a tight control loop (20-30 Hz bandwidth) on the gimbal rate. However, this effort involves development work and the performance goals have not yet been proven. For the initial MRO analysis, minimal improvements have been assumed from the gimbal development efforts. The articulation models have instead been developed based upon in-flight observations of articulation disturbances from 2001 Mars Odyssey spacecraft data.[10] Although the MRO motor design is quite different from Odyssey's motor design (Odyssey uses stepper motors), the harmonic drive design is similar. Therefore, the gimbal model used in the MRO pointing stability analysis is quite conservative.

A Mars orbital simulation was constructed in ADAMS/Simulink to analyze the low-frequency, transient dynamics resulting from solar array and high-gain antenna gimbal motion, as well as from CRISM and MCS gimbal motion. The simulation incorporates moderately high-fidelity models, including a full reaction wheel control model for the spacecraft, a full gimbal control model for the solar array and high-gain antenna, and representative CRISM and MCS disturbance profiles. The spacecraft model consists of an FEM structural model, including structural modes for the solar arrays, compliant gimbals for the solar array and the high-gain antenna, and Odyssey-like harmonic-drive disturbances.

Some simulation results from this model are presented in Figure 13, showing significant stability excursions above HiRISE requirements. Because of these simulation results, an operational constraint has been imposed to stop solar array gimbal motion during high-resolution imaging. Although stopping motion of the high-gain antenna could produce some additional stability gains, HGA motion does not violate the stability requirements and the desire to maintain communications throughout high-resolution imaging outweighs the desire for improved stability. The simulation results have also led to some minor structural design changes to improve HiRISE pointing stability. The gimbal support deck was modified to take disturbances down struts to core cylinder, thereby reducing the transmissibility of vibration from the gimbals to HiRISE.

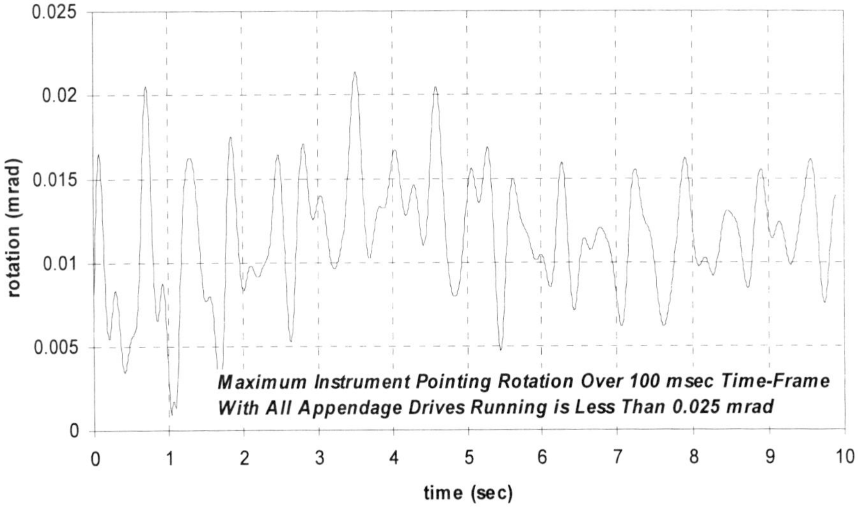

Figure 13. MRO Pointing Stability Simulation Results (Maximum Deviation over 100 msec) including Solar Array, HGA, CRISM, and MCS Gimbal Motion

Reaction Wheels and Inertial Measurement Unit (IMU)

Both the reaction wheels and the inertial measurement unit (IMU) can be significant vibration sources, and can adversely affect pointing stability for high-resolution imaging. Each reaction wheel has its own static and dynamic imbalances that produce vibration when spinning. Because the vibration magnitude is proportional to the square of the speed, better pointing stability can be achieved by operating the reaction wheels at lower speeds. Large momentum-capacity reaction wheels and frequent desaturation events will optimize pointing stability. For the MRO spacecraft, the Navigation team has imposed a maximum desaturation frequency of one desaturation event every 48 hours to meet orbit prediction accuracy requirements, so design flexibility is only available in reaction wheel sizing and balancing. A large capacity off-the-shelf wheel size (100 Nms) has been selected for MRO which provided the best trade-off between momentum capacity and wheel balancing. Extra balancing activities are required for MRO to minimize reaction wheel vibration over the operational speed range (0-2500 RPM). To further limit the effect of wheel vibration on HiRISE stability, the wheels have been placed as far away from the HiRISE optical bench as possible.

The IMU selected for use on MRO contains three ring laser gyros (RLGs), which are dithered to improve performance at low rotation rates. The dither frequencies are in the range of 400 to 700 Hz. The IMU can either be procured with the RLGs internally isolated from the chassis, or without internal isolation. Without internal isolation, relative alignment knowledge between the star tracker and the gyros can be improved by more than an order of magnitude. This improves pointing accuracy when slews to a surface target are required. However, removing the internal isolation greatly increases the dither vibration observed at the IMU mounting interface, which can degrade pointing stability. For MRO, this has been resolved by imposing an operational constraint on targeted imaging. Each off-nadir targeted slew must have time allocated to star acquisition following the slew, which removes the need for extremely tight IMU to star tracker alignment knowledge. This allows MRO to include internal isolation in the IMU, which improves pointing stability.

Simulation results of crosstrack jitter are shown in Figure 14. This figure shows time-history details over 0.3 sec. Included in the simulation results are reaction wheel speeds of 1000, 2000, and 3000 RPM, with static and dynamic imbalance numbers discussed above. The HGA is assumed to be articulating, but the solar arrays are frozen in a nominal orientation. The CRISM cryocooler is assumed to be operating. Finally, the IMU is assumed to be operating without any internal isolation. Although this is not the current baseline configuration, it was analyzed this way to assess a worst-case scenario. As can be seen in the figure, reducing the reaction wheel speeds from 3000 RPM (blue) to 2000 RPM (green) significantly improves the jitter characteristics.

A detailed statistical analysis was performed on the jitter simulation results. The simulation was run for a number of cases; each case represents 10 sec of simulated motion, and results were output at 0.5 msec time steps. Various time windows were examined, with the most stressing time window being 12 msec. The 12 msec time window was passed over the 10 sec of simulation data to provide 20,000 distinct windows of 25 data points each. The mean (in microrad) for each window was computed, as well as the distance of each sample from the mean. Statistics for all 20,000 windows were then computed from these results.

A stressing case in which all three-reaction wheels were spinning at 3000 RPM is shown in Figure 15. The left side of Figure 15 shows the jitter distribution, and the right side shows the cumulative distribution. From this figure we see that the jitter is less than 0.9 microrad approximately 94% of the time for this stressing case. The jitter is less than 0.5 microrad for approximately 72% of the time. These numbers are significant, because the HiRISE pixel width is 1.0 microrad. This means that, even for this bad stack up of error sources and for the longest possible HiRISE integration time, a point source would stay within single TDI column 72% of the time.

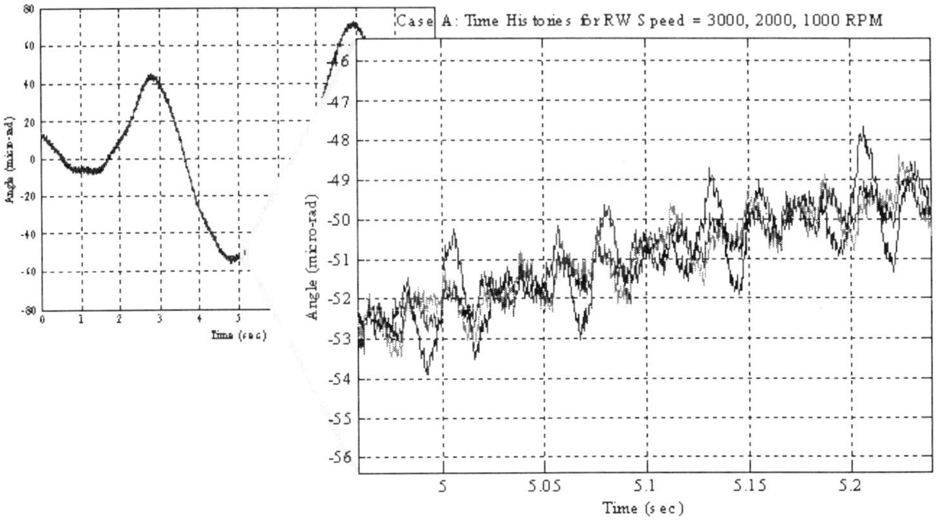

Figure 14. Simulated Time History of Crosstrack Jitter Response at HiRISE Mounting Interface‡

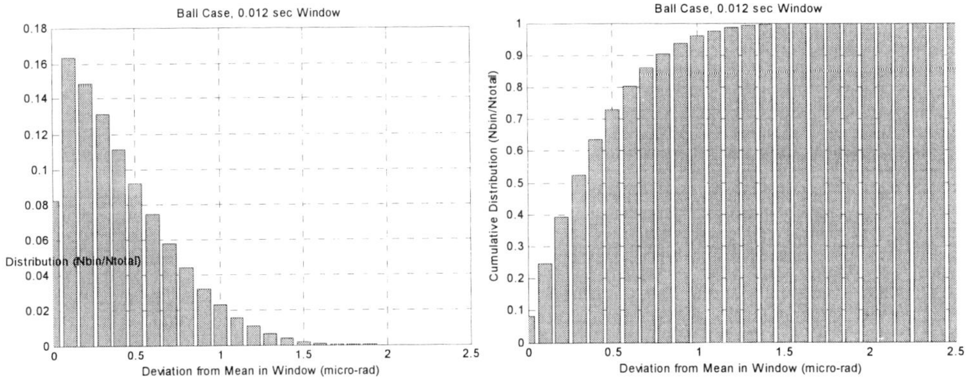

Figure 15. Jitter Distribution [left] and Cumulative Jitter Distribution [right] for 12 msec Imaging Period (RWA speed of 3000 RPM)

OVERALL ESTIMATED PERFORMANCE

A moderate-fidelity simulation was constructed to assess the MRO pointing accuracy and stability characteristics under the above assumptions. Pointing stability results of this simulation are shown in Figure 16. The startup transients die out within the first 100 sec of simulation time; the remainder of the simulation represents realistic pointing stability characteristics. As can be seen in the figure, the solar array motion produces unacceptably large disturbances for high-resolution pointing, even when the MCS gimbal platform is frozen. When the solar array is stopped (approximately 350 sec into the simulation), a transient is observed in the pointing stability followed by very good pointing stability. Without solar array motion, the simulation predicts that the

‡ Red, green, and blue lines correspond to reaction wheel speeds of 1000, 2000, and 3000 RPM, respectively.

HiRISE stability requirements will be met with significant margin. The simulations indicate that the pointing strategies will support the ambitious MRO science objectives, and that the various MRO science goals are achievable. Higher-fidelity simulation models will become available as the MRO design matures, and the pointing accuracy and stability characteristics will continue to be closely scrutinized.

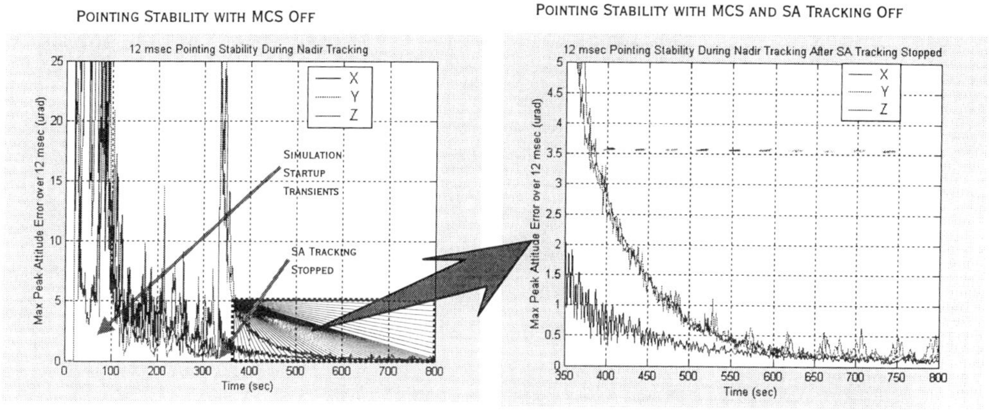

Figure 16. 12 msec Stability Initial Results for HiRISE

CONCLUSION

In 2005 the Mars Reconnaissance Orbiter spacecraft will embark on its ambitious journey to explore Mars. The spacecraft will carry an impressive array of instruments including HiRISE, a high-resolution imager that will obtain stunning new images of Mars with a resolution of 30 cm/pixel. Producing such images, however, requires extremely tight target-relative pointing accuracy and stability capabilities; the precision required is unmatched by any previous interplanetary spacecraft mission. Although the nature of the MRO mission does not allow the spacecraft design to be optimized for high-resolution imaging, the design compromises discussed in this paper still produce an extraordinarily-capable vehicle. Analysis and simulation results indicate the MRO design will meet performance goals the vast majority of the time, with exceedances only under the most challenging combinations of error sources. The resulting design will allow exciting new discoveries as we continue the to expand our knowledge of the planet Mars.

ACKNOWLEDGMENTS

This work was performed at the Jet Propulsion Laboratory, California Institute of Technology, at Lockheed Martin Space Systems, and at Ball Aerospace and Technologies Corp., under NASA contract NAS7-1407. Many thanks go to the many engineers and scientists at these institutions who participated in the MRO pointing stability special study. Special acknowledgement is given to Dr. Phillip Good and Jason Wynn at Lockheed Martin, who worked diligently in developing the MRO stability analysis techniques, as well as Jim Bergstrom, Bill Meiers, Barry Dunn, and Dennis Gallagher, members of the Ball HiRISE team, who were major contributors to the analysis of the HiRISE performance. Inspiration for the HiRISE camera came from Professor Alfred McEwen and his science team at the University of Arizona. Many others contributed in various ways through MRO program workshops and tabletop reviews and helped developed a better understanding between instrument operations and spacecraft performance. The authors gratefully acknowledge all those who participated.

REFERENCES

[1] "Mars Reconnaissance Orbiter Request for Proposal," California Institute of Technology, Jet Propulsion Laboratory, RFP No. JPL-MRO-2005, Apr 2001.

[2] Zurek, R., "Mars Exploration Strategy: Follow the Water," Keynote Presentation, *24th Annual AAS Guidance and Control Conference*, Breckenridge, CO, Feb 2001.

[3] Spath, S. R., and Eckart, D. F., "Fuel Optimization During Mars Global Surveyor Aerobraking," AAS 00-072, in *Guidance and Control 2000*, R. D. Culp and E. M. Dukes, eds., AAS *Advances in the Astronautical Sciences*, Vol. 104, pp. 615-627, 2000 (paper presented at the 23rd Annual AAS Guidance and Control Conference, Breckenridge, CO, Feb 2000).

[4] Chapel, J. D., Johnson, M. A., Sidney, W. P., Willcockson, W. H., Wynn, J. A., Gulick, D., "Aerobraking Safing Approach for 2001 Mars Odyssey," AAS 02-011, in *Guidance and Control 2002*, R. D. Culp and S. D. Jolly, eds., AAS *Advances in the Astronautical Sciences*, Vol. 111, pp. 95-109, 2002 (paper presented at the 25th Annual AAS Guidance and Gontrol Conference, Breckenridge, CO, Feb 2002).

[5] E. J. Lefferts, F. L. Markley, and M. D. Shuster, "Kalman Filtering for Spacecraft Attitude Estimation," *Journal of Guidance*, Sep-Oct, 1982, Vol 5, No 5, p. 417.

[6] Sietzen Jr., F., "Advanced Imagery Raises the Ante," *Aerospace America*, Sep 2001.

[7] "DigitalGlobe Successfully Launches QuickBird Imaging Satellite," *SpaceDaily.com*, Oct 18, 2001.

[8] "French Imaging Spacecraft Lofted into Polar Orbit by Ariane 4," *Space.com*, May 3, 2002.

[9] Ross, R., "Critique of the Mars '05 CRISM Spectrometer Cryocooler Application—An Update," JPL Internal Memorandum, Feb 2002.

[10] Gehling, R., and Lynch, S., "Appendage Deployment Verification for the 2001 Mars Odyssey Spacecraft Using Spacecraft Attitude Telemetry," AAS 03-075, in this volume (paper presented at the 26th Annual AAS Guidance and Control Conferencc, Breckenridge, CO, Feb 2003).

Section VI:
RECENT EXPERIENCES IN GUIDANCE AND CONTROL

SESSION VI

Joint Session Chairperson:	Darrell Zimbelman NASA Goddard Space Flight Center
Joint Session Charperson:	David Haley Johns Hopkins University / Applied Physics Laboratory
Local Session Chairperson:	Eileen Dukes Interplanetary Horizons

The following paper was not available for publication:

AAS 03-076 "Development of FUSE Onboard SW to Accommodate Operations with Limited Gyro and Reaction Wheel Capability," by M.J. White, L. Fisher, T. Englar, JHU/APL(Paper Withdrawn)

The following paper numbers were not assigned:

AAS 03-078 to -099

AAS 03-071

SATELLITE GN&C ANOMALY TRENDS

Brent Robertson and Eric Stoneking*

On-orbit anomaly records for satellites launched from 1990 through 2001 are reviewed to determine recent trends of un-manned space mission critical failures. Anomalies categorized by subsystems show that Guidance, Navigation and Control (GN&C) subsystems have a high number of anomalies that result in a mission critical failure when compared to other subsystems. A mission critical failure is defined as a premature loss of a satellite or loss of its ability to perform its primary mission during its design life. The majority of anomalies are shown to occur early in the mission, usually within one year from launch. GN&C anomalies are categorized by cause and equipment type involved. A statistical analysis of the data is presented for all anomalies compared with the GN&C anomalies for various mission types, orbits and time periods. Conclusions and recommendations are presented for improving mission success and reliability.

INTRODUCTION

A study of past on-orbit anomalies was undertaken to assess how future satellite program resources might be best spent to ensure mission success. The requirements for future spacecraft include advances in reliability, particularly for deep space missions and long duration Earth observing platforms. The American industry has typically tried to achieve this goal by incorporating redundancy, robustness and fault tolerance in spacecraft designs, and verifying those designs with a thorough test program. This approach has had varying degrees of success; it is not clear whether the benefit outweighs the cost for all cases. Although Launch Vehicle failures have contributed to the majority of past space mission losses, this trend may be changing; in 1998 the bulk of satellite losses were caused by failures of on-orbit satellites. In the last four years, space insurance rates for Geosynchronous Earth Orbit (GEO) commercial communications satellites has risen by 129% and major on-orbit anomalies have risen by 146%[1]. Analysis of on-orbit anomalies can provide greater insight into design and process improvements and help in devising more effective verification methods during integration and test.

Mission critical on-orbit anomalies were analyzed to determine the number and characteristics of mishaps. A mission critical failure is defined as a premature loss of a satellite or loss of its ability to perform its primary mission during its design life. On-orbit anomaly data used in this study was taken from a number of sources in the public record. Data was taken from databases available on the web including The Satellite Encyclopedia (TSE)[2], Satellite News Digest[3], Mission and Spacecraft Library[4], Airclaims Space Trak[5], and Encyclopedia Astronautica[6]. Data was also taken from the Space Systems Engineering Database[7] (SSED) and the Mission Failure Analysis for NASA Ames Research Center[8]. All anomaly reports collected

* NASA Goddard Space Flight Center, Code 570, Greenbelt, MD 20771.

were unclassified, and almost all were written, archived and reference-able. Wherever possible, records of anomalies were corroborated between multiple sources. Anomalies associated with Guidance, Navigation and Control (GN&C) was scrutinized further, with every attempt made to obtain mishap reports, visit pertinent web sites, or to contact people with insight into the problem. A statistical analysis of the data was undertaken for all anomalies compared with GN&C anomalies, for various mission types, orbits and time periods.

In this study, GN&C is defined as including the on-orbit Attitude Control System (ACS), the on-orbit Propulsion System and all ground operations and software associated with satellite flight dynamics (trajectory planning and determination, navigation and attitude determination).

DATA COLLECTION

Table 1 contains the satellite classifications and categories used for analysis of anomaly trends. Only on-orbit anomalies involving un-manned spacecraft were investigated; launch and upper stage failures were not recorded. Anomalies were investigated for all on-orbit satellites launched in the years 1990 through 2001. The analysis only investigated satellites that were designed and built by companies or government organizations in the United States, Europe, Canada or Japan. An anomaly was recorded only if it occurred prior to the mission fulfilling its design life. Anomalies that occur during an extended mission were not recorded.

Table 1 Anomaly Classification

Classification	Category
Satellite	Name
Mission	Space Science, Earth Science, Deep Space, Communications, Military, Technology, Other
Orbit	Low Earth Orbit (LEO), High Earth Orbit (HEO), Geosynchronous Earth Orbit (GEO), Heliocentric, Planetary
Launch Date	1990 through 2001
Design Life	
Anomaly Date	
Impact	Total Loss, Partial Loss, Mission Interruption, Shortened Life, Performance Loss
Failure Subsystem	Attitude Control System, Propulsion, Electrical Power System, Command & Data Handling, Communications, Mechanical, Software, Payload, Operations, Unknown

Missions were categorized as space science (including astronomy, observation of the Sun, and Sun-Earth connection) Earth science (including meteorology, remote sensing, observations of the Earth atmosphere, ionosphere and magnetosphere, and geodesy), deep space (Mars, Lunar and asteroids), communications (including telecommunications and Direct TV), military, technology, and other (including test, burial, materials processing and life sciences).

Orbits were categorized as Low Earth Orbit (LEO), High Earth Orbit (HEO), GEO, heliocentric, and planetary. LEO orbits were defined as Earth centered orbits having apogee less than 1500 km. HEO orbits were defined as Earth centered orbits with apogee greater than 1500 km.

The anomaly date is defined as the date that the anomaly first occurred. In some cases, a failure can take a significant time period to fully affect the mission. In general, the date of first occurrence of the anomaly was recorded.

The impact was categorized as total loss, partial loss, mission interruption, shortened life, or performance loss. A total loss resulted if a satellite failed so that it could no longer perform the mission, or if it was deactivated or taken out of service due to an anomaly. A shortened life is equivalent to a total loss, although the loss has not yet occurred. The analysis did not include anomalies that resulted in loss of redundancy, or a minor interruption in the mission due to an operational work-around or software patching. Anomalies that occurred within the planned satellite operational checkout phase were not recorded unless they had an effect on the ability to perform the subsequent mission. A mission interruption was only recorded if it occurred for a substantial time period (e.g., more than a week). In general, an anomaly was recorded if an insurance claim was paid.

The subsystem categories were made up of ACS, Propulsion, Electrical Power Subsystem (EPS), Command & Data Handling (C&DH), communications, mechanical, software, payload, operations, and unknown. Mechanical was assumed to include structures, mechanisms and thermal. ACS was assumed to include all sensor, actuator, and ACS software involved with on-board attitude and orbit control. GN&C was assumed to include ACS, propulsion and all ground operations and software involved with flight dynamics. A number of recent anomalies have been associated with crystalline growth on tin relays that short circuit the spacecraft processor; these were categorized as C&DH. EPS anomalies included solar array, battery, bearing and power transfer assembly, DC-DC converter and power regulator problems. Transponder anomalies on communications satellites were categorized as payload anomalies. In a few instances, an anomaly was attributed to two subsystems.

All of the GN&C anomalies were scrutinized further. Table 2 illustrates data that was collected for GN&C anomalies, in order to discern trends specific to GN&C anomalies. GN&C anomalies were categorized as caused by problems in design, hardware, software, verification, operations or the environment. Design includes analysis and models, processes or misapplication of hardware. The design category was applied at the system level; problems with the design of components typically procured (e.g., reaction wheels, gyros) were categorized as hardware. The hardware category includes materials, parts, workmanship and design of components. The verification category was applied where the test effort should have caught the design or process error (e.g. ACS polarity errors not found in test). The environment category was applied where on-orbit events such as magnetic storms caused the anomaly.

Table 2 GN&C Anomaly Classification

Classification	Category
Cause	Design, Hardware, Software, Verification, Operations, Environment
Equipment Type	Wheel, Gyro, GPS Receiver, Earth Sensor, Thruster, Tank, Pyrovalve, Processor, Nutation Damper

The GN&C equipment category includes wheel (reaction wheel or momentum wheel), gyro, GPS Receiver, Earth sensor, thruster, tank (fuel tank or pressurant tank), pyrovalve, processor and nutation damper. The equipment did not necessarily have to fail, but was at a minimum involved in the anomaly (e.g. a pyrovalve was fired that induced a propulsion system explosion), or the equipment failure caused the anomaly (e.g. the second of four reaction wheels failed causing the mission to be interrupted for a significant amount of time).

To the extent possible, the subjectivity of anomaly reports was removed to prevent bias in analysis results. Multiple sources were used wherever possible to corroborate events. Every

GN&C anomaly was scrutinized by each of the two authors to ensure consistency of reporting. Despite these precautionary efforts, a bias in analysis results does exist. The accuracy of anomaly reports has a direct bearing on the trends observed. Many anomalies are not reported in the public domain, particularly those involving military missions. Satellite insurance claims are not highly publicized due to their proprietary nature. The probability that an anomaly is reported can depend on the severity of the anomaly (i.e. a greater percentage of anomalies causing total loss are reported than those causing performance loss). Many other anomalies are reported, but the details are scarce or the cause of the anomaly is not reported. Being employed by NASA, the authors have greater insight into anomalies on NASA missions than anomalies on other types of missions. In some cases the impact of the anomaly is diminished by the reporting organization. Anomalies were recorded only if they occurred within the satellite design life; the reported design life of a satellite can be subjective. It was assumed that the mission started as soon as the satellite was launched. No value was placed on any of the satellites or the severity of the anomalies. An anomaly that resulted in the loss of a half billion-dollar asset was counted the same as an anomaly that resulted in loss of a University-built micro-satellite. The impact of an anomaly can be a matter of luck; an ACS polarity problem can result in total loss of mission (which would be recorded) or result in the ground catching and uploading software to fix the problem within the planned checkout phase (where it would not be recorded). Finally, the interpretation of an anomaly report and the resulting categorization can be subjective.

DATA ANALYSIS

Figures 1, 2 and 3 illustrate the total number of on-orbit satellites investigated in this study. Every on-orbit satellite that originated from the United States, Europe, Canada or Japan is shown by year launched from 1990 through 2001, mission type and orbit. Satellites that were not placed in the proper orbit by the Launch Vehicle or Upper Stage were not recorded. A total of 764 satellites were recorded. It can be seen from Figure 1 that the launch rate for those satellites investigated peaked in 1998, mostly due to the launch of many LEO communication satellites during that time. Figure 2 shows that the majority of satellites launched were communication satellites, followed by those designed for military purposes. Figure 3 shows that the majority of satellites were launched into LEO orbits, with GEO orbits receiving the next largest number of launches. Very few missions were launched into heliocentric or planetary orbits.

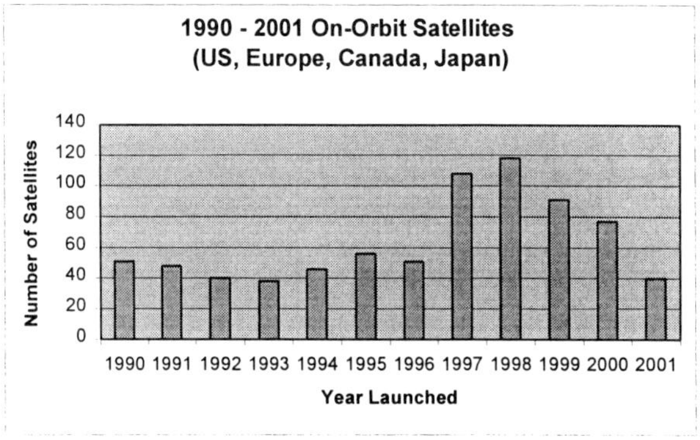

Figure 1 On-Orbit Satellites vs. Year Launched.

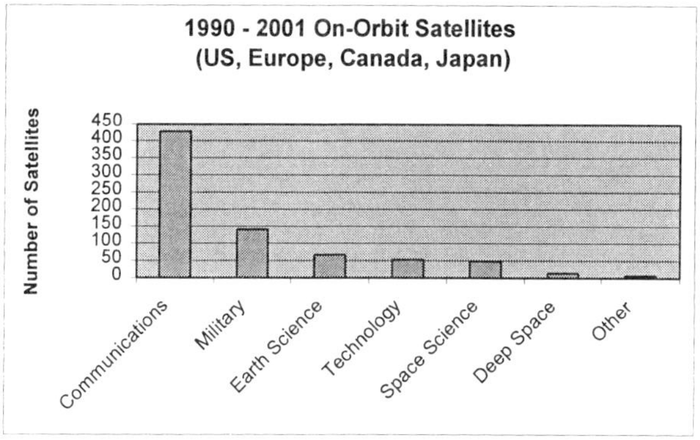

Figure 2 On-Orbit Satellites vs. Mission Type.

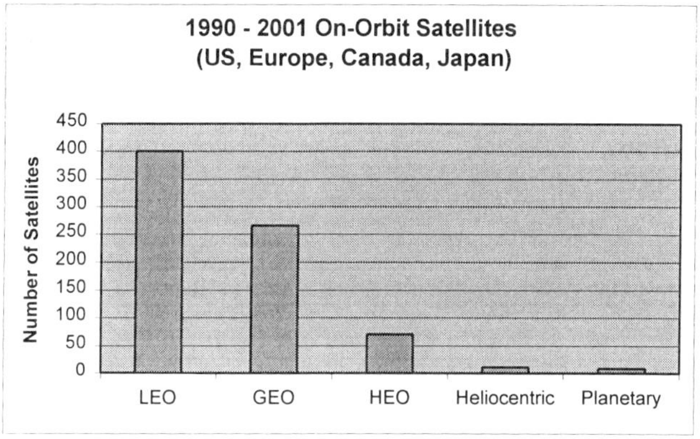

Figure 3 On-Orbit Satellites vs. Orbit.

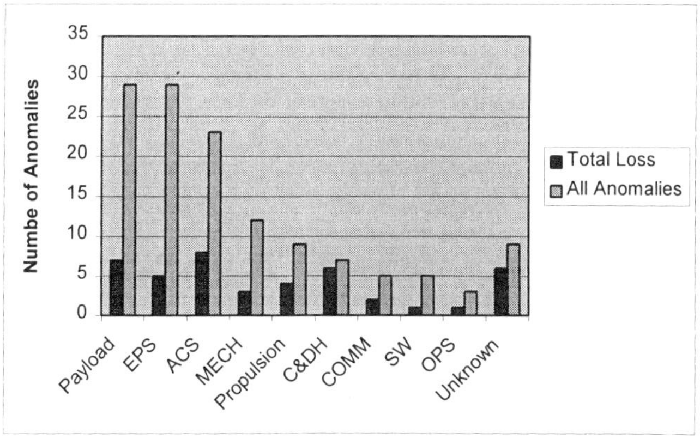

Figure 4 Subsystem Anomalies.

Figure 4 illustrates on-orbit satellite anomalies recorded for satellites launched from 1990 through 2001 relative to the distribution by satellite subsystem. All anomalies are shown, as well as those that resulted in total mission loss. It can be seen that payload, EPS and ACS have a large contribution to reported anomalies. A total of 35 GN&C (ACS, propulsion, and ground operations and software involving flight dynamics) anomalies were reported during the time period investigated, which represents 29% of all anomalies recorded. The GN&C contribution to anomalies that result in total loss is higher, with 13 GN&C anomalies reported representing 37% of all anomalies resulting in total loss.

Table 3 lists those GN&C anomalies recorded for satellites launched from 1990 through 2001.

Table 3 GN&C Anomalies

Satellite	Launch Date	Mishap Date	Impact	Cause
Anik E2	4/5/1991	1/20/1994	Mission Interruption	Magnetic storm destroyed ACS
Aurora 2 (Satcom C5)	5/29/1991	6/1991	Shortened Life	Motor fault
Clementine	1/25/1994	5/1/1994	Partial Loss	Software error caused spin up and loss of fuel
Deep Space 1	10/24/1998	7/1999	Partial Loss	Target tracking problem due to software
Early Bird	12/24/1997	12/28/97	Total Loss	GPS unit shorted to bus draining batteries
Echostar 5	9/23/1999	7/1/2001	Mission Interruption	One of three momentum wheels fails
FUSE	6/1/1999	12/1/2001	Mission Interruption	Second of four reaction wheels fails
Galaxy 4	6/25/1993	5/19/98	Total Loss	Catastrophic attitude control failure due to SCP malfunctions
Galaxy 8i	12/8/1997	9/1/2000	Shortened Life	Three of four xenon ion thrusters fail.
GFO 1	2/10/1998	3/1998	Mission Interruption	GPS receivers fail to maintain nav state; ground-based workaround implemented
Goes 9	5/23/1995	7/7/1998	Total Loss	Taken out of service due to noisy pointing caused by lubrication starvation of momentum wheels.
GPS BII-07	3/26/1990	5/21/1996	Total Loss	3-Axis stabilization failure due to a second reaction wheel failure
Hotbird 2	11/21/1996	12/31/1996	Shortened Life	Fuel tank leak; Apogee transfer anomaly
HST	4/1/1990	11/1/1999	Mission Interruption	Fourth of six gyros fails
IMAGE	3/25/2000	3/25/2000	Mission Interruption	Nutation damper liquid immobilized by surface tension
Intelsat 801	3/1/1997	3/1997	Mission Interruption	Ground command error caused uncontrollable spin
Iridium	6/18/1997	9/1/1997	Total Loss?	Attitude control and propulsion system failure
Iridium	12/8/1997	7/17/1998	Total Loss?	Attitude control and propulsion system failure
Iridium	6/18/1997	11/2/2000	Total Loss?	Failure in orbit – fuel depletion
Iridium 5	5/5/1997	5/5/1997	Mission Interruption	Faulty wheel electronics.
Iridium 11	6/18/1997	6/18/1997	Mission Interruption	Faulty wheel electronics.
Iridium 27	9/14/1997	9/14/97	Total Loss	Thruster anomaly depleted operational fuel
Iridium 42	12/8/1997	12/8/1997	Mission Interruption	Wheel tachometer failure
Landsat 6	10/5/1993	10/5/1993	Total Loss	Satellite exploded when propulsion system pyrovalve was fired, igniting adjacent mixture.
Lewis	8/23/1997	8/26/1997	Total Loss	Design error in ACS; failure to monitor spacecraft during initial operations
Mars Climate Orbiter	12/11/1998	9/23/1999	Total Loss	Failure to use metric units in ground software trajectory models
Mars Observer	9/1/1992	8/1/1993	Total Loss	Probably due to Propulsion System rupture or power short, induced by oxidizer leaking past check valves.
NEAR	2/17/1996	12/1998	Mission Interruption	Main engine fuel burn malfunction due to on-board software limits being exceeded
Nozomi	7/3/1998	12/20/1998	Mission Interruption	Consumed more fuel than expected during Earth swingby due to thruster valve stuck partially open.
Solar A	8/30/1991	12/15/2001	Mission Interruption	Safe mode during solar eclipse, unexpected spin, loss of control
STEP 0	3/13/1994	7/19/1994	Mission Interruption	IMU (gyro) fails
STEP 2	5/19/1994	5/19/1994	Performance Loss	Noisy earth sensor affects pointing accuracy
Telstar 402	9/9/1994	9/9/1994	Total Loss	Propulsion System pyrovalve firing caused explosion
Terriers	5/18/1999	5/18/1999	Total Loss	ACS polarity error controlling magnetic torquer coil
TOMS-EP	7/2/1996	7/2/1996	Mission Interruption	Coarse Sun Sensors miswired; magnetic torque rod polarity error

The time of occurrence of satellite anomalies within their design life was investigated. Figure 5 illustrates when satellites were observed to fail for all anomalies, and those involving GN&C. In this figure, an anomaly was counted only if the satellite completed its design life by the end of 2002, or would have if the anomaly did not occur. It can be seen that the majority of anomalies (51% of all, 50% of all GN&C) occur within 10% of the mission design life.

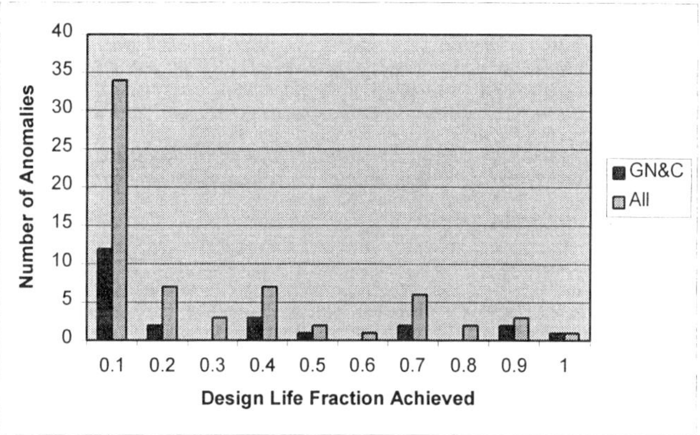

Figure 5 Anomaly Occurrence in Satellite Design Life.

Figure 6 shows the time relationship of cumulative anomalies as a function of mission day, for both all anomalies and GN&C anomalies. It is interesting to note that the slope is not equal to 1, which would be expected if satellites exhibit constant anomaly failure rates. It can be seen that a large number of anomalies are observed on the first day of the mission. After reaching 10% of a satellite design life, the anomaly failure rate declines precipitously and continues to decline thereafter.

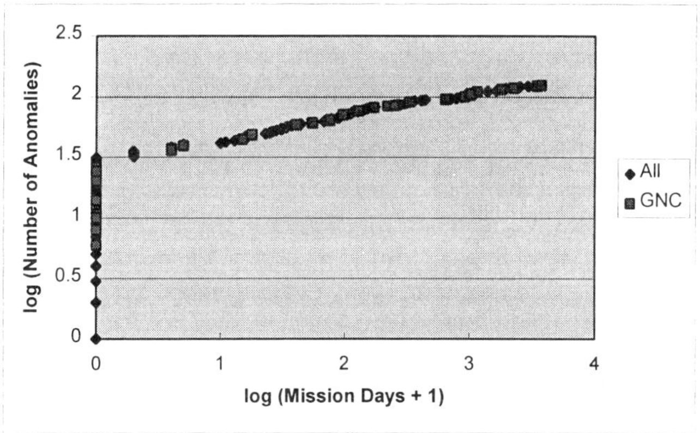

Figure 6 Cumulative Anomalies vs. Mission day.

Mishap rates as a function of launch year, mission category and orbit category were investigated. Mishap rate was defined as the number of anomalies per satellites launched. Mishap rates shown in the following figures are for all anomalies recorded, using the total mission population whether or not the mission has completed its design life. Figure 7 illustrates the mishap rate distributed by year launched. The mishap rate over 1990 through 2001 appears to be fairly constant, especially when considering that the satellites launched later have not yet achieved their design life, making their status as success or failure as yet undetermined. The mishap rate due to GN&C anomalies is fairly constant as well, although it appears that GN&C anomalies have fallen recently, with few GN&C anomalies reported for satellites launched in 2000 or 2001. Although mishap rates observed are fairly constant, the number of reported anomalies has increased dramatically since 1997 due to the large number of satellites launched in 1997 through 2000.

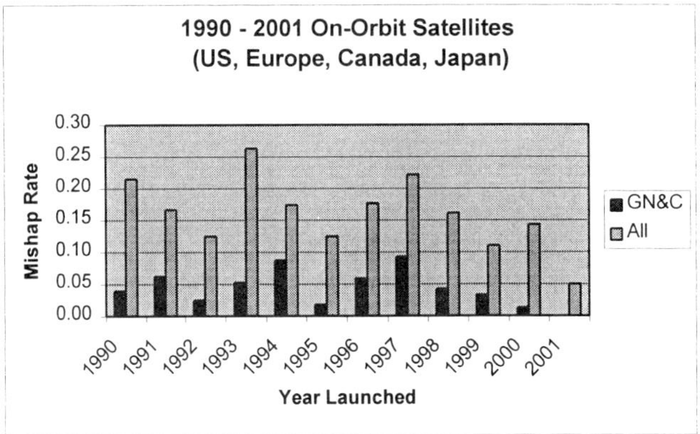

Figure 7 Mishap Rate vs. Year Launched.

Figure 8 shows the mishap rate distributed by mission type. It can be seen that a large variation in mishap rate exists for various mission types. In general, mission types with a large number satellites (communications and military) have a lower mishap rate than missions types with few satellites launched (deep space), which is as expected. Deep space missions have very high mishap rates when compared to other mission types, and the GN&C contribution to these anomalies is very high. The high mishap rate may be explained by the inherent complexity of deep space missions. Deep space missions require high levels of autonomy and have limited communication time with the ground. In addition, the deep space environment is not as well characterized as for other mission types. The deep space thermal environment is extreme, and solar electric power generation may be problematic. Military missions have slightly lower mishap rates than expected when comparing to other mission types; this can probably be explained by under-reporting of anomalies for military missions. Technology missions are seen to have a low GN&C mishap rate when compared to their total mishap rate; this may be explained by the fact that technology missions by their nature have higher mishap rates due to their usually complex technology payload rather than due to any satellite bus subsystem.

Mishap rates distributed by orbit also show variations depending upon the orbit category. Figure 9 shows that GN&C anomalies make up a rather large percentage (roughly one third) of LEO anomalies compared with GEO anomalies; this can be attributed in part to the large number of EPS problems that GEO communication satellites have exhibited recently. The average power level of GEO communication satellites launched from 1996 through 2001 has tripled[1], which has resulted in an increase in EPS technical complexity. It can also be seen that the GN&C contribution to anomalies in heliocentric and planetary orbits is very large; this may be partly explained by the fact that heliocentric and planetary missions cannot rely on magnetic control and require a propulsion system, whereas this is not the case for LEO missions. The large percentage of GN&C anomalies can be also attributed to the difficulty of getting to heliocentric and planetary orbits. Missions in heliocentric and planetary orbits have the highest observed mishap rate.

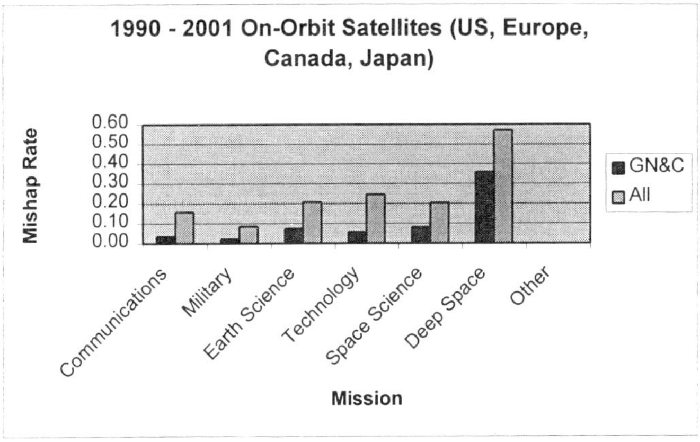

Figure 8 Mishap Rate vs. Mission Type.

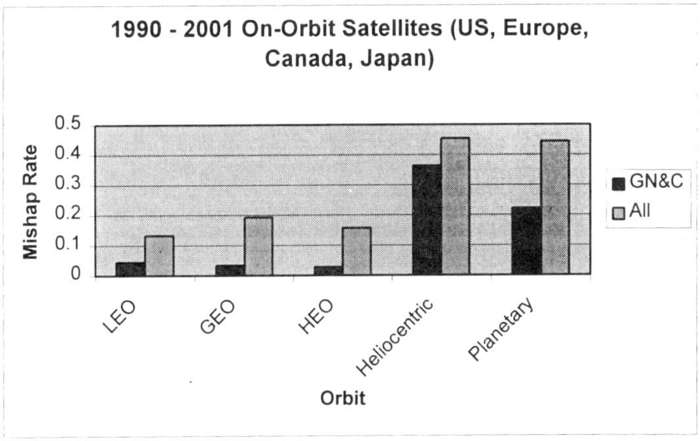

Figure 9 Mishap Rate vs. Orbit.

Figure 10 shows the GN&C failure categories observed. It can be seen that the largest number of GN&C anomalies can be attributed to hardware problems. Design, software and operations problems are the next major cause categories for GN&C anomalies, although more design problems result in total loss. All anomalies categorized as verification were ACS polarity errors.

Figure 11 shows GN&C anomalies categorized by equipment type. Not all GN&C anomalies have an obvious associated equipment type (e.g., an ACS polarity problem with the magnetic system was not given an equipment category). It can be seen that the largest number of GN&C anomalies are attributed to wheel failures. Pyrovalve induced anomalies have been a problem for both propulsion system applications as well as payload applications, and tend to result in total loss of mission. Pyrovalves by themselves are reliable, but in these cases the adjacent systems have not been able to withstand the mechanical or electrical shock generated by the pyrovalve. A number of anomalies have been caused by faulty processor operations. Faulty thrusters have caused a number of anomalies. A number of gyro failures have occurred, but many of them have occurred after the satellite design life was reached so were not counted in this analysis. A number of missions have recently developed gyro-less attitude determination software, so gyro failures now tend to result in mission interruption or performance loss instead of total failure. It is interesting to note that no anomaly records were found involving star trackers; it appears that most star tracker anomalies can also be solved with software uploads.

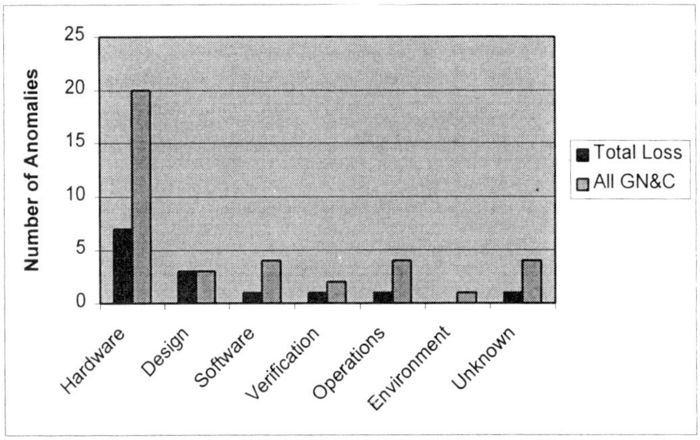

Figure 10 GN&C Anomalies vs. Cause Category.

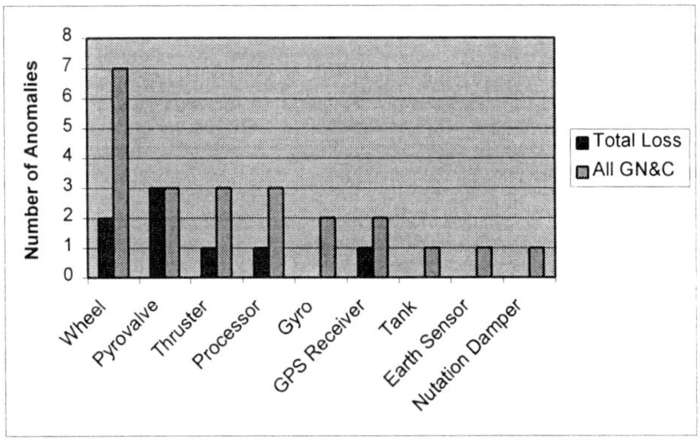

Figure 11 GN&C Anomalies vs. Equipment Type.

CONCLUSION

A study of past on-orbit anomalies was undertaken to assess how future satellite program resources might be best spent to ensure mission success. Spacecraft anomaly trends were surveyed over the last decade, with the hope of learning ways to improve the process of GN&C system development, to reduce the failure rate of future missions. One conclusion that was apparent during the data survey was that industry-wide data is not shared on a routine basis. It is difficult to learn from history if anomaly records are kept out of the public domain.

As expected, most anomalies were observed to occur early in the mission. This indicates that design flaws and latent manufacturing defects have a greater effect on mission success than materials contamination or fatigue/overstress. The standard spacecraft integration and test process already invests significant effort to expose design flaws and physical defects before launch. The fact that some mission critical faults get through shows that this effort is not excessive. Is more testing needed? Not necessarily; other studies have drawn a correlation between parts failures and stress due to excessive testing. It is recommended, however, that testing be done in as flight-like a configuration as possible (e.g. flight harnesses for polarity tests), and that all test results be understood. This may seem obvious, but history shows that test anomalies have been overlooked, only to be found in review of test data after an on-orbit failure.

An exception to the trend of anomalies occurring early in the mission is wheel anomalies. Wheel anomalies tend to occur later in the mission, which suggests that mechanical wear-out is an increasingly significant factor. Since wheel anomalies are the most common GN&C failures, the most significant reduction in GN&C failures may be realized by improving wheel reliability over time. The satellite industry may benefit from investing in more robust life test programs for wheels.

Aside from wheels, the most troublesome GN&C components appear to be pyrovalves. While the pyrovalves themselves are reliable, they are apparently easy to misuse, leading to catastrophic damage of other components. This is counted as a design flaw, not of the pyrovalve but of the system that employs it. The mechanical and electrical interactions of the pyrovalves with surrounding systems must be thoroughly understood.

Many recent anomalies have been caused by EPS problems associated with solar array and battery anomalies. It is apparent that the increase in technical complexity and design lifetime for GEO communication satellites has increased the risk of failure in this subsystem.

Deep space missions and missions in planetary and heliocentric orbits have very high mishap rates when compared to other types, and the GN&C contribution to these anomalies is very high. The GN&C effort for these missions is inherently more technically complex due to the complex orbit trajectory requirements, the need for high levels of autonomy and their dependence on propulsion systems. A higher level of funding, testing and oversight of the GN&C effort for these missions is warranted.

ACKNOWLEDGMENT

The authors would like to acknowledge the help of Dr. Paul Cheng of The Aerospace Corporation, and Walt Thomas and Michael Sampson of NASA/GSFC, each of which contributed to the anomaly data used in the analysis.

REFERENCES

1. Futron Corporation, Satellite Insurance Rates On the Rise – Market Correction or Overreaction, http://www.spacedaily.com/news/futron-insure-2002.pdf, July 10, 2002.

2. Tag's Broadcasting Services Internet, *The Satellite Encyclopedia (TSE)*, http://www.TBS-satellite.com/tse/, V0.58.001, 39th edition - ISSN 1630-277X, August, 2001.

3. Satellite News Digest, http://www.sat-nd.com/.

4. NASA JPL, Mission and Spacecraft Library, http://msl.jpl.nasa.gov/home.html.

5. Encyclopedia Astronautica, http://www.fplib.ru/partners/mwade/spaceflt.htm

6. Airclaims Space Trak, http://www.airclaims.co.uk/products/spacetrak.htm.

7. Aerospace Corporation Space Operations Support Office, Space Systems Engineering Database (SSED), http://www.aero.org/sopso/index.html.

8. Space Launch Projects, *Mission Failure Analysis for NASA Ames Research Center*, Aerospace Report No. TOR-2002(2153)-1402, October, 2001.

INTERNATIONAL SPACE STATION CONTROL MOMENT GYRO FAILURE

Richard R. Burt*

The US Attitude Control system of the International Space Station includes four Control Moment Gyros (CMGs) that provide a micro-gravity environment for science payloads. On June 8th 2002 a CMG suffered a hard failure after approximately 1.5 yrs in operation. This paper presents an overview of the on-orbit configuration, operation, failure, and corresponding root cause analysis of an ISS CMG. A review of telemetry data from the units is presented with events leading up to the time of failure. A description of suspected modes of failure and corresponding analysis are also given.

INTRODUCTION

The International Space Station (ISS) is currently the largest man-made object to ever orbit the Earth and represents one of the greatest engineering and integration efforts the National Aeronautics and Space Administration (NASA) has ever undertaken. The Guidance, Navigation, and Control (GN&C) system is composed of both a US non-propulsive attitude control system and a Russian thruster attitude control system. Nominal operations are conducted under US control using its four CMGs, shown in Figure 1, with the Russian system providing momentum desaturation through thruster assists. When configured for completely non-propulsive control, the CMGs provide the ISS the micro-gravity environment that is required for science payloads.

The US GN&C system uses a two-axis Torque Equilibrium Attitude (TEA) seeker controller to minimize the amount of momentum required to maintain attitude control. Due to thermal issues with external components, several of the assembly stages require a biased attitude to minimize sun exposure to specific surface regions. Analysis of the momentum necessary to maintain the required attitude envelope has shown that the momentum of four CMGs is required for much of the assembly phase. On June 8th of 2002, one of the CMGs suffered a bearing failure and is no longer operational. The failure poses a major obstacle to the ISS program and GNC subsystem as we approach a critical yearlong stage of assembly of very asymmetric configurations.

This paper describes the configuration of the ISS CMGs, their operations, the on-orbit failure, and corresponding root cause analysis.

* The Boeing Company, Mail Code HM5-20, 13100 Space Center Blvd., Houston, Texas 77059.

CMG CONFIGURATION

The four CMGs operate as momentum storage devices that exchange momentum with the ISS through induced gyroscopic torques created by a motor-driven constant-speed momentum wheel mounted in two orthogonal gimbals. Both gimbals have torquer motors and position resolvers mounted on the rotational axis and move by use of a gear train system.

The momentum wheel is mounted inside the inner gimbal and is supported by bearings mounted on both sides of the spin axis. This configuration is shown in Figure 2a below.

Figure 1: CMGs 1-4 Mounted on Z1 Truss.

The Spin Motor is a low torque motor mounted on the spin axis with a control loop operating at 440Hz to maintain the nominal wheel-speed of 6600 rpm. The wheel speed resolver is mounted on the opposite side of the spin axis and provides digital wheel-speeds for downlink telemetry. The Spin Motor Commanded Current (SMCC) ranges from 0 to 1.6 Amps and nominally averages approximately 0.37 Amps. The SMCC varies from viscosity changes in the spin bearing lubricant due to thermal variations and high gimbal rates that induce radial loads on the spin bearings. The SMCC is used as a "barometer" of spin bearing health and status.

An accelerometer is mounted to the inner gimbal of each CMG to measure spin bearing vibrations in the radial direction. The accelerometer is used to track the unbalance of the momentum wheel during operations. Values seen on on-orbit since operations began have been between 0 to 0.04 G.

The Spin Bearings (SB), shown in Figures 2a and 2b, are single roller bearings with 15 steel balls mounted in a retainer that maintains consistent ball motion. The spin bearings use a petroleum lubricant in a active dynamic oiler system. Due to the large influence of temperature on lubricant viscosity, the spin bearings must maintain an 18°C to 50°C temperature range for operations. This is accomplished using pairs of small heaters mounted in the outer race of each bearing.

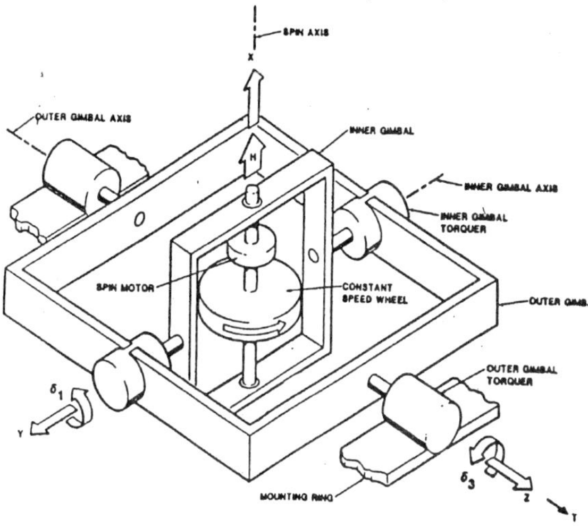

Figure 2a: CMG Gimbal and Wheel Assembly.

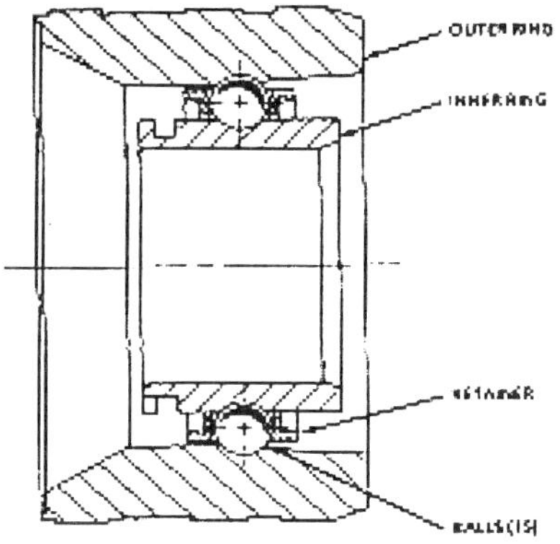

Figure 2b: CMG Spin Bearing Assembly.

OPERATIONS

The ISS currently maintains its attitude with respect to two reference frames due to thermal and power concerns. Local Vertical Local Horizontal (LVLH) is the primary attitude reference frame and will be the nominal attitude used after the ISS assembly is complete. When maintaining the ISS attitude with respect to the LVLH frame, the CMG gimbal motion remains within a 20° range with gimbal rates typically below 0.04 deg/sec. This reference frame keeps the aft face of the Z1 truss out of the sun for the majority of each orbit so the SB heaters cycle constantly to keep the SBs in the required temperature range. The CMG SMCC responds to viscosity changes from the heater cycles and varies in a range of 0.02 Amps.

X-axis Perpendicular to Orbital Plane (XPOP) is a quasi-inertial reference frame that is currently used at high Solar Beta angles ($|S\beta A| > \pm 37°$) to provide sufficient electrical power prior to the installation of additional joints that allow for any-attitude pointing of the solar arrays. This orientation exposes the aft face of the Z1 truss, where the CMGs are mounted, to direct sunlight for the majority of each orbit causing heating of the CMGs. As the magnitude of the SβA increases, the sunlit portion of the orbit increases, which in turn increases the temperatures of the CMGs. CMG heaters are not activated in this orientation; however, small cyclic variations in SMCC are noted at orbit period due to viscosity changes from the sun heating. Large variations in SMCC current can be seen when trended over the entire solar beta range and are shown in Figure 3 below. The CMG gimbals rotate at approximately orbit rate to maintain the same momentum state in this reference frame. Gimbal rates are similar to those seen during LVLH periods.

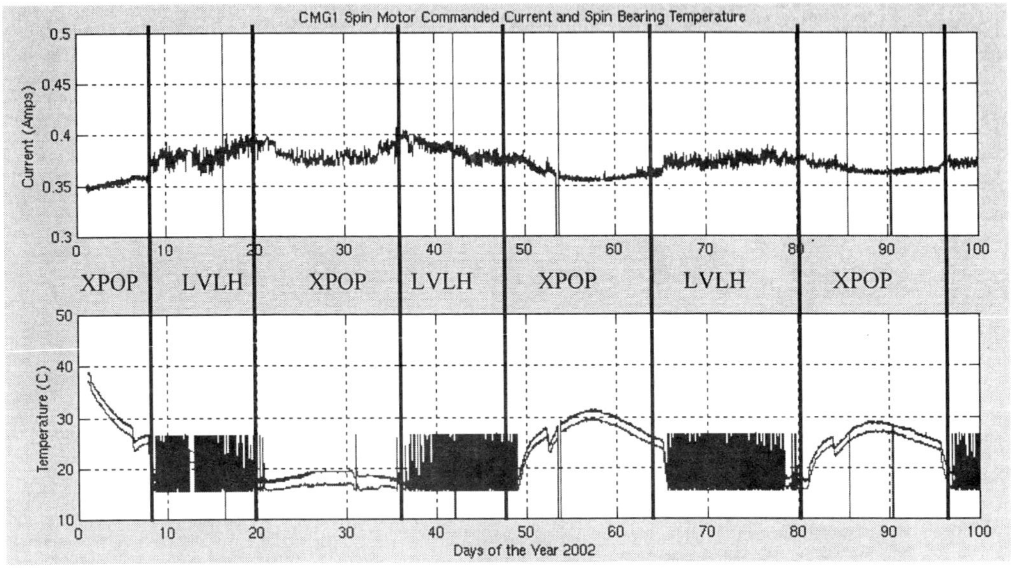

Figure 3: CMG Spin Bearings Response to Flight Attitudes.

Attitude Hold (AH) is a control mode that is used to maintain an attitude for the purpose of dockings and propulsive events. The AH controller has tight roll, pitch, and yaw attitude deadbands and it commands large torques from the CMGs to maintain a fixed attitude, which results in large CMG gimbal rates. Thruster desaturation firings can occur frequently, during which the CMGs return to a lower momentum state. These desaturation operations also result in high gimbal rates.

With the reduction of Shuttle flights to ISS, Russian thrusters are the primary means of conducting reboosts to higher altitudes. The increased use of ISS propellant for reboosts may lead to a future shortage, as a projected decline in the number of Progress refueling flights cannot replace the amount being used.

Prior to April 2002, all attitude reorientation maneuvers were conducted using Russian Thruster control. Attitude maneuver controllers have since been designed to allow the CMGs to maneuver the ISS in concert with Russian thruster desaturations firings, resulting in a 60% saving of maneuver propellant (mostly due to lower attitude maneuver rates). Due to the large torques required to reorient the ISS, the CMG gimbals are commanded at high rates.

ON-ORBIT FAILURE

At the end of May 2002, prior to the launch of shuttle mission UF-2; the ISS was in a high solar beta XPOP period. The ISS attitude was transitioned from XPOP to LVLH on June 7th, several hours prior to shuttle docking. [This docking was reported to have one of the highest velocity docking contact rates to date.] Several hours after docking, the flight control team noted a flag that indicated the accelerometer on CMG1 had measured 0.5 Gs. The CMG1 SMCC was elevated and one spin bearing temperature was beginning to rise. GN&C personnel determined that the CMG would be monitored but no immediate action would be taken. The event repeated itself twice during the next 16 hours. Then, following a 30-minute period with no ground communication, telemetry showed that the CMG1 SMCC was at its maximum value, 1.6 Amps, and the wheel-speed was decreasing. After several seconds of maximum current and wheel-speed reduction, Fault Detection and Isolation software declared the CMG failed and automatically shutdown the spin motor. Following motor shutdown, normal SB drag would result in the wheel speed decaying to 0 rpm in about 12 hours. During this event, the wheel stopped 72 minutes after spin motor shutdown. Data was retrieved from the on-board data recorders to fill in the missing telemetry for analysis purposes. The timeline of events is given with SMCC and SB temperatures shown in Figures 4, 5, and 6.

ROOT CAUSE ANALYSIS

A team of experts in the field of CMGs, roller bearings, and tribology was assembled to review the data and determine the root cause of the failure. A fault tree was developed and sectioned into different failure modes given below:

- Debris Impact
- Loss of Bearing Pre-load
- Loss of Lubrication
- Retainer Instability
- Thermal Clamping
- Loss of Ball Control/Ball Skid

The corresponding analysis for each failure scenario is provided in the following sections. Several separate scenarios were presented that dealt with design and manufacturing issues that were investigated by the vendor. None were found to be a likely cause of failure and are out of the scope of this paper.

Debris Impact

The ISS is impacted regularly by small micrometeoroids and orbital debris that result in damage to the outer surfaces. The ISS is outfitted with shielding to protect critical items and structures from such impacts. The CMGs shielding consists of a Whipple shield or dual walled aluminum shield covered by multi-layer insulation (MLI) to protect the momentum wheel and bearings from damage.

The GN&C team requested that the program task the crewmembers on one of the UF-2 extra-vehicular activities to inspect the area around the CMG for micrometeoroid/orbital debris impacts. The inspection was completed and revealed no perforations of the MLI covering the CMG. It was concluded from the ballistic limit of the CMG shielding that any impact with enough kinetic energy to perforate into the inner assembly would create a significant hole in the MLI and would be visible to the naked eye.

Analysis of the bearing design showed that the probability of debris entering the roller bearings from inside the inner gimbal cover was remote due to a cover on the spin bearing that limits the gap between the spin axis and the outer race of the bearings.

Figure 4: Failure Timeline GMT 158 1200 to 2400.

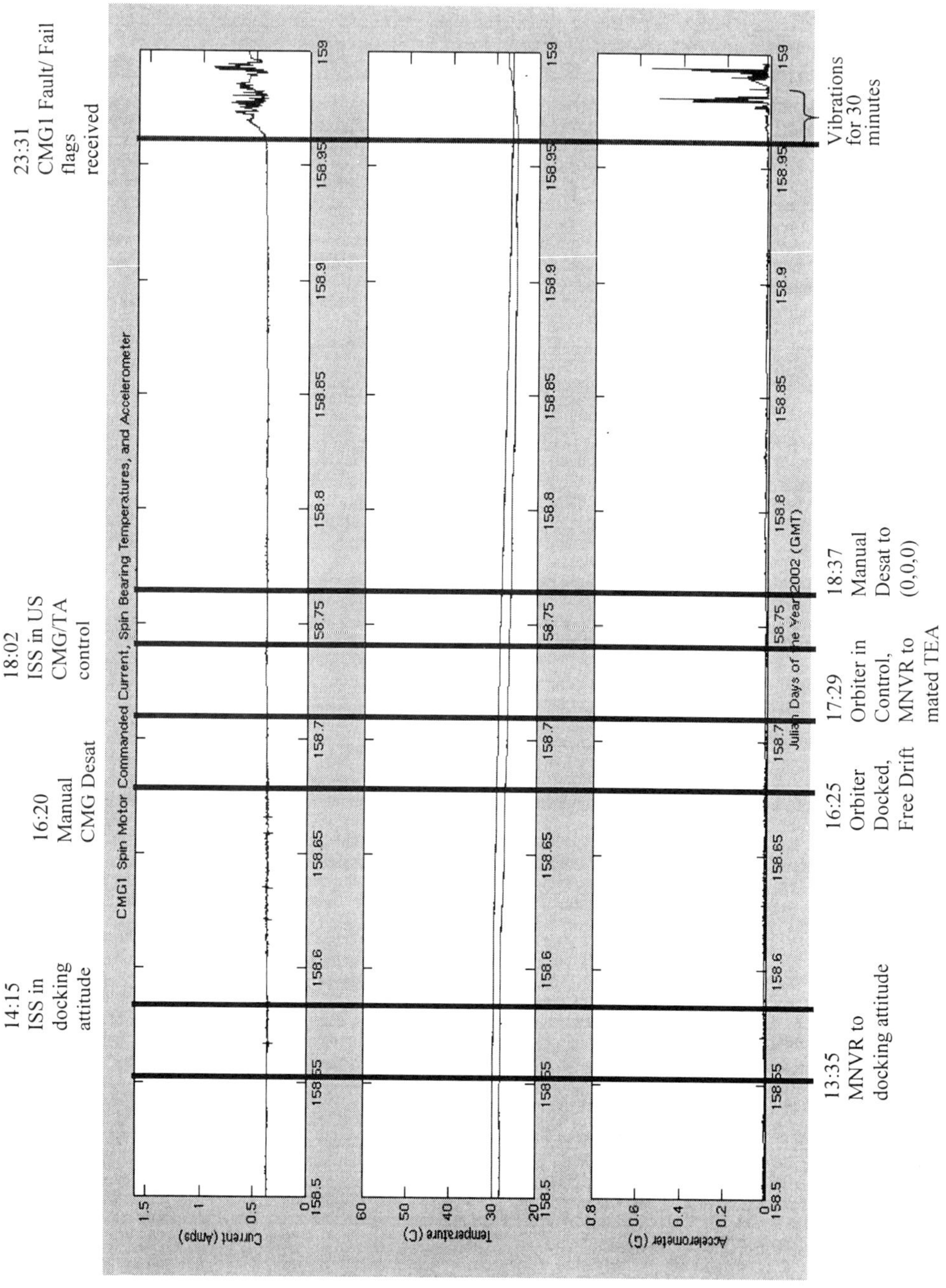

LEGEND: BLUE – ISS EVENTS RED – CMG1 FAILURES/ ANOMALIES
GREEN – CMG1 CLEARING FAULTS

Figure 5: Failure Timeline GMT 159 0000 to 1200.

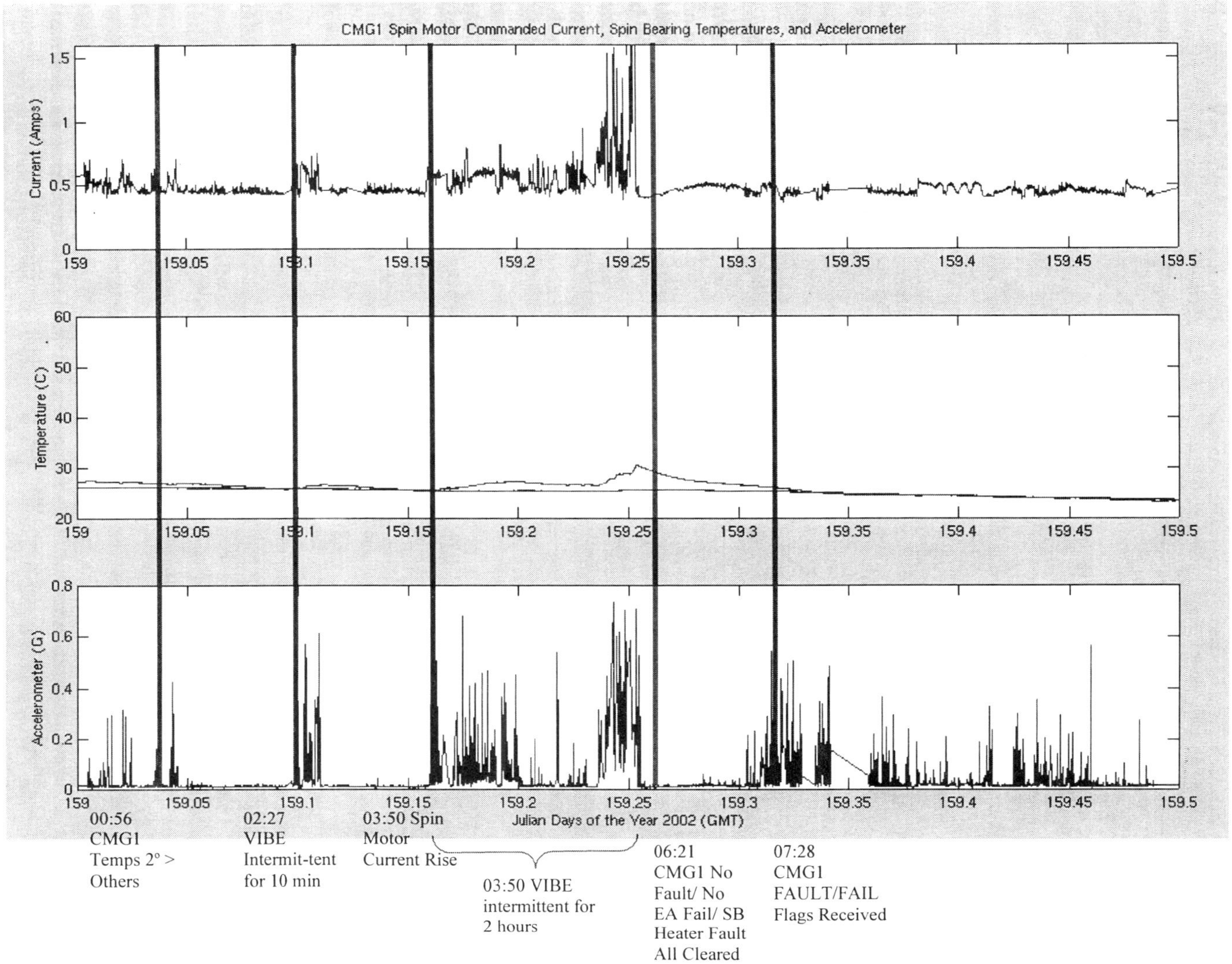

Figure 6: Failure Timeline GMT 159 1200 to 1912.

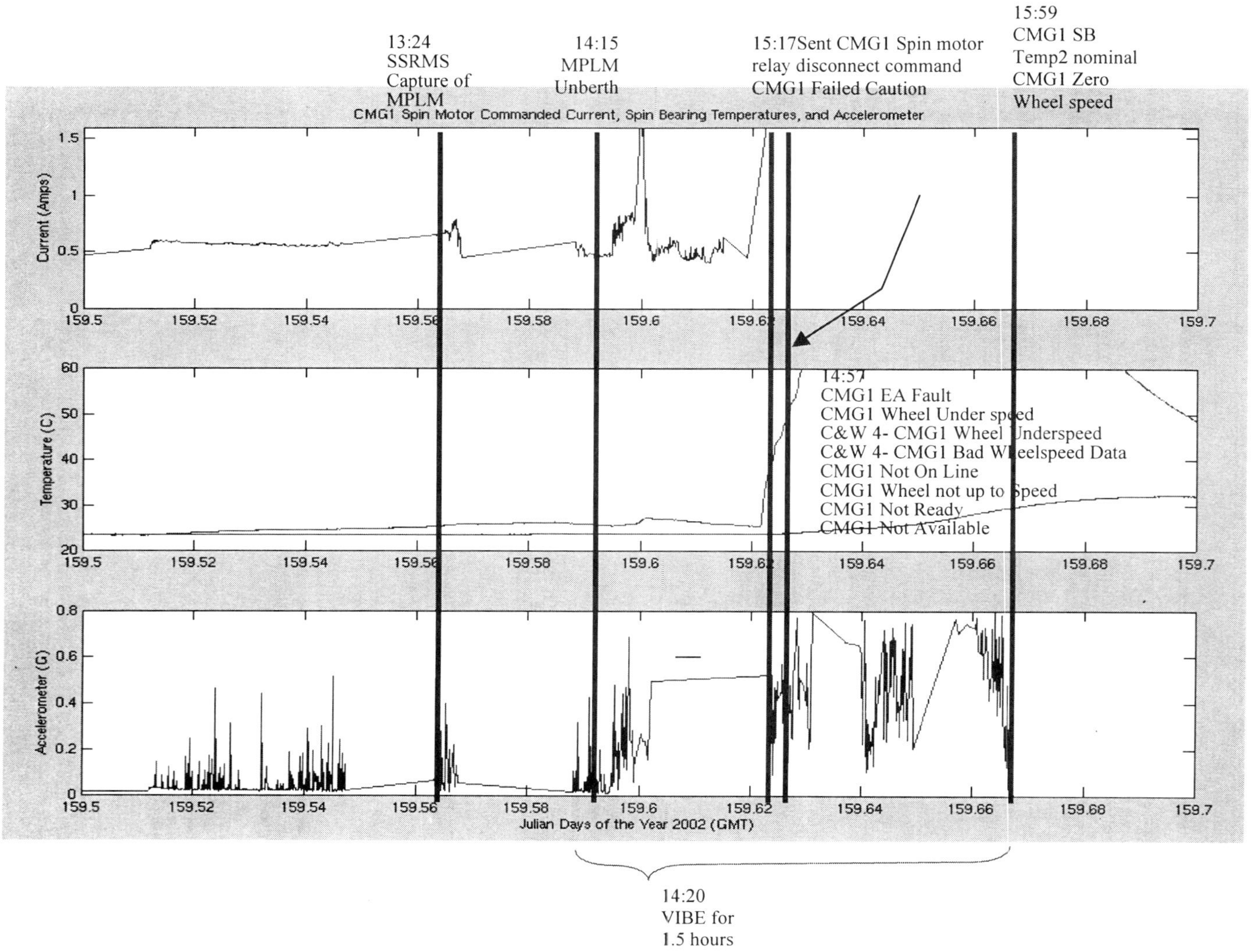

Loss of Bearing Pre-Load

Roller bearings are designed with a pre-load which loads the bearings in the spin axis direction to control the ball contact angle on the race. Pre-load is applied to the CMG spin bearings by use of a belleville washer that applies a spring load. Loss of pre-load would allow the inner and outer race to become unaligned resulting in a hard failure. Excessive loads in the spin axis direction or failure of the bearing belleville washer can cause loss of preload. The shuttle docking was examined to determine if the docking force was enough to cause a loss of pre-load and is shown in Figure 7. It was determined that such small loads incurred by docking would not be enough to cause a loss of pre-load.

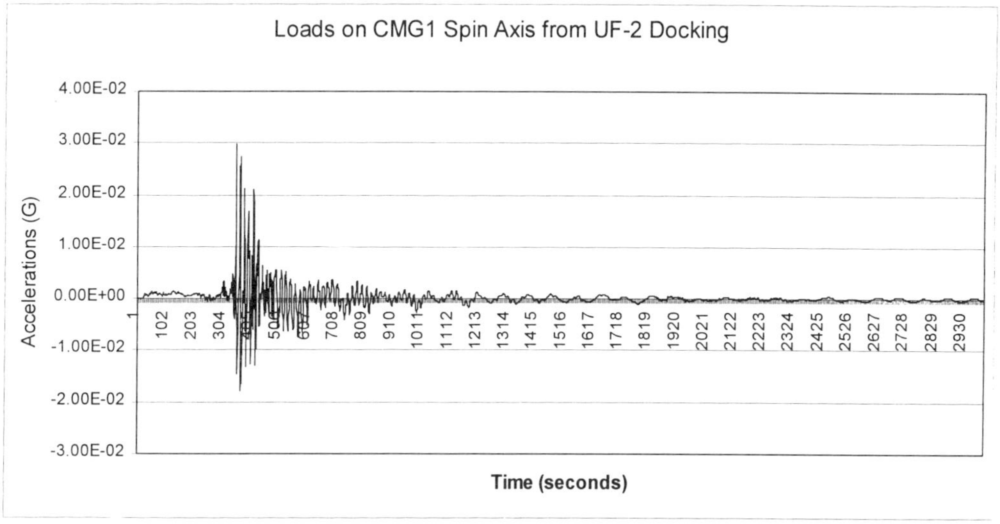

Figure 7: Loads incurred on CMG1 Spin Axis from UF-2 Docking.

A rise in steady state SMCC was seen on CMG2 after the transition from XPOP to LVLH and is shown in Figure 8. This is suspected to be indicative of a small change (~10%) in pre-load due to thermal contraction. Belleville washers are known for fretting, or creating small ridges on the surface that they contact over time. When the CMGs heat and cool, the belleville washer will expand and contract with respect to the surrounding materials. Over time this motion will wear small ridges in the contact surface. It is possible that bearing contraction will be limited as it encounters a small ridge. This would result in small changes in pre-load to the bearing until it is relieved. The drop in SMCC seen for CMG2 corresponded to high vibrations seen during the CMG1 failure providing a release. Such a small change in pre-load alone is however not thought to be a credible cause for failure.

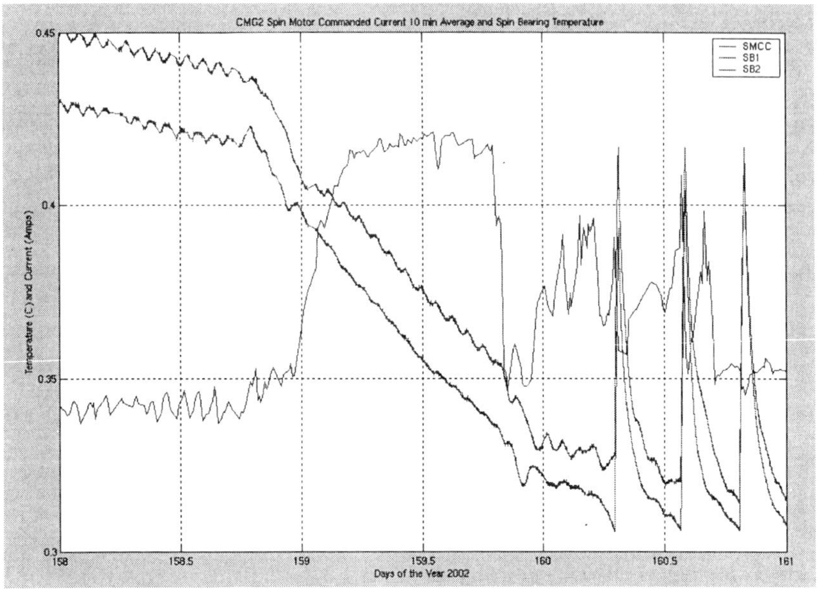

Figure 8: CMG2 Change in SMCC due to Spin Bearing Preload.

Lubrication Starvation/Dry Retainer Instability

The normal mode of spin bearing failure is due to a lack of adequate lube between the ball and the cage pockets, resulting in very high energy at very high frequency. Loss of lubrication has been seen in CMGs and Momentum Wheels during life testing on the ground and has known signatures. Initially, dry retainer instability will be evident from the drag torque and subsequently the SMCC increasing for short periods with an abrupt beginning and departure. Over long periods this will show a downward trend in steady state SMCC as lubrication is depleted. Dry retainer events will increase in frequency and length over this period. Once the lubrication is depleted, the steady state current will begin to rise as the bearing accumulates scarring from non-lubricated metal-on-metal contact, followed by bearing failure. This process could take weeks, to months, to years before a hard failure occurs.

Retainer or cage instability is a high frequency anomaly that results when the retainer contacts the race and initiates a cyclic motion similar to a resonance frequency, which persists until it is relieved by large bearing loads. Retainer instability is not a well-understood phenomenon but it has never been known to cause wheel failure when it first occurs. This is due to the characteristic of the instability; the balls change their path and pick up additional lube resident on the raceway just outside of the normal ball path. The progression to failure occurs because eventually there is no longer any extra lube available.

For this bearing design, the resonance motion would be at approximately 10 times the wheel rotational speed or 1100 Hz. From the telemetry available, the 0.1Hz-downlink frequency is not adequate to determine if this phenomenon was present. From the historic data reviewed, the small temporary rises in SMCC due to retainer instability was never seen on CMG1. Figure 9 shows the SMCC for 2002 before the failure. For the period leading up to failure, the small cyclic variations in SMCC are still apparent in CMG1 as shown in Figure 10. This indicates that at least one of the CMG1 spin bearings was lubricated prior to failure. With the data available, it is impossible to determine if both were properly lubricated.

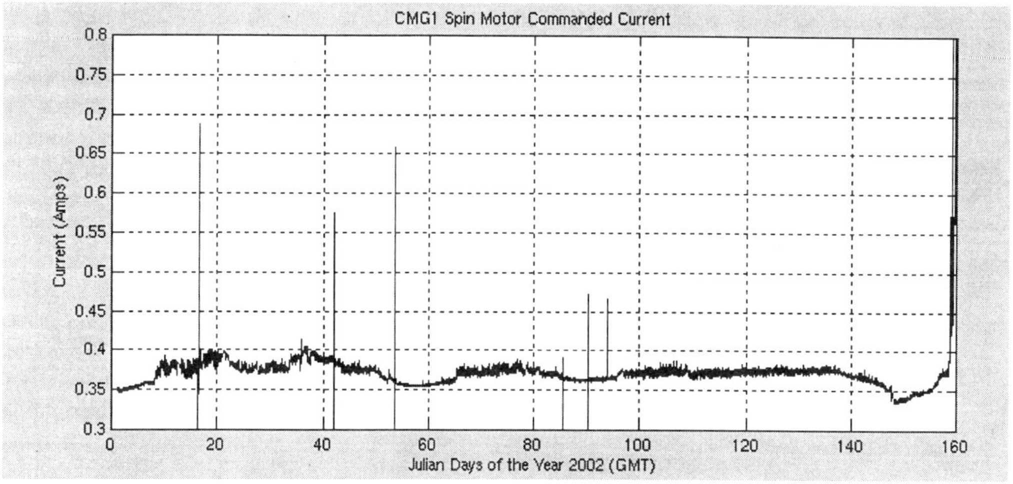

Figure 9: CMG1 SMCC for 2002.

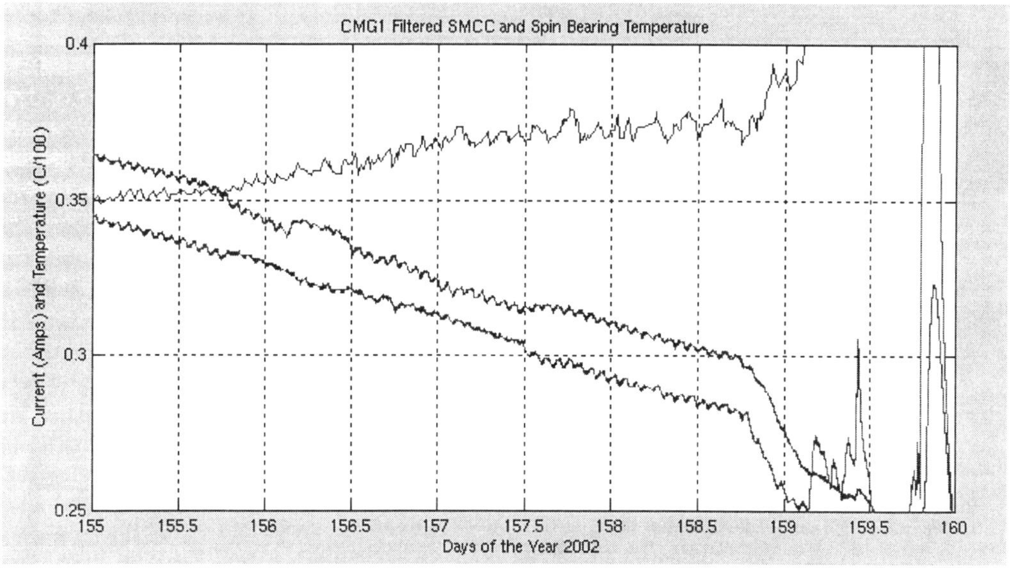

Figure 10: CMG1 SMCC Variation Due to Orbital Thermal Cycle (SMCC-Blue, SB1 Temp-Green, SB2 Temp-Red)

Thermal Clamping

During initial CMG design for Space Station Freedom, LVLH was designated as the primary flight attitude and the orbital inclination was planned to be 28.5°. The CMGs were designed to operate primarily in that thermal environment, which precluded full sun exposure across an entire orbit. The SB design was conducted with the assumption that the thermal power distribution between the inner and outer races was 70%/30%. During LVLH flight periods, the heater cycles the outer race temperatures with little thermal transfer to the inner race. This creates a temperature differential within the bearing as the outer race is warmer and inner race is cooler, causing the gap between the bearings to increase slightly. However, in the XPOP flight attitude, the CMGs are heated evenly and both races maintain approximately the same temperature. As the ISS transitions from XPOP to LVLH, the temperatures drop dramatically as the assembly sheds thermal energy. The outer races conduct thermal energy to the inner and outer gimbals with large surface areas to radiate heat. The inner race transfers the heat down the spin axis into the momentum wheel that acts as heat sink. It has been shown that with a large enough temperature differential, the outer race could clamp down on the balls and cause the bearings to seize. As the bearings rub on the outer race, heating from friction will increase the temperature of the outer race and it will expand. With the initial damage to the bearings, failure would occur quickly after the first instance.

Further analysis determined that the assumption of a 70/30-power distribution was found to be reasonable, and on-orbit conditions would not lead to especially large race temperature differentials. Therefore, the ball clearance between the races was adequate for the environmental thermal variations that were occurring.

Ball Control/Ball Skidding

Roller bearings are designed with bearings being controlled by either the inner or outer race. This control is a function of the rotational speed of the bearing and contact angle of the balls which is a result of the pre-load force. The ISS CMGs were designed with inner race control. It was shown through analysis that during periods of high gimbaling or radial loading, one or more balls in the bearing could become unstable and be controlled by either race.

Ball skidding is the phenomena when one or more balls will ride on the opposite race from the others. This results in the ball attempting to rotate at a different speed, which causes it to slide along the race and add stress to the retainer. If properly lubricated, ball skid will not cause damage. If the ball skid occurs on a dry surface, it will cause smearing which will degrade the surface and lead to premature failure.

Ball skidding is a low frequency event over several seconds and would be plainly visible in downlink telemetry. Figure 11 shows a period of CMG2 SMCC in which ball skid is believed to be occurring. This is also the period of a possible change in bearing pre-load due to thermal contraction for CMG2.

A leading theory of CMG1 failure is that the combination of a change in pre-load on the bearing from thermal contraction and induced ball skidding from high gimbal rates was produced during the same period on CMG1 spin bearing #2. However, due to either less lubrication or more extensive ball skid, the event caused catastrophic failure. This theory is still under investigation and has not been confirmed or dismissed.

CONCLUSIONS

Analysis work is still on going as several distinct failure methods have neither been substantiated nor rejected. However, the knowledge gained during the root cause analysis will benefit the program during both the post-mortem analysis of the failed CMG in 2003, and during future CMG anomalies.

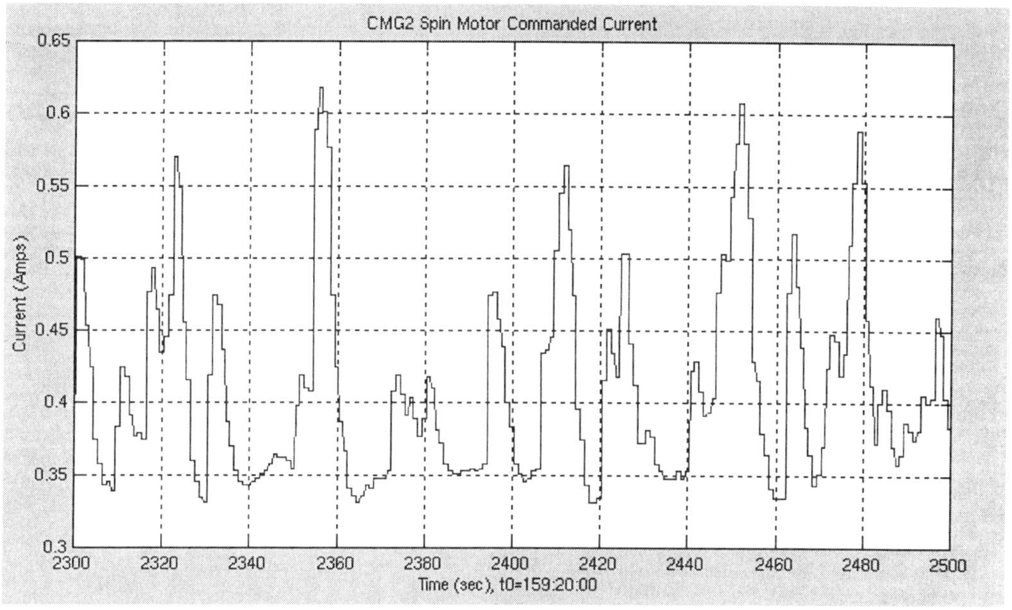

Figure 11: CMG2 Ball Skid.

Micrometeoroid and Orbital Debris impacts were dismissed as a cause of failure after the visual inspection. The root cause team has identified possible small changes to bearing pre-load and ISS flight operations have implemented mitigation procedures to relieve future events. On-orbit data has shown that loss of lubrication trends were not apparent prior to the CMG failure and are not considered being a primary cause of failure. Retainer instability signatures were not seen prior to failure and have never been known to cause a hard failure on the first occurrence; therefore it is also not considered a primary cause of failure. Through further analysis, thermal environment changes have been determined not to cause a significant bearing temperature differential and are not considered a primary cause of failure. Ball skidding is believed to be occurring due to bearing operations and the ISS has incorporated limitation on the amount of time spent in modes that produce frequent periods of high gimbal rates. One operational constraint is a prohibition on the performance of large maneuvers with CMGs.

CMG1 is scheduled to return to Earth in March 2003 after replacement with the flight spare. A detailed teardown and analysis will follow which should provide additional data. Each failure scenario will be compared against the new data and a detailed picture of the failure should become apparent.

ACKNOWLEGEMENTS

The author would like to thank the dedicated team at L-3 Communications in Teterboro, NJ including Charles Gurrisi, Dennis Francia, Raymond Seidel, and Mark Glowacki. The work of specialist is the fields of tribology, CMGs, and failure analysis: Dr. Alan Leveille, Jack Murphy of Boeing Satellite Services, and Stuart Randolph of GB Tech for their contributions. Thanks to John Harduvel and Michael Begley of the Boeing Company for their guidance and assistance. Special thanks to Rick Loffi, Kelly Murdock, and the NASA/Boeing GN&C team for providing a very significant portion of this entire root cause effort, including fault tree development, root cause analysis, tireless effort for months, and overall direction.

REFERENCES

1. Barrows, D., *Technical Description Document for the PG-1 Guidance, Navigation, and Control System*, MDC 95H0223, January 17, 1997.

2. Friesen, L., Whitney, J., and Christiansen, E., *Hypervelocity Impact Development Testing of Control Moment Gyro (CMG) Shielding: Part I – Close-out for Space Station Freedom CMG Impact Tests*, JSC 26925, April 1995.

3. Leveille, A., *CMG Root Cause Meeting at L3 July 30 –Aug 1*, August 5, 2002.

AAS 03-073

PERFORMANCE OF THE STAR TRACKER LIGHT SHADES ON THE EARTH OBSERVING SATELLITE (EOS) AQUA

Thomas Kenney,* Albert Lee,† Michael Donnelly,*
Michael Schroder‡ and Mark McNally**

The TRW-built EOS Aqua spacecraft uses two Ball Aerospace CT-602 star trackers to provide attitude updates to the 3-axis, zero momentum, and controller. Two months prior to the scheduled launch of Aqua, Ball Aerospace reported an error in the model of the star tracker light shades. The light shades, which had been designed specifically for the EOS Common spacecraft, were not expected to meet the stray light rejection requirements of the mission, thus impacting the overall spacecraft pointing performance.

What ensued was an effort to characterize the actual performance of the existing light shade design, determine what could be done within the physical envelope available, and incorporate changes to meet requirements. Changes were made based on this review activity and Aqua was successfully launched on May 4, 2002. The spacecraft is meeting all of its science pointing requirements. Reported here are the light shade design predictions, the solution trade space, and the measured on-orbit performance of these light shades.

INTRODUCTION

The EOS Aqua spacecraft is the 3rd major EOS Observatory, after Terra and Landsat-7. After successfully launching on May 4th, 2002, Aqua is continuing NASA's Earth Science Enterprise leadership in Earth science investigation. The observatory is comprised of six state-of-the-art instruments flying on a new spacecraft bus: AIRS – Atmospheric Infrared Sounder, AMSU-A1 & A2 – Advanced Microwave Sounding Unit A1 & A2, HSB – Humidity Sounder for Brazil, AMSR-E – Advanced Microwave Scanning Radiometer for EOS, MODIS – Moderate Resolution Imaging Spectroradiometer, and CERES – Clouds and the Earth's Radiant Energy System.

Originally scheduled to launch in December 2000, Aqua experienced and overcame numerous technical issues that confronted the Project office. The following technical issues were identified:

- Failures of similar flight hardware on other programs – 4th generation transponder failure on X-38
- Flight hardware glitches – unqualified battery cell crimps

* NASA Goddard Space Flight Center, Greenbelt, Maryland 20771.

† TRW.

‡ Lockheed Martin.

** Ball Aerospace, P.O. Box 1062, Boulder, Colorado 80306.

- Instrument I&T problems – operation, of HSB, with the purge line improperly connected, resulting in bearing lubrication issues
- Launch vehicle concerns – redesign of automatic destruct system (ADS) and fairing air (humidity and temperature) monitoring
- Late hardware deliveries due to paperwork issues
- Other

One of the last issues that arose was the stray light attenuation in the single-stage light shade for the Ball Aerospace CT-602 star trackers. During the launch vehicle Pre-Vehicle On-Stand (Pre-VOS) review February 19th, Aqua Project management was informed by TRW that the star tracker light shade model was in error. It was reported that the star tracker would not meet the specified performance with the sun 35 degrees off the bore sight. The construction of new light shades would require weeks. Aqua was, at the time, scheduled to ship to Vandenberg Air Force Base (VAFB) on the 25th of February and launch on the 18th of April. The immediate question confronting the Project was – whether to ship to VAFB in less than one week. The long-term question was whether the launch date was now in jeopardy.

On February 23rd the following options were presented to Aqua Project management:

- Implement a two-stage light shade with a 35 degree second stage
- Redesign the single-stage light shade
- Modify the existing single-stage light shade
- Implement flight software modifications

With the Project's driving requirement of not impacting the launch readiness date, several of these options were immediately discounted. Implementing a two-stage light shade with a 35 degree second stage angle was not possible. Aqua barely fit in the Delta II 10-foot composite fairing with the single-stage design. A two-stage design would require significant fairing modifications.

In the Project's opinion, redesigning the single-stage light shade would mean sacrificing the launch date as well. The solution that remained was modifying the existing light shade and/or implementing flight software changes to resolve the stray light attenuation problem and launch on time. As a side note, the second EOS Common Spacecraft, Aura, used the same light shade, so the Project could test and modify the Aura light shades without risking the Aqua shades. Having established that there was a viable option, the Aqua Project shipped the spacecraft from TRW's Space Park facility to VAFB. The spacecraft arrived early on the morning of the 25th of February – 52 days away from the scheduled launch.

As the stray light attenuation issue continued to be worked, other items were brought to the forefront: launch vehicle ADS modifications, new pre-MECO loads from the Delta II series 50 RS-27 main engine, and battery cell crimp integrity. As these items were addressed, key personnel continued to vigorously work the light shades. By March 20th, the light shade issue was sufficiently under control and it was dropped from the Project's list of 'major issues'.

DISCUSSION — THE ISSUE

The original star tracker specification called for full performance operation when the tracker bore sight was as close as 35 degrees to the sun line. A similar requirement called for operation as close as 25 degrees to the moon. The original stray light analysis (see "Original Analysis" in Figure 1) predicted that at 35 degrees, the sun would generate approximately 87 counts of background in the CCD image data, well within the limit to ensure that the 5 arc seconds noise equivalent angle (NEA) requirement would be met. The tracker software was designed to set the "Background High" flag, if the background exceeded 60 counts. Ball Aerospace intended this as a warning to the user that background was increasing. The data would be usable down to 35 degrees. However, TRW's flight software was designed to ignore a tracker's position data when this flag was set.

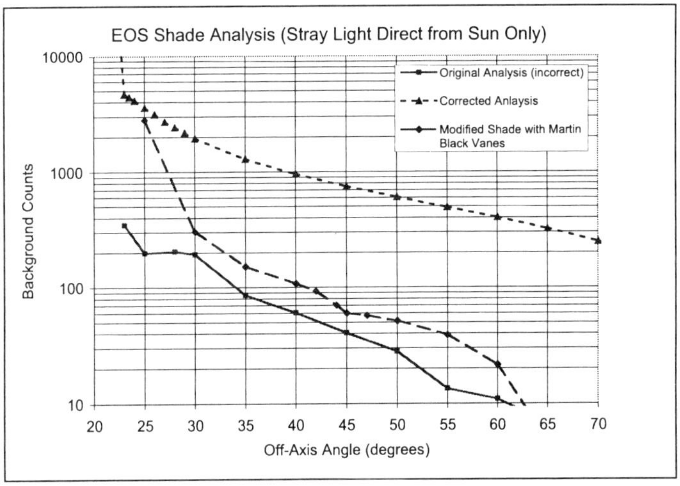

Figure 1: EOS Light Shade Stray Light Analysis.

On February 14th, 2002 Ball Aerospace notified TRW that there was an error in the design of the Aqua light shades and these performance requirements could not be met. The performance of the original design had been determined using the APART design tool from Breault Research Corporation (BRO). An error was discovered when Ball Aerospace compared results from the APART tool for models of the same light shade executed in the newer ASAP™ tool (also from BRO). Ball Aerospace later determined that a function "DELTA" had been applied incorrectly in their APART model. Ball then corrected the model for this error. The revised analysis indicated that the light shade would not meet the required sun performance (see "Corrected Analysis" in Figure 1).

Under normal two-tracker operation, a single tracker's data becoming temporarily invalid would not be a problem for Aqua, since only one tracker is required for full performance. Aqua's geometry is such that there is no chance that in normal flight the sun would interfere with both trackers. However, if one tracker should experience a hard failure, this vulnerability could contribute to periods of larger than expected attitude errors. Although Aqua uses gyro data for position information, periodic updates to the Kalman filter from the tracker are required to compensate for various sources of drift.

THE APPROACH

Given the short time frame to resolve the problem, all parties had to identify where margin existed in the various aspects of the mission so that all options could be considered in the solution trades. The challenge was to find an approach that could be implemented with no impact to the launch vehicle, and as little impact to the spacecraft as possible. Therefore, the trade space included the size of the light shade, the tracker performance requirements, the light shade's vane configuration, the materials used to make the light shade, and the star tracker software.

Project mechanical engineers examined the clearances between the STA light shade and the launch vehicle fairing. From this study they determined that the light shade could increase by at most 2.0 inches in length or 6.0 inches in diameter and maintain sufficient dynamic clearance from the fairing envelope. This restriction eliminated the ideal solution, a true 35 degree 2-stage design, but it did present the possibility of modifying the design to regain some performance.

TRW and GSFC revisited the orbit and spacecraft configuration studies to determine if there was any flexibility in the sun encroachment specification. The Aqua spacecraft is in the standard sun-synchronous polar orbit utilized by Earth resource satellites: 705 km, 98-degree inclination. The ascending node passes over the equator at approximately 1330 local time. The satellite body frame (SBF see Figure 2) is such that the +Z-axis (yaw) is along the local vertical in the nadir direction. The +X-axis (roll) is in the direction of flight. The +Y-axis (pitch) completes the right hand set and is normal to the side opposite the solar array. The two star trackers (STA) are mounted on the zenith side of the vehicle. The STA's boresight is its +Z-axis. The STA frame is defined as a 1-2-3 rotation (f,q, then y) from the SBF.

$$T_{SBF2STA} = \begin{bmatrix} \cos(\psi) & \sin(\psi) & 0 \\ -\sin(\psi) & \cos(\psi) & 0 \\ 0 & 0 & 1 \end{bmatrix} \cdot \begin{bmatrix} \cos(\theta & 0 & -\sin(\theta) \\ 0 & 1 & 0 \\ \sin(\theta) & 0 & \cos(\theta) \end{bmatrix} \cdot \begin{bmatrix} 1 & 0 & 0 \\ 0 & \cos(\phi) & \sin(\phi) \\ 0 & -\sin(\phi) & \cos(\phi) \end{bmatrix}$$

Where,

Table 1
SBF to STA Rotations

Axis	STA1	STA2
ϕ	62.763695°	62.763695°
θ	138.358854°	-138.358854°
ψ	-80°	80°

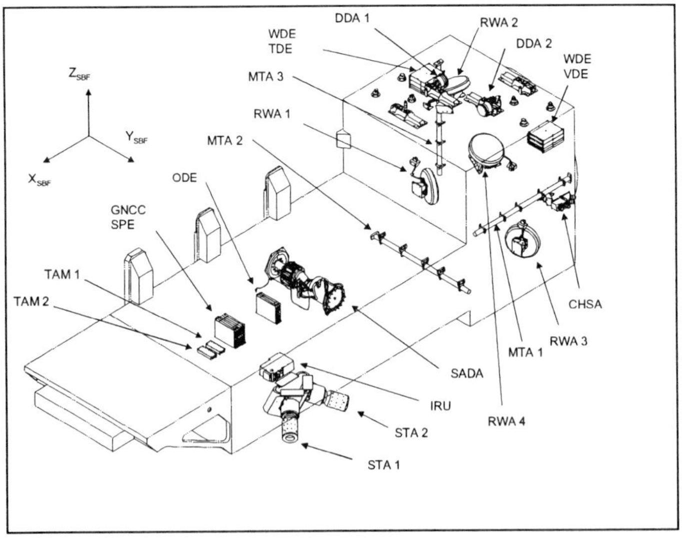

Figure 2: Aqua Attitude Control Component Layout.

As a result of this combined orbit and tracker geometry it was expected that there would be regular close approaches of the sun and moon. However, GSFC Flight Dynamics predicted that at the minimum beta angle, the sun would never be closer than 57 degrees to each STA's boresight (see Figure 3). This result suggested that there might be margin (up to 22 degrees) in the original 35 degrees encroachment specification.

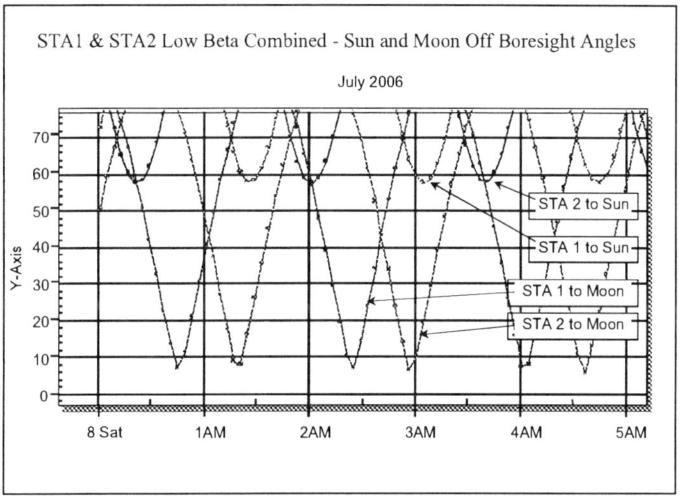

Figure 3: Predicted Sun and Moon Encroachments.

To determine whether the encroachment specification could be eased, TRW examined what spacecraft elements might intrude on the light shades fields of view. Figure 4 shows the impingements. The AMSR-E instrument antenna was responsible for the violation of STA-1's light shade FOV. This large rotating dish (40 RPM) would enter and exit the light shade FOV based on its rotation angle. The AIRS instrument's Earth shield intruded into the STA-2 light shade FOV.

Both of the instruments were covered in thermal blankets so they were both potential sources of reflected light. Based upon data provided by the prime contractor, Ball Aerospace estimated the contribution these surfaces would make to the overall stray light analysis, the results of which are shown in Figure 5. This Figure shows an approximately 30-count increase in background for the modified light shade at 45 degrees relative to the sun-only intrusion analysis of Figure 1.

At the spacecraft level, TRW's simulation of Aqua's science pointing mode predicted that the spacecraft would meet all specifications without a star tracker attitude update for at least 23 minutes. After this interval performance would gradually degrade due to gyro drift.

The STA set the "High Background" flag at 60 counts. The attitude determination software interprets this flag being set as indicating invalid star position data. Figure 6 shows the corrected NEA estimates as a function of sun angle and star intensity, where 160 counts of background occur with the sun at 35 degrees from the STA bore sight. TRW had found that the filters in the flight software made it relatively insensitive to NEA. Given the NEA specification remained below 5 arc seconds for stars as dim as 6.5 MI it was concluded that a higher background than the original 60 counts could be tolerated.

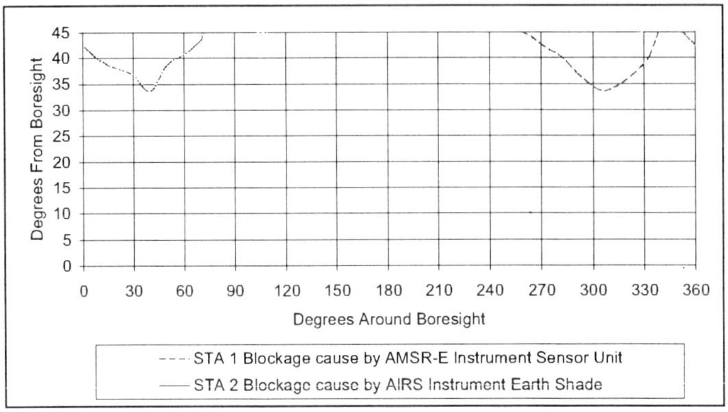

Figure 4: STA 1&2 FOV Blockage by Spacecraft.

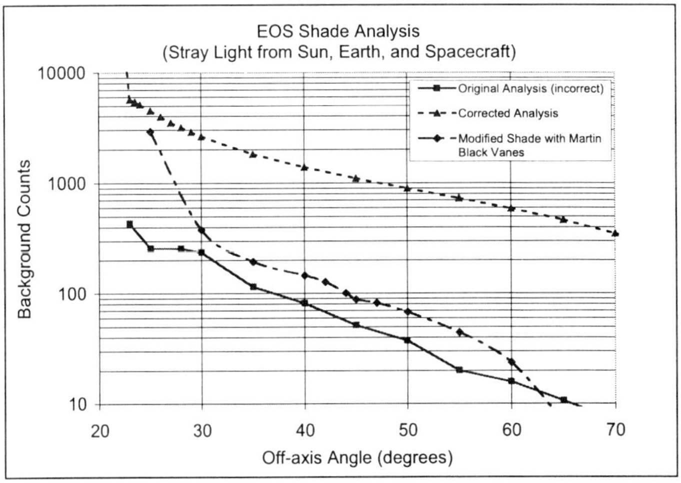

Figure 5: EOS Stray Light Analysis with Secondary Sources.

The original design had the entire light shade interior coated with an optical black paint. This coating was selected for its optical properties and its relatively rugged ground handling characteristics. Less reflective coatings were available, but they were more subject to ground handling damage. This led to an end of life performance model of 4% reflective to account for degradation and contamination. Martin Black, however, is less than 1% reflective at beginning of life and was modeled as 2% reflective at end of life.

THE SOLUTION

Based on these studies and the available trade space, it was obvious that no single change could provide the solution. In fact, the only possible approach was to make a number of modifications throughout the system. Relaxation of specifications, as well as hardware and software changes, were implemented to produce a system that could achieve the revised performance objectives.

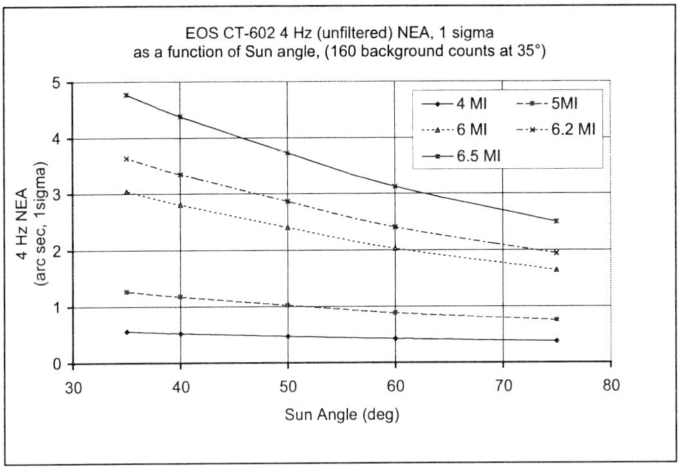

Figure 6: Predicted NEA Performance The Solution.

From the light shade FOV studies it was determined that the sun encroachment specification could be relaxed to 45 degrees. This was done knowing that as a result, more spacecraft structure would be within that angle from the bore sight of both light shades. However, no direct sun would be closer than 57 degrees. The estimated contribution to background from these secondary sources was in the range of 30 counts.

Ball Aerospace recommended that the best use of the available volume was to increase the length of the light shade and to add an additional baffle. This modification provided an incomplete 2nd stage by reducing direct illumination of the original single stage baffles at 45 degrees and establishing a cutoff angle of 66 degrees. The baffles, which had been coated with a black paint, were upgraded to Martin Black. Ball Aerospace updated the light shade model to reflect these changes. The results for the upgraded design are shown in Figure 5.

The tracker had been designed to set a flag in its output data to indicate when the background threshold of 60 counts was exceeded. The latest system analysis indicated that the NEA requirement could be met when the background was as high as 160 counts. To provide the option of increasing the tolerance of background light and thereby providing additional margin on orbit, a software patch was designed that would increase this threshold to 120 counts. This modification also necessitated a change in the trackers charge coupled device (CCD) readout scheme via another software patch. These patches were held in reserve for possible implementation after the satellite was on-orbit. The design goal remained to meet the original 60-count limit.

GROUND VERIFICATION

Ball Aerospace attempted to perform an end-to-end test of the STA. Ideally the test would have included a sun simulator, the flight light shade, and a flight star tracker. The test would have been performed in the flight configuration, the STA exposed to an artificial sun at various angles off the bore sight, and the resulting background at the CCD measured. At the time, the available sun-source simulator was a small fraction of the actual solar intensity. With this input there was a very large quantization error in scaling the results to a full solar intensity, which rendered the experiments inconclusive. Since then, a new facility has been constructed with a brighter simulated sun. The laboratory upgrades have resulted in recent flight data closely matching laboratory test results. These improvements will be helpful for future mission light shade validation prior to flight.

The effects of the software patches were verified via tests at Ball Aerospace. These tests confirmed that both patches would function as designed. The tests also demonstrated that the procedures to install, activate, and deactivate would function.

On March 26, 2002 Ball Aerospace personnel hand carried the modified light shades to VAFB for installation, resulting in no impact to the launch schedule.

FLIGHT RESULTS

Shortly after its launch in May, Aqua achieved stable fine point mode. Soon thereafter background data from both trackers were available. Figures 7 and 8 show typical data collected during one orbit on day 142 (May 22, 2002) for STA-1 and STA-2 respectively. The performance of STA-1 reveals the effects of the rotating AMSR-E antenna. In normal operation, the STA-1 light shade looks toward the sun through the rotating AMSR antenna as the spacecraft emerges from eclipse. The antenna moves in and out of the light shade FOV, obstructing the sun, and actually aids light shade performance by reducing background. Figure 8 shows what is probably more representative of the true light shade behavior. In this case the change in background over time is relatively smooth. Although the background counts for both the entry and exit paths are fairly similar, there is a distinct difference in the behavior prior to the sun entering the FOV and after it exits. Before the rapid rise the background is about 12 counts. After passage the background falls to about 2 counts. This difference is attributed to the effects of the AIRS Earth shield impingement. Since STA-2 performance is, in general, the worst case, the rest of this discussion will be limited to its performance.

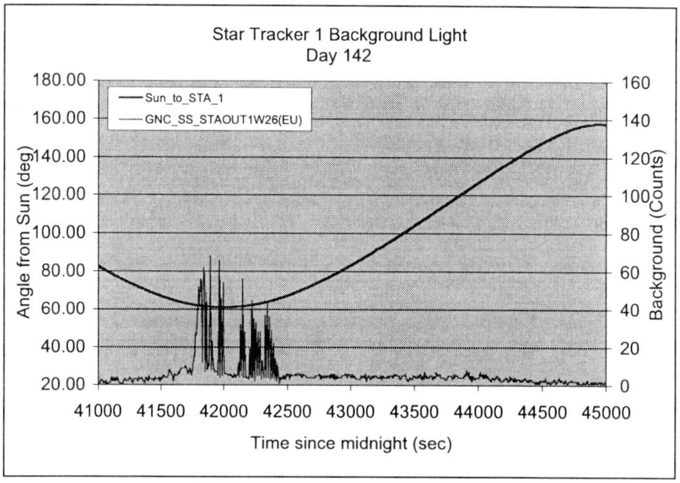

Figure 7: STA-1 Sun Approach Background Telemetry.

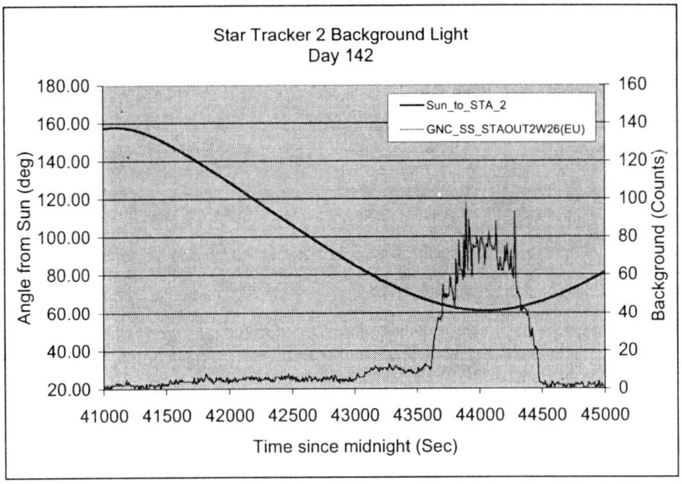

Figure 8: STA-2 Sun Approach Background Telemetry.

From Figure 8 it is clear that the light shade attenuation performance was not as predicted. In fact the 60-count limit predicted to occur at 45 degrees was being violated when the sun was 62 degrees from the bore sight. This resulted in the tracker data being ignored for roughly 470 seconds. The software patches had not yet been loaded. Both trackers were functional and this violation did not affect spacecraft performance. In fact, even if one tracker had failed the system could have tolerated an outage of this duration. Figure 9 shows the background counts as a function of sun angle on day 142 and a line representing the pre-launch predictions. This further illustrates that the predicted attenuation, as a function of sun angle, was off by approximately a factor of 4. Figure 10 shows similar results for day 144 with a slightly lower minimum sun angle. Both Figures show the effect of the sun line trajectory and impingements as a "hysteresis" in the entry and exit paths.

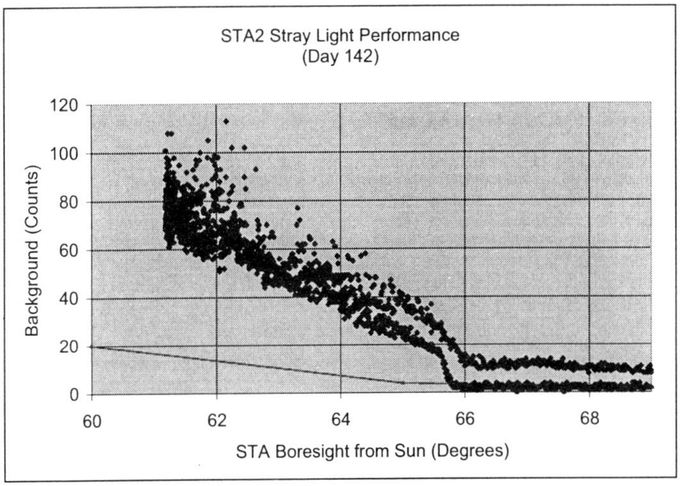

Figure 9: STA-2 Background vs. Sun Angle.

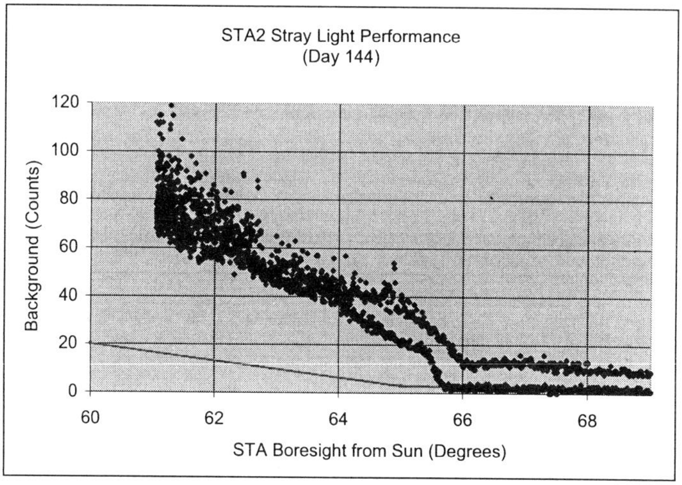

Figure 10: STA-2 Background vs. Sun Angle.

On May 24th (day 144) a series of moon intrusions occurred. In Figure 11, data that includes a lunar crossing is plotted. The outage due to the moon (background > 60 counts) is approximately 585 seconds. From Figure 12 it can be seen that the light shade is performing better than its lunar exclusion requirement. In fact performance is met down to 14 degrees from the moon, compared to the required 25 degrees.

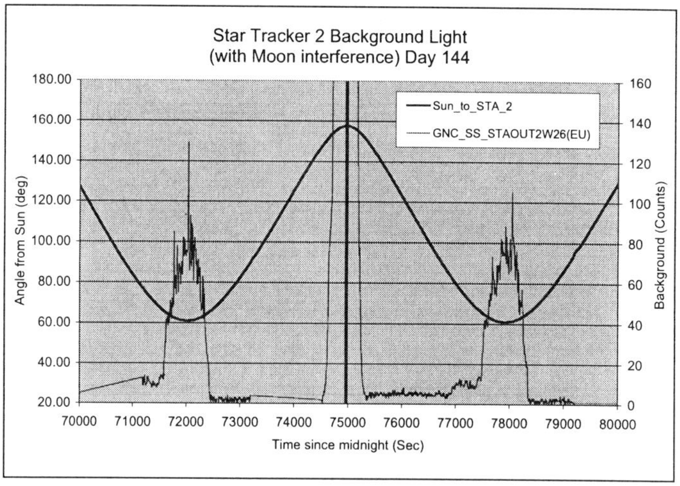

Figure 11: STA-2 Sun and Moon Passage.

On day 216 the software patches were loaded which increased the allowable background to 120 counts in both trackers. The plan was to use the star tracker measurement residuals from the Kalman filter to judge the impact of the patches on performance. Since the residual is the difference between the actual measurement and the predicted measurement it was expected that it would increase if the higher background were negatively impacting the measurements. Each star tracker is capable of tracking up to five stars. For each of these "virtual" trackers there are residuals for both the X and Y measurements. So a total of twenty points were monitored. The residuals are in the telemetry stream and ground processing allows for basic statistics to be calculated for each telemetry point over a window. For Aqua, reference data were collected for a full day (209) and the daily average values for all of the points were averaged. After the patch was applied the residuals were compared to pre-patch values. This was repeated for days 216, 219, and 225. There was no deleterious effect on performance.

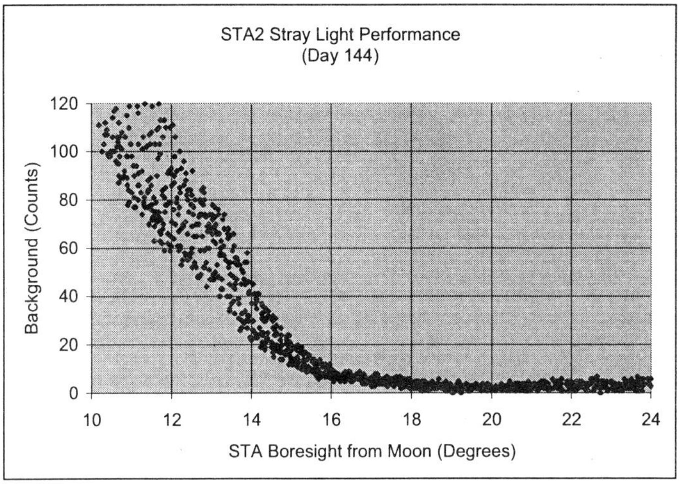

Figure 12: STA-2 Background vs. Moon Angle.

Today Aqua is meeting its pointing requirements. One can construct a marginal performance scenario by postulating a single tracker operational, and worst-case solar beta angle with consecutive sun and moon intrusions. Analysis indicates events like this could result in short periods of degraded attitude knowledge during an average of 6 of the 440 orbits per month.

CONCLUSION

Aqua successfully launched and is operating nominally. In a very short time frame, a star tracker light shade retrofit was devised, analyzed, implemented, and installed on the pad at Vandenburg, enabling Aqua to launch without delay and to accomplish its science mission unimpeded. This was possible only through the cooperative efforts of multiple companies and government agencies.

Prior to launch, the Aqua light scattering model was reviewed by outside experts, and the hardware was tested to the capabilities of the facility. To date, data indicate that the sun-induced background, with respect to the model predictions made prior to launch, is slightly higher than expected. We continue to study this issue for Aqua's sister ship Aura, which has an identical bus. Aura's instrument complement has fewer intrusions into the STA light shade fields of view and is in a slightly better orbit with its ascending node at 1345.

Historical precedents relied on orders of magnitude of margin provided by two-stage light shades for stray light attenuation, at the expense of considerable spacecraft and launch vehicle volume. One can argue that a minimum factor often design margin is still warranted based on these flight results. As further improvements are made to the models, based upon ground and flight data, and a permanent ground test capability is continuously applied to new sensor designs, it may be possible to decrease those safety factors. Aqua quantitatively tested, on-orbit, a modern single-stage light shades in the presence of spacecraft structure impingements, as well as sun and moon intrusions. As such, it represents a leap forward in verifying our understanding of the complex stray light modeling required to design and use small single-stage light shades compatible with smaller launch vehicle envelopes.

ACRONYMS

ADS – Automatic Destruct System
AIRS – Atmospheric Infrared Sounder
AMSU - Advanced Microwave Sounding Unit
AMSR-E – Advanced Microwave Scanning Radiometer for EOS
ASAP™ – Advanced Systems Analysis Program (™Breault Research Corporation)
BRO – Breault Research Corporation
CCD – Charge Coupled Device
CERES – Clouds and the Earth's Radiant Energy System
EOS – Earth Observing System
FOV – Field of View
GSFC – Goddard Space Flight Center
HSB – Humidity Sounder for Brazil
MECO – Main Engine Cut-Off
MODIS – Moderate Resolution Imaging Spectroradiometer
NASA – National Aeronautics and Space Administration
NEA – Noise Equivalent Angle
PSR – Pre-Ship Review
RPM – Revolutions Per Minute
SBF – Satellite Body Frame
STA – Star Tracker Assembly
VAFB – Vandenburg Air Force Base
VOS – Vehicle On-Stand

AAS 03-074

RECENT G&C EXPERIENCES OF THE TIMED SPACECRAFT

Wayne Dellinger, Hongxing Stella Shapiro, Courtney Ray and Tom Strikwerda*

The Thermosphere, Ionosphere, Mesosphere, Energetics and Dynamics (TIMED) spacecraft was launched December 7, 2001. Designed and built for NASA by The Johns Hopkins University Applied Physics Laboratory, the TIMED mission is studying the influences of the sun on the Mesosphere and Lower Thermosphere/Ionosphere (MLTI) region of the Earth's atmosphere.

TIMED is a 3-axis stabilized, nadir-pointing spacecraft with very large solar wings. It is controlled with reaction wheels and torque rods, and uses a ring-laser gyro and two star trackers for attitude estimation. TIMED suffered several problems immediately after launch, however since that time the spacecraft has been performing very well. This paper presents some of the recent G&C experiences of the TIMED spacecraft. These include initial problems encountered at separation, and current performance.

INTRODUCTION

As part of NASA's Solar Connections Program, the Thermosphere Ionosphere Mesosphere Energetics and Dynamics (TIMED) mission has the primary objective of investigating and understanding the energetics and dynamics of the Mesosphere and Lower Thermosphere/Ionosphere (MLTI) region. Encompassing altitudes from 60 to 180 kilometers, this region is the least explored and least understood of Earth's atmosphere. TIMED mission goals include determining the temperature, density, and wind structure of the MLTI, and determining the relative importance of various radiative, chemical, electrodynamical, and dynamical sources and sinks of energy in the MLTI region. Launched at 15:07:35 GMT on Friday, December 7, 2001, the TIMED spacecraft was built, and is operated, for NASA by The Johns Hopkins University Applied Physics Laboratory.

The launch of TIMED yielded great excitement in many ways. With an original launch date of May, 2000, the many hours of hard work had finally come to fruition. (TIMED's launch was co-manifested with JASON I. Delivery problems with JASON delayed launch until December 2001.) The excitement of launch quickly gave way to the excitement of on-orbit performance deficiencies. Finally, the excitement of a science-producing spacecraft settled in.

This paper presents the recent experiences of the TIMED Guidance and Control (G&C) system. Some of these experiences were less than pleasant; however, the end result has been a very successful mission. A brief description of the TIMED spacecraft and its G&C system is presented in the next section. A discussion of the problems encountered immediately post-launch

* The Johns Hopkins University, Applied Physics Laboratory, 11100 Johns Hopkins Road, Laurel, Maryland 20723-6099. E-mail: Wayne.Dellinger@jhuapl.edu.

is then given, followed by a presentation of those things that have gone well. Lastly, summary remarks are presented.

TIMED MISSION

TIMED is a 600-kilogram spacecraft carrying four primary instrument payloads: the Solar EUV Experiment (SEE), the TIMED Doppler Interferometer (TIDI), the Global Ultraviolet Imager (GUVI), and the Sounding of the Atmosphere using Broadband Emission Radiometry (SABER) payload. Each instrument is controlled from a Payload Operations Center (POC) located remotely at the sites of the respective Principal Investigator's institutions. Measurements from these four instruments, combined with simultaneous ground observations, provide the core subset of measurements necessary to define the basic states of the MLTI region.

Launched into a 625 kilometer circular orbit with an inclination of 74.1 degrees, TIMED has a planned mission lifetime of two years. Shown in Figure 1, TIMED consists of a rectangular bus structure and two very large solar wings. Vehicle axes definition uses a right-handed orthonormal triad with the +**z** axis towards the "bottom" in the direction of GUVI and the +**y** axis towards the side of the spacecraft containing SABER. The +**x** axis completes the triad in the direction $+\mathbf{y} \times +\mathbf{z}$.

System Requirements

The G&C subsystem is responsible for the attitude control of the spacecraft during all mission phases. After separation from the launch vehicle the spacecraft must achieve a Sun-pointing "safe" attitude. This provides a power-positive orientation in which the solar wings can generate maximum power and recharge the batteries, if necessary. The next and presumably final phase is the acquisition and sustaining of an operational attitude, described as follows.

During normal operation TIMED is required to maintain a nadir-pointing attitude with the +**z** axis in the direction of the Earth's geocentric center, the +**y** axis in the direction of orbit normal, and the +**x** axis generally in the ram or wake direction. The Sun beta-angle, defined as the angle between the Sun and orbit normal vectors, dictates the **x**-axis direction as follows. SABER contains a thermal cooler for instrument temperature regulation and requires the spacecraft to avoid pointing the +**y** vehicle axis toward the Sun. Orbit precession thus necessitates a π-radian rotation, called the "yaw maneuver," about the **z**-axis approximately once every sixty days.

Architecture

Figure 2 illustrates the TIMED G&C subsystem architecture. The architecture consists of two star trackers, redundant 3-axis Inertial Reference Units (IRUs), redundant 3-axis magnetometers, redundant Sun sensors on each of the +**y** and -**y** sides, three torque rods containing redundant coils, four reaction wheel assemblies, and a pair of redundant processors. A MIL-STD-1553B data bus links the two-processor system. A second MIL-STD-1553B data bus connects the G&C subsystem to the Command and Data Handling (C&DH) system, which sends commands to, and receives data from, G&C. These two data links are the G&C and C&DH buses, respectively.

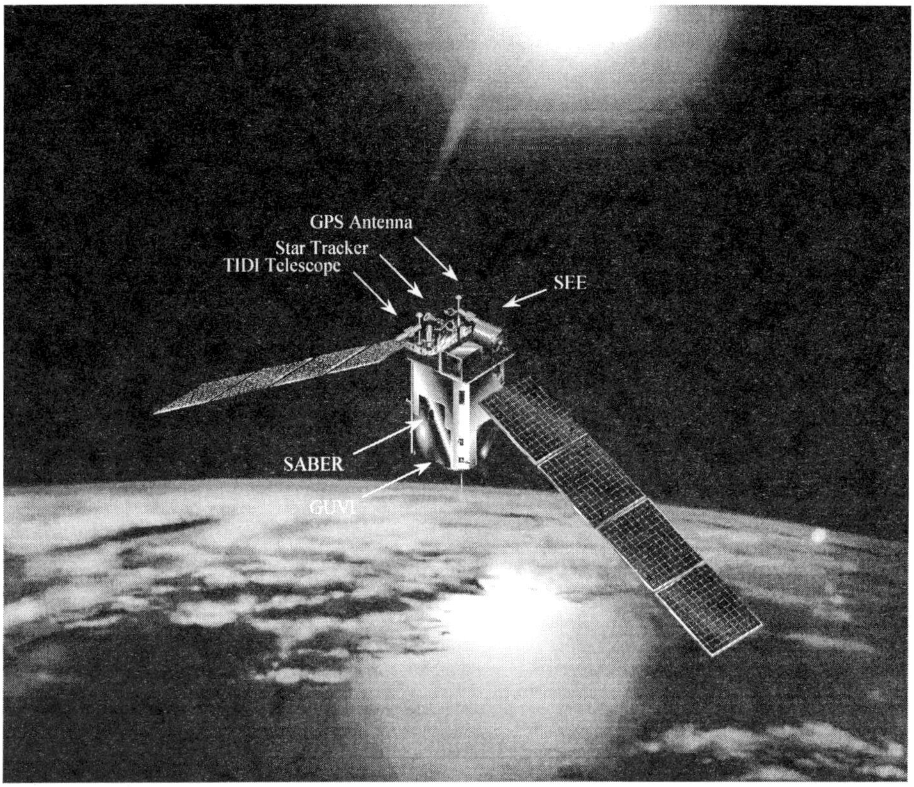

Figure 1 TIMED Spacecraft.

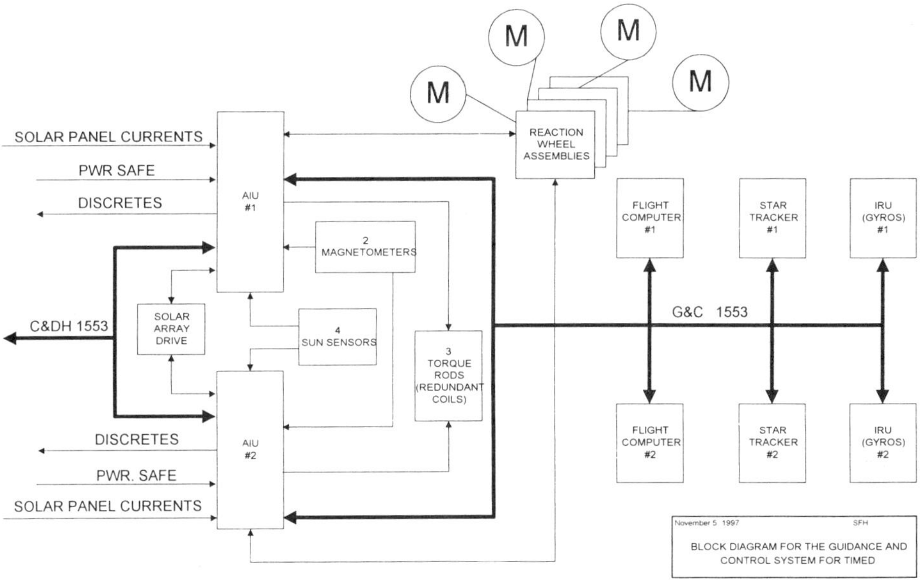

Figure 2 G&C Architecture.

The Attitude Interface Unit (AIU) uses an RTX-2010 as its processor. As the name implies, the AIU is an interface linking the Attitude Flight Computer (AFC) with the rest of the G&C hardware and the C&DH. The AIU is the G&C bus controller and is responsible for sending commands to the actuators and reading sensor data. The AFC uses a 32-bit MONGOOSE V RISC processor. This is the primary computer for the G&C subsystem and is responsible for performing all functions necessary to meet the guidance and control performance requirements.

TIMED also carries a redundant Guidance and Navigation System (GNS). GNS uses Global Positioning Satellite data to calculate a position, velocity, and time state vector at a rate of once per second. The C&DH system passes these data to the G&C subsystem.

INITIAL ON-ORBIT PROBLEMS

Momentum Control

One of the worst fears for a G&C engineer is to have a sign error creep into the design, with a compensating sign error elsewhere to hide it. Try as one might, it is still possible to have this error occur; however, there should be practices and procedures in place to minimize the chance. Clearly on TIMED there were not.

By design, the momentum storage capacity of the reaction wheels was not large enough to absorb the momentum imparted to the spacecraft by the launch vehicle at separation. TIMED thus uses magnetic torque rods, a magnetometer, and an IRU to control and maintain momentum to levels which the wheels can accommodate. This was the first task of the TIMED spacecraft. Real-time telemetry received in the Mission Operations Center (MOC) by way of TDRSS allowed the operations team to watch, and react to, the spacecraft from the moment of separation. All subsystem lead engineers/analysts were present as part of the operations team during the first many hours after launch.

At the time of separation the G&C lead analyst had a real-time plot of system momentum to watch the performance of the dumping process. It was abundantly clear after about five minutes that there was a significant problem as the momentum steadily increased rather than decreased. TIMED does not have a propulsion system, and there is nothing else on-board that could vent to create an external torque to increase momentum; there had to be a sign error in the rod controller logic. Unfortunately there was not a simple means of changing a sign in the code, nor was there a simple way to turn the rods off. A simple unity gain parameter for this purpose, in the path of the commanded rod polarities, did not exist.

The first order of business was to disable rod commanding, which was not as straightforward as one would hope. The G&C subsystem was intended to operate separately from the C&DH subsystem in some cases, notably certain extreme cases of safe mode. While G&C had the ability to send on/off/polarity commands to the rods, and torque commands to the reaction wheels, it could not send power commands to the wheels or magnetometers. In addition, not all data available to C&DH were available to G&C. One particular piece of data was a direct notification of separation; there was a desire to not rely on the C&DH 1553 at this critical time. Hence, a back door method was devised. C&DH has access to the separation switches, and it also has control of powering the magnetometers. When the spacecraft separated from the launch vehicle, C&DH would detect the separation and power the magnetometers. G&C would detect the powering of the magnetometers and proceed with the momentum dumping. Once the magnetometers were powered they were to never be powered off again. As such, the momentum control logic looked for valid magnetometer data to proceed with the torque rod control. If the magnetometer data were not valid the rods would be commanded off. Hence, the means for disabling the torque rod commanding was to turn off the magnetometers. Not an elegant solution, but functional and expedient.

The second order of business was to correct the sign error. Inverting the signs on the magnetometer biases and scale factors made a temporary correction. In effect, the magnetic field direction was switched. The momentum decreased as expected after the magnetometer parameter changes went into effect. It took approximately 80 minutes from discovery of the problem to implementation of the temporary fix. In this time the spacecraft rates did not get above 43.6 mrads/sec (2.5 deg/sec) and the solar wings were charging the battery. The spacecraft was never in a state of near danger; however, several operations personnel, as well as others, suffered minor premature aging.

The root cause of the sign error was discovered several days after launch when there was time to think through the ground testing process. During integration and test (I&T), several tests were performed to check the polarity of the rods. These tests were performed under ground control as well as under flight software control. In all cases the polarity was verified by use of a compass. A rod was turned on, and the polarity checked with the compass. Note this is a *discovery* test rather than a *verification* test. Unfortunately the engineer with the compass believed a compass pointed to the north magnetic pole, when in fact it points to the south (since the Earth's south magnetic pole is located in the geographic north). This introduced a sign error that was implemented in both the flight software as well as the simulated dynamics model, creating compensating errors. This information should be, needs to be, and will be passed on in the future through simple, written, test procedures.

Sun Sensor Orientation

It was known that TIMED would complete momentum dumping about the time it entered an eclipse period. At that point the reaction wheels would be powered on by the C&DH. The controller would command them to absorb the remaining spacecraft momentum, and, having no sun information with which to point the spacecraft, control only rate until eclipse was over. Having just completed approximately 90 minutes of heart pounding excitement, what else could go wrong?

When coming out of eclipse and seeing the sun for the first time, the spacecraft was to reorient to point a vector 611 mrads (35 deg) off the –**z**-axis, lying in the **y**-**z** plane, toward the sun. Depending upon the initial attitude, this maneuver could take up to 20 minutes. At the end of the eclipse the spacecraft began to reorient itself as expected, however after 20 minutes it showed no signs of settling out; commanded wheel torques showed a strange gyration occurring. Something else had gone wrong.

Safe mode on TIMED uses sun sensors and an IRU for attitude/rate knowledge, and wheels for control. In addition, the magnetometers and torque rods are used for momentum control. Upon examining raw sun sensor measurements and solar wing currents, all of which were available in telemetry, it was determined that the spacecraft had settled into what appeared to be a quasi-stable attitude with the –**x**-axis generally pointed at the sun. It appeared that the spacecraft was attempting to point this axis at the sun rather than the intended pointing axis, and it was not doing a very good job at that. The spacecraft would pitch, roll, and yaw as it moved the –**x**-axis around. Once again, hearts were pounding.

Fortunately for all involved, including the spacecraft, the particular attitude into which it settled pointed one of the large solar wings at the sun, thus providing plenty of power. In fact, the battery continued to charge and TIMED remained power positive orbit after orbit. Additionally, this attitude was a comfortable one thermally. TIMED had settled into a pseudo safe-mode; it was not what was intended, but it was in no danger.

At this time many of the engineers from the G&C team were gathered together to diagnose the problem. Software code, hardware drawings, and I&T photographs were examined. In parallel, a decision had to be made regarding the safety of the spacecraft. Given the acceptable power and thermal state of the spacecraft, it was decided the leave TIMED in the hands of the normally scheduled operations team, with everyone's pager and cell phone numbers nearby, and let others go home and get some much needed sleep. In the meantime, several G&C engineers continued to diagnose the problem.

Having examined the software code and hardware drawings, it was determined that they were in agreement. Each particular solar cell, on each of the four sun sensors, was properly mapped into software, given the G&C team's understanding of how the sensors were mounted, and tested, on the spacecraft. However, an I&T photo taken at the launch site revealed the problem; the sensors were not mounted as the G&C team had thought they were. In particular, the two sensors on the 'hot' side, that side which was to be pointed at the sun during safe mode, were mounted ninety degrees from what was expected. This revelation was discovered approximately 19 hours after launch. Subsequent analysis of the raw sensor data revealed that the orientation of the 'hot' side sensors were in fact incorrect; however, the 'cold' side sensors were correct.

With the discovery that the 'hot' side sun sensors were rotated from the expected orientation, two questions had to be answered: how do we fix it, and how did it happen? The immediate order of business was correcting the situation. Since the sun sensors were very coarse in their knowledge capabilities, and to save memory in the AIU, the software did not have parameters for their alignments. Thus a code change was required. This change was incorporated into new code, tested, verified, approved, and loaded to the spacecraft one week after launch. On-orbit tests of the new code were verified during spacecraft checkout.

Having corrected the problem the focus changed to its root cause. It was disturbing, again, to have performed specific phasing tests during I&T and yet still have an error. In this particular case the problem was with the configuration of the spacecraft when the G&C team performed their tests. The two 'hot' side sun sensors are mounted to a panel on the –y side of the spacecraft. This panel was also the main panel through which the inside of the spacecraft was accessed; the panel was removed during most of the I&T build and the two sensors were temporarily hung off to the side. Unbeknownst to the G&C team, the orientation in which they were hung did not agree with the orientation they would have in flight. The G&C team did not ask about the orientation, the I&T team did not volunteer the information, and there was no documentation *specifically* denoting the desired orientation. Thus the tests were performed in an unintended *non-flight* configuration.

Control-Structure Interaction

TIMED has basically two modes of operation: sun-pointing and nadir-pointing. The former is for spacecraft safing operations, while the latter is the science attitude. When the spacecraft failed to attain the nominal sun-pointing attitude it was decided to promote to the nadir-pointing attitude. If this mode worked it would be a safer attitude than the current one. In addition, a higher downlink telemetry rate could be had with a nadir-pointing spacecraft. Before promoting, however, there was the question of what would happen if TIMED had to return to safe mode; where would it end up?

A modified safe mode was devised in which a non-zero rate would be commanded such that the arrays would be sufficiently exposed to the sun as the spacecraft rolled. This mode was tested over several orbits on Saturday afternoon to verify its viability. While not the ideal mode of operation, this mode would work, and buy some time, if it was needed. The spacecraft was then successfully commanded to nadir-pointing mode late Saturday evening. This success allowed many to return home for a day of rest on Sunday.

The week after launch was consumed with fixing safe mode. There was, however, the obvious need to evaluate the performance of the newly attained nadir attitude. After all, that is the science attitude. As part of this evaluation an unexpected 0.1 Hz oscillation in the 1 Hz real-time rate and wheel torque data was observed. The high-rate telemetry, which is stored on the solid-state recorder at 10 Hz and down-linked after-the-fact, revealed the true frequency to be 2.1 Hz; aliasing due to 1 Hz sampling had created the fictitious 0.1 Hz. Review of the structural modal frequencies confirmed that one of the solar wing structural modes was being excited.

Early in the controller design-phase the structural modes were briefly examined with regard to controller stability. The lowest structural mode frequency was 0.25 Hz while the controller bandwidth was 0.01 Hz and so, with the decade or more separation between them, no further analysis was performed. A formal frequency domain analysis was not undertaken, and a suitable fidelity flexible-mode model was deemed unnecessary for the existing time domain simulation. These were serious omissions. Low-pass filtering of the rate data was recommended during reviews, and this was implemented in the AIU for safe mode but not in the AFC for the nadir-pointing modes. The latter decision was based on the desire to avoid filtering the gyro data prior to its use in the Kalman filter. It was an oversight not to filter the data prior to use in the controller.

Analysis of this problem after launch through fine-tuning of a NASTRAN model with flight data revealed a significant modal gain from the 2.1 Hz mode, a mode that actually changes in frequency from 2.0 to 2.6 Hz over the $\pi/2$ radian (90 deg) solar wing range. Frequency domain analysis demonstrated the unstable nature of this mode for the given controller gain set, and incorporation of the NASTRAN model in the time-domain simulation yielded results similar to on-orbit data. A complete analysis of the flexible modes during the controller design would have exposed the problem and it could have been eliminated pre-launch.

A low-pass Butterworth filter was implemented in the software and the code loaded to the AFC. After detailed frequency domain analyses, with time domain simulation concurrence, the filter coefficients and controller gains have been chosen to alleviate the oscillations.

Interesting Discoveries

Another interesting discovery during the checkout phase was the relatively rapid rate at which the momentum was building up. Based on the momentum control parameters and the expected momentum buildup, the G&C team expected to dump momentum approximately once per day. We were instead getting around ten. Analysis has revealed what appears to be a significant spacecraft residual dipole. The calculated external torques track the magnetic field over an orbit proving the torque is clearly magnetic related. See Figure 3. The estimated dipole is primarily in the +y axis direction with a magnitude of approximately 10 A-m^2. While nothing more than an annoyance for TIMED, an actual measurement of the spacecraft residual dipole would have been beneficial, given time and money.

Lastly, from the problem department, TIMED has had four unexpected demotions to safe mode. The first was the result of sun glint on one of the 'cold' side sun sensors that caused the AIU to think there was sun on the SABER side of the spacecraft. This happened just prior to the first planned yaw maneuver during the noon-midnight orbit, where the 'cold' side sensors actually see the sun. Altering the sun keep-out angle during these orbits has alleviated the problem.

The second demotion to safe mode occurred during the AFC code load that added the low-pass filter. Due to some unknown timing problems in the C&DH that only manifest themselves during times of large packet uplinks, the C&DH lost track of the AIU heartbeat. Thinking the AIU had ceased to function, the autonomy software forced a switch to the redundant AIU and the mode demotion.

Safe mode was again autonomously commanded during the second planned yaw maneuver when the maneuver failed to complete in the allotted time. Yaw maneuvers are commanded during eclipse periods to avoid sun in the fields-of-view of some of the instruments, thus the maneuver should complete within the time of a noon-midnight orbit eclipse, about 35 minutes. Ground simulations had predicted maneuver times of 15-20 minutes, and the on-board parameter was set to 25 minutes. Safe mode was commanded when this particular maneuver was within two minutes of completion. In fact, the spacecraft was completely turned around and was just settling the rates to within the 0.0001309 rads/sec (0.0075 deg/sec) specification. Allowing the maneuver more time to complete has kept this from occurring again.

The final safe mode occurred as a result of an AFC reset. This reset has been traced to an apparent error in the processor operating system. Additional telemetry packets are now routinely collected in hopes of catching a reset, should it occur again, and finalizing the diagnosis.

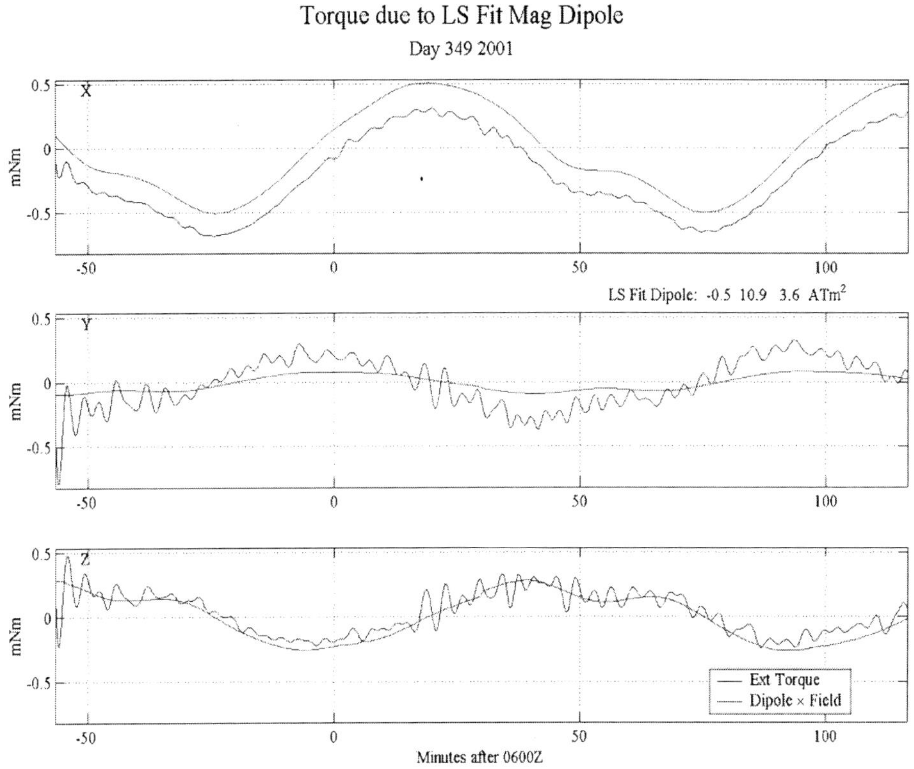

Figure 3 Residual Dipole.

ON-ORBIT PERFORMANCE

Navigation

TIMED has an on-board Guidance and Navigation System (GNS) utilizing the Global Positioning System. The GNS maintains spacecraft time, as well as providing position, velocity, and sun vectors to the ground and the rest of the spacecraft, principally G&C. Position, velocity, and time are used within G&C to determine the commanded nadir attitude and rate. Time is also used for time-tagging sensor measurements and telemetry. The sun vector knowledge, with spacecraft attitude knowledge, allows G&C to report when the sun may be encroaching on the SABER side of the spacecraft and thereby forcing a demotion to safe mode.

GNS has requirements of reporting the spacecraft position to within 300 meters, and velocity within 25 cm/sec. Performance has been verified by comparison with trajectories generated by the GSFC Flight Dynamics Facility based on TDRSS data, thus providing verification against an independent trajectory solution. Results show the GNS system providing position to within 20 meters RMS, with 51 meters maximum error, and velocity to within 2 cm/s, with 5 cm/s maximum error[1]. These values are well within requirements.

Pointing Performance

Pointing requirements for the TIMED spacecraft are fairly modest at 8.7 milliradians (0.5 degrees) in attitude and 0.1309 mrad/sec (0.0075 deg/sec) in rate. Due to the flexible mode issues, the controller gains were 'softened' to avoid exciting the solar wings, resulting in less than stellar performance. Immediately after launch, when the flexible mode issue was first discovered, the control system could not meet the requirements in attitude or rate. Attitude excursions of greater than 17.5 mrads (1 degree) and rate errors greater than 1.0 mrad/sec (0.06 deg/sec) were common. Reduction of the gains helped, albeit for a short period of time. Excitation was clearly wing angle dependent; see Figure 4.

Implementation of the low-pass filter improved performance markedly. Attitude errors are less than 0.8723 mrads (0.05 degrees) with the exception of during some momentum unloads. During some unloads the error will increase to 3.5 mrads (0.2 degrees), but still within the requirement. Rate errors are less than 0.05 mrad/sec (0.003 deg/sec) except during momentum unloads, where the rates exceed the 0.1309 mrad/sec requirement for approximately three minutes during one dump per day. New gains for better control are currently ready for implementation, awaiting final program approval.

Knowledge Performance

TIMED has an attitude knowledge requirement of 524 μrad 3-σ (0.03 deg). Illustrated above in Figure 2, TIMED has two Lockheed-Martin AST-201 star trackers. Their orientation is such that they point up and away from the SABER instrument (see Figure 1), and their bore-sights are approximately $\pi/2$ radians (90 deg) apart. During the noon-midnight orbit, when the sun passes directly over the top deck, the sun-line passes within $\pi/6$ radians (30 deg) of the bore-sights. The chosen orientations were a trade-off between getting too close to the sun and getting sun-glint off the top deck into the trackers. Orienting the tracker bore-sights orthogonally also minimizes the effects of the individual tracker roll errors on the knowledge performance.

Figures 5 and 6 show the filter residual performance for trackers 1 and 2, in the individual tracker frames, over an 11+ hour time span. Performance is as expected given the advertised noise-equivalent-angles of 2.7, 2.7, and 32 μrad for the **x**-, **y**-, and **z**-axes, respectively. It is interesting to note the orbital characteristic in the **x**-axis plot, and the twice-orbital rate characteristic in the **y**-axis plot. While not fully analyzed at this time, it appears that thermal variations are creating these effects.

Thus far the star trackers have behaved very well. Loss of track is common with the moon in the field-of-view, however the orientation of the trackers precludes both being blinded at the same time. No sun interference has been noted. As mentioned before, the sun line gets to within $\pi/6$ radians of the tracker bore-sights. The sunshade design is advertised to allow operation up to approximately $\pi/9$ radians (20 deg) of the bore-sight. While this limit has not been tested, performance has been nominal since launch.

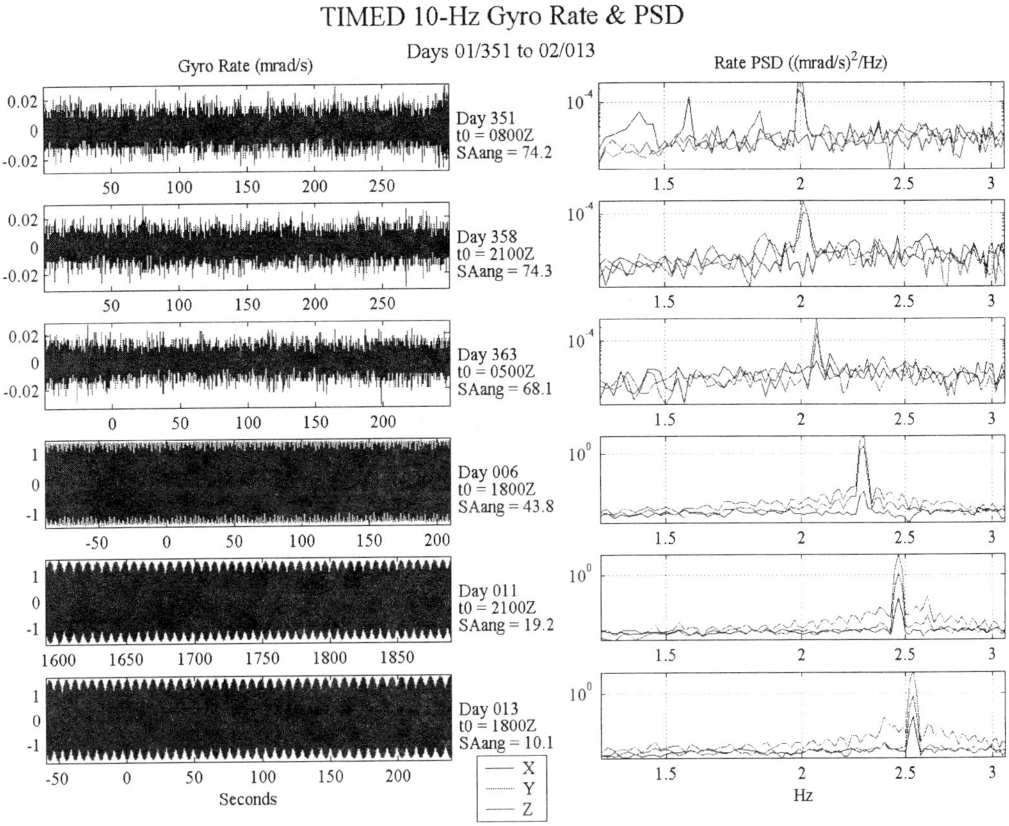

Figure 4 Gyro Rates and PSDs.

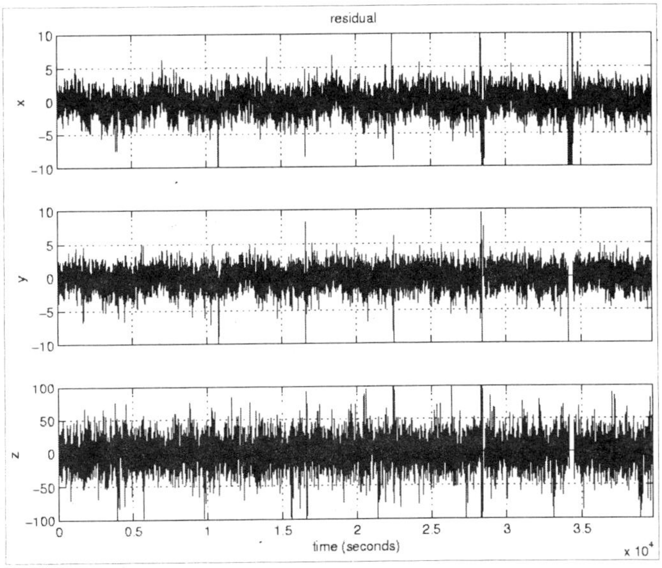

Figure 5 Filter Residuals for Tracker 1 (arcseconds).

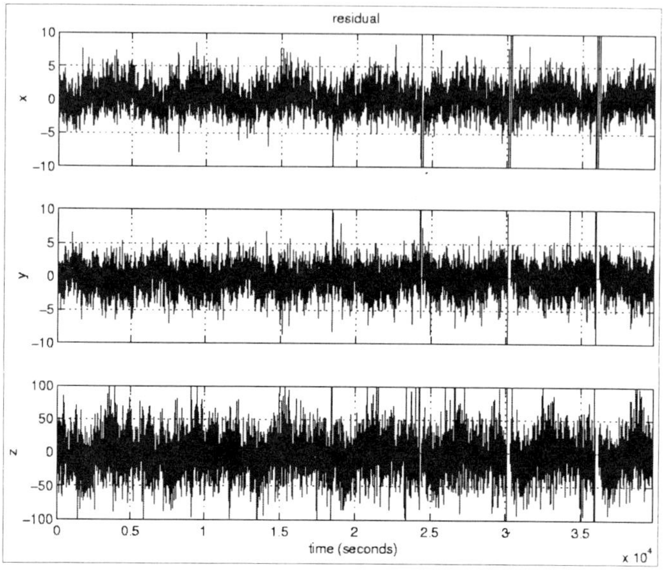

Figure 6 Filter Residuals for Tracker 2 (arcseconds).

Sensor Alignment and Gyro Calibration

Sensor alignment and gyro calibration on TIMED involved estimating alignments of the two trackers relative to the IRU, non-orthogonal gyro axis misalignment, and symmetric scale factors of the gyro. This was accomplished with an alignment calibration algorithm and a composite calibration algorithm developed by Pittelkau.[2,3] The alignment calibration algorithm estimates the alignments from a given maneuver using a Kalman filter. The composite calibration algorithm then uses a least squares approach to form a composite calibration from several maneuvers.

TIMED is, in general, a non-maneuvering spacecraft, and as a result the normal data collection does not provide observability into all of the desired quantities to be estimated. However the yaw maneuver, while not ideal, does provide some observability into the alignments due to increased rates, and rates about all three body axes. (Observability of gyro calibration parameters is not sufficient to allow appropriate estimation.) The alignment calibration was performed using data from two yaw maneuvers.

Results from the calibration are interesting, and the source of numerous discussions. Comparison of the optically measured alignments, measured on the ground, with the alignment calibration estimates show a large rotation primarily about the **z** body axis.[4] In fact, tracker 1 shows a –3.46 mrad (–0.1985 deg) rotation while tracker 2 a –5.7 mrad (-0.3267 deg) rotation. These numbers are considered huge for components that are glued and bolted to the bench. One could infer from the data that a common shift occurred in both trackers, either from shifting of both trackers, shifting of the optical bench, or shifting of the IRU. Currently there is no definitive explanation.

While the instruments on TIMED cannot provide a direct measurement of attitude, one of the instruments, TIDI, can provide scant evidence about the **x** axis. In this regard TIDI is in agreement with the new alignments.

Clearly the above results indicate a strong need to perform this type of calibration on-orbit. While each particular mission may dictate the maneuvering capability available, and the required fidelity of the alignment calibration, the process must be planned for and executed.

Maneuvers

Since the initial two yaw maneuver failures mentioned above, TIMED has successfully completed the five subsequent maneuvers. All instrument calibration maneuvers, sixteen thus far with commanded roll angles up to 0.5136 radians (30 degrees), have been successful.

SUMMARY

The TIMED spacecraft is currently into its second year of operation, more than half-way through a planned two year mission. Even though TIMED sustained some early close-calls in its handling of attitude control, the recovery was quick and complete.

Post-event analysis of the problems encountered have shown the advantages of a tight, documented hardware test program. Most notably, reviewed documentation regarding G&C hardware test procedures, tests that are for verification and not discovery, and improved communication regarding test configurations. Tests must be performed in final flight configuration, and questions must be asked to assure this configuration.

Finally, having passed the initial "bumps in the road," the TIMED G&C system is working very well. Performance of the spacecraft is significantly better than required, and indications are that this trend will continue.

REFERENCES

1. M. S. Asher, "Comparison of GNS Navigation Solution with FDF TDRSS Trajectory, with Historical Perspective," Internal APL Memo SEA-2002-066.

2. M. E. Pittelkau, "Kalman Filtering for Spacecraft System Alignment Calibration," *AIAA Journal of Guidance, Control, and Dynamics*, Vol. 24, No. 6, 2001, pp. 1187-1195.

3. M. E. Pittelkau, "Everything is Relative in System Alignment Calibration," *AIAA Journal of Spacecraft and Rockets*, Vol. 39, No. 3, 2002, pp. 460-466.

4. M. E. Pittelkau and W. F. Dellinger, "Attitude Sensor Alignment and Calibration for the TIMED Spacecraft," AAS 03-153, *Proceedings of the 13th Annual AAS/AIAA Space Flight Mechanics Meeting*, Ponce, Puerto Rico, 2003.

AAS 03-075

APPENDAGE DEPLOYMENT VERIFICATION FOR THE 2001 MARS ODYSSEY SPACECRAFT USING SPACECRAFT ATTITUDE TELEMETRY

Russ Gehling* **and Sue Linch***

The 2001 Mars Odyssey Spacecraft was required to execute several mission critical appendage deployments and articulations along the way to its final Mars-mapping configuration. These events included Solar Array, High Gain Antenna, and Gamma Ray Spectrometer Instrument Boom deployments and positioning of the Solar Array for aerobraking passes. Verification of these events was necessary in order to proceed with nominal mission plans. This verification was accomplished through use of high sample rate (200 samples per second) spacecraft attitude telemetry data compared to analytic model predictions. The use of attitude data eliminated the need for any position monitoring sensors (limit switches, potentiometers, accelerometers etc.) other than the spacecraft Inertial Measurement Unit (IMU) itself, yet provided high-fidelity information regarding the mechanical performance of the spacecraft deployments and articulations. This paper summarizes the flight and simulation results from Mars Odyssey, demonstrating the ability to thoroughly verify spacecraft deployment and articulation activities without the need of additional sensors given the availability of IMU data acquired at sufficiently high sample rates.

INTRODUCTION

During the Mars Global Surveyor mission, it became apparent that accurate knowledge of the spacecraft appendage dynamic characteristics could be inferred from spacecraft body-rate telemetry if acquired at a high enough sample rate. Also, given the availability of higher bandwidth and high accuracy body rate data, appendage deployments and spacecraft appendage configuration could be adequately verified, thereby eliminating the need for deployment position sensors. Fortunately, the inertial measurement units (IMU) selected for use on subsequent spacecraft produced by Lockheed Martin for NASA/JPL Mars missions had the capability to provide the necessary bandwidth (>10 Hz). Further, the sensitivity required by the attitude control system (ACS) provides very high fidelity data for evaluation of the spacecraft structural dynamic characteristics and response.

Given the ACS IMU capabilities, the Mars '98 and the 2001 Mars Odyssey deployment verifications were done through the use of spacecraft body-rate telemetry. This required ensuring that adequate on-board data storage was allocated for engineering telemetry and that the required software code was loaded to store high-rate IMU data. With this in mind, no deployment sensors were included on the spacecraft thereby simplifying the design, testing and verification of the mechanism subsystem as well as conserving onboard data acquisition channels.

This paper summarizes the results of the appendage deployment and positioning verification effort performed for the 2001 Mars Odyssey spacecraft. As shown in Figure 1, the Odyssey spacecraft has three deployed appendages: Solar Array, High Gain Antenna (HGA) and Gamma-Ray Spectrometer (GRS) boom. The solar array was deployed immediately after launch, the HGA after completion of aerobraking, and the GRS boom after several months of initial science operations.

* Lockheed Martin Space Systems, Denver, Colorado. All correspondence should be sent to Russ Gehling, Lockheed Martin Space & Strategic Missiles, P.O. Box 179, M/S S8350, Denver, CO 80201. E-mail: russ.n.gehling@lmco.com.

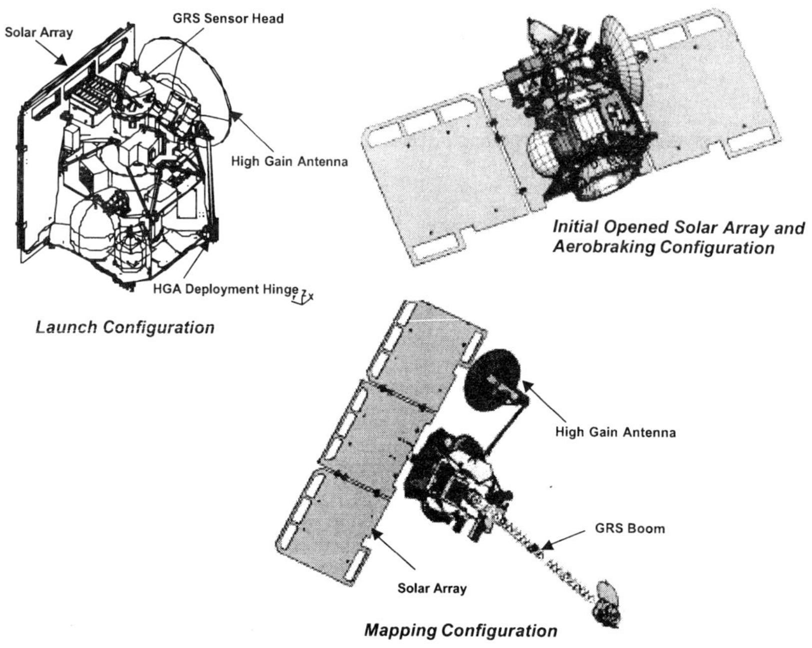

Figure 1. Odyssey Spacecraft

IMU DESCRIPTION AND DATA

The Odyssey spacecraft IMU is based on the standard Miniature Inertial Measurement Unit (MIMU) fromHoneywell (Figure 2). The unit employs three GG1320 ring laser gyros and three ISO-2000 accelerometers. The IMU performance is summarized in Figure 3. Data is output over an RS-422 interface at 200 samples per second (SPS) in the form of angle and translational velocity. For use by the attitude control system (ACS), 20 samples are averaged every 0.1 seconds. However, for appendage deployment events, the raw 200 SPS data was stored and transmitted to Earth. The storage allocation allowed approximately 22 minutes of 200 SPS IMU data to be recorded. Ground processing of the IMU measured angles involved bias correction, filtering the data with a forward-backward difference at 10 Hz, and computation of angular rate by simple differencing.

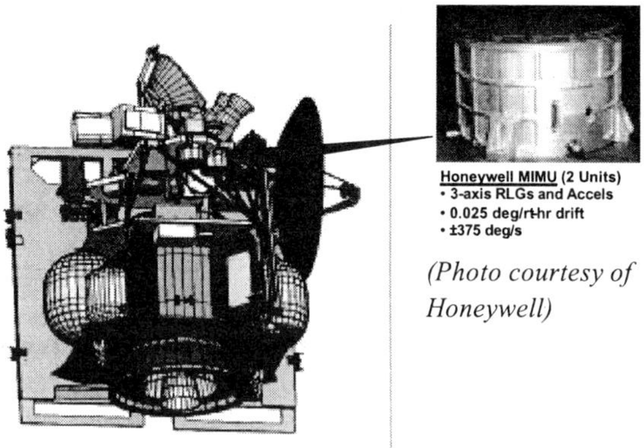

Figure 2. IMU Installation on Spacecraft

- Gyro Performance
  - — Drift < .12 deg/hr (3 sigma over 1 hr)
  - — Random Walk < 0.025 deg/rt-hr,
  - — Scale Factor Error< 250 ppm (3 sigma, 12 months)
  - — G-sens Drift < 0.2 deg/hr/g 3 sigma
  - — Bandwidth >15 Hz
  - — Range ±375 deg/sec BOL (±315 deg/sec EOL) with sign still valid above limit
  - — Performance same as MSP'98
- Accelerometer Performance
  - — Bias Stability < 180 micro g (3 sigma, 12 month)
  - — Scale Factor Error < 240 ppm
  - — Range ±25 g
  - — Performance same as MSP'98

Figure 3. IMU Performance Summary

MODELING

Response simulations were performed using ADAMS software with flexible body appendage models imported from NASTRAN finite element models (Figure 4). Flexible body modeling retained modes up to 10 Hz for the solar array panels, HGA boom, and GRS boom. The spacecraft core vehicle was modeled as a rigid body. Since the attitude control system was disabled for appendage deployments, no controller modeling was necessary for the deployment simulations. Typical simulation time steps were 0.01 seconds. When available, models were tuned to gimbal, boom, and solar array panel stiffness and damper test results. Mass properties were taken from measurements and best estimates.

In some cases, such as the solar array, high fidelity pre-test simulations were performed for the actual deployment event in order to provide rapid determination that the deployment went as expected and was successful. In other cases (GRS boom deployment), a more qualitative assessment of the deployment dynamics was used with more focus on the post-deployed dynamic characteristics of the spacecraft.

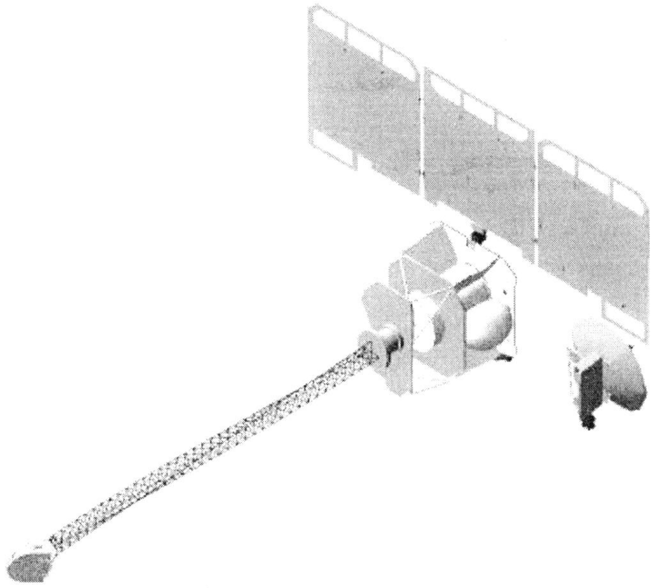

Figure 4. Odyssey Deployment Simulation Model

SOLAR ARRAY DEPLOYMENT

Verification of proper deployment and latching of the solar array was necessary to plan for array articulation and spacecraft aerobraking at Mars. This verification was accomplished by comparing spacecraft body rates recorded during hinge-line opening to analytic predictions.

Solar array release and opening (Figure 5) immediately followed separation from the launch vehicle upper stage and spacecraft rate damping, with autonomous activation of four burn-wire release nuts. The attitude control system was idled for a period of eight minutes to allow the solar array hinge lines to fully open and latch. The eight-minute period was based on the longest predicted deployment time with cold dampers, high frictional losses, and only 75% spring torque.

During release of the solar array separation nuts, spacecraft body rate data was recorded and stored at 200 samples per second (sps). Following acquisition of signal, the stored data was transmitted for post-deployment analysis. At the 200 sps rate, dynamic vehicle response to the solar array deployment was compared to ground testing and simulation results with very high fidelity, thereby allowing evaluation and verification of hinge-line latch-up.

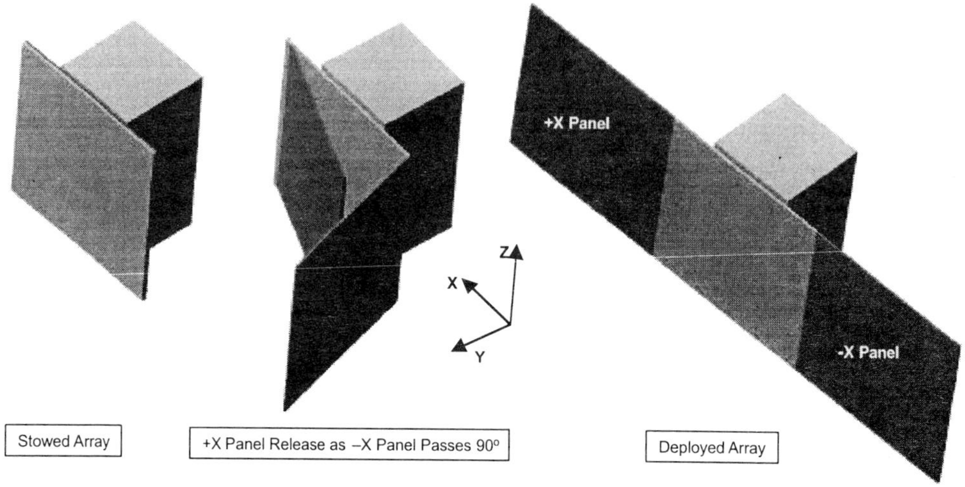

Figure 5. Odyssey Solar Array Panel Opening Sequence

Since the array hinge-lines are parallel to the S/C Z-axis, the most informative body rate information is ω-z. An annotated comparison of observed and predicted spacecraft ω-z is shown in Figure 6 and demonstrates that the array opened as predicted. The actual panel latch-up events are indicated by the transient oscillations identified in Figure 6. Agreement of the vibration frequency (1.5 Hz observed versus 1.4 Hz predicted) indicates positive latch engagement.

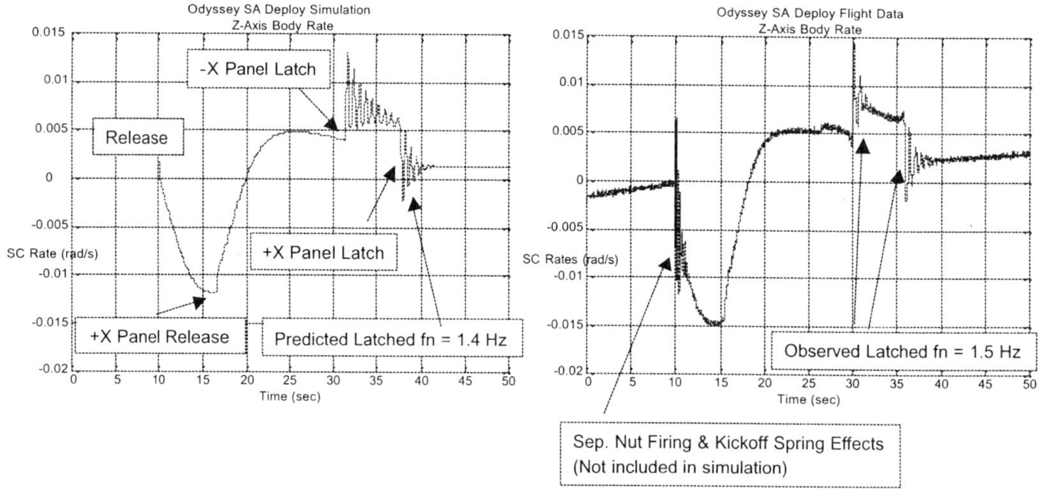

SIMULATION NOTES:

Run with 20 °C Damper Properties & As-Measured Net Deployment Spring Torque

Separation Nut Firing and Kickoff Springs Not Included in Model

Latch Friction Not Included in Model

Figure 6. Spacecraft ω–z During Array Release Event, Observed and Predicted

Comparison of observed and predicted Spacecraft ω-z body rate response during the Odyssey Solar Array release event clearly shows successful panel opening and positive latching of both the –X and +X panels.

SOLAR ARRAY POSITIONING FOR AEROBRAKE PASSES

In order to insure that the array did not back-drive the gimbal during aerobraking, the solar array was required to be accurately positioned to place wheels on the array in front of restraint tabs on the spacecraft body (see Figure 7). During cruise, proper positioning was verified by commanding the gimbal to drive the array wheel against the tab to stall the gimbal. Spacecraft body rate was recorded along with gimbal encoder output to determine the indicated encoder position when contact with the tab occurred. Contact was determined by the sudden cessation of oscillation about the spacecraft x-axis. As shown in Figure 8, the contact occurred between 88.86 and 88.77 degrees (array was positioned at 89 degrees for aeropasses). This verified proper actual array position against the gimbal encoder indicated position.

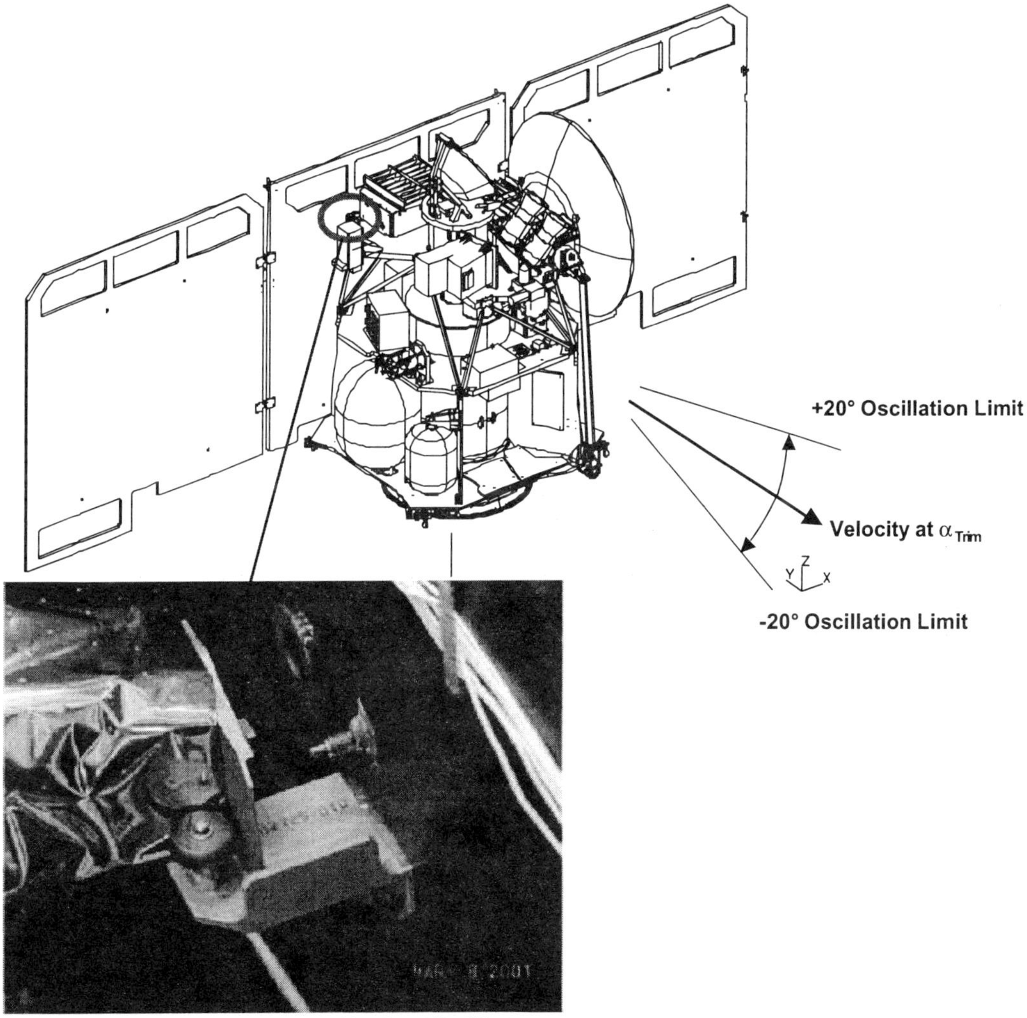

Passive Restraint Wheel & Tab, (0,89) Position, Second Test Side A

Figure 7. Solar Array Aerobraking Restraint

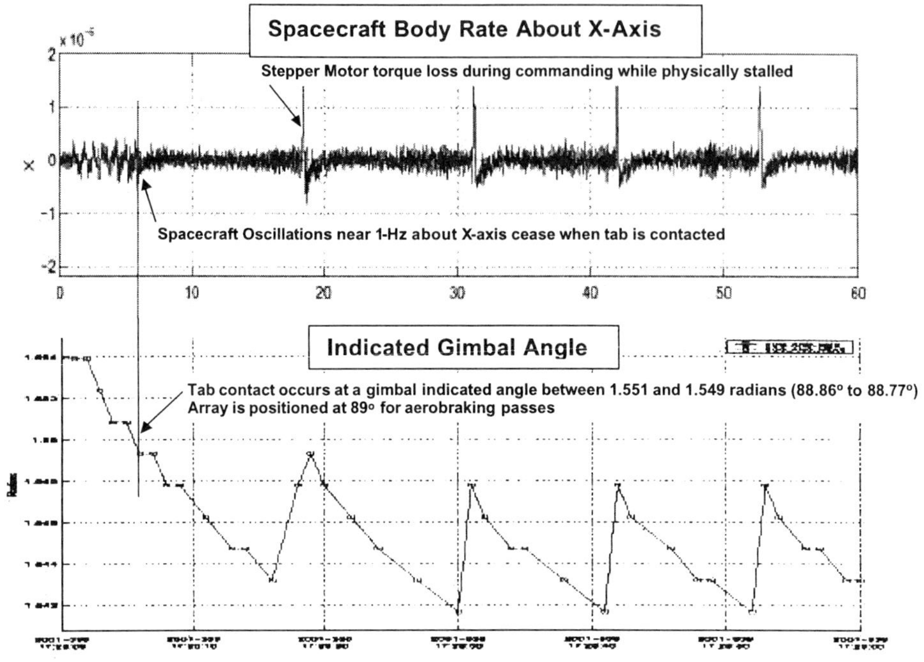

Figure 8. Solar Array Aerobraking Restraint Calibration Data

HGA DEPLOYMENT

Following completion of aerobraking and attainment of its Mars mapping orbit, Odyssey's HGA was deployed. The spacecraft was taken to a safe inertial hold orientation prior to deployment. The deployment sequence is summarized in Figure 9. Figure 10 shows the pre-event prediction of spacecraft body-rate about the spacecraft Y-axis. Note the significant duration of free wheeling motion. The simplified pre-event modeling assumed the three kick-off springs acted together through their center of action. Figure 11 shows the measured spacecraft response during deployment along with a more refined simulation in which the kickoff springs actions were modeled separately. This produced a much smaller free-wheeling motion, but still over-predicted the free-wheeling. This difference in initial motion is likely caused by assuming all the potential energy stored in the kickoff springs was effective with no losses. None-the-less, the good qualitative agreement with the pre-event simulation allowed immediate determination of a successful deployment once the telemetry was received.

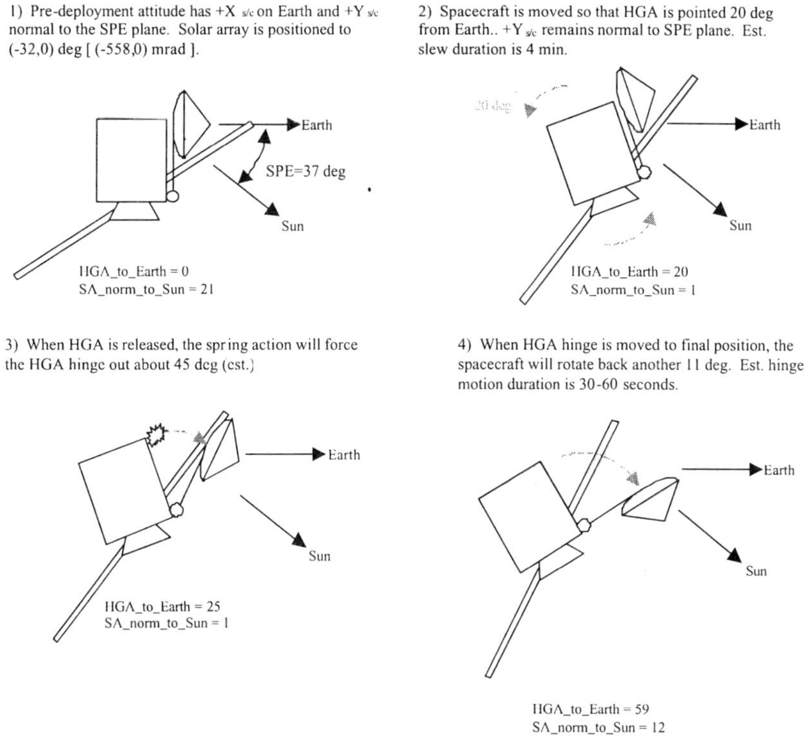

Figure 9. HGA Deployment Sequence

- **Kick-off spring energy included (no losses assumed)**
- **Flexible Solar Array, HGA Boom, and HGA Gimbal included**
- **Motor & Hinge losses (friction) included**

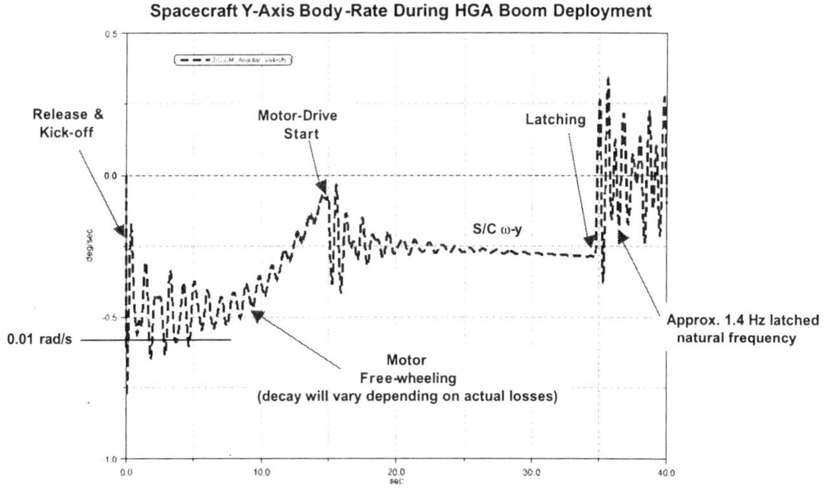

Figure 10. HGA Deployment Pre-Event Simulation Result

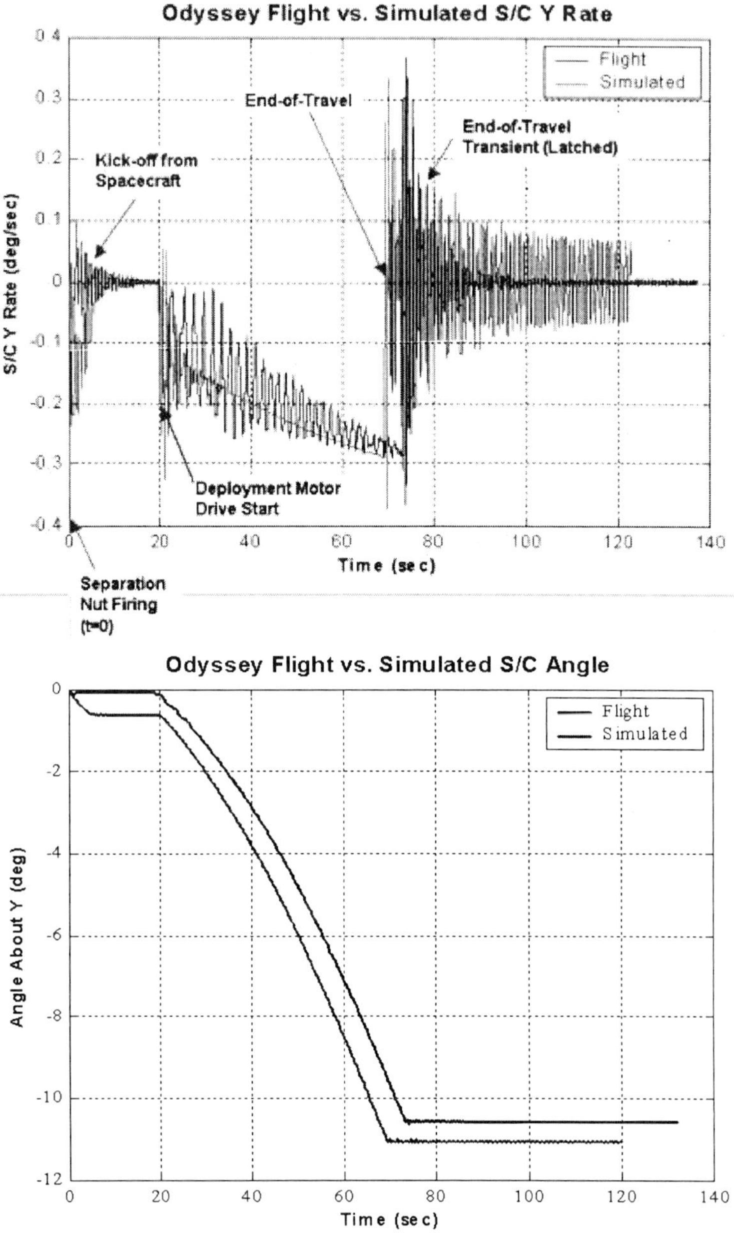

Figure 11. HGA Deployment Data and Post-Event Refined Simulation Result

GRS BOOM DEPLOYMENT

The final deployment event for Odyssey was release of the GRS boom as shown in Figure 12. This section summarizes a detailed evaluation of the Odyssey Spacecraft body-rate data acquired during GRS boom deployment. The purpose of this evaluation was to verify correct and full deployment of the GRS boom. After the deployment data was obtained, some refinements to the deployment modeling and predictions were required to provide a more thorough verification of the deployment event. With these refinements, all significant differences between the pre-deployment predictions and the actual flight data were explained by simplifications made to the modeling and test data interpolations used to arrive at the original predictions. Figure 13 shows the initial predicted and observed spacecraft body rate response to GRS boom deployment. Figure 14 shows the initial deployment time predictions. Note that the purpose of the pre-deployment time-history predictions was to give a qualitative idea of what to expect during the initial flight deployment data evaluation. All characteristics seen in the measured flight data qualitatively matched expected behavior. The significant differences in timing and response are due to testing and modeling simplifications.

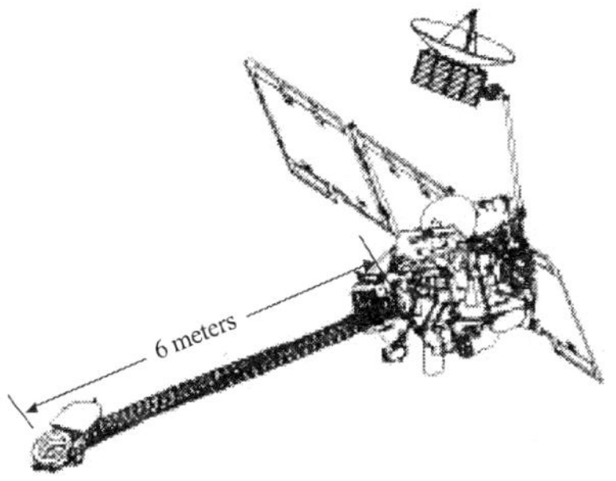

Figure 12. GRS Boom Deployed

The refined evaluation found no anomalies in the deployment event and verified the deployment occurred as planned with no unexpected boom dynamics. Further, the data verified full boom extension and lock-out to the configuration depicted in Figure 12, and proper function of the GRS boom components. Refinements included HFTL (High Force Thermal Latch) release time, boom extension time, peak spacecraft rate response, and partial deployment characteristics.

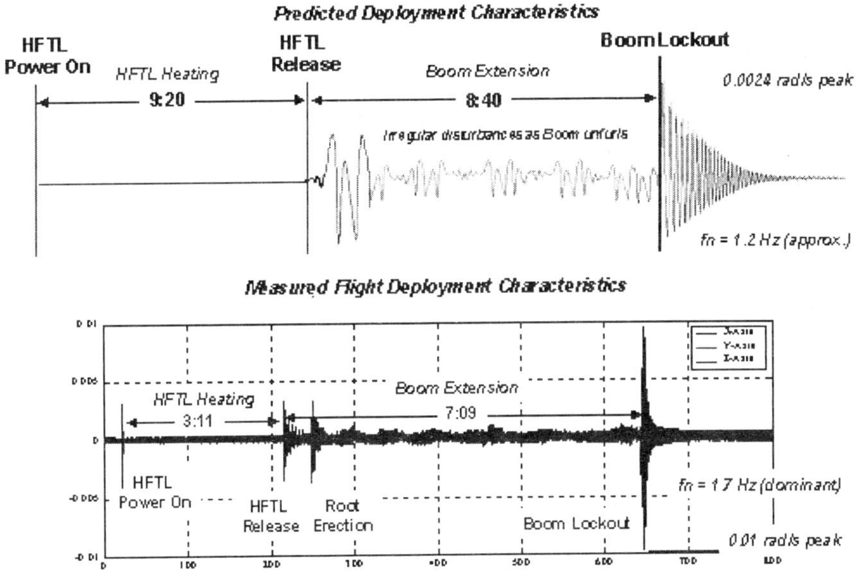

Figure 13. GRS Boom Deployment Time History Characteristics

Summary for GRS Boom deploy at -40C

Bus Voltage V	HFTL Time min:sec	Boom Time min:sec	Total Time min:sec
26	11:33	8:35	20:08
28	10:26	8:35	19:01
30	**9:19**	**8:35**	**17:54**

Summary for GRS Boom deploy at -50C

Bus Voltage V	HFTL Time min:sec	Boom Time min:sec	Total Time min:sec
26	12:07	9:09	21:16
28	11:00	9:09	20:09
30	9:53	9:09	19:02

Figure 14. GRS Boom Deployment Time Data

HFTL Release Time - The HFTL release time was much faster than predicted. Extrapolating from test data, at the actual 32V voltage at the time of deployment, the HFTL predicted heating time to release would have been about 8:00, while the actual in-flight release time was 3:11. Power telemetry indicated the HFTL power cut-off micro-switch was actuated 3 to 4 minutes after HFTL power-on. This indicates the HFTL achieved full stroke and that the quick release time was due to rapid HFTL heating rather than a pre-mature mechanical release. Heat-up time predictions were based on testing performed at ambient air pressure. Convection was known to be a significant heat loss but was accepted as a conservative test condition. The primary reason for the quicker HFTL release is most likely the heating in vacuum rather than in an atmosphere.

Boom Extension Time - The boom extension time was also shorter than predicted. The extension time prediction of 8:40 at -40C was based on linear interpolation between two boom deployment test points: a deployment time of 4 minutes was measured at 27C, and a deployment time of 9 minutes 9 seconds was measured at –50C. Review of damper acceptance data shows a non-linear relationship between damping rate and temperature. As shown in Figure 15, the increase in damping rate from room temperature to –40C could be 81% instead of 106% as given by the linear approximation. Using the polynomial curve-fit results in an extension time predict of 7:30 which agrees better with the actual 7:09 extension time. The remaining extension time difference is attributed to testing in 1-G and some uncertainty regarding the exact damper temperature in flight.

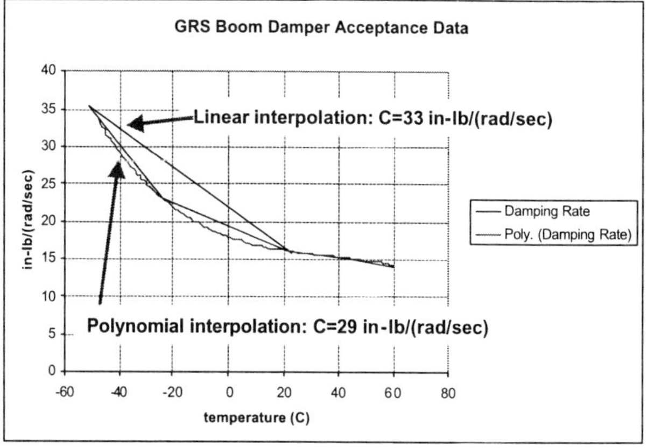

Figure 15. GRS Boom Damper Acceptance Data

Peak Spacecraft Rate Response - The peak spacecraft rate response to boom lock-out is greater than shown in the prediction: 0.006 rad/sec was observed about Z-axis (boom torsional axis) versus a predicted 0.0024 rad/sec. The prediction was based on applying the maximum angular rotation rate about the boom deployment axis (as measured in a boom deployment engineering test) to a rigid spacecraft model. When the same boom angular rate was applied to the spacecraft flexible body model (Figure 16), the predicted response agreed much better with the observed behavior (Figures 17-19).

The flexible model predicted response was 0.0075 rad/sec about the X-axis compared to 0.01 rad/sec measured in the actual flight data. The Y-axis simulation was 0.005 rad/sec compared to the flight data of 0.0025 rad/sec. In the Z-axis, 0.006 rad/sec was measured and predicted in the simulation. The X-axis flexible simulation response peak was largest, agreeing with flight data, and appears to be the result of coupling with Solar Array and HGA dynamics.

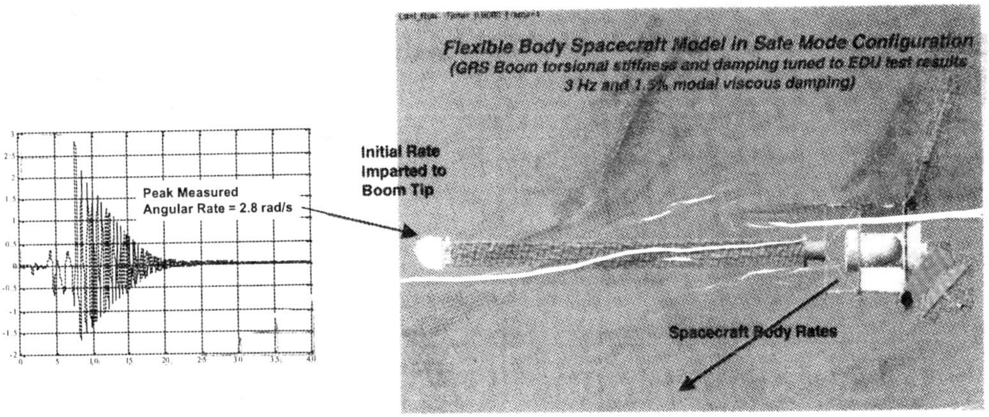

Figure 16. Boom Angular Rates Applied to S/C Flexible Body Model

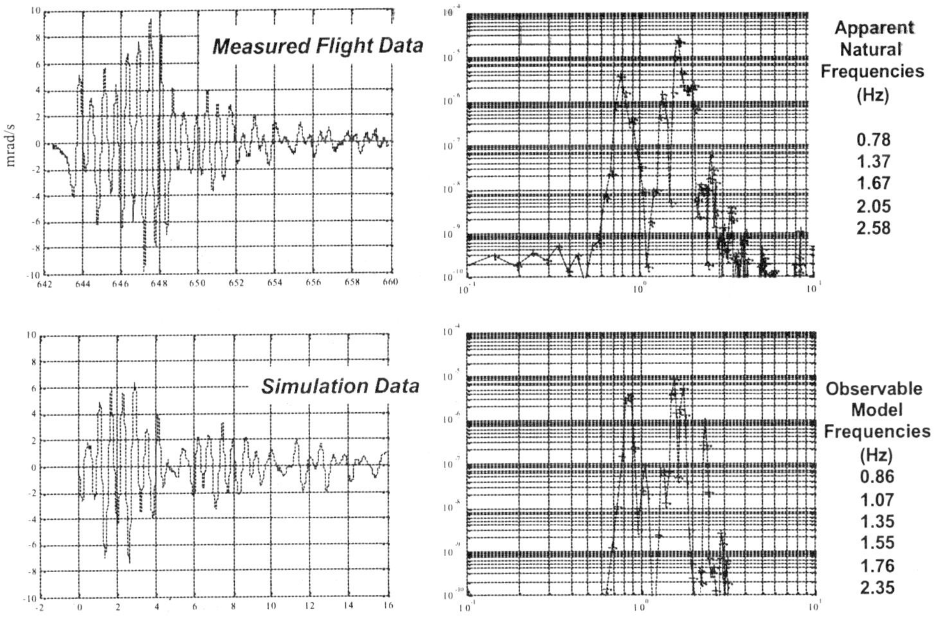

Figure 17. S/C X-Axis Boom Lock-Out Response

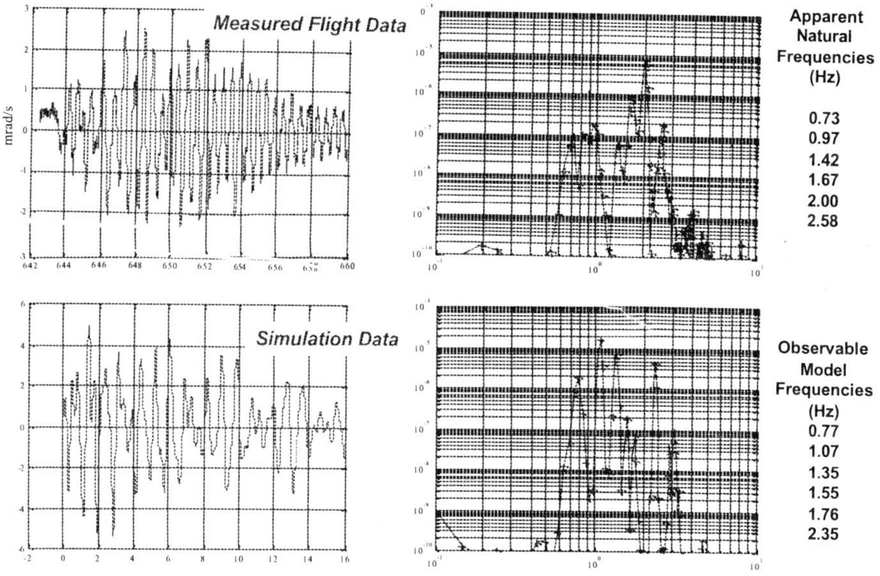

Figure 18. S/C Y-Axis Boom Lock-Out Response

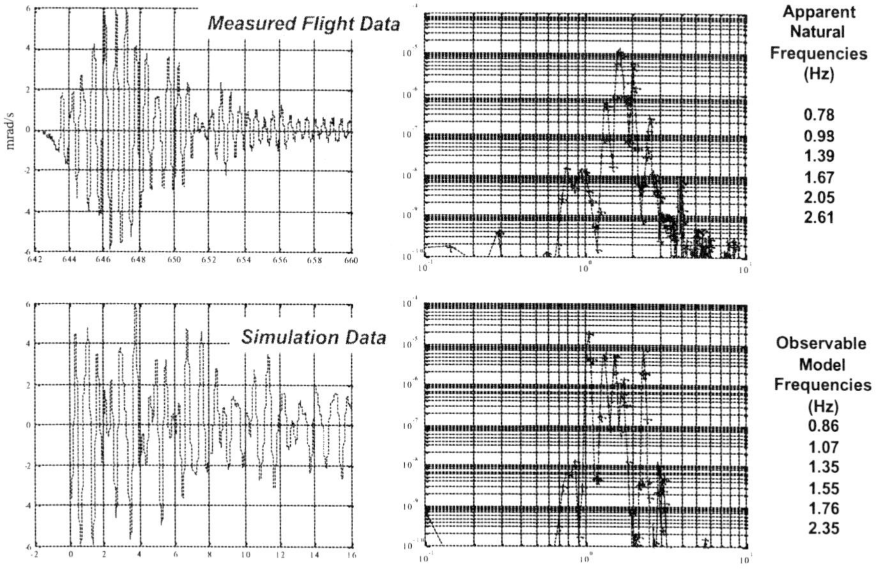

Figure 19. S/C Z-Axis Boom Lock-Out Response

Partial Deployment Characteristics - Based on the natural frequencies inferred from the lockout transient PSD's, the GRS boom was fully deployed. Spacecraft frequencies for the fully deployed GRS boom are shown in Figure 20. Note that a partially deployed boom would show lower natural frequencies.

The measured flight deployment characteristics in Figure 13 were compared to flexible modeling for a partially deployed boom. The initial release prior to root erection demonstrated the softest boom state at 0.23 Hz (Figure 21). The EDU test-correlated model for initial release was 0.31 Hz. After root erection, the natural frequency matches the partial deployment model: 0.4 Hz measured versus 0.41 & 0.45 Hz predicted using the EDU test-correlated model (Figure 22).

Observed Spacecraft Natural Frequencies (Hz)	Model Flex-Body Modes (Hz)	zeta (%)	Mode Description
0.73	0.77	0.4	GRS Boom Bending
0.78	0.86	0.6	GRS Boom Bending
0.98	1.07	0.5	Solar Array
1.39	1.35	0.8	Solar Array
1.67	1.55	1.5	GRS Torsion
2.05	1.76	1.5	Solar Array
2.58	2.35	0.6	HGA

Figure 20. Spacecraft Deployed Frequencies

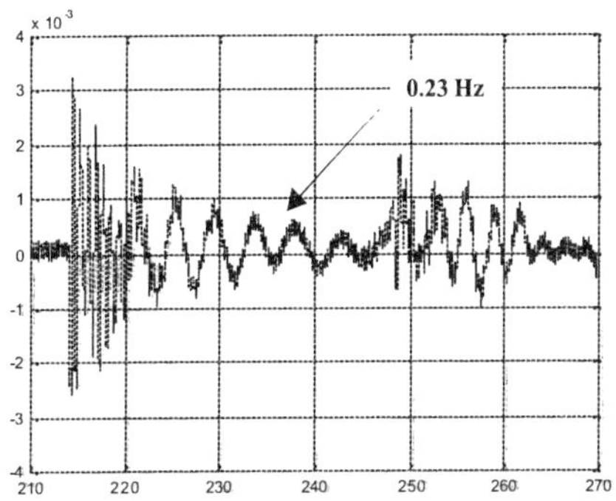

Figure 21. Initial GRS Release S/C Rate

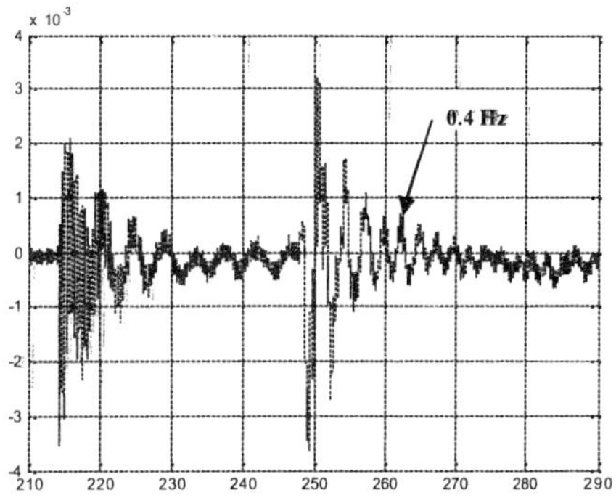

Figure 22. S/C Rate After Boom Root Erection

During extension, observed apparent stiffness varies with erection state of bays. Natural frequencies near 0.6 Hz are seen (Figure 23) that were similar to the EDU partial-deployment testing.

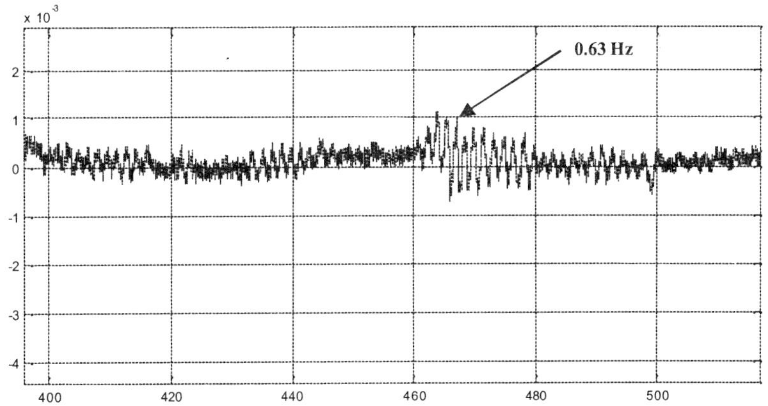

Figure 23. S/C Rotational Rate During Boom Extension

CONCLUSION

All deployment and articulation events were verified using spacecraft body-rate data from the IMU. The planned use of this technique allowed elimination of deployment sensors such potentiometers and limit switches, thereby supporting the low-cost, robust design of the spacecraft. Pre-event simulation predictions allowed nearly immediate declaration of a successful event once the telemetry was received. However, some lingering questions regarding the deployment events could have been answered immediately had high fidelity simulations been performed for all events. Current commercially available modeling and simulation software allow such simulations to be performed very efficiently.

ACKNOWLEDGMENTS

The work described in this paper was performed by Lockheed Martin Space & Strategic Missiles under contract to the Jet Propulsion Laboratory, California Institute of Technology, sponsored by the National Aeronautics and Space Administration (Mars Surveyor Operations Project, Contract number 960794). The Odyssey Attitude Control Group (Jim Chapel, Phil Good, Brad Haack, and Dave Shafter) provided a great deal of assistance obtaining and processing the data presented here. Assistance with much of the mechanical modeling and testing was provided by Katie Oakman of the Odyssey Mechanical Engineering Group.

AAS 03-077

TOPEX/POSEIDON: CONTROLLING THE ATTITUDE OF A 10-YEAR OLD

Paul Sanneman, Beth Lee and Peter Vanderham*

The first joint US-French ocean topography mission, TOPEX/ Poseidon, has been hugely successful in mapping global ocean circulation patterns, including most of what is known about the El Niño condition. The spacecraft and its Attitude Determination and Control System (ADCS) were designed for a primary mission of three years and an extended mission of five years. This paper summarizes the pointing performance, component operations and anomalies associated with attitude control of this satellite during its first ten years of on-orbit life. Specific topics include details regarding long-term parameter trending, star tracker re-incarnation, autonomous maneuver development and testing, alternate attitude determination methods, contingency planning and response, operational changes for mission longevity, reaction wheel faults, and orbit maintenance strategies that significantly reduced propellant usage. The paper concludes with a description of the orbit transfer maneuvers that were conducted during late 2002 to shift the ground track for tandem operations with the JASON mission.

INTRODUCTION

On August 10, 2002, the TOPEX/Poseidon satellite completed ten years of successful on-orbit operation while achieving an incredible science data acquisition return. This mission was initiated as a joint effort between the National Aeronautics and Space Administration (NASA) and the French space agency, Centre National d'Etudes Spatiales (CNES). Data from the TOPEX/Poseidon satellite has provided the capability to significantly increase the understanding of the dynamics of ocean circulation as well as the relationship between ocean circulation and global climate changes. This data was essential to the efforts to understand and predict the El Niño and La Niña weather patterns over the past 10 years.

The TOPEX/Poseidon satellite was designed, built, and tested for a three-year design life and a five-year mission life. In support of the mission, the Attitude Determination and Control Subsystem (ADCS) was closely monitored and maintained in order to meet the extended mission challenges. The information in this paper has been compiled from numerous sources and is intended to provide an overall summary of the long-term on-orbit performance, mission highlights, anomalies, and current status of the ADCS. The key lessons learned will also be presented in order to provide the experience base for similar missions.

* Swales Aerospace, 5050 Powder Mill Road, Beltsville, Maryland 20705.

MISSION AND SPACECRAFT OVERVIEW

Launched on August 10, 1992, the TOPEX/Poseidon satellite represented the first NASA space mission specifically designed and implemented for studying the circulation dynamics of the world's oceans. The primary mission objective is to measure sea-surface topography for the determination of the global ocean circulation. The TOPEX/Poseidon mission data is also used to develop improved large-scale ocean models for extended weather and climate predictions, and for determining the effects on the Earth and its oceans of broad ranging natural and man induced processes.

The TOPEX/Poseidon satellite employs a state-of-the art dual-frequency radar altimeter to measure the precise height of the satellite above the ocean surface. Additionally, the satellite also carries satellite-tracking instrumentation to support the Precision Orbit Determination (POD) process required to determine the satellite's orbit relative to the Earth's reference ellipsoid. Combining these data, the ocean surface height above the local marine geoid, i.e. its topography, is determined for definition of ocean currents (see Reference [1 and 2]). Figure 1 provides an illustration of the spacecraft with callouts for major modules or components. Table 1 provides a summary of the mission and spacecraft parameters.

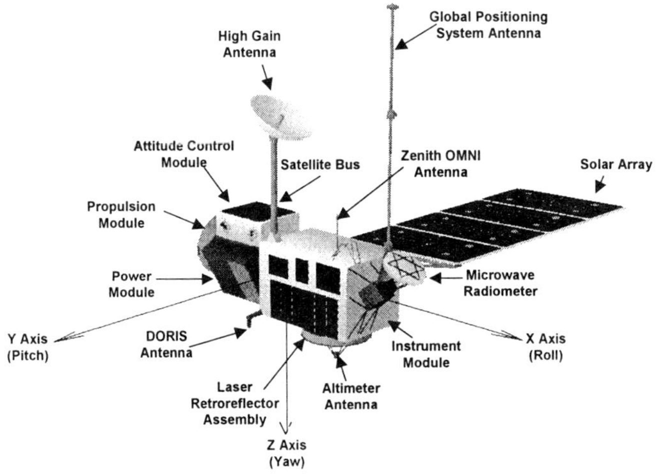

Figure 1 TOPEX/Poseidon Spacecraft.

SUN-NADIR STEERING

Since the TOPEX/Poseidon operational orbit is not Sun-synchronous, the angle between the Sun and the orbit plane, β', varies over the range of +/- 88 degrees during the mission. This β' variation over the year 2002 is shown in Figure 2. In order to satisfy the power requirements, the satellite uses a pitch-axis Solar Array Drive in combination with yaw attitude motion. A continuous yaw axis sinusoidal slew maneuver is performed when the angle between the Sun and the orbit plane increases/decreases past a specified angular limit (nominally at $|\beta'| > 30$ degrees)[1]. This sinusoidal profile, as shown in the lower two plots of Figure 2, allows efficient, smooth command generation while allowing predictable cosine power losses when compared to the yaw position for perfect Sun pointing.

[1] Note that the ADCS was designed to perform Yaw slewing for values of $|\beta'|$ as low as 10 degrees, and that the earlier mission operations utilized a fixed yaw transition threshold of $|\beta'| < 15$ degrees.

Table 1
TOPEX/Poseidon Mission Summary

Science • Measure Sea Level above referenced Ellipsoid to 14 cm • Satellite to Ocean Surface – 4 cm • Satellite to Referenced Ellipsoid – 13 cm	**Power** • Load: 1043 W • SA BOL: 3385 W • Battery: 150 Amp-Hours
Satellite • NASA Multimission Modular Spacecraft (MMS) Bus • Instrument Module (IM)	**Commanded Rate** • 125, 1K or 2K bps **Data Rate** • Telemetry: 16 or 1 Kbps • TDRSS: 512 Kbps
Instruments NASA • Radar Altimeter (ALT) • TOPEX Microwave Radiometer (TMR) • Laser Retroreflector Assembly (LRA) • GPS Demonstration Receiver (GPSDR) • Frequency Reference Unit (FRU) CNES • Doppler Receiver (DORIS) • Solid State Altimeter (SSALT)	**Nadir Pointing** Knowledge (3σ) • Specification: 0.15 deg • Measured: 0.08 deg Control (3σ) • Specification: 0.24 deg • Measured: 0.10 deg
Launch • Vehicle – Ariane 42P • Date – August 10, 1992	**OBC Memory** • 65,536 (18-Bit Words)
Orbit • Altitude – 1336 KM • Inclination – 66.039 degrees • Eccentricity – Less than 0.0001 • Ground Track Repeat in 127 Revs/9.9 days to ±1 KM	**Orbit Control** • Thrusters: 1 N and 22 N • Delta-V: 168 m/sec
Launch Mass • 2495 kg (includes 217 kg fuel)	**Data Storage** • 3 x 5 x 10^8 Bits
Dimensions • Solar Array – 3.3 x 8.7 m • Main Body – 5.5 x 2.8 x 2.8 m	**Precision Orbit Determination** • Center of Mass Error: 1.2 cm (3σ)

The commanded yaw angle is calculated using the following equations:

$$\text{Yaw angle} = 90^\circ + (90^\circ - \beta')\cos(\Omega) \quad \text{for } \beta' \geq 30 \text{ degrees}$$
$$\text{Yaw angle} = -90^\circ - (90^\circ + \beta')\cos(\Omega) \quad \text{for } \beta' \leq -30 \text{ degrees}$$
$$\text{Yaw angle} = 0^\circ \quad \text{for } 0 \leq \beta' < 30 \text{ degrees}$$
$$\text{Yaw angle} = 180^\circ \quad \text{for } -30 < \beta' < 0 \text{ degrees}$$

where the orbit position angle, Ω, is defined to be zero at orbit 6AM, and the spacecraft yaw angle is defined to be zero when the body axes are coaligned with the orbital reference frame. Details regarding the dynamic yaw attitude commanding are given in Reference [3].

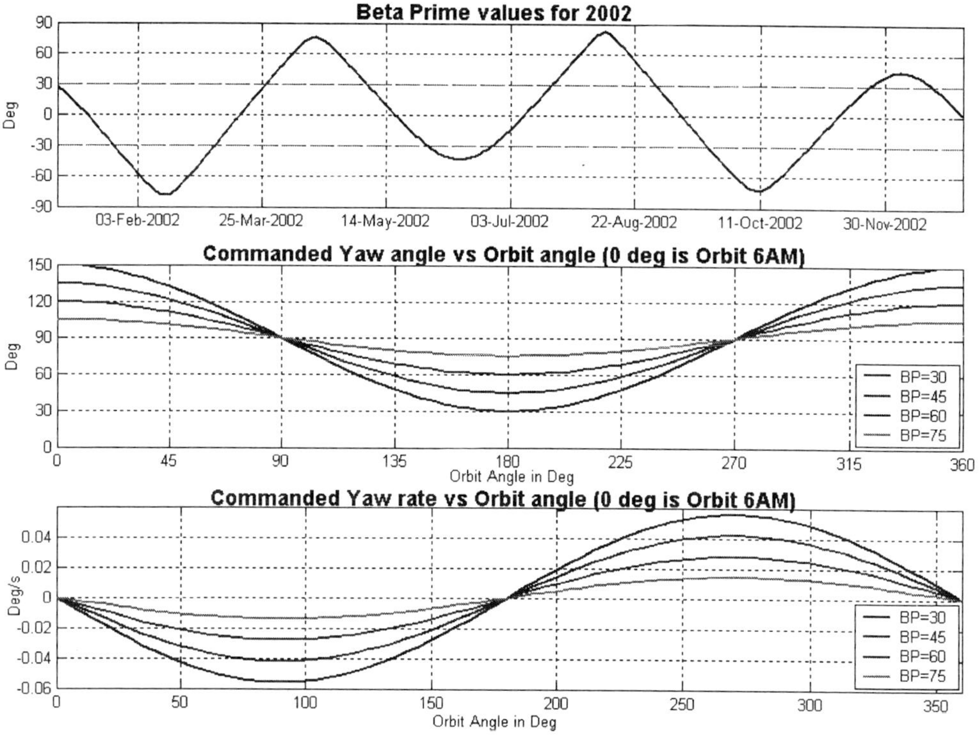

Figure 2 Sinusoidal yaw profile.

ATTITUDE DETERMINATION AND CONTROL SYSTEM

The NASA Multimission Modular Spacecraft (MMS) spacecraft bus design included the Modular Attitude Control System (MACS) for providing accommodation for most of the redundant hardware components of the ADCS (see Reference [4]). The ADCS architecture is a three-axis stabilized, zero-momentum system using three on-axis reaction wheels and a skew reaction wheel. The ADCS hardware components are summarized in Table 2. The ADCS flight software algorithms are hosted in the On-Board Computer (OBC), which includes a NASA Standard Spacecraft Computer (NSSC-1). The OBC acquires gyro data every 64 millisecond and performs the control processing at either 256 millisecond period for thruster-based modes or 512 millisecond for RWA-based modes. The six-state Kalman Filter processing (see Reference [5]) is performed every 32.768 seconds. The OBC control modes include capability for magnetic or thruster despin, Earth acquisition, Normal Mission and delta-v, while the analog Safe Hold Mode includes configurations for Earth, sun, or inertial pointing (see References [6-10]).

MISSION ADCS HIGHLIGHTS SUMMARY

The ten years of orbital operations have included a number of spacecraft and ADCS anomalies and events. Typical of most spaceflight missions, the causes of these unexpected events include hardware component upsets, sensor or actuator degradation, previously undiscovered software ‘features,’ operational procedure errors and even human operator error. On seven occasions, the anomalies caused the spacecraft to enter its Safe Hold Mode of operation – three of these were due to a halt of the OBC. In the case of the star trackers and Solar Array Drive Electronics, the redundant unit was utilized to resume normal mission operations with very little delay. Table 3 provides a summary of these mission highlights.

Table 2
ADCS Component Summary

MACS COMPONENTS	SUPPLIER	UNITS	SIGNIFICANT FEATURES
Dry Rotor Inertial Reference Unit	Teledyne	1	NASA standard with three 2-axis gyros, two rate modes, analog and digital outputs
Reaction Wheel Assemblies	Honeywell	4	Three orthogonal with one skew wheel; 0.2 Nm, 32 Nms@5000 RPM
Magnetic Torquer Bars	Ithaco	4	150 Am^2 each in roll and pitch; two 90 Am^2 in yaw
Advanced Star Trackers	Hughes Danbury	2	8.8° x 9.6° FOV CCD tracker (single star H, V, I) 17 arc-sec, 1-σ @ 10-Hz
Digital Fine Sun sensor	Adcole	1	Two axis sensor with ±32 deg FOV; 60 arc-sec, 3-σ
Three Axis Magnetometer	Schonstedt	2	± 1.0 Gauss range, analog output with 7.8 mG quantization
Remote Interface Unit	Fairchild	2	Provides all command and telemetry for MACS
Attitude Control Electronics	Fairchild	2	ADCS interfaces and analog Safe Hold Mode circuitry
OTHER COMPONENTS	**SUPPLIER**	**UNITS**	**SIGNIFICANT FEATURES**
Earth Sensor Assembly	Ithaco	2	Conical scanners, each with Coarse (±45°) and Fine (±5°), roll and pitch signals
Coarse Sun Sensors	Adcole	22	Redundant set of 6 yaw, 4 pitch and 2 Sun presence
Propulsion Module	Hamilton Standard	1	Four 22N (5 lbf) Delta-V thrusters Four 1N (0.25 lbf) Delta-V/ACS thrusters Eight 1N (0.25 lbf) ACS thrusters
Steering Control Electronics	Honeywell	2	Digital control of High Gain Antenna (HGA) gimbals
Solar Array Drive Electronics	JPL	2	Stepper-motor control with 0.088° counter/potentiometer resolution

POINTING PERFORMANCE TRENDS

Throughout most of the past 10 years, the TOPEX/Poseidon satellite was controlled by the Normal Mission Mode attitude control system. This Proportional-Integral-Derivative-Double Integral controller (see Reference [11]) consistently maintained the required attitude, allowing science data acquisition to continue with minimal interruption. As shown in Table 4, there are four main error sources that contribute to the overall on-orbit pointing performance.

- The portion of the pointing error that can be attributed to the controller is shown in lines 1 and 6 (labeled Attitude Control), and can be directly measured from telemetry data. For the TOPEX/Poseidon satellite, these errors represent the difference between the attitude command (generated from the on-board ephemerides) and the on-board attitude estimate.
- The errors shown in lines 2 and 7 (labeled Attitude Knowledge) represent the error in the on-board estimate of the satellite attitude. These errors cannot be assessed from satellite telemetry, but must be extracted by ground-based post processing. For the TOPEX/Poseidon satellite, JPL personnel developed the Attitude Sensor Calibration (ASCAL) program for this purpose.

- The third major error source is shown in lines 3 and 8 and represents the error in the on-board attitude command. This error is also determined by ground-based processing using both ASCAL and an on-orbit maneuver that is referred to as ABCAL (Altimeter Boresight Calibration).
- The error source shown in line 4 represents the pointing error that is associated with altimeter distortions as the satellite experiences temperature changes throughout the orbit. When the TOPEX/Poseidon satellite is in a full Sun orbit ($\beta' > 57.3$ degrees), the thermal distortions are minimized; conversely, when the Sun is near the orbit plane (β' near 0 degree), the eclipse duration is longest (about 30 minutes). The errors shown in line 4 are based on prelaunch predictions since there is no supporting analysis or measurements to refine those values.

Table 3
Summary of ADCS Mission Highlights

Event	Date	Comment, Action / Response
Safe Hold Mode Entry	28-Aug-92	OBC assembly code had an 'undiscovered' FDC threshold
Safe Hold Mode Entry	25-Sep-92	OBC gyro scale factor error for Low Rate Mode in launch load
ASTRA1B Shutter Closure	26-Nov-92	Reset commands did not fix problem; remains inoperable. Background counts continue to be calculated
Safe Hold Mode Entry	7-Aug-93	Ephemeris file upload procedure change to avoid reoccurrence
ASTRA1-B Probable loss	25-May-94	Monitor status; power/temp change indicate probable loss background counts no longer calculated
ASTRA1-A SEU TEC OFF	13-Oct-94	SEU reset response; hardware commands to recover successful
ASTRA1-A SEU TEC OFF	31-Jul-95	SEU reset response; hardware commands to recover successful
Safe Hold Mode Entry	26-Nov-95	On-Board Computer halt; recovered from Sun Pointed safe hold
Safe Hold Mode Entry	20-Jan-96	On-Board Computer halt; new flight image for rapid recovery
SADE-A Failure	19-Sep-96	Swap to SADE-B; update TMONs for better protection
ESA-1 Turn off	11-Dec-96	To preserve lifetime, due to lubricant issue
ESA-2 Turn off	4-Mar-97	To preserve lifetime, due to lubricant issue. Turned on for OMMs
ASTRA1-B Resurrection	26-Apr-98	Shutter opens and begins tracking stars after power cycling the ASTRAs
ASTRA1-A Shutter Closure	26-Apr-98	Early yaw flip caused shutter to close, background level to high to reopen shutter. CCD Temp rises due to TEC turn off
TAME	1-Dec-98	TOPEX Automated Maneuver Experiment executed
Safe Hold Mode Entry	31-Aug-99	On-Board Computer halt, recovered from Sun Pointed safe hold
Yaw Reaction Wheel Anomaly	1-Sep-99	Tachometer stuck at zero RPM with max torq being applied. While on ACE-B notice wheel frees itself when torq polarity flips
Yaw Wheel Loss of Winding	15-May-00	Increased zero Amp samples, possible loss of motor winding
ASTRA1-B SEU TEC ON	8-Oct-00	Non-terminating software check while shutter was closed. Both ASTRAs power cycled, clearing the self-check
Safe Hold Mode Entry	23-Nov-00	Bad Ephemeris, recovered from Sun Pointed safe hold
Roll Reaction Wheel Anomaly	23-Nov-00	Tachometer stuck at zero RPM with max torq being applied. While on ACE-B notice wheel frees itself when torq polarity flips
IMIU Rollover	26-Apr-01	IMIU clock rolls over, new OBC parameters uploaded
Roll Wheel Loss of Winding	Aug-01	Increased zero Amp samples, possible loss of motor winding; similar to Yaw wheel
TOTM Maneuvers	15-Aug-02 to 16-Sep-02	TOPEX moved into a tandem orbit with JASON via a series of posigrade and retrograde maneuvers

Table 4
TOPEX/Poseidon On-Orbit Satellite Pointing Performance

	Sources of On-Orbit Error	Knowledge			Accuracy		
		Roll	Pitch	Yaw	Roll	Pitch	Yaw
	Random						
1	Attitude Control (3-σ)	0.000	0.000	0.000	0.015	0.019	0.032
2	Attitude Knowledge (3-σ)	0.030	0.042	0.033	0.030	0.042	0.033
3	ABCAL Uncertainty (3-σ)	0.021	0.021	0.000	0.021	0.021	0.000
4	Thermal Distortions (3-σ)	0.030	0.030	0.030	0.030	0.030	0.030
5	RSS Random (Lines 1-4)	0.047	0.056	0.045	0.050	0.059	0.055
	Deterministic						
6	Attitude Control (3-σ)	0.000	0.000	0.000	0.005	0.004	0.055
7	Attitude Knowledge (3-σ)	0.001	0.001	0.003	0.001	0.001	0.003
8	ABCAL Bias Quaternion Error (3-σ)	0.000	0.000	0.000	0.020	0.004	0.000
9	Sum Deterministic (Lines 6-8)	0.001	0.001	0.003	0.026	0.009	0.058
	Total (Sum 5, 9)	0.048	0.057	**0.048**	0.076	0.068	**0.113**
	Yaw Pointing Requirement (3-σ)			**0.070**			**0.440**
	Nadir Pointing Error (3-σ)	**0.075**			**0.102**		
	Nadir Pointing Requirement (3-σ)	**0.150**			**0.240**		

Note: All numbers in degrees, 3-σ

These error sources are further separated as random and deterministic, where a portion of each error is contributed to either knowledge or accuracy.

Attitude Control

As described earlier, the TOPEX/Poseidon satellite follows a commanded yaw attitude that changes with the angle between the Sun and the orbit plane, where the satellite maintains either a fixed yaw attitude or follows a sinusoidal yaw command. This changing yaw attitude profile coupled with the Sunrise/Sunset thermal transients directly affects the attitude errors in all axes. As shown in Table 4, the mean attitude errors are essentially constant, with values of 0.005, 0.004, and 0.055 degree in the roll, pitch, and yaw axes, respectively (line 6). The variation in the attitude error, however, is noticeably reduced when the satellite enters a full Sun orbit ($|\beta'| > 57$ degrees) see Figure 3. During eclipsing orbits, the motion of the large solar array at entrance to or exit from eclipse is the predominant disturbance, causing attitude errors that may exceed 0.2 degree for a short period of time (less than 5 minutes per orbit, see References [12-13]). The standard deviation during eclipsing orbits is therefore not representative of the attitude control performance throughout the orbit. Controller performance in response to random disturbances (line 1) is therefore based on the standard deviation during full Sun orbits.

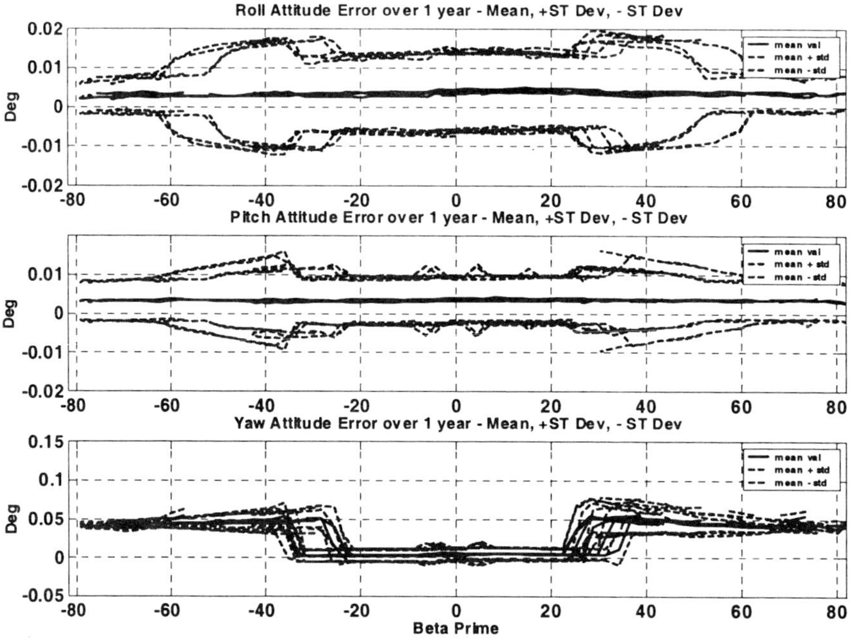

Figure 3 Attitude Errors over a typical year.

Attitude Knowledge

The on-board attitude estimate for the TOPEX/Poseidon satellite is generated by a 6 state Kalman Filter (see Reference [5]). This filter uses celestial measurements from the active star tracker (ASTRA-1B) and the Digital Fine Sun Sensor (DFSS) to periodically correct the attitude estimate that is propagated from the gyro data. The accuracy of this on-board attitude estimate is therefore highly dependent on the accuracy of the information regarding the alignment and scale factors associated with the measurements from the gyro, ASTRA-1B, and the DFSS. The ASCAL program was developed to determine these parameters by post-processing flight telemetry data using a 22-state batch least-squares estimation algorithm. The ASCAL program was used early in the mission to generate new alignment matrices for the gyro, DFSS, and ASTRA-1B. Since there was no noticeable change following launch, prelaunch estimates for the DFSS and ASTRA-1B scale factors were not, and have not, been changed. The gyro scale factors, however, are slowly changing and have been uplinked on a yearly basis.

The on-board Kalman filter attempts to compensate for the errors in sensor scale factors and misalignments via changes to the gyro drift rate bias estimates. For that reason, the on-orbit attitude knowledge error is minimized immediately following the scale factor uplink and slowly increases until the next uplink is performed. This characteristic is most noticeable in the Kalman filter residuals that are the angular difference between where a particular celestial measurement is expected to be found (based on the current attitude estimate) and where it is actually found as measured by the sensor. As shown in Figure 4[2], the residuals are lowest following the uplink,

[2] During the first year of the mission, both star trackers were used for the celestial data. Since late November 1992, one tracker and the DFSS have been used; and data for the last 9 years of the mission was used to generate that figure.

and gradually increase until the next uplink is performed. The largest spikes in that figure are associated with the 180-degree yaw turn that is performed when the Sun crosses through the orbit plane.

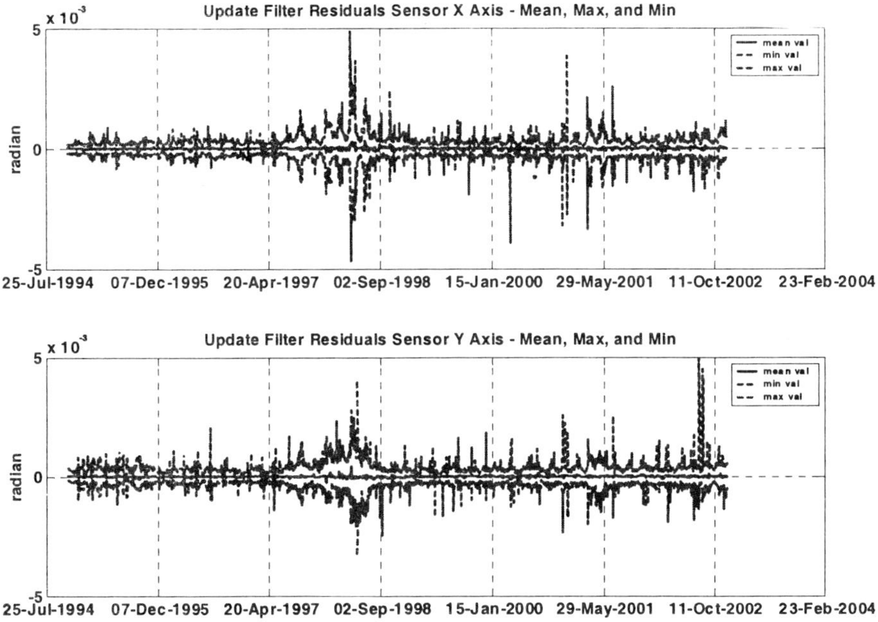

Figure 4 Update Filter Residuals.

The attitude knowledge errors given on lines 2 and 7 of Table 4 were computed using the attitude reconstruction data for numerous ASCAL solutions that were generated during the 10-year mission. The constant bias in the attitude knowledge error is shown in line 7 of the table, while the time-varying component is shown in line item 2.

Altimeter Pointing

On-orbit calibrations of the spacecraft pointing are periodically performed to evaluate the altimeter pointing errors, which may include ADCS-to-Altimeter misalignments, altimeter offsets, and electrical biases. To perform this calibration, a predefined maneuver is executed and then both the ADCS and Altimeter data is processed using ground-based software. The Altimeter boresight roll and pitch axis pointing errors can then be determined and a new attitude bias command uplinked to the ADCS software, if necessary. This entire process is referred to as ABCAL, utilizes the ASCAL software, and was also developed at JPL.

During most of the mission, a calibration slew maneuver that commands the satellite to follow a cross shaped pattern has been used to obtain data for ABCAL processing. This maneuver is executed at a consistent point in the orbit where the altimeter thermal variations are minimized and the ocean surface provides a typically static reference. Telemetry data from this maneuver is used in the ASCAL software to generate the reconstructed attitude estimate. The ABCAL ground-based software then compares this truth attitude estimate with the altimeter signal to determine a measure of the off-nadir attitude. The difference between these two attitude estimates indicates the altimeter boresight misalignment.

Altimeter boresight misalignment was corrected by uplinking a bias pointing command early in the mission. This action essentially removed any structural and thermal distortions that occurred during launch and initial on-orbit operation. Additional ABCAL processing indicates that this altimeter misalignment has not significantly changed throughout the mission; therefore the bias pointing command that was loaded in mid-1994 is still in use today. The error associated with this bias parameter is shown in line 8 of Table 4. Analyses by the Navigation Team indicate that the maximum nadir pointing error due to the ephemeris numerical representation via coefficient tables remains within 0.008 degrees. The random errors associated with altimeter pointing shown in line 3 of Table 4 are the accuracy of the ABCAL process, and include all other uncompensated error sources (such as ephemeris errors, timing errors, data quantization, etc.).

ADCS COMPONENTS OPERATIONAL SUMMARIES

TOPEX utilizes the modular ADCS, similar to that used on Solar Maximum Mission (SMM), LANDSAT-4, LANDSAT-5, and UARS. The ADCS contains all the equipment required for attitude stabilization, attitude determination, attitude acquisition, attitude control, and thruster control during all phases of the mission. The ADCS functions are supported by a suite of sensors and actuators as identified in Table 2. This section reviews the component operational histories.

Star Trackers

TOPEX launched with two Advanced Star Trackers (ASTRAs) that were developed by Hughes Danbury Optical Systems (subsequently became Raytheon Optical Systems and currently is Goodrich). Each ASTRA consists of a sunshade, optics, focal plane, focal plane electronics, and data processing electronics. The tracker uses a Charge Coupled Device (CCD), with a Field of View (FOV) of approximately 6.8 by 9 degrees to detect and track a single star (data output includes horizontal and vertical coordinates and the detector magnitude of intensity). A life-limiting factor in the CCD is the increase of dark current over the operational life of the detector. In order to improve long-term detector performance, the ASTRA is equipped with a thermoelectric cooler (TEC) that maintains the CCD temperature between –2 and 0 degrees Celsius. When a bright object is within 20 Degrees of the boresight, a shutter will automatically close to protect the CCD. The shutter has pinholes to let a small amount of light through while the shutter is closed; and once the background level is below a pre-defined limit (~10000 counts), the shutter reopens.

During the first few months of the mission, both ASTRA1-A and ASTRA1-B were used together as the absolute attitude measurements for the Kalman Filter. However, during most of the mission, only one ASTRA has been operable as explained further below (see References [14 and 15]).

ASTRA1-A

ASTRA1-A was the prime attitude sensor following the suspected loss of ASTRA1-B early in the mission. The observed detector background level began increasing in mid-1997; circuit analyses indicated the cause may have been related to readout circuitry and not CCD degradation. The background increase presented a problem with the hysteresis logic used in the shutter operation; namely, when the level increased to over 10,000 counts in April 1998, any subsequent shutter closure due to moon or sun impingement would trigger a closure that would be prevented from reopening. Due to this situation, ADCS engineers designed and executed Lunar Avoidance Maneuvers (LAM) using the yaw attitude slews to prevent the Moon from closing the shutter. Sadly, an early Yaw Flip maneuver in April 1998 caused an unexpected solar reflection off an interior surface of the baffle closing the shutter forever.

ASTRA1-B

The shutter on ASTRA1-B closed shortly after launch (November 1992) due to an SEU-induced latchup that caused the tracker detector readout A/D converter to change from 'True Binary' to an 'Inverted Offset Twos Complement.' This caused a reverse video effect, which forced the shutter closure. After several orbits, the update filter swapped to the backup sensor mode using ASTRA1-A and the Digital Fine Sun Sensor (DFSS). In April 1998 when ASTRA1-A was lost, a ground command to power cycle both trackers was performed. This action cleared the latchup in ASTRA1-B, and it began tracking stars almost immediately. Since its resurrection, ASTRA1-B has performed nominally. The background level of the CCD remains low and varies with the Sun angle from the orbit plane and yaw slewing profile (see Figure 5). Minimum background level during fixed yaw periods continues to rise slowly (~100 counts per year [3]), as illustrated in Figure 5. Shutter operation is nominal, where closures are mostly caused by the moon passing through the field of view. ASTRA1-B typically tracks 10-25 different stars per day, and no problems associated with the on-board star identification have been observed. The tracker detector magnitude error is shifting at a rate of 0.0117 DM per year, implying continued stable CCD operation.

Inertial Reference Unit

The Dry Rotor Inertial Reference Unit-II (DRIRU-II) uses three two-axis gyros to provide analog and digital outputs for three orthogonal axes (manufactured by Teledyne). Given the history of the DRIRU-II on other space missions in the 1980s and 1990s, two parameters that are closely monitored include the motor current for all three 'gyro' channels (see Figure 6) and the estimated gyro bias (see Figure 7). The current configuration of the gyro channels is provided in Table 5.

Table 5
Gyro Configuration

Gyro Channel	Measurements
A	Primary Roll, Backup Yaw
B	Backup Pitch, Backup Roll
C	Primary Yaw, Primary Pitch

[3] Fixed yaw provides the best representation of the CCD background level. A background level above 10,000 counts would not allow the shutter to reopen following a closure.

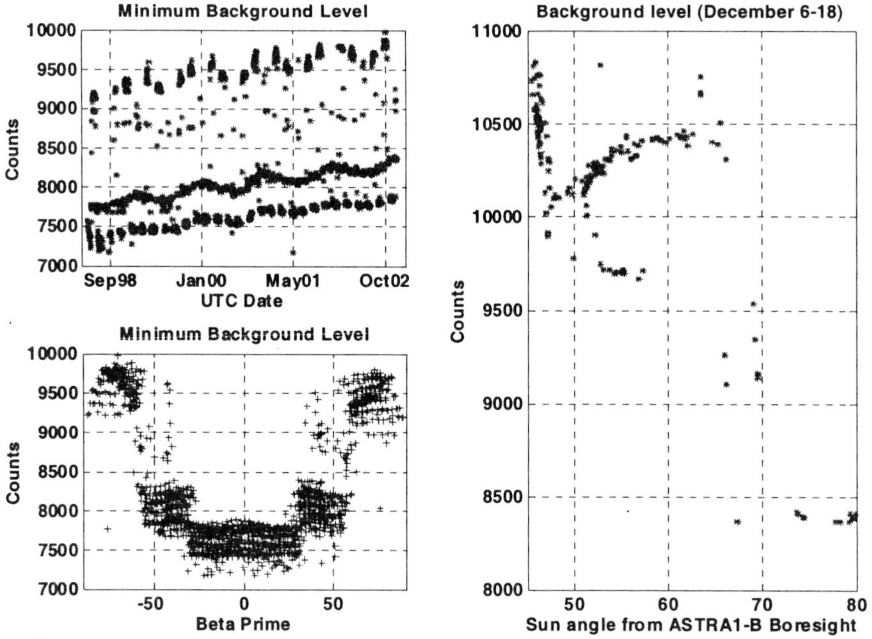

Figure 5 ASTRA1-B Background levels.

Gyro Channel A

Channel A provides primary roll measurements. The large shift in the mean of the motor current was analyzed by Teledyne and determined to be a change in the relative bearing position with respect to its contact with the retainer ring, a not uncommon situation observed on other missions. This new position requires more torque and subsequently more power to maintain the synchronous operating speed. Large motor current spikes have also been observed and correlate with an increased drift in the roll axis gyro bias (see Figure 7) as well a related increase to the update filter residuals. Engineers from Teledyne stated that these spikes are likely due to minor ball bearing retainer instabilities and are not considered to be life threatening.

Gyro Channel B

Channel B provides backup measurements for pitch and roll. Similar to Channel A, gyro spikes have been observed along with increased motor current; however, since the Channel B measurements have never been used in the on-board update filter, no estimates of gyro bias are available.

Gyro Channel C

Channel C provides primary measurements for yaw and pitch. Since launch, the Channel C motor current has been very steady, and no gyro spikes or change in the noise characteristics have been observed.

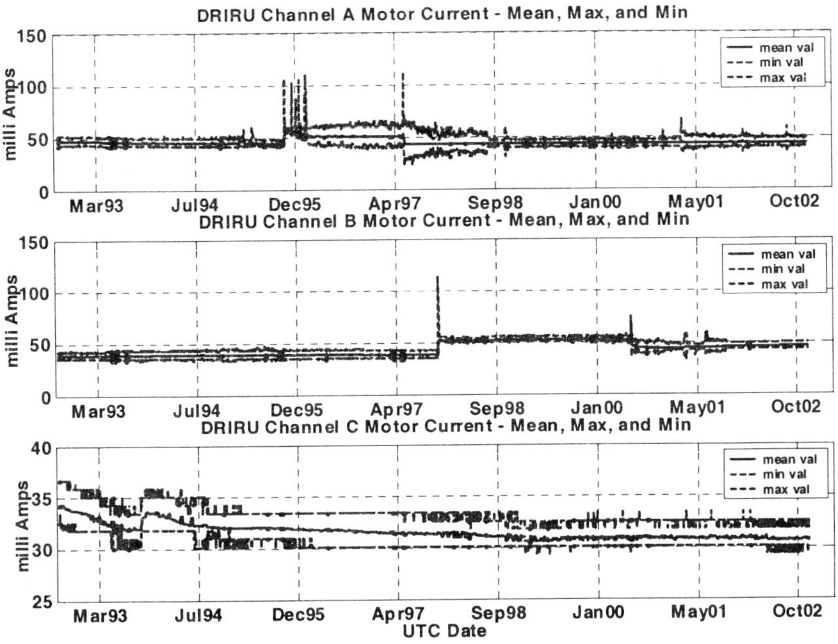

Figure 6 Gyro Motor Current

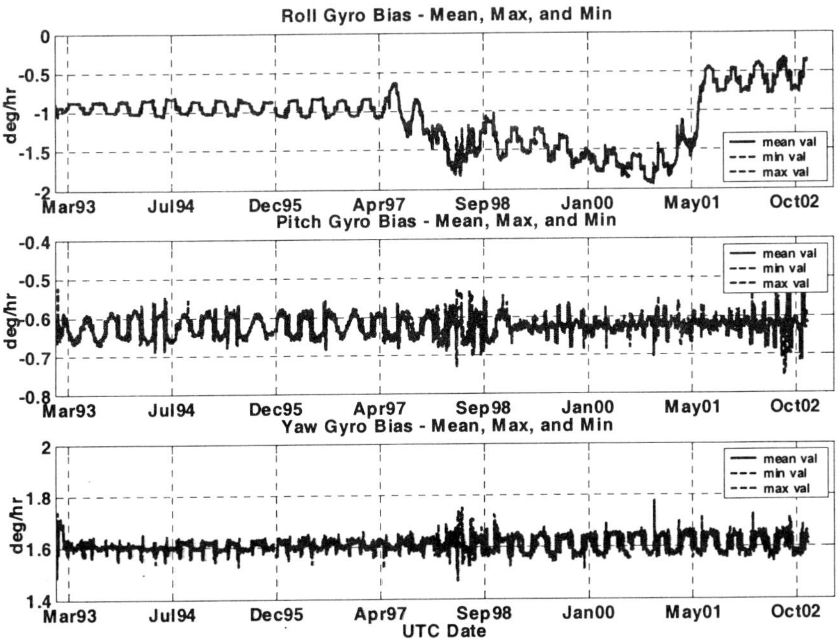

Figure 7 Gyro Bias

Digital Fine Sun Sensor (DFSS)

The DFSS is a non-redundant two-axis Sun sensor manufactured by Adcole Corporation and is used for attitude initialization and as a celestial sensor to support the Update Filter since late 1992. Based on the orbit design and the satellite yaw slewing profile, the Sun position in the DFSS field of view can vary from full orbit availability to a limited region during fixed yaw periods. The DFSS FOV is +/- 32 degrees in each of two orthogonal axes. Early in the mission, solar reflections from the propulsion module prompted development of a software patch that identifies the false, reflected Sun exposure and prevents the usage of the corrupted data. In February 1993, the Update Filter residuals exhibited erratic behavior with residuals repeatedly reaching 150 arcseconds. Analyses indicated the likely cause to be degradation of the DFSS Y-axis reticle, and a software patch was uploaded to the spacecraft to prevent the usage of the data during solar passage through this region of the DFSS.

Earth Sensors

Each of the two independent infrared conical scanners, manufactured by Ithaco Space Systems (now part of Goodrich), produces satellite roll and pitch axis off-nadir measurements. These ESAs demonstrated numerous Single Event Upsets (SEUs) that manifested themselves in two ways (see Reference [16]). In one case, data spikes on the measurement outputs are seen simultaneously in both channels of a given ESA. Analysis of high rate data revealed that the shape of the ESA output was consistent with a low pass filter employed in the ESA electronics. However, since sampled peak values exceeded 5 degrees, the on-board thresholds for Fault Detection and Correction were adjusted to prevent a false trigger. SEUs also caused indications of Wide Angle Mode (WAM) status as single-cycle bit flips. During the early mission, all satellite subpoints during SEUs were plotted on a geographic map and generally correlated with the South Atlantic Anomaly (SAA) as shown in Figure 8. In late 1995, ESA problems and failures on other missions (such as Landsat- 4, Landsat-5) prompted the manufacturer to stop an on-going life-test and perform an analysis on the ESA bearing assembly. The ESAs bearings were lubricated with Braycote 815Z oil, which reacted with some of the filings from the bearing material caused by a tight mechanical tolerance. Based on close coordination with Ithaco, the ADCS team recommended turning off the ESAs to preserve their operational lifetimes. Since the ESAs were only used for FDC during normal operations, minor software code changes were required; however, the operational impact was to force entry to a sun-pointing Safe Hold Mode. ESA-1 was powered off on December 11, 1996, and has only been turned on a few times during safehold events and during the TOTM maneuvers.

ESA-2 was powered off on March 4, 1997, and has been turned on during numerous Solid State Altimeter (SSALT) cycles, Orbit Maintenance Maneuvers (OMMs) and various spacecraft anomalies. Figure 9 shows typical ESA measurements for a one-day period plotted versus latitude. At the TOPEX altitude of 1335 km, the maximum difference between the geodetic Earth (which is the TOPEX nadir target) and a spherical Earth is about 0.16 deg. As shown in Figure 9, the ESA chord width measurement (E-Channel) demonstrates this magnitude of variation.

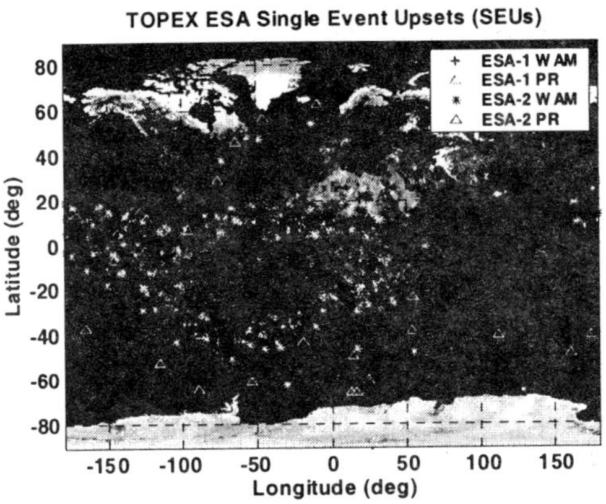

Figure 8 ESA Single Event Upset Locations.

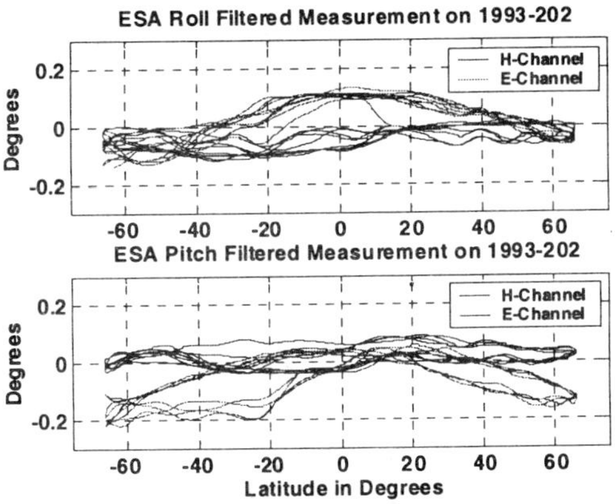

Figure 9 ESA Typical Day.

Reaction Wheel Assemblies (RWA)

The four RWAs (manufactured by Honeywell Defense & Space Systems) are the primary actuators for the ADCS and each consists of a three-phase motor, flywheel, thermistor, redundant tachometers, and drive electronics enclosed in a sealed container. During nominal mission operations, the wheels operate at a mean speed of 2000 RPM for the on-axis wheels and –3300 RPM for the Skew wheel to provide a zero-momentum control regime. Throughout most of the mission, the RWA motor current and mean commanded torques during fixed-yaw periods have provided trends that indicate good motor and bearing performance. However, in the last few years, some degradation related to motor electrical operation has been observed, as discussed below.

Yaw RWA

During a safe hold event in August 1999, the yaw wheel would not respond to the drive commands during a zero-speed crossing while in three-wheel control; the wheel responded only when the torque voltage polarity changed signs. In May 2000, the yaw wheel motor current began showing increasing zero amp telemetry samples and a changing characteristic of the commanded torque to the wheel, see Figure 10. Following analysis of the wheel data and discussions with Honeywell, it was concluded that one of the three motor phase drive circuits was in an open state. Given this understanding, if the wheel is allowed to come to a complete stop, there is a 33% chance that it will not be able to restart in at least one direction. It may be possible to restart the wheel by performing a power cycle or by alternately applying positive and negative maximum torque commands to the wheel. The ADCS team has developed a software patch that will detect the "stuck" wheel and apply a maximum torque or "kickstart" command in the opposite polarity. As long as the ADCS remains in the four-wheel mode (with the Yaw wheel near 2000 RPM), it continues to function nominally.

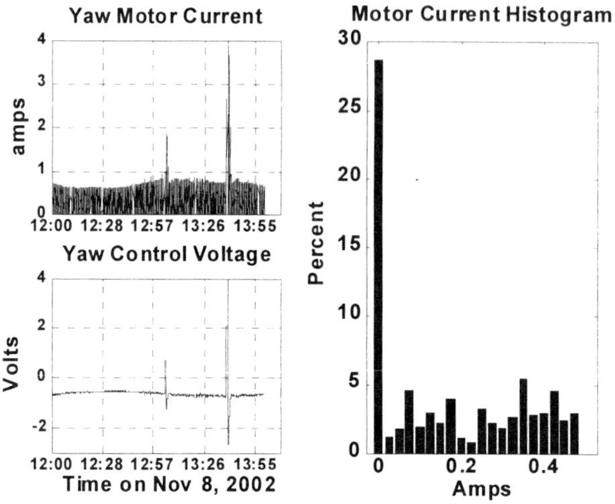

Figure 10 Yaw Reaction Wheel Parameters.

Pitch RWA

The pitch RWA continues to perform nominally with no signs of anomalous behavior, see Figure 11.

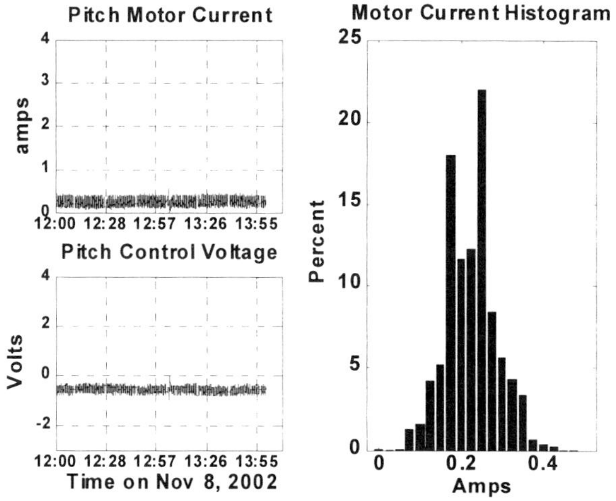

Figure 11 Pitch Reaction Wheel Parameters.

Roll RWA

During the Thanksgiving safe hold in 2000, the roll RWA did not respond to drive commands during a zero-speed crossing while in three-wheel control. The wheel eventually recovered during the transition to four-wheel control. The roll wheel motor current telemetry began exhibiting an increased percentage of zero amp telemetry samples in late 2001, and it is believed that the roll RWA has a similar fault as discussed in the Yaw RWA section above.

Skew RWA

Following the transition from fixed yaw-to-yaw steering in September 2002, the skew wheel motor also began to show an increase in zero amp telemetry samples. This change in telemetry characteristics is similar to the yaw and roll wheels but not identical. The increase in zero amp samples seems to correlate with an increase in RWA temperature while yaw steering, and it could be caused by increased noise in the skew motor current telemetry read-out channel. While TOPEX is in a fixed yaw profile, the wheel temperature decreases and the skew wheel motor current returns to a pre-anomalous state.

Propulsion Subsystem

The Propulsion Module (PM), manufactured by Hamilton Standard, includes a main tank and an Auxiliary Tank Kit (ATK) with a combined hydrazine fuel capacity of 217.3 Kg with an initial pressure of 300 PSI. Figure 12 shows the fuel consumption over the entire mission. The PM also includes four Rocket Engine Modules (REMs), each with four thrusters (one 5lb (22N) and three 0.25lb (1N)), oriented to provide for delta-v and attitude control. Throughout the mission, the propulsion system has performed well with excellent delta-v performance. Unfortunately, there is a lifetime degradation issue with the 4N49 opto-couplers used in the command and telemetry electronics. In January 1996, the opto-couplers associated with thruster telemetry failed due to total dose and the low drive current used in the circuit design. Based on these failures, the analyses were revisited and the opto-coupler circuits used in the thruster and latch valve commands were estimated to have an 8–10 year radiation degradation lifetime. Following an orbit maneuver in May 2001, the command to close a latch valve failed – after 8.75 years on-orbit. The remaining latch valves were all opened at that time with the double-seat feature of the thrusters serving as the only leak protection.

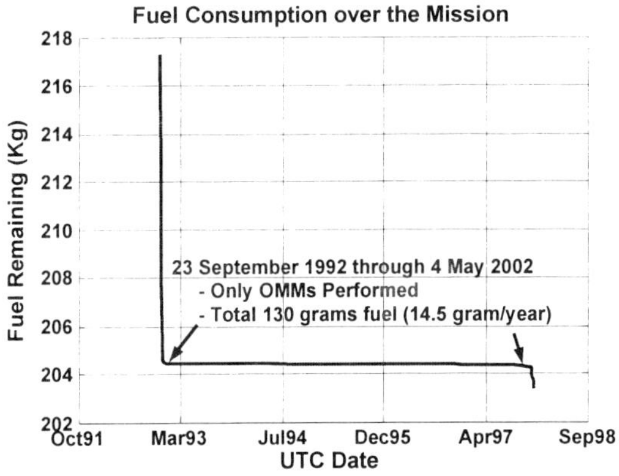

Figure 12 Fuel Consumption.

SPECIAL EVENTS OR ACTIVITIES

TOPEX Autonomous Maneuver Experiment

The TOPEX Autonomous Maneuver Experiment (TAME) was initiated in 1994 as part of NASA / JPL developments in autonomous planning of propulsive maneuvers (see References [17-19]). The demonstration of highly autonomous, goal-seeking control algorithms can enable future deep-space missions and substantiate the possible mission operations cost savings. Since propulsive maneuver development is a multidiscipline activity, a major goal for TAME was to perform the attitude maneuver planning and constraint checking on-board the spacecraft. The TAME team adapted existing path planning and constraint modeling algorithms for upload to the T/P GPS Demonstration Receiver (GPSDR). Due to on-orbit experience with reloading the software to the GPSDR 1750A processor, the capability existed to host the TAME algorithms in the GPSDR in a temporary manner. The existing hardware and software interfaces between the GPSDR and the OBC were also redesigned to support the experiment. The availability of the SDTVB and an engineering model GPSDR unit enabled a complete validation of the algorithm implementation, interfaces, command/telemetry, and operational procedures. The implementation included numerous 'check valves' in the execution and process for ground-based validation of the TAME processing and planned maneuver sequence prior to enabling the plan for on-board execution. TAME was successfully executed on December 1, 1998, during OMM10.

Clock Rollover

The spacecraft internal timing employs a Mission Elapsed Time counter that serves as the on-board clock reference. This counter was reset on the launch pad and included a sufficient number of bits for the extended 5-year mission. However, based on the extended mission success, the clock counter eventually became as significant as the Earth-bound Y2K event: special efforts were required to investigate and plan for the clock rollover. Fortunately, the time distribution and synchronization functions had been designed to accommodate the rollover with a temporary change in ADCS control mode and uploadable OBC parameters. This activity was successfully performed during April 2001 and included the update of the on-board star catalog to correct for the precession of inertial coordinate frame.

TOPEX Orbit Transfer Maneuver

After many years of planning and development, the 'replacement' spacecraft for ocean topography – the French-built JASON-1 mission – was launched by a US Delta-II rocket on December 7, 2001 (see Reference [20]). After orbit maneuvers were completed on JASON-1 to match its orbit in formation with TOPEX/Poseidon, the science observations were correlated and calibrated with respect to the nine-year TOPEX/POSEIDON history. Following this correlation, the JASON-1 science data became the prime source for science observations and the science working group expressed interest in moving TOPEX/POSEIDON to a 'tandem' orbit to allow acquisition of stereo data for improved feature resolution. During mid-2002, the TOPEX/POSEIDON mission operations team developed, planned, and executed a sequence of orbital changes named the TOPEX Orbit Transfer Maneuver (TOTM). These delta-v maneuvers needed to be designed around a number of system constraints including:

- Minimum speed constraints for the roll and yaw RWAs due to known motor faults
- Avoidance of attitude control thruster actuation due to unknown operability of coupled pairs
- Avoidance of solar exposure of the remaining star tracker's field of view
- Thermal and power constraints due to required burn attitudes
- Burn centroid location, burn duration, and spacing for navigation requirements

Due to the numerous operational constraints and concerns, the sequence of seven delta-v maneuvers was performed using only two 1-N thrusters as summarized in Table 6.

Table 6
TOPEX Orbit Transfer Maneuvers Performance

Maneuver	Date	UTC Time	β' (deg)	Duration (sec)	Planned Delta-V Mm/sec	Delta-V Accuracy
TOTM-CAL	2002-227	23:13:29	+71.20	117.584	100.000	+1.64%
TOTM-D231	2002-231	9:03:00	+60.40	186.176	159.000	+2.45%
TOTM-D235	2002-235	12:44:00	+49.20	176.576	153.000	-0.11%
TOTM-D253A	2002-253	17:25:44	-1.00	150.582		
TOTM-D253B	2002-253	20:07:44	-1.00	150.582	260.000 (A+B)	-2.48% (A+B)
TOTM-D256	2002-256	20:17:02	-9.00	127.440	100.000	-0.16%
TOTM-D259	2002-259	19:25:23	-18.00	73.072	57.000	+0.10%

LESSONS LEARNED

Even though the TOPEX/Poseidon baseline mission was just three years, the satellite is continuing to provide high quality reliable data after more than 10 years on orbit. This success can be attributed to the prelaunch efforts by the engineers at both JPL and Fairchild Space, and for the management philosophy and staff dedication that has continued through all the on-orbit operations. Most of the ADCS components still continue to support all phases of the mission even though the high radiation environment has caused some degradation and single event upsets. Since the ADCS components are fully redundant, failures in some of the primary components (such as the failure of one of the star trackers and the solar array drive electronics) has not affected the science data quality.

In the opinion of the authors of this document, however, the JPL mission managers and their management philosophy are the most significant reasons for the success of the TOPEX/Poseidon satellite. These managers realized that continued involvement of the development team was crucial for long-term maintenance and anomaly response. The importance of a flight software maintenance facility was recognized, and for that reason, JPL obtained a copy of the Software Development and Verification Testbed (SDVTB) prior to launch. Throughout the mission, all major command sequences were tested in the SDVTB prior to uplink. The SDVTB was also used to support numerous anomaly and safe hold recovery training exercises. Detailed contingency plans were developed, periodically reviewed, and then modified whenever component performance changed. In some cases, serious possible failures were identified, and detailed software patches were developed and tested before they were needed for on-orbit operations. The JPL managers fostered a team approach to operations, and as a result, all ground personnel have been, and continue to be, dedicated to mission success.

SDVTB and Flight Software (FSW) Maintenance

The SDVTB was developed prior to launch to support real-time testing of all flight software. The SDVTB is comprised of a non-flight version of the OBC, telemetry/command interfaces, and FORTRAN-based simulation of the satellite, its dynamic response and orbital environment. By providing the SDVTB with links to the TOPEX/Poseidon control center, it is possible to verify all critical command sequences prior to uplink. The SDVTB also provided the capability for prelaunch exercises that prepared the ground personnel for on-orbit operations. Early in the mission, the SDVTB was used to support additional activities associated with flight software debugging. When code errors were discovered, patches were developed and tested on the SDVTB prior to loading them in the satellite. All patches were tracked, and after about 5 years on-orbit, a new flight image was generated and tested in the SDVTB. This new image significantly reduced the command activities associated with reloading the flight code.

Contingency Plans and FSW Patch Validation

As with all missions, the operations team performs many contingency planning activities, and the TOPEX/POSEIDON management encouraged development and SDVTB testing of the command sequences and activities associated with any recovery scenario. In some cases, code patches were developed and tested, but not loaded; this approach allowed analysts to isolate and correct a code error, but not use the patch until it was needed. An example is the response to the fault in one of the star trackers, which stopped tracking stars during the first year of operation. Since the ability to recover this tracker was uncertain, all personnel were concerned about the reliability of the second tracker. If both trackers failed, it would not be possible to continue accurate on-board attitude determination. For that reason, the ADCS and FSW team developed an Alternate Attitude Determination (AAD) approach that replaced star tracker data to the Kalman Filter with Earth Sensor measurements. Even though it was not yet needed for on-orbit operations, code implementation for this modified algorithm was developed and extensively tested on the SDVTB. That code was documented and approved for uplink in March 2001, but as of this date, has not been loaded.

The SDVTB was also used to support numerous training activities. In some cases, the training session involved normal operations, while in other situations, a component failure was emulated and the ground personnel were required to identify and respond to that failure. To enhance these training activities, the code to emulate safe hold mode operation was added to the SDVTB. This modification provided the capability to investigate satellite behavior during the transition to and from safe hold mode and to fully test all safe hold mode recovery command sequences; the SDVTB therefore became more useful for contingency training activities.

Operational Changes

Prior to launch and during the early part of the mission, the satellite remained in fixed yaw when the Sun was within 15 degrees of the orbit plane. Later in the mission, the fixed yaw period was extended to 30 degrees, allowing the large solar array to be used to provide increased drag or boost to help maintain the orbital ground track, reducing the need for propulsive maneuvers (see References [21-26]). As component behavior changed during the mission, the ground personnel found new ways to accommodate or eliminate the problem. When the High Gain Antenna gimbals began to overheat during full Sun orbits, changes to the TDRS scheduling shaded the gimbals. Due to the reaction wheel anomalies associated with zero speed crossings, all maneuvers are now designed to keep roll and yaw wheel speeds above 500 RPM. To preserve operational lifetime, both Earth Sensors were powered off and the Fault Detection and Correction (FDC) software was modified to command the satellite to Sun Pointing Safe Hold mode.

Long-Term Trending

Trend data for all of the major ADCS component and performance parameters have been generated and examined throughout the mission. Each day, a ground-system task processes flight telemetry data to compute and record the mean, minimum, maximum, and standard deviation for each parameter. This data is concatenated in a file and is available for analyses. As a minimum, the data for each parameter is examined on a monthly basis, while some parameters are examined on a daily basis. Given the varying orbit-sun geometry, the trend analyses include correlation between component performance and the β' angle.

In addition to the long-term trending, the Spacecraft Performance Analysis Team (SPAT) tracks life-limiting items. In most cases, it is not possible to really measure degradation; however, it is possible to note when the current operational approach may be causing unnecessary degradation to a particular component. This situation was apparent with the High Gain Antenna, where it was placed in a "park" position at the end of each track sequence. Since this approach added unnecessary gimbal cycles to each track, parking the antenna was discontinued.

Documentation

Since the mid-1980s, the ADCS design, analyses, simulations, and implementation details have been documented in hundreds of internal technical memos and numerous conference papers. While the additional time and effort to create, review, and manage these memos is not insignificant, they have been invaluable references throughout the mission. The most often referred to document is the 'flight load memo' that provides the background and/or source of every parameter (gains, limits, thresholds, coefficients, etc.) value in the on-board ADCS FSW.

CONCLUSION

The TOPEX/Poseidon mission has been hugely successful and has made tremendous contributions to our understanding of ocean dynamics and its effects on world climate. The spacecraft and its ADCS have far exceeded their design life due largely to diligent maintenance and designed redundancy.

LIST OF ACRONYMS

ACE	Attitude Control Electronics	MMS	Multimission Modular Spacecraft
AAD	Alternate Attitude Determination	MTB	Magnetic Torquer Bars
ABCAL	Altimeter Boresight Calibration	NASA	National Aeronautics and Space Administration
ADCS	Attitude Determination Control Subsystem	OBC	On-Board Computer
ASCAL	Attitude Sensor Calibration	PM	Propulsion Module
ASTRA	Advanced Star Trackers	POD	Precision Orbit Determination
CCD	Charged Coupled Device	REM	Rocket Engine Module
CNES	Centre National d'Etudes Spatiales	RIU	Remote Interface Unit
CSS	Coarse Sun Sensors	RWA	Reaction Wheel Assemblies
DFSS	Digital Fine Sun sensor	SADE	Solar Array Drive Electronics
DRIRU	Dry Rotor Inertial Reference Unit	SCE	Steering Control Electronics
ESA	Earth Sensor Assembly	SDVTB	Software Development and Verification Testbed
FOV	Field of View	SPAT	Spacecraft Performance Analysis Team
FSW	Flight Software	T/P	TOPEX/Poseidon
MACS	Modular Attitude Control Subsystem	TAM	Three Axis Magnetometer

ACKNOWLEDGEMENTS

The authors would like to acknowledge the dedicated JPL program management of Charles Yamarone, the JPL mission operations leadership of Rod Zeiger, Mark Fujishin, and Parag Vaze, contributions of JPL ADCS engineers [Tooraj Kia, Gene Hanover, Bob Campbell,

Keui Shen, HoSen Lin] and Fairchild ADCS engineers [Raymond Welch, Chuck Gay, Neil Dennehy, Joe Bolek, Darrell Zimbelman, Bob Williams, Dennis Lee]. This work was performed for the Jet Propulsion Laboratory, California Institute of Technology, sponsored by the National Aeronautics and Space Administration.

REFERENCES

1. Hall, *TOPEX/Poseidon Satellite: Enabling a Joint U.S.-French Mission for Global Ocean Study*, 41st Congress of the International Astronautical Federation, Oct 1990, IAF 90-101.
2. Whitacre, *On the Determination of Ocean Circulation by Satellite Radar Altimetry*, 1990.
3. Zimbelman, Lee, Welch, *Dynamic Attitude Command and Control of the TOPEX/POSEIDON Spacecraft*, 1991, AIAA 91-703.
4. Williams, Dennehy, Lee, Welch, *An Augmented MMS MACS for the TOPEX/Poseidon Mission and Beyond*, AAS 90-013, AAS *Advances in the Astronautical Sciences*, Vol. 72, pp. 67-82, 1990.
5. Welch, Dennehy, Kia, Ha, *On-Board Attitude Determination for the TOPEX Satellite*, AIAA Guidance, Navigation and Control Conference, Aug 1989, AIAA 89-3622.
6. Dennehy, Williams, Lee, Welch, *Design of the TOPEX Earth Pointing Safe Hold Mode*, AAS 89-005, AAS *Advances in the Astronautical Sciences*, Vol. 68, pp. 69-100, 1989 (Guidance & Control Conference, Jan. 1989).
7. Sanneman, Lee, Welch, *An Extended Gyrocompassing Controller for the TOPEX/Poseidon Spacecraft*, AAS 91-479, AAS *Advances in the Astronautical Sciences*, Vol. 76, pp. 691-704, 1992.
8. Welch, Lee, Sanneman, *Dot Product Control Law for Non-Orthogonal Thrust Vectors on TOPEX/Poseidon*, AIAA Conference, 1992.
9. Sanneman, Lee, *On-Orbit Performance of the Extended Gyrocompass Controller for the TOPEX/POSEIDON Spacecraft*, AIAA Guidance, Navigation, and Control Conference, 1993.
10. Dennehy, Lee, Welch, *Controls/Structures Interaction Analyses and Simulation for the TOPEX/Poseidon Satellite NMM Digital Controller*, Dynamics of Flexible Structures in Space Conference, May 1990.
11. Dennehy, Welch, Kia, *Attitude Determination and Control Subsystem for the TOPEX Satellite*, AIAA Guidance, Navigation, and Control Conference, Aug 1988, AIAA 88-4129-CP.
12. Dennehy, Zimbelman, Welch, *Sunrise/Sunset Thermal Shock Disturbance Analysis and Simulation for the TOPEX Satellite*, AIAA Guidance, Navigation, and Control Conference, Jan 1990, AIAA 90-0470.
13. Zimbelman, Welch, Born, Optimal Temperature Estimation for Modeling the Thermal Elastic Shock Disturbance Torque, *Journal of Spacecraft and Rockets*, Aug 1991.
14. Kia, Hanover, On-orbit Performance of TOPEX/POSEIDON Star Trackers, 1996, SPIE V.2810.
15. Flynn, Fowski, Kia, *Flight Performance of TOPEX/POSEIDON Star Trackers*, 1993, SPIE V.1949.
16. Gay, Welch, Selby, *Living With On-Orbit Anomalies in the TOPEX/Poseidon Earth Sensors*, AAS 93-044, AAS *Advances in the Astronautical Sciences*, Vol. 81, pp. 361-373, 1993 (Guidance & Control Conference, Feb 1993.)
17. Kia, Aljabri, Goddard, Munson, Kissel, Lin, Vaze, *TOPEX/Poseidon Autonomous Maneuver Experiment (TAME)*, AAS 96-002, AAS *Advances in the Astronautical Sciences*, Vol. 92, pp. 21-37, 1996.
18. Kia, Mellstrom, Klumpp, Munson, Vaze, *TOPEX/Poseidon Autonomous Maneuver Experiment (TAME) Design and Implementation*, AAS 97-004, AAS *Advances in the Astronautical Sciences*, Vol. 94, pp. 41-56, 1997.
19. Kia, Mellstrom, Klumpp, Lin, Sanatar, Shen, Vaze, Nachman, *TOPEX/Poseidon Autonomous Maneuver Experiment (TAME) in-Flight Execution,* AAS 98-001, *Adv. in the Astronautical Sci.*, Vol. 98, pp. 3-19, 1998.
20. Salama, Soroosh, Martin-Mur, Paredes, *TOPEX/Poseidon and Jason-1 Coordinated Navigation,* AAS 01-424, AAS *Advances in the Astronautical Sciences*, Vol. 109, pp. 1769-1782, 2002.
21. Frauenholz, Hamilton, Shapiro, Bhat, *The Role of Anomalous Satellite-Fixed Accelerations in TOPEX/Poseidon Orbit Maintenance*, AAS 93-570, AAS *Advances in the Astronautical Sciences*, Vol. 85, pp. 83-102, 1984.
22. Bhat, Frauenholz, Cannell, *TOPEX/Poseidon Orbit Maintenance Maneuver Design*, AAS 89-408, AAS *Advances in the Astronautical Sciences*, Vol. 71, pp. 645-670, 1990.
23. Bhat, *TOPEX/Poseidon Orbit Acquisition Maneuver Design*, AAS 91-514, AAS *Advances in the Astronautical Sciences*, Vol. 76, pp. 1131-1150, 1992 (AAS/AIAA Astrodynamics Specialists Conference, Aug 1991).
24. Frauenholz, Shapiro, *The Role of Predicted Solar Activity in TOPEX/Poseidon Orbit Maintenance Maneuver design*, AAS 91-515, AAS *Advances in the Astronautical Sciences*, Vol. 76, pp. 1151-1169, 1992.
25. Vincent, *Eccentricity and Argument of Perigee Control for Orbits with Repeat Ground Tracks*, AAS 91-516, AAS *Advances in the Astronautical Sciences*, Vol. 76, pp. 1171-1177.
26. Marshall, Luthcket, *Modeling Radiation Forces Acting on TOPEX/Poseidon for Precision Orbit Determination*, AIAA Conference, revised Feb 1993.

APPENDICES

PUBLICATIONS OF THE AMERICAN ASTRONAUTICAL SOCIETY

Following are the principal publications of the American Astronautical Society:

JOURNAL OF THE ASTRONAUTICAL SCIENCES (1954 -)
Published quarterly and distributed by AAS Business Office, 6352 Rolling Mill Place, Suite #102, Springfield, Virginia 22152. Back issues available from Univelt, Inc., P.O. Box 28130, San Diego, California 92198.

SPACE TIMES (1986 -)
Published bi-monthly and distributed by AAS Business Office, 6352 Rolling Mill Place, Suite #102, Springfield, Virginia 22152.

AAS NEWSLETTER (1962 - 1985)
Incorporated in *Space Times*. Back issues available from AAS Business Office, 6352 Rolling Mill Place, Suite #102, Springfield, Virginia 22152.

ASTRONAUTICAL SCIENCES REVIEW (1959 -1962)
Incorporated in *Space Times*. Back issues still available from Univelt, Inc., P.O. Box 28130, San Diego, California 92198.

ADVANCES IN THE ASTRONAUTICAL SCIENCES (1957 -)
Proceedings of major AAS technical meetings. Published and distributed for the American Astronautical Society by Univelt, Inc., P.O. Box 28130, San Diego, California 92198.

SCIENCE AND TECHNOLOGY SERIES (1964 -)
Supplement to *Advances in the Astronautical Sciences*. Proceedings and monographs, most of them based on AAS technical meetings. Published and distributed for the American Astronautical Society by Univelt, Inc., P.O. Box 28130, San Diego, California 92198.

AAS HISTORY SERIES (1977 -)
Supplement to *Advances in the Astronautical Sciences*. Selected works in the field of aerospace history under the editorship of R. Cargill Hall. Published and distributed for the American Astronautical Society by Univelt, Inc., P.O. Box 28130, San Diego, California 92198.

AAS MICROFICHE SERIES (1968 -)
Supplement to *Advances in the Astronautical Sciences*. Consists principally of technical papers not included in the hard-copy volume. Published and distributed for the American Astronautical Society by Univelt, Inc., P.O. Box 28130, San Diego, California 92198.

Subscriptions to the *Journal of the Astronautical Sciences* and the *Space Times* should be ordered from the AAS Business Office. Back issues of the *Journal* and all books and microfiche should be ordered from Univelt, Incorporated.

ADVANCES IN THE ASTRONAUTICAL SCIENCES SERIES
(1957-)

ISSN 0065-3438,

LIBRARY OF CONGRESS CARD NO. 57-43769

Proceedings of Major AAS Technical Meetings

Vol. 1 Third Annual AAS Meeting, Dec. 6-7, 1956, New York, NY, 1957, 184p., ed. Norman V. Petersen, Microfiche only, $20 (ISBN 0-87703-002-2)

Vol. 2 Fourth Annual AAS Meeting, Jan. 29-31, 1958, New York, NY, 1958, 440p., eds. Norman V. Petersen, Horace Jacobs, Microfiche only, $20 (ISBN 0-87703-003-0)

Vol. 3 First Western National AAS Meeting, Aug. 18-19, 1958 530p., eds. Norman V. Petersen, Horace Jacobs, Microfiche only, $20 (ISBN 0-87703-004-9)

Vol. 4 Fifth Annual AAS Meeting, Dec. 27-31, 1958, Washington, D.C., 1959, 462p., ed. Horace Jacobs, Microfiche only, $20 (ISBN 0-87703-005-7)

Vol. 5 Second Western National AAS Meeting, Aug. 4-5, 1959, Los Angeles, CA, 1960, 364p., ed. Horace Jacobs, Microfiche only, $20 (ISBN 0-87703-006-3)

Vol. 6 Sixth Annual AAS Meeting, Jan. 18-21, 1960, New York, NY, 1961, 968p., eds. Horace Jacobs and Eric Burgess, Hard Cover $45 (ISBN 0-87703-007-3)

Vol. 7 Third Western National AAS Meeting, Aug. 4-5, 1960, Seattle, WA, 1961, 464p., eds. Horace Jacobs and Eric Burgess, Microfiche only, $20 (ISBN 0-87703-008-1)

Vol. 8 Seventh Annual AAS Meeting, Jan. 16-18, 1961, Dallas, TX, 1963, 602p., ed. Horace Jacobs, Microfiche only, $20 (ISBN 0-87703-009-X)

Vol. 9 Fourth Western Regional AAS Meeting, Aug. 1-3, 1961, San Francisco, CA, 1963, 910p., ed. Eric Burgess, Hard Cover $45 (ISBN 0-87703-010-3)

Vol. 10 Manned Lunar Flight (AAS/AAAS Symposium) Dec. 19, 1961, Denver, CO, 1963, 310p., eds. George W. Morgenthaler and Horace Jacobs, Hard Cover $35 (ISBN 0-87703-011-1)

Vol. 11 Eighth Annual AAS Meeting, Jan. 16-18, 1962, Washington, D.C., 1963, 808p., ed. Horace Jacobs, Hard Cover $45 (ISBN 0-87703-012-X)

Vol. 12 Scientific Satellites - Mission and Design (AAS/AAAS Symposium), Dec. 27, 1962, Philadelphia, PA, 1963, 262p., ed. Irving E. Jeter, Hard Cover $25 (ISBN 0-87703-013-8)

Vol. 13 Interplanetary Missions, 9th Annual AAS Meeting, Jan. 15-17, 1963, Los Angeles, CA, 1963, 690p., ed. Eric Burgess, Hard Cover $45 (ISBN 0-87703-014-6)

Vol. 14 Second AAS Symposium on Physical and Biological Phenomena under Zero G Conditions, Jan. 18, 1963, Los Angeles, CA, 1963, 382p., eds. Elliot T. Benedikt and Robert W. Halliburton, Hard Cover $30 (ISBN 0-87703-015-4)

Vol. 15 Exploration of Mars Symposium, Jun. 6-7, 1963, Denver, CO, 1963, 634p., ed. George W. Morgenthaler, Hard Cover $45 (ISBN 0-87703-016-2)

Vol. 16 Space Rendezvous, Rescue, and Recovery Symposium, Sept. 10-12, 1963, Edwards, CA 1963, 1408p., ed. Norman V. Petersen, Hard Cover, Part 1, 1028p., $45 (ISBN 0-87703-017-0); Part 2, 380p., $30 (ISBN 0-87703-018-9)

Vol. 17 Bioastronautics - Fundamental and Practical Problems (AAS/AAAS Symposium), Dec. 30, 1963, Cleveland, OH, 1964, 128p., ed. William C. Kaufman, Microfiche only, $10 (ISBN 0-87703-019-7)

Vol. 18 Lunar Flight Programs, 10th Annual AAS Meeting, May 4-7, 1964, New York, NY, 1964, 630p., ed. Ross Fleisig, Hard Cover $45 (ISBN 0-87703-020-0)

Vol. 19 Unmanned Exploration of the Solar System Symposium, Feb. 8-10, 1965, Denver, CO, 1965, 1000p., eds. George W. Morgenthaler, Robert G. Morra, Hard Cover $45 (ISBN 0-87703-021-9)

Vol. 20 Post Apollo Exploration, 11th Annual AAS Meeting, May 3-6, 1965, Chicago, IL, 1966, 1220p., ed. Francis Narin, Microfiche only, Part I, 572p., $30 (ISBN 0-87703-022-7); Part 2, 648p., $35 (ISBN 0-87703-023-5)

Vol. 21 Practical Space Applications Symposium, Feb. 21-23, 1966, San Diego, CA, 1967, 508p., ed. Lawrence L. Kavanau, Microfiche Only $40 (ISBN 0-87703-024-3)

Vol. 22 The Search for Extraterrestrial Life, 12th Annual AAS Meeting, May 23-25, 1966, Anaheim, CA, 1967, 388p., ed. James S. Hanrahan, Microfiche only $30 (ISBN 0-87703-025-1); Microfiche Suppl. (Vol. 1 AAS Microfiche Series) $12 (ISBN 0-87703-132-0)

Vol. 23 Commercial Utilization of Space, 13th Annual AAS Meeting, May 1-3, 1967, Dallas, TX, 1968, 512p., eds. J. Ray Gilmer, Alfred M. Mayo, Ross C. Peavey, Hard Cover (ISBN 0-87703-026-X); plus Microfiche Suppl. (Vol. 3 AAS Microfiche Series) $60 (ISBN 0-87703-216-5)

Vol. 24 Exploitation of Space for Experimental Research, 14th Annual AAS Meeting, May 13-15, 1968, Dedham, MA, 1968, 363p., ed. Harry Zuckerberg, Hard Cover $30 (ISBN 0-87703-027-8)

Vol. 25 Advanced Space Experiments, Sept. 16-18, 1968, Ann Arbor, MI, 1969, 530p., eds. O. Lyle Tiffany and Eugene M. Zaitzeff, Hard Cover $40 (ISBN 0-87703-028-6)

Vol. 26 Planning Challenges of the 70"s in Space, 15th Annual AAS Meeting, Jun. 17-20, 1969, Denver, CO, 1970, 470p., eds. George W. Morgenthaler and Robert G. Morra, Hard Cover $35 (ISBN 0-87703-053-7); Microfiche Suppl. (Vol. 14 AAS Microfiche Series) $20 (ISBN 0-87703-130-4)

Vol. 27/28 Space Stations (v27) and Space Shuttles and Interplanetary Missions (v28), 16th Annual AAS Meeting, Jun. 8-10, 1970, Anaheim, CA, 1970, Vol. 27, eds. Lewis Larmore and Robert L. Gervais, 606p., Hard Cover $45 (ISBN 0-87703-054-5); Vol. 28, eds. Lewis Larmore and Robert L. Gervais, 488p., Hard Cover $35 (ISBN 0-87703-055-3)

Vol. 29 The Outer Solar System, 17th Annual AAS Meeting, Jun. 28-30, 1971, Seattle, WA, 1971, ed. Juris Vagners, Part 1 (on Microfiche Only), 618p., $40 (ISBN 0-87703-059-6); Part 2, Hard Cover, 740p., $45 (ISBN 0-87703-060-X)

Vol. 30 International Congress of Space Benefits, 19th Annual AAS Meeting, Jun. 19-21, 1973, Dallas, TX, 1974, 528p., ed. Francis S. Johnson, Hard Cover $40 (ISBN 0-87703-065-0)

Vol. 31 The Skylab Results, 20th Annual AAS Meeting, Aug. 20-22, 1974, Los Angeles, CA, 1975, 1174p., eds. William C. Schneider and Thomas E. Hanes, Microfiche only (ISBN 0-87703-072-3); Plus Microfiche Suppl. (Vol. 22 AAS Microfiche Series) $60 (ISBN 0-87703-143-6)

Vol. 32 Space Shuttle Missions of the 80's, 21st Annual AAS Meeting, Aug. 26-28, 1975, Denver, CO, 1977, 1364p., eds. William J. Bursnall, George W. Morgenthaler, Gerald E. Simonson, Hard Cover, Part 1, 598p., $40 (ISBN 0-87703-078-2); Hard Cover, Part 2, 766p., $55 (ISBN 0-87703-087-1); Microfiche Suppl. (Vol. 25 AAS Microfiche Series $65 (ISBN 0-87703-133-9)

Vol. 33 AAS/AIAA Astrodynamics Conference, July 28-30, 1975, Nassau, Bahamas, 1976, 390p., eds. William F. Powers, Herbert E. Rauch, Byron D. Tapley, Carmelo E. Velez, Hard Cover $35 (ISBN 0-87703-079-0); Microfiche Suppl. (Vol. 26 AAS Microfiche Series) $40 (ISBN 0-87703-142-8)

Vol. 34 Apollo Soyuz Mission Report, 1977, 336p., ed. Chester M. Lee, Hard Cover $95 (ISBN 0-87703-089-8)

Vol. 35 The Bicentennial Space Symposium - New Themes for Space: Mankind"s Future Needs and Aspirations, 22nd AAS Meeting, Oct. 6-8, 1976, Washington, D.C., 1977, 242p., ed. William C. Schneider, Hard Cover $25 (ISBN 0-87703-090-1)

Vol. 36 The Industrialization of Space, 23rd Annual AAS Meeting, Oct. 18-20, 1977, San Francisco, CA, 1978, 1160p., eds. Richard A. Van Patten, Paul Siegler, Edward V.B. Stearns, Hard Cover, Part 1, 610p., $55 (ISBN 0-87703-094-4); Hard Cover, Part 2, 550p.,

$45 (ISBN 0-87703-095-2); Microfiche Suppl. (Vol. 28 AAS Microfiche Series) $15 (ISBN 0-87703-121-5)

Vol. 37 Space Shuttle and Spacelab Utilization, What are the Near-Term and Long-Term Benefits for Mankind?, 16th Goddard Memorial Symposium, 24th Annual AAS Meeting, March 8-10, 1978, Washington, D.C., 1978, 865p., eds. George W. Morgenthaler and Manfred Hollstein, Hard Cover, Part 1, 400p., $40 (ISBN 0-87703-096-0); Hard Cover, Part 2, 465p., $45 (ISBN 0-87703-097-9)

Vol. 38 The Future U.S. Space Program, 25th Anniversary Conference, Oct. 20 - Nov. 2, 1978, Houston, TX, 1979, 880p., eds. Richard S. Johnston, Albert Naumann, Jr., Clay W. G. Fulcher, Hard Cover, Part 1, 444p., $45 (ISBN 0-87703-098-7); Hard Cover, Part 2, 436p., $40 (ISBN 0-87703-099-5); Microfiche Suppl. (Vol. 30 AAS Microfiche Series) $15 (ISBN 0-87703-129-0)

Vol. 39 Guidance and Control 1979, Feb. 24-28, 1979, Keystone, CO, 1979, 492p., ed. Robert D. Culp, Hard Cover $45 (ISBN 0-87703-100-2); Microfiche Suppl. (Vol. 31 AAS Microfiche Series) $10 (ISBN 0-87703-128-2)

Vol. 40 AAS/AIAA Astrodynamics Conference, Jun. 25-27, 1979, Provincetown, MA, 1980, 996p., eds. Paul A. Penzo, Bernard Kaufman, Louis Friedman, Richard Battin, Hard Cover, Part 1, 494p., $45 (ISBN 0-87703-107-X); Soft Cover $35 (ISBN 0-87703-108-8); Hard Cover, Part 2, 502p., $45 (ISBN 0-87703-109-6); Soft Cover $35 (ISBN 0-87703-110-X); Microfiche Suppl. (Vol. 32 AAS Microfiche Series) $20 (ISBN 0-87703-139-8)

Vol. 41 Space Shuttle: Dawn of an Era, 26th Annual AAS Meeting, Oct. 29-Nov. 1, 1979, Los Angeles, CA, 1980, 980p., eds. William F. Rector, III and Paul A. Penzo, Hard Cover, Part 1, 452p., $45 (ISBN 0-87703-111-8); Soft Cover $35 (ISBN 0-87703-112-6); Hard Cover, Part 2, 528p., $55 (ISBN 0-87703-113-4); Soft Cover $40 (ISBN 0-87703-114-2); Microfiche Suppl. (Vol. 33 AAS Microfiche Series) $10 (ISBN 0-87703-136-3)

Vol. 42 Guidance and Control 1980, Feb. 17-21, 1980, Keystone, CO, 1980, 738p., ed. Louis A. Morine, Hard Cover $60 (ISBN 0-87703-137-1); Soft Cover $45 (ISBN 0-87703-138-X)

Vol. 43 Shuttle/Spacelab - The New Transportation System and its Utilization, (3rd DGLR/AAS Symposium), Apr. 28-30, 1980, Hannover, Germany, 1981, 342p., eds. Dietrich E. Koelle and George V. Butler, Hard Cover $45 (ISBN 0-87703-144-4); Soft Cover $35 (ISBN 0-87703-146-0)

Vol. 44 Space-Enhancing Technological Leadership, 27th Annual AAS Meeting, Oct. 20-23, 1980, Boston, MA, 1981, 580p., ed. Lawrence P. Greene, Hard Cover $65 (ISBN 0-87703-147-9); Soft Cover $50 (ISBN 0-87703-148-7); Microfiche Suppl. (Vol. 35 AAS Microfiche Series) $10 (ISBN 0-87703-164-9)

Vol. 45 Guidance and Control 1981, Jan. 31-Feb. 4, 1981, Keystone, CO, 1981, 506p., ed. Edward J. Bauman, Hard Cover $60 (ISBN 0-87703-150-9); Soft Cover $50 (ISBN 0-87703-151-7); Microfiche Suppl. (Vol. 36 AAS Microfiche Series) $15 (ISBN 0-87703-156-8)

Vol. 46 AAS/AIAA Astrodynamics Conference, Aug. 3-5, 1981, North Lake Tahoe, NV, 1982, 1124p., eds. Alan L. Friedlander, Paul J. Cefola, Bernard Kaufman, Walt Williamson, G.T. Tseng, Hard Cover, Part 1, 552p., $55 (ISBN 0-87703-159-2); Soft Cover $45 (ISBN 0-87703-160-6); Hard Cover, Part 2, 572p., $55 (ISBN 0-87703-161-4); Soft Cover $45 (ISBN 0-87703-162-2); Microfiche Suppl. (Vol. 37 AAS Microfiche Series) $40 (ISBN 0-87703-163-0)

Vol. 47 Leadership in Space - For Benefits on Earth, 28th Annual AAS Meeting, Oct. 26-29, 1981, San Diego, CA, 1982, 310p., ed. William F. Rector, III, Hard Cover $45 (ISBN 0-87703-168-1); Soft Cover $35 (ISBN 0-87703-169-X)

Vol. 48 Guidance And Control 1982, Jan. 30 -Feb. 3, 1982, Keystone, CO, 1982, 558p., eds. Robert D. Culp, Edward J. Bauman, W. E. Dorroh, Jr., Hard Cover $65 (ISBN 0-87703-170-3); Soft Cover $50 (ISBN 0-87703-171-1); Microfiche Suppl. (Vol. 38 AAS Michrofiche Series) $10 (ISBN 0-87703-180-0)

Vol. 49 Spacelab, Space Platforms, and the Future, Fourth AAS/DGLR Symposium and 20th Goddard Memorial Symposium, Mar. 17-19, 1982, Greenbelt, MD, 1982, 502p., eds. Peter M. Bainum, Dietrich E. Koelle, Hard Cover $55 (ISBN 0-87703-174-6); Soft Cover $45 (ISBN 0-87703-175-4); Microfiche Suppl. (Vol. 42 AAS Microfiche Series) $15 (ISBN 0-87703-181-9)

Vol. 50 Proceedings on an International Symposium on Engineering Sciences and Mechanics, Dec. 29-31, Tainan, Taiwan, 1983, two parts, 1570p., eds. Han-Min Hsia, Richard W. Longman, You-Li Chou, Hard Cover $120 (ISBN 0-87703-176-2); Microfiche Suppl. (Vol. 43 AAS Microfiche Series) $10 (ISBN 0-87703-215-7)

Vol. 51 Guidance and Control 1983, Feb. 5-9, 1983, Keystone, CO, 1983, 494p., eds. Edward J. Bauman, Zubin W. Emsley, Hard Cover $60 (ISBN 0-87703-182-7); Soft Cover $50 (ISBN 0-87703-183-5); Microfiche Suppl. (Vol. 44 AAS Microfiche Series) $10 (ISBN 0-87703-214-9)

Vol. 52 Developing the Space Frontier, 29th Annual AAS Meeting, Oct. 25-27, 1982, Houston, TX, 1983, 436p., eds. Albert Naumann, Grover Alexander, Microfiche Only $45 (ISBN 0-87703-189-4)

Vol. 53 Space Manufacturing 1983, May 9-12, 1983, Princeton, NJ, 1983, 496p., eds. James D. Burke, April S. Whitt, On Microfiche Only $50 (ISBN 0-87703-188-6)

Vol. 54 AAS/AIAA Astrodynamics Conference, Aug. 22-25, 1983, Lake Placid, NY, 1984, two parts, 1370p., eds. G.T. Tseng, Paul J. Cefola, Peter M. Bainum, David A. Levinson, Hard Cover $120 (ISBN 0-87703-190-8); Soft Cover $90 (ISBN 0-87703-191-6); Microfiche Suppl. (Vol. 45 AAS Microfiche Series) $40 (ISBN 0-87703-192-4)

Vol. 55 Guidance and Control 1984, Feb. 4-8, 1984, Keystone, CO, 1984, 500p., eds. Robert D. Culp, Parker S. Stafford, Hard Cover $60 (ISBN 0-87703-199-1); Soft Cover $50 (ISBN 0-87703-200-9); Microfiche Suppl. (Vol. 48 AAS Microfiche Series $15 (ISBN 0-87703-201-7)

Vol. 56 From Spacelab to Space Station, Fifth DGLR/AAS Symposium, Oct. 3-5, 1984, Hamburg, Germany, 1985, 270p., eds. H. Stoewer, Peter M. Bainum, Microfiche Only $30 (ISBN 0-87703-209-2)

Vol. 57 Guidance and Control 1985, Feb. 2-6, 1985, Keystone, CO, 1985, 618p., eds. Robert D. Culp, Edward J. Bauman, Charles A. Cullian, Hard Cover $65 (ISBN 0-87703-211-4); Soft Cover $50 (ISBN 0-87703-212-2); Microfiche Suppl. (Vol. 50 AAS Microfiche Series) $15 (ISBN 0-87703-213-0)

Vol. 58 AAS/AIAA Astrodynamics Conference, Aug. 12-15, 1985, Vail, CO, 1986, two parts, 1556p., eds. Bernard Kaufman, Joseph J.F. Liu, Robert A. Calico, Felix R. Hoots, Hard Cover $140 (ISBN 0-87703-245-9); Soft Cover $110 (ISBN 0-87703-246-7); Microfiche Suppl. (Vol. 51 AAS Microfiche Series); $60 (ISBN 0-87703-247-5)

Vol. 59 Space Station Beyond IOC, 32nd Annual AAS Meeting, Nov 6-7, 1985, Los Angeles, CA, 1986, 188p., ed. M. Jack Friedenthal, Hard Cover $40 (ISBN 0-87703-252-1); Soft Cover $30 (ISBN 0-87703-253-X)

Vol. 60 Space Exploitation and Utilization, First AAS/JRS Symposium, Dec. 15-19, 1985, Honolulu, HI, 1986, 740p., eds. Gayle L. May, Peter M. Bainum, Kenji Ikeda, Tamiya Nomura, Tatsuo Yamanaka, Ryojiro Akiba, Hard Cover $70 (ISBN 0-87703-254-8); Soft Cover $55 (ISBn 0-87703-255-6); Microfiche Suppl. (Vol. 52 AAS Microfiche Series) $10 (ISBN 0-87703-256-4)

Vol. 61 Guidance and Control 1986, Feb. 1-5, 1986, Keystone, CO, 1986, 460p., eds. Robert D. Culp, John C. Durrett, Hard Cover $60 (ISBN 0-87703-257-2); Soft Cover $50 (ISBN 0-87703-258-0); Microfiche Suppl. (Vol. 53 AAS Microfiche Series) $10 (ISBN 0-87703-259-9)

Vol. 62 Tethers in Space, Proceedings of First International Conference on Tethers in Space (NASA & PSN Sponsors; AIAA, AAS, & AIDAA Co-Sponsors), Sept. 17-19, 1986, Arlington, VA, 1987, 784p., eds. Peter M. Bainum, Ivan Bekey, Luciano Guerriero, Paul A. Penzo, Hard Cover $80 (ISBN 0-87703-264-5); Soft Cover $70 (ISBN 0-87703-265-3)

Vol. 63 Guidance and Control 1987, Jan. 31 -Feb. 4, 1987, Keystone, CO, 1987, 638p., eds. Robert D. Culp, Terry J. Kelly, Hard Cover $75 (ISBN 0-87703-268-8); Soft Cover $60 (ISBN 0-87703-269-6)

Vol. 64 Aerospace Century XXI, 33rd AAS Annual Meeting, Oct. 26-29, 1986, Boulder, CO, 1987, all three parts, Hard Cover $225 (ISBN 0-87703-276-9); Soft Cover $180 (ISBN 0-87703-277-7); Part I, Space Missions and Policy, 686p., eds. George W. Morgenthaler, Gayle L. May, Hard Cover $75 (ISBN 0-87703-279-3); Soft Cover $60 (ISBN 0-87703-282-3); Part II, Space Flight Technologies, 608p., eds. George W. Morgenthaler, W. Kent Tobiska, Hard Cover $75 (ISBN 0-87703-280-7); Soft Cover $60 (ISBN

0-87703-283-1); Part III, Space Sciences, Applications, and Commercial Developments, 724p., eds. George W. Morgenthaler, Jean N. Koster, Hard Cover $75 (ISBN 0-87703-281-5); Soft Cover $60 (ISBN 0-87703-284-X); Microfiche Suppl. (Vol. 54 AAS Microfiche Series) $25 (ISBN 0-87703-278-5)

Vol. 65 AAS/AIAA Astrodynamics Conference, Aug. 10-13, 1987, Kalispell, MT, 1988, two parts, 1774p., eds. John K. Soldner, Arun K. Misra, Robert E. Lindberg, Walton Williamson, Hard Cover $180 (ISBN 0-87703-285-8); Soft Cover $150 (ISBN 0-87703-286-6); Microfiche Suppl. (Vol. 55 AAS Microfiche Series); $70 (ISBN 0-87703-287-4)

Vol. 66 Guidance and Control 1988, Jan. 30 -Feb. 3, 1988, Keystone, CO, 1988, 576p., eds. Robert D. Culp, Paul L. Shattuck, Hard Cover $75 (ISBN 0-87703-288-2); Soft Cover $60 (ISBN 0-87703-289-0); Microfiche Suppl. (Vol. 56 AAS Microfiche Series) $10 (ISBN 0-87703-290-4)

Vol. 67 Space - A New Community of Opportunity, 34th AAS Annual Meeting, Nov. 3-5, 1987, Houston, TX, 1989, 472p., eds. William G. Straight, Henry N. Bowes, Hard Cover $70 (ISBN 0-87703-297-1); Soft Cover $55 (ISBN 0-87703-298-X)

Vol. 68 Guidance and Control 1989, Feb. 4-8, 1989, Keystone, CO, 1989, 708p., eds. Robert D. Culp, Robert A. Lewis, Hard Cover $85 (ISBN 0-87703-299-8); Soft Cover $70 (ISBN 0-87703-300-5)

Vol. 69 Orbital Mechanics and Mission Design, Apr. 24-27, 1989, Greenbelt, MD, 1989, 862p., ed. Jerome Teles, Hard Cover $95 (ISBN 0-87703-311-0); Microfiche Suppl. (Vol. 57 AAS Microfiche Series) $10 (ISBN 0-87703-313-7)

Vol. 70 The 21st Century in Space, 35th AAS Annual Meeting, Oct. 24-26, 1988, St. Louis, MO, 1990, 446p., ed. George V. Butler, Hard Cover $90 (ISBN 0-87703-314-5); Soft Cover $75 (ISBN 0-87703-315-3); Microfiche Suppl. (Vol. 58 AAS Microfiche Series) $10 (ISBN 0-87703-316-1)

Vol. 71 AAS/AIAA Astrodynamics Conference, Aug. 7-10, 1989, Stowe, VT, 1990, two parts, 1472p., eds. Catherine L. Thornton, Ronald J. Proulx, John E. Prussing, Felix R. Hoots, Hard Cover $200 (ISBN 0-87703-317-X); Soft Cover $170 (ISBN 0-87703-318-8); Microfiche Suppl. (Vol. 59 AAS Microfiche Series) $50 (ISBN 0-87703-319-6)

Vol. 72 Guidance and Control 1990, Feb. 3-7, 1990, Keystone, CO, 1990, 676p., eds. Robert D. Culp, Arlo D. Gravseth, Hard Cover $95 (ISBN 0-87703-320-X); Soft Cover $80 (ISBN 0-87703-321-8)

Vol. 73 Space Utilization and Applications in the Pacific, Third (PISSTA) AAS/JRS/CSA Symposium, Nov. 6-8, 1989, Los Angeles, CA, 1990, 764p., eds. Peter M. Bainum, Gayle L. May, Tatsuo Yamanaka, Yang Jiachi, Hard Cover $95 (ISBN 0-87703-325-0); Soft Cover $80 (ISBN 0-87703-326-9)

Vol. 74 Guidance and Control 1991, Feb. 2-6, 1991, Keystone, CO, 1991, 730p., eds. Robert D. Culp, James P. McQuerry, Hard Cover $120 (ISBN 0-87703-334-X); Soft Cover $90 (ISBN 0-87703-335-8)

Vol. 75 AAS/AIAA Spaceflight Mechanics Meeting, Feb. 11-13, 1991, Houston, TX, 1991, two parts, 1354p., eds. John K. Soldner, Arun K. Misra, Lester L. Sackett, Richard Holdaway, Hard Cover $220 (ISBN 0-87703-338-2); Soft Cover $190 (ISBN 0-87703-339-0); Microfiche Suppl. (Vol. 62 AAS Microfiche Series) $50 (ISBN 0-87703-340-4)

Vol. 76 AAS/AIAA Astrodynamics Conference, Aug. 19-22, 1991, Durango, CO, 1992, three parts, 2590p., eds. Bernard Kaufman, Kyle T. Alfriend, Ronald L. Roehrich, Robert R. Dasenbrock, Hard Cover $390 (ISBN 0-87703-347-1); Microfiche Suppl. (Vol. 63 AAS Microfiche Series) $75 (ISBN 0-87703-348-X)

Vol. 77 International Space Year (ISY) in the Pacific Basin, Fourth ISCOPS (formerly PISSTA) AAS/JRS/CSA Symposium, Nov. 17-20, 1991, Kyoto, Japan, 1992, 798p., eds. Peter M. Bainum, Gayle L. May, Makoto Nagatomo, Yoshiaki Ohkami, Yang Jiachi, Hard Cover $120 (ISBN 0-87703-351-X); Soft Cover $90 (ISBN 0-87703-352-8)

Vol. 78 Guidance and Control 1992, Feb. 2-6, 1992, Keystone, CO, 1992, 754p., eds. Robert D. Culp, Richard P. Zietz, Hard Cover $120 (ISBN 0-87703-353-6); Soft Cover $90 (ISBN 0-87703-354-4); Microfiche Suppl. (Vol. 64 AAS Microfiche Series) $20 (ISBN 0-87703-355-2)

Vol. 79 AAS/AIAA Spaceflight Mechanics Meeting, Feb. 24-26, 1992, Colorado Springs, CO, 1992, two parts, 1312p., eds. Roger E. Diehl, Ralph G. Schinnerer, Walton E. Williamson, Daryl G. Boden, Hard Cover $240 (ISBN 0-87703-358-7); Microfiche Suppl. (Vol. 65 AAS Microfiche Series) $40 (ISBN 0-87703-359-5)

Vol. 80 Space Business Opportunities, 37th and 38th AAS Annual Meetings, Nov. 5-7, 1990, Dec. 3-5, 1991, Los Angeles, CA, 1992, 380p., eds. Wayne J. Esser, Don K. Tomajan, Hard Cover $90 (ISBN 0-87703-360-9); Soft Cover $70 (ISBN 0-87703-361-7); Microfiche Suppl. (Vol. 61 AAS Microfiche Series) $50 (ISBN 0-87703-331-5); (Vol. 66 AAS Microfiche Series) $60 (ISBN 0-87703-362-5)

Vol. 81 Guidance and Control 1993, Feb. 6-10, 1993, Keystone, CO, 1993, 648p., eds. Robert D. Culp, George Bickley, Hard Cover $120 (ISBN 0-87703-365-X); Soft Cover $90 (ISBN 0-87703-366-8); Microfiche Suppl. (Vol. 67 AAS Microfiche Series) $20 (ISBN 0-87703-367-6)

Vol. 82 AAS/AIAA Spaceflight Mechanics Meeting, Feb. 22-24, 1993, Pasadena, CA, 1993, two parts, 1454p., eds. Robert G. Melton, Lincoln J. Wood, Roger C. Thompson, Stuart J. Kerridge, Hard Cover $240 (ISBN 0-87703-368-4); Microfiche Suppl. (Vol. 68 AAS Microfiche Series) $15 (ISBN 0-87703-369-2)

Vol. 83 Dynamics of Space Tether Systems (English Language Edition), 1993, 508p., by Vladimir V. Beletsky, Evgenii M. Levin, Hard Cover $120 (ISBN 0-87703-370-6); Soft Cover $90 (ISBN 0-87703-371-4)

Vol. 84 AAS/GSFC International Symposium on Spaceflight Dynamics, Apr. 26-30, 1993, Greenbelt, MD, 1993, two parts, 1450p., eds. Jerome Teles, Mina V. Samii, Hard Cover $240 (ISBN 0-87703-378-1); Microfiche Suppl. (Vol. 69 AAS Microfiche Series) $10 (ISBN 0-87703-379-X)

Vol. 85 AAS/AIAA Astrodynamics Conference, Aug. 16-19, 1993, Victoria, British Columbia, Canada, 1994, three parts, 2750p., eds. Arun K. Misra, Vinod J. Modi, Richard Holdaway, Peter M. Bainum, Hard Cover $390 (ISBN 0-87703-380-3); Microfiche Suppl. (Vol. 70 AAS Microfiche Series) $30 (ISBN 0-87703-381-1)

Vol. 86 Guidance and Control 1994, Feb. 2-6, 1994, Keystone, CO, 1994, 700p., eds. Robert D. Culp, Ronald D. Rausch, Hard Cover $120 (ISBN 0-87703-384-6); Soft Cover $90 (ISBN 0-87703-385-4)

Vol. 87 AAS/AIAA Spaceflight Mechanics Meeting, Feb. 14-16, 1994, Cocoa Beach, FL, 1994, two parts, 1272p., eds. John E. Cochran, Jr., Charles D. Edwards, Jr., Stephen J. Hoffman, Richard Holdaway, Hard Cover $240 (ISBN 0-87703-386-2)

Vol. 88 Guidance and Control 1995, Feb. 1-5, 1995, Keystone, CO, 1995, 600p. eds. Robert D. Culp, James D. Medbery, Hard Cover $120 (ISBN 0-87703-399-4); Soft Cover $90 (ISBN 0-87703-400-1)

Vol. 89 AAS/AIAA Spaceflight Mechanics Meeting, Feb. 13-16, 1995, Albuquerque, NM, 1995, two parts, 1774p., eds. Ronald J. Proulx, Joseph J. F. Liu, P. Kenneth Seidelmann, Salvatore Alfano, Hard Cover $280 (ISBN 0-87703-401-X); Microfiche Suppl. (Vol. 71 AAS Microfiche Series) $15 (ISBN 0-87703-402-8)

Vol. 90 AAS/AIAA Astrodynamics Conference, Aug. 14-17, 1995, Halifax, Nova Scotia, Canada, 1996, two parts, 2270p., eds. K. Terry Alfriend, I. Michael Ross, Arun K. Misra, C. Fred Peters, Hard Cover $290 (ISBN 0-87703-407-9); Microfiche Suppl. (Vol. 72 AAS Microfiche Series) $15 (ISBN 0-87703-408-7)

Vol. 91 Strengthening Cooperation in the 21st Century, Sixth ISCOPS (formerly PISSTA) AAS/JRS/CSA Symposium, Dec. 6-8, 1995, Marina Del Rey, CA, 1996, 1154p., eds. Peter M. Bainum, Gayle L. May, Yoshiaki Ohkami, Kuninori Uesugi, Qi Faren, Li Furong, Hard Cover $145 (ISBN 0-87703-409-5)

Vol. 92 Guidance and Control 1996, Feb. 7-11, 1996, Breckenridge, CO, 1996, 744p. Eds. Robert D. Culp, Marv Odefey, Hard Cover $120 (ISBN 0-87703-412-5); Soft Cover $90 (ISBN 0-87703-413-3)

Vol. 93 AAS/AIAA Spaceflight Mechanics Meeting, Feb. 12-15, 1996, Austin, TX, 1996, two parts, 1776p., eds. G. Edward Powell, Robert H. Bishop, John B. Lundberg, Robert H. Smith, Hard Cover $280 (ISBN 0-87703-414-1); Microfiche Suppl. (Vol. 73 AAS Microfiche Series) $10 (ISBN 0-87703-415-X)

Vol. 94 Guidance and Control 1997, Feb. 5-9, 1997, Breckenridge, CO, 1997, 458p. Eds. Robert D. Culp, Stuart B. Wiens, Hard Cover $120 (ISBN 0-87703-430-3); Soft Cover $90 (ISBN 0-87703-431-1)

Vol. 95 AAS/AIAA Spaceflight Mechanics Meeting, Feb. 10-12, 1997, Huntsville, AL, 1997, two parts, 1168p., eds. Kathleen C. Howell, David A. Cicci, John E. Cochran, Jr., Thomas S. Kelso, Hard Cover $240 (ISBN 0-87703-432-X)

Vol. 96 Space Cooperation into the 21st Century, Seventh ISCOPS (formerly PISSTA) JRS/AAS/CSA Symposium, Jul. 15-18, 1997, Nagasaki, Japan, 1997, 1089p., eds. Peter M. Bainum, Gayle L. May, Makoto Nagatomo, Kuninori T. Uesugi, Fu Bing-chen, Zhang Hui, Hard Cover $145.00 (ISBN 0-87703-438-9)

Vol. 97 AAS/AIAA Astrodynamics Conference, Aug. 4-7, 1997, Sun Valley, ID, 1998, two parts, 2184p., eds. Felix R. Hoots, Bernard Kaufman, Paul J. Cefola, David B. Spencer, Hard Cover $310 (ISBN 0-87703-441-9)

Vol. 98 Guidance and Control 1998, Feb. 4-8, 1998, Breckenridge, CO, 1998, 706p. Eds. Robert D. Culp, David Igli, Hard Cover $120 (ISBN 0-87703-448-6); Soft Cover $90 (ISBN 0-87703-449-4)

Vol. 99 AAS/AIAA Spaceflight Mechanics Meeting, Feb. 9-11, 1998, Monterey, CA, 1998, two parts, 1638p., eds. Jay W. Middour, Lester L. Sackett, Louis A. D"Amario, Dennis V. Byrnes, Hard Cover $280 (ISBN 0-87703-450-8); Microfiche Suppl. (Vol. 78 AAS Microfiche Series) $10 (ISBN 0-87703-452-4)

Vol. 100 AAS/GSFC International Symposium on Space Flight Dynamics, May 11-15, 1998, Greenbelt, MD, 1998, two parts, 1092p., ed. Thomas H. Stengle, Hard Cover $250 (ISBN 0-87703-453-2)

Vol. 101 Guidance and Control 1999, Feb. 3-7, 1999, Breckenridge, CO, 1999, 528p. Eds. Robert D. Culp, Douglas Wiemer, Hard Cover $120 (ISBN 0-87703-456-7); Soft Cover $90 (ISBN 0-87703-457-5)

Vol. 102 AAS/AIAA Spaceflight Mechanics Meeting, Feb. 7-10, 1999, Breckenridge, CO, 1999, two parts, 1600p., eds. Robert H. Bishop, Robert D. Culp, Donald L. Mackison and Maria Evans, Hard Cover $280 (ISBN 0-87703-458-3)

Vol. 103 AAS/AIAA Astrodynamics Conference, Aug. 16-19, 1999, Girdwood, AK, 2000, three parts, 2724p., eds. Kathleen C. Howell, Felix R. Hoots, Bernard Kaufman, K. Terry Alfriend, Hard Cover $450 (ISBN 0-87703-467-2)

Vol. 104 Guidance and Control 2000, Feb. 2-6, 2000, Breckenridge, CO, 2000, 738p. Eds. Robert D. Culp, Eileen M. Dukes, Hard Cover $130 (ISBN 0-87703-468-0); Soft Cover $95 (ISBN 0-87703-469-9)

Vol. 105 AAS/AIAA Spaceflight Mechanics Meeting, Jan. 23-26, 2000, Clearwater, FL, 2000, two parts, 1704p., eds. Craig A. Kluever, Beny Neta, Christopher D. Hall, John M. Hanson Hard Cover $300 (ISBN 0-87703-470-2)

Vol. 106 The Richard H. Battin Astrodynamics Symposium, Mar. 20-21, 2000, College Station, TX, 2000, 492p., Eds. John L. Junkins, K. Terry Alfriend, Kathleen C. Howell, Hard Cover $130 (ISBN 0-87703-471-0); Soft Cover $95 (ISBN 0-87703-472-9)

Vol. 107 Guidance and Control 2001, Jan. 31 - Feb. 4, 2001, Breckenridge, CO, 2001, 738p. Eds. Robert D. Culp, Charles N. Schira, Hard Cover $140 (ISBN 0-87703-479-6); Soft Cover $100 (ISBN 0-87703-480-X)

Vol. 108 AAS/AIAA Spaceflight Mechanics Meeting, Feb. 11-15, 2001, Santa Barbara, CA, 2001, two parts, 2174p., eds. Louis A. D'Amario, Lester L. Sackett, Daniel J. Scheeres, Bobby G. Williams, Hard Cover $330 (ISBN 0-87703-487-7)

Vol. 109 AAS/AIAA Astrodynamics Conference, Jul. 30 to Aug. 2, 2001, Quebec City, Quebec, Canada, 2002, three parts, 2592p., eds. David B. Spencer, Calina C. Seybold, Arun K. Misra, Ronald J. Lisowski, Hard Cover $450 (ISBN 0-87703-488-5)

Vol. 110 Space Development & Cooperation Among all Pacific Basin Countries, Ninth ISCOPS (formerly PISSTA) AAS/JRS/CSA Symposium, Nov. 14-16, 2001, Pasadena, CA, 2002, 444p., eds. Peter M. Bainum, Joan Johnson-Freese, Takashi Nakajima, Kuninori T. Uesugi, Hard Cover $130.00 (ISBN 0-87703-489-3); Soft Cover $100.00 (ISBN 0-87703-490-7)

Vol. 111 Guidance and Control 2002, Feb. 6-10, 2002, Breckenridge, CO, 2002, 544p. Eds. Robert D. Culp, Steven D. Jolly, Hard Cover $140 (ISBN 0-87703-491-5); Soft Cover $100 (ISBN 0-87703-492-3)

Vol. 112 AAS/AIAA Spaceflight Mechanics Meeting, Jan. 27-30, 2002, San Antonio, TX, 2002, two parts, 1570p., eds. Kyle T. Alfriend, Beny Neta, Kim Luu, Cheryl A. Hilton Walker, Hard Cover $330 (ISBN 0-87703-487-7)

Vol. 113 Guidance and Control 2003, Feb. 5-9, 2003, Breckenridge, CO, 2003, 662p. Ed. Ian J. Gravseth, Hard Cover $150 (ISBN 0-87703-502-4); Soft Cover $110 (ISBN 0-87703-503-2)

Order from Univelt, Inc., P.O. Box 28130, San Diego, California 92198 (Web Site: http://www.univelt.com) STANDING ORDERS ACCEPTED

SCIENCE AND TECHNOLOGY SERIES (1964-)

ISSN 0278-4017

A Supplement to *Advances in the Astronautical Sciences*. Proceedings and monographs, most of them based on AAS technical meetings.

Vol. 1 Manned Space Reliability Symposium, Jun. 9, 1964, Anaheim, CA, 1964, 112p., ed. Paul Horowitz, Hard Cover $20 (ISBN 0-87703-029-4)

Vol. 2 Towards Deeper Space Penetration (AAS/AAAS Symposium), Dec. 29, 1964, Montreal, Canada, 1964, 182p., ed. Edward R. Van Driest, Hard cover $20 (ISBN 0-87703-030-8)

Vol. 3 Orbital Hodograph Analysis, 1965, 150p., ed. Samuel P. Altman, Hard Cover $20 (ISBN 0-87703-031-6)

Vol. 4 Scientific Experiments for Manned Orbital Flight, 3rd Goddard Memorial Symposium, Mar. 18-19, 1965, Washington, D.C., 1965, 372p., ed. Peter C. Badgley, Hard Cover $30 (ISBN 0-87703-032-4)

Vol. 5 Physiological and Performance Determinants in Manned Space Systems (AAS/HFS Symposium, Apr. 14-15, 1965, Northridge, CA, 1965, 220p., ed. Paul Horowitz, Microfiche Only $20 (ISBN 0-87703-033-2)

Vol. 6 Space Electronics Symposium (AAS/AES Meeting), May 25-27, 1965, Los Angeles, CA, 1965, 404p., ed. Chung-Ming Wong, Hard Cover $30 (ISBN 0-87703-034-0)

Vol. 7 Theodore von Karman Memorial Seminar, May 12, 1965, Los Angeles, CA, 1966, 140p., ed. Shirley Thomas, Hard Cover $30 (ISBN 0-87703-035-9)

Vol. 8 Impact of Space Exploration on Society, Aug. 18-20, 1965, San Francisco, CA, 1966, 382p., ed. William E. Frye, Hard Cover $30 (ISBN 0-87703-036-7)

Vol. 9 Recent Developments in Space Flight Mechanics, (AAS/AAAS Symposium), Dec. 29, 1965, Berkeley, CA, 1966, 280p., ed. Paul B. Richards, Hard Cover $25 (ISBN 0-87703-037-5)

Vol. 10 Space in the Fiscal Year 2001, 4th Goddard Memorial Symposium, Mar. 15-16, 1966, Washington, D.C., 1967, 458p., eds. Eugene B. Konecci, Maxwell W. Hunter, II, Robert F. Trapp, Hard Cover $35 (ISBN 0-87703-038-3)

Vol. 11 Space Flight Specialist Conference, Jul. 6-8, 1966, Denver, CO, 1967, 618p., ed. Maurice L. Anthony, Microfiche Only (ISBN 0-87703-039-1); Plus Microfiche Suppl. (Vol. 2 AAS Microfiche Series) $60 (ISBN 0-87703-221-1)

Vol. 12 Management of Aerospace Programs Conference, Nov. 16-18, 1966, Columbia, MO, 1967, 392p., ed. Walter K. Johnson, Hard Cover $30 (ISBN 0-87703-040-5)

Vol. 13 Physics of the Moon (AAS/AAAS Symposium), Dec. 29, 1966, Washington, D.C., 1967, 260p., ed. S. Fred Singer, Hard Cover $25 (ISBN 0-87703-041-3)

Vol. 14 Interpretation of Lunar Probe Data, Sept. 17, 1966, Huntington Beach, CA, 1967, 270p., ed. Jack Green, Hard Cover $25 (ISBN 0-87703-042-1)

Vol. 15 Future Space Program and Impact on Range and Network Development Symposium, Mar. 22-24, 1967, Las Cruces, NM, 1967, 588p., ed. George W. Morgenthaler, Hard Cover $40 (ISBN 0-87703-043-X)

Vol. 16 Voyage to the Planets, 5th Goddard Memorial Symposium, Mar. 14-15, 1967, Washington, D.C., 1968, 184p., ed. S. Fred Singer, Hard Cover $20 (0-87703-044-8)

Vol. 17 Use of Space Systems for Planetary Geology and Geophysics Symposium, May 25-27, 1967, Boston, MA, 1968, 623p., ed. Robert D. Enzmann, Hard Cover $45 (ISBN 0-87703-045-6); Microfiche Suppl. (Vol. 5 AAS Microfiche Series) $15 (ISBN 0-87703-135-5)

Vol. 18 Technology and Social Progress, 6th Goddard Memorial Symposium, Mar. 12-13, 1968, Washington, D.C., 1969, 170p., ed. Philip K. Eckman, Hard Cover $20 (ISBN 0-87703-046-4)

Vol. 19 Exobiology - The Search for Extraterrestrial Life (AAS/AAAS Symposium) Dec. 30, 1967, New York, NY, 1969, 184p., eds. Martin M. Freundlich, Bernard W. Wagner, Hard Cover $20 (ISBN 0-87703-047-2)

Vol. 20 Bioengineering and Cabin Ecology (AAS/AAAS Symposium) Dec. 30, 1968, Dallas, TX, 1969, 162p., ed. William Cassidy, Hard Cover $20 (ISBN 0-87703-048-0)

Vol. 21 Reducing the Cost of Space Transportation, 7th Goddard Memorial Symposium, Mar. 4-5, 1969, Washington, D.C., 1969, 264p., ed. George K. Chacko, Microfiche only $25 (ISBN 0-87703-049-9)

Vol. 22 Planning Challenges of the 70"s in the Public Domain, 15th Annual AAS Meeting, Jun. 17-20, 1969, Denver, CO, 1970, 504p., eds. William J. Burnsnall, George K. Chacko, George W. Morgenthaler, Hard Cover $40 (ISBN 0-87703-050-2); Microfiche Suppl. (Vol. 13 AAS Microfiche Series) $20 (ISBN 0-87703-131-2); See also Vols. 15-17, AAS Microfiche Series

Vol. 23 Space Technology and Earth Problems Symposium, Oct. 23-25, 1969, Las Cruces, NM, 1970, 418p., ed. C. Quentin Ford, Hard Cover $35 (ISBN 0-87703-051-0); Microfiche Suppl. (Vol. 12 AAS Microfiche Series) $20 (ISBN 0-87703-134-7)

Vol. 24 Aerospace Research and Development, Jul. 14, 1966, Holloman AFB, NM, 1970, 500p., ed. Ernst A. Steinhoff, Microfiche Only $40 (ISBN 0-87703-052-9)

Vol. 25 Geological Problems in Lunar and Planetary Research, Feb. 17-18, 1969, Huntington Beach, CA, 1971, 750p., ed. Jack Green, Hard Cover $45 (ISBN 0-87703-056-1)

Vol. 26 Technology Utilization Ideas for the 70s and Beyond, Oct. 30, 1970, Winrock, AR, 1971, 312p., eds. Fred W. Forbes, Paul Dergarabedian, Microfiche only $30 (ISBN 0-87703-057-X)

Vol. 27 International Cooperation in Space Operations and Exploration, 9th Goddard Memorial Symposium, Mar. 11, 1971, Washington, D.C. 1971, 194p., ed. Michael Cutler, Hard Cover $20 (ISBN 0-88703-058-8)

Vol. 28 Astronomy from a Space Platform (AAS/AAAS Symposium) Dec. 27-28, 1971, Philadelphia, PA, 1972, 416p., eds. George W. Morgenthaler, Howard D. Greyber, Hard Cover $35 (ISBN 0-87703-061-8)

Vol. 29 Space Technology Transfer to Community and Industry, 10th Goddard Memorial Symposium, 18th Annual AAS Meeting, Mar. 13-14, 1972, Washington, D.C., 1972, 196p., eds. Ralph H. Tripp, John K. Stotz, Jr., Hard Cover $20 (ISBN 0-87703-062-6); on Microfiche $15

Vol. 30 Space Shuttle Payloads (AAS/AAAS Symposium) Dec. 27-28, 1972, Washington, D.C., 1973, 532p., eds. George W. Morgenthaler, William J. Bursnall, Hard Cover $40 (ISBN 0-87703-063-4)

Vol. 31 The Second Fifteen Years in Space, 11th Goddard Memorial Symposium, Mar. 8-9, 1973, Washington, D.C., 1973, 212p., ed. Saul Ferdman, Hard Cover $25 (ISBN 0-87703-064-2)

Vol. 32 Health Care Systems Conference, Nov. 21-22, 1972, Dallas, TX, 1974, 265p., ed. Eugene B. Konecci, Hard Cover $25 (ISBN 0-87703-067-7)

Vol. 33 Orbital International Laboratory, 3rd and 4th IAF/OIL Symposia, Oct. 5-6, 1970, Constance, Germany, Sept. 24-25, 1971, Brussels, Belgium, 1974, 322p., ed. Ernst A. Steinhoff, Hard Cover $30 (ISBN 0-87703-068-5)

Vol. 34 Management and Design of Long-Life Systems, Apr. 24-26, 1973, Denver, CO, 1974, 198p., ed. Harris M. Schurmeier, Hard Cover $20 (ISBN 0-87703-069-3)

Vol. 35 Energy Delta, Supply vs. Demand, (AAS/AAAS Symposium) Feb. 25-27, 1974, San Francisco, CA, 1975, 2nd Printing 1976, 604p., eds. George W. Morgenthaler, Aaron N. Silver, Hard Cover $35 (ISBN 0-87703-070-7); Soft Cover $25 (ISBN 0-87703-082-0); on Microfiche $20

Vol. 36 Skylab and Pioneer Report, 12th Goddard Memorial Symposium, Mar. 8, 1974, Washington, D.C., 1975, 160p., eds. Philip H. Bolger, Paul B. Richards, Hard Cover $20 (ISBN 0-87703-071-5)

Vol. 37 Space Rescue and Safety 1974, 7th International IAA Symposium, Sept. 30 - Oct. 5, 1974, Amsterdam, Netherlands, 1975, 294p., ed. Philip H. Bolger, Microfiche Only $25 (ISBN 0-87703-073-1)

Vol. 38 Skylab Science Experiments, (AAS/AAAS Symposium) Feb. 28, 1974, San Francisco, CA, 1976, 274p., eds. George W. Morgenthaler, Gerald E. Simonson, Microfiche only $20 (ISBN 0-87703-074-X)

Vol. 39 Environmental Control and Agri-Technology, 1976, 346p., ed. Eugene B. Konecci, Microfiche only $20 (ISBN 0-87703-075-8)

Vol. 40 Future Space Activities, 13th Goddard Memorial Symposium, Apr. 11, 1975, Washington, D.C., 1976, 182p., ed. Carl H. Tross, Microfiche only $20 (ISBN 0-87703-076-6)

Vol. 41 Space Rescue and Safety 1975, 8th International IAA Symposium, Sept. 21-27, 1975, Lisbon, Portugal, 1976, 230p., ed. Philip H. Bolger, Hard Cover $25 (ISBN 0-87703-077-4)

Vol. 42 The End of an Era in Space Exploration, From International Rivalry to International Cooperation, 1976, 216p., by J.C.D. Blaine, Hard Cover $25 (ISBN 0-87703-084-7); without volume number (ISBN 0-87703-080-4)

Vol. 43 The Eagle Has Returned, Part I, International Space Hall of Fame Dedication Conference, Oct. 5-9, 1976, Alamogordo, NM, 1976, 370p., ed. Ernst. A. Steinhoff, Hard Cover $30 (ISBN 0-87703-086-3)

Vol. 44 Satellite Communications in the Next Decade, 14th Goddard Memorial Symposium, Mar. 12, 1976, Washington, D.C., 1977, 188p., ed. Leonard Jaffe, Hard Cover $20 (ISBN 0-87703-088-X)

Vol. 45 The Eagle Has Returned, Part 2, International Space Hall of Fame Dedication Conference, Oct. 5-9, 1976, Alamogordo, NM, 1977, 454p., ed. Ernst A. Steinhoff, Hard Cover $35 (ISBN 0-87703-092-8)

Vol. 46 Export of Aerospace Technology, 15th Goddard Memorial Symposium, Mar. 31 - Apr. 1, 1977, Washington, D.C., 1978, 174p., ed. Carl H. Tross, Hard Cover $20 (ISBN 0-87703-093-6)

Vol. 47 Handbook of Soviet Lunar and Planetary Exploration, 1979, 276p., by Nicholas L. Johnson, Microfiche Only $25 (ISBN 0-87703-105-3)

Vol. 48 Handbook of Soviet Manned Space Flight, 2nd Edition, 1988, 474p., by Nicholas L. Johnson, Hard Cover $60 (ISBN 0-87703-115-0); Soft Cover $45 (ISBN 0-87703-116-9)

Vol. 49 Space - New Opportunities for International Ventures, 17th Goddard Memorial Symposium, Mar. 28-30, 1979, Washington, D.C., 1980, 300p., ed. William C. Hayes, Jr., Hard Cover $35 (ISBN 0-87703-124-X); Soft Cover $25 (ISBN 0-87703-125-8); see also Vol. 2 AAS History Series

Vol. 50 Remember the Future - The Apollo Legacy, Jul. 20-21, 1979, San Francisco, CA, 1980, 218p., ed. Stan Kent, Hard Cover $25 (ISBN 0-87703-126-6); Soft Cover $15 (ISBN 0-87703-127-4)

Vol. 51 Commercial Operations in Space 1980-2000, 18th Goddard Memorial Symposium, Mar. 27-28, 1980, Washington, D.C., 1981, 214p., eds. John L. McLucas, Charles Sheffield, Hard Cover $30 (ISBN 0-87703-140-1); Soft Cover $20 (ISBN 0-87703-141-X); Microfiche Suppl. (Vol. 34 AAS Microfiche Series) $10 (ISBN 0-87703-165-7); see also Vols. 2 and 3, AAS History Series

Vol. 52 International Space Technical Applications, 19th Goddard Memorial Symposium, Mar. 26-27, 1981, Washington, D.C., 1981, 186p., eds. Andrew Adelman, Peter M. Bainum, Hard Cover $30 (ISBN 0-87703-152-5); Soft Cover $20 (ISBN 0-87703-153-3); see also Vol. 5, AAS History Series

Vol. 53 Space in the 1980"s and Beyond, 17th European Space Symposium, Jun. 4-6, 1980, London, England, 1981, 302p., ed. Peter M. Bainum, Hard Cover $40 (ISBN 0-87703-154-1); Soft Cover $30 (ISBN 0-87703-155-X)

Vol. 54 Space Safety and Rescue 1979-1981 (with abstracts 1976-1978), Proceedings of symposia of the International Academy of Astronautics held in conjunction with the 30th, 31st, and 32nd International Astronautical Federation Congresses, Munich, Germany, 1979, Tokyo, Japan, 1980, and Rome, Italy, 1981, 1983, 456p., ed. Jeri W. Brown, Hard Cover $45 (ISBN 0-87703-177-0); Soft Cover $35 (ISBN 0-87703-178-9); Microfiche Suppl. (Vols. 39-41 AAS Microfiche Series) $39 (ISBN 0-87703-222-X); (ISBN 0-87703-223-8); (ISBN 0-87703-224-6)

Vol. 55 Space Applications at the Crossroads, 21st Goddard Memorial Symposium, Mar. 24-25, 1983, Greenbelt, MD, 1983, 308p., eds. John H. McElroy, E. Larry Heacock, Hard Cover $45 (ISBN 0-87703-186-X); Soft Cover $35 (ISBN 0-87703-187-8)

Vol. 56 Space: A Developing Role for Europe, 18th European Space Symposium, Jun. 6-9, 1983, London, England, 1984, 278p., eds. Len J. Carter, Peter M. Bainum, Hard Cover $45 (ISBN 0-87703-193-2); Soft Cover $35 (ISBN 0-87703-194-0); Microfiche Suppl. (Vol. 46 AAS Microfiche Series) $15 (ISBN 0-87703-195-9)

Vol. 57 The Case for Mars, Apr. 29 - May 2, 1981, Boulder, CO, 1984, Second Printing 1987, 348p., ed. Penelope J. Boston, on Microfiche only $25 (ISBN 0-87703-198-3)

Vol. 58 Space Safety and Rescue 1982-1983, Proceedings of the International Academy of Astronautics held in conjunction with the 33rd and 34th International Astronautical Congresses, Paris, France, Sept. 27 - Oct. 2, 1982, and Budapest, Hungary, Oct. 10-15, 1983, 1984, 378p., ed. Gloria W. Heath, Hard Cover $50 (ISBN 0-87703-202-5); Soft Cover $40 (ISBN 0-87703-203-3)

Vol. 59 Space and Society - Challenges and Choices, April 14-16, 1982, University of Texas at Austin, 1984, 442p., eds. Paul Anaejionu, Nathan C. Goldman, Philip J. Meeks, Hard Cover $55 (ISBN 0-87703-204-1); Soft Cover $35 (ISBN 0-87703-205-X)

Vol. 60 Permanent Presence - Making It Work, 22nd Goddard Memorial Symposium, Mar. 15-16, 1984, Greenbelt, MD, 1985, 190p., ed. Ivan Bekey, Hard Cover $40 (ISBN 0-87703-207-6); Soft Cover $30 (ISBN 0-87703-208-4)

Vol. 61 Europe/United States Space Activities - With a Space Propulsion Supplement, 23rd Goddard Memorial Symposium/19th European Space Symposium, Mar. 27-29, 1985, Greenbelt, MD, 31st Annual AAS Meeting, Oct. 22-24, 1984, Palo Alto, CA, 1985, 442p., eds. Peter M. Bainum, Friedrich von Bun, Hard Cover $55 (ISBN 0-87703-217-3); Soft Cover $45 (ISBN 0-87703-218-1)

Vol. 62 The Case for Mars II, July 10-14, 1984, Boulder, CO, 1985, 730p., ed. Christopher P. McKay, Hard Cover $60 (ISBN 0-87703-219-1); Soft Cover $40 (ISBN 0-87703-220-3)

Vol. 63 Proceedings of 4th International Conference on Applied Numerical Modeling, Dec. 27-29, 1984, Tainan, Taiwan, 1986, 800p., ed. Han-Min Hsia, You-Li Chou, Shu-Yi Wang, Sheng-Jii Hsieh, Hard Cover $70 (ISBN 0-87703-242-4)

Vol. 64 Space Safety and Rescue 1984-1985, Proceedings of the International Academy of Astronautics held in conjunction with the 35th and 36th International Astronautical Congresses, Lausanne, Switzerland, Oct. 7-13, 1984, and Stockholm, Sweden, Oct. 7-12, 1985, 1986, 400p., ed. Gloria W. Heath, Hard Cover $55 (ISBN 0-87703-248-3); Soft Cover $45 (ISBN 0-87703-249-1)

Vol. 65 The Human Quest in Space, 24th Goddard Memorial Symposium, Mar. 20-21, 1986, Greenbelt, MD, 1987, 312p., ed. Gerald L. Burdett, Gerald A. Soffen, Hard Cover $55 (ISBN 0-87703-262-9); Soft Cover $45 (ISBN 0-87703-263-7)

Vol. 66 Soviet Space Programs 1980-1985, 1987, 298p., by Nicholas L. Johnson, Microfiche Only $45 (ISBN 0-87703-266-1)

Vol. 67 Low-Gravity Sciences, Seminar Series 1986, University of Colorado at Boulder, 290p., ed. Jean N. Koster, Hard Cover $55 (ISBN 0-87703-270-X); Soft Cover $45 (ISBN 0-87703-271-8)

Vol. 68 Proceedings of the Fourth Annual L5 Space Development Conference, Apr. 25-28, 1985, Washington, D.C., 1987, 268p., ed. Frank Hecker, Hard Cover $50 (ISBN 0-87703-272-6); Soft Cover $35 (ISBN 0-87703-273-4)

Vol. 69 Visions of Tomorrow: A Focus on National Space Transportation Issues, 25th Goddard Memorial Symposium, Mar. 18-20, 1987, Greenbelt, MD, 1987, 338p., ed. Gerald A. Soffen, Hard Cover $55 (ISBN 0-87703-274-2); Soft Cover $45 (ISBN 0-87703-275-0)

Vol. 70 Space Safety and Rescue 1986-1987, Proceedings of the International Academy of Astronautics held in conjunction with the 37th and 38th International Astronautical Congresses, Innsbruck, Austria, Oct. 4-11, 1986, and Brighton, England, Oct. 11-16, 1987, 1988, 360p., ed. Gloria W. Heath, Hard Cover $55 (ISBN 0-87703-291-2); Soft Cover $45 (ISBN 0-87703-292-0)

Vol. 71 The NASA Mars Conference, Jul. 21-23, 1986, Washington, D.C., 1988, 570p., ed. Duke B. Reiber, Hard Cover $50 (ISBN 0-87703-293-9); Soft Cover $30 (ISBN 0-87703-294-7)

Vol. 72 Working in Orbit and Beyond: The Challenges for Space Medicine, Jun. 20-21, 1987, Washington, D.C., 1989, 188p., ed. David Lorr, Victoria Garshnek, Hard Cover $45 (ISBN 0-87703-295-5); Soft Cover $35 (ISBN 0-87703-296-3)

Vol. 73 Technology and the Civil Future in Space, 26th Goddard Memorial Symposium, Mar. 16-18, 1988, Greenbelt, MD, 1989, 246p., ed. Leonard A. Harris, Hard Cover $50 (ISBN 0-87703-301-3); Soft Cover $35 (ISBN 0-87703-302-1)

Vol. 74 The Case for Mars III: Strategies for Exploration - General Interest and Overview, July 18-22, 1987, Boulder, CO, 1989, 744p., ed. Carol Stoker, Hard Cover $75 (ISBN 0-87703-303-X); Soft Cover $55 (0-87703-304-8)

Vol. 75 The Case for Mars III: Strategies for Exploration - Technical, July 18-22, 1987, Boulder, CO, 1989, 646p., ed. Carol Stoker, Hard Cover $70 (ISBN 0-87703-305-6); Soft Cover $50 (ISBN 0-87703-306-4)

Vol. 76 Global Environmental Change: The Role of Space in Understanding Earth, 27th Goddard Memorial Symposium, Mar. 8-10, 1989, Washington, D.C., 1990, 178p., ed. Richard G. Johnson, Hard Cover $50 (ISBN 0-87703-322-6); Soft Cover $40 (ISBN 0-87703-323-4); Microfiche Suppl. (Vol. 60 AAS Microfiche Series) $10 (ISBN 0-87703-324-2)

Vol. 77 Space Safety and Rescue 1988 - 1989, Proceedings of the International Academy of Astronautics held in conjunction with the 39th and 40th International Astronautical Congresses, Bangalore, India, Oct. 8-15, 1988, and Málaga, Spain, Oct. 7-12, 1989, 1990, 500p., ed. Gloria W. Heath, Hard Cover $70 (ISBN 0-87703-327-7); Soft Cover $55 (ISBN 0-87703-328-5)

Vol. 78 Leaving the Cradle: Human Exploration of Space in the 21st Century, 28th Goddard Memorial Symposium, Mar. 14-16, 1990, Washington, D.C., 1991, 348p., ed. Thomas O. Paine, Hard Cover $70 (ISBN 0-87703-336-6); Soft Cover $55 (ISBN 0-87703-337-4)

Vol. 79 Space Safety and Rescue 1990, Proceedings of the International Academy of Astronautics held in conjunction with the 41st International Astronautical Congress, Dresden, Germany, Oct. 6-12, 1990, 1991, 232p., ed. Gloria W. Heath, Hard Cover $65 (ISBN 0-87703-341-2); Soft Cover $50 (ISBN 0-87703-342-0)

Vol. 80 Prospects for Interstellar Travel, 1992, 390p., by John H. Mauldin, Hard Cover $50 (ISBN 0-87703-344-7); Soft Cover $27 (ISBN 0-87703-345-5)

Vol. 81 Humans and Machines in Space: The Vision, The Challenge, The Payoff, 29th Goddard Memorial Symposium, Mar. 14-15, 1991, Washington, D.C., 1992, 204p., ed. Bradley Johnson, Gayle L. May, Paula Korn, Hard Cover $50 (ISBN 0-87703-356-0); Soft Cover $35 (ISBN 0-87703-357-9)

Vol. 82 Space Safety and Rescue 1991, Proceedings of the International Academy of Astronautics held in conjunction with the 42nd International Astronautical Congress, Montreal, Canada, Oct. 5-11, 1991, 1993, 270p., ed. Gloria W. Heath, Hard Cover $65 (ISBN 0-87703-372-2); Soft Cover $50 (ISBN 0-87703-373-0)

Vol. 83 Space: A Vital Stimulus to Our National Well-Being, 31st Goddard Memorial Symposium, March 9-10, 1993, Arlington, Virginia, and World Space Programs and Fiscal Reality, 30th Goddard Memorial Symposium, April 9-10, 1992, Alexandria, Virginia, 1994, 334p., ed. Gayle L. May, Saunders B. Kramer, Paula Korn, Leonard David, Barbara Sprungman, Hard Cover $70 (ISBN 0-87703-389-7); Soft Cover $50 (ISBN 0-87703-390-0)

Vol. 84 Space Safety and Rescue 1992, Proceedings of the International Academy of Astronautics held in conjunction with the World Space Congress, Washington, D.C., Aug. 28 to Sept. 5, 1992, 1994, 372p., ed. Gloria W. Heath, Hard Cover $70 (ISBN 0-87703-391-9); Soft Cover $55 (ISBN 0-87703-392-7)

Vol. 85 Civil Space in the Clinton Era, 32nd Goddard Memorial Symposium, March 1-2, 1994, Crystal City, Virginia, and Partners in Space . . . 2001, 41st Annual Meeting, November 14-16, 1994, Crystal City, Virginia, 1995, 292p., ed. Donald R. McConathy, Paula Korn, Hard Cover $70 (ISBN 0-87703-397-8); Soft Cover $50 (ISBN 0-87703-398-6)

Vol. 86 Strategies for Mars: A Guide to Human Exploration, 1996, 644p, ed. Carol R. Stoker, Carter Emmart, Hard Cover $70 (ISBN 0-87703-405-2); Soft Cover $45 (ISBN 0-87703-406-0)

Vol. 87 Space Safety and Rescue 1993, Proceedings of the International Academy of Astronautics held in conjunction with the 44th International Astronautical Congress, Graz, Austria, Oct. 16-22, 1993, 1996, 344p., ed. Gloria W. Heath, Hard Cover $70 (ISBN 0-87703-410-9); Soft Cover $50 (ISBN 0-87703-411-7)

Vol. 88 Space Safety and Rescue 1994, Proceedings of the International Academy of Astronautics held in conjunction with the 45th International Astronautical Congress, Jerusalem, Israel, Oct. 9-14, 1994, 1996, 326p., ed. Gloria W. Heath, Hard Cover $70 (ISBN0-87703-416-8); Soft Cover $50 (ISBN 0-87703-417-6)

Vol. 89 The Case for Mars IV: The International Exploration of Mars"Mission Strategy and Architectures, June 4-8, 1990, Boulder, CO, 1997, 790p., ed. Thomas R. Meyer, Hard Cover $80 (ISBN 0-87703-418-4); Soft Cover $55 (ISBN 0-87703-419-2)

Vol. 90 The Case for Mars IV: The International Exploration of Mars"Considerations for Sending Humans, June 4-8, 1990, Boulder, CO, 1997, 502p., ed. Thomas R. Meyer, Hard Cover $70 (ISBN 0-87703-420-6); Soft Cover $55 (ISBN 0-87703-421-4)

Vol. 91 From Imagination to Reality: Mars Exploration Studies of the Journal of the British Interplanetary Society (Part I, Precursors and Early Piloted Missions), 1997, 388p., ed. Robert M. Zubrin, Hard Cover $70 (ISBN 0-87703-426-5); Soft Cover $45 (ISBN 0-87703-427-3)

Vol. 92 From Imagination to Reality: Mars Exploration Studies of the Journal of the British Interplanetary Society (Part II, Base Building, Colonization and Terraformation), 1997, 376p., ed. Robert M. Zubrin, Hard Cover $70 (ISBN 0-87703-428-1); Soft Cover $45 (ISBN 0-87703-429-X)

Vol. 93 Space Safety and Rescue 1995, Proceedings of the International Academy of Astronautics held in conjunction with the 46th International Astronautical Congress, Jerusalem, Israel, Oct. 2-6, 1995, 1997, 482p., ed. Gloria W. Heath, Hard Cover $80 (ISBN 0-87703-416-8); Soft Cover $55 (ISBN 0-87703-417-6)

Vol. 94 Fluid and Electrolyte Regulation in Spaceflight, 1998, 238p., Carolyn S. Leach Huntoon, Anatoliy I. Grigoriev, Yuri V. Natochin, Hard Cover $60 (ISBN 0-87703-442-7); Soft Cover $40 (ISBN 0-87703-443-5)

Vol. 95 Space Safety and Rescue 1996, Proceedings of the International Academy of Astronautics held in conjunction with the 47th International Astronautical Congress, Beijing, China, Oct. 7-11, 1996, 1998, 362p., ed. Gloria W. Heath, Hard Cover $75 (ISBN 0-87703-446-X); Soft Cover $50 (ISBN 0-87703-447-8)

Vol. 96 Space Safety and Rescue 1997, Proceedings of the International Academy of Astronautics held in conjunction with the 48th International Astronautical Congress, Turin, Italy, Oct. 6-10, 1997, 1999, 412p., ed. Gloria W. Heath, Hard Cover $80 (ISBN 0-87703-454-0); Soft Cover $55 (ISBN 0-87703-455-9)

Vol. 97 The Case for Mars V, May 26-29, 1993, Boulder, CO, 2000, 564p., ed. Penelope J. Boston, Hard Cover $80 (ISBN 0-87703-459-1); Soft Cover $55 (ISBN 0-87703-460-5)

Vol. 98 The Case for Mars VI: Making Mars an Affordable Destination, July 17-20, 1996, Boulder, CO, 2000, 578p., ed. Kelly R. McMillen, Hard Cover $80 (ISBN 0-87703-461-3); Soft Cover $55 (ISBN 0-87703-462-1)

Vol. 99 Space Safety and Rescue 1998, Proceedings of the International Academy of Astronautics held in conjunction with the 49th International Astronautical Congress, Melbourne, Australia, Sept. 28 - Oct. 2, 1998, 2000, 410p., ed. Macgregor S. Reid and Walter Flury, Hard Cover $80 (ISBN0-87703-463-X); Soft Cover $55 (ISBN 0-87703-464-8)

Vol. 100 Space Debris 1999, Proceedings of the International Academy of Astronautics held in conjunction with the 50th International Astronautical Federation Congress, Amsterdam, The Netherlands, Oct. 2-8, 1999, 2001, 494p., ed. Joerg Bendisch, Hard Cover $90 (ISBN 0-87703-473-7); Soft Cover $65 (ISBN 0-87703-474-5)

Vol. 101 Space Access and Utilization Beyond 2000, Based on, but not limited to, the Symposium held at the Annual General Assembly of the American Association for the Advancement of Science, Washington, D.C., Feb. 18, 2000, 234p., ed. Yoji Kondo, Charles Sheffield and Frederick C. Bruhweiler, Hard Cover $70 (ISBN 0-87703-475-3); Soft Cover $50 (ISBN 0-87703-476-1)

Vol. 102 Space Safety, Rescue and Quality 1999-2000, Proceedings of the International Academy of Astronautics held in conjunction with the 50th and 51st International Astronautical Federation Congresses, Amsterdam, The Netherlands, Oct. 2-8, 1999; Rio de Janeiro, Brazil, Oct. 2-6, 2000, 2001, 286p., eds. Macgregor S. Reid and Manola Romero, Hard Cover $80 (ISBN 0-87703-483-4); Soft Cover $55 (ISBN 0-87703-484-2)

Vol. 103 Space Debris 2000, Proceedings of the International Academy of Astronautics held in conjunction with the 51st International Astronautical Federation Congress, Rio de Janeiro, Brazil, Oct. 2-6, 2000, 2001, 368p., ed. Joerg Bendisch, Hard Cover $90 (ISBN 0-87703-485-0); Soft Cover $65 (ISBN 0-87703-486-9)

Vol. 104 ISOLATION: NASA Experiments in Closed-Environment Living - Advanced Human Life Support Enclosed System Final Report, 2002, 470p., eds. Helen W. Lane, Richard L. Sauer and Daniel L. Feeback, Hard Cover $90 (ISBN 0-87703-493-1); Soft Cover $65 (ISBN 0-87703-494-X)

Vol. 105 Space Debris 2001, Proceedings of the International Academy of Astronautics held in conjunction with the 52nd International Astronautical Federation Congress, Toulouse, France, Oct. 1-5, 2001, 2002, 306p., ed. Joerg Bendisch, Hard Cover $90 (ISBN 0-87703-496-6); Soft Cover $65 (ISBN 0-87703-497-4)

Vol. 106 Space Safety, Rescue and Quality 2001, Proceedings of the International Academy of Astronautics held in conjunction with the 52nd International Astronautical Federation Congress, Toulouse, France, Oct. 1-5, 2001, 2003, 224p., eds. Macgregor S. Reid and Manola Romero, Hard Cover $80 (ISBN 0-87703-500-8); Soft Cover $55 (ISBN 0-87703-501-6)

Order from Univelt, Inc., P.O. Box 28130, San Diego, California 92198 (Web Site: http://www.univelt.com) STANDING ORDERS ACCEPTED

AAS HISTORY SERIES

Vol. 1 Two Hundred Years of Flight in America: A Bicentennial Survey, Edited by Eugene M. Emme, 1977, 326p, Third Printing 1981, Hard Cover $35 (ISBN 0-87703-091-X); Soft Cover $25 (ISBN 0-87703-101-0); special price for classroom text or bulk purchase.

Vol. 2 Twenty-Five Years of the American Astronautical Society: Historical Reflections and Projections, 1954-1979, Edited by Eugene M. Emme, 1980, 248p, Hard Cover $25 (ISBN 0-87703-117-7); Soft Cover $15 (ISBN 0-87703-118-5).

Vol. 3 Between Sputnik and the Shuttle: New Perspectives on American Astronautics, 1957-1980, Edited by Frederick C. Durant, III, 1981, 350p, Hard Cover $40 (ISBN 0-87703-145-2); Soft Cover $30 (ISBN 0-87703-149-9).

Vol. 4 The Endless Space Frontier: A History of the House Committee on Science and Astronautics, By Ken Hechler, Abridged and edited by Albert E. Eastman, 1982, 460p, Hard Cover $45 (ISBN 0-87703-157-6); Soft Cover $35 (ISBN 0-87703-158-4).

Vol. 5 Science Fiction and Space Futures: Past and Present, Edited by Eugene M. Emme, 1982, 278p, Hard Cover $35 (ISBN 0-87703-172-X); Soft Cover $25 (ISBN 0-87703-173-8).

Vol. 6 First Steps Toward Space, Edited by Frederick C. Durant, III and George S. James, 1986, 318p, Hard Cover $45 (ISBN 0-87703-243-2); Soft Cover $35 (ISBN 0-87703-244-0).

Vol. 7 History of Rocketry and Astronautics, Edited by R. Cargill Hall, 1986, Part I, 250p, Part II, 502p, sold as a set, Hard Cover $100 (ISBN 0-87703-260-2); Soft Cover $80 (ISBN 0-87703-261-0).

Vol. 8 History of Rocketry and Astronautics, Edited by Kristan R. Lattu, 1989, 368p, Hard Cover $50 (ISBN 0-87703-307-2); Soft Cover $35 (ISBN 0-87703-308-0).

Vol. 9 History of Rocketry and Astronautics, Edited by Frederick I. Ordway, III, 1989, 330p, Hard Cover $50 (ISBN 0-87703-309-9); Soft Cover $35 (ISBN 0-87703-310-2).

Vol. 10 History of Rocketry and Astronautics, Edited by Å. Ingemar Skoog, 1990, 330p, Hard Cover $60 (ISBN 0-87703-329-3); Soft Cover $40 (ISBN 0-87703-330-7)

Vol. 11 History of Rocketry and Astronautics, Edited by Roger D. Launius, 1994, 236p, Hard Cover $60 (ISBN 0-87703-382-X); Soft Cover $40 (ISBN 0-87703-383-8).

Vol. 12 History of Rocketry and Astronautics, Edited by John L. Sloop, 1991, 252p, Hard Cover $60 (ISBN 0-87703-332-3); Soft Cover $40 (ISBN 0-87703-333-1).

Vol. 13 History of Liquid Rocket Engine Development in the United States 1955-1980, Edited by Stephen E. Doyle, 1992, 176p, Hard Cover $50 (ISBN 0-87703-349-8); Soft Cover $35 (ISBN 0-87703-350-1).

Vol. 14 History of Rocketry and Astronautics, Edited by Tom D. Crouch, Alex M. Spencer, 1993, 222p, Hard Cover $50 (ISBN 0-87703-374-9); Soft Cover $35 (ISBN 0-87703-375-7).

Vol. 15 History of Rocketry and Astronautics, Edited by Lloyd H. Cornett, Jr., 1993, 452p, Hard Cover $60 (ISBN 0-87703-376-5); Soft Cover $40 (ISBN 0-87703-377-3).

Vol. 16 Out From Behind the Eight-Ball: A History of Project Echo, by Donald C. Elder, 1995, 176p, Hard Cover $50 (ISBN 0-87703-387-0); Soft Cover $30 (ISBN 0-87703-388-9).

Vol. 17 History of Rocketry and Astronautics, Edited by John Becklake, 1995, 480p, Hard Cover $60 (ISBN 0-87703-395-1); Soft Cover $40 (ISBN 0-87703-396-X).

Vol. 18 Organizing for the Use of Space: Historical Perspectives on a Persistent Issue, Edited by Roger D. Launius, 1995, 234p, Hard Cover $60; Soft Cover $40

Vol. 19 History of Rocketry and Astronautics, Edited by J. D. Hunley, 1997, 318p, Hard Cover $60 (ISBN 0-87703-422-2); Soft Cover $40 (ISBN 0-87703-423-0).

Vol. 20 History of Rocketry and Astronautics, Edited by J. D. Hunley, 1997, 344p, Hard Cover $60 (ISBN 0-87703-424-9); Soft Cover $40 (ISBN 0-87703-425-7).

Vol. 21 History of Rocketry and Astronautics, Edited by Philippe Jung, 1997, 368p, Hard Cover $60 (ISBN 0-87703-439-7); Soft Cover $40 (ISBN 0-87703-440-0).

Vol. 22 History of Rocketry and Astronautics, Edited by Philippe Jung, 1998, 418p, Hard Cover $60 (ISBN 0-87703-444-3); Soft Cover $40 (ISBN 0-87703-445-1).

Vol. 23 History of Rocketry and Astronautics, Edited by Donald C. Elder and Christophe Rothmund, 2001, 566p, Hard Cover $85 (ISBN 0-87703-477-X); Soft Cover $60 (ISBN 0-87703-478-8).

Vol. 24 The Origins and Technology of the Advanced Extravehicular Space Suit, by Gary L. Harris, 2001, 558p, Hard Cover $85 (ISBN 0-87703-481-8); Soft Cover $60 (ISBN 0-87703-482-6).

Vol. 25 History of Rocketry and Astronautics, Edited by Hervé Moulin and Donald C. Elder, 2003, 370p, Hard Cover $85 (ISBN 0-87703-498-2); Soft Cover $60 (ISBN 0-87703-499-0).

INDEX

NUMERICAL INDEX*

---

* Unless otherwise indicated, all papers appear in Volume 113, *Advances in the Astronautical Sciences.*

AAS 03-043 Night Sky Tests and In-flight Results of SED16 Autonomous Star Sensor, Ludovic Blarre, Sylvain Thomas, Philippe Jacob, Thierry Foisneau, Didier Vilaire and Marc Pochard

AAS 03-044 Videometer: A New Rendezvous Sensor, Agnès Mestreau-Garreau, Carole Moussu, Ludovic Blarre, Nicolas Perrimon, Philippe Jacob, Marc Pochard, Sabine Ludwig, Paul Da Cunha and Stein Strandmoe

AAS 03-045 to -048 Not Available

AAS 03-049 Properties of BFI's New Proximity Sensor, The Linear Gap Displacement Transducer (LGDT), Andrew Chouinard

AAS 03-050 Extreme Accuracy Star Tracker in Support of Hyper Precision Cold Atom Interferometry, S. P. Airey, G. Bagnasco, M. Barilli, S. Becucci, G. Cherubini and A. Romoli

AAS 03-051 to -054 Not Available

AAS 03-055 to -060 Not Assigned

AAS 03-061 Hubble Space Telescope Relative Pointing Stability Post-Servicing Mission 3b, A. Bradley, J. L. Kelley, J. H. Wirzburger, E. Nelan, D. Smith, O.G. Franz, L.H. Wasserman, M. Wenz and D. Crain

AAS 03-062 Not Available

AAS 03-063 Precision Pointing of the Thuraya Satellite, Richard A. Fowell and Hanching G. Wang

AAS 03-064 Transient Pointing Error Reduction in Spacecraft Performing Station-keeping Operations With Momentum Wheel Control, Lee Barker and Xen Price

AAS 03-065 Design and Analysis of the ST7 Disturbance Reduction System (DRS) Spacecraft Controller, P. G. Maghami, F. L. Markley, M. B. Houghton and C. J. Dennehy

AAS 03-066 Precision Pointing for the Laser Interferometry Space Antenna Mission, T. T. Hyde and P. G. Maghami

AAS 03-067 Mars Reconnaissance Orbiter Design Approach for High-Resolution Surface Imaging, S. W. Lee, E. D. Skulsky, J. Chapel, D. Cwynar, R. Gehling and A. Delamere

AAS 03-068 to -070 Not Assigned

AAS 03-071 Satellite GN&C Anomaly Trends, Brent Robertson and Eric Stoneking

AAS 03-072 International Space Station Control Moment Gyro Failure, Richard R. Burt

AAS 03-073 Performance of the Star Tracker Light Shades on the Earth Observing Satellite (EOS) Aqua, Thomas Kenney, Albert Lee, Michael Donnelly, Michael Schroder and Mark McNally

AAS 03-074 Recent G&C Experiences of the Timed Spacecraft, Wayne Dellinger, Hongxing Stella Shapiro, Courtney Ray and Tom Strikwerda

AAS 03-075 Appendage Deployment Verification for the 2001 Mars Odyssey Spacecraft Using Spacecraft Attitude Telemetry, Russ Gehling and Sue Linch

AAS 03-076 Not Available

AUTHOR INDEX*

* For each author the paper number is given. The page numbers refer to Volume 113, *Advances in the Astronautical Sciences*.